REVIEWS IN MINERALOGY
VOLUME 18

SPECTROSCOPIC METHODS
IN MINERALOGY AND GEOLOGY

EDITOR: Frank C. Hawthorne

AUTHORS:

GORDON E. BROWN, Jr.[1,2]

MICHAEL F. HOCHELLA, Jr.[1,2]

JONATHAN F. STEBBINS [1]

GLENN A. WAYCHUNAS [2]
 [1] Department of Geology
 [2] Center for Materials Research
 Stanford University
 Stanford, California 94305

GEORGES CALAS

JACQUELINE PETIAU
 Laboratoire de Minéralogie- Cristal-
  lographie, C.N.R.S. A.U. 09
 Universités Paris 6 et 7
 Tour 16, 4, place Jussieu
 75252 Paris Cedex 05, France

SUBRATA GHOSE
 Mineral Physics Group
 Department of Geological Sciences
 University of Washington
 Seattle, Washington 98195

FRANK C. HAWTHORNE
 Department of Earth Sciences
 University of Manitoba
 Winnepeg, Manitoba R3T 2N2 Canada

ANTHONY C. HESS

PAUL F. McMILLAN
 Department of Chemistry
 Arizona State University
 Tempe, Arizona 85287

ANNE M. HOFMEISTER
 Geophysical Laboratory
 2801 Upton Street N.W.
 Washington, D.C. 20009

R. JAMES KIRKPATRICK
 Department of Geology
 Inversity of Illinois
 Urbana, Illinois 61801

GEORGE R. ROSSMAN
 Division of Geological & Planetary Sci.
 California Institute of Technology
 Pasadena, California 91125

SERIES EDITOR: Paul H. Ribbe
 Department of Geological Sciences
 Virginia Polytechnic Institute & State University
 Blacksburg, Virginia 24061

REVIEWS in MINERALOGY

(Formerly: SHORT COURSE NOTES)

ISSN 0275-0279

Volume 18: *Spectroscopic Methods*

in Mineralogy and Geology

ISBN 0-939950-22-7

ADDITIONAL COPIES of this volume as well as those listed below
may be obtained from the MINERALOGICAL SOCIETY of AMERICA,
1625 I Street, N.W., Suite 414, Washington, D.C. 20006 U.S.A.

SPECTROSCOPIC METHODS in Mineralogy and Geology

PREFACE and ACKNOWLEDGMENTS

Both mineralogy and geology began as macroscopic observational sciences. Toward the end of the 19th century, theoretical crystallography began to examine the microscopic consequences of translational symmetry, and with the advent of crystal structure analysis at the beginning of this century, the atomic (crystal) structure of minerals became accessible to us. Almost immediately, the results were used to explain at the qualitative level many of the macroscopic physical properties of minerals. However, it was soon realized that the (static) arrangement of atoms in a mineral is only one aspect of its constitution. Also of significance are its vibrational characteristics, electronic structure and magnetic properties, factors that play an even more important role when we come to consider the *behavior* of the minerals in *dynamic* processes. It was as probes of these types of properties that spectroscopy began to play a significant role in mineralogy.

During the 1960's, a major effort in mineralogy involved the characterization of cation ordering in minerals, and this work began to have an impact in petrology via the thermodynamic modeling of inter- and intra-crystalline exchange. This period saw great expansion in the use of vibrational, optical and Mössbauer spectroscopies for such work. This trend continued into the 1970s, with increasing realization that adequate characterization of the structural chemistry of a mineral often requires several complementary spectroscopic and diffraction techniques.

The last decade has seen the greatest expansion in the use of spectroscopy in the Earth Sciences. There has been a spate of new techniques (Magic Angle Spinning Nuclear Magnetic Resonance, Extended X-ray Absorption Fine-Structure and other synchrotron-related techniques) and application of other more established methods (inelastic neutron scattering, Auger spectroscopy, photoelectron spectroscopy). Furthermore, scientific attention has been focused more on *processes* than on crystalline minerals, and the materials of interest have expanded to include glasses, silicate melts, gels, poorly-crystalline and amorphous phases, hydrothermal solutions and aqueous fluids. In addition, many of the important intereactions occur at surfaces or near surfaces, and consequently it is not just the properties and behavior of the bulk materials that are relevant.

This is an exciting time to be doing Earth Sciences, particularly as the expansion in spectroscopic techniques and applications is enabling us to look at geochemical and geophysical processes in a much more fundamental way than was previously possible. However, the plethora of techniques is very forbidding to the neophyte, whether a graduate student or an experienced scientist from another field. There are an enormous number of texts in the field of spectroscopy. However, very few have a slant towards geological materials, and virtually none stress the integrated multi-technique approach that is necessary for use in geochemical and geophysical problems. I hope that this volume will fill this gap and provide a general introduction to the use of spectroscopic techniques in Earth Sciences.

I thank all of the authors for trying to meet most of the deadlines associated with the production of this volume. It is my opinion that the primary function of this volume (and its associated Short Course) is *instructive*. With this in mind, I also thank each of the authors for the additional effort necessary to write a (relatively) brief but clear introduction to a very complex subject, and for good-humoredly accepting my requests to include more explanation *and* shorten their manuscripts.

We are all indebted to Paul Ribbe, series editor of *Reviews in Mineralogy*, and his assistants Marianne Stern and Margie Sentelle, for putting together this volume, despite our tardiness in supplying them with the necessary copy. I would like to thank Barbara Minich, without whom we would have had a volume but no Short Course. Lastly, I would like to thank the CNR Centro di Studio per la Cristallografia Strutturale and the Istituto di Mineralogia, both at the Università di Pavia. Italy, for help and support during the preparation of this volume.

Frank C. Hawthorne
Pavia, Italy
February 29, 1988

REVIEWS in MINERALOGY, Volume 18

FOREWORD

The authors of this volume presented a short course, entitled "Spectroscopic Methods in Mineralogy and Geology", May 13-15, 1988, in Hunt Valley, Maryland. The course was sandwiched between the first V.M. Goldschmidt Conference, organized by the Geochemical Society and held at Hunt Valley, and the spring meeting of the American Geophysical Union, held in Baltimore. This was the sixteenth short course organized by the Mineralogical Society of America, and this volume is the nineteenth book published in the *Reviews of Mineralogy* series [see list of available titles on the opposite page -- all are currently available at moderate cost from MSA].

The fourteen chapters of this volume were assembled from author-prepared copy. Mrs. Marianne Stern was responsible for most of the paste-up, and Mrs. Margie Sentelle assisted with typing and formatting.

Paul H. Ribbe
Series Editor
Blacksburg, VA

TABLE OF CONTENTS

Chapter 1   G. Calas and F. C. Hawthorne

INTRODUCTION TO SPECTROSCOPIC METHODS

Chapter 2   P. F. McMillan and A. C. Hess

SYMMETRY, GROUP THEORY AND QUANTUM MECHANICS

Chapter 3   **F. C. Hawthorne** and **G. A. Waychunas**

SPECTRUM–FITTING METHODS

Chapter 4   **P. F. McMillan** and **A. M. Hofmeister**

INFRARED AND RAMAN SPECTROSCOPY

Chapter 5   **Subrata Ghose**

INELASTIC NEUTRON SCATTERING

Chapter 6   George R. Rossman

VIBRATIONAL SPECTROSCOPY OF HYDROUS COMPONENTS

Chapter 7   George R. Rossman

OPTICAL SPECTROSCOPY

Chapter 8   Frank C. Hawthorne

MÖSSBAUER SPECTROSCOPY

Chapter 9   R. James Kirkpatrick

MAS NMR SPECTROSCOPY OF MINERALS AND GLASSES

Chapter 10   Jonathan F. Stebbins

NMR SPECTROSCOPY AND DYNAMIC PROCESSES IN MINERALOGY AND GEOCHEMISTRY

Chapter 11

G. E. Brown, Jr., G. Calas,
G. A. Waychunas and J. Petiau

X-RAY ABSORPTION SPECTROSCOPY AND ITS APPLICATIONS IN MINERALOGY AND GEOCHEMISTRY

Chapter 12   Georges Calas

– ELECTRON PARAMAGNETIC RESONANCE

Chapter 13   Michael F. Hochella, Jr.

AUGER ELECTRON AND X-RAY PHOTOELECTRON SPECTROSCOPIES

Chapter 14   Glenn A. Waychunas

LUMINESCENCE, X-RAY EMISSION AND NEW SPECTROSCOPIES

Chapter 1   G. Calas and F. C. Hawthorne
INTRODUCTION TO SPECTROSCOPIC METHODS

In the past 10 years, there has been an explosion in the number of new types of spectroscopy that have been developed. This has been accompanied by great increases in efficiency and convenience in the more well-established spectroscopic techniques. Accompanying this technical development is the growing realization that an adequate understanding of geological and geophysical processes requires proper characterization of the constituent materials. Mechanistic descriptions of such processes usually involve the interaction of atoms and molecules in the solid, liquid and gaseous states, and thus we need to know the spatial and energetic characteristics of these materials at the atomic scale. Traditionally, the principal fields of mineralogy and petrology have involved (mixtures of) crystalline minerals. However, the current thrust in these areas is now more concerned with processes, and the materials of interest have expanded to include not just crystalline minerals, but also amorphous minerals, glasses, melts and fluids. In addition, not only are the bulk properties and behaviour of these materials of interest. Many processes of geological importance occur at mineral surfaces, and hence the behaviour of surfaces and interfaces has become of increasing interest to us.

Our traditional methods of diffraction and chemical analysis remain of great importance, but many of the material characterization problems involved with current scientific thrusts are not amenable to such techniques. It is here that spectroscopy has become of such importance in characterizing both the static <u>and</u> dynamic properties of both crystalline <u>and</u> non-crystalline geological materials. However, the one term "spectroscopy" covers a large number of techniques, and the trick is to use the particular method or methods that are sensitive to the problem you wish to consider. One can view the different spectroscopies, together with other analytical and scattering techniques, as a series of tools that one uses to solve or examine a problem of interest; a single tool is generally not sufficient for ones needs - you cannot drive a nail <u>and</u> drill a hole just with a hammer. <u>One should regard these techniques as complementary</u>, and use them in combination to solve the specific problem at hand.

RADIATION

Spectroscopic techniques are concerned with the interaction between radiation and matter, using radiation appropriate to distance and time scales relevant to the microscopic study of atoms, molecules and solids. A wide variety of radiations can be used in this regard:

    (1) electromagnetic radiation
    (2) elementary particles (as electrons, protons or neutrons)
    (3) nuclei
    (4) ions
    (5) ultrasonic waves

Spectroscopic methods are dominated by the first two categories, and so we will consider these in more detail.

Electromagnetic radiation

According to the wave model, electromagnetic radiation consists of oscillating (i.e. time dependent) electric and magnetic fields. The frequency ν (or $\omega=2\pi\nu$) of the radiation is related to the wavelength λ by the relation

$$c = \nu\lambda = \frac{\omega}{2\pi}\lambda \qquad [1]$$

where c is the velocity of propagation in vacuum. The propagation of the radiation is characterized by a wave vector **k**, which has a modulus $k = 2\pi/\lambda$. The duality principle assigns to this radiation the usual parameters of particle dynamics:

$$\mathbf{p} = \hbar\mathbf{k} \qquad [2]$$
$$E = \hbar\omega \qquad [3]$$

both of which satisfy the conservation rules for an isolated system. The electromagnetic spectrum is shown in Figure 1; as indicated, most regions of the spectrum are of use for spectroscopic purposes.

Particle beams

With $m_e = 9.11 \times 10^{-31}$ kg, the electron is a light particle compared to the proton or neutron; it has an elementary electric charge with a spin of $\frac{1}{2}$. The energy of an electron beam is given by

$$E = \frac{\hbar^2}{2m_e}k^2 \qquad [4]$$

where k is the modulus of the wavevector. The energies available with current electron guns vary from some hundred of eV (Low Energy Electron Diffraction - LEED) to several hundreds of keV. For energies spanning 15-200 keV, the characteristics of the associated radiation are: $0.01 \leq \lambda \leq 0.1$ Å and $10^{14} \leq \nu \leq 10^{17}$ Hz.

Available neutron sources produce neutrons with energies spanning 10^{-4} - 1 eV and the following characteristics: $0.3 \leq \lambda \leq 20$ Å and $10^{10} \leq \nu \leq 10^{14}$ Hz. The difference between the energy of electron and neutron beams may be attributed to the much greater mass of the neutron.

UNITS

A number of different units are commonly used in spectroscopy, a circumstance that can be very confusing even to the expert. However, as the quantities of interest can vary by about 15 orders of magnitude (see Fig. 1), standardization to one particular set of units is not very convenient.

The most frequently used wavelength units are micrometers (μm), nanometers (nm), Angstroms (Å), and meters (m):

$$1.0 \ \mu m = 10^3 \ nm = 10^4 \ \text{Å} = 10^{-6} \ m$$

The fundamental frequency unit is the Hertz (Hz), one cycle per second. However, as is apparent from Figure 1, this unit is too small to be convenient, and even in the radio-frequency region, the GHz is used. Compared to the processes that give rise to (spectroscopic) transitions, the second is a long time (see Figure 3). A more appropriate time unit may be defined as the time required by light to travel 1 cm in vacuum (approx. 33.4 psec); this is the wavenumber, with units cm^{-1}. Some authors use the term kayser for

Figure 1.　　The electromagnetic spectrum.

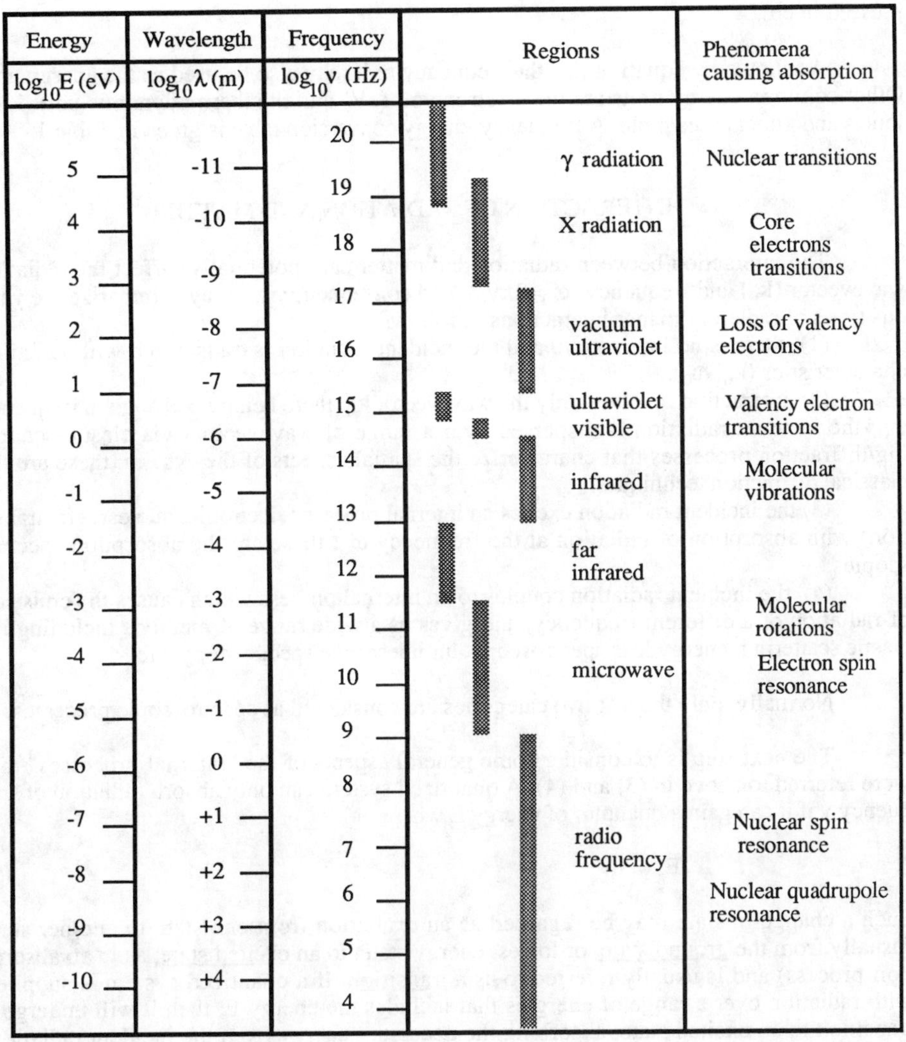

Energy $\log_{10}E$ (eV)	Wavelength $\log_{10}\lambda$ (m)	Frequency $\log_{10}\nu$ (Hz)	Regions	Phenomena causing absorption
5	-11	20	γ radiation	Nuclear transitions
4	-10	19	X radiation	Core electrons transitions
3	-9	18		
2	-8	17	vacuum ultraviolet	Loss of valency electrons
1	-7	16		
0	-6	15	ultraviolet visible	Valency electron transitions
-1	-5	14	infrared	Molecular vibrations
-2	-4	13	far infrared	
-3	-3	12		Molecular rotations
-4	-2	11	microwave	Electron spin resonance
-5	-1	10		
-6	0	9		
-7	+1	8	radio frequency	Nuclear spin resonance
-8	+2	7		
-9	+3	6		Nuclear quadrupole resonance
-10	+4	5		
		4		

Table 1. Conversion table for frequency and energy units used in spectroscopy

	1 Hz $\equiv$	1 cm^{-1} $\equiv$	1 J mol^{-1} $\equiv$	1 eV $\equiv$
Hz	1	2.9979×10^{10}	2.5053×10^{9}	2.4182×10^{14}
cm^{-1}	3.3356×10^{-11}	1	8.3567×10^{-2}	8.0663×10^{3}
J mol^{-1}	3.9915×10^{-10}	11.9660	1	9.6522×10^{4}
eV	4.1353×10^{-15}	1.2397×10^{-4}	1.0360×10^{-5}	1

the cm^{-1} unit, but this is not common. The wavenumber is a common unit used in place of frequency, and has the advantage that it is equal to the reciprocal of the wavelength expressed in cm.

As shown by equation (3), the frequency unit may also be used as the energy unit. Other common energy units are the electron volt (eV) and the thermodynamic units kilojoules and kilocalories/mole. A frequency-energy conversion table is given in Table 1.

INTERACTION OF RADIATION AND MATTER

The interaction between radiation and matter can (potentially) affect the radiation wavevector (k_0) and frequency ($\omega_0 = 2\pi\nu_0$), and consequently we may summarize the various types of radiation-matter interactions as follows:

(1) there is no interaction, and the incident radiation is transmitted with its initial characteristics (k_0, ω_0).

(2) interaction involves only the wavevector k_0, there being no change in frequency ω_0; the incident radiation is dispersed over a range of wavevectors via elastic scattering/diffraction processes that characterize the spatial aspects of the system; these are the classical diffraction techniques.

(3) the incident radiation excites an internal process (electronic, nuclear, etc. transition) with absorption of radiation at the frequency ω_0; these are the absorption spectroscopies.

(4) the incident radiation couples to an internal process which causes the emission of radiation of a different frequency, and gives us a wide range of methods including inelastic scattering, energy loss spectroscopy, luminescence spectroscopy, etc.

Normally, only the last two categories are considered as spectroscopic processes.

The next step is to consider some general aspects of the "internal processes" that were referred to above in (3) and (4). A quantized system can only absorb radiation of frequency ν if it can gain a quantum of energy E where

$$E = h\nu \qquad (5)$$

Such a change in state may be regarded as an excitation from one state to another state (usually from the ground state, or lowest energy state, to an excited state, as in an absorption process) and is usually referred to as a transition. If a quantized system is supplied with radiation over a range of energies that includes the energy E, then it will undergo a transition to an excited state, absorbing the necessary energy from the incident radiation. Consequently the absorption energy E can be measured and used as a probe of the system with regard to the factors that affect the magnitude of E. Alternatively, subsequent emitted radiation on decay from the excited state can also be measured and used in the same way, as a probe of the factors affecting the internal absorption/emission processes of the quantized system.

A BRIEF SURVEY OF SPECTROSCOPIC METHODS

Before we take a look at this, we have to come to grips with the plethora of acronyms that can baffle even the experienced spectroscopist. Table 2 translates the acronyms into normal speech. The different spectroscopic methods can be grouped together according to the physical processes on which they rely.

Atomic resonance spectroscopy

Electromagnetic radiation passes through matter without any modification of intensity, except where the frequency corresponds to that of a transition between ground and excited states of the system. Consequently, by varying the frequency of the incident radiation, a series of electronic transitions can be excited without any modification of the energy of this radiation. Depending on the energy of the incident radiation, the electronic transitions concern core levels or external incompletely-filled electronic shells (transition d- and f- elements). The former are related to energies which span the range from the near infrared through the visible to the ultraviolet and are therefore often associated with color; they may be studied by optical (electronic) absorption spectroscopy. The latter correspond to the x-ray region of the electromagnetic spectrum and can be investigated by x-ray absorption spectroscopy. The complementary methods use either emission, which corresponds to transitions to the ground state (see below under relaxation processes), or reflectance, which is a combination of both absorption and emission. Although the transitions involved are between nuclear rather than electronic states, Mössbauer spectroscopy can also be included here, as here is no modification of the energy of the incident beam in the absorption process.

Vibrational spectroscopy

These correspond to the interaction between electromagnetic waves and molecular states or vibration modes of a crystal. As these interactions are slower than for electronic transitions, the energies concerned are smaller, involving medium and far infrared radiations.

Infrared absorption occurs at one of the eigenfrequencies of the molecular groups if the vibrational mode modifies the length of the equivalent dipole. The interaction between radiation and vibrational processes can also occur by scattering processes, which obey selection rules. Two types of scattering occur, depending on the modification of the energy of the incident beam. Elastic scattering (Rayleigh scattering) emits in all directions a radiation which is in phase with the incident wave at all frequencies because of the forced vibrations of the electronic charges: Rayleigh scatter emerges with exactly the same energy as the incident light. Inelastic scattering is a consequence of the Doppler effect through the interaction with vibrating atoms at the frequency ω_0. Both directions of Doppler shift give rise to two scattered lines, called the Stokes ($\omega - \omega_0$) and anti-Stokes ($\omega + \omega_0$) lines, where ω_0 is an eigenfrequency of the system, and ω is the frequency of the incident radiation. The selection rules here are due to the cumulative effect of the polarization of the medium to observe the absorption. Inelastic scattering is usually much less intense than Rayleigh scattering. If the implied vibrations belong to acoustic phonons, it is Brillouin scattering; for optical phonons, it is Raman scattering. Both correspond to an exploration of the center of the Brillouin zone (large wavelengths).

Energy loss spectroscopy

Brillouin and Raman spectroscopies are typically of this type. Electron Energy Loss Spectroscopy (EELS) is used for studying core-electron transitions under the high resolution electron microscope. Neutron (and to lesser extent X-ray) inelastic scattering is used to explore phonons in a wide range of the Brillouin zone.

Electronic resonances followed by relaxation processes

As a photon is absorbed, a hole is created by the expulsion of an electron following the absorption; this subsequently relaxes by various processes to bring the system back to

Method	Particles used		Information obtained about:
	In	Out	
Photoelectron	Photon	Electron	Filled levels
Inverse photoelectron	Electron	Photon	Empty levels
X-ray emission	–	Photon	Filled levels
X-ray absorption	Photon	–	Empty levels
Visible/UV absorption	Photon	–	Band gap; defects
Electron energy loss	Electron	Electron	Conduction electrons

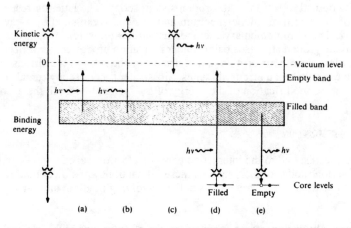

Figure 2. Electron spectroscopy techniques: (a) optical absorption in the visible/UV range; (b) photoelectron spectroscopy; (c) inverse photoelectron spectroscopy; (d) X-ray absorption; (e) X-ray emission. Electrons with energies above the vacuum level can enter or leave the solid; in techniques (b) and (c) the scale shows the kinetic energy measured in a vacuum outside the solid. Modified from Cox (1987).

its ground state, with the emission of a photon or an electron. Detection of these emissions is an alternative way to study electronic transitions, and is widely used in analytical techniques because of its sensitivity.

The emission of photons after absorption by valence electrons gives rise to emission in the optical range; involvement of core electrons gives rise to X-ray emission. If the emission is strictly restricted to the excitation period, it is called fluorescence. If the emission exceeds the excitation period, it is called phosphorescence.

Another type of relaxation (after absorption by core electrons) can be observed through the expulsion of an electron from the outer shells (thus of lower energy); this is the Auger effect, which is thus the equivalent of X-Ray emission. It is also possible to directly analyze the energy of the photoelectrons emitted by the absorption process. This energy is the difference between that of the incident X-Ray and the initial binding energy of the ejected electron. This spectroscopy is called X-ray Photoelectron spectroscopy (XPS, also called ESCA, Electron Spectroscopy for Chemical Analysis).

XPS ionizes electrons from solids, and hence shows the energy of occupied orbitals. This experiment may be reversed: the solid is exposed to an electron beam of known energy, and emits a photon. The electrons undergo transitions to empty states in the conduction band, emitting a photon and giving a picture of the conduction band energies. This is Inverse PhotoElectron Spectroscopy (IPES), also called Bremsstrahlung spectroscopy. The various electron spectroscopies are diagrammatically represented in Figure 2.

Table 2. Spectroscopy acronyms revealed!

AES	Auger Electron Spectroscopy
ARPES	Angle-Resolved PhotoElectron Spectroscopy
ATR	Attenuated Total Reflectance spectroscopy
BIS	Bremsstrahlung Isochromat Spectroscopy
CARS	Coherent Anti-Stokes Scattering
CEMS	Conversion Electron Mössbauer Spectroscopy
CHEXE	CHanneling-Enhanced X-ray Emission spectroscopy
CL	CathodoLuminescence spectroscopy
DCEMS	Depth-resolved Conversion Electron Mössbauer Spectroscopy
DRS	Diffuse Reflectance Spectroscopy
EAPFS	Extended Appearance Potential Fine Structure
EAS	Electronic Absorption Spectroscopy (OAS)
EELFS	Extended Energy Loss Fine Structure
EELS	Electron Energy Loss Spectroscopy
ENDOR	Electron-Nuclear DOuble Resonance spectroscopy
EPR	Electron Paramagnetic Resonance spectroscopy
ESCA	Electron Spectroscopy for Chemical Analysis (XPS, UPS)
ESR	Electron Spin Resonance spectroscopy
EXAFS	Extended X-ray Absorption Fine Structure spectroscopy
EXBIFS	Extended X-ray Bremsstrahlung Isochromat Fine Structure
EXELFS	EXtended Electron (energy) Loss Fine Structure
EXFAS	EXtended Fine Structure in Auger Spectra
FLN	Fluorescence Laser Narrowing spectroscopy
FMR	FerroMagnetic Resonance
FTIR	Fourier Transform InfraRed spectroscopy
HREELS	High-Resolution Electron Energy Loss Spectroscopy
INS	Ion Neutralization Spectroscopy
IPES	Inverse PhotoElectron Spectroscopy
IR	InfraRed spectroscopy
IRS	Internal Reflectance Spectroscopy
ISS	Ion Scattering Spectroscopy
LRS	Laser Raman Spectroscopy
MAS NMR	Magic Angle Spinning Nuclear Magnetic Resonance spectroscopy
NMR	Nuclear Magnetic Resonance spectroscopy
NQR	Nuclear Quadrupole Resonance spectroscopy
OAS	Optical Absorption Spectroscopy (EAS)
PAS	PhotoAcoustic Spectroscopy
PEOL	Proton Excited Optical Luminescence spectroscopy
PES	PhotoElectron Spectroscopy
PIXE	Proton Induced X-ray Emission spectroscopy
RefleEXAFS	Reflection Extended X-ray Absorption Fine Structure spectroscopy
SEELFS	Surface Extended Energy Loss Fine Structure
SEXAFS	Surface Extended X-ray Absorption Fine Structure spectroscopy
SIMS	Secondary Ion Mass Spectroscopy (spectrometry)
SRS	Specular Reflectance Spectroscopy
SXAPS	Soft X-ray Appearance Potential Spectroscopy
TL	ThermoLuminescence (TSL) spectroscopy
TSL	Thermally Stimulated Luminescence (TL) spectroscopy
UMER	Ultrasonically Modulated Electron Resonance spectroscopy
UPS	Ultraviolet Photoelectron Spectroscopy
XANES	X-ray Absorption Near Edge Structure spectroscopy
XAS	X-ray Absorption Spectroscopy
XEOL	X-ray Excited Optical Luminescence spectroscopy
XES	X-ray Emission Spectroscopy
XPS	X-ray Photoelectron Spectroscopy
XRF	X-Ray Fluorescence spectroscopy (spectrometry)

Finally, EXAFS (Extended X-ray Absorption Fine Structure) and XANES (X-ray Absorption Near Edge Structure) can also be included in this type of spectroscopy. The emitted photoelectron is partially backscattered by the neighbouring atoms before the relaxation occurs; interference effects thus modulate the absorption coefficient. They can be described as local electronic scattering: EXAFS corresponds to single scattering, whereas XANES can be described using a multiple scattering model. Both methods probe only the local environment because of the low energy of these photoelectrons (some tens of eV for XANES and up to several hundreds of eV for EXAFS). Unlike LEED, it is chemically selective, as it specifically concerns the absorbing atom.

Magnetic resonance

According to the type of spin involved, magnetic resonance can be nuclear or electronic. Both methods can be described simply as the precession of the spin around a magnetic field $\mathbf{B_0}$, where this precession frequency corresponds to that of the oscillating electromagnetic radiation applied to the sample: the radiation is then absorbed, and this is easily detected. The same fundamental relation holds for both types of resonances:

$$\omega_0 = \gamma B_0 = g\frac{e}{2M} B_0 \qquad [6]$$

where e is the elementary charge,
g is the gyromagnetic factor (of the order of unity)
M is the mass of the electron or nucleus, depending on the type of spin involved.

For a conventional magnetic field $\mathbf{B_0}$ (corresponding to 10^2 - 10^4 G), the radiation frequency is in the range of MHz for Nuclear Magnetic Resonance (NMR) and of GHz for Electron Paramagnetic Resonance (EPR, also called ESR, Electron Spin Resonance); this frequency difference results from the 3 orders of magnitude difference between the masses of the electron and the nucleus. The absorption is caused by a change in the energy of the precessing spins, which can be classically described by a modification of the angle between $\mathbf{B_0}$ and the spin direction.

The evolution of structure can be studied, as the method integrates static and dynamic fluctuations of the magnetic field (in time and space). For EPR and particularly for NMR, the relevant energy levels are quite closely spaced. This results in slow relaxation of the excited spins towards the ground state imposed by the thermodynamical equilibrium of the system. NMR is then a spectroscopy of low frequencies and slow processes.

TIMESCALES

Although spectroscopic methods study energy transfer between various states of the system and give direct access to its evolution in time, it is possible to get static information (structural studies, chemical analysis) as well as data on the dynamics of the system. This arises from the relative magnitude of the measurement time and the duration of the observed process. If the interaction time between radiation and matter is short, the observer has an instantaneous image of the system. In the case of a long interaction time, it is only possible to get an average picture. This interaction time obeys the Heisenberg uncertainty principle, and it is given qualitatively by the linewidth ($\tau \sim \hbar/\Delta E$) or the radiation energy ($\tau \sim \hbar/E$). In the case of magnetic resonance methods, the measurement time depends on the type of relaxation involved; it is often determined by the attainment of the thermodynamical equilibrium of spin states (spin-lattice relaxation time).

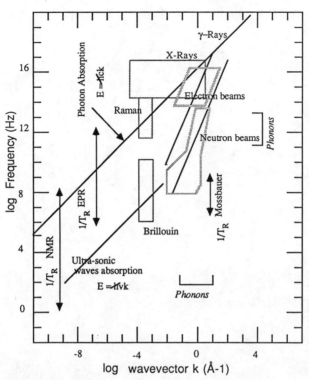

Figure 3.   Frequencies and wavevector moduli in the various regions of the electromagnetic spectrum. The lines indicate the dependence of radiation frequency (energy) on the modulus of wavevector k in the case of absorption processes concerning photons, electron and neutron beams and ultrasonic waves. The domains delimited by dotted lines indicate inelastic scattering processes. Depending on the spectroscopic method used, the observation timescales are qualitatively related to the inverse of radiation frequency or to the relaxation time (T_R) indicated on the drawing. Modified from Janot and George (1986).

The dynamical aspects of a system can be studied if the correct time scale is used to observe the process: infrared spectroscopy, Raman scattering or inelastic neutron scattering for lattice vibrations; NMR, Mössbauer effect or quasi-elastic neutron scattering for atomic transport processes; electronic absorption spectroscopies (optical, X-Rays) for valence fluctuation, etc. The relaxation times and frequencies determining time scales of the various techniques are given in Figure 3.

REFERENCES

Cox, P.E. (1987) The electronic structure and chemistry of solids. Oxford Science Publications, Oxford.

Janot, C. and George, B. (1986) Présentation générale des différentes spectroscopies. In: Méthodes spectroscopiques appliquées aux minéraux. G. Calas, ed., Société française de Minéralogie-Cristallographie, Paris, p. 3-20.

Chapter 2

P. F. McMillan and A. C. Hess

SYMMETRY, GROUP THEORY AND QUANTUM MECHANICS

QUANTUM MECHANICS

Introduction

By the end of the nineteenth century, the concepts of classical mechanics were firmly established, and had been successfully applied to a wide range of problems in physics and chemistry, ranging from the motion of planetary bodies to thermodynamics and the kinetic theory of gases. Despite the obvious successes of the theory, there remained a number of observations which could not be explained, such as the line spectra of atoms and molecules in several regions of the spectrum. The classical theory suggested a continuous radiation or absorption of energy from the electromagnetic field, which was not observed. Further inadequacies with the classical theory arose with the heat capacities of molecules, with the observation of the photoelectric effect, and with the observed intensity distribution of the radiation emitted from a heated body. It was this last problem that led to Planck's (1901) postulation of the original quantum theory, which in its various refinements, has revolutionized the fields of physics and chemistry, and led to our current understanding of atomic and molecular spectroscopy.

The problem posed was that of understanding the intensity distribution of different wavelengths of light emitted by a black body, such as a heated iron bar. It was known that as the temperature was increased, the intensity maximum would shift to shorter wavelengths, and the bar would glow first red hot (at around $600^{o}C$), then white hot (around $1200^{o}C$), before passing into the ultraviolet regions of the spectrum. The classical theory demanded that the internal energy of the "oscillators" in the bar be partitioned equally among all possible wavelengths, which would result in emission of light as ultraviolet, X- and gamma-rays for any arbitrary temperature. This "ultraviolet catastrophe" obviously did not occur, hence the classical concept of equipartition of energy could not be applied to this problem. Planck's solution was to suggest that the oscillators in the bar could only change their energy in discrete increments proportional to their oscillation frequency ν

$$\Delta E = h\nu ,\qquad\qquad (1)$$

where the proportionality constant h became known as <u>Planck's</u> <u>constant</u>, and has a magnitude of $h = 6.621 \times 10^{-34}$ Js. This quantization condition allowed Planck to correctly calculate the statistical distribution of thermal energy among the oscillators at a given temperature T, and successfully reproduced the intensity distribution of the emitted light.

Until this time, light had been regarded almost exclusively as a continuous waveform, described as oscillations in the electromagnetic field. This formed the basis for both the classical theory of optics and Maxwell's synthesis of the laws of electrodynamics. Now light was also to be regarded as having a particulate nature, with each quantum "jump" in energy of an oscillator in the blackbody giving rise to emission of a light "particle" with energy ΔE. These light particles later became known as "<u>photons</u>". Planck himself was unhappy with the consequences of this suggestion, and considered that the light "particles" only existed in the immediate vicinity of the emitting oscillators, but that light should be continuous everywhere else in space. A few years later, Einstein (1905) examined these results, and concluded that a dual wave-particle description was necessary for light at all points in space, and used the quantum theory to rationalize the photoelectric effect observed by Franck, Hertz and others.

By this time, both chemists and physicists had formed some ideas as to the internal structure of the atom. The α-particle scattering experiments of Geiger and Marsden and their analysis by Rutherford (1911) gave the presently accepted model of the nuclear atom, with negatively charged electrons in "orbit" around a positively charged nucleus. Both atoms and molecules were known to have characteristic lines in their absorption and emission spectra, but no model was available for these. Bohr (1913) applied the quantum theory to the problem of the spectral lines of the hydrogen atom, with spectacular success. Bohr took Rutherford's nuclear atom with Coulombic attraction between the electron and proton, and derived the classical equation of motion for the electron in its orbit. At this point, he assumed that the angular momentum values of the electronic orbits were quantized, and calculated a set of electronic energy levels E_n, where the integer n=1,2,3,4... corresponded to quantized circular orbits of successively larger radius. The observed spectral lines were then due to emission or absorption of light as the electron passed from one orbit to another. Bohr's model gave remarkable agreement with the experimental spectrum of the hydrogen atom, lending further support to the budding quantum theory. Later improvements on Bohr's theory by Sommerfeld gave rise to additional quantum numbers l and m_l, which referred to the eccentricity of non-circular orbits, while Uhlenbeck and Goudsmidt suggested a quantum number m_s for the electron spin, in order to explain small splittings in some spectral lines.

At this point, there was growing dissatisfaction with Bohr's theory, mainly due to its inability to treat the spectra of complex atoms, and the arbitrary introduction of the quantization condition on the electronic orbits. A breakthrough was provided by de Broglie (1924) who suggested that, since light could have both wave-like and particle-like descriptions, so could electrons. Ascribing a wave-like motion to the electrons in orbit around the atomic nucleus gave an automatic justification for Bohr's quantization condition, since stable orbits must be associated with standing waves where the orbit circumference must be an integral multiple of the wavelength. Schrodinger (1926a) developed this wave theory into the new wave mechanics, which supplanted Bohr's earlier quantum theory for atomic and molecular structure.

The terminology and structure of Schrodinger's wave equation can be easily developed from the general wave equation. In general, a wave is any periodic disturbance in space-time, and we can choose to describe the electronic motion (or any other atomic or molecular phenomenon such as rotation or vibration) as a wave. The amount of disturbance (u) is a function both of time t and the space coordinates. In one dimension (x), the general wave equation is

$$\frac{\delta^2 u}{\delta x^2} = \frac{1}{v^2} \frac{\delta^2 u}{\delta t^2} \, , \qquad (2)$$

where v is the velocity of propagation of the wave ($v = \delta u / \delta t$). If we assume that x and t are independent variables (i.e., carry out a non-relativistic treatment), we can separate this equation:

$$u(x,t) = X(x) \ e^{2\pi i \nu t} \quad , \tag{3}$$

where ν is the frequency of the wave motion and $i = \sqrt{(-1)}$. $X(x)$ is a function of the space coordinate alone. Substituting this separated equation into (2) gives

$$\frac{d^2X}{dx^2} + \frac{4\pi^2\nu^2}{v^2} = 0 \quad . \tag{4}$$

The time information for the wave motion is now contained in the frequency (ν) and the velocity (v), which are constant and independent of the space coordinate.

Now we use de Broglie's wave-particle duality for the electron. The total energy E is given by the sum of kinetic (T) and potential (V) energies:

$$E = T + V \tag{5}$$

with

$$T = \frac{1}{2} mv^2 = \frac{p^2}{2m} \tag{6}$$

(m, v and p are the mass, velocity, and momentum (p=mv) of the electron). This expression is known as the Hamiltonian form of the kinetic energy). From (5) and (6), we obtain

$$p = \sqrt{2m(E-V)} \quad . \tag{7}$$

By further utilizing de Broglie's relation between the wavelength and the momentum

$$\lambda = h/p \tag{8}$$

(h is Planck's constant), we find that the wavelength may be expressed in the following manner

$$\lambda = \frac{h}{\sqrt{(2m(E-V))}} \quad . \tag{9}$$

Since the frequency of a wave is related to its velocity by

$$\nu = v/\lambda \quad , \tag{10}$$

the frequency may be expressed as

$$\nu^2 = \frac{2mv^2(E-V)}{h^2} \quad . \tag{11}$$

Upon substituting this result into (4), obtain

$$\frac{d^2X}{dx^2} + \frac{8\pi^2m}{h^2} (E - V) \ X = 0 \quad . \tag{12a}$$

More commonly, the time-independent wave function $X(x)$ is given the symbol ψ:

$$\frac{d^2\psi}{dx^2} + \frac{8\pi^2m}{h^2} (E - V) \ \psi = 0 \quad . \tag{12b}$$

The problem is easily generalized to three dimensions:

$$\left[\frac{\delta^2 \psi}{\delta x^2} + \frac{\delta^2 \psi}{\delta y^2} + \frac{\delta^2 \psi}{\delta z^2}\right] + \frac{8\pi^2 m}{h^2}(E - V)\psi = 0 \quad . \tag{13}$$

The partial differentials ($\delta^2/\delta x^2 + \delta^2/\delta y^2 + \delta^2/\delta z^2$) which <u>operate</u> on the wave function ψ are known as the <u>Laplacian operator</u>, and are given the symbol ∇^2 (del squared):

$$\nabla^2 \psi + \frac{8\pi^2 m}{h^2}(E - V)\psi = 0 \quad . \tag{14}$$

This wave equation can be re-written

$$\left[-\frac{h^2}{8\pi^2 m}\nabla^2 + V\right]\psi = E\psi \quad . \tag{15a}$$

The group of terms on the left side of this equation operating on the wave function ψ are known as the <u>Hamiltonian operator</u> H, and the <u>time independent Schrodinger wave equation</u> is usually written symbolically as

$$\hat{H}\psi = E\psi \quad . \tag{15b}$$

In order to apply Schrodinger's wave mechanics to a physical problem, it is necessary to express the potential energy V in terms of the coordinates q_i of the system, i.e., to have an analytical form for $V(q_i)$. For many problems, such an analytic potential function is not known, and a model approximate potential function is often used. When this potential function is substituted into the Schrodinger equation (15), a second order differential equation is obtained which relates the wave function ψ to the coordinates q_i. This differential equation is then solved for ψ, and for the energy E.

In the special case where the problem is not bound (i.e., the potential energy is constant over all space, and there are no restrictions on the coordinates), the solution for the wave function ψ is simply a travelling wave, which may have any energy E defined only by the initial energy of the system. In all other cases where the system is bound (i.e., the motion is restricted to some regions of space with some space-dependent potential function), ψ and E are found to have only discrete solutions (analogous to the standing waves found for the vibrations of a string fixed at both ends). These discrete <u>quantized states</u> are known as <u>eigenfunction</u> solutions ψ_n with associated energy <u>eigenvalues</u> E_n, and the integers n are known as the <u>quantum numbers</u> for the system.

In general, once a suitable potential function $V(q_i)$ has been chosen, the major obstacle to applying quantum mechanics to a physical problem lies in the mathematical solutions of the wave equation (15). This accounts for the often formidable mathematical expressions which appear in most treatments of quantum mechanics. However, there has been considerable research over the past few hundred years into the properties of different types of differential equations, and the solutions to at least some of the differential equations generated by the application of quantum mechanics to physical problems are known. As examples, we can consider the solutions to three "standard" problems in quantum mechanics: (a) the problem of a particle in a three-dimensional box, which forms a basis for discussion of electron motion in metals and other important physical problems; (b) the hydrogen atom, which serves as a starting point for discussion of the electronic structure of more complex atoms and molecules; and (c) the simple harmonic oscillator, which provides a model for the vibrations of molecular systems. The solutions to these and similar problems are discussed in detail in introductory quantum mechanics texts, such as Pauling and Wilson (1963), or Levine (1983).

The particle in a box

Consider a particle of mass m (which could be an electron, mass 9.11×10^{-31} kg) confined in a three-dimensional rectangular box with sides of length a, b and c. We wish to describe the motion of the particle by its wave function $\psi(x,y,z)$. Since the particle is constrained to be within the box, we have the conditions that $\psi(x,y,z) = 0$ for $x,y,z < 0$ and for $x > a$, $y > b$, $z > c$. These conditions are met if the potential energy $V(x,y,z)$ is infinite everywhere outside the box. For the simplest case, we assume that the potential $V(x,y,z)$ is constant everywhere inside the box. The Schrodinger equation for the particle inside the box can then be expressed as

$$\frac{\delta^2 \psi}{\delta x^2} + \frac{\delta^2 \psi}{\delta y^2} + \frac{\delta^2 \psi}{\delta z^2} + \frac{8\pi^2 m}{h^2}(E - V)\psi = 0 \quad . \tag{16}$$

This equation can be separated by the substitution

$$\psi(x,y,z) = X(x)Y(y)Z(z)$$

into three independent differential equations in x, y and z. The equation involving x is

$$\frac{d^2 X}{dx^2} + \frac{8\pi^2 m}{h^2}(E_x - V_x)X = 0 \quad . \tag{17}$$

The equations involving y and z are analogous. Using the boundary conditions for the box, the solution to this differential equation is

$$X_{n_x}(x) = \sqrt{\frac{2}{a}} \sin \frac{n_x \pi x}{a} \quad , \tag{18}$$

where n_x can take integer values $1, 2, 3, \ldots$. This is the quantum number for the X(x) equation. The associated energies are given by

$$E_{n_x} = \frac{(n_x h)^2}{8ma^2} \quad . \tag{19}$$

These constitute the energy levels for the system. The quantized energy levels and wave functions for this system are shown schematically in Figure 1. Since similar solutions are found for the Y(y) and Z(z) equations, the total wave function is

$$\psi_{n_x n_y n_z}(x,y,z) = \sqrt{\frac{8}{abc}} \sin \frac{n_x \pi x}{a} \sin \frac{n_y \pi y}{b} \sin \frac{n_z \pi z}{c} \tag{20}$$

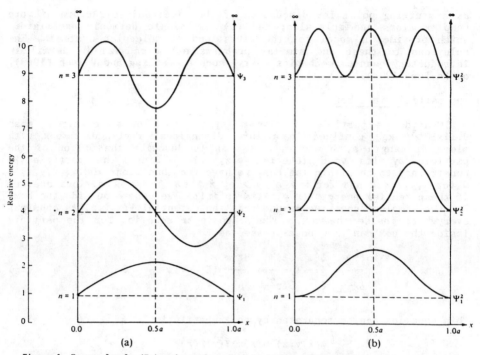

Figure 1. Energy levels (E_n) and wavefunctions ($\psi_n(x)$) for the first three quantum numbers ($n = 1, 2, 3$) for a particle in a one-dimensional box ($0<x<a$). Also shown at right are the squares of the wavefunctions (ψ_n^2), which can be interpreted as the probability of finding the particle between x and x+dx (from Huheey, 1983, p. 15).

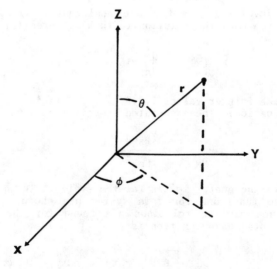

Figure 2. Relationship between Cartesian (x,y,z) and spherical polar coordinates (r, θ, ϕ).

and the total energy is

$$E = E_{n_x} + E_{n_y} + E_{n_z} = \frac{h^2}{8m} \left[\frac{n_x^2}{a^2} + \frac{n_y^2}{b^2} + \frac{n_z^2}{c^2} \right] \ . \tag{21}$$

This simple model is used as a starting point for understanding the electronic structure of metals. In the electron gas model, the conduction electrons are assumed to move independently in the average potential field provided by the atomic cores. The solutions generally appear slightly different to (20) and (21) because a box of infinite dimensions is usually assumed, and the wave-like solutions for the electronic wave function are running waves, not standing waves. These plane waves are characterized by their wave vector k, a vector in the direction of propagation of the electron waves, with magnitude $2\pi/\lambda$, where λ is the wavelength. There is a dependence of the electronic energy on k, and the electronic energy levels are grouped into bands. The independent electron gas model can be refined by recognizing that the electronic motions are correlated through electron-electron repulsion. In further refinements, interactions between the electrons and the atomic cores can be modelled by forcing the wave function to vary strongly in the region of the cores. This is achieved by augmenting the electronic plane wave functions (APW method) with functions which oscillate strongly in the neighbourhood of the cores. Many other methods have been devised for calculating the electronic band structures of metals and other crystals. Introductions to the application of quantum mechanics to the electronic structures of solids can be found in solid state physics texts, such as those by Ashcroft and Mermin (1976), Harrison (1979), and Seitz (1987).

The hydrogen atom

The H atom problem consists of an electron with charge $-e$ in motion relative to a proton of charge $+e$ and mass $M = 1.673 \times 10^{-27}$ kg. The reduced mass of the system is $\mu = mM/(m+M) = 9.105 \times 10^{-31}$ kg. The potential energy function is a Coulombic attraction between the electron and the nucleus

$$V = - e^2/r \ , \tag{22}$$

where e is the electronic charge in c.g.s. units (e = 4.803 esu), and r is the distance between the electron and the nucleus.

The Schrodinger wave equation (15) for the hydrogen atom is then

$$-\nabla^2 \psi + \frac{8\pi^2\mu}{h^2} (E + \frac{e^2}{r}) = 0 \ . \tag{23}$$

Since the Coulombic potential (22) is a simple function of r, it is more convenient to discuss the electronic motion in spherical polar coordinates instead of Cartesian coordinates (Fig. 2):

$$\begin{aligned} x &= r \sin \theta \cos \phi \\ y &= r \sin \theta \sin \phi \\ z &= r \cos \theta \end{aligned} \tag{24}$$

and the wave function ψ becomes a function of r, θ and ϕ. Since the potential energy is only a function of r and not of θ or ϕ (23), this wave function may be separated into component functions:

$$\psi(r,\theta,\phi) = R(r)Y(\theta,\phi) \quad . \tag{25}$$

R(r) is known as the <u>radial wave function</u>, and $Y(\theta,\phi)$ is known as the <u>angular part</u> of the wave function. The Schrodinger wave equation can then be separated into radial and angular parts:

Radial:
$$\frac{1}{r^2} \frac{d}{dr}(r^2 \frac{dR}{dr}) + \frac{8\pi^2\mu}{h^2} (E + \frac{e^2}{r} - \frac{X}{r^2}) = 0 \tag{26a}$$

Angular:
$$\frac{1}{\sin\theta} \frac{\delta}{\delta\theta} (\sin\theta \frac{\delta Y}{\delta\theta}) + \frac{1}{\sin^2\theta} \frac{\delta^2 Y}{\delta\phi^2} + XY = 0 \quad , \tag{26b}$$

where X is an arbitrary constant introduced by the mathematical technique of separation of variables. Note that this constant X will now appear in the solutions to the separated differential equations. Due to the mathematical manipulations which follow, X is usually written in the form $l(l+1)$, where l is another arbitrary constant at this stage; however, l becomes a quantum number when the differential equations are solved. Note also that since the energy E appears only in the radial equation, the quantized electronic energy levels (E_n) arise from solution of the radial wave equation, but are independent of the solutions for the angular equation. (This is not true for more complex atoms or molecules, or when an electric or magnetic field is applied to the system).

The solutions to differential equations of the types found in the radial and angular equations were known many decades before the development of Schrodinger's wave mechanics. The radial wave equation is usually solved by a method involving polynomials (a standard way of solving many types of differential equation). The appropriate polynomials in this case are known as the <u>Laguerre polynomials</u>. The Laguerre polynomial of <u>degree</u> y in some variable x is written as

$$L_y(x) = e^x \frac{d^y(x^y e^{-x})}{dx^y} \quad . \tag{27}$$

If this expression is then differentiated z times with respect to x, we obtain the <u>associated Laguerre polynomial</u> of <u>order</u> z and <u>degree</u> y-z:

$$L_{y-z}^{z}(x) = \frac{d^z}{dx^z} L_y(x) \quad . \tag{28}$$

Solution of the radial wave equation (26a) by the polynomial method gives solutions for R in terms of associated Laguerre polynomials of order 2l+1 and degree n-l-1,

$$R_{nl}(\rho) = N_{nl} \rho^l L_{n-l-1}^{2l+1}(\rho) e^{-\rho/2} \tag{29}$$

where the reduced variable ρ is related to r by

$$\rho = 2\alpha r \quad : \quad \alpha^2 = - \frac{8\pi^2\mu E}{h^2} \quad . \tag{30}$$

N_{nl} is a normalization coefficient, and the quantum numbers n and l can take the values

$$n = 1,2,3,4,\ldots\ldots\infty$$
$$l = 0,1,2,\ldots, n - 1 \quad.$$

The associated solutions for the energies are given by

$$E_n = -\frac{2\pi^2\mu e^4}{n^2 h^2} \quad \text{(in Joules)}. \tag{31}$$

These are the electronic energy levels for the hydrogen atom. The solution of the hydrogen atom problem is discussed in detail in many physical chemistry texts, such as Moore (1972), Berry et al. (1980), or Atkins (1982), or introductory treatments of quantum mechanics , such as Pauling and Wilson (1963) or Levine (1983).

The quantum numbers n and l are known respectively as the principal and azimuthal quantum numbers, and are usually given physical interpretations by analogy with the older quantum mechanical model of Bohr: n defines the average radius of the electron orbit, hence its distance from the nucleus, while l determines the angular momentum of the electron orbit for a given value of n. Different values of l are given symbols derived from spectroscopic terminology:

$$l = 0 : s$$
$$l = 1 : p$$
$$l = 2 : d$$
$$l = 3 : f \ldots$$

An electron with n=1 and l=0 is said to be in a 1s orbital, while an electron with n=3 and l=2 is in a 3d orbital.

Plots of the radial wave equation R(r) for various orbitals are shown in Figure 3, along with the radial distribution functions D(r) = $4\pi R^2 r^2$. This function can be interpreted as the probability of finding the electron in the space between r and r+δr, and provides a useful way of depicting the radial wave function.

The energy E depends only on the quantum number n (Eq. 31), and orbitals with a given value of n but different values of l are degenerate (i.e., have the same energy), for example, the 3p orbitals have the same energy as the 3s. This is only true for the H atom with no perturbing field. If an external electric or magnetic field is applied to the H atom, the degeneracies are lifted (Stark and Zeeman effects). In more complex atoms, the additional electrons provide internal electric and magnetic fields which spontaneously lift the orbital degeneracies. A set of orbitals with a given principal quantum number n is termed a shell:

$$n = 1 : \quad K \text{ shell}$$
$$n = 2 : \quad L \text{ shell}$$
$$n = 3 : \quad M \text{ shell} \ldots$$

Solutions for the angular wave equation Y(θ,ϕ) (26b) are also known, and form a set of functions called the surface spherical harmonics. These functions give the shapes of the s, p, d etc. orbitals, and depend on two quantum numbers, the azimuthal quantum number l and the "magnetic" quantum number m_l.

20

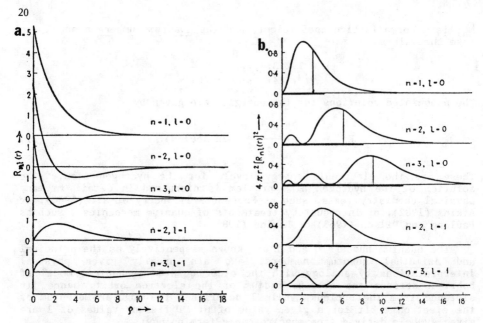

Figure 3. (a) Radial wave functions R(r) for several values of the principal (n) and azimuthal (l) quantum number, obtained from solution of the radial wave equation for the hydrogen atom. (b) Radial distribution functions $D(r) = 4\pi R^2(r)$ for the hydrogen atom (from Pauling and Wilson, 1963, p. 142-143).

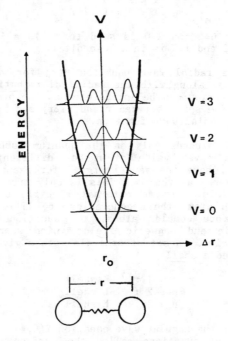

Figure 4. The harmonic oscillator model for molecular vibration. The heavy solid parabola indicates the potential energy function within which the vibration takes place. The light horizontal lines indicate the vibrational energy levels E_v obtained by solution of the vibrational wave equation. The square of the corresponding vibrational wave function ψ_v^2 is shown above each energy level.

It has not yet been possible to obtain exact solutions to the Schrodinger wave equation for more complex atoms or molecules, due to the mathematical difficulties in treating the instantaneous electron-electron repulsions in many-electron systems. However, useful results have been obtained by using the results for the hydrogen atom as a first approximation. For more complex atoms, the nuclear charge has a value Z+. Electrons with low quantum numbers lie in shells close to the nucleus, and these inner electrons shield the outer electrons from the full nuclear potential. Within a given electronic shell, orbitals with different values of l have different radial distribution functions (Fig. 3), hence penetrate the inner shells to different degrees. Because of increased penetration, an s orbital "feels" more of the nuclear charge than a p orbital of the same shell, hence the s orbital is lowered in energy relative to the p orbital. The combined effects of shielding and penetration serve to remove the degeneracy of the s, p, d etc. orbitals within a given shell. The resulting energy level patterns with increasing atomic number can be used to rationalize differences and similarities between different atoms, and provide a quantum mechanical basis for the structure of the Periodic Table.

For more quantitative treatments of the electronic structure of complex atoms, a wide range of mathematical techniques have been developed. In these, a first approximation to the electronic wave function is constructed from linear combinations of hydrogen-like wave functions, known as a basis set. Variable parameters are then adjusted to give the lowest possible energy solution, which corresponds to the best approximation to the true wave function with the chosen basis (the variational principle). In the Hartree-Fock self-consistent field (SCF) method, the many-electron problem is modelled by considering one electron at a time, and considering the motion of that electron in the average field of the nucleus and all the other electrons (Szabo and Ostlund, 1982). Using the variation principle, this gives a better approximation to the wave function for that one electron. This is carried out for all electrons in turn, then the entire process repeated until the average electronic potential function shows no further change, i.e., the field is self-consistent. This process takes account of electron-electron interactions in an average way, but ignores instantaneous electron-electron repulsive terms, or dynamic electron correlation. This can be modelled either by taking the Hartree-Fock solution as a first approximation and applying the methods of perturbation theory (see below) to obtain corrections to the energy, or to allow excited state electronic configurations to "mix" with the ground state basis set. This method is known as configuration interaction (CI).

The electronic wave functions and energies of molecules are conveniently calculated using the method of linear combination of atomic orbitals (LCAO). Appropriate atomic basis sets are constructed for the constituent atoms from hydrogen-like wave functions. Approximate molecular wave functions are then obtained by taking linear combinations of these atomic basis sets. The coefficients in these linear combinations are adjusted to give the lowest molecular energy via the variation principle. This process is then carried out for several possible internuclear distances. That which gives the lowest electronic energy is the equilibrium configuration of the molecule. These results can then be refined by carrying out a perturbation or CI treatment. This type of approach can be applied to crystalline systems, where it is known as a tight-binding method (Ashcroft and Mermin, 1976; Harrison, 1979).

The simple harmonic oscillator

This problem usually forms the starting point for a discussion of molecular vibrations. In the model, two masses m_1 and m_2 are connected by an ideal spring (Fig. 4). At rest, the masses are separated by a distance r_0. On extension or compression of the spring (Δr), the masses experience a restoring force proportional to the displacement

$$F = -k\Delta r \qquad , \qquad (32)$$

where the proportionality constant k is the force constant of the spring. This expression leads directly to the potential energy function for the system

$$V = \frac{1}{2} k\Delta r^2 \qquad (33)$$

which is substituted into the Hamiltonian operator. The Schrodinger equation (15) becomes

$$\frac{d^2\psi}{d\Delta r^2} + \frac{8\pi^2\mu}{h} (E - \frac{1}{2} k\Delta r2) \psi = 0 \qquad , \qquad (34)$$

where μ is the reduced mass for the system

$$\mu = \frac{m_1 m_2}{m_1 + m_2} \qquad . \qquad (35)$$

(This is introduced in order to treat the problem in centre of mass coordinates, and thus ignore translation of the system as a whole).

Like the radial equation for the hydrogen atom, the differential equation (34) can be solved exactly using a power series method. In this case, the solutions for the vibrational wave function ψ are written in terms of the Hermite polynomials H(y), defined by

$$H_v(y) = (-1)^v e^{y^2} \frac{d^v}{dy^v} e^{-y^2} \qquad . \qquad (36)$$

Here v is an integer known as the vibrational quantum number (v = 0,1,2,3,4...), and the variable y is related to Δr by

$$y^2 = \sqrt{\frac{4\pi^2\mu k}{h^2}} \Delta r^2 \qquad . \qquad (37)$$

The first few Hermite polynomials are:

$$H_0(y) = 1$$

$$H_1(y) = 2y$$

$$H_2(y) = 4y^2 - 2$$

$$H_3(y) = 8y3 - 12y \quad .... \qquad (38)$$

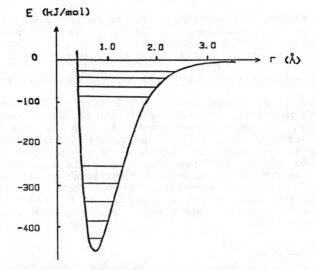

Figure 5. The anharmonic potential function for a typical diatomic molecule. The spacing between vibrational levels decreases with increasing quantum number.

In terms of these polynomials, the solution to the vibrational wave equation is

$$\psi_v = N_v H_v(y) \, e^{-y^2/2} \quad , \tag{39}$$

where N_v is a normalization coefficient.

The vibrational energies are given by

$$E_v = (v + 1/2) \, \frac{h}{2\pi} \sqrt{\frac{k}{\mu}} \tag{40a}$$

or

$$E_v = (v + 1/2) \, h\nu_0 \quad , \tag{40b}$$

where ν_0 (= $1/2\sqrt{(k/\mu)}$) is the frequency of the classical harmonic oscillator with force constant k and reduced mass μ. The first few vibrational energy levels and wave functions for the harmonic oscillator are shown in Figure 4. The energy of the harmonic oscillator with $v = 0$ (the vibrational <u>ground state</u>) is known as the <u>zero point energy</u>, and has magnitude $E_0 = 1/2 \, h\nu_0$.

The simple harmonic oscillator provides a useful first approximation for the vibrational motion of a diatomic molecule. However, it is known that true intermolecular potential functions $V(r)$ are not harmonic, but have a form of the type shown in Figure 5. The vibrational wave equation can be solved using an <u>anharmonic potential function</u> $V(r)$ in the wave equation (15), to give the energy levels and wave functions for the anharmonic oscillator. Unlike the result for the harmonic oscillator, the spacing between these anharmonic levels decreases slightly with increasing vibrational quantum number.

The vibrations of polyatomic molecules can be analyzed by an extension of the simple harmonic oscillator model. Each vibrational normal mode q_i (see Chapter 4) is treated as an independent oscillator, and the vibrational wave equation is solved to give a set of vibrational wave functions $\psi_{v(i)}$ and energy levels $E_{v(i)}$, each set associated with one normal mode. The total vibrational energy for a polyatomic molecule is written as the sum of the individual oscillator energies

$$E_{vib} = (v_1+1/2) \ h\nu_1 + (v_2+1/2) \ h\nu_2 + \ldots (v_i+1/2) \ h\nu_i \ , \qquad (41)$$

and the zero point energy is $E_0 = 1/2 \ \Sigma h\nu_i$. If anharmonicity is included, the individual mode oscillators are no longer independent, and the discussion becomes much more involved, even for relatively simple molecules (Herzberg, 1945, 1950). For crystals, the vibrational modes are expressed in terms of lattice waves with the translational periodicity of the lattice (see Chapter 4). In a quantized model of crystal lattice vibrations, a unit of vibrational excitation (involving a transition between the vibrational ground state and higher excited quantized states) is termed a phonon, which leads to a useful description of many dynamic processes in crystals (Ashcroft and Mermin, 1976).

Spectroscopic transitions

In the above examples, we have applied Schrodinger's wave mechanics to simple systems which serve as models for the electronic and vibrational behaviour of atoms, molecules and crystals. In general, these problems consist of finding a suitable potential energy function for the system, substituting this into the Hamiltonian operator in the Schrodinger wave equation, and finding exact or approximate solutions to the resulting differential equation. In general, the allowed solutions for Ψ form a set related by integer values, which are known as the quantum numbers for the system. These integers arise from the mathematical derivation of the solutions for the differential equations. Each of these eigenfunctions Ψ_n has an associated energy, or eigenvalue, E_n, which also appears in the solution of the differential equations. These energies are usually shown in graphical form as the quantized energy levels for the system.

In a spectroscopic experiment, such as UV/visible spectroscopy for electronic transitions, infrared and Raman spectroscopy for probing vibrational properties, or microwave and radio wave spectroscopy for examining electron and nuclear spin state changes, we are concerned with transitions between quantized states, from some initial state Ψ_n to another state Ψ_m. The energy associated with the transition is the difference between the energies of the two states, $\Delta E = E_m - E_n$, and the intensity of the observed line is related to the probability of the transition $n \rightarrow m$. These probabilities are described by the Einstein transition probabilities for absorption (B_{nm}), spontaneous emission (A_{mn}), and induced emission (B_{mn}). The Einstein coefficient for absorption describes the case where a system is initially in state n and absorbs a quantum of energy from an applied radiation field to undergo a transition to the higher energy state m. This is the case considered in most of the spectroscopic experiments described in this volume.

In the case of infrared absorption (for vibrational transitions) or UV/visible spectroscopy (for electronic transitions), the oscillating electromagnetic field of the incident light constitutes a time-dependent perturbation on the system in its initial state Ψ_n, and this perturbation is responsible for the transition to the higher energy state Ψ_m. Electron spin and nuclear magnetic resonance (ESR and NMR) experiments differ in that the system is perturbed by an applied static magnetic field before the interaction with radiation takes place. These experiments can be considered as (a) a time-independent, or static perturbation followed by (b) a time-dependent perturbation. In a Raman scattering experiment, the system is initially perturbed by the incident light beam before the transition takes place. This can be considered as two consecutive time-dependent perturbations to the system.

The Einstein probability coefficient for absorption B_{nm} in these cases can be calculated using wave mechanics and the mathematical methods of perturbation theory (Schrodinger, 1926b). It is always found that the transition probability is maximized when the energy of the radiation corresponds to ΔE_{mn}, the energy difference between the states E_m and E_n. The set of probabilities for transitions between sets of levels n and m for a system are known as the selection rules for spectroscopic transitions in the system. These are determined by the form of the wavefunctions Ψ_n and Ψ_m (electronic, vibrational, rotational, etc.), and by the nature of the perturbation involved (oscillating electric or magnetic field, interacting with electric or magnetic dipoles, quadrupoles, etc.). The elements of the Einstein probability coefficients are often termed the matrix elements for the transition. This refers to an alternative mathematical treatment of atomic problems due to Heisenberg, Born and Jordan (1926) known as matrix mechanics, which was originally developed to describe the energies and intensities of spectral transitions.

Infrared absorption

In an infrared absorption experiment, the system absorbs a quantum of light with energy in the infrared region of the spectrum to pass from a vibrational state with quantum number v_n to one with quantum number v_m. The interaction process can be described by time-dependent perturbation theory, relating the wave functions in states Ψ_n and Ψ_m through the time-dependent perturbation on the system due to the radiation. The perturbation can be described as an interaction between the oscillating electric field vector E of the light, and the instantaneous dipole moment vector μ of the molecule. For a diatomic molecule, the dipole moment is defined by

$$\mu = Qr , \qquad (42)$$

where Q is the charge difference between the atom centres, and r (magnitude of the vector r) is the distance between them. When $r = r_0$ (the equilibrium bond distance), μ_0 is the permanent molecular dipole moment. During a vibration, the atoms undergo small displacements Δr relative to each other, and the magnitude of the instantaneous dipole moment can be written as

$$\mu(r) = \mu_0 + (\frac{d\mu}{d\Delta r})\Delta r + (\frac{d^2\mu}{d\Delta r^2})(\Delta r)^2 + \ldots \qquad (43)$$

Since the displacements Δr are small, usually terms in $(\Delta r)^2$ and higher are ignored.

From perturbation theory, the value of the transition moment M_{nm} (proportional to the Einstein coefficient for absorption) is given by

$$M_{nm} = \int \psi_n^* \mu(r) \psi_m \, dr = \mu_0 \int \psi_n^* \psi_m \, dr + \int \psi_n^* (\frac{d\mu}{d\Delta r}) \Delta r \psi_m \, dr \qquad (44)$$

(ψ^* is the complex conjugate of the wave function ψ). We can evaluate these integrals from a knowledge of the vibrational wavefunctions ψ_n

and ψ_m, which are given in terms of Hermite polynomials (see discussion above for the harmonic oscillator). The first term in (44), an expression involving $\int \psi_n^* \psi_m dr$, is always zero, due to the orthogonality properties of the wavefunctions ψ_n and ψ_m; therefore, vibrations are infrared active only when the second term in (42) is non-zero. This term has a non-zero value if two conditions are met: (a) if $d\mu/d\Delta r$ is not zero (i.e., there is a dipole moment change during the vibration), and (b) if $n = m \pm 1$, i.e., if the vibrational quantum number v changes by one unit (this follows from the mathematical properties of the Hermite polynomial functions). These two conditions constitute the selection rules for infrared absorption of the harmonic oscillator.

These selection rules allow a number of predictions for the infrared spectra of diatomic molecules. Homonuclear diatomic molecules (i.e., in which both atoms are the same) possess no permanent dipole moment, and this does not change during the vibration. From this observation, homonuclear diatomic molecules should have no infrared absorption or emission spectrum, which is true to a first approximation. (A weak infrared spectrum may be observed for molecules perturbed by some external field, or for unperturbed molecules interacting with radiation through electric multipole or magnetic effects). Secondly, for heteronuclear diatomic molecules, only a single absorption line should be observed corresponding to a transition between two adjacent quantum states. If the lower state is the vibrational ground state (v = 0), this is the fundamental absorption line, from v = 0 to v = 1. The intensity of this transition is proportional to the magnitude of the dipole moment change during the vibration ($d\mu/d\Delta r$). Again, this selection rule is obeyed to a good first approximation for diatomic molecules: a strong infrared absorption is observed at the frequency of the fundamental line, and weak overtone absorptions are observed corresponding to $\Delta v = \pm 2, \pm 3$, etc. These overtones are weakly allowed because the true molecular potential function is not truly harmonic, but is an anharmonic function, for which the selection rules are slightly modified.

These selection rules can be extended to polyatomic molecules and condensed phases. In general, in order for a vibration to be infrared active, the vibrational motion must cause a change in dipole moment, i.e., $d\mu/dq_i \neq 0$, where q_i is the vibrational normal coordinate (see Chapter 4). This requirement may be satisfied even when the molecule has no permanent dipole moment, for example CO_2 has two infrared active vibrations. For crystals, an additional "selection rule" is introduced by the translational symmetry of the crystal. The vibrational normal modes are cooperative lattice distortions. If the mode results in a dipole change within the unit cell, the result is an electric dipole wave within the crystal with a well defined wavelength λ and wave vector $k = 2\pi/\lambda$ in the direction of propagation. This dipole wave can only interact with light when its wavelength is comparable with that of infrared radiation, i.e., when λ is very large or $k \to 0$ (see Chapter 4).

Absorption due to electronic transitions: UV/visible spectroscopy

In the case of UV/visible spectroscopy, we wish to consider a change in the electronic state of the system accompanied by absorption or emission of radiation. Like infrared absorption, we consider the system in its initial and final states, perturbed by the time-dependent field of the incident light. Unlike the infrared experiment, in electronic absorption spectroscopy, we are concerned with interactions between both the electric and magnetic fields of the light beam and the electronic wave function of the system. The selection rules for UV/visible spectroscopic experiments are generally more complicated than those for infrared spectroscopy.

We can examine the form of the selection rules for a simple case. For the hydrogen atom discussed above, we found that three quantum numbers n,l and m_l were required to specify the quantum state of the one electron system. If we consider an interaction with the electric field vector of the light and the electronic wave function of hydrogen (such electric dipole transitions usually give rise to the strongest electronic absorption lines), the probability for a radiative transition in the hydrogen atom between two quantum states is proportional to the integral

$$\int_{-\infty}^{\infty} \psi_{n,l,m_l}^{*} \; u \; \psi_{n',l',m_l'} \; du \; , \tag{45}$$

where u represents the x, y or z directions in space. Since the electronic wave functions of the hydrogen atom are known in terms of associated Laguerre polynomials and surface spherical harmonics (see previous discussion), such transition integrals can be evaluated. When the analysis is carried out, it is found that only those transitions are allowed in which (a) the orbital angular momentum (l) changes by $^+_-1$ and (b) the magnetic quantum number (m_l) changes by $^+_-1$ or 0. The selection rules for radiative electronic transitions in the hydrogen atom are thus

$$\Delta l = {}^+_-1 \; ; \; \Delta m_l = 0, {}^+_-1 \; . \tag{46}$$

The total quantum number, n, is subject to no special selection rules except that its change be integer. This means that electronic transitions such as $1s \rightarrow 2p_z$ or $3p_x \rightarrow 3d_{z^2}$ would be allowed, and would give strong absorption spectra in the UV/visible region, but transitions like $1s \rightarrow 3d_{z^2}$, or transitions between d orbitals, would be forbidden.

When applied to the electronic spectra of polyelectronic atoms or molecules, these selection rules are modified by the Pauli exclusion principle (no two electrons can have the same four quantum numbers, or the electronic wave function must be antisymmetric to exchange of two electrons), and consideration of the electron spin quantum number m_s, but the general form of the selection rule is generally retained. For this reason, the crystal field absorption bands of first row transition metals are generally weak absorption bands, since these involve transitions between electronic levels derived from the metal d orbital manifold (formally forbidden for hydrogen: $\Delta l = 0$). (See the next section for a discussion of this selection rule in terms of the symmetries of the upper and lower electronic states). In contrast, the strong charge transfer bands of transition metal complexes involve transitions between levels derived from metal d orbitals, and generally p or π type levels centered on the ligands ($\Delta l = {}^+_-1$) (see chapter by G. Rossman).

In addition to the above electric dipole selection rules, additional selection rules can be derived for the generally much weaker absorption lines due to magnetic dipole and electric quadrupole and higher multipole transitions.

Magnetic resonance

Electron spin resonance (e.s.r.) and nuclear magnetic resonance (n.m.r.) experiments both depend on the fact that electrons and some nuclei possess magnetic moments which are capable of interacting with applied magnetic fields. The electron has an intrinsic spin described by its spin quantum number $m_S = \pm 1/2$. Any system with unpaired electrons will have a net electronic spin. Since the electrons are charged particles, the spin angular momentum gives rise to a magnetic moment

$$\mu e = -g_e \beta S \quad , \tag{47}$$

where S is the total electronic spin angular momentum, β is the electronic Bohr magneton (9.274 x 10^{-24} Am^2), and g_e is a dimensionless parameter known as the electronic g factor.

If a permanent magnetic field H is applied to a system with unpaired electrons, the field interacts with the magnetic moment due to the electron spin, and results in a static perturbation to the electronic spin wave function. This perturbation can be described by the Hamiltonian

$$\hat{H} = -\mu_e \cdot H \quad , \tag{48a}$$

which becomes

$$\hat{H} = g_e \beta H S_z \tag{48b}$$

if the field is in the z-direction. (S_z is the magnitude of the z component of the spin angular momentum). Because $S = 1/2$ for a single electron, there are two allowed orientations of the spin of a single unpaired electron in such an applied magnetic field: either aligned with the field ($m_S = -1/2$) or against the field ($m_S = +1/2$). In the absence of an external field, the two cases $m_S = \pm 1/2$ are degenerate. Application of the field H as a static perturbation removes the degeneracy, and results in two states with different energies (Fig. 6). The lowest energy state with $m_S = -1/2$ will normally be populated at low temperature. Now application of an oscillating magnetic field (electromagnetic radiation in the microwave or radiofrequency part of the spectrum) results in a time-dependent perturbation of the system, and causes a transition between the lower and upper spin states. The transition probability can be calculated using wave mechanics, as described above for infrared and UV/visible spectroscopy, one the spin Hamiltonian for the system has been determined. The transition frequency ν, corresponding to the position of the absorption line, is determined by the energy separation between the spin states

$$h\nu = g_e \beta H \quad . \tag{49}$$

The magnitude of the electronic g factor, determined from an electron spin resonance experiment, gives detailed information on the electronic wave function at the unpaired electron site, and can be used to study site symmetries, nature of nearest neighbors, electron delocalization, and other related phenomena (see chapter by G. Calas).

Nucleons (protons and neutrons) also each possess an intrinsic spin of $\pm 1/2$. Depending on the numbers of protons and neutrons in a given nucleus, the isotope may possess a net nuclear spin I, which gives rise to a nuclear spin angular momentum vector $Ih/2\pi$. As for electrons, this results in a nuclear magnetic moment

$$\mu_n = \gamma_n h I/2\pi = g_n \beta_n I \tag{50}$$

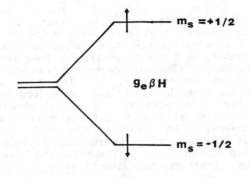

$m_s = +1/2$

$g_e \beta H$

$m_s = -1/2$

field off   **field on**

Figure 6. The behavior of degenerate electronic spin states when an external magnetic field is applied to the system. The transition frequency ν in the ESR experiment is given by $h\nu = g_e\beta H$, where g_e is the electronic g factor to be determined, β is the electronic Bohr magneton, and H is the applied field strength. (After Carrington and McLachlan, 1979, p. 5).

where γ_n is the <u>gyromagnetic ratio</u> of the nucleus, g_n is the <u>nuclear g factor</u>, and β_n is the nuclear magneton (5.05×10^{-22} Am^2). The spin properties of a given nucleus are specified by the magnitude of I, and either by γ_n, or in terms of β_n and its nuclear g factor.

As discussed above for electrons, applying a static magnetic field H results in a perturbation to the nuclear spin system, described by a Hamiltonian

$$\hat{H} = -\mu_n \cdot H \ . \tag{51}$$

If the applied field direction is defined as the z axis, this Hamiltonian becomes

$$\hat{H} = -\gamma_n H I_z h/2\pi = -g_n\beta_n H I z \quad , \tag{52}$$

where I_z is the allowed component of the nuclear spin in the z direction, $\pm n/2$ where n is an integer. For n=1 (i.e., I = 1/2, which is the case for ^{1}H or ^{29}Si), there are two allowed directions for I_z; aligned with the external field ($m_I = +1/2$), corresponding to the lower energy state, and aligned against the field ($m_I = -1/2$), giving a higher energy state (note the sign change compared with the electronic case, due to the difference in charge between electrons and protons).

Once the magnetic field has removed the degeneracy of the nuclear spin states through this static perturbation of the spin system, transitions are induced between the perturbed spin states via a time dependent perturbation applied by an oscillating magnetic field (electromagnetic radiation in the radiofrequency region of the spectrum). The number, positions and intensities of these transitions give valuable information of the detailed environment of the nucleus probed, and details of the site geometry and the charge density near the nucleus can be elucidated (see chapters by R.J. Kirkpatrick and J. Stebbins).

Raman scattering

The final type of spectroscopy considered here consists of an initial time-dependent perturbation on the system, followed by a second time-dependent perturbation which provides a probe of the system. In Raman scattering, a beam of visible light is passed through a sample and the energy of the scattered light is analyzed. In general, both elastic (Rayleigh) and inelastic (Raman) scattering is observed. In most applications of Raman spectroscopy, the inelastic Raman scattering occurs via interaction with the vibrational wave function of the system, and Raman scattering is usually a type of vibrational spectroscopy. Raman scattering can also be used to study a wide range of electronic and magnetic phenomena. The treatment of these is similar to that described here, but appropriate Hamiltonian functions must be chosen to describe each interaction, and different wave functions must be substituted for the vibrational wave functions.

For vibrational Raman scattering, the scattering mechanism can be described in terms of the instantaneous dipole moment μ_{ind} induced in the system by the incident light beam:

$$\mu_{ind} = \alpha E = \alpha E_0 \cos 2\pi\nu t \quad . \tag{53}$$

E is the oscillating electric field vector of the radiation with frequency ν and amplitude E_0, and α is the molecular polarizability which expresses the deformability of the electron density by the radiation field. Since μ_{ind} and E are not necessarily collinear, α is a second rank tensor:

$$\alpha = \begin{bmatrix} \alpha_{xx} & \alpha_{xy} & \alpha_{xz} \\ \alpha_{yx} & \alpha_{yy} & \alpha_{yz} \\ \alpha_{zx} & \alpha_{zy} & \alpha_{zz} \end{bmatrix} \quad . \tag{54}$$

Since the polarizability will in general change during a molecular vibration, it is commonly expanded in a Taylor series of the form

$$\alpha = \alpha_0 + \left(\frac{d\alpha}{d\Delta r}\right) \Delta r + \ldots \tag{55}$$

where α_0 is the equilibrium polarizability. In this description, the action of the incident light beam in creating the instantaneous induced dipole moment constitutes the time-dependent perturbation on the system. In the second step of the analysis, the vibrational wave functions corresponding to the initial and final states of the system are allowed to interact, modulated by the induced dipole moment. This treatment results in the selection rules for the vibrational Raman effect. The total instantaneous dipole moment for the system is given by

$$\mu = \mu_0 + \mu_{ind} \quad , \tag{56}$$

where μ_0 is the permanent molecular dipole moment. The Raman intensity for a transition between vibrational states n and m is proportional to the square of the transition moment M_{nm}, given by

$$M_{nm} = \int \psi_n^* \mu(r) \psi_m \, dr = E \int \psi_n^* \left(\frac{d\alpha}{d\Delta r}\right) \Delta r \, \psi_m \, dr \tag{57}$$

after substituting from equations (52), (54) and (56), and simplifying using the orthogonality properties of the vibrational wave functions ψ_n

and ψ_m. This integral is only non-zero when (a) the vibrational quantum number changes by one unit ($\Delta v = {}^+1$) between states n and m, and (b) the term in ($d\alpha/d\Delta r$) is not zero. The first selection rule is relaxed for anharmonic molecular vibrations, allowing overtone bands to appear in the Raman spectrum as for infrared absorption. The second part of the selection rule implies that, in order for a vibrational mode to be Raman active, there must be a change in molecular polarizability associated with the vibration. This condition can be difficult to ascertain by simply visualizing the vibrational mode of interest, and it is often difficult to predict the Raman activity of particular vibrations of even simple molecules by inspection. However, as discussed below, the methods of symmetry and group theory provide powerful techniques for predicting the infrared and Raman activities of all vibrational modes of even complex molecules and crystals.

Summary

As we have seen, quantum mechanics provides a framework for a detailed understanding of the structure and dynamics of atoms, molecules and solids, and forms a basis for interpretation of the transitions involved in the various atomic and molecular spectroscopies. In principle, quantum mechanical methods can be used to calculate the transition energies and intensities for any given spectroscopic experiment, if the initial and final wavefunctions for the system are known. In practice, this process is generally laborious and highly technical in nature. In many cases, severe approximations must be made to the true Hamiltonian of the system to reduce the problem to manageable proportions. These approximations can reduce the range of applicability of the calculations, and may remove some of the essential characteristics of a given problem.

It is usually possible to greatly reduce the magnitude of a problem without affecting its rigorous solution by recognizing and exploiting the underlying symmetry of the system, and using the algebraic methods of group theory to manipulate the symmetry properties. Even when a quantitative result is not required, the application of symmetry and group theory can be applied in a systematic way to provide qualitative information, such as the number and spectroscopic activity of expected absorption bands. In the next section we will discuss the methods of symmetry and group theory, with associated applications to quantum mechanics and spectroscopy.

SYMMETRY AND GROUP THEORY

Introduction

As we have described above, in spectroscopic experiments on molecular systems (including crystalline and non-crystalline solids), transitions from one quantum state to another are induced when the molecule interacts with incident radiation. The classification of the quantum states (which may be electronic, vibrational, rotational, etc.) and the subsequent description of the spectroscopic interaction are greatly simplified by exploiting the <u>symmetry</u> possessed by the molecule. In general, the molecular symmetry is described in terms of a set of conventionally chosen <u>symmetry elements</u>, which express certain spatial relations between different parts of the molecule. For any molecular system, the set of <u>symmetry operations</u> demonstrating the symmetry of the molecule forms a mathematical <u>group</u> (a special type of <u>set</u>, satisfying particular combination relations between the elements of the set). <u>Group theory</u> is the mathematical framework within which quantitative descriptions of the symmetry possessed by a structure are constructed. In this section, we will provide an introduction to the basic concepts of symmetry and group theory, as applied to molecular structure and spectroscopy. An excellent introduction to the use of symmetry in molecular spectroscopic problems is given by Cotton (1971), while Bouckaert et al. (1936), Koster (1957) and Birman (1984) give detailed discussions of the application of symmetry to electronic and vibrational problems in crystals.

Symmetry elements and operations

We begin by first developing the nomenclature and conventions used to describe the symmetry of finite objects. This is done by introducing a set of symbols that represent the symmetry possessed by such objects. It is important to make the distinction between symmetry elements and symmetry operations. A symmetry element is a geometric entity possessed by an object, such as a line, a point or a plane, about which a symmetry operation can be performed. The corresponding symmetry operation is the movement of the body with respect to that symmetry element, such that after the movement all points in the new orientation are coincident with equivalent points in the initial orientation.

In general, the symmetry of molecules and crystals is described in terms of the following five symmetry elements:

  a) the identity element: I (or E)
  b) a proper axis of rotation: C_n
  c) a mirror plane: σ
  d) an improper axis of rotation: S_n
  d) an inversion center: i.

Here we have used the Schonflies symbols for the symmetry elements. These symbols are those most commonly used in spectroscopy and in molecular chemistry. Another set of symbols, the Hermann-Mauguin or International symbols, are more common in crystallography. These symmetry elements and their associated symmetry operations are most easily described by the use of examples.

Identity element. The identity element is possessed by all finite objects, and involves the operation of doing nothing, in the sense that no movement of the body has been carried out. All points in the "new" orientation are equivalent to those in the "old" orientation. The identity "operation" is introduced to satisfy the algebraic requirements of the group, to be defined below.

Proper rotation axis. The operation corresponding to a C_n proper rotation axis is a rotation of $2\pi/n$ about that axis. For example, the molecule NH_3 possesses a three-fold rotation axis C_3 (Fig. 7). In this figure, each of the atom centers has been given a label. This is a bookkeeping device for us to discuss the object before and after the symmetry operation, and does not imply that the H atoms are distinguishable. When the molecule is rotated through an angle of $\phi = 120°$ ($2\pi/3$) in a clockwise sense, an equivalent orientation (II) is obtained. This operation is designated $C_3{}^1$ (it is conventional to express both the symmetry element and the symmetry operation by the same symbol). Rotation of NH_3 about the axis through an angle of $\phi = 240°$ ($4\pi/3$: $C_3{}^2$) also results in an equivalent orientation, III (Fig. 7). Finally, if the molecule is rotated through an angle $\phi = 360°$ ($C_3{}^3$), we not only obtain an equivalent orientation but one identical to the original. This operation is then equivalent to the identity operation

7.

8.

9.

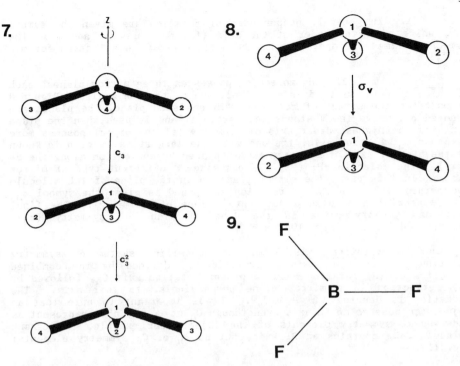

Figure 7. The three-fold rotation axis of NH_3 and the rotation operations C_3^1 and C_3^2. (Nitrogen is the central atom 1; the hydrogen atoms are numbered 2-4).

Figure 8. The effect of reflection in a σ_v mirror plane of NH_3. The mirror plane is perpendicular to the plane of the paper, and contains atoms 1 and 3.

Figure 9. The planar molecule BF_3 also contains a horizontal mirror plane in the plane of the paper, σ_h, which is perpendicular to the principal C_3 axis.

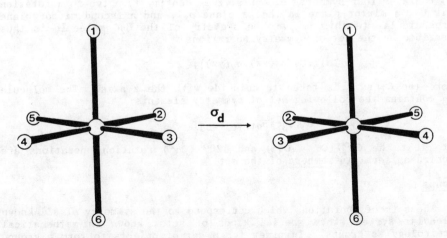

Figure 10. The effect of reflection in one of the dihedral σ_d planes of SF_6 (S is the central atom). The dihedral mirror planes bisect the FSF angles. The σ_d plane in question is perpendicular to the plane of the paper.

($C_3^3 \equiv I$). In general, proper axes of rotation are given the symbol C_n^m, where n is the <u>order</u> of the axis (for NH_3, n = 3) and m is the number of times the operation has been performed. In all cases for m = n, $C_n^n \equiv I$.

<u>Mirror plane</u>. The NH_3 molecule possesses three mirror planes, each of which contain the C_3 axis and bisect one of the HNH angles. Figure 8 illustrates the effect of reflection in one such plane. The planes are denoted σ_v, where the v stands for <u>vertical</u>, and is used when the plane contains the highest order axis of symmetry (if the object possess more than one C_n rotation axis, the one with the largest value of n is known as the highest axis of symmetry and is placed by convention along the z-axis). The molecule BF_3 (Fig. 9) possesses a different type of mirror plane. In this case, the mirror plane is in the plane of the molecule and perpendicular to the C_3 rotation axis, and is given the symbol σ_h. (The subscript h denotes a <u>horizontal</u> plane of symmetry). The final type of symmetry plane is the <u>dihedral</u> plane of symmetry (σ_d), illustrated for SF_6 (Fig. 10).

<u>Improper rotation axis</u>. In the Schonflies system of symmetry notation, an improper axis of rotation is defined by the combined operation of (a) rotation about a proper rotation axis C_n, followed by (b) reflection in a mirror plane perpendicular to that axis. The operation is denoted $S_n^m = \sigma_h C_n^m$. It is important to note that an object can possess an S_n axis even though C_n or σ_h may not be present as independent symmetry elements of the object (for example, the ethane molecule C_2H_6 contains an S_6 axis, but no σ_h or C_6 symmetry elements: Fig. 11).

<u>Inversion center</u>. The operation associated with the inversion centre (i) corresponds to reflection in a point. For example, in the molecule SF_6 (Fig. 12), there is an inversion center at the S atom, which takes F_1 into F_4, F_2 into F_5, and F_3 into F_6.

Sets of symmetry operations

Any molecule (or unit cell of a crystal) contains some <u>set</u> of one or more of these symmetry elements. For example, the molecule H_2O (Fig. 13) contains four symmetry elements; the identity I, a two-fold rotation axis C_2, a mirror plane in the xz plane σ_{xz}, and a second mirror plane perpendicular to this, σ_{yz}. The symmetry of the H_2O molecule is thus described by the <u>set</u> of symmetry operations

$$\{I, C_2, \sigma_v(xz), \sigma_v(yz)\},$$

where the C_2 axis is taken to coincide with the z axis. The molecule NH_3 contains the following set of symmetry elements:

$$\{I, C_3, C_3^2, \sigma_{v1}, \sigma_{v2}, \sigma_{v3}\}.$$

Note that the $60°$ ($C_3 \equiv C_3^1$) and $120°$ (C_3^2) rotation operations are counted as separate <u>members</u> of the set.

Groups

The sets of operations which correspond to the symmetry of all known molecular systems form special kinds of sets, known in mathematical terminology as <u>groups</u>. In order for a set of objects to form a group, the members of the set must satisfy a certain number of conditions concerning the combination properties of members of the set. To show

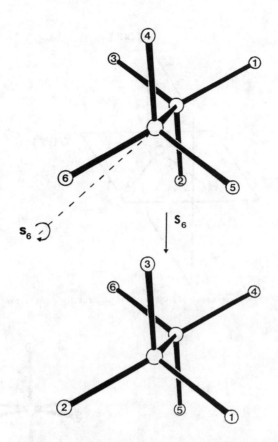

Figure 11. The S_6 improper rotation axis of the staggered conformotaion of ethane (the central atoms are carbon; hydrogen atoms are numbered 1-6).

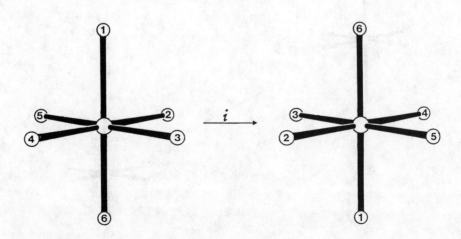

Figure 12. The inversion center of SF_6 (S is the central atom).

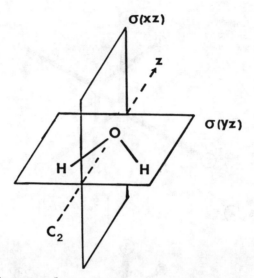

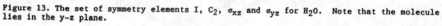

Figure 13. The set of symmetry elements I, C_2, σ_{xz} and σ_{yz} for H_2O. Note that the molecule lies in the y-z plane.

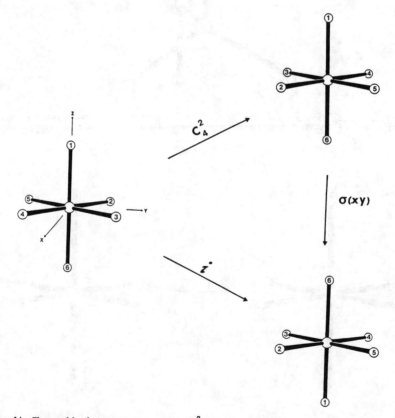

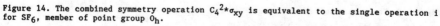

Figure 14. The combined symmetry operation $C_4{}^2 \star \sigma_{xy}$ is equivalent to the single operation i for SF_6, member of point group O_h.

that a set constitutes a group, it is necessary to define a law of combination for the set. For sets of symmetry operations, we define the combination law as follows: if M and N are two symmetry operations, the combined operation M*N (the "product") corresponds to performing operation N first, then performing operation M, on the molecule. For example, for the molecule SF_6, the combined operation $C_4^2(z)*\sigma_{xy}$ corresponds to first reflecting the molecule in the σ_{xz} mirror plane, then performing a 180° rotation of the molecule about the z axis $(C_4^2(z))$ (Fig. 14). The resulting configuration of the molecule is the same as if we had performed the single operation corresponding to the inversion center, i. We can express this algebraically as $i = C_4^2(z)*\sigma_{xy}$. Note that, in general, the product M*N does not equal the product N*M. (Those groups for which the commutative law of combination does hold are known as <u>Abelian</u> groups).

Using this definition of a combination law, the conditions necessary for a set to constitute a group are:

1) For every element M and every element N in a set, the product M*N is also in the set, i.e., the set is <u>closed</u>.
2) For any three elements of the set, $\overline{M,N}$ and P, the <u>associative law</u> of combination is obeyed, i.e., M*(N*P) = (M*N)*P.
3) The set must contain one element known as the identity element, I, such that it <u>commutes</u> with every other element of the set; N*I = I*N = N.
4) For every element M in the set, there must be an element N such that their product M*N = N*M = I. In this case, N is said to be the <u>inverse</u> of M, written $N = M^{-1}$.

As noted above, the sets of symmetry operations for all known molecules are found to obey these rules, and hence constitute groups.

In order to avoid writing down the set of operations for each individual case, such as

$$\{I, C_3, C_3^2, C_{2(1)}, C_{2(2)}, C_{2(3)}, \sigma_h, S_3, S_3^2, \sigma_{v1}, \sigma_{v2}, \sigma_{v3}\}$$

for BF_3, or $\qquad \{I, C_2, \sigma(xz), \sigma(yz)\}$

for H_2O, we use a symbolic notation. For molecules like H_2O, the set of operations is given the symbol C_{2v}, while for BF_3, the group symbol is D_{3h}. These symbols, known as the <u>Schonflies symbols</u>, are constructed in a systematic way to denote the particular set of elements present for the symmetry group. (An alternative system of group symbols is used within the <u>Hermann-Mauguin (International)</u> system). Throughout this text, we use the Schonflies nomenclature, because this is more commonly encountered in spectroscopic applications.

Combination of symmetry operations: multiplication tables

In order to investigate the combination properties of a set of operations, it is usual to construct a <u>multiplication table</u>. For example, the multiplication table for the set of elements which constitute the group C_{2v},

$$\{I, C_2, \sigma_v(xz), \sigma_v(yz)\} ,$$

is formed by taking all possible pairwise products of the four symmetry operations (Table 1). The entries in this table are conventionally arranged: the first member of a product pair appears as a column entry on the left, while the second operation appears on the top row. It can be seen that the product of all members of the set are indeed members of the set, therefore the set is closed under the chosen combination law "followed by", and axiom 1 is satisfied. The associative law is also obeyed (axiom 2): for example, $C_2(\sigma(yz)*\sigma(xz)) = (C_2*\sigma(yz))*\sigma(xz)$. Each member of the set is its own inverse (see Table 1), therefore the set obeys axiom 4. Finally, the set contains the identity element. All axioms are thus satisfied, and therefore the set forms a group. In this particular case, the group is Abelian: elements are symmetric across the diagonal of the multiplication table, indicating that every member commutes with every other.

As a second example, the multiplication table for the group C_{3v} (e.g., NH_3) is shown in Table 2. The most striking difference from C_{2v} is that the group C_{3v} is non-Abelian, i.e., the commutative law is not obeyed, and the order in which operations are carried out is important.

In these molecular symmetry groups, we are not concerned with translational symmetry operations, and at least one point within the molecule remains unchanged ("invariant") under all operations of the group. For this reason, these groups are known as point groups. (When translational operations are considered explicitly, as in crystallography, the group is known as a space group). The number of symmetry operations in the group is said to be the order (h) of the point group; for example, the order of point group C_{2v} is four, for C_{3v}, h = 6.

Matrix representations of symmetry operations

In the above discussion, the interrelations between members of a point group were described algebraically, after we had defined a combination law. It is also useful to express these relationships numerically, via the use of transformation matrices. This provides a framework for a quantitative, rather than an intuitive, use of molecular symmetry. We can establish the general form of the matrix representations for the five symmetry operations, C_n, S_n, σ, I and i, by investigating the effect that each of these operations has on a general point P(x,y,z) relative to three Cartesian axes.

Proper rotations C_n. Figure 15 shows a clockwise rotation of Θ degrees about the z-axis of a general point, P, in a Cartesian coordinate system. The relationship between the "new" coordinates (x',y',z') and the original set (x,y,z) is:

$$x' = x \cos \Theta + y \sin \Theta$$
$$y' = -x \sin \Theta + y \cos \Theta$$
$$z' = z \ . \tag{58}$$

The matrix representation of this system of linear equations is:

$$\begin{bmatrix} x' \\ y' \\ z' \end{bmatrix} = \begin{bmatrix} \cos \Theta & \sin \Theta & 0 \\ -\sin \Theta & \cos \Theta & 0 \\ 0 & 0 & 1 \end{bmatrix} \begin{bmatrix} x \\ y \\ z \end{bmatrix} \ . \tag{59}$$

Since a C_n^m rotation operation corresponds to a rotation of $\Theta = (360xm)^\circ/n$ about the rotation axis, the matrix

$$C_n^m = \begin{bmatrix} \cos m\Theta & \sin m\Theta & 0 \\ -\sin m\Theta & \cos m\Theta & 0 \\ 0 & 0 & 1 \end{bmatrix} \tag{60}$$

<u>Table 1: Multiplication table for C$_{2v}$</u>

C$_{2v}$	I	C$_2$	σ(xz)	σ(yz)
I	I	C$_2$	σ(xz)	σ(yz)
C$_2$	C$_2$	I	σ(yz)	σ(xz)
σ(xz)	σ(xz)	σ(yz)	I	C$_2$
σ(yz)	σ(yz)	σ(xz)	C$_2$	I

<u>Table 2: Group multiplication table for C$_{3v}$</u>

C$_{3v}$	I	C$_3$	C$_3{}^2$	σ$_{v1}$	σ$_{v2}$	σ$_{v3}$
I	I	C$_3$	C$_3{}^2$	σ$_{v1}$	σ$_{v2}$	σ$_{v3}$
C$_3$	C$_3$	C$_3{}^2$	I	σ$_{v3}$	σ$_{v1}$	σ$_{v2}$
C$_3{}^2$	C$_3{}^2$	I	C$_3$	σ$_{v2}$	σ$_{v3}$	σ$_{v1}$
σ$_{v1}$	σ$_{v1}$	σ$_{v2}$	σ$_{v3}$	I	C$_3$	C$_3{}^2$
σ$_{v2}$	σ$_{v2}$	σ$_{v3}$	σ$_{v1}$	C$_3{}^2$	I	C$_3$
σ$_{v3}$	σ$_{v3}$	σ$_{v1}$	σ$_{v2}$	C$_3$	C$_3{}^2$	I

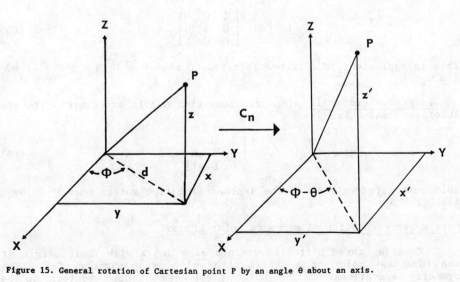

Figure 15. General rotation of Cartesian point P by an angle θ about an axis.

corresponds to a general rotation operation C_n^m about the z axis.

Improper rotation axis S_n. An improper rotation operation, S_n^m is defined as an m-fold rotation about a C_n axis, followed by reflection in a plane perpendicular to this axis. Taking the rotation axis along the z direction, the matrix representation for this operation is

$$\begin{bmatrix} x' \\ y' \\ z' \end{bmatrix} = \begin{bmatrix} \cos\Theta & \sin\Theta & 0 \\ -\sin\Theta & \cos\Theta & 0 \\ 0 & 0 & -1 \end{bmatrix} \begin{bmatrix} x \\ y \\ z \end{bmatrix} , \qquad (61)$$

and the matrix representation of the S_n^m symmetry operation is

$$S_n^m = \begin{bmatrix} \cos m\Theta & \sin m\Theta & 0 \\ -\sin m\Theta & \cos m\Theta & 0 \\ 0 & 0 & -1 \end{bmatrix} . \qquad (62)$$

Reflection in a mirror plane σ. The relationship between old and new Cartesian coordinates on reflection of point P in the xy plane is

$$\begin{aligned} x' &= x \\ y' &= y \\ z' &= -z . \end{aligned} \qquad (63)$$

The matrix representation for this operation is:

$$\sigma(xy) = \begin{bmatrix} 1 & 0 & 0 \\ 0 & 1 & 0 \\ 0 & 0 & -1 \end{bmatrix} . \qquad (64)$$

Note that this may be obtained from the general matrix representation for an improper rotation axis about z, if $\Theta = 360^\circ$; i.e., $\sigma(xy) \equiv S_1^1(z)$.

Identity operation I. Clearly the matrix representation of the identity element is

$$I = \begin{bmatrix} 1 & 0 & 0 \\ 0 & 1 & 0 \\ 0 & 0 & 1 \end{bmatrix} . \qquad (65)$$

This is equivalent to a proper rotation C_n when n = 1 ($\Theta = 360^\circ$), i.e., $I \equiv C_1^1$.

Inversion center i. The transformation matrix associated with the inversion center is

$$i = \begin{bmatrix} -1 & 0 & 0 \\ 0 & -1 & 0 \\ 0 & 0 & -1 \end{bmatrix} . \qquad (66)$$

This matrix corresponds to the improper rotation matrix when $\Theta = 180^\circ$, i.e., $i \equiv S_2^1$.

Matrix representation of symmetry point groups

Consider three Cartesian vectors $\underline{x}$, $\underline{y}$ and $\underline{z}$ with their origin at the invariant point of a system with point group symmetry C_{2v}. The symmetry operations of the point group cause these vectors to be transformed to a new set, $\underline{x}'$, $\underline{y}'$ and $\underline{z}'$, which can be expressed using the matrix notation described above. For example, the transformation matrix corresponding to the identity operation is

$$I = \begin{bmatrix} 1 & 0 & 0 \\ 0 & 1 & 0 \\ 0 & 0 & 1 \end{bmatrix} , \qquad (65)$$

so that

$$\begin{bmatrix} x' \\ y' \\ z' \end{bmatrix} = \begin{bmatrix} 1 & 0 & 0 \\ 0 & 1 & 0 \\ 0 & 0 & 1 \end{bmatrix} \begin{bmatrix} x \\ y \\ z \end{bmatrix} . \qquad (67)$$

Likewise, the matrix representation for the C_2 operation is

$$C_2 = \begin{bmatrix} -1 & 0 & 0 \\ 0 & -1 & 0 \\ 0 & 0 & 1 \end{bmatrix} , \qquad (68)$$

so that

$$\begin{bmatrix} x' \\ y' \\ z' \end{bmatrix} = \begin{bmatrix} -1 & 0 & 0 \\ 0 & -1 & 0 \\ 0 & 0 & 1 \end{bmatrix} \begin{bmatrix} x \\ y \\ z \end{bmatrix} . \qquad (69)$$

The transformation matrices corresponding to the $\sigma(xz)$ and $\sigma(yz)$ mirror plane reflections are:

$$\sigma(xz) = \begin{bmatrix} 1 & 0 & 0 \\ 0 & -1 & 0 \\ 0 & 0 & 1 \end{bmatrix} \quad ; \quad \sigma(yz) = \begin{bmatrix} -1 & 0 & 0 \\ 0 & 1 & 0 \\ 0 & 0 & 1 \end{bmatrix} . \qquad (70)$$

Since these matrix transformations are equivalent to the symbols

$$\{I, C_2, \sigma(xz), \sigma(yz)\}$$

which define the point group C_{2v}, we can write the set of matrices

$$\left\{ \begin{bmatrix} 1 & 0 & 0 \\ 0 & 1 & 0 \\ 0 & 0 & 1 \end{bmatrix} , \begin{bmatrix} -1 & 0 & 0 \\ 0 & -1 & 0 \\ 0 & 0 & 1 \end{bmatrix} , \begin{bmatrix} 1 & 0 & 0 \\ 0 & -1 & 0 \\ 0 & 0 & 1 \end{bmatrix} , \begin{bmatrix} -1 & 0 & 0 \\ 0 & 1 & 0 \\ 0 & 0 & 1 \end{bmatrix} \right\}$$

as a <u>matrix</u> <u>representation</u> of point group C_{2v}. The three Cartesian vectors <u>x</u>, <u>y</u>, and <u>z</u> are said to form a <u>basis</u> for this particular matrix representation. In general, a representation of a group is given the symbol Γ, so this particular representation of point group C_{2v} in terms of the <u>basis</u> <u>set</u> $\{\underline{x}, \underline{y}, \underline{z}\}$ would be $\Gamma_{(x,y,z)}$.

Irreducible representations

We could also choose to represent the point group by considering the [1x1] matrix representations for each symmetry operation of the point group associated with each of the individual basis vectors, <u>x</u>, <u>y</u>, and <u>z</u>, separately:

C_{2v}:	{I	C_2	$\sigma(xz)$	$\sigma(yz)$}
Basis $\underline{x}$: Γ_x = { 1 ,		−1 ,	1 ,	−1 }
Basis $\underline{y}$: Γ_y = { 1 ,		−1 ,	−1 ,	1 }
Basis $\underline{z}$: Γ_z = { 1 ,		1 ,	1 ,	1 } .

In doing this, we have <u>reduced</u> the dimensionality of the original three-dimensional basis representation $(\underline{x}, \underline{y}, \underline{z})$, to express the group representation $\Gamma_{(x,y,z)}$ in terms of three one-dimensional bases <u>x</u>, <u>y</u> and

These particular one-dimensional representations $\{1,-1, 1,-1\}$ (Γ_x), $\{1,-1,-1, 1\}$ (Γ_y) and $\{1, 1, 1, 1\}$ (Γ_z) can not be reduced any further by any different choice of basis vectors, and are known as <u>irreducible representations</u> of the group.

The number of possible irreducible representations for any point group is fixed by the properties of the group. For the point group C_{2v}, there are four irreducible representations:

$$\Gamma_1 = \{1, 1, 1, 1\}$$
$$\Gamma_2 = \{1, 1,-1,-1\}$$
$$\Gamma_3 = \{1,-1, 1,-1\}$$
$$\Gamma_4 = \{1,-1,-1, 1\} \; .$$

The representations Γ_x, Γ_y and Γ_z constitute three of these four (Γ_4, Γ_3 and Γ_1 respectively). In the present case, the numbers 1 and -1 could be thought of as expressing the "symmetry character" of the x, y and z basis vectors under the symmetry operations of the point group $\overline{C}_{2v}$, if we define +1 to denote "symmetric" with respect to each operation (i.e., the vector remains the same), and -1 for "antisymmetric" behavior (the vector changes direction). This description gives a physical interpretation for the four possible irreducible representations of point group C_{2v}. No matter what basis is chosen, (a) all members of any basis must be symmetric to the identity (I) symmetry operation, and (b) it is not possible to be simultaneously antisymmetric with respect to the C_2 axis and both mirror planes, hence $\{1,-1,-1,-1\}$ does not constitute a valid irreducible representation; from this Γ_1 through Γ_4 are the only possible irreducible representations.

In discussions of molecular chemistry and spectroscopy, these irreducible representations are given conventional symbols. For the point group C_{2v}, the symbols used for the irreducible representations are

$$\Gamma_1 \equiv A_1$$
$$\Gamma_2 \equiv A_2$$
$$\Gamma_3 \equiv B_1$$
$$\Gamma_4 \equiv B_2 \; .$$

Reduction of matrix representations

In the above discussion, we chose a set of three Cartesian vectors (x, y, z) centered at the origin to provide a basis for the matrix representation of the point group C_{2v}. When dealing with molecular symmetry, it is usual to encounter bases which are much larger. Consider the molecule H_2O, which belongs to point group C_{2v}. We can constitute a basis by setting Cartesian vectors on each of the three atoms H(1), H(2) and O(3) (Fig. 16). With this nine-dimensional basis $[x_1,y_1,z_1,x_2,y_2,z_2,x_3,y_3,z_3]$, the symmetry operations of C_{2v} constitute a set of [9x9] matrices. For example the identity operation corresponds to the [9x9] unit matrix:

$$I = \begin{bmatrix} 1 & 0 & 0 & 0 & 0 & 0 & 0 & 0 & 0 \\ 0 & 1 & 0 & 0 & 0 & 0 & 0 & 0 & 0 \\ 0 & 0 & 1 & 0 & 0 & 0 & 0 & 0 & 0 \\ 0 & 0 & 0 & 1 & 0 & 0 & 0 & 0 & 0 \\ 0 & 0 & 0 & 0 & 1 & 0 & 0 & 0 & 0 \\ 0 & 0 & 0 & 0 & 0 & 1 & 0 & 0 & 0 \\ 0 & 0 & 0 & 0 & 0 & 0 & 1 & 0 & 0 \\ 0 & 0 & 0 & 0 & 0 & 0 & 0 & 1 & 0 \\ 0 & 0 & 0 & 0 & 0 & 0 & 0 & 0 & 1 \end{bmatrix} \; . \qquad (71)$$

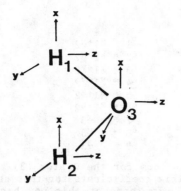

Figure 16. General Cartesian displacement vectors for H_2O (point group C_{2v}).

The C_2 operation causes atoms 1 and 2 to exchange positions, and reverses the direction of all x and y vectors, so the matrix representation of C_2 is:

$$C_2 = \begin{bmatrix} 0 & 0 & 0 & -1 & 0 & 0 & 0 & 0 & 0 \\ 0 & 0 & 0 & 0 & -1 & 0 & 0 & 0 & 0 \\ 0 & 0 & 0 & 0 & 0 & 1 & 0 & 0 & 0 \\ -1 & 0 & 0 & 0 & 0 & 0 & 0 & 0 & 0 \\ 0 & -1 & 0 & 0 & 0 & 0 & 0 & 0 & 0 \\ 0 & 0 & 1 & 0 & 0 & 0 & 0 & 0 & 0 \\ 0 & 0 & 0 & 0 & 0 & 0 & -1 & 0 & 0 \\ 0 & 0 & 0 & 0 & 0 & 0 & 0 & -1 & 0 \\ 0 & 0 & 0 & 0 & 0 & 0 & 0 & 0 & 1 \end{bmatrix} . \tag{72}$$

Now the matrix representation of the point group C_{2v} consists of a set of four [9x9] matrices, corresponding to the transformation matrices for I, C_2, $\sigma(xz)$ and $\sigma(yz)$ operating on the nine basis vectors of the H_2O molecule. If we had chosen to describe a larger molecule with C_{2v} symmetry, such as CH_2Cl_2, the associated matrix representation of the point group would be even larger (15x15). These large matrix representations are unwieldy to work with. However, we stated above that all of these matrix representations can be expressed in terms of a number of elementary, irreducible representations.

In the above example, we found that the irreducible representations corresponding to x (Γ_3 or B_1), y (Γ_4 or B_2) and z (Γ_1 or A_1) could easily be written down by examining the column matrix of basis vectors

$$\begin{bmatrix} x \\ y \\ z \end{bmatrix}$$

and the coefficients of the (diagonal) transformation matrices corresponding to the symmetry operations of the point group. In the present case for H_2O, we cannot immediately write down irreducible representations corresponding to the basis vectors x_1, y_1, z_1, x_2, y_2, z_2, x_3, y_3 and z_3, since the transformation matrices are not diagonal. However, it is possible to express the large, unreduced [9x9] matrix representation for H_2O in terms of elementary irreducible representations, if we choose a different set of basis vectors. This new set is chosen so that all of the terms in the transformation matrices lie along the matrix diagonals. To construct the new <u>basis set</u>, we simply take linear combinations of the Cartesian vectors for atoms (1) and (2), to give the new basis as

$$\begin{aligned}
\xi_1 &= x_1 + x_2 \\
\xi_2 &= x_1 - x_2 \\
\xi_3 &= y_1 + y_2 \\
\xi_4 &= y_1 - y_2 \\
\xi_5 &= z_1 + z_2 \\
\xi_6 &= z_1 - z_2 \\
\xi_7 &= x_3 \\
\xi_8 &= y_3 \\
\xi_9 &= z_3 \ .
\end{aligned} \tag{73}$$

Note that the basis vectors for the O atom (3) remain unchanged, since the transformation matrix coefficients for this center already lie along the matrix diagonals. In terms of this new basis, the transformation matrices for I, C_2, $\sigma(xz)$ and $\sigma(yz)$ are all diagonal:

$$I = \begin{bmatrix}
1 & 0 & 0 & 0 & 0 & 0 & 0 & 0 & 0 \\
0 & 1 & 0 & 0 & 0 & 0 & 0 & 0 & 0 \\
0 & 0 & 1 & 0 & 0 & 0 & 0 & 0 & 0 \\
0 & 0 & 0 & 1 & 0 & 0 & 0 & 0 & 0 \\
0 & 0 & 0 & 0 & 1 & 0 & 0 & 0 & 0 \\
0 & 0 & 0 & 0 & 0 & 1 & 0 & 0 & 0 \\
0 & 0 & 0 & 0 & 0 & 0 & 1 & 0 & 0 \\
0 & 0 & 0 & 0 & 0 & 0 & 0 & 1 & 0 \\
0 & 0 & 0 & 0 & 0 & 0 & 0 & 0 & 1
\end{bmatrix}$$

$$C_2 = \begin{bmatrix}
-1 & 0 & 0 & 0 & 0 & 0 & 0 & 0 & 0 \\
0 & 1 & 0 & 0 & 0 & 0 & 0 & 0 & 0 \\
0 & 0 & -1 & 0 & 0 & 0 & 0 & 0 & 0 \\
0 & 0 & 0 & 1 & 0 & 0 & 0 & 0 & 0 \\
0 & 0 & 0 & 0 & 1 & 0 & 0 & 0 & 0 \\
0 & 0 & 0 & 0 & 0 & -1 & 0 & 0 & 0 \\
0 & 0 & 0 & 0 & 0 & 0 & -1 & 0 & 0 \\
0 & 0 & 0 & 0 & 0 & 0 & 0 & -1 & 0 \\
0 & 0 & 0 & 0 & 0 & 0 & 0 & 0 & -1
\end{bmatrix}$$

$$\sigma(xz) = \begin{bmatrix}
1 & 0 & 0 & 0 & 0 & 0 & 0 & 0 & 0 \\
0 & -1 & 0 & 0 & 0 & 0 & 0 & 0 & 0 \\
0 & 0 & -1 & 0 & 0 & 0 & 0 & 0 & 0 \\
0 & 0 & 0 & 1 & 0 & 0 & 0 & 0 & 0 \\
0 & 0 & 0 & 0 & 1 & 0 & 0 & 0 & 0 \\
0 & 0 & 0 & 0 & 0 & -1 & 0 & 0 & 0 \\
0 & 0 & 0 & 0 & 0 & 0 & 1 & 0 & 0 \\
0 & 0 & 0 & 0 & 0 & 0 & 0 & -1 & 0 \\
0 & 0 & 0 & 0 & 0 & 0 & 0 & 0 & 1
\end{bmatrix} \tag{74}$$

$$\sigma(yz) = \begin{bmatrix}
-1 & 0 & 0 & 0 & 0 & 0 & 0 & 0 & 0 \\
0 & -1 & 0 & 0 & 0 & 0 & 0 & 0 & 0 \\
0 & 0 & 1 & 0 & 0 & 0 & 0 & 0 & 0 \\
0 & 0 & 0 & 1 & 0 & 0 & 0 & 0 & 0 \\
0 & 0 & 0 & 0 & 1 & 0 & 0 & 0 & 0 \\
0 & 0 & 0 & 0 & 0 & 1 & 0 & 0 & 0 \\
0 & 0 & 0 & 0 & 0 & 0 & -1 & 0 & 0 \\
0 & 0 & 0 & 0 & 0 & 0 & 0 & 1 & 0 \\
0 & 0 & 0 & 0 & 0 & 0 & 0 & 0 & 1
\end{bmatrix} \ .$$

Since these transformation matrices are diagonal, we can now write the transformation coeffients associated with each basis element as before.

Table 3: Individual basis representations for H_2O

$$\{ I \quad , \quad C_2 \quad , \quad \sigma(xz) \quad , \quad \sigma(yz)\}$$

$$
\begin{aligned}
\Gamma(\xi_1): & \quad \{ \ 1 \ , \ -1 \ , \ \ 1 \ , \ -1 \ \} \ \equiv \ \Gamma_3 \ (B_1) \\
\Gamma(\xi_2): & \quad \{ \ 1 \ , \ \ 1 \ , \ -1 \ , \ -1 \ \} \ \equiv \ \Gamma_2 \ (A_2) \\
\Gamma(\xi_3): & \quad \{ \ 1 \ , \ -1 \ , \ -1 \ , \ \ 1 \ \} \ \equiv \ \Gamma_4 \ (B_2) \\
\Gamma(\xi_4): & \quad \{ \ 1 \ , \ \ 1 \ , \ \ 1 \ , \ \ 1 \ \} \ \equiv \ \Gamma_1 \ (A_1) \\
\Gamma(\xi_5): & \quad \{ \ 1 \ , \ \ 1 \ , \ \ 1 \ , \ \ 1 \ \} \ \equiv \ \Gamma_1 \ (A_1) \\
\Gamma(\xi_6): & \quad \{ \ 1 \ , \ -1 \ , \ -1 \ , \ \ 1 \ \} \ \equiv \ \Gamma_4 \ (B_2) \\
\Gamma(\xi_7): & \quad \{ \ 1 \ , \ -1 \ , \ \ 1 \ , \ -1 \ \} \ \equiv \ \Gamma_3 \ (B_1) \\
\Gamma(\xi_8): & \quad \{ \ 1 \ , \ -1 \ , \ -1 \ , \ \ 1 \ \} \ \equiv \ \Gamma_4 \ (B_2) \\
\Gamma(\xi_9): & \quad \{ \ 1 \ , \ \ 1 \ , \ \ 1 \ , \ \ 1 \ \} \ \equiv \ \Gamma_1 \ (A_1)
\end{aligned}
$$

As shown at the right of this table, the one-dimensional group representations which correspond to each of the basis vectors $\Gamma(\xi_i)$ are the irreducible representations for the point group C_{2v}. By changing the basis for the group representation, we have succeeded in reducing the [9 x 9] matrix representation into elementary, one-dimensional representations. We could write this result in symbolic form for the irreducible representations:

$$\Gamma = 3A_1 + A_2 + 2B_1 + 3B_2 . \tag{75}$$

This type of analysis can be applied to many types of physical problems. For example, if we were interested in describing the vibrational modes of the water molecule, we would choose as our basis Cartesian displacement vectors centered on each atom, as in the above discussion. Reduction of the matrix representation of the point group with this basis would result in the irreducible representations above, which would represent the symmetries of the Cartesian degrees of freedom for the water molecule. These degrees of freedom include both internal (vibrational) and external (rotations and translations of the entire molecule) degrees of freedom. To deduce the symmetries of the vibrational modes, we must subtract the symmetry species (irreducible representations) corresponding to the three translational (T_x, T_y, T_z) and the three rotational motions (R_x, R_y, R_z) of the molecule:

$$
\begin{aligned}
\Gamma_{trans} &= A_1(T_z) + B_1(T_x) + B_2(T_y) \\
\Gamma_{rot} &= A_2(R_z) + B_1(R_y) + B_2(R_x).
\end{aligned}
\tag{76}
$$

This leaves the symmetries of the vibrational modes as:

$$
\begin{aligned}
\Gamma_{vib} &= \Gamma_{total} - \Gamma_{trans} - \Gamma_{rot} \\
&= 2A_1 + B_2 .
\end{aligned}
\tag{77}
$$

We could then use symmetry to deduce the infrared and Raman activity of these vibrational modes, and hence predict or interpret the vibrational spectra. If instead we had been interested in the electronic structure of the H_2O molecule, we could have used sets of s and p orbitals centered on the H and O atoms as the basis for the group representation, instead of sets of Cartesian vectors.

Similarity transformations

When we wish to apply symmetry to a given problem, we first choose a basis, which is equivalent to choosing a coordinate system for the problem. We then construct a matrix representation for the symmetry point group in terms of that basis. The next step is to diagonalize this matrix representation, in order to express the result in terms of irreducible representations. In the above example, we did this by inspection, by constructing appropriate linear combinations of the original basis vectors. This can can be unreliable and time-consuming for more complex systems, and much more efficient methods for reduction of group representations have been developed, based on the algebraic relations of group theory.

In general, once a basis set is defined, there is an associated transformation matrix A_R for each symmetry operation R of the point group:

$$A_R = \begin{bmatrix} a_{11} & a_{12} & a_{13} & a_{14} & \cdots a_{1n} \\ a_{21} & a_{22} & a_{23} & \cdots\cdots\cdots \\ a_{31} & a_{32} & \cdots\cdots\cdots\cdots \\ a_{41} & \cdots\cdots\cdots\cdots\cdots \\ \cdots\cdots\cdots\cdots\cdots\cdots \\ a_{n1} \cdots\cdots\cdots\cdots\cdots a_{nn} \end{bmatrix} . \tag{78}$$

The set of [n x n] matrices constitute the unreduced matrix representation for the point group for the problem. The object is to find some transformation of basis vectors which will simultaneously diagonalize all of the matrices A_R, and thus allow us to express the group in terms of irreducible representations. This can be expressed mathematically as finding the transformation matrix S which will simultaneously reduce all matrices A_R to block-diagonal form (D_R) via the similarity transformation

$$D_R = S^{-1}A_R S , \tag{79}$$

where S^{-1} is the inverse of matrix S.

Doubly- and triply-degenerate representations

For some point groups like C_{2v}, the irreducible representations are all one dimensional or singly degenerate; i.e., they all form sets of [1x1] matrices. Complete reduction of the matrix representation leads to fully diagonal matrices D_R with ± 1 along the diagonal. Other point groups for which this occurs include C_1, C_s, C_i, C_2, C_{2h}, D_2 and D_{2h}. This is not the case for the other point groups. In these, complete reduction of the original representation leads to block-factored matrices D_R, which contain [2x2] or [3x3] blocks along the leading diagonal that cannot be further reduced by any transformation. For example, consider the molecule NH_3 which possesses C_{3v} symmetry. We choose as our basis the three principal Cartesian axes and place the C_3 rotation axis coincident with the z-axis. We can immediately write down the matrix representations for the symmetry operations since we have already determined their general form in terms of the rotation angle Θ:

$$I = \begin{bmatrix} 1 & 0 & 0 \\ 0 & 1 & 0 \\ 0 & 0 & 1 \end{bmatrix} \qquad C_3^1 = \begin{bmatrix} -1/2 & -\sqrt{3}/2 & 0 \\ \sqrt{3}/2 & -1/2 & 0 \\ 0 & 0 & 1 \end{bmatrix}$$

$$C_3^2 = \begin{bmatrix} -1/2 & \sqrt{3}/2 & 0 \\ -\sqrt{3}/2 & -1/2 & 0 \\ 0 & 0 & 1 \end{bmatrix} \qquad \sigma_{v1} = \begin{bmatrix} 1 & 0 & 0 \\ 0 & -1 & 0 \\ 0 & 0 & 1 \end{bmatrix}$$

$$\sigma_{v2} = \begin{bmatrix} -1/2 & -\sqrt{3}/2 & 0 \\ -\sqrt{3}/2 & 1/2 & 0 \\ 0 & 0 & 1 \end{bmatrix} \qquad \sigma_{v3} = \begin{bmatrix} -1/2 & \sqrt{3}/2 & 0 \\ \sqrt{3}/2 & 1/2 & 0 \\ 0 & 0 & 1 \end{bmatrix} .$$

The representations for the individual basis vectors are:

$$\Gamma_z = \{\ 1\ ,\ 1\ ,\ 1\ ,\ 1\ ,\ 1\ ,\ 1\}$$

$$\Gamma_{x,y} = \{\ \begin{bmatrix} 1 & 0 \\ 0 & 1 \end{bmatrix}\ ,\ \begin{bmatrix} -1/2 & -\sqrt{3}/2 \\ \sqrt{3}/2 & -1/2 \end{bmatrix}\ ,\ \begin{bmatrix} -1/2 & \sqrt{3}/2 \\ -\sqrt{3}/2 & -1/2 \end{bmatrix}\ ,$$

$$\begin{bmatrix} 1 & 0 \\ 0 & -1 \end{bmatrix}\ ,\ \begin{bmatrix} -1/2 & -\sqrt{3}/2 \\ -\sqrt{3}/2 & 1/2 \end{bmatrix}\ ,\ \begin{bmatrix} -1/2 & \sqrt{3}/2 \\ \sqrt{3}/2 & 1/2 \end{bmatrix}\ \}. \qquad (80)$$

The representation for z is one-dimensional, or <u>singly degenerate</u>. This corresponds to one of the irreducible representations for point group C_{3v}, and is given the symbol A_1. The x and y vectors are interchanged by the symmetry operations, and there exists no similarity transformation that will further diagonalize their transformation matrices. When such a situation occurs, the resulting two-dimensional irreducible representation is said to be <u>doubly degenerate</u>, and is given the symmetry symbol E. (In the cubic point groups, all three directions in space are equivalent, and <u>triply degenerate</u> representations occur. These are given the symmetry symbol T or F).

There are three irreducible representations for point group C_{3v}; these can be written in tabular form as

Table 4: Irreducible representations for C_{3v} in matrix form

C_{3v}:	I	C_3^1	C_3^2	σ_{v1}	σ_{v2}	σ_{v3}
A_1	[1]	[1]	[1]	[1]	[1]	[1]
A_2	[1]	[1]	[1]	[-1]	[-1]	[-1]
E	$\begin{bmatrix} 1 & 0 \\ 0 & 1 \end{bmatrix}$	$\begin{bmatrix} -1/2 & -\sqrt{3}/2 \\ \sqrt{3}/2 & -1/2 \end{bmatrix}$	$\begin{bmatrix} -1/2 & \sqrt{3}/2 \\ -\sqrt{3}/2 & -1/2 \end{bmatrix}$	$\begin{bmatrix} 1 & 0 \\ 0 & -1 \end{bmatrix}$	$\begin{bmatrix} -1/2 & -\sqrt{3}/2 \\ -\sqrt{3}/2 & 1/2 \end{bmatrix}$	$\begin{bmatrix} -1/2 & \sqrt{3}/2 \\ \sqrt{3}/2 & 1/2 \end{bmatrix}$

Point group character tables

When two- and three-dimensional representations are present, instead of giving the full matrix form for irreducible representations, it is more convenient to give only the <u>characters</u> $\chi(A_R)$ of the matrices corresponding to each symmetry operation R. The character (or trace) of a square matrix is the sum of the elements along its leading diagonal. For example, for the matrix

$$A = \begin{bmatrix} a_{11} & a_{12} & a_{13} \\ a_{21} & a_{22} & a_{23} \\ a_{31} & a_{32} & a_{33} \end{bmatrix},$$

the character is

$$\chi(A) = a_{11} + a_{22} + a_{33}. \qquad (81)$$

The character of a matrix is not affected by a similarity transformation; i.e., if

$$A' = S^{-1}AS, \qquad (82)$$

the character is unchanged:

$$\chi(A') = \chi(A) \ . \tag{83}$$

This means that for matrix representations of groups, the characters of the transformation matrices are unaffected by any change of basis, including reduction of the representation. In terms of matrix characters, the irreducible representations for point group C_{3v} can be written as

Table 5a: Character table for point group C_{3v}

C_{3v}	I	$C_3{}^1$	$C_3{}^2$	σ_{v1}	σ_{v2}	σ_{v3}
A_1	1	1	1	1	1	1
A_2	1	1	1	-1	-1	-1
E	2	-1	-1	0	0	0

This type of table is known as a <u>point group character</u> table, and is the conventional way of tabulating symmetry information for molecular point groups. Character tables and related symmetry information are found in many books on molecular symmetry, such as Cotton (1971), or as reference sets of tables (Salthouse and Ware, 1972). In the above character table, it can be seen that certain sets of symmetry elements have exactly the same characters; for example $C_3{}^1$ and $C_3{}^2$ always behave in the same way in the three symmetry species A_1, A_2 and E, while σ_{v1}, σ_{v2} and σ_{v3} all behave in the same way. Each of the subsets of symmetry elements $\{C_3{}^1, C_3{}^2\}$ and $\{\sigma_{v1}, \sigma_{v2}, \sigma_{v3}\}$ are said to form a <u>class</u> within the point group, and their characters can be grouped together in a shortened form of the character table:

Table 5b: Point group character table for C_{3v} (short form)

C_{3v}	I	$2C_3$	$3\sigma_v$		
A_1	1	1	1	z	x^2+y^2, z^2
A_2	1	1	-1	R_z	
E	2	-1	0	(x,y), (R_x,R_y)	(x^2-y^2,xy), (xz,yz)

One useful result which can be derived from the mathematical theory of groups is that there are always just as many irreducible representations for a group as there are classes. In the above example, there are three classes of symmetry operations, $\{I\}$, $\{C_3{}^1, C_3{}^2\}$ and $\{\sigma_{v1}, \sigma_{v2}, \sigma_{v3}\}$, and three irreducible representations A_1, A_2 and E.

Also shown in the point group character table for C_{3v} above (in the first column to the right of the character table) are the symmetry species corresponding to x, y and z translations of the origin, and the symmetry species for rotations (R_x, R_y and R_z) about these axes. This useful information is usually tabulated along with the characters of the matrix representations in point group tables (e.g., Cotton, 1971; Salthouse and Ware, 1972). The infrared active vibrations of a molecule with C_{3v} symmetry would belong to symmetry species A_1 and E, the same species as those for the z and (x,y) translation vectors. This is because the condition for infrared activity of a vibration is that there must be a change in dipole moment $\Delta\mu$ during the vibration. Since the dipole moment change is a vector, this can be expressed in terms of Cartesian components $\Delta\mu_x$, $\Delta\mu_y$ and $\Delta\mu_z$, which belong to the same symmetry species as Cartesian translations of the origin, x, y and z. On the far right of the point group character table are shown the quadratic combinations of x, y and z (x^2, y^2, z^2, xy, xz, yz). These correspond to the symmetry species of Raman active vibrational modes of a molecule with point group symmetry C_{3v}. This is because the condition for Raman activity of a vibrational mode is that there must be a change in polarizability α during the vibration. This polarizability change can be expressed in terms of its tensor elements α_{xx}, α_{yy}, α_{zz}, α_{xy}, α_{xz} and α_{yz}, which transform in the same way as the quadratic combinations of x, y and z within the point group. Once the symmetry species for vibrational modes of a molecule have been determined, these observations allow immediate prediction of the number and type of infrared and Raman active vibrations for the molecule or crystal.

Reduction of a group representation

As an example of the application of these symmetry relations to a molecular problem, consider the vibrational motions of a molecule with N nuclei. The vibrational, translational and rotational degrees of freedom of this molecule can be described as linear combinations of Cartesian displacements (x_i, y_i, z_i) of each atom (these displacement vectors constitute the basis for the problem). There will be 3N degrees of freedom, so the matrix representation for the molecular motions within the point group for the molecule will be a set of [3Nx3N] matrices, one for each symmetry operation R. In general, this matrix representation will be reducible, to give a set of R block-diagonal matrices, where each set of [1x1], [2x2] or [3x3] blocks corresponds to an irreducible representation. The representation for the molecular motions can be expressed as a sum of these irreducible representations, which can then be separated into internal (vibrational) and external (translations and rotations) degrees of freedom, and the infrared and Raman activities of the vibrational modes deduced.

Let the original, unreduced representation (set of [3Nx3N] matrices) be Γ; χ_R^Γ is the character of the matrix within this representation corresponding to the symmetry operation R. The point group contains γ symmetry species, or distinct irreducible representations, and the characters of the matrices corresponding to each operation R within these are χ_R^1, χ_R^2, $\chi_R^3 \ldots \chi_R^\gamma$. These irreducible representations will appear a certain number of times in the completely reduced form of Γ, i.e.,

$$\chi_R^\Gamma = a_1 \chi_R^1 + a_2 \chi_R^2 + a_3 \chi_R^3 + \ldots + a_\gamma \chi_R^\gamma \quad . \tag{84}$$

There are γ unknowns a_γ in this expression, and there will be one equation of this type for each class of symmetry operation of the point group. Since the number of classes is always equal to the number of irreducible representations γ, the coefficients a_γ can be determined uniquely. Once the coefficients are known, the original matrix representation can be expressed as the direct sum of irreducible representations:

$$\Gamma = a_1\Gamma_1 + a_2\Gamma_2 + a_3\Gamma_3 + \ldots + a_\gamma\Gamma_\gamma \quad , \tag{85}$$

where Γ_1, Γ_2, etc. are usually given their appropriate symmetry symbols A_1, A_2, E, etc. within the molecular point group.

Example: the vibrational modes of water. It is useful to illustrate this process with a simple example. We have already discussed the vibrational modes of the water molecule H_2O earlier, but not in a systematic way. The molecule belongs to point group C_{2v}, and contains three atoms, hence there are nine degrees of freedom. If we set Cartesian displacement vectors $(x_1,y_1,z_1,x_2,y_2,z_2,x_3,y_3,z_3)$ on each atom as before (Fig. 16), the matrix representation of the identity operation is

$$
I = \begin{bmatrix}
1 & 0 & 0 & 0 & 0 & 0 & 0 & 0 & 0 \\
0 & 1 & 0 & 0 & 0 & 0 & 0 & 0 & 0 \\
0 & 0 & 1 & 0 & 0 & 0 & 0 & 0 & 0 \\
0 & 0 & 0 & 1 & 0 & 0 & 0 & 0 & 0 \\
0 & 0 & 0 & 0 & 1 & 0 & 0 & 0 & 0 \\
0 & 0 & 0 & 0 & 0 & 1 & 0 & 0 & 0 \\
0 & 0 & 0 & 0 & 0 & 0 & 1 & 0 & 0 \\
0 & 0 & 0 & 0 & 0 & 0 & 0 & 1 & 0 \\
0 & 0 & 0 & 0 & 0 & 0 & 0 & 0 & 1
\end{bmatrix} . \tag{86}
$$

The character of this matrix (which forms part of the unreduced representation) is $\chi_I{}^\Gamma = 9$. The unreduced matrix representation for the C_2 operation is

$$
C_2 = \begin{bmatrix}
0 & 0 & 0 & -1 & 0 & 0 & 0 & 0 & 0 \\
0 & 0 & 0 & 0 & -1 & 0 & 0 & 0 & 0 \\
0 & 0 & 0 & 0 & 0 & 1 & 0 & 0 & 0 \\
-1 & 0 & 0 & 0 & 0 & 0 & 0 & 0 & 0 \\
0 & -1 & 0 & 0 & 0 & 0 & 0 & 0 & 0 \\
0 & 0 & 1 & 0 & 0 & 0 & 0 & 0 & 0 \\
0 & 0 & 0 & 0 & 0 & 0 & -1 & 0 & 0 \\
0 & 0 & 0 & 0 & 0 & 0 & 0 & -1 & 0 \\
0 & 0 & 0 & 0 & 0 & 0 & 0 & 0 & 1
\end{bmatrix} . \tag{87}
$$

In this case ($R = C_2$), $\chi_R{}^\Gamma = -1$. Note that the only terms which contribute to the character of the unreduced matrix representation are those which lie along the leading diagonal, i.e., those which correspond to atom centers (in this case oxygen, atom 3) which are unshifted by the symmetry operation (i.e., those atoms which lie on the symmetry element being considered). Further, the value of the character contributed by each unshifted atom is easily obtained from the general matrix form for symmetry transformations in terms of a rotation angle Θ (see above). The character for a proper rotation $C_n{}^m$ is

$$
\chi(C_n{}^m) = 2 \cos m\Theta + 1 \tag{88}
$$

while the character for an improper rotation $S_n{}^m$ is

$$
\chi(S_n{}^m) = 2 \cos m\Theta - 1 , \tag{89}
$$

where $\Theta = 360°/n$. Note that the character associated with a mirror plane σ (which is equivalent to $S_1{}^1$) is $\chi_\sigma = 1$; the character for the identity (per unshifted atom) is $\chi_I = 3$, and the character for an inversion center is $\chi_i = -3$.

Using these observations, we can write down the number of unshifted atoms and their contributions to the total character for all four symmetry operations of point group C_{2v}:

	I	C_2	$\sigma(xz)$	$\sigma(yz)$
Number of unshifted atoms	3	1	1	3
Contribution to χ^Γ per atom	3	–1	1	1
Total character χ_R^Γ	9	–1	1	1

The characters of the irreducible representations in point group C_{2v} are:

	I	C_2	$\sigma(xz)$	$\sigma(yz)$
A_1	1	1	1	1
A_2	1	1	–1	–1
B_1	1	–1	1	–1
B_2	1	–1	–1	1

Now we can set up the simultaneous equations

$$\chi_R^\Gamma = a_1 \chi_R^{A_1} + a_2 \chi_R^{A_2} + a_3 \chi_R^{B_1} + a_4 \chi_R^{B_2} : \qquad (90)$$

$$
\begin{aligned}
I: \quad 9 &= a_1 + a_2 + a_3 + a_4 \\
C_2: \quad -1 &= a_1 + a_2 - a_3 - a_4 \\
\sigma(xz): \quad 1 &= a_1 - a_2 + a_3 - a_4 \\
\sigma(yz): \quad 3 &= a_1 - a_2 - a_3 + a_4 .
\end{aligned}
$$

Solution of these simultaneous equations gives the coefficients: $a_1 = 3$, $a_2 = 1$, $a_3 = 2$, $a_4 = 3$. The direct sum of the irreducible representations for the Cartesian degrees of freedom for H_2O is then:

$$\Gamma = 3A_1 + A_2 + 2B_1 + 3B_2 . \qquad (91)$$

Table 6: Point group character table for C_{2v}

C_{2v}	I	C_2	$\sigma(xz)$	$\sigma(yz)$		
A_1	1	1	1	1	z	x^2, y^2, z^2
A_2	1	1	–1	–1	R_z	xy
B_1	1	–1	1	–1	x, R_y	xz
B_2	1	–1	–1	1	y, R_x	yz

We can then examine the point group character table for H_2O (Table 6) to find the symmetry species of the translational and rotational external degrees of freedom, Γ^{trans} ($A_1 + B_1 + B_2$) and Γ^{rot} ($A_2 + B_1 + B_2$). These can be subtracted from the total Cartesian degrees of freedom to give the vibrational mode symmetries as

$$\Gamma^{vib} = 2A_1 + B_2 . \qquad (92)$$

We can further examine the character table as described above to predict the infrared and Raman activities of these vibrations: in the case of H_2O, all three modes are active in both the Raman and the infrared spectra.

Orthogonality

One of the most useful results of formal group theory to be applied to molecular symmetry are the statements concerning <u>orthogonality</u> properties of group representations (for a detailed treatment, consult Hammermesh (1962)). The orthogonality relations give rise to simple algorithms for reducing group representations, which are essential in applying symmetry to molecular problems. We can illustrate the orthogonality relations by examining various sums and products of the irreducible representations of the group C_{3v} in matrix form (Table 4).

First we define a nomenclature for each matrix element: for a symmetry element R in an irreducible representation γ, subscripts i and j denote the row and column of the matrix element $R_{ij}(\gamma)$. For example, within the E representations, if $R = \sigma_{v2}$, $R_{11}(E) = -1/2$, and $R_{12}(E) = -\sqrt{3}/2$ (see Table 4). Note that for $\gamma = A_1$ or A_2, all of the $R_{ij} = \pm 1$. Now for a given symmetry operation R, choose any two matrix elements $R_{ij}(\mu)$ and $R_{kl}(\nu)$, where at least one of the following conditions is true: $i \neq k$, $j \neq l$, or $\mu \neq \nu$. We multiply $R_{ij}(\mu)$ and $R_{kl}(\nu)$ together, and take the sum over all such pairs for all of the symmetry operations of the group:

$$\sum R_{ij}^{(\mu)} R_{kl}^{(\nu)} \ .$$

The number of terms in this sum is equal to the order of the group h (for C_{3v}, $h = 6$). For example, in the present case, let $\mu = A_2$ and $\nu = E$, and take $ij = 11$ and $kl = 22$. We will form the sum of $R_{11}(A_2)R_{22}(E)$ over all operations R:

R	$R_{11}^{(A2)}$	$R_{22}^{(E)}$	$R_{11}^{(A2)} \cdot R_{22}^{(E)}$
I	1	1	1
C_3^1	1	$-1/2$	$-1/2$
C_3^2	1	$-1/2$	$-1/2$
σ_{v1}	-1	-1	1
σ_{v2}	-1	$1/2$	$-1/2$
σ_{v3}	-1	$1/2$	$-1/2$.

It is obvious from this table that $\Sigma R_{11}^{(A_2)} R_{22}^{(E)} = 0$.

This sum of products of like matrix elements turns out be zero for all such pairs, i.e.,

$$\sum R_{ij}^{(\mu)} R_{kl}^{(\nu)} = 0 \tag{93}$$

if at least one of $i \neq j$, $k \neq l$ or $\mu \neq \nu$ is true. This states that two different irreducible representations μ and ν, are <u>orthogonal</u> to each other, and that within a two- or three-dimensional representation, different matrix elements ij and kl are also orthogonal.

Now examine the case where <u>all</u> of $i = j$, $k = l$ <u>and</u> $\mu = \nu$ <u>is</u> true, i.e.,

$$\sum R_{ij}^{(\mu)} R_{ij}^{(\mu)} = \sum (R_{ij}^{(\mu)})^2 . \tag{94}$$

For example, choose $\mu = E$ and $ij = 12$ (see Table 4 for numerical values):

$$\sum (R_{12}^{(E)})^2 = 0^2 + (\frac{-\sqrt{3}}{2})^2 + (\frac{\sqrt{3}}{2})^2 + 0^2 + (\frac{-\sqrt{3}}{2})^2 + (\frac{\sqrt{3}}{2})^2 = 3 . \tag{95}$$

As a second example, choose $\mu = A_1$:

$$\sum (R_{11}^{(A_1)})^2 = 1^2 + 1^2 + 1^2 + 1^2 + 1^2 + 1^2 = 6 . \tag{96}$$

Note that the order of the group is 6, and that the degeneracies (n) of representations E and A_1 are respectively: $n_E = 2$ and $n_{A1} = 1$. In general,

$$\sum (R_{ij}^{(\mu)})^2 = \frac{h}{n_\mu} , \tag{97}$$

where h is the order of the group, and n_μ is the degeneracy of the representation of interest. Finally, the above two cases can be summarized in "<u>the</u> <u>great</u> <u>orthogonality</u> <u>relation</u>" for group representations:

$$\sum R_{ij}^{(\mu)} R_{kl}^{(\nu)} = \delta_{ik} \delta_{jl} \delta_{\mu\nu} \frac{h}{n_\mu} \tag{98}$$

where the δ's are the <u>Kronecker delta</u> symbol; for example $\delta_{ik} = 0$ for $i \neq k$, but $\delta_{ik} = 1$ for $i = k$.

Orthogonality properties of matrix characters

If we express the irreducible representations in terms of the characters of their matrix representations, the orthogonality relations become even simpler. The character χ_R of matrix R is defined as the sum of its leading diagonal elements. For triply degenerate species,

$$\chi_R^\mu = R_{11}^\mu + R_{22}^\mu + R_{33}^\mu . \tag{99}$$

(For singly and doubly degenerate species, $R_{22} = R_{33} = 0$ and $R_{33} = 0$ respectively). The square of the character is

$$(\chi^{\mu}_{R})^{2} = (R^{\mu}_{11})^{2} + (R^{\mu}_{22})^{2} + (R^{\mu}_{22})^{2} + 2\ (R^{\mu}_{11}R^{\mu}_{22} + R^{\mu}_{22}R^{\mu}_{33} + R^{\mu}_{33}R^{\mu}_{11})\ . \quad (100)$$

Now consider the sum of squares of the characters over all operations in the group:

$$\Sigma_{R}(\chi^{\mu}_{R})^{2} = \Sigma_{R}(R^{\mu}_{11})^{2} + \Sigma_{R}(R^{\mu}_{22})^{2} + \Sigma_{R}(R^{\mu}_{22})^{2} + 2\Sigma_{R}(R^{\mu}_{11}R^{\mu}_{22} + R^{\mu}_{22}R^{\mu}_{33} + R^{\mu}_{33}R^{\mu}_{11})\ . \quad (101)$$

From the great orthogonality relation stated above (equation (98)), the summations in $R_{11}R_{22}$, $R_{22}R_{33}$ and $R_{33}R_{11}$ are all equal to zero, while the sums of squares are

$$\Sigma_{R}(R^{\mu}_{11})^{2} = \Sigma_{R}(R^{\mu}_{22})^{2} = \Sigma_{R}(R^{\mu}_{33})^{2} = \frac{h}{n_{\mu}}\ . \quad (102)$$

However, for $n_{\mu} = 1$ (i.e., a singly degenerate species), $R_{22} = R_{33} = 0$, and for $n_{\mu} = 2$ (doubly degenerate), $R_{33} = 0$, so in general

$$\Sigma_{R}(\chi^{\mu}_{R})^{2} = h\ . \quad (103)$$

If instead of taking the sums of squares of characters, the products $\chi^{\mu}\chi^{\nu}$ are taken and summed over all operations of the group, we obtain

$$\Sigma_{R}\ \chi^{\mu}_{R}\chi^{\nu}_{R} = 0\ . \quad (104)$$

These two results, (101) and (102), can be combined into the great orthogonality relation for group characters:

$$\Sigma_{R}\ \chi^{\mu}_{R}\chi^{\nu}_{R} = \delta_{\mu\nu}\ h\ . \quad (105)$$

This form of the orthogonality relation is the most convenient for reducing a general group representation.

An alternative method for reduction of group representations

Above, we presented a method for reducing group representations by solving sets of simultaneous equations. This method is completely general, but can be cumbersome for more complicated systems. There is an alternative and more convenient way to reduce a group representation, which relies on the orthogonality properties of group representations. The expression for the sum of characters for a given symmetry operation R was

$$\chi^{\Gamma}_{R} = a_{1}\chi^{1}_{R} + a_{2}\chi^{2}_{R} + a_{3}\chi^{3}_{R} + \ldots + a_{\gamma}\chi^{\gamma}_{R}\ . \quad (84)$$

We can multiply both sides by χ_{R}^{1}, and take the sum over all operations R:

$$\Sigma_{R}\chi^{1}_{R}\chi^{\Gamma}_{R} = a_{1}\Sigma_{R}\chi^{1}_{R}\chi^{1}_{R} + a_{2}\Sigma_{R}\chi^{1}_{R}\chi^{2}_{R} + \ldots + a_{\gamma}\Sigma_{R}\chi^{1}_{R}\chi^{\gamma}_{R}\ . \quad (106)$$

From the orthogonality relation for matrix characters (103), this becomes

$$\Sigma \chi_R^1 \chi_R^\Gamma = a_1 \Sigma \chi_R^1 \chi_R^1 = a_1 h \qquad (107)$$

i.e.,

$$a_1 = \frac{1}{h} \Sigma \chi_R^1 \chi_R^\Gamma$$

and similarly for the other irreducible representations:

$$a_2 = \frac{1}{h} \Sigma \chi_R^2 \chi_R^\Gamma \;, \quad a_3 = \frac{1}{h} \Sigma \chi_R^3 \chi_R^\Gamma \;,$$

and in general,

$$a_\gamma = \frac{1}{h} \Sigma \chi_R^\gamma \chi_R^\Gamma \;. \qquad (108)$$

We illustrate the use of this reduction formula for the vibrational modes of another simple molecule, methane (CH_4).

Example: the vibrational modes of methane

Methane is a tetrahedral molecule, and belongs to point group T_d. There are five atoms, hence fifteen Cartesian degrees of freedom. As before, we construct a table of the number of atoms lying on each symmetry element, and the contribution to the total character of the unreduced Cartesian representation per unshifted atom:

T_d	I	$8C_3$	$3C_2$	$6S_4$	$6\sigma_d$
Number of unshifted atoms	5	2	1	1	3
Contribution to χ_R^Γ per atom	3	0	-1	-1	1
χ_R^Γ	15	0	-1	-1	3

The character table for T_d ($h = 24$) is:

T_d	I	$8C_3$	$3C_2$	$6S_4$	$6\sigma_d$
A_1	1	1	1	1	1
A_2	1	1	1	-1	-1
E	2	-1	2	0	0
F_1	3	0	-1	1	-1
F_2	3	0	-1	-1	1

Using the reduction formula for group characters given above, the coefficients of the irreducible representations for the Cartesian degrees of freedom of methane can be easily obtained:

$$aA_1 = 1/24[(1x15) + 8(1x0) + 3(1x-1) + 6 (1x-1) + 6(1x3)] = 1$$

$$aA_2 = 1/24[(1x15) + 8(1x0) + 3(1x-1) + 6(-1x-1) + 6(-1x3)] = 0$$

$$aE = 1/24[(2x15) + 8(-1x0) + 3(2x-1) + 6(0x-1) + 6(0x3)] = 1$$

$$aF_1 = 1/24[(3x15) + 8(0x0) + 3(-1x-1) + 6(1x-1) + 6(-1x3)] = 1$$

$$aF_2 = 1/24[(3x15) + 8(0x0) + 3(-1x-1) + 6(-1x-1) + 6(1x3)] = 3 ,$$

so the total degrees of freedom are

$$\Gamma^{tot} = A_1 + E + F_1 + 3F_2 . \tag{109}$$

From tabulated forms of the character table (e.g., Salthouse and Ware, 1972), it is found that the x-, y- and z-translations of the entire molecule are triply degenerate and belong to symmetry species F_2, while the three rotations (R_x, R_y and R_z) belong to species F_1. This leaves the vibrational degrees of freedom for CH_4 as

$$\Gamma^{vib} = A_1 + E + 2F_2 . \tag{110}$$

The infrared active modes belong to the same symmetry species as the x-, y- and z- translation vectors, so the infrared modes for methane are

$$\Gamma^{IR} = 2F_2 . \tag{111}$$

The Raman-active vibrations belong to the same symmetry species as the quadratic combinations of x, y and z. These are A_1 ($x^2+y^2+z^2$), E ($2z^2-x^2-y^2$, x^2-y^2), and F_2 (xy, xz, yz), so the Raman active modes of methane are

$$\Gamma^R = A1 + E + 2F_2 . \tag{112}$$

From this analysis, we can predict that the vibrational spectrum of methane should contain two infrared active modes and four Raman active vibrations. Further, two of the Raman modes should coincide with the two IR bands. This prediction is experimentally confirmed (e.g., Herzberg, 1945).

Infrared and Raman active vibrations of crystals

In an ordered crystal structure, only those vibrational modes for which all unit cells vibrate in phase can give rise to an infrared or Raman spectrum (see Chapter 4), so it is only necessary to consider the unit cell symmetry to determine the number and species of infrared and Raman active modes. For simple structures, the symmetry methods described above can be easily used. For example, MgO has the rock salt structure with space group Fm3m. The associated point group is m3m in terms of the Hermann-Mauguin symbols, or O_h in the Schonflies notation. There are two atoms in the primitive unit cell: one Mg and one O on special positions (0,0,0) and (1/2,1/2,1/2) respectively. We can treat this cell as a molecule, and examine the behavior of x-, y- and z-translations of Mg and O within the unit cell point group, or alternatively determine the character for the Mg and O atoms, as described above for H_2O or CH_4. After reduction of the group representation, the symmetry species for Mg and O displacements are

$$\Gamma = 2F_{1u} . \tag{113}$$

Included in these are the x-, y- and z- translations of the entire crystal, which belong to species F_{1u} within point group O_h. The vibrational species for MgO when all unit cells vibrate in phase (this is termed the <u>Brillouin</u> <u>zone</u> <u>center</u>, or at wave vector $\mathbf{k} = 0$: see Chapter 4) are then simply

$$\Gamma_{vib} = F_{1u} \quad . \tag{114}$$

From examination of the point group character table for O_h, we find that since x-, y- and z- vectors transform like F_{1u} species, these are the infrared active species. We then predict that MgO (and any crystal with the rock salt structure) should have one infrared absorption band of F_{1u} symmetry, which will appear for x-, y- and z- polarizations of the incident radiation with respect to the crystal axes, but no Raman spectrum.

This type of analysis can be carried out for any crystal, but can become difficult and laborious for complex crystal structures. A much more convenient method for determining the symmetry of optical modes of crystals is to determine the species of x-, y- and z- translations for each atom within the <u>site</u> <u>group</u> for each atom in the unit cell, then <u>correlate</u> these site group species with the symmetry species for the point group of the unit cell. Correlation tables used for this purpose are often found along with tables of point group symmetry properties (Wilson et al., 1955; Cotton, 1971; Salthouse and Ware, 1972; Fateley et al., 1972; Decius and Hexter, 1977).

As an example, consider the vibrational modes of ilmenite, $FeTiO_3$. The space group is R3, or unit cell point group C_{3i} ($\equiv S_6$). There are two formula units of $FeTiO_3$ within the primitive unit, or 10 atoms, so we expect $(3\times10)-3 = 27$ vibrational modes. Within the structure, both Fe and Ti occupy sites with symmetry C_3, while the oxygen atoms lie on general positions with symmetry C_1. From point group tables, the symmetry species for Cartesian displacements of the metal atoms belong to species $A(z) + E(x,y)$ within point group C_3. The total contribution from all four metal atoms is then $4A + 4E$ (C_3 sites). From group correlation tables, A and E species of point group C_3 correlate into symmetry species A_g, E_g, A_u and E_u of point group C_{3i}. The Fe and Ti vibrations of ilmenite then give rise to

$$\Gamma_{(Fe,Ti)} = 2A_g + 2E_g + 2A_u + 2E_u \tag{115}$$

symmetry species of the crystal. The twelve oxygen atoms on sites with symmetry C_1 likewise give rise to

$$\Gamma_{(O)} = 3A_g + 3E_g + 3A_u + 3E_u \tag{116}$$

symmetry species. These are then combined to give the total symmetry species for atomic displacement in the ilmenite crystal:

$$\Gamma_{tot} = 5A_g + 5E_g + 5A_u + 5E_u \quad . \tag{117}$$

Finally, from the point group character table, the symmetry species corresponding to translation of the entire crystal are $A_u + E_u$, so the symmetry species of the true vibrational modes are:

$$\Gamma_{vib} = 5A_g + 5E_g + 4A_u + 4E_u \quad . \tag{118}$$

Again from the point group character tables, the A_g and E_g modes will be Raman active. Both A_g and E_g modes will be present in parallel polarization experiments (x^2, y^2 and z^2 diagonal products appear in both A_g and E_g representations), but only E_g modes will appear in crossed polarized experiments (xy, xz and yz off-diagonal products). The A_u and E_u modes will be infrared active. The E_u modes will appear in spectra with the incident light polarized normal for to the c axis (x- and y-polarizations), while the A_u modes will be polarized parallel to z. Finally, because of the center of symmetry, we expect no coincidences between the peaks observed in the infrared spectrum, and those observed in the Raman.

This type of symmetry analysis is extremely useful in interpretation of observed spectra. In the above example for ilmenite, displacements of both metal atoms and oxygen contribute to all symmetry species, hence all infrared and Raman bands may contain mixed components of metal and oxygen vibrations. However, in some cases, displacements of particular atoms in the structure may not contribute to certain symmetry species, which can aid in describing the spectra in terms of atomic vibrations. For example, in the orthorhombic Pbnm perovskites ABX_3 (including $MgSiO_3$ perovskite), the B cations occupy sites on centers of inversion, so that displacements of the B cations cannot contribute to the Raman spectrum, only to the infrared spectrum.

A further extension of this type of analysis can be applied when "molecular" groups can be identified within the structure. For example, in the case of calcite and other carbonates, the CO_3^{2-} anions can be identified as distinct structural species. The free CO_3^{2-} anion is trigonal planar, with symmetry D_{3h}. The symmetry species of the internal modes of the CO_3^{2-} group are

$$
\begin{aligned}
\nu_1 &- A_1' \\
\nu_2 &- A_2'' \\
\nu_3 &- E' \\
\nu_4 &- E' \ ,
\end{aligned}
$$

where ν_1, ν_2, ν_3, and ν_4 correspond to symmetric stretching, symmetric out-of-plane deformation, asymmetric stretching, and OCO bending vibrations respectively (White, 1974). The frequencies of these vibrations are known from work on aqueous solutions of alkali carbonates. In the calcite structure, there are two carbonate units in the primitive unit cell (Z=2). The site symmetry of each anion in the crystal is reduced to D_3, so the internal mode symmetries must be correlated from point group D_{3h} for the free anion to D_3:

D_{3h} species		D_3 species
A_1'	--------	A_1
A_2'	--------	A_2
E'	--------	E
A_1''	--------	A_1
A_2''	--------	A_2
E''	--------	E

The unit cell symmetry is R3c, corresponding to point group D_{3d}. The symmetry species of the internal vibrations of both carbonate anions must then be correlated from point group D_3 to D_{3d}:

D_3 species		D_{3d} species
A_1	-------	$A_{1g} + A_{1u}$
A_2	-------	$A_{2g} + A_{2u}$
E	-------	$E_g + E_u$.

From these correlations, we can deduce that the internal modes of the CO_3^{2-} anions give rise to internal crystal modes within the calcite structure with symmetries:

$$\nu_1 \dashrightarrow A_{1g} + A_{1u}$$
$$\nu_2 \dashrightarrow A_{2g} + A_{2u}$$
$$\nu_3 \dashrightarrow E_g + E_u$$
$$\nu_4 \dashrightarrow E_g + E_u \ .$$

For point group D_{3d}, A_{1g} and E_g modes are Raman active, and A_{2u} and E_u modes are infrared active. The A_{2g} and A_{1u} modes are spectroscopically inactive. We then predict that one Raman peak of A_{1g} symmetry should arise from the ν_1 symmetric stretch of the CO_3^{2-} anions in the calcite structure, one infrared band should be derived from ν_3, and two Raman and infrared bands would be associated with bending and deformation modes of the CO_3^{2-} groups. From the knowledge of the infrared and Raman spectra of free CO_3^{2-}, these modes can be easily identified in the spectra of calcite, and can be used to make definitive structural statements about calcite and related carbonate structures (e.g., White, 1974; Scheetz and White, 1977; Bischoff et al., 1985).

This type of symmetry analysis for large molecules and crystals can provide detailed structural information from interpretation of the observed spectra, and is a standard method of analyzing infrared and Raman data. A number of worked examples relevant to mineralogy can be found in Fateley et al. (1972), Farmer (1974), Decius and Hexter (1977) and McMillan (1985).

Selection rules for electronic transitions

Symmetry arguments can also be used to predict or rationalize the relative intensity and polarization dependence of absorption bands associated with transitions between electronic states. The total electronic state for an isolated atom or ion in the gas phase is described by an atomic term symbol. For example, the ground state of gaseous atomic O (outer electronic configuration $2s^2 2p^4$) is a ^3P state, and low-lying excited electronic states are ^1D and ^1S states. These symbols as written refer to the total orbital angular momentum L (an S state has L=0, a P state has L=1, and a D state has L=2), and the total electron spin multiplicity 2S+1 (singlet, triplet, etc.), and transitions between these states are determined by the selection rules discussed earlier. For polyelectronic atoms, the selection rules become $\Delta L = \pm 1$ and $\Delta S = 0$. These selection rules, based on the orbital angular momentum and electronic spin, can be used to interpret the absorption spectra of isolated atoms and ions.

If an ion is placed in a crystalline environment, it is usual to correlate the free ion spectroscopic states with the symmetry of the crystalline environment. For example, the electronic ground state of a free Fe^{2+} (d^6) ion has term symbols ^5D. The spectroscopic symbol for orbital angular momentum can alternatively be given a symmetry symbol within the point group for a sphere (point group O(3)): $D \equiv D_g^{(2)}$ (degeneracy 5). This symmetry symbol can then be correlated into the symmetry species for an octahedral point group O_h:

Sphere: point group O(3)   Octahedron: point group O_h

$$D_g^{(2)} \ \text{----------------} \ E_g + T_{2g}$$

This shows that the application of an octahedral field to a free Fe^{2+}

ion results in splitting the lowest electronic state of the free ion into two states of E_g and T_{2g} symmetry. It is found that the the T_{2g} state is lower in energy. We can now use symmetry arguments to decide whether or not a transition between these states is allowed.

The strongest interaction with light occurs via the oscillating electric field of the electromagnetic radiation and the electric dipole field of the electronic wave function. The transition probability for an electric dipole transition between electronic states m and n is proportional to the square of the transition moment integral (see (44) and (45))

$$M_{nm} = \int \psi_n \mu \psi_m \, d\tau \,, \tag{119}$$

where ψ_m and ψ_n represent the wave functions for the intial and final states of the system, and μ is the electric dipole moment operator, which can be expressed in terms of its Cartesian component vectors μ_x, μ_y and μ_z, aligned along the three Cartesian axes. In order for this integral to be non-zero by symmetry, the combined function $\psi_n \mu \psi_m$ must contain a component which will span the totally symmetric representation of the molecular point group (in the case of an ion in an octahedral field, this would be the A_{1g} representation). The reason for this is that the interaction with light cannot change the symmetry properties of the electronic wave function. Considering the example of Fe^{2+} in an octahedral field, we can determine whether or not a transition between the T_{2g} and E_g states is allowed by symmetry. The direct product of the symmetry species of the initial and final electronic states is $T_{2g} \times E_g$ = $T_{1g} + T_{2g}$. The x, y and z dipole moment vector components belong to symmetry species T_{1u} within point group O_h, and the product

$$T_{1u} \times (T_{1g} + T_{2g}) = A_{1u} + A_{2u} + 2T_{1u} + 2T_{2u} \,. \tag{120}$$

Tables of such direct products are given along with many point group character tables (e.g., Salthouse and Ware, 1972). This final product does not contain an A_{1g} species, hence a transition between the T_{2g} and E_g electronic states of Fe^{2+} in an octahedral complex or crystal site would not be allowed; i.e., we would expect no absorption of visible light due to a transition between d electrons centered on the Fe^{2+} ion. Such transitions are in fact observed (see chapter by G. Rossman) due to vibronic interaction (an interaction between vibrational and electronic degrees of freedom: the vibrational motion distorts the symmetry of the electronic wave function), but are very weak compared to other, symmetry-allowed electronic transitions.

REFERENCES

Ashcroft, N.W. and Mermin, N.D. (1976) "Solid State Physics", Holt-Saunders, Philadelphia.

Atkins, P.W. (1982) "Physical Chemistry", 2nd. ed., W.H. Freeman, San Francisco.

Berry, R.S., Rice, S.A. and Ross, J. (1980) "Physical Chemistry", John Wiley and Sons, New York.

Birman, J.L. (1984) "Theory of Crystal Space Groups and Lattice Dynamics - Infrared and Raman Optical Processes of Insulating Crystals", Springer-Verlag, Berlin.

Bischoff, W.D., Sharma, S.K. and Mackenzie, F.T. (1985) Carbonate ion disorder in synthetic and biogenic magnesian calcites: a Raman spectral study. Am. Mineral., 70, 581-589.

Bohr, N. (1913) Constitution of atoms and molecules. Phil. Mag., 26, 1-

25.

Born, M., Heisenberg, W. and Jordan, P. (1926) On quantum mechanics. Z. Phys., 35, 557-615.

Bouckaert, L.P., Smoluchowski, R. and Wigner, E. (1936) Theory of Brillouin zones and symmetry properties of wave functions in crystals. Phys. Rev., 50, 58-67.

de Broglie, L.V. (1923) Ondes et quanta. C. R., Acad. Sci. Paris, 177, 507-510; Waves and quanta. Nature 112, 540.

Carrington, A. and McLachlan, A.D. (1979) I"ntroduction to Magnetic Resonance", John Wiley and Sons, New York.

Cotton, F.A. (1971) "Chemical Applications of Group Theory". 2nd. ed., Wiley-Interscience, New York.

Decius, J.C. and Hexter, R.M. (1977) "Molecular Vibrations in Crystals", McGraw-Hill, New York.

Einstein, A. (1905) Uber einen die erzeugung und verwandiwadlung des lichtes betreffenden heuristischen gesichtspunkt. Ann. Phys., 17, 132-148.

Farmer, V.C. (1974), ed., "The Infrared Spectra of Minerals", Mineralogical Society, London.

Fateley, W.G., Dollish, F.R., McDevitt, N.T. and Bentley, F.F. (1972) "Infrared and Raman Selection Rules for Molecular and Lattice Vibrations: The Correlation Method", Wiley-Interscience, New York.

Hammermesh, M. (1962) "Group Theory", Addison-Wesley, Reading, MA.

Harrison, W.A. (1979) "Solid State Theory", Dover Press, New York.

Herzberg, G. (1945) "Molecular Spectra and Molecular Structure. II. Infrared and Raman Spectra of Polyatomic Molecules", Van Nostrand Reinhold, New York.

Herzberg, G. (1950) "Molecular Spectra and Molecular Structure. I. Spectra of Diatomic Molecules", Van Nostrand Reinhold, New York.

Huheey, J.E. (1983) "Inorganic Chemistry", 3rd. Ed., Harper and Row, New York.

Koster, G.F. (1957) Space groups and their representations. Solid State Phys., 5, 173-256.

Levine, I.N., (1983) "Quantum Chemistry", Allyn and Bacon.

Moore, W.J. (1972) "Physical Chemistry", Wiley-Interscience, New York.

Pauling, L. and Wilson, E.B. (1963) "Introduction to Quantum Mechanics with Applications to Chemistry", Dover Press, New York.

Planck, M. (1901) Uber das gesetz der energieverteilung im normal spectrum. Ann. Phys., 4, 553-563.

Rutherford, E. (1911) Scattering of α and β particles by matter and the structure of the atom. Phil. Mag., 21, 669-688.

Salthouse, J.A. and Ware, M.J. (1972) "Point Group Character Tables and Related Data", Cambridge University Press, Cambridge.

Scheetz, B.E. and White, W.B. (1977) Vibrational spectra of the alkaline earth double carbonates. Am. Mineral., 62, 36-50.

Schrodinger, E. (1926a) Quantisierung als Eigenwertproblem. Ann. Phys., 79, 361-376.

Schrodinger, E. (1926b) Quantisierung als Eigenwertproblem: Storungstheorie, mit Anwendung auf den Starkeffekt der Balmerlinien. Ann. Phys., 80, 437-490.

Seitz, F. (1987) "The Modern Theory of Solids", Dover Press, New York.

Szabo, A. and Ostlund, N.S. (1982) "Modern Quantum Chemistry - Introduction to Advanced Electronic Structure Theory", MacMillan, New York.

White, W.B. (1974) The carbonate minerals. In "The Infrared Spectra of Minerals", ed. V.C. Farmer, p. 227-284. Mineralogical Society, London.

Wilson, E.B., Decius, J.C. and Cross, P.C. (1955) "Molecular Vibrations", McGraw-Hill, New York.

Chapter 3 F. C. Hawthorne and G. A. Waychunas

SPECTRUM–FITTING METHODS

INTRODUCTION

When one records a spectrum, in the simplest case one gets an analogue representation of signal intensity as a function of energy. From this, one can approximately measure band positions and estimate qualitative intensities. However, a major thrust in the spectroscopy of minerals is to measure site-occupancies and/or chemical composition. For this one needs derivative results more precise (and hopefully more accurate) than those obtained by qualitative processing of analogue data. One needs digitized data and an objective method of deriving quantitative information from that data. Details of data acquisition are technique dependent, but much of the data reduction necessary to obtain quantitative results is common to many spectroscopic techniques. This chapter will provide a general background in the numerical methods and general philosophy of "curve-fitting" techniques.

There is often a tendency to treat curve-fitting as a "black-box" procedure. This is extremely dangerous. It is very easy to make mistakes, particularly when one lacks a basic understanding of the general principles. This has happened with distressing frequency in the mineralogical literature and has damaged the credibility of the technique, when it is the user rather than the technique that is at fault. Consequently, all spectroscopic practitioners should be aware of certain general principles and technique associated with this aspect of spectroscopy. For this, some knowledge of statistics is essential; you may refresh your memory at the end of this chapter, where Appendix A gives definitions of the common statistical quantities used here.

GENERAL PHILOSOPHY

One may summarize the general principles of curve-fitting very briefly:
(i) From one's physical/chemical knowledge of the experiment, one sets up a mathematical model that will describe the raw data of the experiment;
(ii) One then changes the variable parameters of the model to minimize the deviations between the calculated "data" and the observed data:
(iii) If the "fit" or agreement between the calculated data and the observed data is statistically acceptable, then the (mathematical) model is taken as being a possible description of the experimental situation.

Although this sounds very straightforward, it is necessary to define very exactly what we mean by many of the terms used in this description; this will be done in the following sections. Before we do this, there is a little matter of terminology. This procedure of curve-fitting is frequently referred to by the term deconvolution. This is wrong; deconvolution is a specific mathematical operation in Fourier analysis (that is discussed in detail later in this chapter). Our spectrum consists of a summation of a series of separate curves (spectral bands), which together with a random noise component constitutes our observed <u>envelope</u> of curves. Thus we resolve the observed spectrum into its constituent bands, and refer to the process as spectrum resolution.

SETTING UP THE MODEL

In this section, we use the word model in the most general sense to mean the mathematical model we set up to describe the data. This type of model can be divided into three parts:

(i) <u>data reduction procedures</u> to correct for various experimental factors that are independent of the other two parts of the model; this converts the raw data into a form convenient for the ensuing calculations;

(ii) deciding on an algorithm to <u>model the background</u>, that is the response of the equipment when no spectroscopic signal is being observed;

(iii) deciding on a function that adequately models the digitized spectrum signal, that is the band shape.

Data reduction

This is normally quite a simple procedure in most spectroscopic techniques, often being restricted to assigning weights to each observation (this will be dealt with later on). For some methods, a "split-beam" technique is used, in which the incoming radiation is split into two beams, only one of which passes through the sample, the other passing through a similar path but without undergoing sample absorption. By suitable subtraction or ratioing methods, complex background profiles may be removed easily.

Background modeling

This varies tremendously from one spectroscopic method to another. In some cases, we have a good idea what the background function is, and this can be modeled as part of the fitting procedure. In other cases, there is no <u>a priori</u> ideal shape for the background and a variety of <u>ad hoc</u> methods are used, often at the data reduction stage. We will examine these in increasing order of sophistication.

<u>Linear interpolation</u>: this is diagrammatically illustrated in Figure 1(a). On either side of the band of interest, the background intensity is counted at specific points B(1) and B(2). A straight line is drawn between these two points, and the intensity below this line is taken as the background intensity. This is subtracted from the total integrated intensity between points B(1) and B(2) to get the intensity ($\equiv$ area) under the peak.

This method assumes that the background is a linear function of energy, which is often not the case. The effect of this is shown in Figure 1(b). A concave background can lead to significant underestimation of the peak intensity. Obviously, different nonlinear type backgrounds can lead to different sorts of error.

<u>Non-linear interpolation</u>: the principal problem here is to find a suitable analytical function with which to model background behavior.

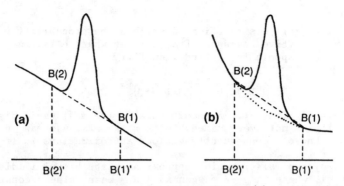

Figure 1.   Background subtraction.   For linear background (a), a linear interpolation
gives the correct background; for a non-linear background (b), linear interpolation gives
the wrong background and hence the wrong integrated peak intensity.

In some cases, there is insufficient data to warrant such a procedure,
and backgrounds must be "drawn by eye".   Obviously, any method that
requires   such   a   procedure   cannot   be   considered   as   completely
quantitative.   On the other hand, a semi-quantitative method is better
than no method at all.

Most   analytical   background functions involve either polynominals
or circular functions (sines and/or cosines), and may be implemented in
two different ways:

(i) the spectroscopic bands are removed from the data, and the
remaining   background   points   are then fit to the background function;
this   background   function   is   then considered fixed.   The background
intensity   for   each   data point in the complete data set is calculated
from   the   background   function   and   then subtracted from the total
intensity at that point.   The area under each band then represents the
intensity   and   energy   of   the   spectroscopic response to the incident
radiation.

(ii) an approximate background function is derived as in (i).   The
subtraction of the background intensity is done in the   actual
curve-fitting process, with certain of the background terms considered
as   variable   parameters.   Thus   the background fitting is iteratively
optimized   throughout   the   fitting   procedure.   This   is   the   most
satisfactory   method,   provided   that   there are enough background data
points to constrain a good fit by the background function.   This latter
point   is   of   considerable   importance as there may be high correlation
between   background   and   band   function   variables   if   there   are
insufficient background data.

Modeling the band shape

Choosing an approximate function   to accurately model the band
shape is one of the   more   difficult   aspects   of   generalized
curve-fitting procedures.   What one is essentially doing is looking at
the   distribution   of   (absorption)   events   as a function of energy.
Consequently, band shapes are normally described by one of the usual
distribution functions.   We will take a look at four examples, but it
is the Lorentzian and Gaussian functions that are of most use in
spectroscopic applications.

The <u>binomial</u> distribution describes the probability P(x,n,p) of observing x positive responses from n tries where p is the probability of a positive response for an individual try:

$$P(x, n, p) = \binom{n}{x} p^x (1-p)^{n-x} \quad , \tag{1}$$

where $\binom{n}{x} = n!/x!(n-x)!$. Although this is the most basic distribution function, it is not very amenable for general use as n and p are usually not known; consequently various approximations to it are used.

The <u>Poisson</u> distribution approximates the binomial distribution for p << 1, where n --> ∞ when np = μ = mean value = constant. Some algebraic juggling gives

$$P(x, \mu) = \frac{\mu^x}{x!} e^{-\mu} \quad . \tag{2}$$

The variance may be evaluated thus:

$$\sigma^2 = \sum_{x=0}^{\infty} \left[(x-\mu)^2 \frac{\mu^2}{x!} e^{-\mu} \right] = \mu \quad . \tag{3}$$

This result is of considerable importance to us with regard to the weighting of observations (see later section). When we make a single observation, consisting of a number of positive responses (i.e., counts), the distribution of possible results should follow a Poisson distribution if the number of counts is small (i.e., p << 1). Thus the variance of the mean value μ is μ (i.e., equation (3)). If x is one value taken from this distribution (i.e., a single experimental result), we can make the approximation

$$x = \mu \quad , \tag{4}$$

and thus the variance of the single observation x is x (and thus the standard derivation of x is $\sqrt{x}$).

The <u>Gaussian</u> distribution approximates the binomial distribution for n --> ∞ and np >> 1. It is probably the most useful distribution function, and many applications have shown that it is an apt description of the distribution of random observations for the conditions stated (n --> ∞, np >> 1):

$$P(x, \mu, \sigma) = \frac{1}{\sigma \sqrt{2\pi}} \exp\left[-\frac{1}{2} \left(\frac{x-\mu}{\sigma} \right)^2 \right] \quad . \tag{5}$$

One extremely useful characteristic of this function is that the most probable estimate of the mean value, μ, is the average of the observations x:

$$\mu = \bar{x} \quad , \tag{6}$$

and the standard deviation, σ, is a variable in the function. Another important characteristic of this curve is its <u>half-width</u>, Γ, the full-width at half-maximum height; this has very important spectroscopic implications. The half-width is defined as the range of x within which the probability P(x,μ,σ) is half its maximum value. A little algebra shows that

$$\Gamma = 2.354 \, \sigma \quad . \tag{7}$$

Another important distribution function is the <u>Lorentzian</u> (or <u>Cauchy</u>) distribution:

$$P(x, \mu, \Gamma) = \frac{1}{\pi} \frac{\Gamma/2}{(x-\mu)^2 + (\Gamma/2)^2} \quad . \quad\quad\quad [8]$$

This function is unrelated to the binomial distribution but has been found appropriate for describing resonance data; hence it is perhaps the most important function from a spectroscopic viewpoint. For purely mathematical reasons, we cannot define a standard deviation for the Lorentzian distribution; instead we can characterize its dispersion by its half-width, Γ, which is defined such that $P(x,\mu,\Gamma) = \frac{1}{2}P(\mu,\mu,\Gamma)$ for $x-\mu = \Gamma/2$; that is, when the deviation from the mean is equal to one-half of the half-width, then the probability function is half of its maximum value (which occurs at the mean value $x = \mu$).

A comparison of the Lorentzian and Gaussian curves is shown in Figure 2, in which both curves have the same half-width. Note that the Gaussian curve has a higher maximum value, whereas the Lorentzian curve has a wider tail. Very frequently one finds that none of the ideal distribution functions adequately models the experimental data. In this case, "mixtures" of curves can be used, whereby one's model curve consists, for example, of A Lorentzian character and (1-A) Gaussian character.

Table 1 summarizes the details of the various distribution functions we have considered. Although they are the most common in spectroscopic applications, they are not the only sorts of curves one can use; and several other functions are also shown.

CRITERION OF "BEST FIT"

Having set up a mathematical model, the next step is to "fit" the model to the experimental data. This normally involves varying the parameters of the model until the model shows the best agreement with the observed data. This raises the question of what do we mean by the best agreement or best fit. Intuitively one expects the best agreement when some function of the deviations between the observed values and the corresponding values calculated from the model is minimized. However, what is this function?

Consider that our model (Fig. 3) is of the form

$$y = mx + c \quad . \quad\quad\quad [9]$$

Intuitively one might expect that the best fit occurs when the sum of the deviations from the observed data are minimized. However, this is not a good measure, as positive and negative deviations tend to cancel each other out. This can be overcome by minimizing the sum of the magnitudes of the deviations, but this produces numerous practical problems with regard to the minimization procedure.

Least-squares method

There is no unique method for defining a correct criterion of best fit. However, if we assume a Gaussian distribution of

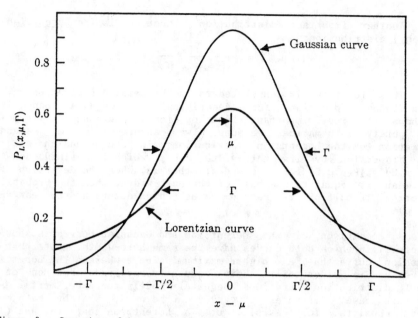

Figure 2. Comparison of Gaussian and Lorentzian curves of equal half-width. Note that the Gaussian curve has a higher maximum whereas the Lorentzian curve has a wider tail; after Bevington (1969).

Table 1. Various distribution functions

---

(1) Binominal distribution: $P(x,n,p) = \binom{n}{x} p^x (1-p)^{n-x}$

(2) Poisson distribution: $P(x,\mu) = \frac{\mu^x}{x!} e^{-\mu}$

(3) Gaussian distribution: $P(x,\mu,\sigma) = \frac{1}{\sigma\sqrt{2\pi}} \exp\left[-\frac{1}{2}\left(\frac{x-\mu}{\sigma}\right)^2\right]$

(4) Lorentzian distribution: $P(x,\mu,\Gamma) = \frac{1}{\pi} \frac{\Gamma/2}{(x-\mu)^2 + (\Gamma/2)^2}$

---

(1) $P(x,n,p)$ = probability of observing x positive responses from n tries, where p = probability of response for an individual try

(2) approximates (1) for p << 1, where n $\longrightarrow \infty$ when np = μ = mean value = constant

(3) approximates (1) for n $\longrightarrow \infty$ and np >> 1; σ = standard deviation

(4) unrelated to (1). Γ = half-width, a measure of the dispersion of the distribution

---

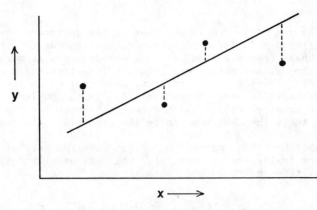

Figure 3. A linear model (y = mx + c) to the four data points shown in the graph. We wish to minimize some function of the deviations of the points from the line for it to be the "best fit".

probabilities (see equation (5)), we can derive a useful and robust method: the <u>method of</u> <u>least-squares</u>.

Suppose we wish to fit our experimental observations thus:

$$y = f(x) \tag{10}$$

Where our experimental observations are (x_i, y_i), the discrepancy δy_i between the observed and calculated value is given by

$$\delta y_i = y_i - f(x_i) \tag{11}$$

Let us write the actual relationship (which we cannot know) as

$$y(x) = f_o(x) \tag{12}$$

For any value x_i of x, the probability of making the observed measurement y_i is given by

$$P_i = \frac{1}{\sigma_i \sqrt{2\pi}} \exp\left[-\frac{1}{2}\left(\frac{y_i - y(x_i)}{\sigma_i}\right)^2\right] \tag{13}$$

The probability of making the observed set of N observations of y_i is the product of the N probabilities of equation (3):

$$P(f_o) = \prod_{i=1}^{N} P_i = \prod_{i=1}^{N} \frac{1}{\sigma_i \sqrt{2\pi}} \exp\left[-\frac{1}{2}\sum_{i=1}^{N}\left(\frac{y_i - y(x_i)}{\sigma_i}\right)^2\right] \tag{14}$$

For any <u>estimated</u> function f, the probability that we will make the observed set of measurements is

$$P(f) = \prod_{i=1}^{N} \frac{1}{\sigma_i \sqrt{2\pi}} \exp\left[-\frac{1}{2}\sum_{i=1}^{n}\left(\frac{\delta y_i}{\sigma_i}\right)^2\right] \tag{15}$$

The important point about equations (12) to (14) is that we do not know $f_o(x)$ and hence cannot evaluate the probabilities in equations (13) and (14). However, what we can do is use the method of maximum likelihood to allow us to overcome our ignorance. We make the assumption that the observed set of measurements is more likely to come from the actual parent distribution (equation (12)) rather than any other distribution with different numerical parameters for f(x). Thus the probability of equation (12) is the maximum probability possible with equation (14), and the best estimates for the parameters of f(x) are the values which maximize the probablity of equation (15).

In equation (15), maximizing P(f) requires minimizing the summation term inside the exponential, which is usually designated χ^2, and sometimes referred to as the residual:

$$\chi^2 \equiv \sum_{i=1}^{N} \left(\frac{\delta y_i}{\sigma_i}\right)^2 = \sum_{i=1}^{N} \frac{1}{\sigma_i^2} \left[y_i - f(x_i)\right]^2 \qquad (16)$$

Thus the "best fit" to the data for a specific model is the one that minimizes the weighted sum of the squares of the deviations between the observed and 'calculated' data. The method by which this fit is found is called the least-squares method, and will be discussed later.

MINIMIZATION METHODS

We have seen how we decide what is the criterion of minimization (optimal fit); now we need to examine the methods by which such minimization is done. These fall into three broad groups, each of which uses a different general philosophy:
   (i) Pattern search methods,
   (ii) Gradient methods,
   (iii) Analytical solution methods.
Pattern search methods are generally rather crude, and are usually used only in conjunction with one of the other two methods. However, it is very instructive to work through some very simple pattern search examples, as they give one a feeling for the convergence process in general (and what can go wrong with it), something one does not easily get from the less visual methods of (ii) and (iii).

Pattern search

The general idea of these methods is very simple, one just makes a search throughout parameter space such that one always moves to lessen the residual, without using any of the properties of the algorithm used to model the spectrum.

The simplest method is to divide up parameter space into a network and calculate the residual at nodes of this network. One picks an arbitrary starting point and determines by inspection of the surrounding grid points, which is the best way to move. Having done this, the process is iterated until one cannot move any more, at which time one is, hopefully, at or near the minimum value of the residual.

A very simple 2-dimensional example is shown in Figure 4. We start at point (1,1) and move down the first column until we can move

Figure 4.   X²-values at network nodes in parameter space; lines connecting nodes show paths for various starting points.

	1	2	3	4	5	6	7	8	9	10
1	547	527 -	483 -	432	403 -	317 -	286 -	253 -	201	233
2	486	517	446	397 -	352	333	277	221	156	201
3	444	483	443	373	306	386	261	153	83	163
4	367	399	376	351	256	401	274	143	[72]	136
5	329	365	333	321	243	302	255	216	117	200
6	283	313	290	262	196	171	125	153	137	222
7	241 -	234	228	200	132 --	84 ---	[4]	73	165	247
8	276	222 -	168 -	143	161	96	38	113	193	264
9	331	296	199	127 -	101 --	74 --	57	186	261	307
10	374	342	247	168	193	111	88	203	290	356

no further at row 7 in Figure 4.   Then we inspect one point each way along row 7, and move in the direction defined by the lesser value until we can move no further along this row.   The process is repeated until one cannot move at all [point (7,7)]; at this stage, it is (again, hopefully) at or close to the minimum (that is, the best fit). If one wishes to be more precise, then one can construct another such network on a larger scale close to the end-point of the first search, and repeat the procedure.

Let us repeat this example, but start at point (1,2) instead. This time we move along row 1 and follow a very different path; we arrive at the minimum point much more quickly.   This illustrates two important points concerning fitting procedures in general:
(i)   <u>the path taken to a minimum can be dependent on one's starting point</u>;
(ii) <u>some paths to a minimum are much shorter than others</u>.
Consequently, it is of importance to try and optimize one's path to the minimum to reduce computation time.

Let us repeat the example again, starting at point (1,5).   We move along row 1, eventually converging at point (9,4).   Note that this is a different end-point from the previous two paths. We are still at a minimum, that is we cannot move according to our criterion of moving along a row or column to a lower residual.   However, the value of the residual is greater than that on the previous paths; we have converged to a false minimum, it is only a local minimum and not a global minimum.   Thus
(iii) <u>convergence at a false (local) minimum can be a major problem in all fitting procedures</u>,
and there is no intrinsic way of identifying whether or not one has converged only at a local minimum.

The example we have been considering is only a 2-variable problem. It serves to indicate some of the features of convergence problems, but is not a practical method.   Each of the parameters is varied independently (i.e., along a row or down a column); consequently, with

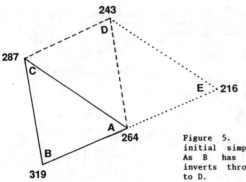

Figure 5. A two-variable simplex: ABC is the initial simplex with the χ^2 values indicated. As B has the largest χ^2 value, this vertex inverts through the center of the opposite edge to D.

a large number of variable parameters, the path taken through variable space is extremely convoluted and inefficient.

Simplex method

A more sophisticated pattern search approach is the <u>simplex method</u>, which varies all parameters simultaneously, exploiting the geometric properties of a simplex. A simplex is a geometrical figure with (n+1) vertices, where n is the number of variable parameters; Figure 5 shows the analogous simplex for a 2-dimensional problem. Optimization proceeds in the following manner. The values of χ^2 are calculated at the vertices of the simplex (points A,B and C in Fig. 5). The vertex with the largest χ^2 value (B in Fig. 5) is inverted through the centre of the opposite edge (to point D in Fig. 5). Next, χ^2 is evaluated at point D, and the process is repeated to form ADE. Iteration of this procedure moves the simplex around parameter space until it arrives in the vicinity of the minimum, where it cannot move.

There are various modifications of the simplex method, whereby the amount of movement during inversion is variable; this allows the simplex to expand or contract depending on how close it is to the minimum, and can greatly improve the rate of convergence. Simplex methods can be very useful when inadequate starting models are available for analytical least-squares methods.

Gradient method

Continuing with our 2-dimensional example developed in the previous section, a more efficient path down to the minimum would be down the steepest slope (the analogy of a mountain stream is a good one), and the method is often referred to as the <u>method of steepest descent</u>.

The gradient $\nabla\chi^2$ is that vector the components of which are equal to the rate at which χ^2 increases in that direction:

$$\nabla\chi^2 = \sum_{j=1}^{N} \left(\frac{\partial\chi^2}{\partial a_j} \, \hat{a}_j \right) \quad , \tag{17}$$

where $\hat{a}_j$ is a unit vector. The gradient may be obtained exactly by evaluation the partial derivatives of the χ^2 function, or approximated from the observed variation of χ^2 for small incremental changes in the variables:

$$\left(\nabla\chi^2\right)_j = \frac{\partial\chi^2}{\partial a_j} \sim \frac{\chi^2(a_j + w\delta a_j) - \chi^2 a_j}{w\delta a_j} \quad , \tag{18}$$

where δa_j is an incremental step in a_j and w is a weight (≈ 0.1). A combination of the two-gradient evaluation methods is optimal, as the analytical calculation of the partial derivative is slow, but the approximation is imprecise. Most efficient is to initially calculate the derivatives exactly, and then increment all variables in the optimum direction, either approximating χ^2 every (few) increments, or continuing until χ^2 begins to rise, whereupon χ^2 is calculated exactly again. This method does have problems with irregular surfaces (those containing curving valleys) and is not good close to the minimum. Consequently, it has tended to be replaced by analytical solutions.

Analytical solution methods

All of the previous methods miminize χ^2 and hence are classed as least-squares methods. However, common usage often reserves the use of this term to analytical solution methods.

Linear functions. Let us first consider the case in which our dependent variable, $y(x)$ is a linear function of the coefficients, a_j, of the fitting function:

$$y(x) = \sum_{j=0}^{n} a_j f_j(x) \quad . \tag{19}$$

Remember that here we are talking about the function being linear in the coefficients; thus the equation $y = ae^{bx}$ is linear in a, but not in b; similarly $y = ax + bx^2$ is linear in both a and b.

At the minimum point of equation (19), the derivative of χ^2 with respect to each variable is zero:

$$\frac{\partial\chi^2}{\partial a_i} = 0 \qquad (i = 1,n) \quad . \tag{20}$$

Writing χ^2 as

$$\chi^2 = \sum_{i=1}^{N} \left[\frac{1}{\sigma_i^2} \left(y_i - y(x_i)\right)^2 \right] = \sum_{i=1}^{N} \left[\frac{1}{\sigma_i^2} \left(y_i - \sum_{j=0}^{n} a_j f_j(x_i)\right)^2 \right] \tag{21}$$

and applying the conditions of equation (20) we get

$$\frac{\partial\chi^2}{\partial a_k} = \frac{\partial}{\partial a_k} \left[\sum_{i=1}^{N} \left[\frac{1}{\sigma_i^2} \left(y_i - \sum_{j=0}^{n} a_j f_j(x_i)\right)^2 \right] \right] \quad (k = 1,n) \quad . \tag{22}$$

There are n of these equations, one for each coefficient a_k in the function of equation (22). Note that the subscript k must be used in the derivative, as this subscript is independent of j in the function expression itself; however, they are the same coeficients.

Taking the derivative and using the relation

$$\frac{\partial}{\partial a_k} \sum_{j=0}^{n} a_j \ f_j(x_i) = f_k(x_i)$$

together with a little algebraic manipulation gives

$$\sum_{i=1}^{N} \left[\frac{1}{\sigma_i^2} y_i \ f_k(x_i) \right] = \sum_{j=0}^{n} \left[a_j \sum_{i=1}^{N} \left[\frac{1}{\sigma_i^2} f_j(x_i) \ f_k(x_i) \right] \right] \quad (k = 0,n) \qquad . \ (23)$$

The (n+1) unknowns in this series of equations are a_j (j=0,n); as there are (n+1) equations of this form (i.e., (k=0,n) in equation (23)), then we may solve these equations for a_j, the (fitting) parameters of interest in the model.

These equations could be solved by successive subtraction (elimination of variables) as one does with simple simultaneous equations. Of course, this is far too clumsy for most uses, and matrix methods are normally used. Equation (23) can be much more compactly represented in matrix form:

$$R_k = \sum_{j=0}^{n} a_j \ M_{jk} \quad \text{or} \quad R = a \ M \quad , \qquad (24)$$

where R_k is a column vector (the <u>normal vector</u>) given by

$$R_k = \sum_{i=1}^{N} \left[\frac{1}{\sigma_i^2} y_i \ f_k(x_i) \right] \quad , \qquad (25)$$

and M_{jk} is a symmetric n x n matrix (the <u>curvature matrix)</u> given by

$$M_{jk} = \sum_{i=1}^{N} \left[\frac{1}{\sigma_i^2} f_j(x_i) \ f_k(x_i) \right] \quad . \qquad (26)$$

The set of equations (24) are often called the <u>normal equations</u>. Multiplying both sides of equation (24) by M^{-1}, the inverse of the curvature matrix M is

$$aMM^{-1} = RM^{-1} = a \quad . \qquad (27)$$

Hence one may write the solution to the normal equations as

$$a_j = \sum_{k=0}^{n} R_k \ M_{jk}^{-1} \quad , \qquad (28)$$

and the principle labor involves the inversion of the curvature matrix; hence this is sometimes known as the <u>matrix inversion method</u>.

It is instructive to briefly consider the evaluation of the inverse matrix, as it is here that the least-squares method can give problems. The inverse of a matrix is defined as the adjoint of the matrix divided by its determinant:

$$M^{-1} = \frac{M^A}{|M|} \quad . \qquad (29)$$

If the determinant is zero (i.e., $|M| = 0$), the inverse of the matrix is indeterminate and the matrix is <u>singular</u>. When this is the case, one cannot find a solution to one's problem, the fitting process failing (diverging) when this happens. This occurs generally because one of the constituent equations of equation (23) can be expressed (either exactly or approximately) as a linear combination of some of the other equations.

<u>Linearization of non-linear functions</u>. Here we will consider the case in which there is a non-linear relationship between our dependent variable $y(x)$, and the coefficients of our fitting function.

We may recognize two cases, one trivial and one not. Consider for example, the function

$$y(x) = ae^{bx} \quad . \tag{30}$$

This is linear in a, but not linear in b. However, we may take logarithms

$$\ln y(x) = \ln a + bx \quad , \tag{31}$$

whereupon $\ln y(x)$ is linear in $\ln$ a and b, and we may proceed with our least-squares method as above.

In the second case, linearization is not achieved so easily. For example, in the case of fitting a Lorentzian curve plus a quadratic background, we can write

$$y(x) = \frac{1}{\pi} \frac{\frac{a_1}{2}}{(x-a_2)^2 + (\frac{a_1}{2})^2} + a_3 + a_4 x + a_5 x^2 \quad , \tag{32}$$

and we cannot separate out all of the a_j parameters to form an expression involving simple summation. In this case, we may use Taylor's expansion, which approximates a function $y(x)$ around a point $x = a$ by the following expression:

$$y(x) = y(a) + y'(a) (x-a) + \ldots + \frac{y^{n'}(a)}{n!} (x-a)^n + \ldots \quad . \tag{33}$$

If $(x-a)$ is small, then we may ignore terms involving $(x-a)^n$, $n>1$ and writing $(x-a)$ as δa, we get

$$y(x) = y_o(x) + \sum_{j=1}^{n} \left[\frac{\partial y_o(x)}{\partial a_j} \delta a_j \right] \quad . \tag{34}$$

This function is now linear in the parameter increments δa_j, and we can write χ^2 directly as a function of δa_j:

$$\chi^2 = \sum_{i=1}^{N} \left[\frac{1}{\sigma_i^2} \left(y_i - y_o(x_i) - \sum_{j=1}^{n} \left[\frac{\partial y_o(x_i)}{\partial a_j} \delta a_j \right] \right)^2 \right] \quad . \tag{35}$$

As before, taking the derivative and setting it equal to zero (the minimization criterion), we get

$$\sum_{i=1}^{N} \left[\frac{1}{\sigma_i^2} \left(y_i - y_o(x_i) \right) \right] = \sum_{j=1}^{n} \left[\delta a_j \sum_{i=1}^{N} \left(\frac{1}{\sigma_i^2} \frac{\partial y_o(x_i)}{\partial a_j} \cdot \frac{\partial y_o(x_i)}{\partial a_k} \right) \right] \quad (k=1,n) \quad . \quad (36)$$

Writing this in matrix form

$$\mathbf{R} = \delta a \mathbf{M} \quad , \tag{37}$$

where

$$R_k = \sum_{i=1}^{N} \left[\frac{1}{\sigma_i^2} \left[y_i - y_o(x_i) \right] \right] \tag{38}$$

and

$$M_{jk} \approx \sum_{i=1}^{N} \left[\frac{1}{\sigma_i^2} \frac{\partial y_o(x_i)}{\partial a_j} \cdot \frac{\partial y_o(x_i)}{\partial a_k} \right] \quad . \tag{39}$$

Thus we now have a set of n linear equations by which we can calculate δa_j, the parameter shifts necessary to move the initial parameter values to their values for the minimum X^2. This set of equations may be solved by matrix inversion methods as outlined above.

SOME ASPECTS OF SPECTRUM REFINEMENT

Tactics

Here we will consider some of the tactics involved in least-squares refinement, as judicious use of these can greatly increase refinement efficiency (and decrease frustration for the spectroscopist).

One starts with a model function to represent the spectrum. This usually consists of a background function and several line-shape functions. Generally the model function is linear in some parameters and nonlinear in other parameters. The linear parameters will converge to the neighborhood of their correct values whatever their starting values, and thus should be refined _first_, with the "nonlinear" parameters held constant; of course, it is sensible to use starting values close to the final values just for efficiency of computation. The nonlinear parameters in one's model have normally been linearized by Taylor expansion of the initial nonlinear equations. We assume that δa_j (= x-a$_j$) is small and hence the nonlinear terms in the Taylor expansion can be ignored. There are two important points with regard to this:

(i) if δa_j are too large (that is we are far from our minimum value of X^2), then the approximation breaks down and the process probably will not converge.

(ii) the values we get for a$_j$ are also only an approximation, and we have to iterate through the procedure until $\delta a_j \approx 0$.

Thus it is important to have a reasonably accurate starting estimate for the parameters. Generally one's model function will be linear in some parameters but not in others. It is usually good practice to first refine the important linear parameters (background, perhaps a scaling constant), and then start to refine the nonlinear parameters. If one's starting parameters are close to the true values, then one can immediately refine all variable parameters simultaneously. More often, one is not that close and a more cautious approach may be necessary, gradually increasing the number of variable parameters as χ^2 decreases. When the starting model is not good, then immediate refinement of all variables can cause oscillation and slow convergence or even divergence. In spectroscopic applications, a common tactic is to initially fix peak positions and half-widths, and refine peak areas (or even peak area ratios).

Convergence is attained when the least-squares refinement procedure calculates parameter shifts that are much less than the standard deviations of the variable parameters. At this stage, it is IMPERATIVE that all variable parameters be refined simultaneously. If this is not done, covariance terms will be missing from the dispersion (variance-covariance) matrix, and derivative parameter standard deviations will be wrongly calculated (i.e., underestimated by omission of covariance terms in equation A(12), see Appendix A).

Correlation and constraints

A spectrum contains a specific amount of information, and it is not possible to get out of the spectrum refinement procedure more information than the spectrum contains. However, what we can do is input _external_ information into the fitting process such that we can extract the information in the spectrum in a form that we want. This procedure is very common in spectrum fitting, and involves the use of _linear constraints_ in the fitting process. In the simplest case, a variable, a_k, is constrained to be related to a set of other variables:

$$a_k = \sum_{i=1}^{n} b_i \, f(a_i) \qquad\qquad (i \neq j) \qquad . \qquad (40)$$

The set of functions $b_i f(a_i)$, $(i=1,n; i \neq j)$ is the external information put into the fitting process. Usually what this does is to greatly _decrease variable correlation_ in the fitting process; this leads to more precisely determined variables, but it should be realized that their accuracy depends on the correctness of the constraint equations used. An example of this is shown in Figure 6. Mossbauer spectra of a series of minerals was used to determine Fe^{3+}/Fe^{2+} ratios, using the constraint that all the half-widths of the Fe^{2+} peaks were equal. Figure 6 shows that the half-widths of the Fe^{3+} doublet is strongly correlated with the Fe^{3+}/Fe^{2+} ratio, with weak Fe^{3+} doublets showing very wide half-widths. It seems probable that the half-width constraint for the Fe^{2+} peaks is not exact, and that the Fe^{3+} doublet is absorbing the error associated with this; the weaker and less well defined the Fe^{3+} doublet, the more it can widen and absorb error from the Fe^{2+} doublets. Thus one has to be very careful in the use of such constraints.

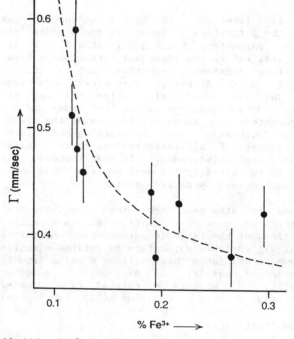

Figure 6.   Half-width of Fe^{3+} doublet as a function of $Fe^{3+}/(Fe^{3+}+Fe^{2+})$ for a series of isostructural minerals examined by Mössbauer spectroscopy.

In spectrum fitting procedures, constraints can involve

(i) constraining peak half-widths,
(ii) constraining peak intensities,
(iii) constraining peak positions,
(iv) constraining peak shapes,
(v) constraining peak splittings.

Often, the first two types involve constraining parameters to be equal (equal half-widths is a very common constraint); however, it is sometimes advantageous to constrain such parameters to have constant ratios.

The use of constraints in spectrum fitting is often essential to get any result at all. However, one must be very careful about the results; if the constraints are not appropriate, then the results will be wrong _despite_ (possibly) having high precision.

False minima and model testing in least-squares fitting

It was noted earlier that there is no mathematical way to test least-squares minima for global validity. Rather, the general method is to attempt convergence from a variety of starting points in parameter space. This is frequently not an option if many constraints must be used for convergence, or if the fit is extremely sensitive to the values of particular parameters.

Another strategy involves making assumptions about the nature of false minima. In least-squares refinement, false minima occur because of the existence of a relative minimum for a particular subset of the data set. If these data could be given zero weight in the refinement procedure, the true global minimim would be obtained. Although the subset responsible cannot be determined by analytic tests, it's effect can be overwhelmed or circumvented. The former is achieved by increasing the number of independent data. This effectively dilutes the possibility of finding a relative minimum in parameter space. The latter can be done (sometimes) by separating the data at random into multiple sets and refining each set separately. It would be most unlikely that each data subset would lead to the same minimum in the same region of parameter space unless the minimum were truly global.

Another consideration in the detection of false mimina is the reasonableness of the refined parameter values. If these can be subjected to model tests that independently affirm likelihood, or if constraints can be used, the possibility of false minima are reduced. It is also possible to put a restriction on refined parameters such that they will not adopt unreasonable values, (e.g., negative bond lengths). In the latter case, the refinement is said to be restrained (Prince, 1982).

Signal-to-noise effects in fitting

Up to now, we have considered only one goodnes-of-fit parameter, χ^2. This allows us to determine the probability that the model we fit is more (or less) representative of the data than some other fit. This is done by reference to $P_x(\chi^2, v)$, the integral of the χ^2 probability distribution function (Bevington, 1969). However, it will not allow us to compare fits to data of differing signal/noise ratios. This is because χ^2 is sensitive to the absolute size of the residuals. Fits of the same model to indentical data sets with different ratios of signal/noise will show a strong dependence of χ^2 on the signal/noise ratio (Waychunas, 1986). The easiest way around the problem is to attempt all trial models on all data sets and compare the χ^2 values obtained only for a given data set. Because this is time consuming or impractical, one must either make allowances in χ^2 or develop a new goodness-of-fit parameter insensitive to signal/noise variations. One such parameter was derived by Ruby (1973) and is called MISFIT.

Variations on χ^2 goodness-of-fit parameter

The MISFIT goodness-of-fit parameter is defined as the ratio of two values calculated from a test fit, $M = D/S$, where D is the discrepancy or distance of the fit from the actual data, and S is the signal or total spectrum above a baseline. MISFIT is thus a fractional assessment of the fit quality, and in an ideal case, should be independent of the magnitude of S.

The quantity D is derived from χ^2, and has the formulation

$$D = \sum_{I=1}^{N} \left[\left(\frac{Y_c(I) - Y_d(I)}{\sqrt{Y_d(I)}} \right)^2 - 1 \right], \qquad [41]$$

where $Y_d(I)$ are the experimental data points,
$Y_c(I)$ are the calculated model data points,
N is the total number of data points.

S is defined as

$$S = \sum_{I=1}^{N} \left[\left(\frac{(Y_o(I) - Y_d(I))}{\sqrt{Y_d(I)}} \right)^2 - 1 \right] \quad , \tag{42}$$

where Y_o is the baseline value.

Unlike χ^2, MISFIT is not compared with a distribution function, but with its own uncertainty, ΔMISFIT. This is a simple comparison which has obvious meaning. If ΔMISFIT is small relative to MISFIT, then we know the percentage error in the fit. We then seek to minimize MISFIT by adopting improved models. If ΔMISFIT is significant relative to MISFIT, then we are not testing the model very well (i.e. the data are too poor). This use of ΔMISFIT is critical for MISFIT to have any utility over χ^2. ΔMISFIT is defined as

$$\Delta M = \left(\frac{1}{S} \right) \sqrt{\left[n(1 + M^2) + 4D(1 + M) \right]} \quad . \tag{43}$$

Some confusion in the use of MISFIT may occur, as Ruby (1973) actually devised two formulations based on differing assumptions. Either form works fairly well, but both may diverge to large values at very low S values and large data point variances. These deviations have been discussed by Waychunas (1986).

Another function which measures goodness-of-fit is the crystallographic R-factor and similar functions. The R-factor is defined (Hamilton, 1964) as

$$R = \left[\frac{\sum_{I=1}^{N} w_i \, (|F|_i^{obs} - |F|_i^{calc})^2}{\sum_{I=1}^{N} w_i \, (F_i^{obs})^2} \right] \quad , \tag{44}$$

where $|F|_i^{obs}$ and $|F|_i^{calc}$ refer to the observed and calculated values of some function $|F|$ at data point i, having weight w_i.

In the limit of a perfect fit, R goes to zero. The R-factor is scaled by the size of the observed points, so that it is relatively insensitive to the magnitude of differing data sets. R-factors calculated for fit models can be tested for significance by evaluating the ratio R

$$R = \frac{R_1}{R_0} = \left[\frac{p}{n-p} * F_{p,n-p,\Omega} + 1 \right]^{\frac{1}{2}} \quad , \tag{45}$$

where p is the number of fitted parameters,
  n is the number of data points,
  F is the well-known probability distribution,
  Ω is the significance level (1-confidence level to which the
    R-factors are to be tested.
For example, suppose one has 70 data points and must fit 10 parameters to a confidence level of 95%. From tables of the F distribution, $F_{10, 60, 0.05} = 1.9926$. R is thus found to be 1.1541. If a satisfactory R value obtained from fitting procedures is 10.50%,

then a value of 12.12% represents a poorer fit in 95% of the possible cases. Similarly, an R value of 9.10% represents a better fit in 95% of all cases. A similar application of the F distribution can be used with X^2 values (Bevington, 1969).

It is also possible to devise weighting schemes for goodness-of-fit parameters, such that some data points do not contribute significantly to the fit parameter regardless of their true variance. This effectively results in selective fitting of some partition of the spectral data set. Such action would damage the utility of a statistical analysis of the goodness-of-fit parameter magnitude, but may aid least-squares fitting by improving chances for minimization of the fit parameter.

Variations in spectral line shape

When treating any aspect of mineralogical spectroscopy, we will frequently be concerned with the spectral line shape. We will need to understand the significance of broader lines in the spectrum of one material over another, of asymmetry in line shapes, and the types of line shape variations that are the result of particular physical processes in the sample or in the spectrometer. The ability to assume a particular line shape allows us to constrain fitting procedures and extract parameters that are much simpler to manipulate and compare. The basic experimental line shapes were noted earlier, viz. Gaussian and Lorentzian or Cauchy. These lines are symmetric about their centroids, and represent the distributions of energy emitted or absorbed by a particular atom or atomic system in the spectroscopic process. It is usually possible to calculate the minimum possible line width. Such a line width is never actually observed because of the effects of the spectrometer itself, or of non-ideal conditions in the sample. The line width and shape are affected by:

Excited-state lifetime. The natural line width for a spectroscopic line is determined by the lifetime of the excited state via the Heisenberg uncertainty principle: $T*\delta E \geq h/2\pi$, where T is the lifetime of the excited state, h is Planck's constant and δE is the energy uncertainty. The longer lived the excited state, the more sharply defined is the transition energy.

Döppler and collision effects. These effects are primarily relevant to the spectroscopy of independent molecules rather than atoms and molecules in solids. However, Gaussian line-broadening can occur due to the Döppler shifts created by rapid thermal vibrations at high temperatures.

Saturation effects. The line shape can be altered if excitation of a system is so powerful that the excited state is saturated (i.e., the population of excited atoms or electrons is equal to that in the ground (unexcited) state). Under such conditions, the absorption coefficient of the sample is dependent on the intensity of the incident radiation, instead of being an inherent characteristic of the sample.

Relaxation effects. The excited states of a system will decay with a characteristic rate that depends on the physical processes affecting the particular state. Optical spectra usually record electronic states with lifetimes of about 10^{-8} s, so that a fast relaxation process occurs. In this process, energy from the excited state is removed and delivered to the crystal structure as vibrations, or to other electronic centers such as excited electrons or excitons. Much slower relaxation rates may be found in NMR and certain far-IR excited states. Saturation broadening is increasingly more likely for slower relaxation rates.

Besides leading to a change in the line shape of a given excited state transition, relaxation processes can create "mixing" of several excited (or ground) states, or averaging of physical parameters. The latter effects can create very complex line shapes, such as the partial evolution of multiline spectral features into singlets. In Figure 7, the effect of averaging nuclear magnetic hyperfine spin states is shown in the Mössbauer spectrum of FeF_3. The low-temperature spectrum consists of a six line pattern representing allowed transitions to the ground state from each of the hyperfine magnetic states in a static magnetic field (see Chapter 7). This static field is due to magnetic ordering of the Fe^{3+} electronic spins. As the temperature is increased toward the magnetic disordering point (the Néel point), the electronic spins begin to decouple and spin fluctuations occur. The time averaged field seen by the Mössbauer hyperfine states is weaker and varies in direction. This reduces the splitting of these states, and the spectrum begins to collapse. Above the Néel point, the electronic spins are not coupled and there is no separation of the various hyperfine states. Temperature changes near the Néel point have dramatic effects on the observed Mössbauer spectra, but there is little effect over other temperature ranges.

In very dilute Fe^{3+} minerals and compounds, a related effect occurs but usually over a larger temperature range. In such materials, there is no definite magnetic ordering point, but at low temperatures the electronis spins on each individual Fe^{3+} will couple to produce hyperfine state splitting. As the individual ions are well separated in a dilute system, there is little relaxation due to spin-spin exchange and the relaxation rate is slow. Raising the temperature increases fluctuation of the Fe^{3+} spins, but there is no sudden loss of coupling as would occur with long-range magnetic order. Hence there is a gradual change in the observed Mössbauer spectrum (Fig. 8).

Relaxation due to fluctuations in chemical shifts can affect both NMR and Mössbauer spectra. One example of this in the latter is electron hopping. If near-neighbor Fe^{2+} and Fe^{3+} ions exchange the 6th 3d electron between them, the oscillating valence is manifested on the Mössbauer nuclei as a fluctuating chemical shift (isomer shift). At slow oscillation rates, discrete spectral lines due to each valence state are observed, but these smear into one another as the rate increases. At hopping rates faster than the lifetime of the Mössbauer excited state, only one valence-averaged set of features remains (see Chapter 8). The NMR analog could be an atom whose chemical shift fluctuates as it diffuses through a crystal or liquid. The relaxation spectra for these hopping and diffusion cases might look like the idealized spectra in Figure 9.

A very complete treatment of relaxation effects in spectroscopy can be found in Poole and Farach (1971).

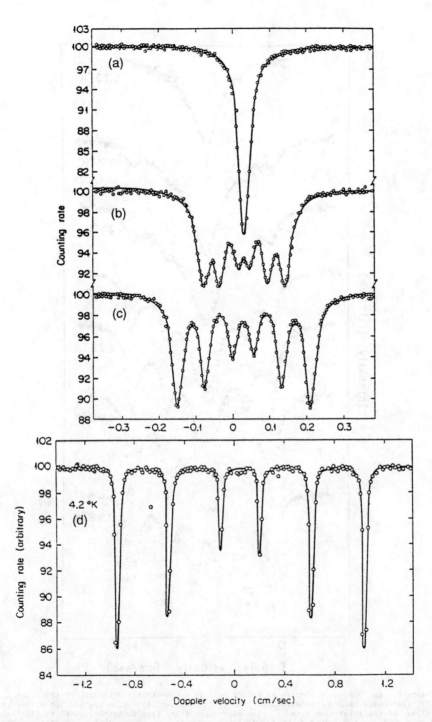

Figure 7. Mössbauer spectrum of FeF₃ at (a) 363.4 K, (b) 362.7 K, (c) 361.5 K, (d) 4.2 K. There is a tremendous change in the spectrum near the Néel temperature of 363.1 K. Note in particular how the spacing of the outmost lines changes; after Wertheim et al. (1968).

84

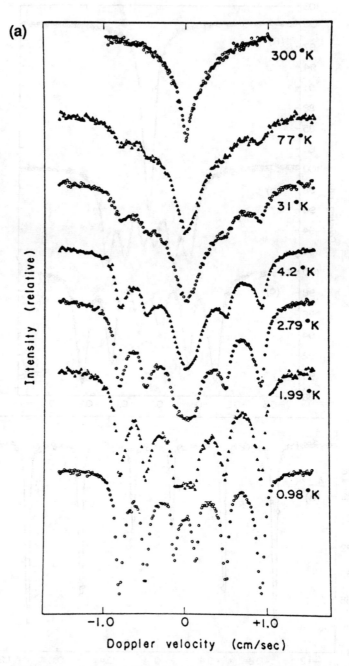

Figure 8. (a) Mössbauer spectra of dilute Fe^{3+} in a nonmagnetic material, Ferrichrome A. (b) [following page]   Calculations   of Fe^{3+} Mössbauer relaxation spectra for differing values of the relaxation rate t.   The values for t are (top to bottom) 10^{-12} s, 10^{-9} s, 2.5×10^{-9} s, 5.0×10^{-9} s, 7.5×10^{-9} s, 2.5×10^{-8} s, 7.5×10^{-8} s and 10^{-6} s.   The lifetime of the ^{57}Fe excited state is about 10^{-8} by comparison.   Relaxation rates faster than the lifetime result in averaged spectra; after Wickman (1966).

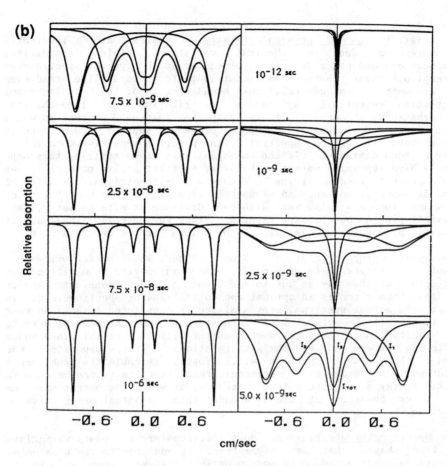

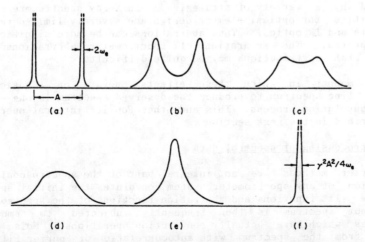

Figure 9.   Collapse of a doublet spectrum due to increasing rate of fluctuations in the abcissa parameter: (a) through (f) represent an increase in the rate of 500 fold.   Note that for individual doublet lines, the width is proportional to the fluctuation rate, w_e, but for the averaged singlet (f), the width is inversely proportional to w_e.   The latter effect is sometimes termed "motional narrowing"; after Poole and Farach (1971).

Distribution of physical states. Samples are inherently inhomogeneous due to chemical variations, grain boundaries, dislocations and other defects. Depending on the type of spectroscopy attempted, these features will contribute to spectral line broadening or asymmetry. In general, the broadening is of the Gaussian type, suggesting essentially a random distribution of line-shifting perturbations. However, it is easy to imagine a context in which broadening and asymmetry are correlated. Suppose a spectroscopic measurement reveals a spectral line position sensitive to a metal-oxygen bond distance. If the potential well characterizing this bond has a hard repulsive potential but very soft attractive potential, we would expect random strains to create a non-symmetric distribution of bond distances, and hence an asymmetric line shape. The probability of a transition may also vary with bond distance or site geometry, and possibly in a non-linear fashion. Thus the observed line shape is actually a convolution over all of the "micro" states of the system.

Spectrometer resolution. Even the best spectrometer system has resolution limits. In many cases, these may derive from diffraction criteria, or they may be due to electronic considerations. An example of the former occurs in optical and optical-analog spectrometers, in which the actual spectrum being analyzed is convoluted by various slit functions. If the natural line-width and homogenous broadening effects are small, then the spectrometer output will be spectral lines whose width is determined by the slit functions. If the opposite is the case, the output spectrum will closely resemble the true sample spectrum. An example of electronic resolution limits can be seen in Si(Li) energy dispersive detector spectra, in which the output spectrum has lines hundreds of times broader than the actual X-ray emission spectrum.

Spectrometer aberrations. Most spectrometers introduce variations in line shapes that are asymmetric, in addition to the broadening effects already noted. In modern optical systems, these effects are minimized by a variety of strategies. In X-ray spectrometers, the possibilities for optical element design are severely limited relative to visible and IR optics. Thus aberrations can be more significant in X-ray spectra. The separation of spectrometer aberrations from physical state distributions may be quite difficult.

The manner in which these effects operate on a basic delta function line spectrum to produce the observed spectrum is one example of the convolution process. This and other Fourier integral operations are described in the next section.

Fourier processing of spectral data

Fourier methods are an integral part of the spectroscopic art. The action of any spectrometer system convolutes the initial spectrum with the slit functions and aberration functions of the spectrometer. The output spectrum is then frequently subjected to smoothing procedures which are actually convolution operations. Noise can be removed from the spectrum with autocorrelation or Fourier filtering procedures.

Attempts can also be made to enhance the apparent resolution of a spectrum via deconvolution and maximum entropy Fourier procedures. In the former case, we attempt to correct (i.e., remove) the spectrometer or other convolution effects, and in the latter case, we make approximations to get around the limited data set (much smaller than from negative to positive infinity) observed.

Fourier transforms and integrals

The Fourier transform of the function $f(x)$ is usually defined as

$$F(s) = \int_{-\infty}^{\infty} f(x)\ e^{-i2\pi xs} dx \quad . \tag{46}$$

Substitution of $F(s)$ for $f(x)$ into the same integral will result in the original function. This is the cyclical nature of the Fourier transform. If $F(s)$ is the Fourier transform of $f(x)$, then $f(x)$ is the Fourier transform of $F(s)$. The key relation between the transform and the original function is the change of variable, from x to s. These variables have reciprocal dimensions. In harmonic analysis and electronics, the two variables are time and frequency; in vibrational systems, distance and wavelength; in diffraction analysis, they can be vectors in real and reciprocal space; in quantum mechanical operations, position and momentum. The usual formulae for Fourier transformation are written slightly differently:

$$F(s) = \int_{-\infty}^{\infty} f(x)\ e^{-i2\pi xs} dx \quad , \tag{47}$$

$$f(x) = \int_{-\infty}^{\infty} F(s)\ e^{i2\pi xs} ds \quad . \tag{48}$$

These forms differ in the exponent, so that they are <u>not</u> identical and may operate differently on particular functions. To keep them straight, we may refer to the first as a <u>forward transform</u> and the second as a <u>back transform</u> (but the names can be reversed). Combining the two results in the Fourier integral theorem,

$$f(x) = \int_{-\infty}^{\infty} \left[\int_{-\infty}^{\infty} f(x)\ e^{-i2\pi xs} dx \right] e^{i2\pi xs} ds \quad . \tag{49}$$

For conditions under which this integral can be evaluated, the reader should consult a text on Fourier methods, (e.g., Bracewell, 1986). In particular, discontinous functions create problems in evaluation. Some examples of $f(x)$ and $F(s)$ pairs are shown in Figure 10.

Properties of Fourier transforms

A few of the theorems important to Fourier analysis are stated here without proof; these can readily be found in most references on Fourier transforms.

<u>Similarity theorem</u>. If $f(x)$ has the Fourier transform $F(s)$, then $f(ax)$ has the transform $|a|^{-1} F(s/a)$. This theorem allows scaling of the area under transforms.

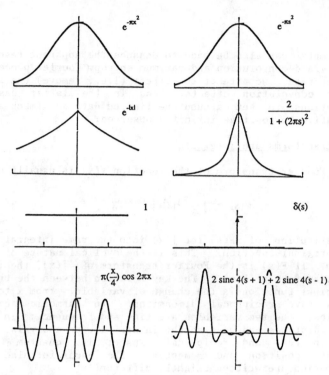

Figure 10. Some examples of Fourier transform pairs. From top to bottom: Gaussian --> Gaussian, exponential --> Lorentzian, constant --> delta function at position s, truncated cosine function --> peaks with finite widths and side lobes at position s; after Bracewell (1986).

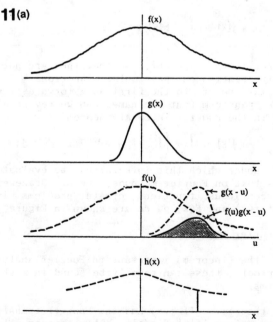

Figure 11. Graphical depiction of convolution: (a) the value of $h(x) = f(x)*g(x)$ calculated at one point from $f(u)g(x-u)$, and the form of the integral $h(x)$; (b) decomposition of small segments of $f(x)$ into the characteristic shape of $g(x)$, then superimposed to obtain part of the convolution integral; after Bracewell (1986).

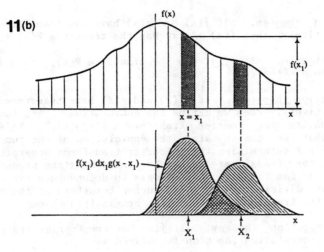

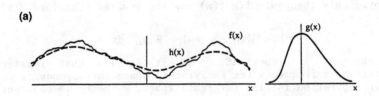

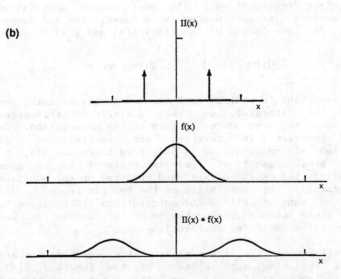

Figure 12.   Examples of convolution; (a) smoothing effect of convolution; (b) convolution of delta function spectrum with Gaussian to obtain Gaussian line spectrum; after Bracewell (1986).

<u>Addition theorem</u>. If $f(x)$ and $g(x)$ have the transforms $F(s)$ and $G(s)$, respectively, then $f(x) + g(x)$ has the transform $F(s) + G(s)$.

<u>Shift theorem</u>. If $f(x)$ has the transform $F(s)$, then $f(x-a)$ has the transform $e^{-i2\pi as}F(s)$.

<u>Convolution theorem</u>. If $f(x)$ and $g(x)$ have transforms $F(s)$ and $G(s)$ respectively, then we define the convolution of the functions as $f(x)*g(x)$ having the Fourier transform $F(s)G(s)$. This theorem indicates that we can obtain the convolution of one function with another by first determining their Fourier transforms separately, then multiplying the transforms and back transforming the product. As a corollary, we can imagine a process to deconvolute two convoluted functions by division of the Fourier transform of the convoluted function by the transform of one of the original functions.

<u>Autocorrelation theorem</u>. If $f(x)$ has the Fourier transform $F(s)$, <u>then</u> its autocorrelation function is defined as

$$\int_{-\infty}^{\infty} f*(u)\ f(u + x)\ du \quad , \tag{50}$$

which has the Fourier Transform $|F(s)|^2$.

<u>Rayleigh's theorem</u>. If $f(x)$ has the Fourier transform $F(s)$, then

$$\int_{-\infty}^{\infty} |f(x)|^2\ dx = \int_{-\infty}^{\infty} |F(s)|^2\ ds \quad . \tag{51}$$

This theorem is important, as it implies that least-squares minimization in frequency (or reciprocal) space corresponds to the same type of operation in time (or real) space. Hence we can operate in either variable space, whichever is more convenient.

<u>Convolution-deconvolution</u>. The most common application of the Fourier transform in spectroscopy is in convolution and deconvolution procedures. The convolution of functions $f(x)$ and $g(x)$ is defined as

$$f(x)*g(x) = \int_{-\infty}^{\infty} f(u)\ g(x-u)\ du \quad . \tag{52}$$

Here it is important to realize that the two functions being convoluted are within the integral, so that a simple multiplication of the functions does not reproduce the form of the convolution. Secondly, the second function is reversed and shifted. A graphical interpretation of convolution is depicted in Figure 11a, in which each small area segment of $f(x)$ is replaced by an equal area element having the shape of $g(x)$ and centered on the position of the original segment. The convolution of the two functions, $h(x)$, is then equal to the sum of all such contributions at the given position. Figure 11b gives another interpretation of the convolution, and Figure 12 gives several examples of convolutions.

One well-known type of convolution is a smoothing operation. It consists of replacing each element of the function $f(x)$ with a Gaussian, rectangular (boxcar), or other function of the same area. Various types of polynomial smoothing functions are frequently used to approximate Gaussian smoothing.

Deconvolution procedures are most often used to reduce spectral line width due to spectrometer broadening. However, when line shape analysis is crucial, such as in Mössbauer spectroscopy, the source line shape can be removed by deconvolution (Vincze, 1982; Lin and Preston, 1974). If all broadening functions and the natural line width can be removed from a spectrum, what remains is the distribution of physical parameters in the sample. As an example, the distribution of isomer shift and quadrupole splitting can be obtained in this way for Mössbauer spectra (Window, 1971). Various numerical approximations to the Fourier transform method have also been evaluated (Wivel and Morup, 1981). An example of a deconvoluted Mössbauer spectrum in which sample thickness effects have been removed (Ure and Flinn, 1971) is shown in Figure 13.

Fourier filtering

In most spectra, the background consists of a variety of differing noise signals. The usual noise is due to counting statistics (i.e., noise with a random of Gaussian distribution). Electronic effects may add periodic noise. Incorrect values in computer memory due to data transfer or transcription problems may introduce data spikes. Fourier filtering is a type of deconvolution in which the Fourier transform of the original spectrum is multiplied by a window function. Typically, the window function has non-zero values only in the vicinity of frequency (or reciprocal) space in which the signals of interest occur. The other frequencies are thus set to zero in the product of the transforms. Back Fourier transformation then yields a filtered spectrum; most statistical noise, as it contains a wide distribution of frequencies, will be removed along with any other periodic noise. Spikes in the raw spectrum will also be removed. An example of Fourier filtering is shown in Figure 14.

The advantage of filtering procedures is that often some sense can be made out of spectral data that would otherwise be too noisy for further analysis. An additional advantage, used in the analysis of EXAFS spectra and seismic frequency spectra, is that certain frequency components can be separated for detailed analysis procedures without handling the full spectrum. This is of particular value when one frequency range in the spectrum is subject to less uncertainty than others.

A major disadvantage of Fourier filtering is the inability to use normal goodness-of-fit parameters when least-squares fitting a filtered spectrum. The filtering removes the statistical variations in the spectrum, so that variances become small or zero. The way around this problem is to calculate some estimate of fit-quality, such as a crystallographic R-factor type parameter, in order to compare fitting models. Then these models can be applied to the original unfiltered spectrum with the usual χ^2 parameter.

Correlation functions

The autocorrelation function has been defined above. Another type of correlation function is the cross correlation function defined as

$$\int_{-\infty}^{\infty} f*(u)\ g(u+x)\ du \qquad (53)$$

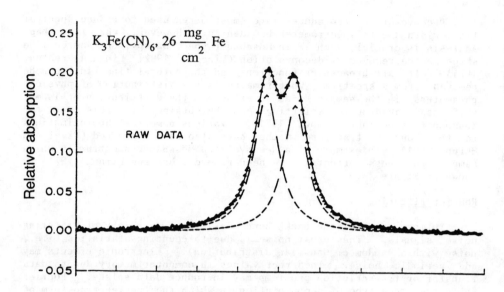

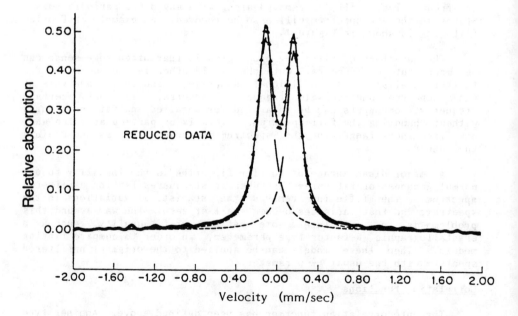

Figure 13. Deconvolution of blackness broadening in the Mössbauer spectrum of potassium ferricyanide. Top: raw data taken with a $^{57}Co:Cr$ source having a line width of 0.13 mm/s. Overall spectral line width is about 0.32 mm/s. Bottom: spectrum after deconvolution process. Spectral line width is now about 0.18 mm/s; after Ure and Flinn (1971).

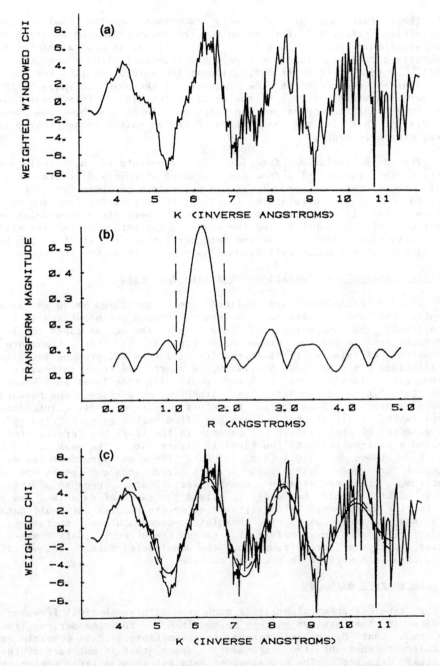

Figure 14. Fourier filtering of EXAFS spectrum; (a) original weighted EXAFS spectrum as as function of k (Å^{-1}); (b) Fourier transform of (a) to R space (Å); (c) Back transform of (b) over the range 1.2<R<2.0 Å compared to original data (solid lines). Calculated fit to back transform is given by the dashed line. Note evidence of higher frequency components in the raw spectrum not present in the filtered spectrum, (e.g. oscillations between 4 and 6 Å^{-1} are due to the second small peak in R space at about 2.8 Å). After Waychunas et al. (1986).

These functions are extremely important in the analysis of diffraction patterns. For example, the autocorrelation function of the electronic charge density in a crystal is equivalent to the generalized Patterson function (for a complete derivation, see Cowley, 1981). Note that from the definition of the autocorrelation function, its Fourier transform is the square of the absolute value of the Fourier transform, $F(s)$, of the original function. This means that the autocorrelation transform can carry no information about the phase of $F(s)$. This is one description of the so-called phase problem in X-ray crystallography.

The autocorrelation function is a measure of the correlation between the values of a function evaluated at points differing by x. This aspect can be used to remove noise from periodic functions or spectra. The autocorrelation function of a periodic function with random noise is equivalent to the sum of the separate autocorrelation functions of the function and the noise. The periodic function will give rise to a periodicity in the autocorrelation from which it can be evaluated, but the noise will produce only a smooth background curve.

Fourier "ripples" -- limitations of finite data sets

Fourier transforms are defined over the range $-\infty$ to $+\infty$; as a result, the use of data sets any shorter causes perturbations in the transform. The endpoints of the data set act as rapid impulse functions and generate sinusoidal ripples in the transform. Transforms of single lines in spectra will thus not produce infinite oscillations in $\underline{s}$ space. Similarly, a short wave train in frequency space will transform to a broad peak with side lobes in $\underline{x}$ space. Thus for finite data sets in either $\underline{x}$ or $\underline{s}$ space, the forward Fourier transform - back Fourier transform process will introduce complicating oscillations. These are often called Fourier "ripples". An example of such ripples can be seen in the EXAFS Fourier transform (called a pair-correlation function) in Figure 14b. The peak at 1.6 Å has side lobes at 1.0 Å and 2.1 Å. These annoying features may interact in phase with side lobes from adjacent peaks in a spectrum, creating strange non-physical beats. Because of this, every effort should be made to extend the range of data in $\underline{x}$ and $\underline{s}$ for Fourier operations. The existance of severe ripples for small data sets may preclude use of convolution-deconvolution operations. However, it may be possible to extend the effective data range by making particular assumptions about the uncollected data. One way of doing this is described in the next section.

Maximum entropy methods

A type of spectral analysis much used with geophysical frequency spectra is the maximum entropy method (MEM). The name derives from the fact that by making reasonable assumptions, it is possible to reconstruct some of the frequency space that is not part of the original data set. The new enhanced data set spans a large region in frequency space, and thus its Fourier transform will have narrower lines and generally enhanced resolution over an original spectrum.

The assumptions in the MEM are mainly that in any limited data set, the unmeasured frequency components are unlikely to be all zero, or all have perfect periodicity to infinity. The actual case lies somewhere in between, and can be estimated by generating a spectral distribution function with the maximum entropy. This function must be constrained to agree with that part of the spectral distribution function which can be measured. The MEM analysis has been applied to various forms of spectra with very short data ranges, with mixed results. Misapplication of the technique can result in spectra with high resolution but spurious features. A good review of progress in MEM with geophysical applications is that of Ulrych and Bishop (1975).

REFERENCES

Bevington, P.R. (1969) Data Reduction and Error Analysis for the Physical Sciences. McGraw-Hill, New York, 336 p.

Bracewell, R. (1986) The Fourier Transform and its Applications. Second revised edition. McGraw-Hill, New York, 474 p.

Champeney, D.C. (1973) Fourier Transforms and their Physical Applications. Academic Press, London, 256 p.

Cowley, J. (1981) Diffraction Physics. Elsevier (North-Holland), Amsterdam, 430 p.

Hamilton, W.C. (1964) Statistics in Physical Science. Ronald Press, New York, 230 p.

Lin, T.M. and Preston, R.S. (1974) Comparison of techniques for folding and unfolding Mössbauer spectra and data analysis. In Gruverman, I.J., Seidel, C.W. and Dieterov, D.K., eds., Mössbauer Effect Methodology 9, 205-233.

Poole, C.P. and Farach, H.A. (1971) Relaxation in Magnetic Resonance. Academic Press, New York, 392 p.

Prince, E. (1982) Mathematical Techniques in Crystallography and Materials Science. Springer-Verlag, New York, 192 p.

Ruby, S.L. (1982) Why MISFIT when you already have X^2? In Gruverman, I.J. and Seidel C.W., eds., Mössbauer Effect Methodology 8, 263-276.

Ulrych, T.J. and Bishop, T.N. (1975) Maximum entropy spectral analysis and autoregressive decomposition. Rev. Geophys. Space Phys. 13, 183-200.

Ure, M.C.D. and Flinn, P.A. (1971) A technique for removal of the "blackness" distortion of Mössbauer spectra. In Gruverman, I.J., ed., Mösbauer Effect Methodology 7, 245-262.

Vincze, I. (1982) Fourier evaluation of broad Mössbauer spectra. Nucl. Spectra. Nucl. Instr. Methods 199, 247-262.

Waychunas, G.A. (1986) Performance and use of Mössbauer goodness-of-fit parameters to spectra of varying signal/noise ratio and possible misinterpretations. Am. Mineral. 71, 1261-1265.

-----, Brown, G.E.,Jr. and Apted, M.J. (1986) X-ray K-edge absorption spectra of Fe minerals and model compounds: II. EXAFS. Phys. Chem. Minerals 13, 31-47.

Wertheim, G.K., Guggenheim, H.J. and Buchanan, D.N.E. (1958) Sublattice magnetization in FeF_3 near the critical point. Phys. Rev. 169, 465-470.

Wickman, H.H. (1966) Mössbauer paramagnetic hyperfine structure. In Gruverman, I.J., ed., Mössbauer Effect Methodology 2, 39-66.

Window, B. (1971) Hyperfine field distributions from Mössbauer spectra. J. Phys. E: Sci. Instrum. 4, 401-402.

Wivel, C. and Morup, S. (1981) Improved computational procedure for evaluation of overlapping hyperfine parameter distributions in Mössbauer spectra. J. Phys. E: Sci. Instrum. 14, 605-610.

APPENDIX A: SOME STATISTICAL DEFINITIONS

ACCURACY: a measure of how close the experimental result is to the "true" value.

PRECISION: the measure of how exactly the result is determined (i.e reproducibility) without any reference to a "true" value.

RANDOM ERROR: indefiniteness of result due to finite precision of the experiment.

SYSTEMATIC ERROR: reproducible inaccuracy caused by faulty experimental technique or a faulty model.

PARENT POPULATION: hypothetical infinite set of "data" points of which the experimental data points are assumed to be a random sample.

PARENT DISTRIBUTION: probability distribution controlling the (random) sample data assumed to be drawn from the parent population.

EXPECTATION VALUE: denoted by < >, it is the weighted average of a function f(x) over all values of x:

$$\langle f(x) \rangle = \lim_{N \to \infty} \left[\frac{1}{N} \sum_{i=1}^{N} f(x_i) \right] = \sum_{j=1}^{n} \left[f(x_j) P(x_j) \right] = \int_{-\infty}^{\infty} f(x) P(x) dx \quad , \quad A(1)$$

where P(x) is the probability function that defines the probability of obtaining a specific value of x in any random experiment.

MEAN VALUE: for a series of N observations, the sample mean value is the average of the observations, x; for the parent distribution (see above), the parent mean value, μ, is the limit of x as N -> ∞; thus

$$\mu \approx \bar{x} = \frac{1}{N} \sum_{i=1}^{N} x_i \quad . \qquad A(2)$$

MEDIAN VALUE: for the parent population, the median $\mu_{\frac{1}{2}}$ is that value of xi for which the probability of any observation being less than the median is equal to the probability of it being greater than the median.

$$P(x_i \leq \mu_{\frac{1}{2}}) = P(x_i \geq \mu_{\frac{1}{2}}) = \frac{1}{2} \quad . \qquad A(3)$$

MOST PROBABLE VALUE: for the parent population, the most probable value μ_{max} is that value of x for which the parent distribution has its greatest value:

$$P\left(\mu_{max}\right) \geq P\left(x \neq \mu_{max}\right) \quad . \qquad A(4)$$

Note that for a symmetrical parent distribution function, all of these parent values are coincident.

AVERAGE DEVIATION: this is defined as the average of the magnitudes of the deviations from the mean of the parent distribution (cf., equation A(2)):

$$\alpha \equiv \lim_{N\to\infty} \left[\frac{1}{N} \sum_{i=1}^{N} |x_i - \mu| \right] \quad , \qquad\qquad A(5)$$

about the mean value.

VARIANCE: like the average deviation, the variance, σ^2, is a measure of the dispersion of the observations, defined as

$$\sigma^2 \equiv \lim_{N\to\infty} \left[\frac{1}{N} \sum_{i=1}^{N} (x_i - \mu)^2 \right] = \lim_{N\to\infty} \left[\frac{1}{N} \sum_{i=1}^{N} x_i^2 \right] - \mu^2 \quad . \qquad A(6)$$

This is a very convenient measure (more so than the average deviation) as the expression $(x_i - \mu)^2$ occurs in several distribution functions (e.g., Gaussian function). We can rewrite it as

$$\sigma^2 = \left\langle (x_i - \mu)^2 \right\rangle = \left\langle x^2 \right\rangle - \mu \quad . \qquad\qquad A(7)$$

For a finite set of observations, the sample variance is defined as

$$\sigma^2 \approx \frac{1}{N} \sum_{i=1}^{N} (x_i - \bar{x})^2 \quad . \qquad\qquad A(8)$$

This is normally modified (Bevington, 1969, p. 19) to the form

$$\sigma^2 \approx s^2 \equiv \frac{1}{N-1} \sum_{x=1}^{N} (x_i - \bar{x})^2 \quad , \qquad\qquad A(9)$$

where s^2 is called the sample variance.

COVARIANCE: this is defined by analogy with variance as

$$\sigma_{ij}^2 = \lim_{N\to\infty} \frac{1}{N} \sum_{i=1}^{N} \left[(x_i - \bar{x})(y_i - \bar{y}) \right] \quad , \qquad\qquad A(10)$$

where x and y are different variables. If the deviations in x and y are random, then $\sigma_{ij}^2 = 0$. However, if the deviations are correlated, then $\sigma_{ij}^2 \neq 0$ and is a measure of the degree of correlation between x and y.

STANDARD DEVIATION: the standard deviation is defined as the square root of the variance

$$\sigma = \sqrt{\sigma^2} \quad . \qquad\qquad A(11)$$

Thus it is the root-mean-square of the deviations, and is a measure of the uncertainty of a result assuming random error only.

The standard deviation is a very important quantity because it allows us to perform hypothesis tests to determine the significance (or otherwise) of a result.

PROPAGATION OF ERROR: let $x_i (i=1,n)$ be a set of experimental results, each with an associated variance σ_i^2, and let y be some parameter that is related to the experimental results by the function

$$y = f(x_1, x_2 ... x_n) \qquad . \qquad\qquad A(12)$$

When calculating y, we must assign a standard deviation or we cannot assess the significance of the result. If σ_y^2 is the variance of y,

$$\sigma_y^2 = \sum_{i=1}^{n} \sum_{j=1}^{n} \sigma_{ij}^2 \frac{\partial f}{\partial x_i} \frac{\partial f}{\partial x_j} \qquad , \qquad\qquad A(13)$$

where σ_{ii}^2 and σ_{jj}^2 are the variances of x_i and x_j, and σ_{ij}^2 is the covariance of x_i and x_j.

Chapter 4   P. F. McMillan and A. M. Hofmeister
INFRARED AND RAMAN SPECTROSCOPY

INTRODUCTION

Vibrational spectroscopy involves the use of light to probe the vibrational behaviour of molecular systems, usually via an absorption or a light scattering experiment. Vibrational energies of molecules and crystals lie in the approximate energy range 0 - 60 kJ/mol, or 0 - 5,000 cm^{-1}. This corresponds to the energy of light in the infrared region of the electromagnetic spectrum, and direct absorption of light by molecular vibrations is an infrared absorption experiment. Molecular vibrations may also be studied by a light scattering experiment, in which the energy of an incident light beam, usually in the visible region of the spectrum, is slightly raised or lowered by inelastic interaction with the vibrational modes. This experiment is known as Raman scattering (or combination scattering in many Russian publications) after one of its discoverers (Raman and Krishnan, 1928; Landsberg and Mandelstam, 1928). A related experiment, known as Brillouin scattering, is used to study the very low frequency vibrations which are sound waves in crystals and other condensed phases.

Both infrared and Raman spectroscopies give rise to a set of absorption or scattering peaks as a function of energy. These constitute a vibrational spectrum. Individual peaks in the spectrum correspond to energies of vibrational transitions within the sample, or in more common terminology, to the frequencies of its vibrational modes.

One of the most obvious applications of vibrational spectroscopy is in qualitative and quantitative analysis. This has been one of the major applications of infrared spectroscopy in mineralogy and geochemistry. A vibrational spectrum is characteristic of a given sample. Furthermore, individual peaks may be associated with the presence of particular structural groups within the sample, hence the spectra may be used to infer the presence of particular phases or molecular groups. The analysis can be made quantitative with suitable calibration. Vibrational spectroscopy can be a useful supplement or alternative to analysis by normal chemical or diffraction methods, since vibrational spectra can be easily obtained for crystalline or amorphous solids, liquids or gases, and can be easily applied to systems involving elements of low atomic weight.

Since a vibrational spectrum is dependent on the interatomic forces in a particular sample, it is a sensitive probe of the microscopic structure and bonding within the material. Vibrational spectroscopy can provide much useful structural information from the positions, symmetries and relative intensities of observed peaks, and any changes in these with variables such as temperature, composition, and applied pressure. In principle, a detailed analysis of the vibrational spectrum should give a wealth of information on the nature of interatomic interactions, and many of the vibrational studies in the literature have been carried out with this in mind. Unfortunately, however, it is not straightforward to unravel the interatomic force field of even the simplest molecules from vibrational data alone. Recently it has become possible to calculate the structures and dynamic properties of relatively large molecules and simple crystals using first-principles or

semi-empirical quantum mechanical methods. Vibrational spectra provide
sensitive tests and calibrations for these calculations, and can in this
way lead to a better microscopic understanding of interatomic forces.

As noted above, vibrational transitions of molecules lie in the
energy range 0-5000 cm^{-1} (0-60 kJ/mol), corresponding to temperatures
(kT) of approximately 0-6000 K. This range covers those temperatures
most commonly measured in the laboratory, or pertinent to most
geochemical and geophysical studies. The thermal excitation of
vibrations is largely responsible for changes in the internal energy of
most materials, and the heat capacity of a substance may be calculated
from a knowledge of its vibrational spectrum (Kieffer, 1985). This
direct link to useful thermodynamic properties has provided a further
impetus for the measurement of vibrational spectra.

From the above discussion, it is evident that vibrational
spectroscopy is a useful tool for sample characterization and analysis,
for structural studies and as a probe of interatomic forces, and for
calculation of thermodynamic and elastic properties. All of these have
been applied to problems in mineralogy and geochemistry, and the field
of mineral vibrational spectroscopy is an active and fascinating area of
research.

VIBRATIONAL THEORY OF MOLECULES AND CRYSTALS

Classical models

The theory of molecular and crystal vibrations and their interaction
with light is well established, and is described in a large number of
texts. An introduction to these topics has recently been presented by
McMillan (1985), and only a brief summary is given below.

The simplest description of vibrations of molecules and crystals is
in terms of a classical mechanical model. Nuclei are represented by
point masses, and the interatomic interactions (bonding and repulsive
interactions) by springs. The atoms are then allowed to undergo small
vibrational displacements about their equilibrium positions, and their
equations of motion are analyzed in terms of classical (Newtonian)
mechanics. If the springs are assumed ideal in that the restoring force
is directly proportional to displacement (Hooke's law), then the
vibrational motion is harmonic, or sinusoidal in time. The
proportionality constant which relates the restoring force to
vibrational displacement is termed the <u>force</u> <u>constant</u> of the spring.
Solution of the equations of motion for the system allows a set of
vibrational <u>frequencies</u> ν_i to be identified. Each frequency corresponds
to a particular atomic displacement pattern, known as a <u>normal</u> <u>mode</u> of
vibration (Fig. 1). In many vibrational studies, the object is to
deduce the form of the normal modes associated with particular
vibrational frequencies, in order to use vibrational spectroscopy as a
structural tool.

Quantum mechanical models

The classical model allows a description of the basic features of
vibrational motion, but does not give any insight into why vibrational
spectra are line spectra rather than continuous absorptions, nor into
the interaction of vibrations with light. For this, it is necessary to
construct a quantum mechanical model. Schrodinger's wave equation (see
Chapter 2) is constructed in terms of the vibrational displacement
coordinates q_i, and an appropriate potential energy function $V(q_i)$ is

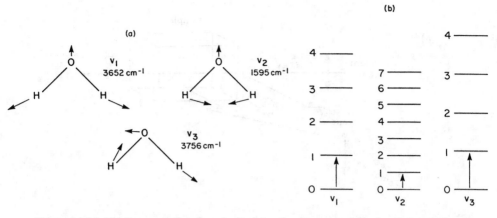

Figure 1. (a) The three normal modes of vibration for the water molecule. Each normal mode is associated with a classical vibrational fequency, ν_1, ν_2, or ν_3. (b) The quantized vibrational energy levels for H_2O. The quantum numbers v_1, v_2 and v_3 are associated with the three vibrational modes, and each can take integer values 0,1,2,3... The fundamental frequencies ν_1, ν_2 and ν_3 correspond to transitions between vibrational states with $v_i=0$ and $v_i=1$ as shown by arrows.

assumed, to give a set of partial differential equations in the vibrational wave equation. Solution of these differential equations gives a set of vibrational wave functions Ψ_i, each describing a vibrational normal mode, and a set of associated vibrational energies. These wave functions and energies are quantized, in that they can take only discrete values determined by a vibrational quantum number v_i, which can have integer values 0,1,2,3, etc.. The quantized energies are usually shown on an energy level diagram as the vibrational energy levels for the system (Fig. 1). In a vibrational spectroscopic experiment, the system undergoes a transition between vibrational levels with quantum numbers v_i and v_j, and light is absorbed or emitted with an energy ($\Delta E=h\nu$) corresponding to the separation between the levels. This separation is generally on the order of 0-60 kJ/mol, corresponding to 0-5000 cm^{-1}, and the light is in the infrared region of the electromagnetic spectrum.

Interaction with light: infrared and Raman spectroscopy

In infrared absorption spectroscopy, infrared radiation is passed through a sample and the intensity of the transmitted light is measured as a function of its wavelength. Absorption of light occurs at frequencies corresponding to the energies of vibrational transitions, and a set of absorption maxima is observed.

In a Raman scattering experiment, visible light (usually a monochromatic beam from a laser) is passed through the sample. Most of the light exits from the sample with no change, but a small fraction (around 10^{-3} of the incident intensity) is scattered by the atoms. A small proportion of this scattered light (around 10^{-6} of the incident light) interacts with the sample in such a way as to induce a vibrational mode. When this occurs, the energy of the scattered light is reduced by an amount corresponding to the energy of the vibrational transition. This type of inelastic scattering is known as Raman scattering, while elastic light scattering with no change in energy is known as Rayleigh scattering (Fig. 2). In Raman spectroscopy, the energy of the scattered light is analyzed using a spectrometer, and Raman lines appear as weak peaks shifted in energy from the Rayleigh

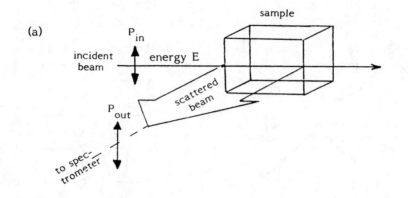

Figure 2. <u>Raman scattering</u>. (a) The incident laser beam with energy E is passed through the sample, and the scattered light is sent to the spectrometer. (b) The Raman spectrum consists of a strong central line at the incident laser energy E corresponding to Rayleigh scattering, and weaker Raman-shifted lines at $E \pm e_i$, where $e_i = h\nu_i$ correspond to the energies of vibrational transitions in the sample. Energies are usually expressed in wavenumbers (cm^{-1}). The incident laser frequency is taken as 0 cm^{-1}, and Stokes Raman-shifted frequencies (E-e) are given as positive wavenumber values. (c) The energy level diagram for Rayleigh and Raman scattering. Two vibrational levels with $v_i = 0$ and $v_i = 1$ are shown. These are separated by an energy $e = h\nu$, where ν is the vibrational frequency. The incident laser photon (energy E) excites the molecule to a short-lived ($\approx 10^{-14}$ s) electronic "virtual state", which decays with release of a photon. When the final vibrational state of the molecule is higher than that of the initial state, the released photon energy is E-e, and Stokes Raman scattering has occurred. When the final state is lower, the released photon has energy E+e, and anti-Stokes Raman scattering has occurred. When the initial and final states are the same, Rayleigh scattering has taken place, and the incident and released photons have the same energy E.

line (which has the energy of the incident beam). The positions of
these Raman peaks relative to the incident laser line correspond to the
frequencies of Raman-active vibrations in the sample.

Different vibrational modes have different relative intensities in
infrared and Raman spectra; some modes are active in one and not the
other, and some modes are not observed at all. The infrared and Raman
activities of particular modes are determined by the quantum mechanical
selection rules for the vibrational transition (see Chapter 2), and by
the mode symmetry. In a simple model, the selection rules can be
rationalized by considering the interaction between the oscillating
electric field vector of the light beam and a changing molecular dipole
moment associated with the vibration. In an infrared experiment, the
light interacts directly with an oscillating molecular dipole, so for a
vibrational mode to be infrared active, it must be associated with a
changing dipole moment. In general, asymmetric vibrations tend to give
stronger infrared absorption than symmetric species, since they are
associated with larger dipole moment changes. Similarly highly polar
("more ionic") molecules and crystals have stronger infrared spectra
than non-polar samples.

In Raman scattering, the light beam induces an instantaneous dipole
moment in the molecule by deforming its electronic wave function. The
atomic nuclei tend to follow the deformed electron positions, and if the
nuclear displacement pattern corresponds to that of a molecular
vibration, the mode is Raman active. The magnitude of the induced
dipole moment is related to the ease with which the electron cloud may
be deformed, described by the molecular polarizability α. The Raman
activity of a given mode is related to the change in polarizability
during the vibration. In general, molecules containing easily
polarizable atoms (such as iodine, sulphur or titanium) have very strong
Raman spectra, while similar molecules with less polarizable atoms
(silicon, carbon, oxygen) have much weaker spectra. In contrast to
infrared spectra, the most symmetric modes tend to give the strongest
Raman signals as these are associated with the largest changes in
polarizability.

Infrared and Raman activities of particular vibrational modes for
simple molecules and crystals can often be deduced by inspection, but
this becomes unreliable or impossible for more complicated cases. The
problem is greatly simplified by use of the molecular or unit cell
symmetry, and the methods of group theory (Cotton, 1971; Woodward, 1972;
Fateley et al., 1972; McMillan, 1985: see Chapter 2).

Crystal lattice vibrations

The number of vibrational modes observed for a molecule is equal to
the number of classical degrees of vibrational freedom, 3N-6, where N is
the number of atoms in the molecule (3N-5 for a linear molecule). In
the case of a crystal, N is very large (on the order of 10^{23}), but most
of the modes are not observed in infrared or Raman spectroscopy. This
is due to the translational symmetry of the atoms in the crystal. The
vibration of each atom about its equilibrium position is influenced by
the vibrational motion of its neighbours. Since the atoms are arranged
in a periodic pattern, the vibrational modes take the form of
displacement waves travelling through the crystal; these are known as
lattice vibrations. These lattice waves may be described as
longitudinal, where the nuclear displacements are parallel to the wave
propagation direction, or transverse, where the displacements are
perpendicular to the propagation direction (Fig. 3).

104

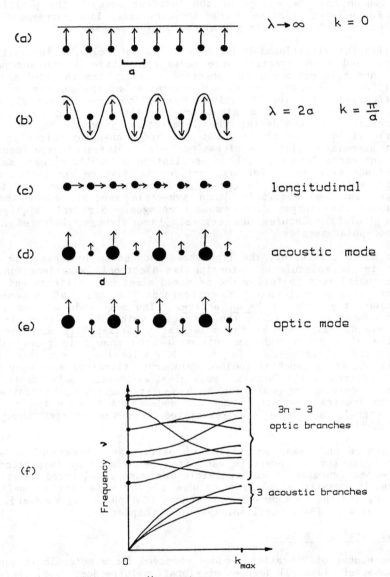

Figure 3. Various types of crystal lattice vibration. (a) The wavelength of this lattice mode is long compared with the crystal lattice constant a. This mode is said to lie at the Brillouin zone centre (k=0). (b) This mode has wavelength λ=2a, and lies at the edge of the Brillouin zone (k=π/a). The waves in (a) and (b) represent transverse lattice vibrations for a monatomic chain of atoms. (c) This illustrates a longitudinal lattice vibrationa for the same monatomic chain. (d) For any crystal, there are three lattice vibrations where all three atoms in the unit cell move in phase in the same direction. These are the acoustic modes. For crystals with more than one atom in the primitive unit cell, there are modes where atoms in the unit cell move in opposing directions (illustrated for a diatomic chain). These motions can generate a changing dipole moment and hence interact with light: these are termed optic modes. (f) A typical dispersion diagram in one direction in reciprocal space for a crystal, in this case with n=4 atoms in its unit cell. Only long wavelength lattice vibrations (near k=0) can be infrared or Raman active due to the long wavelength of light compared with crystal lattice spacings. These points are marked with dots.

The nuclear displacements give rise to an oscillating dipole moment, which interacts with light in a spectroscopic experiment. The frequency of this oscillating dipole wave is defined by the oscillation frequency of individual atoms about their equilibrium positions, and its wavelength is defined by that of the associated lattice vibration. In order for the lattice vibration to interact with light, it must have a wavelength comparable to that of the light. In a typical infrared experiment, light has a wavelength of approximately 5×10^4–5×10^6 Å (2000–20 cm^{-1}), whereas for Raman scattering, the wavelength used is usually in the blue-green region of the visible spectrum, or 10^3–10^4 Å. These light wavelengths are much larger than the dimensions of crystalline unit cells (10–100 Å), hence only very long wavelength lattice modes can interact with light in an infrared or Raman experiment. In these long wavelength lattice vibrations, the vibrations within adjacent unit cells are essentially in phase, so the number of vibrational modes which may be observed in infrared or Raman spectroscopy is equal to $3n-3$, where n is the number of atoms in the primitive unit cell. These $3n-3$ vibrations which can interact with light are termed the "optic modes". Transverse and longitudinal optic modes are termed TO and LO modes for convenience.

Crystal lattice vibrations are usually described in terms of a wave vector, k. The direction of k is the direction of propagation of the lattice wave, and the magnitude of k is $2\pi/\lambda$ where λ is the wavelength of the lattice wave. The relationship between the frequency of a particular normal mode and the wavelength of its propagation through the lattice is known as a dispersion relation, and is usually represented graphically as a dispersion curve $\nu(k)$. Each normal mode is associated with a "branch" of the dispersion diagram. In any particular crystallographic direction in reciprocal space, there are 3n branches, where n is the number of atoms in the primitive unit cell. Three of these are the acoustic branches, responsible for propagation of sound waves through the lattice. At infinite wavelength (i.e., at k=0), the three acoustic modes have zero frequency, and correspond to translations of the entire crystal. The remaining $3n-3$ branches are known as the optic branches, since these can give rise to infrared and Raman active vibrations for long wavelength modes (k≈0) (Fig. 3). The shortest wavelength for lattice vibrations is defined by the lattice constant a, with adjacent unit cells vibrating exactly out of phase. The wavelength of the lattice wave is then 2a, corresponding to k=π/a. The phase relations between vibrations in adjacent unit cells define a region in reciprocal space between k=-π/a and k=π/a, known as the first Brillouin zone. Long wavelength lattice vibrations with k=0 are said to lie at the centre of the first Brillouin zone (Fig. 3).

As for molecules, crystal lattice vibrations are more completely described by a quantum mechanical model, and the vibrational spectra of crystals correspond to transitions between vibrational states. The "unit" of vibrational excitation in a crystal is known as a phonon, by analogy with the term "photon" for a quantized unit of light energy (Chapter 2).

INSTRUMENTATION FOR INFRARED AND RAMAN EXPERIMENTS

In general, a spectroscopic experiment requires a light source, a means of providing energy resolution of the light before or after interaction with the sample, and a detection system. A number of light sources are used in infrared experiments to cover the wide range of wavelengths involved, including heated blackbodies and gas discharge lamps. Early Raman scattering experiments used a high intensity discharge lamp, but since the early 1960's, the laser has been the optical source of choice. Most modern infrared spectroscopy is carried out with an interferometer, while grating spectrometers are used in Raman spectroscopy and some older infrared instruments. A wide range of thermoelectric and photoelectric detectors are used in infrared spectroscopy, whereas photomultiplier or diode array detectors sensitive in the visible region of the spectrum are used in Raman spectroscopy.

Infrared sources

For infrared spectroscopy, it is necessary to distinguish between the near-IR ($10,000-4,000$ cm^{-1}: $1-2.5$ μm), the mid-IR ($4,000-400$ cm^{-1}: $2.5-25$ μm), and the far-IR ($400-10$ cm^{-1}: $25-1000$ μm). The near-IR region is used to study overtone and combination bands, frequently due to vibrations involving hydrogen, which are 10 to 1000 times weaker than mid-IR bands. The mid-IR contains strong fundamental absorptions such as the Si-O stretching and bending vibrations of silicate minerals. The far-IR also contains fundamental absorptions, commonly 2-20 times weaker than the mid-IR bands, and usually associated with metal-oxygen vibrations and complex deformations of polymeric units.

Various infrared sources are used, depending on the wavelength region of interest. For most measurements, a heated silicon carbide "Globar", essentially a black body source, is used. This gives useful intensity between approximately 9000 and 50 cm^{-1}. Some instruments use a Nernst filament, consisting of a mixture of rare earth oxides. For far-IR studies of silicates, a Globar is usually sufficient, but below $100-50$ cm^{-1} more intensity can be obtained from a mercury arc lamp encased in quartz. Near-IR experiments at wavenumbers greater than around 9000 cm^{-1} require a tungsten filament lamp as source. For further details, see Mohler and Rothschild (1971), Griffiths and deHaseth (1986), or Hollas (1987).

Raman sources - lasers

Good introductions to the basic priciples of laser operation and descriptions of the various types of laser are given by Siegman (1971), Svelto (1976) and Demtroder (1982). Most Raman spectroscopy is carried out using a laser in the visible region of the spectrum, and the most popular excitation source is the argon ion gas laser. This is a high power continuous laser which is extremely stable. The most widely used lasing transitions are the 488 nm and the 514.5 nm lines of the Ar^{+} ion, as these provide the most power (1-2 W output for a 4-5 W laser). There are also a number of other weaker lasing lines in the blue-green and ultraviolet regions of the spectrum; these are useful for spectroscopy, especially when working with samples which fluoresce in the red end of the visible region. Other continuous gas lasers used for Raman spectroscopy include the krypton and helium-neon lasers. The He-Ne laser has a lasing transition (of Ne^{+}) in the red region at 632.8 nm which is occasionally used for Raman work, but is rarely of high enough power to obtain high signal to noise ratios with conventional Raman instruments. The high power krypton laser has a number of lasing lines of reasonable intensity ranging from the yellow to the ultraviolet. These are useful for studies in which the sample absorbs or fluoresces in the blue-green region of the spectrum (such as Fe-containing minerals), resulting in problems for Raman spectroscopy with the argon laser. The same degree of spectral coverage can be obtained by using a dye laser. In this case, an organic dye is made to fluoresce over a broad band of visible wavelengths by excitation ("pumping") with a visible laser, often a high power argon laser. Individual laser wavelengths are selected by a diffraction grating or etalon system, and the output laser energy can be tuned continuously over a wide range of wavelengths by rotating the grating. This type of tuning is essential in resonance Raman experiments, where the Raman signal is examined as the laser wavelength is scanned across an electronic transition of the sample. Modern commercial ring dye lasers pumped by 15-20 W argon lasers can achieve on the order of 1-2 W of power from the yellow to the UV (260 nm) by the use of a range of dyes and frequency doubling crystals for different spectral ranges. Some Raman experiments require extremely high power densities which are difficult to achieve with continuous gas lasers, or require time resolution. These conditions are met by pulsed lasers. The most common for Raman spectroscopy is the Nd-YAG pulsed laser, where the lasing transition is the $^4F_{3/2}-^4I_{11/2}$ transition of Nd^{3+} ions in a yttrium aluminum garnet matrix. This transition lies in the infrared, at 1.06 µm, but may be frequency-doubled to give a laser line at 503 nm, in the green region of the spectrum.

Spectrometers and interferometers

A monochromator is a device for providing energy resolution of incident light, and generally consists of an entrance slit, a dispersing element which may be a prism or diffraction grating, and an exit slit. A spectrometer for infrared or Raman studies may consist of one or more monochromators mounted in series or parallel. In a prism spectrometer, as used in early infrared experiments, the dispersing elements were alkali halide prisms, and dispersion of the light took place via the refractive index of the prism material. Such instruments are preserved in only a few laboratories. Raman spectrometers generally are grating instruments, in which dispersion takes place via selective reflection off the grating surface due to constructive and destructive interference caused by the regularly ruled surface. Most modern IR instruments are interferometers, in which the signal is collected in the time domain, then converted to an energy scale via a Fourier transform. Grating IR spectrometers are still present in many laboratories, but are now only marketed by one company (Perkin-Elmer), and are likely to be phased out over the next five to ten years. Fourier transform infrared (FTIR) instruments have many advantages over conventional scanning infrared spectrometers. More recently, Fourier transform Raman experiments have been carried out, using laser sources in the near infrared (Chase, 1987).

A schematic drawing of a double monochromator used for Raman spectroscopy is shown in Figure 4. Light enters the spectrometer through the entrance slit S_1. Usually, the light from the sample is focused by a collection lens, mounted before in front of this slit and must be collimated by a parabolic mirror (C_1) before reaching the grating (G_1). Some spectrometers contain curved gratings, which obviates the need for the mirror. Older gratings were ruled mechanically using a ruling engine, and often periodic errors in the grating spacing would give rise to spurious reflections, or "grating ghosts" in the spectrum. Modern gratings are prepared by photolithography using laser holography ("holographic gratings"), and the problem of grating ghosts has disappeared, although spurious lines may still appear due to stray reflections within the spectrometer, or diffraction off the edges of slits, mirrors and other optical elements within the spectrometer. Constructive and destructive interference of light from the grating surface determines a relation between the reflection angle and the wavelength reflected (the "grating equation": $m\lambda = d(\sin \alpha + \sin \beta)$; where m is the order of the reflection, d is the

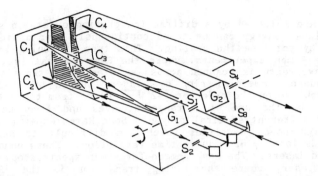

Figure 4. Schematic of a Raman double monochromator system. This drawing shows the Instruments S.A. Ramanor U-1000 system in use at Arizona State University. Light enters the spectrometer through the entrance slit S_1 and falls on the collimating mirror C_1 to be sent to the first holographic grating G_1. The dispersed light is refocussed by the second mirror C_2, then exits the first monochromator through slit S_2. The light then enters a coupling region (between baffles) designed to physically separate the two spectrometers, before entering the second monochromator at S_3. After mirror C_3, grating G_2, and mirror C_4, the light exits the double monochromator through slit S_4 to fall on the detector. Both grating turn on the same axis as shown for wavelength selection. (Adapted from Instruments S.A.'s schematic of the U-1000 spectrometer system).

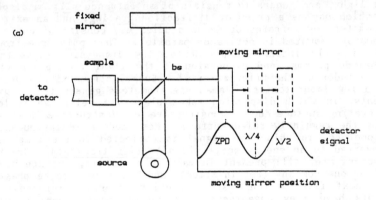

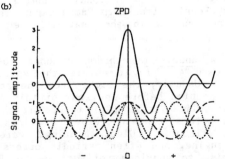

Figure 5. The principles of FTIR spectroscopy. (a) The Michelson interferometer. Light from the source is split by the beamsplitter (bs). One half is reflected by a fixed mirror, the other by a moving mirror. The two beams recombine at the beam splitter, and are sent through a sample area to the detector. With no sample, for a pure single frequency incident beam with wavelength λ, the output signal is a sine wave with maximum amplitude when the difference in distances of the mirrors from the beam splitter is $n\lambda/2$ where n is an integer. Fourier transformation of this signal gives a delta function at the frequency of the signal. The absorption spectrum of the sample modifies the resultant interferogram observed at the detector, and is recovered by an inverse Fourier transform. (b) The resultant interferogram for three pure frequencies. The three sine waves add constructively only at the ZPD (zero path difference) point, which corresponds to the point of zero displacement (n=0) of the moving mirror. (Adapted from Nicolet Corporation's operating manual for their 7199 FTIR instrument).

spacing of grooves on the grating, and α and β are the angles made by incident and reflected light to the grating normal). The light reflected from the grating G_1 is then sent to a second parabolic mirror C_2 to be focused into the second monochromator, then through the exit slit (S_4) to the detector. Rotation of the grating about an axis parallel to the long axis of the slit causes different wavelengths to be incident on the slit, and energy resolution of the light beam is achieved. The width of the exit slit determines the spectral bandpass $\Delta\lambda/\lambda$, hence plays an important role in determining the resolution of observed bands in the spectroscopic experiment.

In a dispersive (double beam) infrared spectrometer (Hadni, 1967), the beam produced by the source is chopped and passed through two compartments, one containing the sample, and the other a reference (usually a metal screen of known absorbance, or a blank KBr disc). The chopped beam allows alternate sampling of the sample and reference beams. The modulated beam is directed through a variable size slit (which determines the resolution), then dispersed by a grating (or a prism in older instruments) on to the detector. One instrument will generally have a set of different gratings mounted on a grating wheel, each operating over a particular wavelength range, and the gratings are mechanically rotated as the spectrum is scanned. Since a wide range of wavelengths is covered, the slit width must be constantly changed to obtain a constant energy resolution $\Delta\lambda/\lambda$. The scan time is determined by the time taken for the instrument to view each resolution element, multiplied by the number of these elements in the spectrum.

Dispersive instruments are rapidly losing popularity in comparison to Fourier transform infrared (FTIR) spectrometers, due to the decreasing cost of FTIR instruments coupled with the inherent disadvantages of dispersive spectrometers:

- No internal reference exists for frequency calibration, hence the instrument must be calibrated against external standards.
- Only a small proportion of the energy is available at the detector at any time due to the monochromating slit, although the sample receives all of the incident energy, and may thermally degrade.
- The time to acquire a spectrum is long: a single scan generally takes on the order of 30 minutes.
- Obtaining high resolution scans is limited by the gratings and slits, and resolution is generally gained at the expense of spectral intensity.
- Stray light can interfere with the measurements.

Fourier transform infrared spectrometers are based on the principle of the Michelson interferometer (Fig. 5a). Light from the infrared source is passed through a beam splitter which directs half of the incident intensity to a fixed mirror, and half to a moving mirror. The two beams are recombined at the beam splitter, so that constructive or destructive interference occurs, depending on the instantaneous position of the moving mirror relative to the fixed mirror. The result is an interferogram, plotted as light intensity vs. time. For a pure single frequency source such as a laser, the result is a sine wave. This is the Fourier transform of a delta function at the laser frequency, when intensity is plotted against light energy. In a Fourier transform spectroscopic experiment, light intensity is measured in the time domain, then converted via a Fourier transform (using appropriate computer software) to intensity vs. energy (usually expressed in units of wavelength, frequency, or wavenumber).

When the source emits a range of light frequencies, all the resulting sine waves add constructively only when the moving mirror is the same distance from the beam splitter as the fixed mirror. The result is shown for three sine waves in Figure 5b. Note that every data point of the resulting interferogram contains information over the entire infrared region covered by the source. This signal is modified by the particular IR absorbance signature of the sample before reaching the detector. During the FTIR experiment, a reference laser line (usually from a He-Ne laser) is also passed through the beam splitter. The laser signal is monitored by a separate detector, and is used to precisely determine the position of the moving mirror relative to the fixed mirror, and also serves as an internal frequency standard.

A schematic of a Fourier transform infrared spectrometer is shown in Figure 6. This type of spectrometer has several distinct advantages over the grating instrument:

- A single scan of the moving mirror (usually taking 0.25-1 second) contains all of the spectral information. Desired signal-to-noise ratios are then obtained by averaging multiple scans until a satis-factory spectrum is obtained. This is the so-called "multiplex advantage".
- The energy reaching the detector is limited only by the source output and sample absorbance, not by instrumental slits (the Jaquinot advantage). This is important with small or weakly absorbing samples.
- The laser provides an internal frequency standard with an accuracy of approximately 0.01 cm^{-1}; this is particularly important for performing spectral subtractions.
- The resolution is determined by the number of data points sampled during the distance travelled by the moving mirror. This only perceptibly affects the time required to obtain a spectrum at resolutions below around 0.5 cm^{-1}.
- Stray light does not affect the measurements because it is not part of the interferogram.

Infrared detectors

Infrared detectors can be separated into two classes: thermal detectors and quantum detectors. Thermal detectors operate by sensing recharge of temperature of an absorbing material (the temperature changes are related to the infrared radiation coming from the sample). These detectors have slow response times (time constants of 0.01-0.1 s), but generally have a wide wavelength range of operation. Quantum detectors rely on the interaction of incoming IR photons with the electronic wavefunction of the solid detector material. These detectors have fast response times, but only cover narrow wavelength ranges.

For the mid-IR (5000-400 cm^{-1}), the most common detectors are liquid-nitrogen-cooled MCT (HgCdTe) or triglycine sulfate (TGS) detectors. For far-IR work (below 500 cm^{-1}), a TGS detector is often used, although these have very low sensitivity. Liquid-helium-cooled bolometers with Si or Ge elements are now commercially available that cover the far-IR range with a hundred-fold increase in sensitivity. The main disadvantages of these detectors are their cost and the difficulties associated with using liquid helium as coolant. For the near-IR (9000-2000 cm^{-1}), liquid-nitrogen-cooled InAs or InSb detectors are available.

Thermal detectors include the TGS, DTGS (deuterated TGS) and MCT

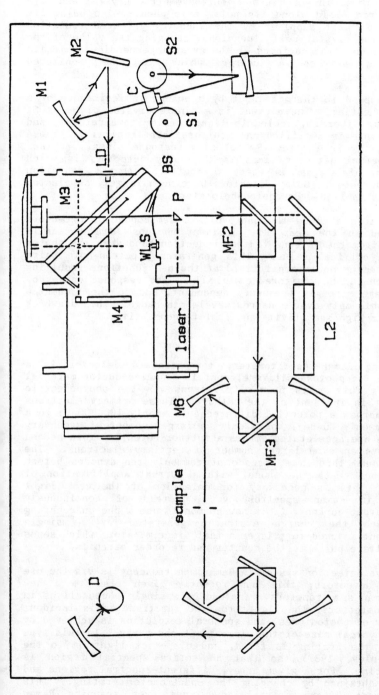

Figure 6. Schematic of the Nicolet 7199 Fourier transform infrared spectrometer in use at the Geophysical Laboratory. Infrared light from one source (S2) is collimated (C) then reflected (mirrors M1,M2) to the beam splitter (BS). The beam is split and directed to the fixed (M3) and moving (M4) mirrors, recombined, and directed into the sample chamber by mirrors MF2, MF3 and M6. After the sample, other mirrors direct the light to the detector (D). A reference laser beam (laser) is collinear with the IR beam. Its beam is removed from the IR detector path by a prism (P), and the laser signal is detected at LD. A white light source (WLS) which is also used as a reference is monitored by an off-axis beam splitter/detector system (dashed lines). A second laser (L2) can also be used for alignment of the optics. (Adapted from Nicolet Corporation's schematic of the 7199 instrument).

detectors commonly used in far- to mid-IR studies. These are pyroelectric detectors, in which a non-centrosymmetric crystal exhibits an internal electric field along its polar axis when cooled below its Curie temperature. Heat supplied by the incoming infrared radiation thermally alters the crystal cell dimensions, changing the value of the electric polarization. The surfaces of the crystal perpendicular to its polarization axis develop a charge which can be monitored electronically.

Bolometers respond to thermal input by a change in resistance of a semiconductor on heating. The element is a single crystal of Ge or Si doped with Cu, As, Ga or Sb, cooled to liquid-helium temperatures, and the resistance change is amplified and monitored electronically. These have high sensitivity in the far-IR, but slow response times. Another member of the thermal detector class is the Golay detector, in which thermal expansion of a gas is used to monitor infrared radiation incident on the detector. Golay detectors have been largely supplanted by recent advances in liquid He-cooled bolometers.

Quantum detectors include the photovoltaic semiconductor detectors (InAs, InSb) used in the near-IR. Incident photons in the near-IR excite electrons from the valence to the conduction band when the photon energy exceeds a critical value. This generates electron-hole pairs which are separated by an internal field at the p-n junction, resulting in a voltage which can be monitored externally. The response curves of these detectors are strongly frequency dependent, and so these quantum detectors are usable only over a narrow wavelength range. This drawback is offset by their high sensitivity and fast response time.

Raman detectors

In a scanning Raman spectrometer, the standard detector is a photomultiplier. The photosensitive element is a semiconductor material (commonly GaAs). This emits primary electrons via the photoelectric effect when light is incident on the cathode. These primary electrons are accelerated across a potential difference (commonly 100-200 V) to a metal dynode (commonly Cu-Be), which emits a larger number of secondary electrons. These are accelerated across a further potential to a second dynode which emits an even larger number of tertiary electrons. The process is continued through a cascade of commonly ten dynodes before the final electrons reach the anode, with a current amplification of around 10^5-10^7. This is necessary for detection of the weak light signals output in Raman spectroscopy. Instead of continuously monitoring the output current, it is now common to use a photon counting method, in which the charge avalanche generated by a single photoelectron event is used to trigger a fast discriminator, which sends a square wave voltage pulse to the counter and recorder circuits.

An alternative detection system in Raman spectroscopy is via the use of diode array detectors; this has recently given rise to a new generation of Raman spectrometers. Instead of a single photocell as in photomultiplier detectors, the light from the spectrometer is incident on a linear array of photodiodes, and spectral resolution is achieved by the number and physical size of the elements of the array. In this type of spectrometer, the grating is fixed, and disperses light on to the diode array (Hemley, 1987), so that an entire spectral region is observed at one time. For a given array, different spectral regions and resolutions are obtained by changing the spectrometer grating. This type of instrument offers several advantages over scanning Raman spectrometers. The most obvious is that diode array spectrometers allow collection of an entire Raman spectrum in a fraction of the time required for a normal scan. This permits considerable signal averaging, giving a multiplex advantage over conventional scanning instruments. For this reason, diode array Raman spectrometers are likely to become the instrument of choice for studies in the diamond anvil cell, which involve small quantities of weakly scattering sample. Diode array instruments can also be used more easily for kinetic studies, in which the Raman spectrum is followed as a function of time.

Infrared windows, cell material and dispersing media

For the mid- to near-IR, KBr or other alkali halides can be used as window materials. These can also be ground with powdered samples to form a sample support when pressed into a disc. CaF_2 or BaF_2 can be used for near-IR and some mid-IR measurements. For the far-IR, various plastics are used as window material. The most common is polyethylene, with a useful range of 10-500 cm^{-1}. Other useful materials for this range are polystyrene, TPX (a transparent commercial plastic), or quartz. For far-IR studies, petroleum jelly, polyethylene or nujol (a liquid paraffin) are the most common dispersing media. Useful specifications on these and other materials can be found in IR trade catalogues. For high pressure work in the diamond anvil cell, type II diamonds are useable over the entire IR range, whereas type I diamonds can only be used in the far-IR (due to intense absorption from nitrogen).

For Raman experiments, powdered samples are usually run without support. Samples with high refractive index can give considerable parasite scattering of the laser beam when run as powders, due to reflection at the many sample/air interfaces. These samples may be embedded in a support material of similar refractive index, such as an alkali halide, to give a more massive sample which the laser can traverse with little or no scattering at the sample/support interface. The only window required in most Raman experiments is that of the detector housing. Quartz is transparent over the visible range, while silica glass or CaF_2 may be used for UV applications. For Raman experiments in the diamond anvil cell, it is important to choose low-fluorescence diamonds, otherwise the weak Raman signal is swamped by a fluorescence background from the diamond.

Infrared beam splitters and polarizers

In the mid-IR, a Ge-coated KBr beam splitter may be used from 5800 to 400 cm^{-1}, while Si or CaF_2 beam splitters may be used in the near-IR, from 10,000 to 1800 cm^{-1}. One of the major nuisances in current far-IR spectroscopy is the narrow range covered by the Mylar beam splitters used. For most experiments, a 6 μm Mylar (polyethylene terephthallate) sheet allows measurement from 550 to 70 cm^{-1}. Below this, thicker sheets cover increasingly smaller wavenumber ranges; for example a 25 μm sheet can be used from 100 to 20 cm^{-1}. Griffiths and deHaseth (1986) discuss the characteristics of various beam splitters and materials, and the wavelength ranges of each are listed in IR trade catalogs.

In both Raman and infrared experiments, it is often useful to obtain polarized spectra in order to assign mode symmetries, and determine relative orientations of vibrating groups. In Raman spectroscopy, a sheet of Polaroid plastic is used to select one polarization of the scattered radiation after the sample. (The incident laser beam is already plane polarized). If necessary, the polarization of the incident or scattered beams may be rotated by inserting a mica 1/4-wave plate, or by using a quartz wedge polarization rotator. If a circular polarized beam is required, this may be obtained by using a quartz polarization scrambler. For far-IR experiments, the polarizing element most commonly used is a wire grid, typically gold on a polyethylene substrate, which covers the range 550-10 cm^{-1}. Common near-IR polarizers are gold wire grids on AgBr covering roughly 5000 to 300 cm^{-1}, or KSR-5 (thallium bromoiodide) which covers 10000-300 cm^{-1}. Near-IR measurements above 4000 cm^{-1} can be optimized by using a LiIO$_3$ substrate.

Instrument calibration

In any type of spectroscopic experiment, two types of calibration are important: wavelength and intensity. The latter is usually at least an order of magnitude more difficult than the former. Secondly, it is also important to decide whether or not an absolute or a relative calibration is necessary for a particular experiment.

In Raman spectroscopy, wavelength calibration is fairly simple: the emission lines of simple gas discharge lamps (Hg, Ne, for example) are known with great accuracy, and can be used to calibrate the instrument over the entire visible spectrum. It is important to note, and either correct or calibrate, any non-linearities in the wavelength readout of scanning Raman instruments. During a Raman scan, non-lasing lines from the exciting laser ("laser plasma lines") may be observed in the spectrum, and can provide useful internal standards. It is also common practice to keep a few samples with well-known spectra as secondary standards, for routine checks on the instrument calibration.

Intensity calibration is generally much more difficult. The throughput of the spectrometer may be calibrated as a function of wavelength using a standard intensity lamp. However, even if the spectrometer throughput is known, it is usually extremely difficult to characterize the laser power density at the sample giving rise to Raman scattering. For these reasons, there have been very few measurements of absolute Raman scattering cross sections. Intensity measurements in Raman spectroscopy are usually relative to an internal or external standard. If the intensity standard is external, care must be taken to ensure that the experimental conditions are exactly reproduced. Due to the difficulties involved, Raman spectroscopy has been a much less popular technique for quantitative analysis than IR spectroscopy. Wopenka and Pasteris (1986) and Pasteris et al. (1987) have recently discussed in detail theoretical and practical aspects of quantitative Raman spectroscopy as applied to fluid inclusion analysis.

Like Raman spectrometers, dispersive IR spectrometers require both wavelength and intensity calibration. The recorded wavelengths may differ from true values by an unspecified amount due to mechanical slippage or misalignment of the gratings. This can be corrected by obtaining a spectrum of a standard material, such as a thin polystyrene film, and comparing the observed peak positions to the accepted standard values. Large corrections (more than a few wavenumbers) usually require realignment of the instrument. The recorded absorbances are calibrated by measuring the absorbance of wire screens of various mesh densities, and constructing a calibration curve for true vs. measured absorbance. Fourier transform instruments do not require wavelength calibration because the He-Ne laser provides an internal standard to an accuracy of +0.01 cm^{-1}. Errors in absolute absorbance do occur in FTIR instruments, but these are generally small (Hirschfeld, 1979; Griffiths and deHaseth, 1986).

Quantitative analysis via IR spectroscopy relies on the Bouguer-Beer-Lambert law. This states that the absorbance is proportional to the concentration of the absorbing species for reasonably dilute substances:

$$A(\nu) = I(\nu)Ct$$

where $A(\nu)$ is the measured absorbance at frequency ν, $I(\nu)$ is the molar absorptivity of the species at that frequency, t is the path length of the IR beam through the sample, and C is the concentration of the species. $I(\nu)$ is generally established from measured absorption spectra of standard samples of known concentration and path length. Deviations from Beer's law may occur for many reasons, including stray light, variable scan speed, insufficient resolution, phase correction errors, or physical and chemical properties of the sample (Hirschfeld, 1979; Griffiths and deHaseth, 1986), but these are usually small for modern FTIR instruments and mineralogical problems.

EXPERIMENTAL INFRARED AND RAMAN SPECTROSCOPY

Sample preparation and measurement techniques

Raman and infrared spectra of most gases and liquids at ambient conditions are readily obtained in gas cells or cuvettes, using appropriate long path lengths or multiple pass cells to amplify weak signals. However, most experiments in mineralogy and geochemistry are carried out on solid samples. When the samples are large (100 μm or more for Raman; a few tens of mm across for infrared) single crystals and glass samples are usually oriented and polished for polarized infrared reflectance or absorption and Raman scattering using normal macroscopic methods. (Oriented spectra may be obtained for much smaller samples using the newly developed microbeam methods, decribed below).

Raman spectroscopy on powders or polycrystalline samples is carried out either by packing the powder in a glass capillary, by mounting in a refractive index-matched medium, or most simply by striking a smoothed surface of the powder with the incident beam, and attempting to minimize stray light entering the spectrometer. One problem with powder methods is that, if the sample is sensitive to the laser beam, transformation of the sample may occur within a few grains at the focus of the beam, and it may be impossible to characterize this later (see Akaogi et al. (1984) and McMillan and Akaogi (1987) for an example of this in β-Mg_2SiO_4).

Powder infrared spectroscopy is carried out by dispersing the finely ground sample in an inert matrix (alkali halide disc, nujol mull, or petroleum jelly slurry) or supporting it on a glass or plastic film, and passing the beam through the sample and matrix material. This is the most common type of infrared spectroscopy carried out in the mineral sciences, but has a number of inherent disadvantages which should be understood before any interpretation of the spectra is made.

Infrared spectra of powders can also be obtained by photoacoustic spectroscopy (PAS). This technique requires essentially no sample preparation and results in an infrared absorption spectrum with little reflection component (Vidine, 1980). The sensitivity of PAS for infrared spectroscopy is low, and requires use of a Fourier transform instrument. In the PAS technique, modulated light is absorbed by a sample housed in a closed compartment. This results in heating of the sample surface and the air immediately surrounding the sample. Because the sample is confined, a modulated pressure wave is generated, and this "sound" wave can be detected by a microphone. The main disadvantages to PAS for routine infrared analysis of powdered samples are that it is essentially a surface technique, and that it is currently limited to the mid-IR and visible regions.

A related type of spectroscopy is photothermal beam deflection spectroscopy (Low and Tascom, 1985), which also requires no sample preparation. A modulated IR beam strikes the surface of a sample. Selectively absorbed IR radiation heats the sample and the air around it, resulting in a refractive index gradient in the surrounding air. A laser beam passing through this gradient is deflected, and the beam position is monitored and transformed to give an absorption spectrum of the sample. The results are similar to those obtained for samples in KBr pellets (Low and Tascom, 1985).

Infrared absorption spectroscopy of minerals

Pure infrared absorption spectra of minerals are obtained by passing an infrared beam through a thin film of sample. For silicates, the appropriate thickness depends on the material and on the strength of the bands of interest. For beryl (Hofmeister et al., 1987a) and olivine (Hofmeister, 1987; Hofmeister et al., 1987b), 10 to 15 µm is appropriate for far-IR studies and 0.5-2 µm for mid-IR studies. For silicate spinels (Hofmeister et al., 1986a), far-IR studies of the weakest band require a 30 µm slab. In general, thinner samples are better for mid-IR studies: at thicknesses of a few microns, LO modes begin to interfere with the TO absorption spectrum. The development of high sensitivity FTIR microscopes has rendered such absorption studies feasible. The very thin films necessary for these studies can be obtained in two ways. One involves compressing a powdered sample in a diamond anvil cell, and the spectrum measured after the pressure is released. The resulting spectrum appears to be free of LO components (Hofmeister et al., 1986a,b; 1987b). The second method is to cleave or ion-thin a bulk sample and use an FTIR microscope to view an area of appropriate thickness (Hofmeister, 1987; Hofmeister et al., 1987b). Polarization properties and absolute peak intensities can be measured with this latter technique, and the wavelength range is only limited by the spectrometer. The major disadvantages lie in the non-trivial sample preparation, requiring orientation, polishing and ion-thinning of microscopic samples.

Infrared reflection spectroscopy

An infrared reflectance spectrum is generally more difficult to interpret than a Raman or infrared absorption spectrum for a number of reasons, and it is important to understand the interaction of infrared radiation with a solid in order to correctly interpret the features observed. The infrared reflectance of crystals and other condensed phases can be understood by combining Maxwell's expressions for the propagation of light through dense media, with the classical theory of lattice dynamics. When an infrared beam is incident on a surface, parts of the beam may be reflected, transmitted or absorbed. An infrared reflectance spectrum $R(\nu)$ of a crystal has a characteristic form (Fig. 7) which can be analyzed to give the optical constants (refractive index n and dielectric behaviour ε) as a function of the wavelength λ (or frequency ν) of the incident light, and the frequencies of infrared active transverse and longitudinal optic modes.

Consider first the hypothetical case where light resonates with an infrared frequency of the crystal, but no energy is absorbed. The

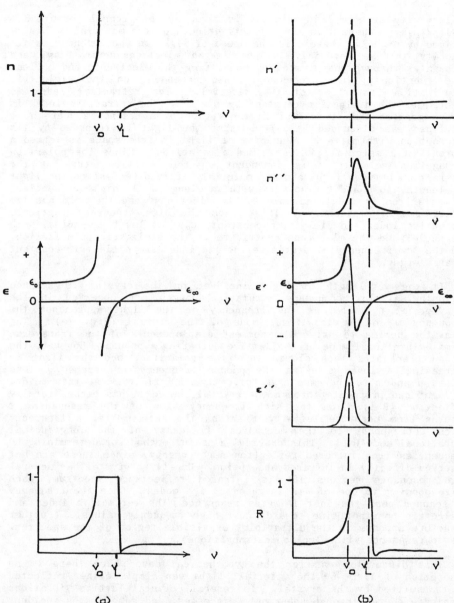

Figure 7. <u>Optical constants and infrared reflectance</u>. (a) The hypothetical for no absorption by the material. The refractive index n(ν) and the dielectric constant ε(ν) tend to infinity in the region of a resonance (ν_0). Immediately following the resonance, the dielectric constant rises from −∞ to cross the axis (ε=0) at the frequency of the longitudinal optic mode (ν_L). The refractive index is undefined between ν_0 and ν_L. The infrared reflectance R(ν)=1 between ν_0 and ν_L. (b) A "real" case with infrared absorption through anharmonic processes. The refractive index and dielectric constant are now both complex. The imaginary part of the dielectric constant ε"(ν) shows a maximum at the TO frequency ν_0, while the LO frequency is identified from the point at which the real part (ε'(ν)) crosses the frequency axis. The reflectivity curve does not rise perfectly to R=1, and there may be some structure in the "forbidded" region ν_0 to ν_L.

reflectivity at normal incidence R is given by the <u>Fresnel</u> formula $R = |n-1|^2/|n+1|^2$, where n is the refractive index of the <u>material</u>. This is defined by $n = c/v$, where c is the speed of light in vacuum and v is its velocity in the material. Light is an electromagnetic beam, and propagates through the crystal by modulating the electronic and nuclear wave functions; this process is then dependent on the dielectric polarization properties of the material. The refractive index is related to the dielectric constant by $\varepsilon = n^2$. The refractive index is always greater than 1 (except in the region of a resonance where n is undefined, as discussed below), because the light is <u>retarded</u> by its interaction with matter. A quantum of light in free space is termed a <u>photon</u>: in a material, it is termed a <u>polariton</u>. Since the polariton propagation velocity v is dependent on the precise nature of the interaction between light and the material, it is dependent on the light wavelength, so n and ε also vary with wavelength. At high frequencies, the polariton is modulated only by the electrons, and the more massive nuclei remain stationary. This gives the high frequency (optical) refractive index and dielectric constant, ε_∞. At low frequencies, both the nuclei and the electrons contribute to the dielectric polarization, and the low frequency or static refractive index and dielectric constant ε_0 are larger.

If the wavelength of light incident on the crystal is scanned beginning at low frequency, dramatic changes are observed in the behavior of n and ε as the frequency of the light approaches the frequency ν_0 of a vibrational mode of the crystal (Fig. 7a). The polariton begins to drive the nuclear displacements of the transverse normal vibrational mode of the lattice, creating a phonon. However, the phonon propagates much slower than the polariton, and the light is increasingly slowed down as the phonon frequency is reached. This causes n and ε to increase asymptotically to infinity. At this point, no light can propagate through the crystal, hence it has become totally reflecting (R=1). On the high frequency side of the resonance, ε returns from negative infinity to cross the axis ε=0 at a frequency which corresponds to the mechanical frequency of the longitudinal vibrational mode (ν_L). This provides a useful method for determining LO frequencies from infrared reflection spectroscopy, since these are not observed directly in infrared absorption. Usually a crystal has several such resonance regions in its infrared reflectance spectrum, each corresponding to one of its infrared active modes. Once the frequency is scanned past the last infrared resonance, the refractive index and dielectric constant rise toward their high frequency values. (Further structure appears in the ultraviolet or visible region of the spectrum, due to resonance with electronic transitions).

This discussion was for the hypothetical case where there is no absorption of light by the material: light was simply either reflected or transmitted by the crystal. In general, crystal lattice vibrations dissipate energy through anharmonic processes, and some of the incident light energy is absorbed by the crystal. This is termed <u>dielectric loss</u> behaviour. Absorption is included in the treatment by allowing the optical constants to become complex: $n = n' + in''$ and $\varepsilon = \varepsilon' + i\varepsilon''$, so that $\varepsilon' = n'^2 - n''^2$ and $\varepsilon'' = 2n'n''$. The <u>dielectric loss function</u> $\varepsilon''(\nu)$ has a maximum at the transverse optical frequency ν_{TO}, and the absorption coefficient is determined by the magnitude of n'' at this point. Plots of these functions are shown schematically in Figure 7b. In this case, the (measured) infrared reflection coefficient R is given in terms of the complex refractive index by

$$R = \frac{(n'-1)^2 + (n'')^2}{(n'+1)^2 + (n'')^2} \quad .$$

There are two major methods for extracting the optical constants from the measured reflectivity $R(\nu)$. If we define $r=\sqrt{R}$ as the real part of the reflectivity, the real (r) and imaginary (r^*) parts of the reflectivity may be related by the expression

$$r^* = r\, e^{i\theta}$$

where θ is the phase shift between real and imaginary parts. The value of θ is proportional to the magnitude of the absorption coefficient, defined by $A(\nu) = 4\pi n''\nu$, where ν is the optical frequency. The task of a Kramers-Kronig analysis is to extract the phase shift $\theta(\nu)$ from the measured reflection spectrum $R(\nu)$, and hence determine the infrared absorption spectrum $A(\nu)$ and other optical properties of the crystal. The form of the equation used in Kramers-Kronig analysis is

$$\theta(\nu_i) = \frac{2\nu_i}{\pi} \int_0^\infty \frac{\ln r(\nu) - \ln r(\nu_i)}{\nu_i{}^2 - \nu^2}\, d\nu \quad .$$

Once $\theta(\nu_i)$ has been determined (and hence r and r^*), the optical constants n' and n'' (and hence ε' and ε'') can be calculated from the Fresnel equations:

$$n' = \frac{(1-r^2)}{1 + r^2 - 2r\cos\theta} \quad ; \quad n'' = \frac{2r}{1 + r^2 - 2r\cos\theta} \quad .$$

The TO and LO mode frequencies can then be established from the resulting plots of ε' and ε'' versus frequency ν (Fig. 7b). The TO frequency occurs at the maximum in the $\varepsilon''(\nu)$ curve, whereas the LO frequency occurs at the point at which the $\varepsilon'(\nu)$ curve crosses the ν axis. The Kramers-Kronig equations have been discussed by many authors (Born and Huang, 1954; Hadni, 1967; Piriou, 1974; Turrell, 1972; Decius and Hexter, 1977).

The problem with the Kramers-Kronig method is that the use of the integral in the expression for $\theta(\nu_i)$ strictly requires knowledge of the measured reflectivity over all frequencies ($\nu=0$ to $\nu=\infty$). In practice, R is known only over a finite range (usually 4000 to 50 cm^{-1} if far-IR experiments have been carried out). The remaining values are estimated by fitting synthetic "wings" to the reflectivity function. The method most commonly used for estimating these wings is due to Anderman et al. (1965), who assumed the lowest and highest energy peaks to be due to undamped simple harmonic oscillators. Their theoretical reflectance spectrum is then calculated and joined smoothly to the low- and high-frequency ends of the experimentally determined region. This approach gives reasonable results for ε', ε'', n' and n'' over the region of measurement, provided the reflectance spectra are smooth and continuous. If the limiting values of the dielectric constants ε_0 and ε_∞ are known from complementary experimental measurements (capacitance measurements for ε_0, and refractive index measurements for ε_∞), these can be used as additional constraints on the synthetic wings (Spitzer et al., 1962).

An alternative method for reducing experimental reflectivity data is known as <u>classical oscillator</u> (classical dispersion) <u>analysis</u> (Piriou and Cabannes, 1968; Gervais et al., 1972; Piriou, 1974). In this method, the crystal is approximated as a system of damped harmonic oscillators with appropriate frequencies and dipole moments. The complex dielectric function is given as

$$\varepsilon(\nu) = \varepsilon_\infty + \sum_j f_j \, \nu_j^2 \, / (\nu_j^2 - \nu^2 + i\nu\Gamma_j) \quad ,$$

where f_j is the <u>oscillator strength</u> of the jth oscillator (related to the dipole moment change of that vibration), ν_j is its vibrational frequency, and Γ_j is its <u>damping coefficient</u>. The width of the peak in the $\varepsilon''(\nu)$ curve is 2Γ. This expression is used to fit the measured reflectivity

$$R = \frac{|\sqrt{\varepsilon}-1|}{|\sqrt{\varepsilon}+1|} \quad .$$

For analysis of a given spectrum, a Kramers-Kronig analysis (as described above) can be used to estimate the initial parameters for classical dispersion analysis. The reflectivity spectrum is then calculated and compared to experiment. Successive trials and adjustments of the parameters are then made until a reasonable fit to the data is obtained. This type of analysis is successful if the number of oscillators used is the minimum required by the data, and if the dispersion parameters are uniquely determined within reasonable uncertainties. Very good fits to the infrared reflectivity of silicate minerals have been obtained using the classical oscillator method (Piriou and Cabannes, 1968; Gervais et al., 1972; 1973a,b,c; Piriou, 1974; Gervais and Piriou, 1975).

Infrared powder transmission spectra

The most common method of obtaining infrared spectra of minerals has been via a powder transmission experiment, in which the powdered sample is dispersed in a pressed KBr disc, a nujol mull or a petroleum jelly slurry, and the IR beam is passed through the powder and support material. The resulting spectrum shows minima in infrared transmission usually taken to correspond to bulk absorption peaks of the sample (these are often re-plotted on a relative absorbance scale for publication). Obviously, any polarization information on absorption bands is lost, but there are several other major problems with this method which are not always recognized. These include (a) effects arising from interference in the infrared beam; (b) shifting of TO (absorption) peak positions toward the frequency of the LO component; (c) splitting of a single absorption band into separated TO and LO components; (d) appearance of additional surface modes. These effects can cause major changes in the appearance of a true absorption spectrum both in the number and shapes of the bands observed, and can be dependent on the shape and size of the sample particles.

In an infrared powder experiment, the particle size is usually on the order of 1-100 μm, comparable to the wavelength of infrared light, and diffraction may occur. Such interference effects are size and shape dependent, and may give rise to spurious structure in far- and mid-IR bands (Born and Huang, 1954; Tuddenham and Lyon, 1967; Luxon and Summitt, 1969; Luxon et al., 1969; Sherwood, 1972). Interference fringes may also appear in thin film absorption studies. In principle, features due to this type of diffraction can be removed from the interferogram in FTIR studies, but commonly the side bands due to the interference are so close to the main spectral spike that any subtraction modifies the spectrum (i.e., information is lost from the wings of the interferogram, leading to indeterminate shifts in band position and alteration of intensities) (Hirschfeld, 1979).

When the incident IR beam strikes a given particle, it is partly absorbed and partly reflected, depending on the geometry of incidence and the optical characteristics of the sample and dispersing medium. The resulting spectrum contains components of both pure absorption (with a single peak at the TO frequency) and reflection (with a reflection band extending between the TO and LO frequencies). If the LO–TO splitting of a given band is small (which is the case for most weak IR bands, and is especially true in the far-IR), then the transmission minimum observed lies close to the true TO frequency. If the LO–TO splitting is moderate (common for most mid-IR bands of silicate minerals), the transmission minimum may be shifted to higher frequency (towards the frequency of the LO component). This shift can be on the order of 50–100 cm^{-1} (Piriou, 1974). For very large LO–TO splittings, as encountered in many titanium-bearing samples, the TO and LO components may appear as separate minima in the transmission spectra.

Finally, the large surface-to-volume ratio of the powdered particles lends itself to the excitation of surface modes. These surface mode frequencies lie between the TO and LO frequencies of the bulk sample, and can give rise to strong infrared absorptions in the powder spectrum. Serna et al. (1982) showed that the frequencies of surface modes for corundum structure oxides are highly dependent on particle shape, particularly the axial ratio, and that this results in apparent shifts of fundamental vibration frequencies. In certain cases, surface mode absorptions have been observed to dominate the powder transmission spectrum (Sherwood, 1972).

All of these effects may be present to varying degree in the powder infrared spectra of minerals, and a given effect may be more important for particular bands within a spectrum. In general, the problems seem to be more pronounced for minerals with simpler spectra. For example, the powder infrared spectra of simple cubic minerals such as rock salt oxides and halides, perovskites and spinels are rendered almost uninterpretable by these effects, but the powder spectra of cubic garnets with a large number of bands can be useful. For non-cubic minerals, accidental degeneracies and near-degeneracies between bands of different symmetries can further complicate the powder spectra, and polarized IR measurements are necessary to reliably observe the entire spectrum. Despite these problems, powder IR spectroscopy has been applied successfully to a wide range of mineralogical problems over the past years (Lazarev, 1972; Farmer, 1974a; Karr, 1975), and may continue to be useful if the above potential problems are recognized. However, with the advent of micro-IR spectroscopy coupled with FTIR techniques, it is likely that powder studies will be replaced by single crystal absorption and reflection studies using the IR microscope.

Far-infrared spectroscopy

Infrared spectra are so much more difficult to measure in the far-IR region due to instrumental and intrinsic sample problems, that far-IR spectroscopy may be considered an entirely different experiment to mid- or near-IR measurements. Because of this, far-IR spectra of minerals have been much less well studied than the mid- and near-IR regions. The problems associated with far-IR measurements include weak sources, generally less sensitive detectors, low energy resolution which varies

strongly with wavelength, the fact that many absorption bands in the
far-IR are generally weaker than those in the mid-IR region, and
interference from strong rotational bands of even trace amounts of H_2O
vapour. The first four problems become more severe with decreasing
wavenumber. However, the multiplexing advantage of Fourier transform
methods overcomes the weakness of far-IR sources and absorption bands,
and the recent development of a silicon detector for liquid-helium-
cooled bolometers has dramatically increased their sensitivity. The
most intense, sharp water vapour peaks occur at 305, 280, 250, 230, 205,
150, 140-120, and 100 cm^{-1} (Mohler and Rothschild, 1971), and it is
important not to incorrectly assign these to sample absorptions in
mineral spectra. Several FTIR instruments are now available which
operate under vacuum, or with a sample compartment purged with dry
nitrogen (the rest of the IR optical bench is evacuated); this
alleviates the problems associated with atmospheric water vapour.

Polarized infrared and Raman spectroscopy

The symmetries of infrared and Raman active vibrational modes can be
deduced from polarized spectroscopic experiments on oriented single
crystals. If the point group character table is examined, the symmetry
species of infrared active modes (obtained by factor group analysis)
transform in the same way as one or more of the Cartesian translation
vectors x, y or z (Chapter 2). A crystalline sample may be cut, mounted
and polished normal to one of these directions and an infrared
reflection or transmission experiment carried out with the beam normal
to the cut face. If the infrared beam is polarized, dipole moment
changes parallel to one of the two crystal axes contained within the
plane will be selected. If these correspond to different symmetry
species, then only normal modes belonging to that species will be
excited. The spectrum of the other symmetry species may be obtained by
rotating the crystal. This type of polarized infrared spectroscopy
allows assignment of the symmetry types of individual infrared active
modes. Furthermore, as infrared activity is dependent on a changing
dipole moment, the orientation of the dipole moment change of a given
vibration relative to the crystal axes may be determined, which can give
information on the crystal structure and the nature of the vibrational
modes. This type of analysis has been particularly useful in the
analysis of O-H stretching vibrations in minerals (see chapter by G.
Rossman).

A similar experiment can be carried out with Raman scattering. The
Raman active vibrations correspond to symmetry species which transform
in the same way as the Cartesian products xx, yy, zz, xy, xz or yz. In
this case, both the incident and the scattered beams may be polarized.
If their polarizations are parallel (VV or HH geometries), diagonal
elements of the scattering tensor xx,yy or zz are selected, while if the
polarizations are perpendicular (VH or HV), off diagonal elements xy,xz
or yz appear in the spectrum. This allows a depolarization ratio ρ =
I_{VH}/I_{VV} to be measured, which can be used to determine the symmetry type
of the vibration. Most Raman scattering experiments are carried out
with the scattered beam detected at 90° to the incident laser, and clean
faces are required for entrance and exit of the laser beam and
observation of the Raman scattering.

There are a number of practical considerations in polarized infrared
or Raman spectroscopy. The crystal must be properly oriented, usually
by optical or X-ray methods, or incorrect mode symmetries will be
identified. Small errors in orientation or non-planarity of polished
faces are difficult to avoid, and may result in leakage, where strong

modes of one symmetry may appear weakly in the spectrum for another orientation. In particular, in infrared spectroscopy, longitudinal components of one mode may appear in the spectrum of a transverse mode propagating at right angles to the first mode. In the case of IR measurements, a few per cent misorientation from normal incidence does not appreciably affect the results. Care must be taken to identify these problems, otherwise the wrong number of modes might be attributed to particular symmetry species, resulting in an incorrect vibrational analysis. Especially in Raman experiments, it is important to have a good optical polish on the laser entrance and exit faces, and to try to direct the beam through parts of the sample free from cracks, inclusions or other imperfections, to avoid excess parasite scattering of the laser beam. Finally, when carrying out polarization measurements, it is important to account for the polarizing effect of the spectrometer gratings, which can have a large effect on the results. This can be avoided by mounting a quartz polarization scrambler at the entrance to the spectrometer.

Micro-Raman spectroscopy

Over the past decade, the development of microbeam techniques for Raman spectroscopy has had a major impact on vibrational studies in the earth sciences. In micro-Raman spectroscopy, the incident laser beam is directed into an optical microscope and focussed through the objective on to the sample. The scattered light is collected back through the same objective (near 180° scattering geometry), and sent into the spectrometer. The focus of the incident laser beam forms an approximate cylinder, whose dimensions are dictated by the wavelength of the laser light and the optical characteristics of the objective. The diameter of the cylinder is fixed by the diffraction limit of light in air, and is approximately one μm for an objective with numerical aperture near 0.9 and laser light in the blue-green region of the spectrum. The length (depth) of the scattering cylinder is approximately three μm for the same objective and laser light. This shows that micro-Raman spectroscopy has better lateral spatial resolution than depth resolution. The depth resolution can be improved by adding diaphragms near the back focal plane of the microscope, before the spectrometer entrance slit.

Micro-Raman instruments have proved extremely useful in mineralogy and geochemistry, allowing oriented single crystal studies on samples only a few μm across. These techniques have revolutionized the study of fluid inclusions (Pasteris et al., 1987). The laser beam can be focused several tens or hundreds of μm within samples transparent to the laser light, to obtain Raman spectra in-situ of the contents of micron-sized fluid inclusions. This permits unambiguous, non-destructive analysis of the molecular species in the inclusion, and the analyses can be made quantitative by suitable calibration.

A further application of micro-Raman spectroscopy is for in-situ studies at high pressures in the diamond anvil cell. In these studies, the incident laser beam usually enters the cell at ~135° to the normal axis of the diamond, in order to minimize problems of fluorescence from the diamond along the laser beam path. In conventional (non-micro) diamond cell Raman experiments, there is no control over position within the sample, and the spectra may averaged over a large pressure gradient, and any assemblage of phases present in the cell. With the micro-beam techniques, individual phases may be studied at well-defined points within the cell. Recently micro-Raman spectroscopy has been used to calibrate the pressure gradient across a diamond cell using observed pressure shifts in the Raman line of diamond (Sharma et al., 1985). In these diamond cell experiments, the excitation volume at the sample is extremely small, hence the Raman scattering is usually weak. The recent development of diode array spectrometers has revolutionized these studies, and it is now possible to obtain useful spectra from even weakly scattering samples (such as amorphous silicates) at high pressure, using micro-Raman with diode array detection in a diamond anvil cell (Hemley, 1987; Hemley et al., 1986a; 1987a).

Micro-infrared techniques

In contrast to micro-Raman spectroscopy, micro-sampling in the infrared is in its infancy. Within the past four years, commercial instruments have been developed which focus the IR beam on to the area of the sample viewed by the user. Two types of IR microscopes are currently available: one based on lenses and the other based on reflecting optics. Both are suitable only for IR interferometry (i.e., are single beam instruments), and both have intrinsic limits on the area sampled. If the diameter of the sample approaches the wavelength of the infrared radiation, the incident beam is diffracted. In principle, to measure far-IR spectra down to 100 cm^{-1} (the lower limit for IR-active vibrations of most silicates), sample size must exceed 100 μm, whereas near-IR measurements can be made on samples only a few microns across. In practice, the commercially available IR microscopes are limited to sample diameters of 10 μm.

Micro-infrared spectroscopy is only beginning to be applied to problems of mineralogical and geochemical interest. Polarized single crystal reflection spectra can be obtained down to the diffraction limit (Hofmeister, 1987), and a new technique involving measurement of absorption spectra from ion-thinned samples has revealed IR bands not resolvable by previous studies (Hofmeister et al., 1987a; Hofmeister, 1987).

Fourier transform Raman spectroscopy

The most common problems in conventional Raman spectroscopy concern:
- the presence of a background (usually from sample fluorescence);
- thermal decomposition of the sample under high power laser excitation;
- the long scan times required for signal averaging;
- the need for external frequency calibration.

Some of these problems are alleviated by the new diode array detection systems described earlier. A second approach has been the recent development of Fourier transform Raman instruments (Chase, 1987), which are especially appropriate for strongly fluorescent or thermally unstable samples. For sample excitation, the FT-Raman instrument uses a Nd-YAG laser operating at 1.06 μm, which is well below the threshold for most electronic fluorescence transitions. The weak Raman cross-section at this wavelength is compensated by the multiplexing (i.e., multiple rapid scanning) ability of the FT spectrometer, and high resolution spectra may be obtained in a fraction of the time required for the same experiment with a conventional scanning instrument. One current limitation is that the incident (elastically scattered) laser line must be removed from the spectrum by a set of optical filters, which limits the lowest frequency observable to around 200 cm^{-1}; however, this may soon be improved (Chase, 1987). This technique will certainly find wide application in mineral spectroscopy, many of which (e.g., iron-bearing phases) show strong fluorescence in conventional Raman spectroscopy.

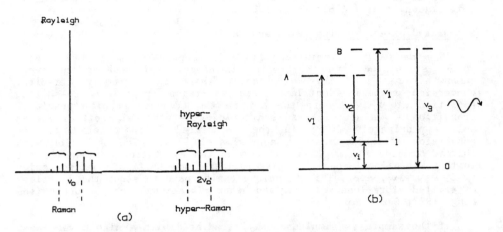

Figure 8. (a) In Raman spectroscopy, the Stokes and anti-Stokes Raman lines are distributed around the Rayleigh line at the incident laser frequency ν_0. In hyper-Raman spectroscopy, the hyper-Raman lines are shifted from the first harmonic of the laser frequency, $2\nu_0$. The selection rules for Raman and hyper-Raman spectroscopy are different, so different numbers and relative intensities of lines are generally observed. (b) Schematic energy level diagram for the coherent anti-Stokes Raman effect (CARS).

Resonance Raman spectroscopy

In normal Raman scattering, the incident laser energy is chosen to lie far from any electronic transitions of the sample, to avoid absorption, fluorescence or other electronic effects. However, if the laser frequency is instead tuned to coincide with an electronic absorption, this can cause large intensity enhancement of Raman-active modes associated with the absorbing species (see Long, 1977). The resonance Raman effect can be used to study specific <u>chromophores</u> (species with electronic transitions in the visible to near-UV). For example, resonance Raman spectroscopy has been used to identify the color-producing species in ultramarine (synthetic blue lazurite: $Na_8Al_6Si_6O_{24}S_4$) as S^{2-} and S^{3-} complexes (Clark and Franks, 1975).

Hyper-Raman scattering

If a sample is excited by a laser beam of sufficiently high power (usually by a pulse from a ruby or Nd-YAG laser), two incident laser photons (incident energy E_0; frequency ν_0) may be combined within the sample through non-linear optical mixing. As a consequence of this, weak Rayleigh scattering is observed at twice the incident frequency ($2\nu_0$) in addition to the normal Rayleigh line at the incident laser frequency ν_0. This <u>frequency-doubled</u> photon can also interact with the vibrational modes of the sample, and a <u>hyper-Raman</u> spectrum appears at frequencies $\pm\nu_i$ around the hyper-Rayleigh line (Fig. 8a). The interest in hyper-Raman spectroscopy is that the selection rules are different to those for normal Raman scattering. For example, all infrared-active modes of a sample are observed in the hyper-Raman spectrum. This could be extremely useful for obtaining infrared frequencies for samples whose infrared spectra are difficult to measure. Some hyper-Raman results have been reported for SiO_2 glass (Denisov et al., 1978; 1984), and for some oxide perovskites both at ambient pressure and in the diamond anvil cell (Inoue et al., 1985; Vogt, 1986).

Stimulated Raman, inverse Raman and CARS

If a sample is illuminated with a laser pulse of sufficiently high power, stimulated Raman scattering is observed instead of the hyper Raman effect. In this experiment, there is extremely efficient conversion of the incident laser beam (at frequency ν_0) to Raman-shifted radiation at $\nu_0+\nu_i$, where the ν_i are the frequencies of vibrational transitions. In normal Raman scattering, the conversion efficiency is only approximately 10^{-6}; in the stimulated Raman experiment, the conversion efficiency can be approximately 50%. The radiation with shifted frequency $\nu_0+\nu_i$ can now act as the excitation frequencies for further vibrational modes, to give further Raman shifting to frequencies $\nu_0+2\nu_i$, $\nu_0+3\nu_i$, etc. This technique can be used to study the lifetimes of excited vibrational states, and hence vibrational decay process (see Long, 1977; Demtroder, 1981).

If the sample is simultaneously irradiated with a high power laser pulse at frequency ν_0 and a continuum extending from ν_0 to approximately $\nu_0\pm\delta\nu$ (where $\delta\nu$ covers the wavenumber range of Raman frequencies of interest, usually approximately 4000 cm^{-1}), absorptions are observed in the continuum spectrum at Raman-active frequencies ν_i (accompanied by emission of photons at the laser frequency ν_0). This is the inverse Raman effect. Under irradiation by the laser and the continuum, the sample absorbs a quantum of energy ($E=hc(\nu_0+\nu_i)$) from the continuum, undergoes an internal transition between two vibrational levels separated by an energy $hc\nu_i$, and light is re-emitted at the laser frequency ν_0. This technique can be used to study the lifetimes of vibrational states, and also the internal re-distribution of vibrational energy between different vibrational modes.

In the CARS (coherent anti-Stokes Raman scattering) experiment, two lasers with frequencies ν_1 and ν_2 are "mixed" within the sample ($\nu_1>\nu_2$). The frequency of one is held constant (ν_1), and the other is tuned to give frequency differences $\nu_1-\nu_2=\nu_i$, where ν_i corresponds to a vibrational frequency of the sample. When this occurs, coherent emission is observed at a frequency $\nu_3=2\nu_1-\nu_2$ ($=\nu_1+\nu_i$). The energy level diagram describing this process is shown in Figure 8b. The first laser frequency (ν_1) pumps the molecule from its vibrational ground state (with vibrational quantum number v=0) to some virtual excited state A (this virtual state does not correspond to a true excited electronic state of the sample, but corresponds to a transient deformation of the electronic wave function due to the incident photon). Simultaneous irradiation with the second laser (ν_2) instantaneously stimulates a transition from this virtual state back to the first excited vibrational state (v=1). A second photon from the initial laser (ν_1) raises the molecule to a further excited virtual state B. This unstable virtual state then decays back near instantaneously to the vibrational ground state, with emission of a photon of frequency $\nu_3=\nu_1+\nu_i$. In principle, in a CARS experiment, there is no need for a spectrometer. The second laser frequency (ν_2) is tuned across the energy range of interest, to provide differences $\nu_1-\nu_2$ corresponding to the range of vibrational frequencies ν_i of the sample, and a detector placed after the sample detects strong emission of light when the frequency difference $\nu_1-\nu_2$ matches each vibrational frequency ν_i. This could prove useful for spectroscopic applications in which it is difficult to collect the scattered light for analysis by a spectrometer.

The laser techniques briefly described above form part of a family of relatively new laser spectroscopies. At present, most require fairly sophisticated experimental technique, and have not yet been applied to samples of geological interest. However, these spectroscopies provide powerful probes of vibrational and electronic properties, and it is certain that the next few years will see application of these techniques to problems in mineralogy and geochemistry.

Spectroscopy at high pressures and temperatures

There is a range of techniques for obtaining vibrational spectra of solids, liquids and gases at high temperatures and pressures. For low temperatures and pressures (several tens of degrees and several hundred bars), gases and liquids can be sealed in glass capillaries.
For higher pressures (up to a few kilobars) and temperatures, a more robust cell must be constructed, usually of a steel-based alloy with sapphire or quartz windows (Irish et al., 1982; Kruse and Franck, 1982; Sherman, 1984). Both internal and external heating arrangements have been used, and pressure and temperature can be easily monitored and controlled in this type of cell. Cells of this type have been used up to around fifty kilobars (although most are used up to only five or ten kilobars) and several hundred degrees Celsius. This type of cell can also be used for solid samples, if the sample is mounted in the light path and pressure is applied via an inert gas.

Studies at higher pressure (to hundreds or thousands of kilobars), usually use a diamond anvil cell (Jayaraman, 1983; Ferraro, 1984; Hemley et al., 1987a). A number of cell designs are available, and have been used for infrared and Raman spectroscopy. These cells can be heated through the gasket material, but the temperature is limited to around 500-600°C due to the oxidation of diamond. Heating to much higher temperatures is also carried using high energy pulses from a Nd-YAG laser directed at the sample, but the temperature is difficult to measure and control, and it is difficult to carry out spectroscopic experiments. A number of challenges are evident for the future of high pressure-high temperature studies: we must find reliable methods for obtaining spectra at pressures of several hundreds to thousands of kilobars and at over 1000°C for studies applicable to processes deep in the Earth, and we must continue to develop fluid cells capable of working at several tens of kilobars and at up to 1000°C, to obtain direct structural and thermodynamic information on geochemical fluids.

VIBRATIONAL SPECTRA OF MINERALS

Simple structures: diamond, graphite and rock salt.

The diamond structure has space group Fd3m (point group O_h) with two atoms in the smallest volume unit cell. It has one triply degenerate optic mode at k=0 with symmetry F_{2g}. This mode is Raman active, but infrared inactive. Since the mode is not polar, it is not affected by TO-LO splitting, and a single line is observed in the Raman spectrum. For carbon with the diamond structure, this line occurs at 1332 cm^{-1}. Recently, Sharma et al. (1985) have used observed shifts in the frequency of this line to calculate the stress distribution across the diamonds in a diamond anvil cell. Phonon dispersion relations in diamond have been measured along several directions in reciprocal space via inelastic neutron scattering (Bilz and Kress, 1979). These have recently been calculated to within a few per cent of experiment using ab initio pseudopotential theory (Louie, 1982).

128

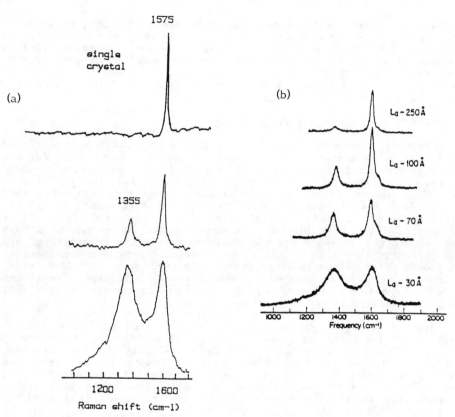

Figure 9. Raman spectra in the C-C stretching region for single crystal and partially ordered graphites. (a) Spectra from Tuinstra and Koenig (1970). The 1355 cm^{-1} band increases in intensity with decreasing crystal order, as evidenced by X-ray diffraction. (b) Spectra from Lespade et al. (1982). The average crystallite size or "correlation length" L_α, calculated from X-ray data, gives a measure of the degree of order of the graphite lattice.

The other common polytype of carbon is the well-known layer structure of graphite, found in a range of terrestrial and extra-terrestrial assemblages (Smith and Buseck, 1981; Buseck and Bo-jun, 1985). Two Raman-active E_{2g} modes and infrared modes of A_{2u} and E_{1u} symmetry are expected for perfectly crystalline graphite. The E_{1u} and one of the E_{2g} vibrations correspond to C-C stretching vibrations within the graphite layer and occur at 1588 and 1582 cm^{-1} respectively. The A_{2u} mode involves out-of-plane bending of the hexagonal carbon layers and occurs at 868 cm^{-1}. The second Raman-active E_{2g} mode is associated with relative displacements of the layers along the c-axis. Since the inter-layer forces are weak compared to the strong intra-layer C-C bonding, this vibration occurs at relatively low frequency (140 cm^{-1}) (Tuinstra and Koenig, 1970; Song et al., 1976). The graphite structure may be disordered in a number of ways, for example by rotating adjacent layers around the c axis, by puckering sheets, or by disordering layer spacings (Rouzaud et al., 1983; Buseck and Bo-Jun, 1987). This disorder has a marked effect on the Raman spectrum. With increasing disorder, the 1582 cm^{-1} band broadens, and moves to higher frequency, while a new broad band appears at 1650 cm^{-1} and increases in intensity (Tuinstra and Koenig, 1970; Lespade et al., 1984; Fig. 9). These effects have been rationalized by considering the phonon dispersion curves of perfectly crystalline graphite (Lespade et al., 1982). If the crystal is uniformly disordered, this is equivalent to considering a larger and larger unit cell, or smaller Brillouin zone, which means that more of the dispersion curve appears in the Raman spectrum (see Lespade et al., 1982). With this level of understanding, the Raman spectrum of graphite can be used to define a degree of disorder within the sample.

The cubic rock salt and cesium chloride structures are taken by a wide range of binary oxides, halides and sulphides. They have a single optic mode at k=0 with F_{1u} symmetry which is infrared active and Raman inactive. This mode involves relative displacement of cations and anions along the three crystallographic directions. Since the mode is polar (i.e., involves a dipole moment change), it is split by TO-LO splitting at the Brillouin zone centre into doubly degenerate transverse components and a longitudinal component, to give a broad reflection band (see Fig. 7). The TO and LO frequencies for a wide range of rock salt and cesium chloride compounds have been measured by infrared reflection spectroscopy and tabulated (Born and Huang, 1954; Farmer, 1974b; Ferraro, 1984), while dispersion curves for many of these have been obtained by inelastic neutron scattering (Bilz and Kress, 1979). Recently, there has been considerable interest in using ab initio methods to calculate static and dynamic properties of these simple oxides, including their phonon dispersion relations, sparked by developments in high pressure research, and interest in chemical and structural modelling of the lower mantle and core (Mehl et al., 1986; Wolf and Bukowinski, 1987a).

There have been a number of studies of the effect of particle size and shape on the infrared spectra of MgO and NaCl (Luxon et al., 1969; T.P. Martin, 1970), while there has also been recent interest in the Raman spectra of small particles of MgO (100-100,000 Å in diameter). These particles show first order Raman lines which are not allowed by normal symmetry selection rules, but result from surface modes of the particles, and bulk modes rendered Raman active by symmetry distortions in the small crystallites (Bockelmann and Sclecht, 1974; Ishikawa et al., 1985).

Fluorite and rutile structures

These common structures occur for a wide range of MX_2 compounds, including oxides, halides and sulphide minerals. Common rutile phases include the type mineral TiO_2, cassiterite (SnO_2) and stishovite (high pressure SiO_2), while the fluorite (CaF_2) structure has been suggested as a possible candidate for a high pressure form of SiO_2 (Liu, 1975, 1982; Carlsson et al., 1984; but see Bukowinski and Wolf, 1986). The infrared and Raman spectra of fluorites are very simple. The fluorite structure MX_2 has point group O_h with one formula unit in its primitve cell (the crystallographic cell Fm3m has Z=4, but contains 4 lattice points). Factor group analysis gives the k=0 vibrational modes as F_{1u} + F_{2g}. The X atoms occupy sites with T_d symmetry in the structure. Symmetry analysis shows that the Raman active vibration (F_{2g}) is entirely due to displacement of these atoms in the x, y and z directions, while the IR active vibration (F_{1u}) involves displacements of both M and X atoms. Polarized Raman and IR reflectance data are available for a number of oxides and fluorides with the fluorite structure (Farmer, 1974b; Ferraro, 1984).

The rutile structure has space group P4/mnm (point group D_{4h}), with two molecules of MX_2 in the primitive unit cell. Factor group analysis gives the k=0 vibrational modes and their activities as:

$$A_{1g}(R) + A_{2g} + B_{1g}(R) + B_{2g}(R) + E_g(R) + A_{2u}(IR) + 2B_{1u} + 3E_u(IR),$$

where R and IR denote Raman and infrared activity (the other modes are not spectroscopically active). Polarized Raman and infrared reflection spectra have been measured for TiO_2, GeO_2, SnO_2 and many other rutile type compounds (Porto et al., 1967; Summitt, 1968; Beattie and Gilson, 1968, 1969; Scott, 1970; Katiyar et al., 1971; Farmer, 1974b; Sharma et al., 1979). Dispersion curves for some of these materials are also known to varying degrees of accuracy and completeness from inelastic neutron scattering measurements and lattice dynamic calculations (Traylor et al., 1971; Katiyar et al., 1971; Bilz and Kress, 1979).

The Raman spectrum of stishovite (natural and synthetic) has only been measured recently (Hemley et al., 1986b,c; Hemley, 1987; Fig. 10), and the four expected modes were observed. A previous study had found a rather different spectrum for a natural sample (Nicol et al., 1980), most likely due to impurities coating the stishovite sample derived from leaching of Coconino sandstone by HF during the separation process (Hemley et al., 1986c; Hofmeister et al., 1986b).

There have been many powder infrared studies of rutile-type compounds, but these have been plagued with problems associated with particle size and shape effects (Tarte, 1963a; Luxon and Summitt, 1969; Farmer, 1974b). From factor group analysis, only four infrared absorption peaks are expected for the rutile structure, but many more features are usually observed in powder transmission spectra. This has been a problem with published infrared studies of stishovite. For example, Lyon's (1962) powder spectrum showed seven distinct peaks and shoulders between 500 and 1000 cm^{-1} (Fig. 10), and Nicol et al. (1980) reported two additional bands at 330 and 260 cm^{-1}, while Kieffer (1979b) found 10-12 bands in the near- and far-infrared. Hofmeister et al. (1986b) have recently carried out infrared reflection measurements for stishovite, and find the expected four TO frequencies at 838, 582 and 460 cm^{-1} (all E_u) and 763 cm^{-1} (A_{2u}) (Fig. 10). The extra peaks and shoulders observed in powder transmission measurements are due to impurities, and to leakage from LO modes into the absorption spectrum.

Much information on the form of the normal modes in rutile structures can be obtained from symmetry analysis of the vibrational displacements (Katiyar, 1970; Traylor et al., 1971; Farmer, 1974b). Since the metal atoms lie on inversion centres in the structure (D_{2h} sites), their displacements do not contribute to the Raman active modes, which are entirely due to oxygen motion. The infrared active modes have components of both oxygen and metal displacements. Since there is only one infrared mode of A_{2u} symmetry, the relative atomic displacements within this mode are uniquely determined by symmetry. This is also true for the Raman active vibrations, but not for the three infrared active E_u modes, which mix with each other. The infrared active A_{2u} mode corresponds to a relative displacement of M and X atoms along the z-axis (Fig. 11). The Raman active B_{2g} vibration can be regarded as derived from the totally symmetric stretching vibration of individual MX_6 octahedral groups in the structure. Since this vibration involves in-phase stretching of six M-X bonds, the B_{2g} vibration is generally the highest frequency mode for rutile structures. The A_{1g} mode is related to the other symmetric stretching mode of the MX_6 octahedra, where two axial M-X bonds contract and four equatorial bonds expand. Because of this, the A_{1g} vibration generally lies at lower frequency than the B_{2g} mode. The B_{1g} mode corresponds to a rotation of the MX_6 octahedra about

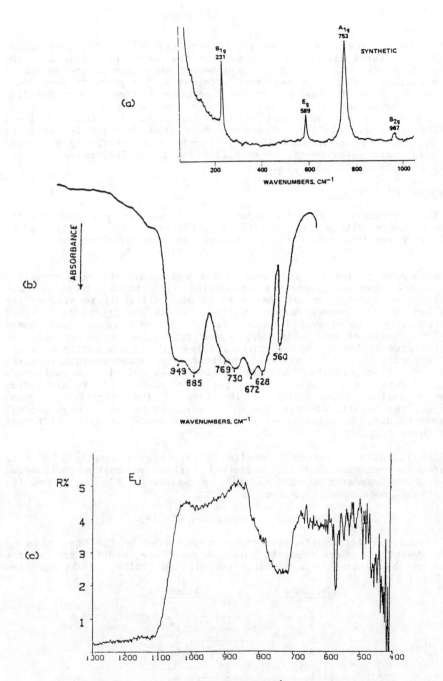

Figure 10. Raman and infrared spectra of stishovite. (a) Raman spectrum of synthetic stishovite from Hemley et al. (1986b). (b) Powder transmission infrared spectrum of natural (Meteor Crater) stishovite from Lyon (1962). (c) Micro-IR reflectance spectrum of synthetic stishovite (Hofmeister et al., 1986b; Hofmeister, Xu and Akimoto, unpublished). Three strong E_u bands are seen at about 1050–850, 700–580 and 570–470 cm^{-1}. The shoulder near 780 cm^{-1} is due to leakage from the weaker A_{1u} polarization. Analysis of the reflectivity data give TO mode frequencies as 838 (E_u), 763 (A_{1u}), 582 (E_u) and 460 (E_u), as expected from symmetry analysis.

their two-fold symmetry axis parallel to z (Fig. 11). This corresponds to the structural distortion necessary to transform the rutile into the $CaCl_2$ structure. Since this is an expected high pressure polymorph of MX_2 structures, there has been much interest in the pressure dependence of the B_{1g} mode in the rutile type structures (Peercy and Morosin, 1973). Pressure shifts have recently been measured to 350 kbar for all but one of the optically active bands of stishovite (Hemley, 1987; Hofmeister et al., 1986b). The Raman data show that the frequency of the B_{1g} mode decreases markedly ("softens") with increasing pressure, which is possible evidence for stabilization of an SiO_2 phase with the $CaCl_2$ structure at higher pressure.

Corundum and ilmenite

The corundum structure (corundum Al_2O_3; hematite Fe_2O_3) contains two formula units within its R3c (D_{3d}) primitive cell. Factor group analysis gives the zone-centre vibrations for this structure as:
$$2A_{1g}(R) + 3A_{2g} + 5E_g(R) + 2A_{1u} + 2A_{2u}(IR) + 4E_u(IR) .$$

Since so many of the IR- and Raman-active modes have the same symmetry, the normal mode displacements associated with these vibrations are likely to be strongly mixed, hence it is more difficult to discuss the detailed form of normal modes for the corundum structure than in the simpler structures discussed above. All of the infrared and Raman modes have components of both metal and oxygen displacements, except for the two infrared active A_{2u} modes, which involve only oxygen motion parallel and perpendicular to the plane of the four metal atoms coordinating the oxygen. Iishi (1978a) has calculated some of the normal mode displacements for corundum using a rigid ion model. This calculation gives a useful visualization of the form of the vibrational modes. However, the results of this type of calculation can be strongly model-dependent, and the actual mode displacements could be quite different from those calculated.

The ilmenite structure (ilmenite $FeTiO_3$, high pressure $MgSiO_3$) is related to corundum, but two different cations ordered on octahedral sites. The symmetry of the ilmenite structure is R3 (C_{3i}), and the vibrational mode symmetries are:

$$5A_g(R) + 5E_g(R) + 4A_u(IR) + 4E_u(IR).$$

The corundum and ilmenite structures are related by the loss of a c-glide symmetry element from the corundum structure, and the zone-centre modes can be related by a symmetry correlation (White, 1974a; McMillan and Ross, 1987):

Corundum		Ilmenite
A_{1g}	----------	A_g
A_{2g}	----------	A_g
E_g	----------	E_g
A_{1u}	----------	A_u
A_{2u}	----------	A_u
E_u	----------	E_u

This symmetry correlation was used by McMillan and Ross (1987) to compare the vibrational spectra of Al_2O_3 corundum and $MgSiO_3$ ilmenite, in order to constrain a heat capacity calculation for the latter phase using Kieffer's (1979a,b,c; 1980; 1985) model.

Raman-active Modes: Rutile Structure

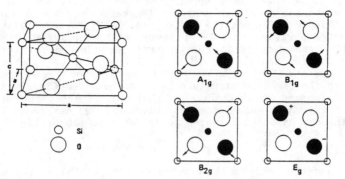

Figure 11. Drawing of the rutile structure of stishovite, and the form of the Raman-active normal modes. After Traylor et al. (1971); taken from Hemley et al. (1986b).

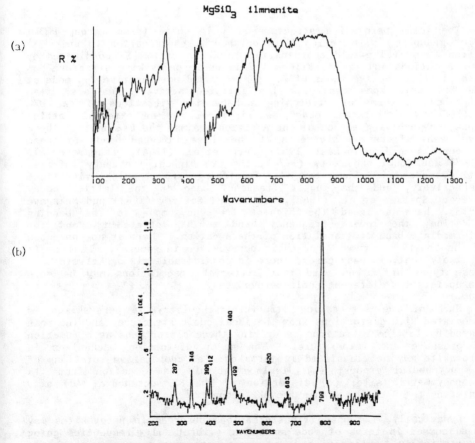

(a)

(b)

Figure 12. Vibrational spectra for $MgSiO_3$ ilmenite. (a) Unpolarized IR reflectance data (Hofmeister and Ito, unpublished). Bands are observed at around 320, 450, 530, 630, 690 and 750-900 (broad) cm^{-1}. A weak band may be present near 200 cm^{-1}, but this could also result from interference fringes, since the sample size was close to the diffraction limit for this wavelength of IR light. A further weak band may be present near 730 cm^{-1}. (b) Unpolarized Raman spectrum from McMillan and Ross (1987). Nine of the ten expected Raman modes are observed.

The infrared and Raman spectra of the most common ilmenites and corundum compounds have been measured (Barker, 1963; Porto and Krishnan, 1967; Phillippi, 1970; Beattie and Gilson, 1970; Farmer, 1974b; Onari et al., 1977; Lucovsky et al., 1977; Baran and Botto, 1978; 1979; Serna et al., 1982), and dispersion curves have been determined for Al_2O_3 by inelastic neutron scattering and a rigid ion model calculation (Kappus, 1975; Bialas and Stolz, 1975). The unpolarized Raman spectrum for $MgSiO_3$ with the ilmenite structure has been measured (Ross and McMillan, 1984; McMillan and Ross, 1987), in which nine of the expected ten bands were observed, while Hofmeister (unpublished data) has recently observed five of the six expected infrared modes for this phase (Fig. 12).

Perovskites

There has been a large literature devoted to the static and dynamic properties of perovskite phases, mainly generated by their unique material properties. There has recently been intense interest in perovskites within the earth sciences, since (Mg,Fe,Ca) silicates are known to transform to the perovskite structure at high pressures and temperatures.

The ideal perovskite structure ABX_3 is cubic (space group Pm3m: point group O_h) with cation B in octahedral coordination to anion X, cation A in twelve-fold coordination to X, and anion X coordinated by four B cations and two A cations. This cubic perovskite structure has three infrared active modes of F_{1u} symmetry and one inactive F_{2u} mode at the Brillouin zone centre. The infrared active bands are often described in terms of stretching and bending vibrations of the BX_6 octahedra (an "internal mode" description), and an external lattice vibration involving motion of the A cation against the $\overline{BX_6}$ units. There have been different assignments of the three infrared bands to these motions. According to Last (1957), Perry et al. (1964), Nakagawa et al. (1967) and Couzi and Huong (1974), the two highest frequency infrared active modes are due to BX_6 stretching and bending vibrations respectively, and the lowest frequency band is the lattice mode. However, Spitzer et al. (1962), Miller and Spitzer (1963) and Nakagawa (1973), have assigned the highest frequency mode to the bending vibration, the lowest frequency band to BX_6 stretching, and the intermediate band to translation of the A cation. There is no consensus on the "best" assignment for cubic perovskites in general. In fact it is likely that, in many cases, there is considerable mixing between these types of motion, and that different descriptions may be more appropriate for different cubic perovskites.

Much of the interesting structural chemistry of perovskites is associated with distortions from the ideal cubic structure, and the role played by lattice vibrations in driving these distortions as a function of pressure and temperature. The major distortion mechanisms in perovskite may be classified in terms of; (a) cooperative rotations of BX_6 octahedral groups; (b) displacement of the A cation from its centrosymmetric site; (c) distortion of the BX_6 octahedra; (d) off-centering the B cation within the BX_6 group.

Dynamically, the phase transformations observed in perovskites may be discussed in terms of mode softening (vibrational frequencies going asymptotically to zero with temperature, pressure or composition) and the appearance of dynamic instabilities (vibrational frequencies becoming imaginary) at the Brillouin zone boundary of the ideal cubic perovskite, giving rise to a wide range of cubic, tetragonal, orthorhombic and trigonal structures with a variety of unit cell sizes,

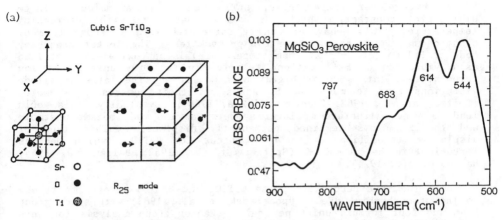

Figure 13. (a) Schematic of the R_{25} lattice vibration in cubic perovskite. The vibration is an octahedral rotation motion as shown at left. Adjacent octahedra along the crystallographic direction [111] vibrate exactly out of phase. This corresponds to the R point of the first Brillouin zone. (b) Powder infrared absorption spectrum of $MgSiO_3$ perovskite (Williams et al., 1987).

related by displacive phase transformations (Cochran and Zia, 1968; Fleury et al., 1968; Shirane and Yamada, 1969; Lockwood and Torrie, 1974). These distortions may occur as a function of temperature, pressure or composition (Nilsen and Skinner, 1968; Alain and Piriou, 1981; Wolf and Jeanloz, 1985). For example, the cubic phase $SrTiO_3$ exhibits a second order transition to a tetragonal structure (I4/mcm) on cooling to 110 K (Fleury et al., 1968; Shirane and Yamada, 1969). This transition can be understood through an elegant analysis of the lattice dynamics of the cubic phase (Fleury et al., 1968). Consider a rotational motion of the TiO_6 octahedra about one C4 axis (Fig. 13a). At the edge of the first Brillouin zone {111}, adjacent unit cells along the [111] direction will vibrate exactly out of phase (Fig. 13a). This point in reciprocal space is known as the R-point for the cubic unit cell (Koster, 1957). Since each octahedron has three unique C_3 axes, this R point vibration is triply degenerate, and is given the symmetry symbol R_{25} (Bouckaert et al., 1936; Cowley, 1964; Boyer and Hardy, 1981). From inelastic neutron scattering studies, this mode shows a large decrease in frequency with temperature (Shirane and Yamada, 1969; Stirling, 1972), consistent with a large anharmonicity for the mode. At the phase transition temperature, the triply degenerate vibration R_{25} becomes unstable; i.e. the classical restoring force for the vibration is less than zero, and the oxygen atoms follow the displacement vectors of the R_{25} mode to a new, lower energy structure. In the case of $SrTiO_3$, the new structure has tetragonal symmetry. Because the unstable mode R_{25} is triply degenerate, the c-axes of the tetragonal phase are randomly distributed in a strain-free environment, and a microscopic domain structure results. A similar condensation of R_{25} modes is responsible for the trigonal symmetry of $LaAlO_3$ perovskite below the transition temperature (Scott, 1969; Axe et al., 1969). The tetragonal phase of $SrTiO_3$ has a larger unit cell than the cubic phase, and has zone centre phonons with symmetries (Fleury et el., 1968):

$$A_{1g}(R) + 2A_{2g} + 2B_{1g}(R) + B_{2g}(R) + E_g(R) + A_{1u} + E_u(IR)$$

so that the 110 K transition of $SrTiO_3$ is accompanied by the abrupt appearance of a first order Raman spectrum, which is not allowed for the cubic phase (Perry et al., 1967; Nilsen and Skinner, 1968).

At even lower temperatures, $SrTiO_3$ shows an anomalously large dielectric constant, which increases markedly with decreasing temperature. This behaviour can be correlated with a second type of dynamic transition noted above for perovskites. The lowest frequency mode of $SrTiO_3$ (80 cm^{-1} at room temperature) is approximately described as a vibration of Sr^{2+} cations relative to the TiO_6 units. At low temperatures, this mode is highly anharmonic, and the large amplitude vibrations tend to move the Sr atoms off their centres of symmetry (Spitzer et al., 1962; Cochran and Zia, 1968). Eventually, this would lead to a new structure of lower symmetry as the mode became unstable and the Sr atoms remained permanently displaced. This type of displacive transition is known to occur for $BaTiO_3$, which becomes ferroelectric below 430 K (Spitzer et al., 1962; Ballantyne, 1964; Harada et al., 1971).

Both the type perovskite phase $CaTiO_3$ and $MgSiO_3$ perovskite belong to the orthorhombic $GdFeO_3$ type (Sasaki et al., 1983) with space group Pbnm and four formula units per cell. Factor group analysis for this orthorhombic structure gives the following optic modes at k=0:

$$7A_g(R) + 7B_{1g}(R) + 5B_{2g}(R) + 5B_{3g}(R)$$

$$+ 8Au + 7B_{1u}(IR) + 9B_{2u}(IR) + 9B_{3u}(IR).$$

These Pbnm orthorhombic perovskite structures are generated from the ideal cubic structures in several steps involving different distortion mechanisms: (a) a condensation of one R_{25} rotational mode of the BX_6 octahedra to generate the I4/mcm tetragonal structure, as for $SrTiO_3$; (b) condensation of M- and X-point BX_6 rotational modes (lattice vibrations at the edge of the cubic Brillouin zone along {110} and {100} respectively) to give an orthorhombic structure with space group Cccm; finally (c) a displacement of the A cations from their centrosymmetric sites to give the Pbnm perovskite structure (Cochran and Zia, 1968; Lockwood and Torrie, 1974).

A number of authors have reported infrared and Raman spectra of Pbnm orthorhombic perovskites (Perry et al., 1964, 1965; Perry, 1971; Pasto and Condrate, 1973; Saine et al., 1981; Balachandran and Eror, 1982; Ross et al., 1986), including the geophysically important $MgSiO_3$ perovskite (Weng et al., 1983; Williams et al., 1987), but in general the spectra are not well understood and individual bands have not been assigned to particular vibrational modes with any certainty. The highest frequency bands (near 800 cm^{-1}) in the infrared spectrum of $MgSiO_3$ perovskite (Fig. 13b) can be correlated with the highest frequency F_{1u} mode of the hypothetical cubic structure, and can be interpreted in terms of asymmetric stretching vibrations of the SiO_6 octahedral groups (Weng et al., 1983; Williams et al., 1987). However, the importance of coupling between this vibration and octahedral deformation vibrations or displacements of the Mg atoms is not yet known. We cannot even begin to assign the lower frequency infrared bands or the observed Raman active modes. In fact, until reliable, parameter-free calculations of the atomic displacement patterns for $MgSiO_3$ and other perovskites are available, we will not be able to attempt a realistic assignment of their infrared or Raman spectra. This type of work has begun in a number of laboratories (Boyer and Hardy, 1981; Wolf and Jeanloz, 1985; Wolf and Bukowinski, 1987b; Hemley et al., 1987b).

Spinel

This is a common structure type for minerals with stoichiometry

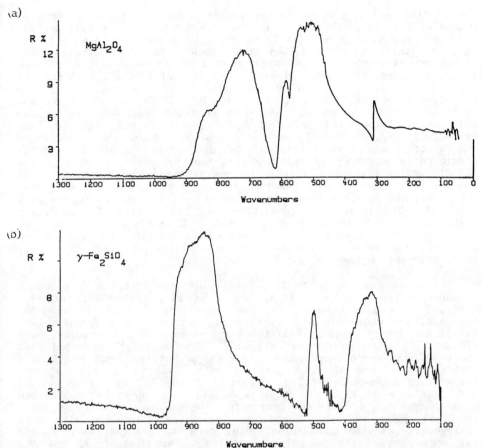

Figure 14. Infrared reflectance spectra of spinels. (a) $MgAl_2O_4$. Two intense broad bands are observed between 775 and 660 cm^{-1} and 560 and 468 cm^{-1}. Weak narrow bands are found at 605-580 cm^{-1} and 310-290 cm^{-1}. These correspond to the four bands expected from symmetry analysis. The origin of the shoulder at 830 cm^{-1} is unknown (Hofmeister, unpublished data). (b) $\gamma-Fe_2SiO_4$. Three bands are obvious at 900-800, 520-480 and 380-300 cm^{-1}. The existence of the fourth weak band at 210-190 cm^{-1} was confirmed by thin film absorption spectroscopy (Hofmeister and Yagi, in prep., Hofmeister et al., 1986a).

AB_2O_4, including the type phase $MgAl_2O_4$, chromite ($FeCr_2O_4$), magnetite (Fe_3O_4) and a high pressure form of $(Mg,Fe)_2SiO_4$. Factor group analysis gives the vibrational mode symmetries of spinel as:

$$A_{1g}(R) + E_g(R) + F_{1g} + 3F_{2g}(R) + 2A_{2u} + 2E_u + 4F_{1u}(IR) + 2F_{2u}.$$

There have been many vibrational studies of spinels, mainly via powder infrared methods (White and DeAngelis, 1967; Preudhomme and Tarte, 1971a,b,c; 1972), but there remain a number of problems with both the experimental spectra and their interpretation. Preudhomme and Tarte (1971c) observed that many of their spinel spectra were dependent on sample preparation conditions, which can be attributed to particle size and shape effects. A further problem in the spectra of spinels is the possibility of cation disorder (White and DeAngelis, 1967; Wood et al., 1986). Infrared powder positions observed for $MgAl_2O_4$ spinel by Preudhomme and Tarte (1971c) (688, 580, 522, 709 cm^{-1}) do not agree with the TO frequencies obtained in the single crystal reflectance study of O'Horo et al. (1973) (670, 485, 428 and 305 cm^{-1}). The synthetic sample measured by O'Horo et al. (1973) is expected to be disordered relative to pure natural spinel (Yamanaka and Takeuchi, 1983; Wood et al., 1986; Finger et al., 1986). Analaysis of the IR reflectivity on natural (ordered) spinel (Fig. 14a) gives TO positions at 670, 585, 485 and 305 cm^{-1}, which agree with neither the results of Preudhomme and Tarte (1971c) nor O'Horo et al. (1973). Jeanloz (1980) and Akaogi et al. (1984) observed a number of subsidiary shoulders and peaks on the two main transmission bands of Mg_2SiO_4, Fe_2SiO_4, Ni_2SiO_4, Co_2SiO_4 and Mg_2GeO_4 spinels, which could well be due to reflectance contributions to the spectra or other particle size and shape dependent effects, unrelated to any symmetry splitting or other structural effects. From the discussion above for $MgAl_2O_4$, the powder transmission minima cannot be trusted to lie close to the true TO mode frequencies. In fact, for $\gamma-Fe_2SiO_4$, the powder transmission bands are shifted up to 100 cm^{-1} away from the TO frequencies determined by thin film IR absorption spectroscopy (Hofmeister et al., 1986a). IR reflectance spectra of these silicate spinels are well-behaved, exhibiting the four bands expected from factor group analysis (Hofmeister and Ito, in prep.; Hofmeister and Yagi, in prep.; Fig. 14b).

The Raman spectrum of $MgAl_2O_4$ spinel has been reported by a number of authors (Fraas et al., 1973; O'Horo et al., 1973; Ishii et al., 1982), but is also a matter of some controversy. From the symmetry analysis, one mode with A_{1g} symmetry is expected. Both Fraas et al. (1973) and O'Horo et al. (1973) found two modes with A_{1g} symmetry, near 770 and 725 cm^{-1}. O'Horo et al. (1973) "tentatively" suggested that their 727 cm^{-1} mode corresponded to a two-phonon band, derived from a combination of the 410 (E_g) and 311 cm^{-1} (F_{2g}) modes, although the temperature dependence of the intensity of the 727 cm^{-1} peak was no different from that of the other peaks observed (normally combination Raman bands show a different temperature dependence of their intensity to first order peaks). There are two further problems with this assignment: (a) within the spinel point group O_h, combination of E_g and F_{2g} modes gives rise to only F_{1g} and F_{2g} representations, not A_{1g} (Wilson et al., 1955, p. 331); (b) the frequency 727 cm^{-1} is 6 cm^{-1} larger than the sum of 410 and 311 cm^{-1}. O'Horo et al. (1973) suggested that this discrepancy could be due to anharmonic effects; however this would only be true for an overtone, not a combination band, and in most molecular systems, anharmonicity causes the overtone frequency to be less than the sum of components, not more (Herzberg, 1945, 1950). Ishii et al. (1982) also observed the same (725 cm^{-1}) mode in the Raman spectrum of a non-stoichiometric spinel $MgO.3Al_2O_3$. The relative intensities of the high frequency modes were similar to those for stoichiometric $MgAl_2O_4$, suggesting that the 725 cm^{-1} peak is not simply related to either normal-inverse disorder or non-stoichiometry in any of the spinels studied by Fraas et al. (1973) or O'Horo et al. (1973). There are further anomalies in the Raman spectra of $MgAl_2O_4$ at lower frequencies. All three studies (Fraas et al., 1973; O'Horo et al., 1973; Ishii et al., 1982) observed an F_{2g} mode near 310 cm^{-1}. Fraas et al. (1973) also reported a weak peak at 225 cm^{-1} with F_{2g} or E_g symmetry, but did not observe a weak feature at 492 cm^{-1} reported by O'Horo et al. (1973). O'Horo et al. (1973) assigned this weak 492 cm^{-1} peak to F_{2g} symmetry, but did not observe the 225 cm^{-1} mode of Fraas et al. (1973). Ishii et al. (1982) observed both the 492 (490) and 225 (223) cm^{-1} peaks for $MgO.3Al_2O_3$ spinel. These authors followed the assignment of O'Horo et al. (1973) for stoichiometric spinel, and suggested that the 223 cm^{-1} peak was associated with the defective nature of the non-stoichiometric spinel lattice. It is obvious from this summary that the Raman spectrum of $MgAl_2O_4$ spinel deserves further study, preferably on well-characterized samples of known stoichiometry and Mg-Al order.

McMillan and Akaogi (1987) have reported the Raman spectrum of Mg_2SiO_4 spinel, which shows the five Raman modes expected for the spinel structure (Fig. 15a). Two modes of A_{1g} and F_{2g} symmetry respectively are expected, derived from the symmetric (ν_1) and asymmetric (ν_3) stretching vibrations of the tetrahedral SiO_4 units. These can be assigned to the strong peaks at 795 and 837 cm^{-1} respectively, since this is a characteristic region for Si-O stretching for tetrahedral orthosilicate structures (Piriou and McMillan, 1983). Guyot et al. (1986) have reported a Raman spectrum of a natural $(Mg,Fe)_2SiO_4$ spinel. This spectrum contained an additional band at 880 cm^{-1}, whose origin is not clear. Yamanaka and Ishii (1986) have recently obtained the Raman spectrum of Ni_2SiO_4 spinel. In this case, the ν_1-derived A_{1g} mode was observed at 849 cm^{-1}, at higher frequency than the ν_3-derived F_{2g} mode at 810 cm^{-1}, which is unusual for an orthosilicate structure. It is not easy to assign the lower frequency Raman modes for these silicate spinels to OSiO bending or metal cation vibrations in these spectra, since these motions are likely mixed. The same is true for the infrared active vibrations, which all have the same symmetry ($4F_{1u}$).

Olivines, garnets and Al_2SiO_5 polymorphs

The expected zone centre modes for minerals with the olivine structure are:

$$11A_g(R) + 11B_{1g}(R) + 7B_{2g}(R) + 7B_{3g}(R)$$

$$+ 10A_u + 9B_{1u}(IR) + 13B_{2u}(IR) + 13B_{2u}(IR).$$

Polarized single crystal IR reflectance and Raman spectra have been obtained for a few olivines (Servoin and Piriou, 1973; Gervais et al., 1973a; Hohler and Funck, 1973; Stidham et al., 1976; Iishi, 1978b; Hofmeister, 1987), while powder IR data are available for many more (Oehler and Gunthard, 1969; Burns and Huggins, 1972; Tarte, 1963b; Toropov et al., 1963; Farmer, 1974c; Handke and Urban, 1982). The infrared and Raman spectra for forsterite have recently been summarized by McMillan (1985). Besson et al. (1982) have recently measured the pressure dependence of high frequency Raman bands of forsterite to 40 kbar, while Hofmeister et al. (1987b) have measured the pressure dependence of the IR bands of Fe-Mg olivine to 450 kbar. Gervais et al. (1973b,c) studied the anharmonicity of vibrational modes in the related phases Be_2SiO_4 (phenacite) and $ZrSiO_4$ (zircon) by high temperature infrared reflectance.

There have been many discussions of the assignment of vibrational modes in olivines. The infrared and Raman modes between 800 and 1000 cm^{-1} fall in the region expected for Si-O stretching modes, and the observed bands may be assigned to combinations of the ν_1 (symmetric) and ν_3 (asymmetric) stretching modes of the SiO_4 groups in the structure. Although the oxygens are also coordinated to other metals such as Ca, Mg, Al or Fe, this internal mode assignment appears to be a useful approximation (Piriou and McMillan, 1983). The SiO_4 groups in the olivine structure are distorted (C_s site symmetry), and both the infrared and Raman spectra show systematic changes with degree of distortion (Tarte, 1963b; White, 1975; Piriou and McMillan, 1983). Bending vibrations of the silicate tetrahedra would be expected at lower frequency, but it is difficult to separate these from vibrations involving the octahedral cations, and it is likely that these modes are all strongly coupled. There have been some isotopic substitution studies on olivines (Paques-Ledent and Tarte, 1973; Handke and Urban, 1982) which have shed some light on the nature of vibrations below 600 cm^{-1}, but more work is needed for unambiguous assignment. There have also been a number of vibrational calculations for olivine structures, using both valence and ionic force field models, which give some insight into the nature of the normal modes (Devarajan and Funck, 1975; Iishi, 1978b; Price et al., 1987), although the results of such calculations can be highly model dependent.

Like the olivines, the garnets also contain "isolated" SiO_4 groups, and the high frequency bands (800–1000 cm^{-1}) in their infrared and Raman spectra may be interpreted in terms of Si–O stretching modes of these groups. The garnet structure has space group Ia3d (point group O_h), with a large number of IR- and Raman-active vibrations:

$$3A_{1g}(R) + 5A_{2g} + 8E_g(R) + 14F_{1g} + 14F_{2g}(R)$$

$$+ 5A_{1u} + 5A_{2u} + 10E_u + 17F_{1u}(IR) + F_{2u}.$$

Tarte and Deliens (1973) and Moore et al. (1971) have obtained Raman and powder infrared spectra for a range of synthetic and natural garnets, and established some relationships between band positions and the garnet crystal chemistry. Like the olivines, the bands in the 800–1000 cm^{-1} region may be interpreted in terms of Si–O stretching vibrations of the SiO_4 tetrahedra, but the bands below 600 cm^{-1} are not well understood. The effects of pressure and temperature on the mid-IR spectra of a ternary garnet ($Al_{45}Py_{41}Gr_{13}$) were determined by Dietrich and Arndt (1982): peak shifts were observed to be linear in pressure and temperature over the ranges studied (0–50 kbar; 25–250 K).

There has recently been considerable interest in the silicate garnet majorite ($Mg_4Si_4O_{12}$), which may be an important mantle mineral. Jeanloz (1981) published a powder infrared spectrum of a natural majorite ($Mg,Fe)_4Si_4O_{12}$ separated from a meteorite sample. Kato and Kumuzawa (1985) obtained a powder infrared spectrum for a synthetic sample of $Mg_4Si_4O_{12}$ garnet prepared at high temperature and pressure. These authors suggested that the $Mg_4Si_4O_{12}$ garnet had tetragonal symmetry, based on X-ray diffraction and peak splittings observed in the infrared spectrum.

Powder mid-IR spectra for the Al_2SiO_5 polymorphs andalusite, sillimanite and kyanite have been published (Farmer, 1974c; Kieffer, 1979b). Polarized Raman and mid-IR reflectance spectra of andalusite and sillimanite were measured by Iishi et al. (1979) and Salje and Werneke (1982), who observed most of the expected IR and Raman bands, while McMillan and Piriou (1982) obtained an unpolarized Raman spectrum for the triclinic polymorph kyanite. Iishi et al. (1979) have carried out a lattice dynamic calculation for andalusite, but some of the vibrational mode assignments suggested by this calculation have been questioned (McMillan and Piriou, 1982), and in general, the vibrations of the Al_2SiO_5 polymorphs are not yet well understood (Farmer, 1974c). Some interesting Raman data have been obtained for natural and synthetic mullites. McKeown et al. (1984) obtained polarized single crystal Raman data for a natural mullite sample, while McMillan and Piriou (1982) reported unpolarized Raman spectra for a number of synthetic compositions. In these studies, a band near 700 cm^{-1} appeared to correlate with excess Al_2O_3 component in the mullites, and could be associated with a vibration of AlO_4 sites in the mullite structure.

Chain and ring silicates

The anhydrous pyrosilicates contain the melilites, rankinite ($Ca_3Si_2O_7$) and thortveitite ($Sc_2Si_2O_7$). These minerals contain the Si_2O_7 structural group, except for gehlenite ($CaAl_2SiO_7$), which has AlOSi linkages between AlO_4 and SiO_4 tetrahedra. There are also a number of hydrous phases, mainly within the epidote group, which also contain Si_2O_7 units. There have been a number of powder infrared studies of pyrosilicate minerals, and a few Raman studies (Lazarev, 1972; Scheetz, 1972; Farmer, 1974c; Sharma and Yoder, 1979; Gabelica-Roberts and Tarte, 1979; Bretheau-Raynal et al., 1979; Conjeaud and Boyer, 1980; Sharma et al., 1983). These phases all show peaks in the 600-700 cm^{-1} region which can be assigned to the symmetric stretching of the SiOSi linkage, and there is a well-developed correlation between the frequency of this band and the SiOSi angle in the pyrosilicate (Lazarev, 1972; Farmer, 1974c). Peaks in the 800-1100 cm^{-1} region may be generally assigned to Si-O stretching vibrations, both of the terminal -SiO_3 groups (with "non-bridging" oxygens) and the SiOSi linkage (Lazarev, 1972; Farmer, 1974c; McMillan 1984). Tarte et al. (1973) and Gabelica-Robert and Tarte (1979) have used isotopic sustitution techniques to make more detailed assignments of the vibrational spectra of a number of pyrosilicates, but many of the modes are complex mixtures of cation and silicate anion vibrations.

At high pressures, some olivines M_2SiO_4 transform to a "β" phase which also contains Si_2O_7 units (Horiuchi and Sawamoto, 1981). The β-phase can also be described as one member of a family of structures related to spinel ("spinelloids") (Horiuchi et al., 1980; Hyde et al. 1982; Price, 1983). In some systems (e.g., $NiAl_2O_4$-Ni_2SiO_4), several spinelloids are commonly found intergrown (Davies and Akaogi, 1983). McMillan and Akaogi (1987) measured the Raman spectrum of β-Mg_2SiO_4 and found no broadening of the observed peaks (Fig. 15a), suggesting the absence of spinelloid stacking disorder in this phase. Jeanloz (1980) and Akaogi et al. (1984) have measured powder IR spectra for β-Co_2SiO_4 and β-Mg_2SiO4, while Williams et al. (1986) reported that the measured powder IR frequencies of these phases showed a linear dependence on pressure to 300 kbar. Recently, Smyth (1987) has suggested that β-Mg_2SiO_4 might constitute a host for water within the mantle. Preliminary Raman data (McMillan et al., 1987; Fig. 15b) indicate that synthetic β-Mg_2SiO_4 can accept structural OH groups supporting Smyth's (1987) suggestion.

There has been relatively little work carried out on ring or chain silicates. Lazarev (1962, 1972) has discussed the powder infrared spectra of catapleite, benitoite and other minerals containing Si_3O_9 rings, while Ignatiev and Lazarev (1972a,b) have obtained spectra for phases $M_3Si_3O_9$ (M = Ca,Sr,Pb). Conjeaud and Boyer (1980) obtained unpolarized Raman spectra for the $CaSiO_3$ phases α-wollastonite (containing Si_3O_9 rings) and β-wollastonite (with pyroxenoid chains). The spectra of these two structurally distinct phases showed only minor differences. The ring silicate beryl ($Be_3Al_2Si_6O_{18}$) has been studied in much more detail. Polarized IR reflectance and Raman spectra were obtained by Adams and Gardner (1974a) and single crystal IR absorption spectra by Hofmeister et al. (1987a). Gervais et al. (1972) reported the temperature dependence of the infrared reflectance spectrum. The structure has a large number of k=0 vibrational modes: 58 expected in the IR and Raman spectra, along with 78 inactive species. All but three of the Raman bands have been observed, and all of the infrared modes can be inferred from combining the data from the three studies.

142

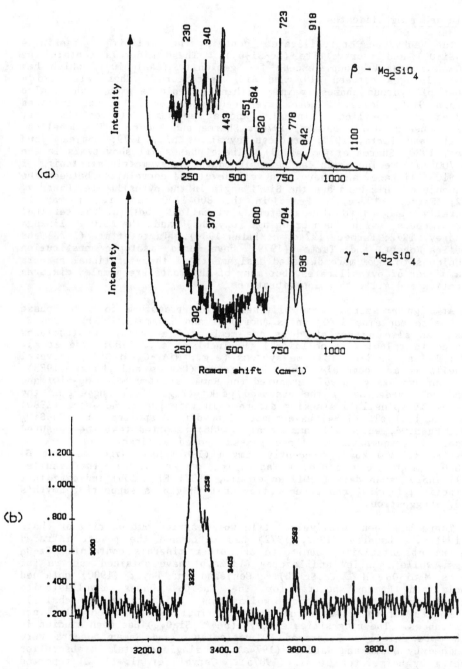

Figure 15. (a) Raman spectra of β-Mg$_2$SiO$_4$ ("modified spinel" and γ-Mg$_2$SiO$_4$ (spinel) from McMillan and Akaogi (1987). (b) Raman spectrum of the OH stretching region for β-Mg$_2$SiO$_4$, showing evidence for structurally bound OH groups in this phase (McMillan, Akaogi and Smyth, in prep.).

The results of a number of early powder IR studies on chain silicates have been summarized by Lazarev (1972). Omori (1971) obtained a powder IR spectrum for diopside, including the far-IR region, while Etchepare (1970) and Zulumyan et al. (1976) carried out polarized single crystal Raman and infrared reflection measurements. White (1975) has presented unpolarized Raman spectra of clinoenstatite and orthoenstatite, and Sharma et al. (1983) have obtained a similar spectrum for Ca-Tschermak's pyroxene ($CaAl_2SiO_6$). Dietrich and Arndt (1982) have measured pressure shifts of high energy IR modes of natural orthopyroxene to 50 kbar. The spectra of the chain silicates show strong infrared and Raman peaks in the 850-1100 cm^{-1} region, which may be generally assigned to Si-O stretching motions, and weak IR bands and strong Raman bands are generally observed in the 600-700 cm^{-1} region, which can be associated with vibrations of the SiOSi linkages.

Amphiboles and micas

Much of the interest in the IR spectra of amphiboles has stemmed from the relation between the number, position and intensities of peaks in the OH stretching region and the M and A site occupancies (Strens, 1966; Burns and Strens, 1966; Rowbotham and Farmer, 1973; Strens, 1974; Hawthorne, 1983a,b; Mottana and Griffin, 1986; Raudsepp et al., 1987). The factors affecting the O-H stretching frequencies are still not well understood, and remain an area of active research for amphiboles and other hydrous minerals. Naummann et al. (1966) have used inelastic neutron scattering to investigate the low frequency vibrations due to hindered rotational and translational motions of the OH groups in a number of amphiboles. There have been few Raman studies of amphiboles: White (1975) presented an unpolarized spectrum of natural anthophyllite, and Blaha and Rosasco (1978) reported a spectrum of tremolite. McMillan et al. (1985) and Sheu et al. (1986) have reported unpolarized Raman spectra for series of synthetic pargasites and richterites. The vibrational spectra of these phases are not well understood at all, and deserve some systematic study.

As for the amphiboles, there has been a large amount of powder IR spectroscopy on hydrated sheet silicates, especially the micas and clay minerals, and especially in the OH stretching region (Farmer, 1974d). Since many of these structures cleave easily, there have also been a number of oriented single crystal absorption studies, which are useful in determining the orientation of the OH group, and in assigning vibrations (Farmer, 1974d). Infrared peaks at 669 and 465 cm^{-1} in the infrared spectrum of talc are assigned to <u>librational</u> (hindered rotation and translation) motions of OH, consistent with the inelastic neutron scattering studies of Naummann et al. (1966). Loh (1973), Blaha and Rosasco (1978) and Haley et al. (1982) have obtained Raman spectra of a number of micas, but the spectra are not easily interpreted. Velde and Couty (1985) have obtained far-IR powder spectra of a number of phyllosilicates which indicate that the low frequency modes are not dominated by alkali ion vibrations, as might have been supposed, but are due to complex lattice vibrations. Powder spectra of the muscovite-celadonite series (Velde, 1978) have shown that ordering is probable in both the octahedral and tetrahedral sites, and that the octahedral species have little effect on the silicate network vibrations. Velde (1980) found small but observable differences between the infrared spectra of 1M and $1M_d$ margarites, suggesting that IR spectroscopy is sensitive to details of the stacking sequence. In contrast, Clemens et al. (1987) have shown that stacking sequence disorder in phlogopite has no effect on the IR or Raman spectra, nor on the enthalpy of solution of this phase. Finally, the infrared spectra of aluminosilicate micas show

144

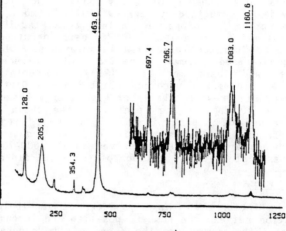

Figure 16. Unpolarized Raman spectrum of α-quartz. The broad band near 206 cm⁻¹ is the "soft mode" observed by Raman and Nedungadi (1940).

a sharp peak near 826 cm^{-1}, which seems to be associated with the presence of tetrahedral Al in the structure (Farmer, 1974d; Jenkins, 1985). Some authors have correlated this and similar bands with the presence of AlOAl linkages, and hence Al/Si disorder, in the mica structure (Farmer and Velde, 1973; Langer et al., 1981; Jenkins, 1985), but other assignments for this peak have also been made (Farmer, 1974d; Velde, 1979).

Silica polymorphs

The vibrational spectrum of α-quartz has received a considerable amount of attention for a number of reasons: it is the simplest silicate, and has served as a model structure for solid state physicists interested in silicates; it is piezoelectric and does not contain an inversion centre, hence polar modes (i.e. TO–LO splitting) are observed in both the infrared and the Raman spectrum; and several modes show complex temperature dependence across the α–β phase transition. The infrared and Raman spectra of α-quartz have been known for many years (the infrared spectrum was first studied in the late 19th century (e.g., Nichols, 1897), and the second published observation of the Raman effect, by Landsberg and Mandelstam (1928), was for α-quartz), and there are a number of now classic infrared and Raman studies of α-quartz at ambient conditions (Spitzer and Kleinmann, 1961; Scott and Porto, 1967; Russel and Bell, 1967; She et al., 1971). Raman and Nedungadi (1940) first measured the Raman spectrum as a function of temperature through the α–β phase transition, and noted that the mode near 207 cm^{-1} at room temperature (Fig. 16) behaved anomalously with temperature. This band became increasingly weak and diffuse as the transition temperature was approached, and Raman and Nedungadi (1940) suggested that the atomic displacements associated with this mode were responsible for the displacive α–β quartz transition. This was in agreement with lattice dynamic calculations carried out by Saksena (1940), which predicted a dynamical instability of this normal mode as quartz transformed from the α– to the β– structure. The relations between dynamic instabilities and displacive phase transitions were later formalized by Cochran (1960), and such dynamically unstable modes are known as "soft modes" (Scott, 1974). Later work showed that the temperature dependence of the 207 cm^{-1}

Raman mode for quartz was not quite so simple. Shapiro and co-workers (1967) suggested that, although the 207 cm^{-1} mode showed some frequency decrease on approaching the transition temperature, a second weak peak near 147 cm^{-1} was in fact the one which showed soft mode behaviour. Scott (1968) considered that this 147 cm^{-1} mode was in fact due to second order Raman scattering, based on its Raman band profile at low temperatures. At present, there are still some gaps in the data, and the Raman spectrum of α-quartz in the 10-300 cm^{-1} range would be worthy of further study as a function of temperature across the α-β transition.

Most recently, it has been shown that the α- and β-phases of quartz are related by a series of incommensurate phases occurring in the transition region, and that the appearance of these incommensurate phases is strongly linked to the lattice dynamics of α- and β-quartz (Dolino, 1986). The interrelation of structural, dynamic and thermodynamic properties of α- and β-quartz through the incommensurate phase region has been studied in an elegant series of experiments by G. Dolino and colleagues (Dolino, 1986; Berge et al., 1986; Bethke et al., 1987; Dolino et al., 1987).

Despite the large number of experimental and theoretical studies of the dynamic properties of α-quartz, the assignment of its vibrational spectrum is still not well understood, in that calculated mode displacements are in only partial agreement with isotopic substitution studies (Sato and McMillan, 1987). Further, non-empirical, calculation of the lattice dynamics of quartz will be needed before its vibrational spectrum, and hence the microscopic origin of the α-β transition, is understood.

The other polymorphs of SiO_2 have received much less attention, and deserve further study. Raman and IR spectra have been obtained for α- and β-cristobalite by Simon and McMahon (1952), Gaskell (1966), Plendl et al. (1967), Bates (1972) and Etchepare et al. (1978). Plendl et al. (1967) and Etchepare et al. (1978) have also presented some infrared and Raman data for samples of tridymite. Raman spectra of natural and synthetic coesites have been measured at atmospheric pressure (von Stengel, 1977; Sharma et al., 1981; Boyer et al., 1985), and as a function of pressure by Hemley (1987). The pressure dependence of vibrational modes in α-quartz has been reported by a number of authors (Asell and Nicol, 1968; Dean et al., 1982; Wong et al., 1985; Hemley, 1987).

Feldspars, cordierite and other framework aluminosilicates

Early work, especially via powder IR spectroscopy, has been summarized by Moenke (1974). von Stengel (1977) obtained single crystal infrared reflection and Raman spectra for a number of alkali feldspars, while Matson et al. (1986) have presented unpolarized Raman data for sanidine and anorthite. McMillan et al. (1982) and Sharma et al. (1983) have also published Raman spectra of polycrystalline anorthite samples. There are considerable differences between these two spectra, which are are almost certainly due to differences in structural state. At present, the relationships between spectra and structure of these and other framework aluminosilicates are not well understood.

R.F. Martin (1970) has studied the powder infrared spectra of alkali feldspars as a function of Al-Si order, while a number of authors have investigated the infrared and Raman spectra of order-disorder series of cordierite (Langer and Schreyer, 1969; White, 1974a; Putnis and Bish, 1983; McMillan et al., 1984). These studies found systematic changes in the spectra with increasing Al-Si order, but these could not be related to specific structural changes, as the detailed assignment of vibrational modes in these complex aluminosilicates is not yet well understood.

In a recent elegant study, Salje (1986) used the temperature dependence of the line widths and intensities of Raman modes of alkali feldspars to follow the evolution of order parameters during the displacive triclinic-monoclinic transition. This type of analysis does not depend on assignment of the vibrational modes, and provides a useful quantitative method for modelling the temperature dependence of the entropy during such displacive transitions (Salje, 1985).

There have been a few scattered studies of other aluminosilicate minerals. Henderson and Taylor (1977, 1979) have obtained powder IR spectra for a wide range of sodalite minerals. Martin and Lagache (1975) have published powder IR spectra for a number of synthetic leucites and pollucites, and Matson et al. (1986) have obtained unpolarized Raman spectra for nepheline and carnegeite. The complex structural relations within these phases (Lange et al., 1986) suggest a need for a more detailed study. Some work at high pressure has been carried out. Couty and Velde (1986) have studied the mid-IR spectra of albite and sanidine feldspars at up to 50 kbar pressure. Velde and Besson (1981) have carried out a similar infrared and Raman study of analcime, while Sharma and Simons (1981) have used Raman spectroscopy to study high pressure phases in the $LiAlSi_2O_6$ system, including spodumene and eucryptite.

Carbonate minerals

Geologically important carbonates (MCO_3) include both single and double carbonates with M=Ca,Mg,Fe,Mn,Sr,Ba in a variety of crystal structures. All of the carbonate structures contain isolated CO_3^{2-} groups. The vibrational spectrum of the "free" trigonal planar CO_3^{2-} ion in aqueous solution is quite well understood (Herzberg, 1945). The free CO_3^{2-} ion (point group symmetry D_{3h}) has six vibrational modes: a symmetric stretching vibration (ν_1), an out-of-plane bend (ν_2), a doubly degenerate asymmetric stretch (ν_3), and a doubly degenerate bending mode (ν_4) (Fig. 17). These modes have symmetries

$$A_1'(R) + A_2''(IR) + E'(R,IR) + E'(R,IR) \quad ,$$

and occur at frequencies 1063, 879, 1415 and 680 cm^{-1} respectively (Herzberg, 1945). Using symmetry analysis, the observed infrared and Raman bands of crystalline carbonates can be assigned to modes derived from these <u>internal</u> vibrations of the carbonate anion, and to <u>external</u> modes involving translation and rotation ("libration") of the carbonate groups and vibrations of the metal cations (Porto et al., 1966; White, 1974b; Bischoff et al., 1985). In general, strong Raman modes in the 1000-1100 cm^{-1} region are due to the symmetric stretching vibration (ν_1) of the carbonate groups, while weak Raman and strong IR peaks near 1400 cm^{-1} are due to the asymmetric stretch (ν_3). Infrared modes near 800 cm^{-1} are derived from the out-of-plane bend (ν_2). Raman peaks are often observed near 1700 cm^{-1} due to the first overtone of this vibration ($2\nu_2$). Infrared and Raman modes in the 700-900 cm^{-1} region are due to the in-plane bending mode (ν_4). This mode is doubly degenerate for undistorted CO_3^{2-} groups. As the carbonate groups become distorted from regular planar symmetry, this mode splits into two components. Infrared and Raman spectroscopy provide a sensitive test of this distortion (White, 1974b; Scheetz and White, 1977; Bischoff et al., 1985).

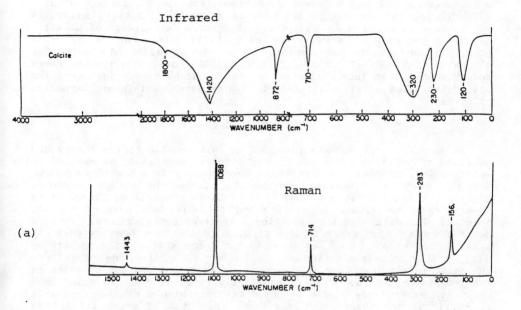

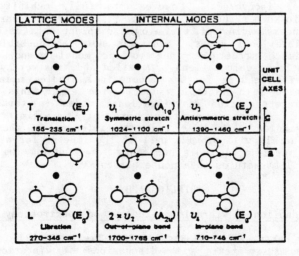

Figure 17. (a) Infrared and Raman spectra of calcite from Scheetz and White (1977). (b) Schematic of principal internal and external vibrational modes of calcite from Bischoff et al. (1985).

The IR and Raman spectra for $CaCO_3$ with the calcite and aragonite structures have been completely determined (White, 1974b; Frech et al., 1980) except for the far-IR spectrum of aragonite. Scheetz and White (1977) have reported Raman and powder IR spectra for a range of alkaline earth carbonates. Yamamoto et al. (1975) have studied dolomite ($CaMgCO_3$), and Rossman and Squires (1974) have obtained IR spectra for alstonite ($CaBa(CO_3)_2$). Bischoff et al. (1985) have recently used Raman spectroscopy in an interesting study, to investigate ordering asociated with orientation of the CO_3^{2-} ions in synthetic and authigenic magnesian calcites.

SUMMARY

In this chapter, we have provided an introduction to the theory and practice of infrared and Raman spectroscopies, especially as applied to mineralogical problems. Vibrational spectroscopy is a useful technique for the study of minerals, and can be applied in a wide range of analytic, thermodynamic and structural studies. For all of these, it is important to constitute a complete and reliable data base of mineral spectra. The data base for vibrational spectroscopy of minerals is very heterogeneous: some minerals and mineral groups have been extensively studied, and others hardly at all. Within these studies, the quality of data is also inhomogeneous: some spectra are definitive, whereas others should be further examined to resolve contradictions and inadequacies in the published spectra. This is especially true for minerals which have only been studied via powder infrared methods which are known to result in spurious peak shapes and positions. Although it seems trivial to state, there is need for many more experimental data on well-defined mineral series. Finally, the field of mineral spectroscopy provides a rich playground for the solid state physicist and structural chemist: natural and synthetic minerals exhibit an entire spectrum (no pun intended) of structural and dynamic behaviour, and many phenomena must remain to be discovered. However, to fully exploit vibrational spectroscopy in such structural studies, it will be necessary to achieve a detailed understanding of the atomic displacements associated with particular vibrational modes. This is not yet possible, even for relatively simple structures, due to the model-dependent nature of current vibrational calculations. Detailed and trustworthy vibrational assignments will require further theoretical studies using ab initio methods, coupled with isotopic substitution experiments, to provide unambiguous interpretation of vibrational modes in terms of atomic displacements. Such calculations are at the limit of current theoretical studies. We hope that the interest in the vibrational spectra of minerals will provide part of the impetus for development of the advanced theoretical tools necessary for studying the dynamics of complex mineral structures in a non-empirical way.

BIBLIOGRAPHY

General introduction to vibrational theory and spectroscopy

Moore (1972); Berry, Rice and Ross (1980); Atkins (1982). Good general physical chemistry texts with discussions of vibrations of small molecules and their interaction with light.

Kittel (1976); Ashcroft and Mermin (1976); Harrison (1979). Solid state physics texts with discussions of lattice vibrations and their interaction with light.

Vibrations of molecules and crystals

Steele (1971); Woodward (1972). Excellent introductions to vibrational theory of small molecules.
Herzberg (1945, 1950). A classic two volume set discussing the theory and practice of vibrational spectroscopy of small molecules; also Huber and Herzberg (1979): an updated extension of Herzberg's (1950) compilation of data on diatomic molecules.
Wilson, Decius and Cross (1955). A classic discussion of vibrational theory of small molecules, and introduction to the matrix method of vibrational calculation.
Bellamy (1958); Dollish et al. (1974). Infrared and Raman spectra of organic molecules (two of many compilations and discussions).
Davies (1963); Jones (1971); Nakamoto (1978). Discussions of the vibrational spectra of inorganic compounds, with much useful data.
Ross (1972); Nakamoto (1978). Compilations of IR and Raman data for inorganic systems.
Sherwood (1972); Turrell (1972); Cochran (1973); Decius and Hexter (1977). Useful texts discussing many aspects of crystal lattice vibrations, infrared and Raman spectroscopy of crystals, inelastic neutron scattering experiments, symmetry analysis, and lattice dynamical calculations.
Born and Huang (1954). The classic treatment of crystal lattice dynamics.

Symmetry and group theory useful for vibrational spectroscopy

Salthouse and Ware (1972). Tables of symmetry properties for molecules and crystals.
Cotton (1971); Woodward (1972); Fateley et al. (1972). Symmetry and group theory for molecules and crystals: applications to the vibrational problem.
Koster (1957) Special symmetry points throughout the Brillouin zone for all crystal lattice types.
Birman (1984). Detailed discussion of crystal lattice dynamics and their symmetry properties (also Decius and Hexter, 1977).

Infrared and Raman spectroscopy

Hadni (1967); Bell (1972); Ferraro and Basile (1982); Steel (1983). Thorough discussions of infrared spectroscopy and instrumentation.
Griffiths and DeHaseth (1986); Hollas (1987). Infrared components, instrumentation and techniques, sample accessory design.
Mohler and Rothschild (1971). Far infrared techniques.
Gilson and Hendra (1970); Tobin (1970); Hayes and Loudon (1970); Long (1977). Excellent discussions of the theory and practice of Raman spectroscopy of gases, liquids and solids.
Demtroder (1982). Laser spectroscopies, spectrometers and interferometers, detectors; also Siegman (1971); Svelto (1976): Introduction to theory and operation of lasers.

Compilations and series

Lazarev (1972); Farmer (1974); Karr (1975). Useful compilations and discussions of mineral spectra.
Mathieu (1973); Ferraro and Basile (1978; 1979; 1982; 1985); Durig (1972-present); Clark and Hester (1973-present).
Also Specialist Periodical Reports of the Chemical Society (London): Spectroscopic Properties of Inorganic and Organometallic Compounds (1968-present: various editors).

REFERENCES

Adams, D.M. and Gardner, I.R. (1974a) Single-crystal vibrational spectra of beryl and dioptase. J. Chem. Soc., Dalton Trans., 74, 1502-1505.

Akaogi, M., Ross, N.L., McMillan, P. and Navrotsky, A. (1984) The Mg_2SiO_4 polymorphs (olivine, modified spinel, spinel)-thermodynamic properties from oxide melt solution calorimetry, phase relations, and models of lattice vibrations. Am. Mineral., 69, 499-512.

Alain, P. and Piriou, B. (1981) Temperature and composition dependence of vibrational modes in $Sm_{1-x}La_xAlO_3$. Solid State Comm., 37, 901-906.

Anderman, G., Caron, A. and Dows, D.A. (1965) Kramers-Kronig dispersion analysis of infrared reflection bands. J. Opt. Soc. Am., 55, 1210-1216.

Asell, J.F. and Nicol, M. (1968) Raman spectrum of α quartz at high pressures. J. Chem. Phys., 49, 5395-5399.

Ashcroft, N.W. and Mermin, N.D. (1976) "Solid State Physics", Holt-Saunders, Philadelphia.

Atkins, P.W. (1982) "Physical Chemistry", 2nd ed., W.H. Freeman, San Francisco.

Axe, J.D., Shirane, G. and Muller, K.A. (1969) Zone-boundary phonon instability in cubic $LaAlO_3$. Phys. Rev., 183, 820-823.

Balachandran, U. and Eror, N.G. (1982) Laser-induced Raman scattering in calcium titanate. Solid State Comm., 44, 815-818.

Ballantyne, J.M. (1964) Frequency and temperature response of the polarization of barium titanate. Phys. Rev., 136, A429-A436.

Baran, von E.J. and Botto, I.L. (1978) Die IR-Spektren einiger Doppeloxide mit Ilmenit-Struktur. Z. anorg. allg. Chem., 444, 282-288.

Baran, von E.J. and Botto, I.L. (1979) Die Raman-Spektren von $ZnTiO_3$ und $CdTiO_3$/einiger. Z. anorg. allg. Chem., 448, 188-192.

Barker, A.S. (1963) Infrared lattice vibrations and dielectric dispersion in corundum. Phys. Rev., 132, 1474-1481.

Bates, J.B. (1972) Raman spectra of α and β cristobalite. J. Chem. Phys., 57, 4042-4047.

Beattie, I.R. and Gilson, T.R. (1968) Single crystal laser Raman spectroscopy. Proc. R. Soc. A, 307, 407-429.

Beattie, I.R. and Gilson, T.R. (1969) Oxide phonon spectra. J. Chem. Soc. A, 2322-2327.

Beattie, I.R. and Gilson, T.R. (1970) Single-crystal Raman spectra of nearly opaque materials. Iron(III) oxide and chromium(III) oxide. J. Chem. Soc. A, 980-986.

Bell, R.J. (1972) "Introductory Fourier Transform Spectroscopy". Academic Press, New York.

Bellamy, L.J. (1958) "The Infrared Spectra of Complex Molecules", 2nd ed., Methuen, London.

Berge, B., Bachheimer, J.P., Dolino, G. and Vallade, M. (1986) Inelastic neutron scattering study of quartz near the incommensurate phase transition. Ferroelectrics, 66, 73-84.

Berreman, D.W. (1963) Infrared absorption at longitudinal optic frequency in cubic crystal films. Phys. Rev., 130, 2193-2198.

Berry, R.S., Rice, S.A. and Ross, J. (1980) "Physical Chemistry", John Wiley and Sons, New York.

Besson, J.M., Pinceaux, J.P. and Anastopoulos, C. (1982) Raman spectra of olivine up to 65 kilobars. J. Geophys. Res., 87, 10,773-10,775.

Bethke, J., Dolino, G., Eckold, G., Berge, B., Vallade, M., Zeyen, C.M.E., Hahn, T., Arnold, H. and Moussa, F. (1987) Phonon dispersion and mode coupling in high-quartz near the incommensurate phase transition. Europhys. Lettes., 3, 207-212.

Bialas, H. and Stolz, H.J. (1975) Lattice dynamics of sapphire (corundum). Part I. Phonon dispersion by inelastic neutron scattering. Z. Physik. B, 21, 319-324.

Bilz, H. and Kress, W. (1979) "Phonon Dispersion Relations in Insulators". Springer-Verlag, New York.

Birman. J.L. (1984) "Theory of Crystal Space Groups and Lattice Dynamics-Infrared and Raman Optical Processes of Insulating Crystals. Springer-Verlag, Berlin.

Bischoff, W.D., Sharma, S.K. and Mackenzie, F.T. (1985) Carbonate ion disorder in synthetic and biogenic magnesian calcites: a Raman spectral study. Am. Mineral., 70, 581-589.

Blaha, J.J. and Rosasco, G.J. (1978) Raman microprobe spectra of individual microcrystals and fibers of talc, tremolite, and related silicate minerals. Anal. Chem., 50, 892-896.

Bockelmann, H.K. and Sclecht, R.G. (1974) Raman scattering from microcrystals of MgO. Phys. Rev. B, 10, 5225-5231.

Born, M. and Huang, K. (1954) "The Dynamical Theory of Crystal Lattices". Clarendon Press, Oxford.

Bouckaert, L.P., Smoluchowski, R. and Wigner, E. (1936) Theory of Brillouin zones and symmetry properties of wave functions in crystals. Phys. Rev., 50, 58-67.

Boyer, H., Smith, D.C., Chopin, C. and Lasnier, B. (1985) Raman microprobe (RMP) determinations of natural and synthetic coesite. Phys. Chem. Mineral, 12, 45-48.

Boyer, L.L. and Hardy, J.R. (1981) Theoretical study of the structural phase transition in $RbCaF_3$. Phys. Rev. B, 24, 2577-2591.

Bretheau-Raynal, F., Dalbiez, J.P., Drifford, M. and Blanzat, B. (1979) Raman spectroscopic study of thortveitite structure silicates. J. Raman Spectr., 8, 39-42.

Bukowinski, M. and Wolf, G. (1986) Equation of state and stability of fluorite-structured SiO_2. J. Geophys. Res. B, 91, 4704-4710.

Burns, R.G. and Strens, R.G.J. (1966) Infrared study of the hydroxyl bands in clinoamphiboles. Science, 153, 890-892.

Burns, R.G. and Huggins, F.E. (1972) Cation determinative curves for Mg-Fe-Mn olivines from vibrational spectra. Am. Mineral., 57, 967-985.

Buseck, P.R. and Bo-Jun, H. (1985) Conversion of carbonaceous materials to graphite during metamorphism. Geochim. Cosmochim. Acta, 49, 2003-2016.

Buseck, P.R. and Bo-Jun, H. (1987) Electron microscope investigation of the structures of annealed carbons. J. Energy Fuels, 1, 105-110.

Carlsson, A.E., Ashcroft, N.W. and Williams, A.R. (1984) Properties of SiO_2 in a high-pressure fluorite structure phase. Geophys. Res. Letts., 11, 617-619.

Chase, B. (1987) Fourier transform Raman spectroscopy. Anal. Chem., 59, 881-889.

Clark, R.J.H. and Franks, M.L. (1975) The resonance Raman spectrum of ultramarine blue. Chem. Phys. Letts., 34, 69-72.

Clark, R.J.H. and Hester, R.E. (1973-present) eds. "Advances in Infrared and Raman Spectroscopy". Heyden Press, London.

Clemens, J.D., Circone, S., Navrotsky, A., McMillan, P.F., Smith, B.K. and Wall, V.J. (1987) Phlogopite: high temperature solution calorimetry, thermodynamic properties, Al-Si and stacking disorder, and phase equilibria. Geochim. Cosmochim. Acta, in press.

Cochran, W. (1960) Crystal stability and the theory of ferro electricity. Adv. Phys., 9, 387-423.

Cochran, W. (1973) "The Dynamics of Atoms in Crystals", Edward Arnold, London.

Cochran, W. and Zia, A. (1968) Structure and dynamics of perovskite-type crystals. Phys. Status Solidi, 25, 273-283.

Conjeaud, M. and Boyer, H. (1980) Some possibilities of Raman microprobe in cement chemistry. Cement and Concrete Res., 10, 61-70.

Cotton, F.A. (1971) "Chemical Applications of Group Theory", 2nd ed., John Wiley and Sons, New York.

Couty, R. and Velde, B. (1986) Pressure-induced band splitting in infrared spectra of sanidine and albite. Am. Mineral., 71, 99-104.

Couzi, M. and Huong, P.V. (1974) Spectres infrarouge et Raman des perovskites. Ann. Chim., 9, 19-29.

Cowley, R.A. (1964) Lattice dynamics and phase transitions of strontium titanate. Phys. Rev., 134, A981-A997.

Damen, T.C., Porto, S.P.S. and Tell, B. (1966) Raman effect in zinc oxide. Phys. Rev., 142, 570-574.

Davies, M. (1963) ed., "Infrared Spectroscopy and Molecular Structure", Elsevier, New York.

Davies, P.K. and Akaogi, M. (1983) Phase intergrowths in spinelloids. Nature, 305, 788-790.

Dean, K.J., Sherman, W.F. and Wilkinson, G.R. (1982) Temperature and pressure dependence of the Raman active modes of vibration of α-quartz. Spectrochim. Acta, 38A, 1105-1108.

Decius, J.C. and Hexter, R.M. (1977) "Molecular Vibrations in Crystals", McGraw-Hill, New York.

Demtroder, W. (1981) "Laser Spectroscopy. Basic Concepts and Instrumentation". Springer-Verlag, Berlin.

Denisov, V.N., Mavrin, B.N., Podobedov, V.B. and Sterin, Kh. E. (1978) Hyper-Raman scattering and longitudinal-transverse splitting of vibrations in fused quartz. Sov. Phys. Solid State, 20, 2016-2017.

Denisov, V.N., Mavrin, B.N., Podobedov, V.B., Sterin, Kh. E. and Varshal, B.G. (1984) Law of conservation of momentum and rule of mutual exclusion for vibrational excitations in hyper-Raman and Raman spectra of glasses. J. Non-Cryst. Solids, 64, 195-210.

Devarajan, V. and Funck, E. (1975) Normal coordinate analysis of the optically active vibrations (k=0) of crystalline magnesium orthosilicate Mg_2SiO_4 (forsterite). J. Chem. Phys., 62, 3406-3411.

Dietrich, P. and Arndt, J. (1982) Effects of pressure and temperature on physical behavior of mantle-relevant olivine, orthopyroxene and garnet: II. Infrared absorption and microscopic Gruneisen parameters. In "High Pressure Researches in Geoscience", ed. W. Schreyer, pp. 307-319. E. Schweizerbart'sche Verlagsbuchhandlung, Stuttgart.

Dolino, G. (1986) The incommensurate phase of quartz. In "Incommensurate Phases in Dielectrics", ed. R. Blinc and A.P. Levanyuk, p. 205-232. Elsevier, Amsterdam.

Dolino, G., Bastie, P., Berge, B., Vallade, M., Bethke, J., Regnault, L.P. and Zeyen, C.M.E. (1987) Stress-induced <3-q>-<1-q> incommensurate phase transition in quartz. Europhys. Letts., 3, 601-609.

Dollish, F.R., Fateley, W.G. and Bentley, F.F. (1974) Characteristic Raman frequencies of organic compounds. John Wiley, New York.

Durig, J. (1972-present) ed. "Vibrational Spectra and Structure". Elsevier, Amsterdam.

Etchepare, J. (1970) Spectres Raman du diopside cristallise et vitreux. C.R. Acad. Sci. Paris, ser. B, 270, 1339-1342.

Etchepare, J., Merian, M. and Kaplan, P. (1978) Vibrational normal modes of SiO_2. II. Cristobalite and tridymite. J. Chem. Phys., 68, 1531-1537.

Farmer, V.C. (1974a) "The Infrared Spectra of Minerals", Mineralogical Society, London.

Farmer, V.C. (1974b) The anhydrous oxide minerals. In "The Infrared Spectra of Minerals", ed. V.C. Farmer. p. 183-204. Mineralogical Society, London.

Farmer, V.C. (1974c) Orthosilicates, pyrosilicates, and other finite-chain silicates. In "The Infrared Spectra of Minerals", ed. V.C. Farmer. p. 285-303. Mineralogical Society, London.

Farmer, V.C. (1974d) The layer silicates. In "The Infrared Spectra of Minerals", ed. V.C. Farmer. p. 331-363. Mineralogical Society, London.

Farmer, V.C. and Velde, B. (1973) Effects of structural order and disorder on the infrared spectra of brittle mica. Mineral. Mag., 39, 282-288.

Fateley, W.G., Dollish, F.R., McDevitt, N.T. and Bentley, F.F. (1972) Infrared and Raman selection rules for molecular and lattice vibrations: the correlation method. Wiley-Interscience.

Ferraro, J.R. (1984) "Vibrational Spectroscopy at High External Pressures". Academic Press, Orlando, Florida.

Ferraro, J.R. and Basile, L.J. (1978) "Fourier Transform Infrared Spectroscopy : Applications to Chemical Systems". Vol. 1, Academic Press, Orlando, Florida.

Ferraro, J.R. and Basile, L.J. (1979) "Fourier Transform Infrared Spectroscopy : Applications to Chemical Systems". Vol. 2, Academic Press, Orlando, Florida.

Ferraro, J.R. and Basile, L.J. (1982) "Fourier Transform Infrared Spectroscopy: Techniques using the Fourier transform". Vol. 3. Academic Press, Orlando, Florida.

Ferraro, J.R. and Basile, L.J. (1985) "Fourier Transform Infrared Spectroscopy : Applications to Chemical Systems". Vol. 4, Academic Press, Orlando, Florida.

Finger, L.W., Hazen, R.M. and Hofmeister, A. (1986) High-pressure crystal chemistry of spinel ($MgAl_2O_4$) and magnetite (Fe_3O_4): comparison with silicate spinels. Phys. Chem. Minerals, 13, 215-220.

Fleury, P.A., Scott, J.F. and Worlock, J.M. (1968) Soft phonon modes and the $110^\circ K$ phase transition in $SrTiO_3$. Phys. Rev. Letts., 21, 16-19.

Fraas, L.M., Moore, J.E. and Salzberg, J.B. (1973) Raman characterization studies of synthetic and natural $MgAl_2O_4$ crystals. J. Chem. Phys., 58, 3585-3592.

Frech, R., Wang, E.C. and Bates, J.B. (1980) The i.r. and Raman spectra of $CaCO_3$ (aragonite). Spectrochim. Acta, 36A, 241-246.

Gabelica-Robert, M. and Tarte, P. (1979) Vibrational spectrum of akermanite-like silicates and germanates. Spectrochim. Acta, 35A, 649-654.

Gaskell, P.H. (1966) Thermal properties of silica. Part 1.-Effect of temperature on infra-red reflection spectra of quartz, cristobalite and vitreous silica. Trans. Faraday Soc., 62, 1493-1504.

Gervais, F. and Piriou, B. (1974) Temperature dependence of transverse- and longitudinal-optic modes in TiO_2 (rutile). Phys. Rev. B, 10, 1642-1654.

Gervais, F. and Piriou, B. (1975) Temperature dependence of transverse- and longitudinal-optic modes in the α and β phases of quartz. Phys. Rev. B, 11, 3994-3950.

Gervais, F., Piriou, B. and Cabannes, F. (1972) Anharmonicity of infrared vibration modes in beryl. Phys. Stat. Sol. B, 51, 701-712.

Gervais, F., Piriou, B. and Servoin, J.L. (1973a) Etude par reflexion infrarouge des modes internes et externes de quelques silicates. Bull. Soc. fr. Mineral. Cristall., 96, 81-90.

Gervais, F., Piriou, B. and Cabannes, F. (1973b) Anharmonicity of infrared vibration modes in the nesosilicate Be_2SiO_4. Phys. Stat. Sol. B, 55, 143-154.

Gervais, F., Piriou, B. and Cabannes, F. (1973c) Anharmonicity in silicate crystals: temperature dependence of A_u type vibrational modes in $ZrSiO_4$ and $LiAlSi_2O_6$. J. Phys. Chem. Solids, 34, 1785-1796.

Gilson, T.R. and Hendra, P.J. (1970) "Laser Raman Spectroscopy". John Wiley ans Sons, New York.

Griffiths, P.R. and de Haseth, J.A. (1986) "Fourier Transform Infrared Spectroscopy". John Wiley and Sons, New York.

Guyot, F., Boyer, M., Madon, M., Velde, B. and Poirier, J.P. (1986) Comparison of the Raman microprobe spectra of $(Mg,Fe)_2SiO_4$ and Mg_2GeO_4 with olivine and spinel structures. Phys. Chem. Minerals, 13, 91-95.

Hadni, A. (1967) "Essentials of Modern Physics Applied to the Study of the Infrared", Pergamon Press.

Haley, L.V., Wylie, I.W. and Koningstein, J.A. (1982) An investigation of the lattice and interlayer water vibrational spectral regions of muscovite and vermiculite using Raman microscopy. J. Raman Spectr., 13, 203-205.

Handke, M. and Urban, M. (1982) IR and Raman spectra of alkaline earth metals orthosilicates. J. Mol. Structure, 79, 353-356.

Harada, J., Axe, J.D. and Shirane, G. (1971) Neutron-scattering study of soft modes in cubic $BaTiO_3$. Phys. Rev. B, 4, 155-162.

Harrison, W.J. (1979) "Solid State Theory". Dover.

Hawthorne, F.C. (1983a) Quantitative characterization of site-occupancies in minerals. Am. Mineral., 68, 287 306.

Hawthorne, F.C. (1983b) The crystal chemistry of the amphiboles. Can. Mineral., 21, 173-480.

Hayes, W. and Loudon, R. (1970) "Scattering of Light by Crystals". John Wiley and Sons, New York.

Hemley, R. (1987) Pressure dependence of Raman spectra of SiO_2 polymorphs: α-quartz, coesite and stishovite. In "High-Pressure Research in Mineral Physics", eds. M.H. Manghnani and Y. Syono, in press.

Hemley, R.J., Mao, H.K., Bell, P.M. and Mysen, B.O. (1986a) Raman spectroscopy of SiO_2 glass at high pressure. Phys. Rev. Letts., 57, 747-750.

Hemley, R.J., Mao, H.-K., Bell, P.M. and Akimoto, S. (1986b) Lattice vibrations of high-pressure SiO_2 phases: Raman spectrum of synthetic stishovite. Physica 139/140 B, 455-457.

Hemley, R.J., Mao, H-K. and Chao, E.C.T. (1986c) Raman spectrum of natural and synthetic stishovite. Phys. Chem. Minerals, 13, 285-290.

Hemley, R.J., Bell, P.M. and Mao, H.K. (1987a) Laser techniques in high-pressure geophysics. Science, 237, 605-612.

Hemley, R.J., Jackson, M.D. and Gordon, R.G. (1987b) Theoretical study of the structure, lattice dynamics, and equations of state of perovskite-type $MgSiO_3$ and $CaSiO_3$. Phys. Chem. Minerals, 14, 2-12.

Henderson, C.M.B. and Taylor, D. (1977) Infrared spectra of anhydrous members of the sodalite family. Spectrochim. Acta, 33A, 283-290.

Henderson, C.M.B. and Taylor, D. (1979) Infrared spectra of alumino-germanate and aluminate sodalites. Spectrochim. Acta, 35A, 929-935.

Herzberg, G. (1945) "Molecular Spectra and Molecular Structure. II. Infrared and Raman Spectra of Polyatomic Molecules", Van Nostrand Reinhold, New York.

Herzberg, G. (1950) "Molecular Spectra and Molecular Structure. I. Spectra of Diatomic Molecules", 2nd ed., Van Nostrand Reinhold, New York.

Hirschfeld, T. (1979) Quantitative FTIR: A detailed look at the problems involved. In "Fourier Transform Infrared Spectroscopy : Applications to Chemical Systems". Vol. 2, p. 193-242. ed. J.R. Ferraro and L.J. Basile, Academic Press.

Hofmeister, A.M. (1987) Single crystal absorption and reflection spectroscopy of forsterite and fayalite. Phys. Chem. Minerals, 14, 499-513.

Hofmeister, A.M., Mao, H.-K. and Bell, P.M. (1986a) Spectroscopic determination of thermodynamic properties of $γ-Fe_2SiO_4$ at mantle pressures. EOS, 67, 395.

Hofmeister, A.M., Xu, J. and Akimoto, S. (1986b) Thermodynamic implications of infrared spectroscopy of stishovite at mantle pressures. Int. Mineral. Assoc. (Stanford) Abstr. Prog., p. 126.

Hofmeister, A.M., Hoering, T.C. and Virgo, D. (1987a) Vibrational spectroscopy of beryllium aluminosilicates: heat capacity calculations from band assignments. Phys. Chem. Minerals, 13, 215-220.

Hofmeister, A.M., Xu, J., Mao, H.-K., Bell, P.M. and Hoering, T.C. (1987b) Thermodynamics of Fe-Mg olivines at mantle pressures: Part I. Mid- and far-infrared spectroscopy at pressure. Am. Mineral., in press.

Hohler, V. and Funck, E. (1973) Vibrational spectra of crystals with olivine structure. I. Silicates. Zeit. Naturfors. B, 28, 125-139.

Hollas, J.M. (1987) "Modern Spectroscopy". John Wiley and Sons.

Horiuchi, H. and Sawamoto, H. (1981) $β-Mg_2SiO_4$: Single-crystal X-ray diffraction study. Am. Mineral., 66, 568-575.

Horiuchi, H., Horioka, K., and Morimoto, N. (1980) Spinelloids: A systematics of spinel-related structures obtained under high pressure conditions. J. Mineral. Soc. Japan, 2, 253-264.

Huber, K.P. and Herzberg, G. (1979) "Molecular Spectra and Molecular Structure. IV. Constants of Diatomic Molecules". Van Nostrand Reinhold, New York.

Hyde, B.G., White, T.J., O'Keeffe, M. and Johnson, A.W.S. (1982) Structures related to those of spinel and the β-phase, and a possible mechanism for the transformation olivine-spinel. Z. Krist., 160, 53-62.

Ignatiev, I.S. and Lazarev, A.N. (1972a) Refinement of the interpretation of the vibrational spectra of $Sr_3Si_3O_9$, $Sr_3Ge_3O_9$ and their solid solutions. Inorg. Mat., 8, 268-279.

Ignatiev, I.S. and Lazarev, A.N. (1972b) Vibrational spectra of silicates and germanates with puckered X_3O_9-type rings. Inorg. Mat., 8, 280-284.

Iishi, K. (1978a) Lattice dynamics of corundum. Phys. Chem. Minerals, 3, 1-10.

Iishi, K. (1978b) Lattice dynamics of forsterite. Am. Mineral., 63, 1198-1208.

Iishi, K., Salje, E. and Werneke, Ch. (1979) Phonon spectra and rigid-ion model calculations on andalusite. Phys. Chem. Minerals, 4, 173-188.

Inoue, K., Akimoto, S. and Ishidate, T. (1985) Study of the hyper-Raman scattering in solids under pressure. In "Solid State Physics Under Pressure", ed. S. Minomura, pp. 87-92. Terra Scientific Publishing Co.

Irish, D.E., Jarv, T. and Ratcliffe, C.I. (1982) Vibrational spectral studies of solutions at elevated temperatures and pressures. III. A furnace assembly for Raman spectral studies to 300°C and 15MPa. Appl. Spectroscopy, 36, 137-140.

Ishii, M., Hiraishi, J. and Yamanka, T. (1982) Structure and lattice vibrations of Mg-Al solid solution. Phys. Chem. Minerals, 8, 64-68.

Ishikawa, K., Fujima, N. and Komura, H. (1985) First-order Raman scattering in MgO microcrystals. J. Appl. Phys., 57, 973-975.

Jayaraman, A. (1983) Diamond anvil ell and high-pressure physical investigations. Rev. Mod. Phys., 55, 65-108.

Jeanloz, R. (1980) Infrared spectra of olivine polymorphs: α, β-phase and spinel. Phys. Chem. Minerals, 5, 327-339.

Jeanloz, R. (1981) Majorite: vibrational and compressional properties of a high-pressure phase. J. Geophys. Res., 86, 6171-6179.

154

Jenkins, D.M. (1985) Assessment of IR criteria for Al-Si order-disorder in trioctahedral micas. Geol. Soc. Am. Abstr. Prog. 17, 619.

Jones, L.H. (1971) "Inorganic Vibrational Spectroscopy", Vol. 1, Marcel Dekker.

Kappus, W. (1975) Lattice dynamics of sapphire (corundum). Part II. Calculations of the phonon dispersion. Z. Physik. B, 21, 325-331.

Karr, C. (1975) ed., "Infrared and Raman Spectroscopy of Lunar and Terrestrial Materials", Academic Press, New York.

Katiyar, R.S. (1970) Dynamics of the rutile structure I. Space group representations and the normal mode analysis. J. Phys. C: Solid State Phys., 3, 1087-1096.

Katiyar, R.S., Dawson, P., Hargreave, M.M. and Wilkinson, G.R. (1971) Dynamics of the rutile structure III. Lattice dynamics, infrared and Raman spectra of SnO_2. J. Phys. C: Solid State Phys., 4, 2421-2431.

Kato, T. and Kumuzawa, M. (1985) Garnet phase of $MgSiO_3$ filling the pyroxene-ilmenite gap at very high temperature. Nature, 316, 803-805.

Kieffer, S.W. (1979b) Thermodynamics and lattice vibrations of minerals: 2. Vibrational characteristics of silicates. Rev. Geophys. Space Phys., 17, 20-34.

Kieffer, S.W. (1985) Heat capacity and entropy: systematic relations to lattice vibrations. In "Microscopic to Macroscopic - Atomic Environments to Thermodynamic Properties", Rev. Mineral., 14, eds. S.W. Kieffer and A. Navrotsky, p. 65-126. Mineralogical Society of America, Washington, D.C.

Kittel, C. (1976) "Introduction to Solid State Physics", 5th ed., John Wiley and Sons, New York.

Koster, G.F. (1957) Space groups and their representations. Solid State Phys., 5, 173-256.

Kruse, R. and Franck, E.U. (1982) Raman spectra of hydrothermal solutions of CO_2 and $KHCO_3$ at high temperatures and pressures. Ber. Bunsenges. Phys. Chem., 86, 1036-1038.

Landsberg, G. and Mandelstam, L. (1928) Eine neue Erscheinung bei der Lichtzerstreuung in Krystallen. Naturwiss., 16, 557.

Lange, R.A., Carmichael, I.S.E. and Stebbins, J.F. (1986) Phase transitions in leucite ($KAlSi_2O_6$), orthorhombic $KAlSiO_4$, and their iron analogues ($KFeSi_2O_6$, $KFeSiO_4$). Am. Mineral., 71, 937-945.

Langer, K. and Schreyer, W. (1969) Infrared and powder X-ray diffraction studies on the polymorphism of cordierite, ($Mg_2Al_4Si_5O_{18}$). Am. Mineral., 54, 1442-1459.

Langer, K., Chatterjee, N.D. and Abraham, K. (1981) Infrared studies of some synthetic and natural $2M_1$ dioctahedral micas. N. Jb. Miner. Abh., 142, 91-110.

Last, J.T. (1957) Infrared-absorption studies on barium titanate and related materials. Phys. Rev., 105, 1740-1750.

Lazarev, A.N. (1962) Vibrational spectra of silicates. IV. Interpretation of the spectra of silicates and germanates with ring anions. Optics and Spectroscopy, 12, 28-31.

Lazarev, A.N. (1972) "Vibrational Spectra and Structure of Silicates", Consultants Bureau, New York.

Lespade, P. Al-Jishi, R. and Dresselhaus, M.S. (1982) Model for Raman scattering from incompletely graphitized carbons. Carbon, 20, 427-431.

Lespade, P., Marchand, A., Couzi, M. and Cruege, F. (1984) Caracterisation de materiaux carbones par microspectrometrie Raman. Carbon, 22, 375-385.

Liu, L. (1975) High pressure phase transformations and compression of ilmenite and rutile. II. Geophysical implications. Phys. Earth. Planet. Interiors, 10, 344-347.

Liu, L. (1982) High pressure phase transformations of the dioxides: implications for structures of SiO_2 at high pressure. In "High Pressure Research in Geophysics", eds. S. Akimoto and M.H. Manghnani, pp. 349-360. Center for Academic Publications Japan, Tokyo.

Lockwood, D.J. and Torrie, B.H. (1974) Raman scattering study of the three structural phases of KMnF3. J. Phys. C: Solid State Phys., 7, 2729-2744.

Loh, E. (1973) Optical vibrations in sheet silicates. J. Phys. C: Solid State Phys., 6, 1091-1104.

Long, D.A. (1977) "Raman Spectroscopy", McGraw-Hill, New York.

Louie, S.G. (1985) Pseudopotentials and total energy calculations: applications to crystal stability, vibrational properties, phase transformations, and surface structures. In "Electronic Structure, Dynamics, and Quantum Structural Properties of Condensed Matter", ed. J.T. Devreese and P. van Camp, p. 335-398. Plenum Press.

Low, M.J.D. and Tascom, J.M.D. (1985) An approach to the study of minerals using photothermal beam deflection spectroscopy. Phys. Chem. Minerals, 12, 19-32.

Lucovsky, G., Sladek, R.J. and Allen, J.W. (1977) ir reflectance spectra of Ti_2O_3: Infrared-active phonons and Ti 3d electronic effects. Phys. Rev. B, 16, 5452-5459.

Luxon, J.T. and Summitt, R. (1969) Interpretation of the infrared absorption spectra of stannic oxide and titanium dioxide (rutile) powders. J. Chem. Phys., 50, 1366-1370.

Luxon, J.T., Montgomery, D.J. and Summitt, R. (1969) Effect of particle size and shape on the infrared absorption of magnesium oxide powders. Phys. Rev., 188, 1345-1356.

Lyon, R.J.P. (1962) Infra-red confirmation of 6-fold coordination of silicon in stishovite. Nature, 196, 266–267.

Mammone, J.F., Sharma, S.K. and Nicol, M. (1980) Raman study of rutile (TiO$_2$) at high pressures. Solid State Comm., 34, 799–802.

Mathieu, J.P. (1973) ed. "Advances in Raman Spectroscopy", vol. 1. Heyden Press, London.

Martin, T.P. (1970) Interaction of finite NaCl crystals with infrared radiation. Phys. Rev. B, 1, 3480–3488.

Martin, R.F. (1970) Cell parameters and infrared absorption of synthetic high to low albites. Contrib. Mineral. Petrol., 26, 62–74.

Martin, R.F. and Lagache, M. (1975) Cell edges and infrared spectra of synthetic leucites and pollucites in the system KAlSi$_2$O$_6$-RbAlSi$_2$O$_6$-CsAlSi$_2$O$_6$. Can. Mineral., 13, 275–281.

Matson, D.W., Sharma, S,K, and Philpotts, J.A. (1986) Raman spectra of some tectosilicates and of glasses along the orthoclase-anorthite and nepheline-anorthite joins. Am. Mineral., 71, 694–704.

McKeown, D.A., Galeener, F.L. and Brown, G.E. (1984) Raman studies of Al co-ordination in silica-rich sodium aluminosilicate glasses and some related minerals, J. Non-Cryst. Solids, 68, 361–378.

McMillan, P. (1984) Structural studies of silicate glasses and melts: applications and limitations of Raman spectroscopy. Am. Mineral., 69, 622–644.

McMillan, P. (1985) Vibrational spectroscopy in the mineral sciences. In "Microscopic to Macroscopic – Atomic Environments to Thermodynamic Properties", Rev. Mineral., 14, eds. S,W. Kieffer and A. Navrotsky, pp. 9–63. Mineralogical Society of America, Washington, D.C.

McMillan, P. and Piriou, B. (1982) The structures and vibrational spectra of crystals and glasses in the silica-alumina system. J. Non-Cryst. Solids, 53, 279–298.

McMillan, P.F. and Ross, N.L. (1987) Heat capacity calculations for Al$_2$O$_3$ corundum and MgSiO$_3$ ilmenite. Phys. Chem. Minerals, 14, 225–234.

McMillan, P.F. and Akaogi, M. (1987) The Raman spectra of β- (modified spinel) and γ- (spinel) Mg$_2$SiO$_4$. Am. Mineral., 72, 361–364.

McMillan, P., Piriou, B. and Navrotsky, A. (1982) A Raman spectroscopic study of glasses along the joins silica-calcium aluminate, silica-sodium aluminate and silica-potassium aluminate. Geochim. Cosmochim. Acta, 46, 2021–2037.

McMillan, P., Putnis, A. and Carpenter, M.A. (1984) A Raman spectroscopic study of Al-Si ordering in synthetic magnesium cordierite. Phys. Chem. Minerals, 10, 256–260.

McMillan, P., Graham, C.M. and Ross, N.L. (1985) Vibrational spectroscopy of F-OH pargasites. EOS, 65, 1141.

McMillan, P., Smyth, J.R. and Akaogi, M. (1987) OH in β-Mg$_2$SiO$_4$. EOS, 68, 1456.

Mehl, M.J., Hemley, R.J. and Boyer, L.L. (1986) Potential-induced breathing model for the elastic moduli and high-pressure behavior of the cubic alkaline-earth oxides. Phys. Rev. B, 33, 8685–8696.

Miller, R.C. and Spitzer, W.G. (1963) Far infrared dielectric dispersion in KTaO$_3$. Phys. Rev., 129, 94–98.

Moenke, H.H.W. (1974) Silica, the three-dimensional silicates, borosilicates, and beryllium silicates. In "The Infrared Spectra of Minerals", ed. V.C. Farmer. p. 365–382. Mineralogical Society, London.

Mohler, K.D. and Rothschild, W.G. (1971) "Far-Infrared Spectroscopy". Wiley-Interscience.

Moore, W.J. (1972) "Physical Chemistry". Wiley-Interscience.

Moore, R.K., White, W.B. and Long, T.V. (1971) Vibrational spectra of the common silicates: I. The garnets. Am. Mineral., 56, 54–71.

Mottana, A. and Griffin, W.L. (1986) Crystal chemistry of two coexisting K-richterites from St. Marcel (Val d'Aosta, Italy). Am. Mineral., 71, 1426–1433.

Nakagawa, I. (1973) Transverse and longitudinal lattice frequencies and interionic potentials in some AMF$_3$ perovskite fluoride crystals. Spectrochim. Acta, 29A, 1451–1461.

Nakagawa, I., Tsuchida, A. and Shimanouchi, T. (1967) Infrared transmission spectrum and lattice vibration analysis of some perovskite fluorides. J. Chem. Phys., 47, 982–989.

Nakamoto, K. (1978) "Infrared and Raman Spectra of Inorganic and Coordination Compounds", 3rd ed., Wiley-Interscience.

Naumann, A.W., Safford, G.J. and Mumpton, F.A. (1966) Low- frequency (OH)-motions in layer silicate minerals. Clays Clay Minerals, 14, 367–383.

Nichols, E.F. (1897) A method for energy measurements in the infra-red spectrum and the properties of the ordinary ray in quartz for waves of great wave length. Phys. Rev. (ser. I), 4, 297–313.

Nicol, M., Besson, J.M. and Velde, B. (1980) Raman spectra and structure of stishovite. In "High Pressure Science and Technology", ed. B. Vodar and Ph. Marteau, pp. 891–893. Pergamon Press.

Nilsen, W.G. and Skinner, J.G. (1968) Raman spectrum of strontium titanate. J. Chem. Phys., 48, 2240–2248.

Oehler, O. and Gunthard, Hs. H. (1969) Low-temperature infrared spectra between 1200 and 20 cm^{-1} and normal-coordinate analysis of silicates with olivine structure. J. Chem. Phys., 51, 4719–4727.

O'Horo, M.P., Frisillo, A.L. and White, W.B. (1973) Lattice vibrations of MgAl$_2$O$_4$ spinel. J. Phys. Chem. Solids, 34, 23–28.

Omori, K. (1971) Analysis of the infrared absorption spectrum of diopside. Am. Mineral., 56, 1607-1616.

Onari, S., Arai, T. and Kudo, K. (1977) Infrared lattice vibrations and dielectric dispersion in α-Fe$_2$O$_3$. Phys. Rev. B, 16, 1717-1721.

Paques-Ledent, M. Th. and Tarte, P. (1973) Vibrational studies of olivine-type compounds-I. The i.r. and Raman spectra of the isotopic species of Mg$_2$SiO$_4$. Spectrochim. Acta, 29A, 1007-1016.

Pasteris, J.D., Wopenka, B. and Seitz, J.C. (1987) Practical aspects of quantitative laser Raman spectroscopy for the study of fluid inclusions. Geochim. Cosmochim. Acta, in press.

Pasto, A.E. and Condrate, R.A. (1973) Raman spectrum of PbZrO$_3$. J. Am. Ceram. Soc., 56, 436-438.

Peercy, P.S. and Morosin, B. (1973) Pressure and temperature dependence of the Raman-active phonons in SnO$_2$. Phys. Rev. B, 7, 2779-2786.

Perry, C.H. (1971) Dielectric properties and optical phonons in para- and ferro-electric perovskites. In "Far Infrared Spectroscopy", eds. K.D. Moller and W.G. Rothschild, pp. 557-591. John Wiley and Sons, New York.

Perry, C.H., Khanna, B.N. and Rupprecht, G. (1964) Infrared studies of perovskite titanates. Phys. Rev., 135, A408-A412.

Perry, C.H., McCarthy, D.J. and Rupprecht, G. (1965) Dielectric dispersion of some perovskite zirconates. Phys. Rev., 138, A1537-A1538.

Perry, C.H., Fertel, J.H. and McNelly, T.F. (1967) Temperature dependence of the Raman spectrum of SrTiO$_3$ and KTaO$_3$. J. Chem. Phys., 47, 1619-1625.

Phillippi, C.M. (1970) Analytical infrared spectra of particulate alpha-aluminas. Dev. Appl. Spectroscopy, 7B, 23-33.

Piriou, B. (1964) Etude des bandes restantes de la magnesie et du corindon, influence de la temperature. Rev. Int. Hautes Temp. Ref., 3, 109-114.

Piriou, B. (1974) Etude des modes normaux par reflexion infrarouge. Ann. Chim. 9, 9-17.

Piriou, B. and Cabannes, F. (1968) Validite de la methode de Kramers-Kronig et application a la dispersion infrarouge de la magnesie. Optica Acta, 15, 271-286.

Piriou, B. and McMillan, P. (1983) The high-frequency vibrational spectra of vitreous and crystalline orthosilicates. Am. Mineral., 68, 426-443.

Plendl, J.N., Mansur, L.C., Hadni, A., Brehat, F., Henry, P., Morlot, G., Naudin, F. and Strimer, P. (1967) Low temperature far infrared spectra of SiO$_2$ polymorphs. J. Phys. Chem. Solids, 28, 1589-1597.

Porto, S.P.S. and Krishnan, R.S. (1967) Raman effect of corundum. J. Chem. Phys., 47, 1009-1011.

Porto, S.P.S., Giordmane, J.A. and Damen, T.C. (1966) Depolarization of Raman scattering in calcite. Phys. Rev., 147, 608-611.

Porto, S.P.S., Fleury, P.A. and Damen, T.C. (1967) Raman spectra of TiO$_2$, MgF$_2$, ZnF$_2$, FeF$_2$, and MnF$_2$. Phys. Rev., 154, 522-526.

Preudhomme, J. and Tarte, P. (1971a) Infrared studies of spinels-I. A critical discussion of the actual interpretations. Spectrochim. Acta, 27A, 961-968.

Preudhomme, J. and Tarte, P. (1971b) Infrared studies of spinels-II. The experimental bases for solving the assignment problem. Spectrochim. Acta, 27A, 845-851.

Preudhomme, J. and Tarte, P. (1971c) Infrared studies of spinels-III. The normal II-III spinels. Spectrochim. Acta, 27A, 1817-1835.

Preudhomme, J. and Tarte, P. (1972) Infrared studies of spinels-IV. Normal spinels with a high-valency tetrahedral cation. Spectrochim. Acta, 28A, 66-79.

Price, G.D. (1983) Polytypism and the factors determining the stability of spinelloid structures. Phys. Chem. Minerals, 10, 77-83.

Price, G.D., Parker, S.C. and Leslie, M. (1987) The lattice dynamics of forsterite. Mineral. Mag., 51, 157-170.

Putnis, A. and Bish, D.L. (1983) The mechanism and kinetics of ordering in Al,Si ordering in Mg cordierite. Am. Mineral., 68, 60-65.

Raman, C.V. and Krishnan, K.S. (1928) A new type of secondary radiation. Nature, 121, 501.

Raman, C.V. and Nedungadi, T.M.K. (1940) The α-β transformation of quartz. Nature, 145, 147.

Raudsepp, M., Turnock, A.C., Hawthorne, F.C., Sherriff, B.L. and Hartman, J.S. (1987) Characterization of synthetic pargasitic amphiboles (NaCa$_2$Mg$_4$M^{3+}Si$_6$Al$_2$O$_{22}$(OH,F)$_2$; M^{3+}=Al,Cr,Ga,Sc,In) by infrared spectroscopy, Rietveld structure refinement, and ^{27}Al, ^{29}Si, and ^{19}F MAS NMR spectroscopy. Am. Mineral., 72, 580-593.

Ross, S.D. (1972) "Inorganic Infrared and Raman Spectra", McGraw-Hill, New York.

Ross, N. and McMillan, P. (1984) The Raman spectrum of MgSiO$_3$ ilmenite. Am. Mineral., 69, 719-721.

Ross, N.L., Akaogi, M., Navrotsky, A., Susaki, J.I. and McMillan, P. (1986) Phase transitions among the CaGeO$_3$ polymorphs (wollastonite, garnet, and perovskite structures): studies by high-pressure synthesis, high-temperature calorimetry, and vibrational spectroscopy and calculation. J. Geophys. Res. B, 91, 4685-4696.

Rossman, G.R. and Squires, R.L. (1974) The occurrence of alstonite at Cave-In Rock, Illinois. Min. Record, 5, 266-269.

Rouzaud, J.N., Oberlin, A. and Beny-Bassez, C. (1983) Carbon films: structure and microtexture (optical and electron microscopy, Raman spectroscopy). Thin Solid Films, 105, 75-96.

Rowbotham, G. and Farmer, V.C. (1973) The effect of "A" site occupancy upon the hydroxyl stretching frequency in clinoamphiboles. Contrib. Mineral. Petrol., 38, 147-149.

Russel, E.E. and Bell, E.E. (1967) Measurement of the optical constants of crystal quartz in the far infrared with the asymmetric Fourier-transform method. J. Opt. Soc. Am., 57, 341-348.

Saine, M.C., Husson, E. and Brusset, H. (1981) Etude vibrationelle d'aluminates et de gallates de terres rares-I. Aluminates de structure perovskite. Spectrochim. Acta, 37A, 985-990.

Saksena, B.D. (1940) Analysis of the Raman and infra-red spectra of α-quartz. Proc. Indian Acad. Sci., 12, 93-138.

Salje, E. (1985) Thermodynamics of sodium feldspar I: order parameter treatment and strain induced coupling effects. Phys. Chem. Minerals, 12, 93-98.

Salje, E. (1986) Raman spectroscopic investigation of the order parameter behaviour in hypersolvus alkali feldspar: displacive phase transition and evidence for Na-K site ordering. Phys. Chem. Minerals, 13, 340-346.

Salje, E. and Werneke, Ch. (1982) The phase equilibrium between sillimanite and andalusite as determined from lattice vibrations. Contrib. Mineral. Petrol., 79, 56-67.

Salthouse, J.A. and Ware, M.J. (1972) "Point Group Character Tables and Related Data", Cambridge University Press, Cambridge.

Sasaki, S., Prewitt, C.T. and Liebermann, R.C. (1983) The crystal structure of CaGeO$_3$ perovskite and the crystal chemistry of the GdFeO$_3$-type perovskites. Am. Mineral., 68, 1189-1198.

Sato, R.K. and McMillan, P.F. (1987) Infrared and Raman spectra of the isotopic species of alpha quartz. J. Phys. Chem., 91, 3494-3498.

Scheetz, B.E. (1972) Vibrational spectra of selected melilite minerals. M.S. thesis, Penn. State University, University Park, PA.

Scheetz, B.E. and White, W.B. (1977) Vibrational spectra of alkaline earth double carbonates. Am. Mineral., 62, 36-50.

Scott, J.F. (1968) Evidence of coupling between one- and two-phonon excitations in quartz. Phys. Rev. Letts., 21, 907-910.

Scott, J.F. (1969) Raman study of trigonal-cubic phase transitions in rare-earth aluminates. Phys. Rev., 183, 823-825.

Scott, J.F. (1970) Raman spectra of GeO$_2$. Phys. Rev. B, 1, 3488-3493.

Scott, J.F. (1974) Soft-mode spectroscopy: experimental studies of structural phase transitions. Rev. Mod. Phys., 46, 83-128.

Scott, J.F. and Porto, S.P.S. (1967) Longitudinal and transverse optical lattice vibrations in quartz. Phys. Rev., 161, 903-910.

Serna, C.J., Cortina, C.P. and Ramos, J.V.G. (1982) Infrared surface modes in corundum-type microcrystalline oxides. Spectrochim. Acta, 38, 797-802.

Servoin, J.L. and Piriou, B. (1973) Infrared reflectivity and Raman scattering of Mg$_2$SiO$_4$ single crystal. Physica Status Solidi, 55, 677-686.

Shapiro, S.M., O'Shea, D.C. and Cummins, H.Z. (1967) Raman scattering study of the alpha-beta phase transition in quartz. Phys. Rev. Letts., 19, 361-364.

Sharma, S.K. and Yoder, H.S. (1979) Structural study of glasses of akermanite, diopside and sodium melilite compositions by Raman spectroscopy. Carnegie Inst. Wash. Ybk. 78, 526-532.

Sharma, S.K. and Simons, B. (1981) Raman study of crystalline polymorphs and glasses of spodumene composition quenched from various pressures. Am. Mineral., 66, 118-126.

Sharma, S.K., Virgo, D. and Kushiro, I. (1979) Relationship between density, viscosity and structure of GeO$_2$ melts at low and high pressures. J. Non-Cryst. Solids, 33, 235-248.

Sharma, S.K., Simons, B. and Yoder, H.S. (1983) Raman study of anorthite, calcium Tschermak's pyroxene, and gehlenite in crystalline and glassy states. Am. Mineral., 68, 1113-1125.

Sharma, S.K., Mammone, J.F. and Nicol, M. (1981) Raman investigation of ring configurations in the structure of vitreous silica. Nature, 292, 714-715.

Sharma, S.K., Mao, H.-K., Bell, P.M. and Xu, J.A. (1985) Measurement of stress in diamond anvils with micro-Raman spectroscopy. J. Raman Spectroscopy, 16, 350-352.

She, C.Y., Masso, J.D. and Edwards, D.F. (1971) Raman scattering by polarization waves in uniaxial crystals. J. Phys. Chem. Solids, 32, 1887-1900.

Sherman, W.F. (1984) Infrared and Raman spectroscopy at high pressures. J. Mol. Struct., 113, 101-116.

Sherwood, P.M.A. (1972) "Vibrational Spectroscopy of Solids". Cambridge University Press.

Sheu, J.L., Welch, M., Graham, C.M. and McMillan, P. (1986) Vibrational spectra of tremolite-richterite amphiboles. EOS, 67, 1270-1271.

Shirane, G. and Yamada, Y. (1969) Lattice-dynamical study of the 110°K phase transition in SrTiO$_3$. Phys. Rev., 177, 177-863.

Siegman, A.E. (1971) "An Introduction to Lasers and Masers". McGraw-Hill, New York.

Simon, I. and McMahon, H.O. (1953) Study of the structure of quartz, cristobalite, and vitreous silica by reflection in infrared. J. Chem. Phys., 21, 23-30.

Smith, P.K. and Buseck, P.R. (1981) Graphitic carbon in the Allende meteorite: a microstructural study. Science, 212, 322-324.

Smyth, J.R. (1987) Beta-Mg_2SiO_4: a potential host for water in the mantle? Am. Mineral., in press.

Song, J.J., Chung, D.D.L., Eklund, P.C. and Dresselhaus, M.S. (1976) Raman scattering in graphite intercalation compounds. Solid State Comm., 20, 1111-1115.

Spitzer, W.G. and Kleinman, D.A. (1961) Infrared lattice bands of quartz. Phys. Rev., 121, 1324-1335.

Spitzer, W.G., Miller, R.C., Kleinman, D.A. and Howarth, L.E. (1962) Far infrared dielectric dispersion in $BaTiO_3$, $SrTiO_3$, and TiO_2. Phys. Rev., 126, 1710-1721.

Steel, W.H. (1983) "Interferometry". Cambridge University Press.

Steele, D. (1971) "Theory of Vibrational Spectroscopy". W.B. Saunders.

Stidham, H.D., Bates, J.B. and Finch, C.B. (1976) Vibrational spectra of synthetic single crystal tephroite, Mn_2SiO_4. J. Phys. Chem, 80, 1226-1234.

Stirling, W.G. (1972) Neutron inelastic scattering study of the lattice dynamics of strontium titanate: harmonic models. J. Phys. C: Solid State Phys., 5, 2711-2730.

von Stengel, M.O. (1977) Normalschwingungen von alkalifeldspaten. Z. Krist., 146, 1-18.

Strens, R.G.J. (1966) Infrared study of cation ordering and clustering in some (Fe,Mg) amphibole solid solutions. Chem. Comm., 519-520.

Strens, R.G.J. (1974) The common chain, ribbon, and ring silicates. In "The Infrared Spectra of Minerals", ed. V.C. Farmer. p. 305-330. Mineralogical Society, London.

Summitt, R. (1968) Infrared absorption in single-crystal stannic oxide: optical lattice-vibration modes. J. Appl. Phys., 39, 3762-3767.

Svelto, O. (1976) "Principles of Lasers". Heyden Press, London.

Tarte, P. (1963a) Applications nouvelles de la spectrometrie infrarouge a des problemes de cristallochimie. Silicates Industriels, 28, 345-354.

Tarte, P. (1963b) Etude infrarouge des orthosilicates et des orthogermanates-II. Structures du type olivine et monticellite. Spectrochim. Acta, 19, 25-47.

Tarte, P. and Deliens, M. (1973) Correlations between the infrared spectrum and the composition of garnets in the pyrope-almandine-spessartine series. Contrib. Mineral. Petrol., 40, 25-37.

Tarte, P., Pottier, M.J. and Proces, A.M. (1973) Vibrational studies of silicates and germanates-V. I.R. and Raman spectra of pyrosilicates and pyrogermanates with a linear bridge. Spectrochim. Acta, 29A, 1017-1027.

Thompson, P. and Grimes, N.W. (1978) Observation of low energy phonons in spinel. Solid State Comm. 25, 609-611.

Tobin, M.C. (1971) "Laser Raman Spectroscopy". Wiley-Interscience.

Toropov, N.A., Fedorov, N.F. and Shevyakov, A.M. (1963) Infrared absorption spectra of orthosilicates of some bivalent elements. Russ. J. Inorg. Chem., 8, 697-699.

Traylor, J.G., Smith, H.G., Nicklow, R.M. and Wilkinson, M.K. (1971) Lattice dynamics of rutile. Phys. Rev. B, 3, 3457-3472.

Tuddenham, W.M. and Lyon, R.J.P. (1960) Infrared techniques in the identification and measurement of minerals. Anal. Chem., 32, 1630-1634.

Tuinstra, F. and Koenig, J.L. (1970) Raman spectrum of graphite. J. Chem. Phys., 53, 1126-1130.

Turrell, G. (1972) "Infrared and Raman Spectra of Crystals". Academic Press.

Velde, B. (1978) Infrared spectra of synthetic micas in the series muscovite-MgAl celadonite. Am. Mineral., 63, 343-349.

Velde, B. (1979) Cation-apical oxygen vibrations in mica tetrahedra. Bull. Mineral., 102, 33-34.

Velde, B. (1980) Cell dimensions, polymorph type, and infrared spectra of synthetic white micas: the importance of ordering. Am. Mineral., 65, 1277-1282.

Velde, B. and Besson, J.M. (1981) Raman spectra of analcime under pressure. Phys. Chem. Minerals, 7, 96-99.

Velde, B. and Couty, R. (1985) Far infrared spectra of hydrous layer silicates. Phys. Chem. Minerals, 12, 347-352.

Vidine, D.W. (1980) Photoacoustic Fourier transform infrared spectroscopy of solid samples. Appl. Spectroscopy, 34, 314-319.

Vogt, H. (1986) Soft-mode spectroscopy in ferroelectrics by hyper-Raman scattering. In "Proc. Tenth Int. Conf. Raman Spectroscopy", eds. W.L. Peticolas and B. Hudson, pp. 11-5-11-6. University Printing Dept., University of Oregon.

Weng, K., Xu, J., Mao, H.K. and Bell, P.M. (1983) Preliminary Fourier-transform infrared spectra data on the SiO_6^{8-} octahedral group in silicate perovskite. Carnegie Inst. Wash. Yrbk., 82, 355-356.

White, W.B. (1974a) Order-disorder effects. In "The Infrared Spectra of Minerals", ed. V.C. Farmer. p. 87-110. Mineralogical Society, London.

White, W.B. (1974b) The carbonate minerals. In "The Infrared Spectra of Minerals", ed. V.C. Farmer. p. 227-284. Mineralogical Society, London.

White, W.B. (1975) Structural interpretation of lunar and terrestrial minerals by Raman spectroscopy. In "Infrared and Raman Spectroscopy of Lunar and Terrestrial Materials", ed. C. Karr, p. 325-358. Academic Press.

White, W.B. and DeAngelis, B.A. (1967) Interpretation of the vibrational spectra of spinels. Spectrochimica Acta, 23A, 985–995.

Williams, Q., Jeanloz, R. and Akaogi, M. (1986) Infrared vibrational spectra of beta-phase Mg_2SiO_4 and Co_2SiO_4 to pressures of 27 GPa. Phys. Chem. Minerals, 13, 141–145.

Williams, Q., Jeanloz, R. and McMillan, P. (1987) The vibrational spectrum of $MgSiO_3$–perovskite: zero pressure Raman and mid-infrared spectra to 27 GPa. J. Geophys. Res. B, 92, 8116–8128.

Wilson, E.B., Decius, J.C. and Cross, P.C. (1955) "Molecular Vibrations", McGraw-Hill, New York. Re-published in 1980 by Dover.

Wolf, G.H. and Jeanloz, R. (1985) Lattice dynamics and structural distortions of $CaSiO_3$ and $MgSiO_3$ perovskites. Geophys. Res. Letts., 12, 413–416.

Wolf, G. and Bukowinski, M.S.T. (1987a) Variational stabilization of the ionic charge densities in the electron-gas theory of crystals: applications to MgO and CaO. Phys. Chem. Minerals, in press.

Wolf, G.H. and Bukowinski, M.S.T. (1987b) Theoretical study of the structural properties and equations of state of $MgSiO_3$ and $CaSiO_3$ perovskites: implications for lower mantle composition. In U.S.-Japan Joint Seminar on High Pressure Research: Applications in Geophysics and Geochemistry, eds. M. Manghnani and S. Akimoto, in press.

Wong, P.T.T., Baudais, F.L. and Moffatt, D.J. (1986) Hydrostatic pressure effects on TO–LO splitting and softening of infrared active phonons in α–quartz. J. Chem. Phys., 84, 671–674.

Wood, B.J., Kirkpatrick, R.J. and Montez, B. (1986) Order- disorder phenomena in $MgAl_2O_4$ spinel. Am. Mineral., 71, 999–1006.

Woodward, L.A. (1972) "Introduction to the Theory of Molecular Vibrations and Vibrational Spectroscopy". Oxford University Press.

Wopenka, B. and Pasteris, J.D. (1986) Limitations to quantitative analysis of fluid inclusions in geological samples by laser Raman microprobe spectroscopy. Appl. Spectroscopy, 40, 144–151.

Yamamoto, A., Utida, T., Mirata, H. and Shiro, Y. (1975) Optically active vibrations and effective charges of dolomite. Spectrochim. Acta, 31A, 1265–1270.

Yamanaka, T. and Takeuchi, Y. (1983) Order/disorder transition in MgAl2O4 spinel at high temperatures up to $1700^{\circ}C$. Z. Krist., 165, 65–78.

Yamanaka, T. and Ishii, M. (1986) Raman scattering and lattice vibrations of Ni_2SiO_4 spinel at elevated temperature. Phys. Chem. Minerals, 13, 156–160.

Zulumyan, N.O., Mirgorodskii, A.P., Pavinich, V.F. and Lazarev, A.N. (1976) Study of calculation of the vibrational spectrum of a crystal with complex polyatomic anions. Diopside $CaMgSi_2O_6$. Optics and Spectroscopy, 41, 622–627.

Chapter 5

INELASTIC NEUTRON SCATTERING

INTRODUCTION

In crystals strong forces exist between neighboring atoms. Hence, if one atom is displaced from its mean equilibrium position, the neighboring atoms also undergo displacements. As a result, the atomic motions are collective rather than individual, which can be analyzed into a spectrum of normal modes of vibrations which travel as waves through the crystal. The energy of these waves is quantized. The "pseudoparticles" associated with these waves are known as "phonons", which carry quantums of energy $\hbar\omega$. Phonons play a very important role in determining the transport and thermodynamic properties of solids, such as electrical resistivity, thermal conductivity, superconductivity and specific heat. Neutrons are ideally suited for the study of phonons, as well as the spin waves (magnons) in magnetic materials.

Basic properties of the neutron

Neutron is a subatomic particle with zero charge, mass m = 1.0087 atomic mass units, spin $\frac{1}{2}$, and magnetic moment, μ_n = -1.9132 nuclear magnetons. These four properties make neutron a very effective tool for the study of condensed matter. The zero mass means the neutron has great penetrability in bulk samples with very little absorption. Depending on their energies, neutrons are classified as ultracold (0.00025 meV), cold (1 meV), thermal (25 meV) and epithermal (1000 meV).

The thermal neutrons are usually obtained by slowing down high energy neutrons obtained from a steady state (reactor) or a pulsed (spallation) source by inelastic collisions in a moderating material containing light atoms. Their characteristics are listed in Table 1.

Table 1. Physical characteristics of thermal neutrons

Quantity	Unit	Definition	Thermal
Energy, E	meV	$\hbar^2 k^2/2m$	25
Temperature, T	K	E/k_B	290
Wavelength, λ	Å	$(h^2/2mE)^{\frac{1}{2}}$	1.8
Wavevector, $\underline{k}$	Å^{-1}	$(2mE)^{\frac{1}{2}}/\hbar$	3.5
Velocity, v	m/s	$(2E/m)^{\frac{1}{2}}$	2200
Frequency, ν	THz	E/h	6.045
Angular frequency, ω	rad/s	$E/\hbar$	3.98×10^{13}
Wavenumber, $\bar{\nu}$	cm^{-1}	E/hc	201.64

Note: 1 meV$\equiv$0.24 THz$\equiv$8.07 cm^{-1}$\equiv$11.71 K; 1 THz$\equiv$10^{12} s^{-1}$\equiv$4.14 meV.

The thermal neutrons (~25 meV) have a wavelength (~1.8 Å) which is comparable to the interatomic spacings in solids. Because the neutron scattering cross-section is independent of the atomic number (unlike x-rays), elastic scattering (neutron diffraction) is extensively used for the location of light elements (hydrogen, deuterium, etc.) in the presence of heavy elements (lead, uranium, etc.) in crystal structures.

The kinetic energy of the thermal neutrons is comparable to those associated with lattice vibrational modes in crystals. This property along with the suitable wavelength make the neutron a powerful tool to probe the detailed lattice dynamics of solids by inelastic scattering. In this respect, inelastic neutron scattering is more advantageous than optical spectroscopy (IR, Raman), because with neutrons the phonon frequencies throughout the Brillouin zone (unit cell in reciprocal space), i.e., phonon dispersion relations can be measured, whereas because of the large wavelengths of light, optical techniques are capable of measuring the phonon frequencies only at the center of the Brillouin zone. In principle, x-rays can also be used to study the phonons in solids, but because of the high energy (~10^3 eV), the energy change associated with the creation or annihilation of a phonon is very small, and is very difficult to detect.

The magnetic moment associated with the neutron means that neutrons are scattered from the magnetic moments associated with transition metal ions with unpaired spins in magnetic samples. Hence, the neutron is a unique tool for the study of spin arrangement (magnetic structure) by elastic scattering (neutron diffraction) as well as spin dynamics by inelastic scattering in magnetically ordered materials.

Neutron scattering cross-section

The neutron scattering cross section, σ, for atoms in a sample which are identical and noninteracting is given by

$$\frac{d\sigma}{d\Omega} = |\underline{b}|^2 \text{ and the total cross section } \sigma = 4\pi|\underline{b}|^2, \tag{1}$$

where $d\Omega$ is the solid angle subtended by the scattered neutrons; $\underline{b}$ is the scattering length, which depends on the atomic number, atomic weight and its spin state relative to that of the neutron. In many cases, the scattering lengths for two spin states $I = \pm\frac{1}{2}$ of the neutron-nucleus system are quite different, giving rise to the "incoherent" scattering. The total scattering cross section σ for the nucleus is the sum of the two terms due to spin combinations $I+\frac{1}{2}$ and $I-\frac{1}{2}$, so that

$$\sigma = \sigma_{coh} + \sigma_{inc} . \tag{2}$$

For hydrogen (proton spin $I = \pm\frac{1}{2}$), the very large incoherent scattering cross section, $\sigma_{inc} = 79.1$ barns (10^{-24} cm^2) far outweighs the coherent scattering cross section, $\sigma_{coh} = 2$ barns. Neutron scattering cross sections for the different elements and their isotopes are listed by Bacon (1975).

Neutron sources

Sources for high flux neutron beams suitable for neutron scattering studies are primarily of two types: (a) steady state sources from nuclear fission reactors, and (b) pulsed spallation sources produced by an accelerated beam of protons or electrons impinging on a heavy metal target, such as U, Pb, W, etc. The high flux reactors designed in the mid-1960s and using water (or D_2O) as coolant have practically reached the limits of feasible values of surface heat flux on the fuel plate. A new high flux reactor at

Oak Ridge National Laboratory using D_2O as coolant and reflector is planned to produce a flux of 4×10^{15} neutrons/cm^2s, which is about the highest we can expect to reach with existing materials and technology. In contrast to the steady sources, pulsed sources operate briefly at high power producing momentarily high flux, then turn off while the heat is being dissipated. Hence, pulsed sources are capable of producing higher fluxes for short times (~1 μs) than those from a nuclear reactor. The currently operating high intensity reactor and pulsed (spallation) sources are listed in Table 2 (Carpenter and Yelon, 1986). The pulsed sources have a component of higher energy neutrons above 100 meV, which are more suitable for a global determination of the phonon spectrum (phonon density of states).

Applications of inelastic neutron scattering

The collective atomic or spin motions in a solid are studied by the <u>coherent</u> inelastic neutron scattering, whereas the individual motions (e.g., water molecules in a zeolite) can be studied by the <u>incoherent</u> neutron scattering. The latter technique can also be used to sample the entire spectrum of the collective atomic motions, viewed as motions of individual atoms.

The coherent neutron scattering technique can be used to determine:

(a) <u>Phonon dispersion relations</u> in a crystal, i.e., the dependence of the angular frequency, ω on the wavevector q, which is numerically equal to $2\pi/\lambda$ for any direction in a crystal. Normally these results are shown in reciprocal space, a unit cell of which is known as the "Brillouin zone".

(b) <u>Phonon density of states</u> which is the frequency distribution function $g(\omega)$, where $g(\omega)d\omega$ is the number of phonons with frequencies lying within the interval from ω to $\omega+d\omega$. The function $g(\omega)$ is what determines the specific heat.

(c) Temperature and pressure dependences of phonon frequency, which yield the <u>mode Grüneisen parameters</u>.

(d) <u>Soft modes</u>, i.e., phonon modes which show a strong frequency dependence on temperature/pressure near a second order phase transition (see Ghose, 1985).

(e) Phonon intensity, peak shapes and lifetimes as a function of temperature/pressure, which give valuable information on the <u>dynamical structure factor, anharmonicity and phonon-phonon interaction</u>.

(f) <u>Electron-phonon interactions</u> (Kohn anomalies) in metals and semiconductors.

(g) <u>Energy level separation in transition metal ions</u> in magnetic materials.

(h) <u>Magnon (spin wave) dispersion relations</u> and magnon density of states in ferro-, ferri- and antiferromagnetic materials.

Table 2a. Highest-flux reactors operating in U.S. and Europe

Reactor	Power (MW)	Ambient moderator Flux (n/cm²s)	Cold sources	Coolant	Reflector	Hot sources	Year
HFBR (Brookhaven)	60	1×10^{15}	L-H_2	D_2O	D_2O	None	1960
HFIR (Oak Ridge)	100	1.5×10^{15}	None	H_2O	Be	None	1960
HFR (Ill, Grenoble)	57	1.2×10^{15}	L-D_2(2)	D_2O	D_2O	2000 K graphite	1972
HFIR II (Oak Ridge)	200	4.0×10^{15}	Not determined	D_2O	D_2O	Not determined	1990(?)

Table 2b. Pulsed neutron spallation sources

Facility	Characteristics	Average fast-neutron production rate (n/s)	Source pulse width (µs)	Pulse repetition frequency (Hz)	Year
	Spallation sources				
IPNS (Argonne, USA)	12-µA, 500-MeV protons, U target	1.5×10^{15}	0.1	30	1981
SNS (Rutherford, UK)	800-MeV protons, U target	4.0×10^{16}	0.27	50	1984
LAMPF-WNR (Los Alamos, USA)	800-MeV protons, W target	$(1-2) \times 10^{15}$	~5	120	1977
WNR/PSR (Los Alamos, USA)	100-µA, 800-MeV protons, W target	1.2×10^{16}	0.27	12	1985
KENS (Tsukuba, Japan)	500-MeV protons, U target	3.0×10^{14}	0.07	15	1980

Quasi-elastic (incoherent) scattering can be used to determine diffusive motions of hydrogen and other organic molecules in solids (e.g., zeolites).

LATTICE DYNAMICS AND NEUTRON SCATTERING

Basic concepts of lattice dynamics

The basic concepts of lattice dynamics can be defined by considering small oscillations around the equilibrium atomic positions, known as the harmonic approximation. We also assume that the electronic wave functions change adiabatically during nuclear motions (Born-Oppenheimer approximation). Any collective atomic motion can be described as a superposition of normal vibrational modes. In a solid, the number of normal modes is equal to the number of vibrational degrees of freedom of the system, $3N - 6 \approx 3N$, since the number of atoms N in this case is very large. The frequencies of normal modes can be determined by diagonalizing the force-constant matrix (see below). For amorphous solids, this involves diagonalizing a $3N \times 3N$ force constant matrix, which is a formidable problem unless N is a tractable number in terms of computer space. On the other hand, for a crystalline solid, the presence of translational symmetry requires that the displacements in different unit cells must be equivalent. This means that the normal modes are propagating vibrational waves with wavevectors q determined by the size and symmetry of the unit cell. The problem reduces to that of determining the motions of r particles in the unit cell; for each value of q, the (circular) frequencies $\omega_j(q)$ are determined by diagonalizing the $3r \times 3r$ force constant matrix. The relation

$$\omega = \omega_j(q) \qquad (j = 1,2,\ldots r) \qquad (3)$$

between the frequency and the wavevector is known as the dispersion relation. The index j, which distinguishes the various frequencies corresponding to the same propagation vector, characterizes the various branches of the dispersion relation. The pattern of motion of the nuclei in the cell is determined by the polarization vectors e_d^j (d = 1,2,...r). A certain branch of the dispersion relation is called longitudinal or transverse if the polarization vectors are parallel or perpendicular to q, respectively.

Within the harmonic approximation, the vibrational modes of a crystal can be expressed as (see Born and Huang, 1954)

$$\frac{1}{\sqrt{NM_d}} A_j(q)e_d^j(q) \exp[i(q \cdot \underline{\ell} - \omega_j(q)t)] \quad , \qquad (4)$$

where $A_j(q)$ = amplitude, which determines the amount of energy carried by this phonon and $e_d^j(q)$ = phonon eigenvector. In the factor $1/\sqrt{NM_d}$, N is the number of unit cells in the crystal and M_d is the mass of the dth atom in the ℓth unit cell.

Low frequency vibrations in a solid are the sound waves with a linear dispersion relation

$$\omega = v_j q \qquad (j = 1,2,3) \qquad , \qquad (5)$$

where v_j is the velocity of sound. Hence, three phonon branches, whose frequencies approach zero as $q \to o$, are known as <u>acoustic branches</u>. The remaining branches are known as <u>optic</u> branches, because some of these modes interact strongly with light.

To determine the allowed values of q, we adopt the <u>periodic boundary conditions</u>, which require that for a parallelopiped with edges $N_1 a_1$, $N_2 a_2$, $N_3 a_3$ ($N_1 N_2 N_3 = N$ is the number of primitive unit cells; a_1, a_2, a_3 = cell edges), the atomic displacements be periodic with periods $N_i a_i$ ($i = 1,2,3$). These conditions are satisfied, if

$$q = \frac{h_1}{N_1} b_1 + \frac{h_2}{N_2} b_2 + \frac{h_3}{N_3} b_3 \qquad (6)$$

where h_1, h_2 and h_3 are integers. Because of translational symmetry, the atomic displacements and frequencies $\omega_j(q)$ are periodic functions of q with the period of the reciprocal lattice

$$\omega_j(q + \tau) = \omega_j(q) \qquad (7)$$

$$e_d^j(q + \tau) = e_d^j(q) \quad , \qquad (8)$$

where τ is a reciprocal lattice vector.

Hence, all physically distinct values of q can be obtained by restricting the allowed values of q in one of the primitive cells of the reciprocal lattice, known as the <u>Brillouin zone</u>. The first Brillouin zone is defined by planes which are perpendicular bisectors of lines joining a reciprocal lattice point to its neighboring points. The Brillouin zone for the face centered cubic lattice is shown in Figure 1, where the symmetry points and lines are labelled following the usual convention (e.g., Δ for the [100] direction, Λ for the [111] direction, Γ for the zone center, etc.).

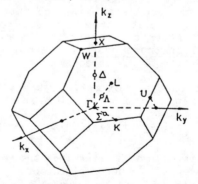

Figure 1. Brillouin zone for a fcc lattice. The symmetry directions are indicated (Stassis, 1986).

The energy $E_j(q)$ and momentum p of the phonon are related to the frequency and wavevector of the corresponding vibrational wave by

$$E_j(\underline{q}) = \hbar\omega_j(\underline{q}) \tag{9}$$

$$\underline{p} = \hbar\underline{q} \quad (\hbar\text{=Planck's constant}/2\pi) \tag{10}$$

The harmonic approximation implies a completely free motion of noninteracting phonons. However, anharmonic interactions are always present and elastic (and inelastic) collisions between the phonons occur, which provide a mechanism for the thermal equilibrium of the phonon gas. The life time of the phonons is thus finite and is determined by the phonon-phonon, as well as phonon-electron, phonon-photon and other elementary exciations in the solid. Since any number of identical phonons can be excited at any given time and this number is determined by the equilibrium condition, the phonon gas obeys Bose-Einstein statistics. Thus, the average number n_j of phonons in a given state at thermal equilibrium is given by

$$n_j = \frac{1}{\exp(E_j/k_BT)-1} \quad , \tag{11}$$

where k_B is the Boltzmann's constant and T is the temperature.

At high temperature ($E_j/k_BT \ll 1$), the number of phonons in a given state is proportional to the temperature and inversely proportional to their energy.

For the calculation of the thermodynamic properties of solids, such as the specific heat, we have to evaluate the frequency distribution or phonon density of states, $g(\omega)$:

$$g(\omega) = \frac{1}{3N} \sum_{j,\underline{q}} \delta[\omega - \omega_j(\underline{q})] \quad . \tag{12}$$

The frequency distribution is normalized, such that

$$\int_0^\infty g(\omega)d\omega = 1 \quad . \tag{13}$$

Since the allowed values of $\underline{q}$ are closely spaced, $g(\omega)$ can also be expressed as

$$g(\omega) = \frac{V}{3N(2\pi)^3} \sum_j \int \frac{dS}{|\nabla\omega_j(\underline{q})|} \quad , \tag{14}$$

where dS is an element of the constant frequency surface corresponding to the frequency ω for the jth branch, and V is the volume of the unit cell.

Analysis for the phonon dispersion curves: Phenomenological models

The experimental determination of phonon dispersion curves and phonon density of states with examples will be described in sections below. In this section, we present a simple model based on Born-von Kármán formalism for the analysis of these data.

In the harmonic approximation, the effective Hamiltonian can be written as

$$H = T + \frac{1}{2} \sum_{\substack{\ell d \alpha \\ \ell' d' \beta}} \Phi^{\ell, \ell'}_{d\alpha, d'\beta} u^{\ell}_{d\alpha} u^{\ell'}_{d'\beta} \tag{15}$$

where, T = kinetic energy, $u^{\ell}_{d\alpha}$ = displacement of the dth atom in the ℓth cell from its equilibrium position along the α-axis, and force constant,

$$\Phi^{\ell, \ell'}_{d\alpha, d'\beta} = \Phi^{\ell', \ell}_{d'\beta, d\alpha} = \frac{\partial^2 \Phi}{\partial u^{\ell}_{d\alpha} \partial u^{\ell'}_{d'\beta}} \tag{16}$$

are evaluated at equilibrium position. The equations of motion can be obtained using Hamilton's equations, where M_d is the mass of the dth atom and $\ddot{u}^{\ell}_d$ is the double derivative with respect to time.

$$M_d \ddot{u}^{\ell}_{d\alpha} = - \sum_{\ell' d' \beta} \Phi^{\ell, \ell'}_{d\alpha, d'\beta} u^{\ell'}_{d'\beta} \quad . \tag{17}$$

From this equation, it is clear that $-\Phi^{\ell, \ell}_{d\alpha, d'\beta}$ is simply the force on the atom ℓ, d along the α axis due to unit displacement of the atom ℓ', d' along the β axis. The <u>force constants</u> $\Phi^{\ell}_{d\alpha, d\beta}$ are given in units of dynes/cm.

The translational symmetry requires that

$$\sum_{\ell' d'} \Phi^{\ell, \ell'}_{d\alpha, d'\beta} = 0 \quad . \tag{18}$$

The crystal symmetry puts additional restrictions on the force constants, which can be analyzed using Group theory. If $\underset{\approx}{S}$ is the orthogonal matrix corresponding to a symmetry element and $\underset{\approx}{\tilde{S}}$ is its transpose,

$$\Phi^{\ell, \ell'}_{d\alpha, d'\beta} = \underset{\approx}{\tilde{S}} \Phi^{\ell_1, \ell'_1}_{d_1, d'_1} \underset{\approx}{S} \quad . \tag{19}$$

These relations considerably reduce the independent force constants needed in a particular force constant model.

After the force constant matrices are determined or parametrized, the phonon frequencies are given by

$$\omega^2(\underline{q}) e_{d\alpha}(\underline{q}) = \sum_{d'\beta} D_{d\alpha, d'\beta}(\underline{q}) e_{d'\beta}(\underline{q}) \tag{20}$$

or in matrix form

$$\omega^2(\underline{q}) \underset{\approx}{e}(\underline{q}) = \underset{\approx}{D}(\underline{q}) \underset{\approx}{e}(\underline{q}) \quad . \tag{21}$$

The <u>dynamical matrix</u> $\underset{\approx}{D}$ is a 3rx3r matrix, and $\underset{\approx}{e}(q)$ is a 3r-component column matrix. The matrix elements of the dynamical matrix are

$$D_{d\alpha, d'\beta}(\underline{q}) = \frac{1}{\sqrt{M_d M_{d'}}} \sum_{\ell} \Phi^{0, \ell}_{d\alpha, d'\beta} e^{i \underline{q} \cdot \underline{\ell}} \quad . \tag{22}$$

The diagonalization of the dynamical matrix yields the frequencies of the vibrational waves of wavevector $\underline{q}$. Because the dynamical matrix is hermitian, the eigen-frequencies are real and the eigenvectors can be chosen as orthonormal.

Phonon dispersion relations: Theory of inelastic neutron scattering

The scattering of a monochromatic beam of neutrons by a crystal is described by the scattering cross section σ, which consists of an incoherent and a coherent part. The coherent one phonon scattering of neutrons from a single crystal involves both the conservation of momentum and energy and is the most powerful technique for the detailed investigation of the lattice dynamics of single crystals.

Coherent one phonon inelastic scattering. The coherent (energy E- and angle Ω- dependent) cross section for scattering of the neutron due to the creation or annihilation of a single phonon of frequency $\omega_j(q)$ is given by

$$\frac{d^2\sigma_{coh}}{d\Omega dE}(\underline{Q},\omega) = \frac{k_1}{k_0} \frac{(2\pi)^3}{2V} \frac{[n_j(\underline{q}) + \frac{1}{2} \pm \frac{1}{2}]}{\omega_j(\underline{q})} |F_j(\underline{Q})|^2 \delta[E_1 - E_0 \pm \hbar \omega_j(\underline{q})] \cdot$$
$$\delta(\underline{Q} \mp \underline{q} - \underline{\tau}) \qquad (23)$$

Here, the scattering vector $\underline{Q} = \underline{k}_1 - \underline{k}_0$, where $\underline{k}_1$ and $\underline{k}_0$ are scattered and incident neutron wavevectors and $[n_j(\underline{q})+\frac{1}{2} \pm \frac{1}{2}]$ is the population factor; the upper and lower + and - signs correspond to phonon creation and phonon annihilation respectively.

The delta functions indicate conservation of energy and momentum during the scattering process, i.e.,

$$\underline{Q} = \pm \underline{q} + \underline{\tau} \qquad (24)$$

$$E_1 - E_0 = \mp \hbar\omega_j(\underline{q}) \quad . \qquad (25)$$

These two conditions enable the phonon dispersion relation $\omega_j(\underline{q})$ to be measured. The vector representation of the momentum conservation relation is shown in Figure 2.

The inelastic structure factor $F_j(\underline{Q})$ is given by

$$F_j(\underline{Q}) = \sum_d M_d^{-\frac{1}{2}} \bar{b}_d \underline{e}_d^j(\underline{q}) \cdot \underline{Q} \ e^{i\underline{Q} \cdot \underline{d}} \ e^{-W_d} \qquad (26)$$

The sum extends over all the atoms of the unit cell. Here, M_d = mass of the dth atom, $\bar{b}_d$ = neutron scattering length averaged over different isotopes, $\underline{d}$ = position vector, and, e^{-2W_d} = Debye-Waller factor of the dth atom.

The dynamical structure factor can be calculated from polarization vectors based on a lattice dynamical model. A prior knowledge of $F_j(Q)$ helps considerably in the suitable choice of the regions of reciprocal space for experimental observations of phonons and their labelling.

Anharmonic effects. The most important consequences of anharmonicity are thermal expansion and thermal resistivity. Hence, the harmonic approximation is not truly valid even at room temperature. The phonon "self-energy" Π contains a frequency shift Δ and a line broadening ωΓ given by:

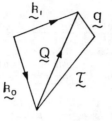

$\underset{\sim}{k_0}$ incident neutron wave vector

$\underset{\sim}{k_1}$ scattered neutron wave vector

$\underset{\sim}{Q}$ scattering vector

$\underset{\sim}{q}$ phonon wave vector

$\underset{\sim}{\tau}$ reciprocal lattice vector

Figure 2. Momentum conservation relation in a coherent inelastic neutron scattering experiment on a single crystal.

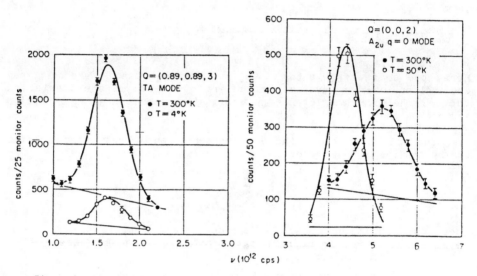

Figure 3. Line widths of two phonon modes in rutile, TiO_2 at two temperatures (Traylor et al., 1971).

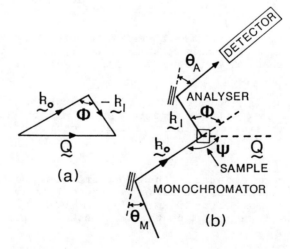

Figure 4. (a) Wavevector diagram showing the scattering vector $\underset{\sim}{Q} = \underset{\sim}{k_1} - \underset{\sim}{k_0}$. (b) Scheme of the scattering geometry on a triple-axis spectrometer (Rao, 1978).

$$\Pi(q_j,T) = \Delta(q_j,T) - i\omega\Gamma(q_j,T) \quad . \tag{27}$$

Usually, the anharmonic effects are large only for some of the phonons in a crystal. Soft modes which are detected close to a second order phase transition invariably show large anharmonicity. Typical temperature dependent line shapes for rutile, TiO_2 at two temperatures are shown in Figure 3 (Traylor et al., 1971). Note that for one mode, the phonon intensity changes, but the frequency remains constant, whereas for the other mode both frequency and intensity change as a function of temperature. The frequency dependence on temperature (or pressure) gives us the mode Grüneisen parameter.

Coherent inelastic neutron spectroscopy

In this technique, scattered neutron groups are observed whenever the energy and momentum conservation laws are simultaneously satisfied at certain ω and q. Neutron spectrometers are designed to measure, E_0 and E_1, the incident and scattered neutron energies and the angles Φ and ψ, which are the scattering angle and the angle defining the orientation of the single crystal respectively, such that ω and q corresponding to the center of the scattered neutron group can be determined.

The wavelength or velocity of the neutrons can be determined by one of two ways:

(a) Bragg scattering from a single crystal used as a monochromator or analyzer (graphite, beryllium, germanium, etc.) from the relation

$$\lambda = 2d_{hk\ell}\sin\theta \quad . \tag{28}$$

(b) Time-of-flight method, where the transit time of neutrons over a finite flight path is electronically measured, from which the neutron velocity can be determined.

Triple-axis spectrometer. For phonon measurements, the crystal spectrometer based on the Bragg law is more convenient and is widely used. In the momentum conservation relation $Q = k_1 - k_0$, the neutron wavevectors, k_1 and k_0 form a closed triangle (Fig. 4a). The corresponding scattering geometry is shown in Figure 4b. By changing the monochromator- and the analyzer-angles, θ_M and θ_A, k_0 and k_1 can be varied. The instrument known as the triple axis spectrometer first designed by Brockhouse (1961) allows variable settings for the angles $\theta_M, 2\theta_M$, ϕ, ψ, θ_A and $2\theta_A$. Figure 5 shows the details of a triple-axis spectrometer, which allows the measurement of the neutron group corresponding to a point on the dispersion relation, ω-q. Two methods are commonly used for this purpose:

(a) Constant Q method, where the momentum transfer is constant, and (b) Constant E method, where the energy transfer is constant.

An example of a measurement of a phonon peak by the constant Q method is illustrated in Figure 6. A low intensity counter (usually a [235]U fission chamber) placed before the sample is used to monitor the

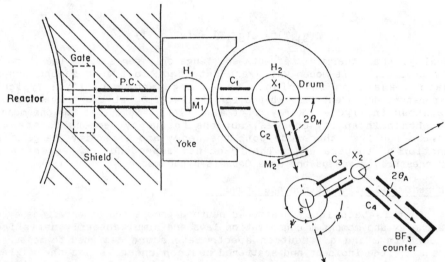

Figure 5. The triple-axis spectrometer. M1, M2: Moderators, C1, C2, C3, C4: Collimators, x1: monochromator, S: sample crystal, x2: analyzer (Brockhouse, 1961).

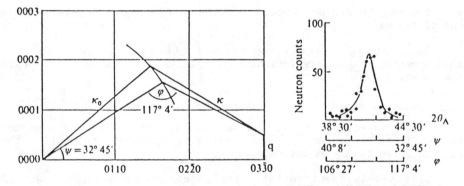

Figure 6. Observation of phonons by the constant Q method in magnesium. The reciprocal lattice construction is shown on the left and the corresponding phonon peak on the right (Iyenger, 1965).

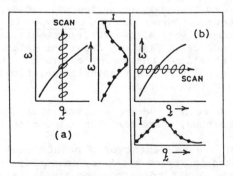

Figure 7. The resolution ellipsoid and focusing of phonon groups during constant Q and constant ω (i.e., E) scans (Rao, 1978).

neutron flux incident on the sample. The counting times are controlled by the monitor counter to compensate for any changes in the incident neutron flux during the experiment. The signal counter is usually a high efficiency $^{10}BF_3$ gas detector, although ^3He gas detectors are also used in some installations.

The overall resolution of the instrument is determined by the collimators, C_1, C_2, C_3, C_4 (Fig. 5) and the mosaic spreads of the monochromator, analyzer and sample crystals. Computer programs for the evaluation of the resolution function are available at all neutron scattering centers. The resolution function can be expressed as a resolution ellipsoid, which is typically quite elongated along $\Delta\omega$, and more elongated along $\Delta Q_{\parallel}$ than along $\Delta Q_{\perp}$. The resolution ellipsoid is responsible for the focusing characteristics of the spectrometer. The orientation of the long axes of the ellipsoid with the dispersion surface determines the widths of the observed neutron groups - the sharpest focusing is obtained when the long axes are parallel to the dispersion surface (Fig. 7) (Rao, 1978).

Experimental results: Phonon dispersion relations in forsterite, Mg_2SiO_4. Forsterite is orthorhombic with space group Pnma ($\underline{a}$ = 10.1902, $\underline{b}$ = 5.9783 and $\underline{c}$ = 47534 A and four formula units per unit cell). The crystal structure consists of two types of edge-sharing $[MgO_6]$ octahedra forming chains, cross linked by isolated $[SiO_4]$ tetrahedra (Fig. 8). Because of the complexity of the crystal structure (28 atoms in the unit cell), a rigid "molecular" ion model was adopted for the lattice dynamical calculations, in which the "molecule" (here the $[SiO_4]$ group) is associated with six degrees of freedom, three translational and three rotational, and two magnesium atoms (Mg1 and Mg2) are associated with three degrees of translational freedom each. The potential energy of the crystal was assumed to contain two body potentials of the form:

$$V(r) = \frac{e^2}{4\pi\varepsilon_0} \frac{Z(Kk)Z(K'k')}{r} + a \exp\left\{-\frac{br}{R(Kk) + R(K'k')}\right\} \qquad (29)$$

where Z_k and R_k are "effective charge" and "effective radius" of the kth ion, and K is the rigid unit, r is the interatomic separation, e is the electron charge and ε_0 is the vacuum permittivity.

Constants a and b were set equal to 18.22 eV and 8.5 respectively. Z_k values were adjusted to maintain charge neutrality. The effective charges and radii were treated as adjustable parameters, which were optimized with respect to the known crystal structure and elastic constants (Rao et al., 1987, 1988). The optimized values used for the phonon dispersion calculations are given in Table 3.

Table 3. *Effective ionic charges and radii used in the rigid 'molecular' ion model calculations*

	Mg1	Mg2	Si	O1	O2	O3
Charge	1.60	1.80	1.0	−1.2	−1.0	−1.10
Radius (Å)	1.15	1.185	0.685	1.062	0.993	1.030

The phonon dispersion relations for the external modes calculated along the three principal symmetry directions are shown in Figure 9

174

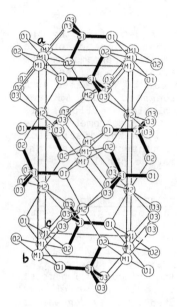

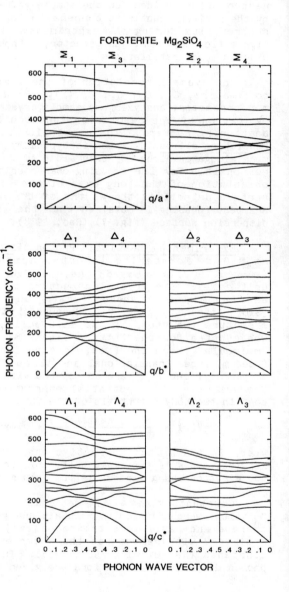

Figure 8 (left). Crystal structure of forsterite, Mg_2SiO_4 (Matsui and Busing, 1984).

Figure 9 (right). Phonon dispersion relations (external modes) in forsterite, Mg_2SiO_4 calculated from a rigid molecular ion model (Rao et al., 1988).

(Rao et al., 1988). One phonon dynamical structure factors and their polarizations were also calculated using the rigid "molecular" ion model and were used as guides for assigning the observed phonon peaks during the inelastic scattering experiment.

Phonon dispersion relations in a single crystal of synthetic forsterite (3.4x1x0.6 cm) were measured at room temperature using a triple axis spectometer in the constant Q mode at the High Flux Beam Reactor of the Brookhaven National Laboratory (Ghose et al., 1987; Rao et al., 1988). A (111)Ge crystal was used as a monochromator and (004) pyrolytic graphite as analyzer. Two different neutron incident energies of 45.30 and 60 meV were used during the phonon scans. Over 150 phonon measurements were made along the three principal symmetry directions in the yx, yz and xz planes. Some of the selected acoustic and optic phonon scans are shown in Figure 10.

The calculated (lines) and measured (circles) phonon branches along the three principal symmetry directions Σ, Δ, Λ in the Brillouin zone are shown in Figure 11. Eight zone center optic phonons were observed. The frequencies of the zone center optic phonons measured by inelastic neutron scattering are compared with those measured previously by IR and Raman scattering in Table 4. The optic phonon at 104 cm^{-1} has been observed for the first time, which is both IR and Raman inactive. It is heartening to note that in spite of the simplicity of the lattice dynamical model used, the agreement between the calculated and experimentally measured phonon branches is very satisfactory. No doubt in the future more sophisticated (ab initio quantum mechanical) models will be used for the lattice dynamical calculations of forsterite. Nonetheless, the experimentally measured phonon frequencies and phonon intensities will serve as valuable tests for such future models. In the meantime, we should note that the availability of the calculated neutron cross sections along the symmetry directions in the Brillouin zone for the phonon branches were extremely helpful as guides during the inelastic scattering experiments and resulted in considerable saving of very expensive neutron beam time.

Table 4. *Frequencies of zone center phonons* (cm^{-1}) *measured by inelastic neutron scattering compared with Raman and infrared measurements under ambient conditions*

INS	Raman	IR
104	–	–
144	142	144
184	183	–
192	192	–
200	–	201
258	260	–
315	318	313
325	324	323

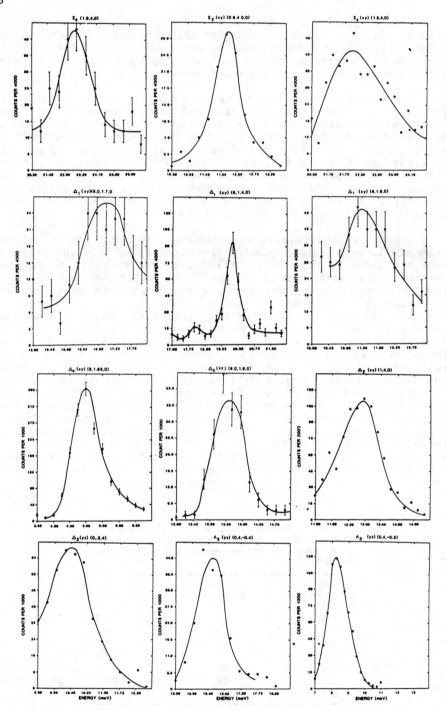

Figure 10. Selected acoustic and optic phonons in forsterite measured by inelastic neutron scattering (Ghose et al., 1987).

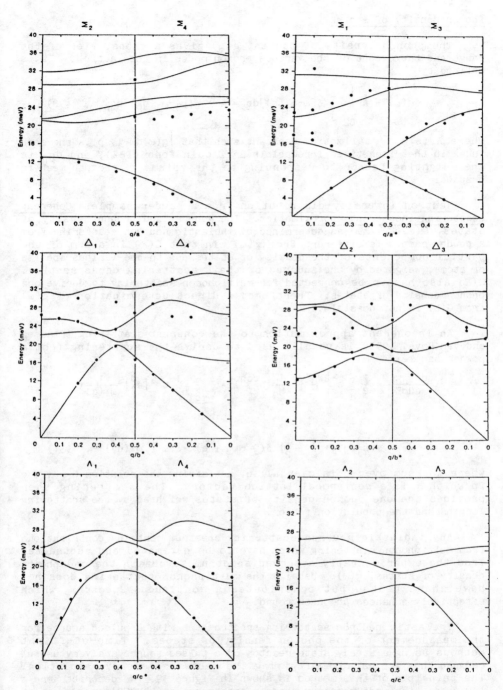

Figure 11. Phonon dispersion relations in forsterite. The full lines are calculated from rigid "molecular" ion model; the filled circles represent measurements by inelastic neutron scattering (Ghose et al., 1987).

Phonon density of states

The phonon density of states, $g(\omega)$ gives a global view of the range and the extent of the various phonon modes in the lattice. It is defined by

$$g(\omega) = A \int_{BZ} \sum_j \delta[\omega-\omega_j(q)]dq = A \sum_{jp} \delta[\omega-\omega_j(q_p)]dq_p \quad , \quad (30)$$

where A is a normalization constant such that $\int g(\omega)d\omega=1$, p is the mesh index in the discretised Irreducible Brillouin Zone (IBZ), and dq_p is the weighting factor corresponding to the volume of the pth mesh in q-space.

Neutron intensity distribution from a powder sample: Coherent scattering-incoherent approximation. The generalized phonon density of states, $G(E)(E = \hbar\omega)$ as determined by inelastic neutron scattering from a powder sample is different from $g(\omega)$, in that $G(E)$ is a sum of the partial components of the density of states due to the various species of atoms, weighted by the squares of their scattering cross section. $G(E)$ also has to be corrected for multiphonon scattering to derive one phonon density of states. Therefore, a direct determination of $g(\omega)$ from $G(E)$ is not possible.

An incoherent approximation to the coherent scattering (Placzek and Van Hove, 1955) is usually used to derive the scattering from a powder sample:

$$\frac{d^2\sigma_{coh}}{d\Omega dE} = \frac{1}{2} \sum_d e^{-2W_d(Q)} \frac{1}{M_d}(b_d^{coh})^2 \cdot \sum_{\tau g j} \frac{1}{3} Q^2 |e_j^d|^2 \frac{1}{\omega_j(q)} \cdot$$

$$\{n_j(q) + \frac{1}{2} \pm \frac{1}{2}\} \delta(Q - \tau \pm q) \cdot \delta\{\omega \pm \omega_j(q)\} \quad , \quad (31)$$

where the sum over q is over all q, satisfying the relation $Q = \tau \pm q$ in which τ is a reciprocal lattice vector. The scattering then provides the one phonon density of states weighted by the scattering lengths and the population factor.

Any inelastic neutron scattering spectrum contains contributions from multiphonon scattering which have to be estimated and subtracted from the experimentally observed spectrum to obtain the one-phonon density of states, $g(\omega)$. Usually the multiphonon scattering does not have sharp peaks, but contributes a continuous spectrum, which effectively enhances the background.

Inelastic neutron scattering spectrometer with a pulsed source for the measurement of the phonon density of states. Time-of-flight methods using a steady state reactor or a pulsed source are very useful because they are capable of giving the necessary energy resolution. The principle of this method is shown in Figure 12. We describe here a low resolution medium energy chopper spectrometer (LRMECS) operating at the Argonne National Laboratory using the Intense Pulsed Neutron Source (Fig. 13). The chopper is phased to the source and gives pulses of

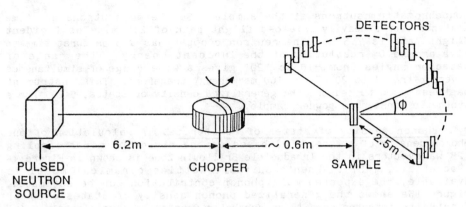

Figure 12. Geometry of the inelastic chopper spectrometer with a pulsed monochromatic beam at the sample. The scattered neutron energy is measured by the time-of-flight method using multi-detectors (Rao et al., 1988).

IPNS Low Resolution Medium Energy Chopper Spectrometer

[Courtesy of D.L. Price, 1988]

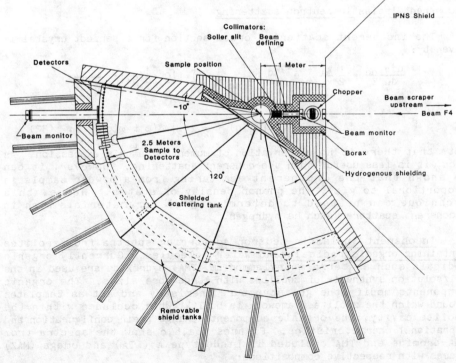

Figure 13. The low resolution medium-energy chopper spectrometer (LRMECS) at the Intense Pulsed Neutron Source (IPNS) at Argonne National Laboratory.

monochromatic neutrons at the sample. Scattered neutrons are time analyzed over a heavily shielded flight path of 2.5 m. The incident flight path is 7.5 m. The neutron chopper has a 5 μs burst time to give around 6% resolution of the incident energy. The range of detector angles from -10 to 120° gives a wide range of simultaneous Q vector from 1 to 20 A^{-1} at 100 meV energy transfers. This instrument has been used to measure the generalized density of states, G(E) from a synthetic forsterite powder sample.

Phonon density of states of forsterite. A calculation of one phonon density of states, $g(\omega)$ based on the rigid ion model sampling 125 wavevectors in the Irreducible Brillouin Zone is shown in Figure 14 (Rao et al., 1987). When a suitable lattice dynamical model is available, the expected multiphonon contribution can be estimated. Figure 15a shows the generalized phonon density of states G(E) of forsterite measured on a 46 gm synthetic forsterite powder sample using the Intense Pulsed Neutron Source at the Argonne National Laboratory by the time-of-flight method. Figure 15b is the G(E) calculated from the rigid-ion model, where resolution broadening and multiphonon contribution have been included. The agreement between the experimental and theoretical curves is very satisfactory. Note that the broad peaks within the 20-100 meV region are principally due to external modes (libration of the silicate tetrahedra coupled to the translation of magnesium atoms), whereas the peak between 100 and 140 meV is from the internal vibrations of the [SiO₄) tetrahedra.

Incoherent inelastic neutron scattering

The incoherent scattering cross section for a perfect crystal is given by:

$$\frac{d^2\sigma_{inc}}{d\Omega dE} = \frac{1}{2}\frac{k_1}{k_0} \sum_d e^{-2W_d} \bar{b}^2_{d\,inc} \frac{1}{M_d} \cdot \sum_{j\mathbf{q}} \{|\mathbf{Q}\cdot\mathbf{e}^j_d(\mathbf{q})|^2 \frac{1}{\omega_j(\mathbf{q})}\cdot$$

$$\left[n_j(\mathbf{q}) + \frac{1}{2} \pm \frac{1}{2}\right] \delta(\omega \pm \omega(\mathbf{q})) \quad . \quad (32)$$

Note that there is no conservation of momentum in this expression. As such, it is less useful than the coherent scattering. However, it can be shown that the incoherent scattering from a powder sample is proportional to $g(\omega)$, the phonon density of states. Hence, this technique can be used to determine $g(\omega)$ for materials rich in incoherent scatterer such as hydrogen.

Incoherent inelastic neutron scattering spectra from zeolites containing organic radicals and water molecules. Currently organic radicals such as tetra-methylammonium (TMA) hydroxide are used in the gel/solution synthesis of zeolites with large pore sizes. The organic components modify the gel structure chemistry and act as templates around which the zeolite framework is built. When occluded within the zeolite cavity, the organic components change their configuration and vibrational characteristics. Figures 16a,b,c show the spectra from TMA-bromide and TMA occluded in Linde type A (LTA) and omega (MAZ) phases with respective compositions:

$$Na_{2.57}TMA_{0.92}Al_{3.49}Si_{8.51}O_{24}.nH_2O$$

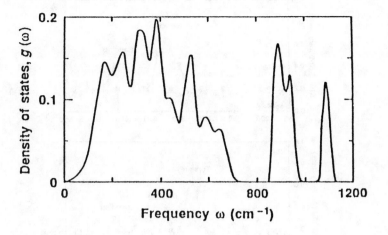

Figure 14. Phonon density of states of forsterite, Mg_2SiO_4, calculated from the rigid-ion model (Rao et al., 1987).

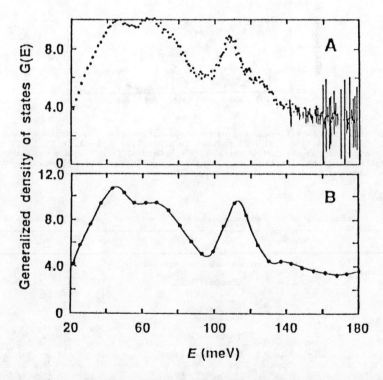

Figure 15. Generalized phonon density of states G(E) of forsterite at 300 K: Comparison of results from experimental determination by inelastic neutron scattering (A) and theoretical calculation based on the rigid-ion model (B). Resolution broadening and multiphonon contributions are included in both cases. In the experimental curve, energies below 20 meV could not be observed because of overlap with elastic scattering (Rao et al., 1987).

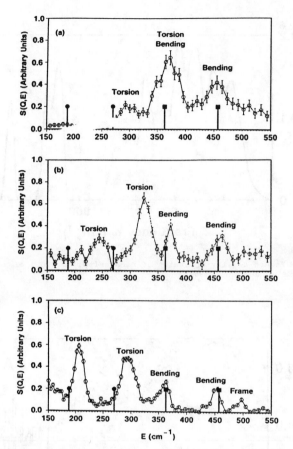

Figure 16. Incoherent inelastic neutron scattering from (a) tetramethylammonium (TMA) bromide at 16 K, (b) TMA occluded in Linde Type A zeolite at 14 K, and (c) TMA occluded in the zeolite omega at 12 K. Vertical lines indicate the positions of scaled torsional (filled circle) and bending (filled square) modes from a calculation for the free ion (Brun et al., 1987).

Table 5. Comparison of the energies in neutron spectra and IR-, FIR- and Raman spectra [meV]. The energies are given in meV

Zeolite	Neutrons			IR	FIR	Raman
Natrolite	a) 87	b) 78	c) Symmetric stretching O-T-O:			88
a) Present work	68	60	Libration of H_2O:	78		66
b) Ref. [8]	38	48		74		61
c) Ref. [7]	25	34		60		
	18	30	Bending O-T-O:	53	50	53
	13	24	Pore opening:	45	46	44
	8	21			44	
	17		Translational H_2O:	39	41	40
	11					37
		8	Optical mode of lattice:	28	29	29
					27	14
					14	12
					12	
			Translational Ca—O:	35		34
				33		
			Na—H_2O:	17		17
				16		
			Translational Na—O:	23		25
				21		20

and

$$Na_{2.31}TMA_{0.64}Al_{2.95}Si_{9.05}O_{24}.nH_2O$$

(Brun et al., 1987). From the spectra it can be seen that the position of the torsional modes (Raman inactive) are very sensitive to the TMA environment.

Nature of the binding and the librational and translational modes of water molecules in zeolites and clay minerals can be determined from the inelastic neutron scattering spectra. Figure 17 shows the generalized phonon density of states from a number of zeolites (Fuess et al., 1986). Only the spectrum of natrolite with ordered water molecules shows well resolved peaks with a sharp separation between translational and librational motions of H_2O. A comparison of the neutron spectra with the IR, FIR and Raman spectra of natrolite is given in Table 5.

INELASTIC MAGNETIC SCATTERING

It is possible to measure the energy level structure both in dilute (isolated), uncoupled atoms as well as for coupled atoms showing cooperative magnetic effects.

Magnetically dilute (uncoupled) systems

A few magnetic ions in a diamagnetic salt represents an uncoupled system. This situation is approximated by the room temperature state of some salt with a very low Curie temperature (~1K), in which case the energy levels of the magnetic ions are determined by the crystalline field from the surrounding ligands. Inelastic neutron scattering in such a system involves energy exchange between the neutron and these energy levels. Wavelength spectrum for neutrons scattered from Ho_2O_3 (Fig. 18) shows peaks at 0.002 eV, 0.010 eV and 0.037 eV; these energies represent the splitting of the lowest J state of the holmium ion by the crystalline field (Cribier and Jacrot, 1960).

Magnetically coupled systems

Coupled systems are represented by ferro-, ferri- or antiferro-magnets below their magnetic ordering (Curie or Néel) temperatures. In the paramagnetic state, the spin system is completely disordered, but the magnetic coupling forces between ions do not disappear, but are dependent on temperature. For such a system, a Gaussian spread of energy in the scattered neutron beam is usually observed. The root mean square change of energy, δE

$$\delta E = 2\{\frac{2}{3}zS(S+1)J^2\}^{1/2} \quad , \tag{33}$$

where S = spin quantum number of the paramagnetic ion, J = exchange integral between near neighbors, and z = number of each type of atom. At lower temperature, the exchange forces become dominant over disordering effects due to thermal motion and the spin system shows a high degree of order. However, the spins may still deviate from perfect order, where the variation of the spin direction takes a waveform. These "spin waves" are known as magnons (Fig. 19), which

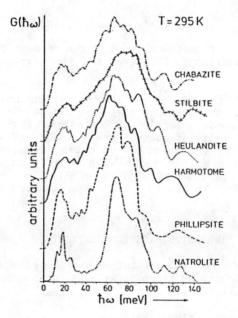

Figure 17. Generalized phonon density of states G(E) of natural zeolites at 295 K (Fuess et al., 1986).

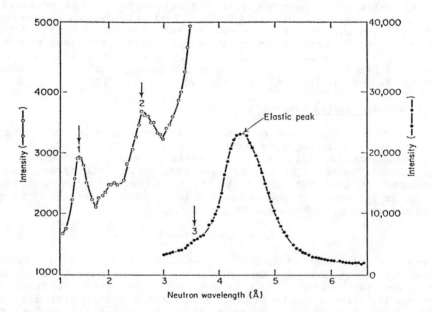

Figure 18. Energy levels of a rare earth ion, Ho³⁺ in Ho₂O₃: Wave length analysis of a neutron beam of initial wavelength 4.35 A after scattering from Ho₂O₃ powder at 300 k (Cribier and Jacrot, 1960).

possess a wavelength and quantized energy. Just like the case of phonons, magnon dispersion relation involves the dependence of the frequency ω with the wavevector q.

Ferromagnets. For a cubic ferromagnet, the dispersion law

$$\hbar\omega = C + 2JS(Z - \sum_{r}\cos q \cdot r \quad , \tag{34}$$

where C = small constant, J = exchange interaction, S = spin quantum number, Z = number of nearest neighbors, and r = vector between an atom and its nearest neighbors.

For a face-centered cubic cell (Z = 12), for magnons in the [111] direction

$$\hbar\omega = C + 12JS\left(1 - \cos\frac{qa}{\sqrt{3}}\right) \quad , \tag{35}$$

where a is the unit cell edge.

The magnon spectrum for the alloy $Co_{0.08}Fe_{0.08}$, shows a parabolic form as expected from which values of J = 16±1.6 meV, and C = 1.3 meV can be derived assuming S = 0.92 (Sinclair and Brockhouse, 1960).

Antiferromagnets and ferrimagnets. For one dimensional spin arrays with nearest neighbor interactions (Kittel, 1971), for small values of q

$$\hbar\omega = 2JSa^2q^2 \qquad \text{for ferromagnetism,} \tag{36}$$

and

$$\hbar\omega = 4JSaq \qquad \text{for antiferromagnetism} \tag{37}$$

where a is the spin separation. The measurement of the dispersion curve for $RbMnF_3$, an antiferromagnet with the perovskite structure (Windsor and Stevenson, 1966) (Fig. 20) confirms this relation, where the Mn^{2+} ions occur at the unit cell corners surrounded octahedrally by six F^- ions. Of the three exchange interactions, J_1, J_2, J_3 for first, second and third neighbors respectively J_1 = 0.30 meV, whereas $J_2, J_3 \approx 0$.

Magnetite, Fe_3O_4 has the inverted spinel structure, with two Fe^{3+} ions in the A-sites, and two Fe^{3+} and two Fe^{2+} ions in the B-sites in the primitive unit cell. The magnetic structure is ferrimagnetic, where the moments on the A sites are aligned together and are directed antiparallel to the aligned moments on B sites. The exchange inter-action between nearest neighbor ions in A and B sites is large and antiferromagnetic, the other interactions being weak (Néel, 1948). The magnon dispersion relations in magnetite for an acoustic and an optic magnon in the [001] directions, measured by Brockhouse and Watanabe (1963) are shown in Figure 21a, which are compared to the theoretical calculations by Kaplan (1952) (Fig. 21b), where the solid lines are calculated on the assumption of nearest neighbor A-B interactions only. The agreement between the calculated and observed magnon branches indicates the essential correctness of the Néel model.

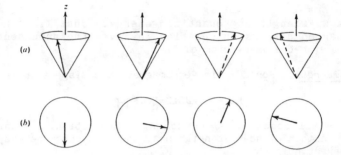

Figure 19. (a) Physical picture of a spin wave. The z axis is the mean direction of the spin vector. (b) Projection of spin S in the xy plane (Squires, 1978).

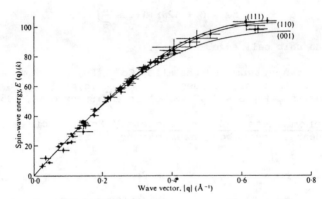

Figure 20. Spin wave (magnon) disperion in cubic RbMnF₃ with the perovskite structure along three cyrstalline directions (Windsor and Stevenson, 1966).

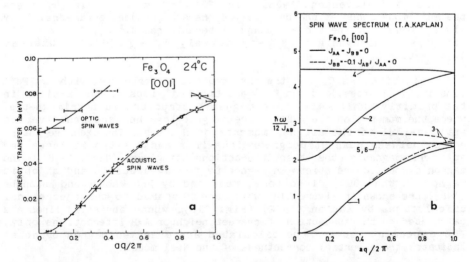

Figure 21. Magnon dispersion curves in magnetite, Fe₃O₄. (a) Experimental curve at room temperature by inelastic neutron scattering (Brockhouse and Watanabe, 1963), (b) Theoretical curve (Kaplan, 1952).

The magnons can be distinguished from phonons by the application of a magnetic field, which affects the magnon intensity but not the phonon intensity. The temperature dependence of the scattered neutron groups may also be different for the phonons and magnons. An important difference between magnons and phonons is that for the magnons, the magnetic form factor due to the large spread of wavefunctions (similar to x-ray form factors) enters the magnon structure factor expression, thereby reducing the scattering cross section substantially at large Q. Magnon density of states can be determined in a way similar to that for phonons.

QUASI-ELASTIC NEUTRON SCATTERING

When neutrons with the same energy interact with nuclei bound in a crystal, the scattered neutron spectrum consists of a sharp line with energy transfer $\hbar\omega = 0$, as well as a contribution due to the inter-action with the atomic vibrations. Physically, the sharp line corresponds to processes where the neutron has exchanged momentum with the crystal as a whole without suffering any energy exchange with the internal quantum states of the crystal. In principle, this line is similar to the Mössbauer line for gamma ray interactions with the crystal. If, however, the scattering particle undergoes translational diffusion in a liquid or in a crystal at higher temperatures, this motion leads to a line broadening, which is called quasi-elastic. Other random motions, such as rotational jumps of molecules in a crystal, also produce a quasi-elastic line around $\hbar\omega=0$. Hence, quasi-elastic scattering can be used to determine the rate and the geometry of such non-periodic motions. For currently available spectrometers (e.g., Quasi-Elastic Neutron Scattering Spectrometer (QENS) at Argonne National Laboratory) with a resolution of 10^{-6} eV, the observable time scale of motion is in the region of 10^{-10} to 10^{-13} sec.

The region of small energy transfers (~1 meV) around the elastic line is analyzed as a superposition of a Gaussian (elastic) and a Lorentzian. A typical example of quasi-elastic (time-of-flight) neutron scattering spectra from solid methane in its high temperature phase is shown in Figure 22 (Kapulla, in Springer, 1972). The rate of jump between 12 possible orientations of the CH_4 molecule in the crystal has been determined to be ~10^{12}/sec.

Because of the large incoherent scattering cross section of hydrogen ($\sigma_{inc} = 79.1$ barns), diffusion of hydrogen or molecules containing hydrogen, such as water and other organic molecules are well suited for quasi-elastic scattering studies.

Diffusive motion of water molecules in clay minerals and zeolites

The quasi-elastic spectra have been analyzed to determine the diffusion coefficient of water molecules between the layers of a clay mineral or channels in zeolites. In the approximation of a simple diffusion, the quasi-elastic line width (ΔE) is given by

$$\Delta E = 2\hbar DQ^2, \tag{38}$$

where, D = molecular translational diffusion coefficient, and Q = momentum transfer.

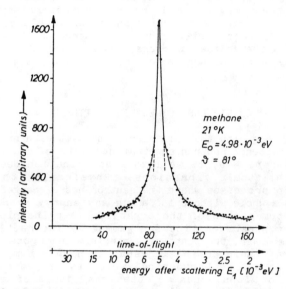

Figure 22. Quasi-elastic (incoherent) neutron scattering spectrum of solid methane in its high temperature phase obtained by the time-of-flight method (Kapulla, 1971).

The quasi-elastic energy broadening as a function of momentum transfer square due to water molecules in lithium montmorillonite with different $\underline{c}$ spacings is shown in Figure 23a (Olejnik and White, 1972). Figure 23b shows the logarithm of the diffusion constant against the inverse of the water layer thickness in montmorillonites and vermiculites.

The diffusion constants, residence times and correlation lengths for water molecules in the zeolite harmotome $Ba_2Al_4Si_{12}O_{32} \cdot nH_2O$ (n = 4,8,12) measured from quasi-elastic scattering spectra are given in Table 6 (Fuess and Stuckenschmidt, 1987).

CONCLUSIONS

Inelastic neutron scattering spectroscopy is a powerful tool to explore lattice vibrations in a crystal throughout the Brillouin zone, which supplements the data obtained through the use of well known optical techniques (IR, Raman and Brillouin scattering). Coherent inelastic neutron scattering provides data on phonon frequency and width, phonon dispersion relations and density of states, mode Grüneisen parameter and anharmonicity from changes in phonon frequency and width due to changes in temperature and pressure, soft modes associated with second order phase transitions, etc. This technique is also very useful for the determination of the energy level splittings in transition metal ions and magnon (spin wave) dispersion and magnon

Li - MONTMORILLONITE

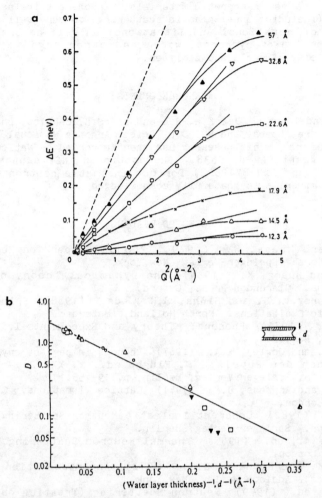

Figure 23a. Quasi-elastic energy broadening in lithium montmorillonite for different c lattice spacings (indicating various thicknesses of intercalculated water layers) (Olejnik and White, 1972).

Figure 23b. Logarithm of diffusion constant of water (D) as a function of inverse water layer thickness in lithium and sodium montmorillonites and vermiculites (Olejnik and White, 1972).

Table 6. Diffusion constant, residence time and correlation length, $\langle l \rangle$

	$\mathbf{D}$ $(10^{-5}cm^2sec^{-1})\tau_0$	10^{12} sec	$\langle l \rangle$, nm
T = 295 K			
$4H_2O$	5.0	5.5	0.41
$8H_2O$	25.0	2.9	0.66
$12H_2O$			
215 K	5.0	5.5	0.41
295 K	8.0	4.2	0.45
320 K	10.0	3.7	0.47

density of states in magnetic materials. Incoherent inelastic neutron scattering (including quasi-elastic scattering) is very well suited for the study of rotational and librational as well as translational diffusive motion of hydrogen, water and other organic molecules occluded in clay minerals and zeolites.

ACKNOWLEDGMENTS

I am indebted to Dr. K.R. Rao and Dr. S.L. Chaplot, Bhabha Atomic Research Centre, Bombay and Dr. D.L. Price, Argonne National Laboratory for discussions. This research has been supported by National Science Foundation grant EAR-8719638. The continuing NSF support (grants EAR-8417767 and EAR-8719638) for our inelastic neutron scattering research on minerals is gratefully acknowledged.

SUGGESTED READING

Bilz, H. and Kress, W. (1979) Phonon Dispersion Relations in Insulators. Springer-Verlag, New York.

Born, M. and Huang, K. (1954) The Dynamical Theory of Crystal Lattices. Clarendon Press, Oxford.

Borovik Romanov, A.S. and Sinha, S.K., eds. (1988) Spin Waves and Magnetic Excitations. North Holland, Amsterdam.

Brüesch, P. (1982) Phonons: Theory and Experiments I. Springer-Verlag, Berlin.

Cochran, W. and Cowley, R.A. (1967) Phonons in perfect crystals. In: Handbuch der Physik. S. Flügge ed., x. XXV/2a. Light and Matter 1a, Springer-Verlag, Berlin, p. 59-156.

Donovan, B. and Angress, J.F. (1971) Lattice Vibrations. Chapman and Hall, London.

Dorner, B. (1982) Coherent Inelastic Neutron Scattering in Lattice Dynamics. Springer-Verlag, Berlin.

Egelstaff, P.A., ed. (1965) Thermal Neutron Scattering. Academic Press, London.

Ghatak, A.K. and Kothari, L.S. (1972) An Introduction to Lattice Dynamics. Addison Wesley, New York.

Kostorz, G., ed. (1979) Neutron Scattering (Treatise on Materials Science and Technology, v. 15). Academic Press, New York.

Lovesey, S.W. (1984) Theory of neutron scattering from condensed matter, volumes 1 and 2. Clarendon Press, Oxford.

_____ and T. Springer, eds. (1977) Dynamics of Solids and Liquids by Neutron Scattering. Springer-Verlag, New York.

Maraduddin, A.A., Montroll, E.W., Weiss, G.H. and I.P. Ipatova (1971) Theory of Lattice Dynamics in the Harmonic Approximation (2nd ed.). Academic Press, New York.

Marshall, W. and Lovesey, S.W. (1971) Theory of Thermal Neutron Scattering. Clarendon Press, Oxford.

Reissland, J.A. (1973) The Physics of Phonons. John Wiley, New York.

Sköld, K. and Price, D.L., eds. (1986) Neutron Scattering, Part A, Methods of Experimental Physics, V. 15. Academic Press, New York.

Springer, T. (1972) Quasielastic Neutron Scattering for the Investigation of Diffusive Motions in Solids and Liquids. Springer-Verlag.

Squires, G.L. (1978) Introduction to the Theory of Thermal Neutron Scattering. Cambridge University Press, Cambridge, England.

Venkataraman, G., Feldkamp, L.A. and Sahni V.C. (1975) Dynamics of Perfect Crystals. MIT Press, Cambridge, Massachusetts.

REFERENCES CITED

Brockhouse, B.N. (1961) Methods for neutron spectrometry. In: Inelastic Scattering of Neutrons in Solids and Liquids. IAEA, Vienna, p. 113-151.

_____ and Watanabe, H. (1963) Spin waves in magnetite from neutron scattering. In: Inelastic Scattering of Neutrons in Solids and Liquids, V. II. Int. Atomic Energy Agency, Vienna, p. 297-308.

Brun, T., Curtiss, L., Iton, L., Kleb, R., Newsam, J., Beyerlein, R. and Vaughn, D. (1987) Inelastic neutron scattering from tetramethylammonium cations occluded in zeolites. In: IPNS Progress Report, 1986. Argonne National Laboratory, Argonne, IL, p. 50-51.

Carpenter, J.M. and Yelon, W.B. (1986) Neutron sources. In: Neutron Scattering. K. Sköld and D.L. Price, eds., p. 99-196.

Cribier, D. and Jacrot, B. (1960) Diffusion des neutrons et effet stark cristallin dans les oxydes de terre rare. C.R. Acad. Sci. Paris 250, 2871-2873.

Fuess, F. and Stuckenschmidt, E. (1987) Arrangement and dynamics of water in natural zeolites. Proc. Int. Conf. Neutron Scattering, Sydney, Australia.

Fuess, H., Stuckenschmidt, E. and Schweiss, B.P. (1986) Inelastic neutron scattering studies of water in natural zeolites. Ber. Bunsenges. Phys. Chem. 90, 417-421.

Ghose, S. (1985) Lattice dynamics, phase transition and soft modes. In: Reviews of Mineralogy, V. 14 Microscopic to Macroscopic: Atomic Environments to Mineral Thermodynamics. S.W. Kieffer and A. Navrotsky, eds., Mineral. Soc. Am., p. 127-163.

_____ , Hastings, J.M., Corliss, L.M., Rao, K.R., Chaplot, S.L. and Choudhury, N. (1987) Study of phonon dispersion relations in forsterite, Mg_2SiO_4 by inelastic neutron scattering. Solid State Comm. 63, 1045-1050.

Iyenger, P.K. (1965) Crystal diffraction techniques. In: Thermal Neutron Scattering. P.A. Egelstaff, ed., Academic Press, London, p. 97-140.

Kaplan, H. (1952) A spin-wave treatment of the saturation magnetization of ferrites. Phys. Rev. 86, 121.

Kapulla, H. (1971) In: Quasielastic Neutron Scattering for the Investigation of Diffusive Motions in Solids and Liquids. T. Springer (1972) Springer Verlag, Berlin.

Kittel, C. (1971) Introduction to Solid State Physics, 4th ed. John Wiley, New York.

Olejnik, S. and White, J.W. (1972) Thin layers of water in vermiculites and montmorillomites - modification of water diffusion. Nature, Phys. Sci. 236, 15-16.

Placzek, G. and Van Hove, L. (1955) Interference effects in total neutron scattering cross-section of crystals. Nuov. Cim. 1, 233-244.

Price, D.L. and Sköld, K. (1986) Introduction to Neutron Scattering. In: Neutron Scattering. K. Sköld and D.L. Price, eds., pp. 1-97.

Rao, K.R. (1978) Thermal neutron scattering and dynamics of solids. In: Lattice Dynamics. K.R. Rao, ed., Indian Physics Assoc., Bombay, p. 323-355.

_____, Chaplot, S.L., Choudhury, N., Ghose, S., Hastings, J.M., Corliss, L.M. and Price, D.L. (1988) Forsterite, Mg_2SiO_4: Lattice dynamics and inelastic neutron scattering. Phys. Chem. Minerals (in press).

_____, Chaplot, S.L., Choudhury, N., Ghose, S. and Price, D.L. (1987) Phonon density of states and specific heat of forsterite, Mg_2SiO_4. Science 236, 64-65.

Sinclair, R.N. and Brockhouse, B.N. (1960) Dispersion relation for spin waves in a fcc cobalt alloy. Phys. Rev. 120, 1638-1640.

Stassis, C. (1986) Lattice dynamics. In: Neutron Scattering, K. Sköld and D.L. Price, eds., Academic Press, New York, p. 369-441.

Traylor, J.G., Smith, H.G., Nicklow, R.M. and Wilkinson, M.K. (1971) Lattice dynamics of rutile. Phys. Rev. B3, 3457-3472.

Windsor, C.G. (1986) Experimental techniques. In: Nuclear Scattering, Pt. A. K. Sköld and D.L. Price, eds., Academic Press, New York, p. 196-257.

_____ and Stevenson, R.W.H. (1966) Spin waves in $RbMnF_3$. Proc. Phys. Soc. London 87, 501-504.

Chapter 6

VIBRATIONAL SPECTROSCOPY OF HYDROUS COMPONENTS

INTRODUCTION

Water molecules, hydroxide ions and fluid inclusions are important components of many natural systems, and are also prominent in a variety of synthetic minerals and related technological materials. Water and OH ions can be both a major component required by the mineral's stoichiometry and an accidental trace component. The trace hydrous species can have a disproportionately important role in the physical, chemical, rheological, electronic and optical properties of the material.

In minerals, hydrogen is most commonly bonded to oxygen. The strongly polar OH groups absorb infrared photons efficiently. Furthermore, if the OH groups are structurally oriented within the host, the amount of incident radiation absorbed can be strongly dependent upon the relative orientation of the OH dipole and the direction of linear polarization of the interrogating light.

General discussions of the vibrational spectra of the water molecule and the OH ion are presented in many general spectroscopy texts such as Nakamoto (1978). The topic of the infrared spectroscopy of OH and H_2O in minerals has been reviewed in Farmer (1974) and more recently by Aines and Rossman (1984a).

GENERAL CONCEPTS

The isolated OH^- ion has a single motion which absorbs infrared radiation. It is the O-H stretching motion which absorbs energy in the 3735 cm^{-1} region. In crystalline solids, the H atom is usually hydrogen bonded to another ion, usually oxide. The exact frequency of the O-H stretch will depend upon the strength of the hydrogen bond. Nakamoto et al. (1955) and Novak (1974) presented the correlation between the frequency of the O-H stretch and the strength of the hydrogen bond as measured by the $O-H\cdots O$ distance (Fig. 1). In general, the frequency of OH stretch occurs at lower energies in systems in which the OH ion is more strongly hydrogen bonded.

Group theory shows that the non-linear triatomic H_2O molecule has C_{2v} symmetry. As such, it has two fundamental stretching motions and one bending motion (Fig.1, Chapter 4). Both the symmetric and asymmetric stretching motions absorb infrared radiation in the 3400-3200 cm^{-1} region. The bending motion absorbs in the 1630 cm^{-1} region. In liquid water, these absorptions are broad because of the continuum of hydrogen-bonding interactions which can occur in bulk water (Fig. 2).

Prominent OH and H_2O absorption positions in the infrared and near-infrared are listed in Table 1. These values are necessarily approximate because the exact OH stretching energies are host dependent.

THE EXPERIMENT

The stoichiometric hydrous components are studied by the

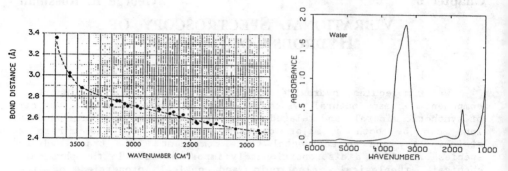

Figure 1 (left). Correlation between the O-H stretching frequency and the strength of hydrogen bonding as measured by the O-H···O distance. Data from Nakamoto et al. (1955).

Figure 2 (right). Infrared spectrum of liquid water showing the H-O-H bending mode near 1630 cm⁻¹, the two, overlapping O-H stretching modes near 3400 and 3200 cm⁻¹, and the much weaker combination stretch+bend mode near 5200 cm⁻¹. Thin film of water approximately 4.5 μm thick.

Table 1. General positions of water and OH vibrational bands in minerals

wavenumber		wavelength	species and motions
1630 cm⁻¹	=	6.1 μm	H_2O bend
3400 cm⁻¹	=	2.9 μm	OH stretch
3500 cm⁻¹	=	2.8 μm	OH stretch
4200 cm⁻¹	=	2.5 μm	M-OH motions
5200 cm⁻¹	=	1.9 μm	H_2O combination mode (bend + stretch)
7100 cm⁻¹	=	1.4 μm	1st OH overtone (stretch)

Table 2. Representative molar absorptivity values for hydrous species

band	ε-value	species	reference
3400 cm⁻¹	81	H_2O (liquid water)	Thompson (1965)
	56	H_2O (in glass)	Newman et al. (1986)
	100	OH (in glass)	Newman et al. (1986)
	206	OH (spessartite)	Rossman et al. (1988)
4500 cm⁻¹	0.76	Si-OH	Scholze (1960)
	1.73	X-OH	Newman et al. (1986)
5200 cm⁻¹	1.14	H_2O	Scholze (1960)
	1.61	H_2O (in glass)	Newman et al. (1986)

conventional methods of infrared spectroscopy (Chapter 4). Most studies have used the KBr pellet method to determine the specific species (OH or H_2O) present, and to use the OH stretching modes as probes of local cation ordering. However, single-crystal studies are preferred because they generally avoid the problem of absorbed water on the fine powders used with the KBr method, they provide information about the orientation of the OH and H_2O units in polarized light, and they provide a known path-length which affords better quantitation. Unfortunately, it has frequently been difficult or impossible to prepare single-crystals thin enough to keep the intense OH bands "on scale" when the "water" content of the mineral is above about two percent.

INTENSITIES

One of the principal uses of the infrared spectroscopy of the hydrous components is for analytical concentration information. Infrared absorpt-ion spectroscopy is probably the easiest and one of the most accurate methods of determining the absolute concentration of minor amounts of OH and H_2O in minerals. Concentrations can be determined through Beer's law calibration based upon:

$$Absorbance = \epsilon \cdot path \cdot concentration,$$

where path is in cm, and concentration is in moles per liter. The molar absorptivity (also referred to as the extinction coefficient), ϵ, is a constant for each system of interest.

Unfortunately, the intensities need independent calibration for accurate results. Few absolute calibrations have been performed on water and OH in minerals. The water absorption in the $3400\ cm^{-1}$ region is sensitive to about 1 μg of water per square cm of sample. The calibration factor varies by about a factor of two in different systems as long as the water is not strongly structurally oriented (Table 2). The calibration of the stretching bands of OH ions is less well determined. Indications are that the intensities of the OH bands are more variable than those of molecular water. More calibration data are available for glasses than are available for crystals.

Paterson (1982) proposed general calibration curves for water in minerals in the OH stretching region, derived largely from glass standards and solutions. He presents them in terms of the integrated absorbance which has a smaller temperature dependence than the molar absorptivity values (ϵ-values). He finds that the intensity of a unit concentration of "water" increases as the strength of the hydrogen bond increases; that is, the intensity increases as the band moves to lower energies.

SPECIFIC APPLICATIONS

Water of hydration and OH are constituents of a large number of minerals. They are recorded, often without comment, in the thousands of spectra routinely taken of minerals. Occasionally, the OH region spectra of a few mineral families have been the objects of detailed study. In particular, zeolites have been the object of extensive study because of their importance to catalytic processes. Most mineralogical studies have been primarily concerned with OH groups which act as probes of the local cation environment. The OH stretching frequency will change in response to the local grouping of cations about a shared OH ion. The amphibole and mica spectra have been the most extensively examined.

Amphiboles

The OH stretching band occurs in the range from 3600-3700 cm^{-1}. Whereas the spectrum of an end-member amphibole consists of a single OH band, the spectrum of most amphiboles consist of multiple bands (Fig. 3 above). They arise from the effects of different local cation environments which in turn result from chemical substitutions at the cation sites. Several attempts have been made to use the intensities of the OH bands to derive cation ordering schemes for the amphiboles using a model developed by Strens (1966) and Burns and Strens (1966) and expanded by Burns and Greaves (1971), Nikitina et al. (1973) and Law (1976). This application was reviewed by Strens (1974) and more recently by Hawthorne (1981), and applied by Burns and Prentice (1968), Rowbotham and Farmer (1973), Maresch and Langer (1976), Hawthorne (1983) and Mottana and Griffin (1986).

The orientational dependence of the OH stretching mode in riebeckite has been presented by Hanisch (1966).

Micas

Mica spectra consist of a series of narrow bands in the OH region. In end-member trioctahedral micas, a single OH stretching band at about 3710 cm^{-1} results from a single OH environment, $Mg_3(OH)$. In chemically more complex micas, multiple bands arise from multiple groupings such as $Mg_2Fe(OH)$ and $MgFe_2(OH)$ $[]Al_2(OH)$, etc. There has been extensive mica OH spectral data presented which has been previously reviewed by Farmer (1974, Chapter 15). The interpretation of these features has been largely concerned with the identification of the individual groupings responsible for each band. Velde (1983) concluded that the average electronegativity of the cluster of octahedral ions which are bound to the OH group gives the best correlation with the mica OH stretching frequency. Octahedral cations produced a shift of about 96 cm^{-1} per electronegativity unit in trioctahedral micas compared to 170 cm^{-1} in the dioctahedral micas. Substitutions in the tetrahedral layer gave much smaller shifts.

The use of the OH bands in the 3800-3200 cm^{-1} range to establish cation site occupancy has assumed that the intensities of the individual OH absorptions are proportional to the concentration of the OH groups in each chemical environment, and that each environment has the same proport-ionality factor. When multiple cation-OH groupings are involved, a number of studies have indicated that, in fact, there are different absorption intensities associated with each particular cation-OH grouping (Rouxhet, 1970; Gilkes et al., 1972; Sanz and Stone, 1983).

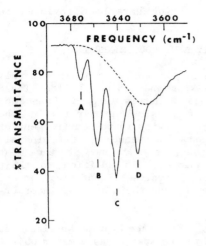

Figure 3. Infrared spectrum of the OH region of powdered actinolite from Cumberland, Rhode Island. The peaks are assigned to the following cation groupings about the OH ion: (A) Mg_3 (B) Mg_2Fe^{2+} (C) $MgFe_2^{2+}$ (D) Fe_2^{2+}. Spectrum from Burns and Greaves (1971).

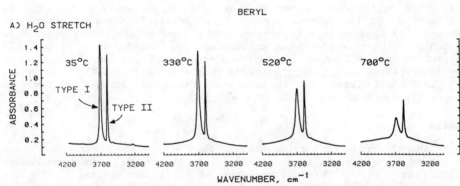

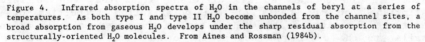

Figure 4. Infrared absorption spectra of H_2O in the channels of beryl at a series of temperatures. As both type I and type II H_2O become unbonded from the channel sites, a broad absorption from gaseous H_2O develops under the sharp residual absorption from the structurally-oriented H_2O molecules. From Aines and Rossman (1984b).

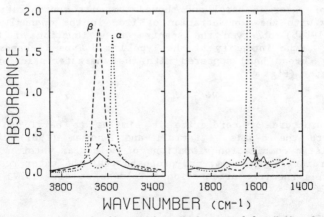

Figure 5. Infrared spectrum of a 25 μm thick cordierite crystal from Haddam, Connecticut, showing the bending mode of Type II H_2O in the channels at 1630 cm^{-1}, and the two stretching modes of Type II H_2O at 3574 cm^{-1} and 3632 cm^{-1}. Weaker absorption in the α direction at 3689 cm^{-1} is from Type I H_2O in the channel. From Goldman et al. (1977).

Additional details of the mica OH region have been reviewed in Rossman (1984)

HYDROUS SPECIES IN NOMINALLY ANHYDROUS MINERALS

Minor amounts of water and OH ion are commonly found in minerals normally considered to be anhydrous (Wilkins and Sabine, 1973). These trace hydrous species can have a disproportionately large influence on the chemical, mechanical, electronic and physical properties of the mineral. They also may reflect the water activity in the environment of formation, or hold a record of processes undergone by the mineral. Infrared spectroscopy presents the easiest and one of the most sensitive means of detecting their presence. In addition to their detection, the study of these components involves identification of their chemical species, their concentration, and their mode of incorporation in the structure.

Many studies casually include evidence for trace hydrous species, but relatively few have addressed the minor hydrous species as a major object of study. An important aspect which needs to be addressed in any study is whether or not the hydrous species is structurally incorporated in the mineral, as opposed to being an accidental, mechanically entrapped phase which could be viewed as a contaminant in the studied phase. Often, this aspect is not addressed. Nevertheless, there are many examples of hydrous species occurring in minerals that are flawless both visually and to the optical microscope. The ones most intensively studied are reviewed below with emphasis on single-crystal studies.

Beryl

The structure of beryl contains rings of six SiO_4 tetrahedra which stack to form channels running parallel to the c-axis. Wood and Nassau (1967) studied the infrared spectrum of beryl and deduced that water resides in the channel at two different sites that can be readily distinuished on the basis of the polarization properties of the spectrum. One site has the H-H vector of the water molecule oriented parallel to the c-axis whereas at the other site, the H-H vector is oriented perpendicular to the c-axis. Goldman et al. (1978) studied the zonation of water in beryl and showed that its concentrations was not correlated with the concentration of iron in the channel. Aines and Rossman (1984b) observed the spectra as a function of temperature; above 300°C, the intensity of the Type I and Type II bands decreased, and a new, broad band appeared with the characteristics of unbonded, gaseous water (Fig. 4).

Cordierite

The structure of cordierite is similar to beryl, with channels parallel to the c-axis. Farrall and Newnham (1967) observed the characteristic combination vibrations of molecular water near 5200 cm^{-1} in the cordierite spectrum, and proposed that the water resided in the channel in a specific orientation. Goldman et al. (1977) studied the

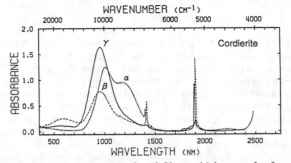

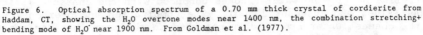

Figure 6. Optical absorption spectrum of a 0.70 mm thick crystal of cordierite from Haddam, CT, showing the H_2O overtone modes near 1400 nm, the combination stretching+ bending mode of H_2O near 1900 nm. From Goldman et al. (1977).

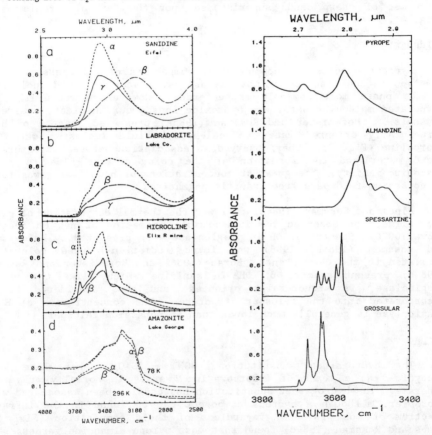

Figure 7 (left). Infrared spectra of H_2O and OH in feldspars. (A) OH in 5.0 mm thick sanidine from the Eifel region, Germany. (B) OH in 5.0 mm thick labradorite from Lake County, Oregon. (C) Molecular H_2O in 0.5 mm thick microcline from San Diego County, California. (D) Fluid inclusion water in 5.0 mm thick microcline from Lake George, Colorado. Also shown is the ice spectrum with the sample at liquid nitrogen temperature. Modified from Hofmeister and Rossman (1985b).

Figure 8 (right). Comparison of the OH region in the liquid nitrogen temperature spectra of four garnets. All features above the baseline are from OH. The pyrope and almandine spectra are for samples about ten times as thick as the spessartite and grossular. From Aines and Rossman (1984d).

spectra of single crystals and determined that, like beryl, water occurred in two distinct orientations in the channel (Fig. 5), one with the H-H vector parallel to the channel (Type I) and the other with the H-H vector perpendicular to the channel (Type II). They associated the Type II spectra with Na-rich cordierites, an observation confirmed by Armbruster and Irouschek (1983) in the case of Na,Be-rich cordierites. Both Aines and Rossman (1984b) and Schreyer (1986) showed that the two types of water could be readily distinguished in the infrared spectra of cordierite powders. Spectra obtained at temperatures up to 900°C show that, like beryl, the water becomes unbonded with the characteristics of a gas (Aines and Rossman, 1984b). The combination and overtones bands of the channel water are readily observed in the near infrared spectrum as a set of sharp bands superimposed upon the Fe^{2+} electronic bands (Fig. 6).

Feldspar

Hydrous species are common minor components of feldspars, and are observed in the spectrum of most feldspars. Water in fluid inclusions is the phase most commonly encountered. However, structurally in-corporated molecular water and hydroxide ions occur in clear, unaltered crystals. Hofmeister and Rossman (1985a) presented spectra of both structurally oriented water molecules and fluid inclusion water in microcline (Fig. 7). They observed an association between structurally bound water and the formation of the color in the blue to green amazonite variety. The greatest concentrations of bound water was found in potassium feldspars from granitic pegmatites.

Aines and Rossman (1984b) followed the thermal dehydration of clear microcline, and observed the irreversible formation of new OH species above 700°C. Spectra of OH in plagioclase were presented by Hofmeister and Rossman (1985b, 1986) who found radiation-induced coloration unrelated to the OH content. Beran (1986) and Hofmeister and Rossman (1985b) present spectra of OH in sanidine which, like the OH in plagioclase, is structurally oriented, and thus considered to be incorporated into the feldspar structure. The concentration of H in plagioclase is generally much lower than in potassium feldspar.

Garnet

The hydrogarnet substitution, $4OH^- = SiO_4^{4-}$, has long been recognized as a means of incorporating "water" in normally anhydrous minerals. The hydrogarnet substitution has usually been associated with grossular, although a hydrous component can be found in the infrared spectrum of most garnets, regardless of their composition (Fig. 8). Aines and Rossman (1984d) found that most pyrope-almandine garnets show OH in their spectra. Aines and Rossman (1984c) also found that OH is in garnets irrespective of whether they are of crustal or mantle origin. Rossman (1986) presented spectra showing the great variability of the grossular OH spectra which contrasts with the consistency of the pyrope-almandine spectra; he also noted that titianian andradite (melanite) garnets and uvarovites are among the most OH-rich found. Rossman and Aines (1986) found strongly anisotropic OH absorption in a birefringent grossular from Asbestos, Quebec, but noted that anisotropic OH is not found in all anisotropic garnets.

Natural glasses

H_2O and OH are common constituents of both commercial and natural glasses. Stolper (1982) has studied the spectra of a series volcanic glasses and synthetics of petrologically relevant composition, and has determined calibration curves for a number of glass compositions. He established that water preferentially enters as OH in glasses of low total "water" content, although molecular water was observed in the spectrum of all glasses of greater that 0.5 wt % total "water" content. At high (>4%) total "water" contents, the concentration of molecular water exceeded the OH content. Newman et al. (1986) have extended the range of calibrations to rhyolitic compositions.

Raman spectra of hydrous glasses of similar composition also record weak bands hear 3600 cm^{-1} due to the OH stretches of both water and OH^-, together with an even weaker band near 1600 cm^{-1} from the H_2O bending motion (McMillan et al., 1983; 1986).

Olivine

A variety of OH defects have been recognized in the spectrum of olivine (Beran and Putnis, 1983; Aines and Rossman, 1984a; Freund and Oberheuser, 1986). Miller et al. (1987) examined the spectrum of olivines from seventeen localities, and found that OH defects were present in all. They identified more than 30 different OH bands and found "water" concentrations which spanned a range of nearly three orders of magnitude. In some cases, the OH spectrum indicated that domains of other minerals were incorporated in the olivine. The OH band of serpentine at 3685 cm^{-1} was observed in the spectrum of a number of crystals. A talc band at 3678 cm^{-1} was seen in one crystal and the titano-humite pattern (Kitamora et al., 1987) was present in others. In each of these cases, the second phase was not visible under the optical microscope. Examination of the OH spectra has also played a role in deformation studies of olivine crystals (Mackwell et al., 1985). These show that OH can be introduced and removed from olivine under controlled pressure, temperature and oxygen fugacity conditions.

Silica

The system most thoroughly studied is quartz. Hydroxide is present in both natural and synthetic quartz. Several percent water can be present in chalcedony and opal. Furthermore, molecular water plays a major role in the dielectric and mechanical properties of synthetic quartz.

Natural quartz spectra usually show a set of narrow bands in the OH region. Kats (1962) showed that these narrow absorptions are OH stretching vibrations. The bands at 3318, 3383, and 3432 cm^{-1} arise from Al^{3+} substitutions charge-compensated by H^+ (Fig. 9). The positions of the bands reported by Kats (1962) plus those of other, more recent studies, are summarized in Aines and Rossman (1984a). Many other bands are associated with specific alkali ions such as Na, Li and K. These

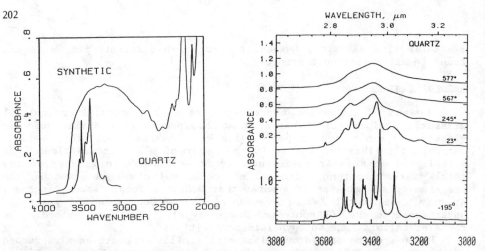

Figure 9 (left). Comparison of the infrared absorption spectrum of synthetic quartz with "broad band" absorption from molecular H_2O centered near 3200 cm^{-1} and natural quartz with a series of sharp OH absorptions in the 3200 - 3600 cm^{-1} region.

Figure 10 (right). Spectrum of a 5.07 mm thick single crystal of quartz from Hot Springs, Arkansas, taken at a series of temperatures from liquid nitrogen temperature up through the α-β transition. The sharp bands due to OH ions merge to a nearly continuous blur at high temperature. From Aines and Rossman (1985).

Table 3. Summary of OH vibrations in low temperature silicas.

2340 cm^{-1}	heat treated	strongly H-bonded OH
3440 broad band	chalcedony	water, H-bonded Si-OH
3585	"	OH at internal structural defects
3740		isolated surface Si-OH
3400 broad band	Opal c	water, H-bonded Si-OH
3650	"	OH groups at internal sites
4560	opal c	weakly H-bonded internal Si-OH
4480 broad band	chalcedony	weakly H-bonded internal Si-OH
4380		strongly H-bonded surface Si-OH
5200		water
5250		monomeric water
5180	78 K spectra	ice

From Graetsch et al. (1985) and references therein.

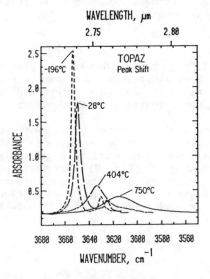

Figure 11. Spectrum of OH^- ions in topaz at various temperatures, showing the wavenumber shift which occurs as the temperature is changed and showing the increased band-width at high temperature. From Aines and Rossman (1985).

absorptions are usually measured at liquid nitrogen temperature because of the strong temperature dependence on the width of these bands. Aines and Rossman (1985) measured the temperature dependence of these bands from 78 K up to 800 K, and found that these bands broaden into a nearly continuous blur by the temperature of the α-β transition (Fig. 9). This behavior contrasts with the high-temperature behavior of other minerals such as muscovite. Chakraborty and Lehmann (1976) discussed OH concentration calibrations for the sharp band absorptions.

In low-temperature forms of natural quartz, as well as opal and some synthetic quartz, the narrow O-H bands are superimposed on a much broader absorption. The broad band absorption (Fig. 10) has been of major scientific and economic interest because a number of physical properties, including the mechanical Q and dielectric properties, are associated with its presence. Most prominent in the geophysics literature are studies of "water" in quartz and its relationship to hydrolytic weakening (Kekulawala et al., 1981; Kronenberg et al., 1986). Patterson (1982) discusses the general problem of water concentration calibration in this system.

Opals, which can contain up to 10 wt % "H_2O", show both molecular water and Si-OH groups in the infrared spectrum (Langer and Flörke, 1974). The near-infrared overtone and combination bands confirmed that molecular water was the major species present. Chalcedony contains both molecular water and Si-OH units. Flörke et al (1982) studied agates and found near-IR absorption at 5200 and 7100 cm^{-1} due to molecular water, and at 4450 cm^{-1} due to Si-OH groups on surfaces and at low-angle grain boundaries. Frondel (1982) found additional OH in chalcedony at structural sites. In an extensive study of water in Brazilian agates, Graetsch et al. (1985) found spectroscopic evidence for water in closed micropores, regions of accumulated defects, and hydrated surfaces. They also found that all surfaces were covered with silanol groups. Additional OH groups were located within the crystallites at structural defects. Table 3 summarizes the positions of OH vibrations in low temperature silicas.

Other minerals

Although the OH overtones and water combination modes in the mid-infrared are routinely recorded and published as part of the study of the optical spectrum of minerals, and the OH portion of the mid-infrared spectrum are routinely reported in powder spectra, comparatively few single-crystal studied have been directed at the OH groups themselves. Examples include axinite (Beran, 1970d), benitoite (Beran, 1974a), datolite (Sahl, 1966) diopside (Beran, 1976), enstatite (Beran and Zemann, 1986), joaquinite (Rossman, 1975), kyanite (Beran, 1970c), lawsonite (Labotka and Rossman, 1974), nepheline (Beran, 1974b), rutile (Beran and Zemann, 1970), sellaite (Beran and Zemann, 1985), sillimanite (Beran et al., 1983), titanite (Beran, 1970a), topaz (Aines and Rossman, 1984b), vivianite (Piriou and Poullen, 1987) and zircon (Aines and Rossman, 1985).

Of particular interest are the single-crystal lawsonite study in which both H_2O molecules and OH groups occur together in the same spectrum, and the topaz spectrum (Fig. 11) which illustrates the magnitude of wavenumber shift which can be encountered in OH bands as a function of temperature. Among studies of powdered minerals which are of interest beyond the routine characterization of the spectrum of the mineral, the spectrum of zoisite by Langer and Lattard (1980) is noteworthy because of the very low energy OH band. Also of interest are the works of Caruba and Iacconi (1983) and Caruba et al. (1985) who studied water in zircon with powdered samples.

SUMMARY

Infrared absorption spectra are extremely sensitive to the OH vibration. As a result, the vibrations of OH and H_2O have been identified in the spectra of numerous minerals where they can be either major constituents or trace components. Furthermore, spectroscopic criteria which distinguish OH ions from H_2O molecules can be applied in many circumstances. With suitable calibration, sensitive analytical determinations of the hydrous component in many minerals can be achieved with detection limits approaching a few micrograms.

REFERENCES

Aines, R.D. and Rossman, G.R. (1984a) Water in minerals? A peak in the infrared. J. Geophys. Research 89, 4059-4071.

_____ and _____ (1984b) The high temperature behavior of water and carbon dioxide in cordierite and beryl. Amer. Mineral. 69, 319-327.

_____ and _____ (1984c) Water content of mantle garnets. Geology 12, 720-723.

_____ and _____ (1984d) The hydrous component in garnets. Pyraspites. Amer. Mineral. 69, 1116-1126.

_____ and _____ (1985) The high temperature behavior of trace hydrous components in silicate minerals. Amer. Mineral. 70, 1169-1179.

Armbruster, T. and Irouschek, A. (1983) Cordierites from the Lepontine Alps: Na + Be -> Al substitution, gas content, cell parameters, and optics. Contrib. Mineral. Petrol. 82, 389-396.

Beran, A. (1970a) Messung des Ultrarot-Pleochroismus von Mineralen. IX. Der Pleochroismus der OH-Streckfrequenz in Titanit. Tschermaks Mineral. Petr. Mitt. 14, 1-5

_____ (1970b) Ultrarotspektroskopischer Nachweis von OH-Gruppen in den Mineralen der Al_2SiO_5-Modifikationen. Österr. Akad. Wiss., Math.-naturwiss. Kl., Anzeiger Jg. 1970, 184-185.

_____ (1970c) Messung des Ultrarot-Pleochroismus von Mineralen. XII. Der Pleochroismus der OH-Streckfrequenz in Disthen. Tschermaks Mineral. Petr. Mitt. 16, 129-135.

_____ (1970d) Messung des Ultrarot-Pleochroismus von Mineralen. XIII. Der Pleochroismus der OH-Streckfrequenz in Axinit. Tschermaks Mineral. Petr. Mitt. 15, 71-80.

_____ and Zemann, J. (1970) Messung des Ultrarot-Pleochroismus von Mineralen. XI. Der Pleochroismus der OH-Streckfrequenz in Rutil, Anatas, Brookit und Cassiterit. Tschermaks Mineral. Petr. Mitt. 15, 71-80.

_____ (1974a) Das Absroptionsspektrum des Benitoits in nahen Ultrarot. Tschermaks Mineral. Petr. Mitt. 21, 47-51.

_____ (1974b) UR-spektroskopischer Nachweis von H_2O in Nephelin. Tschermaks Mineral. Petr. Mitt. 21, 299-304.

_____ (1976) Messung des Ultrarot-Pleochroismus von Mineralen. XIV. Der Pleochroismus der OH-strekfrequenz in Diopsid. Tschermaks Mineral. Petr. Mitt. 23, 79-85.

_____ and Putnis, A. (1983) A model of the OH positions in olivine, derived from infrared-spectroscopic investigations. Phys. Chem. Minerals 9, 57-60.

_____, Hafner, S., and Zemann, J. (1983) Untersuchungen über den Einbau von Hydroxilgruppen im Edelstein-Sillimanit. N. Jahrb. Mineral. Mh. 1983, 219-226.

_____, Sturma, R., and Zemann, J. (1983) Ultrarotspektroskopische Untersuchungen über den OH-Gehalt einiger Granate. Österr. Akad. Wiss., Math.-naturwiss. Kl., Anzeiger Jg. 1983, 75-78.

_____ and Zemann, J. (1985) Polarized absorption spectra of sellaite from the Brumado mine, Brazil, in the near infrared. Bull. Geol. Soc. Finlande 57, 113-118.

_____ and Zemann, J. (1986) The pleochroism of a gem-quality enstatite in the region of the OH stretching frequency, with a stereochemical interpretation. Tschermaks Mineral. Petr. Mitt. 35, 19-25.

_____ (1986) A model of water allocation in alkali feldspar, derived from infrared-spectroscopic investigations. Phys. Chem. Minerals 13, 306-310.

_____, and Götzinger, M.A. (1987) The quantitative IR spectroscopic determination of structural OH groups in kyanites. Mineral. Petrol. 36, 41-49.

Burns, R.G. and Strens, R.G.J. (1966) Infrared study of the hydroxyl bonds in clinoamphiboles. Science 153, 890-892.

_____ and Prentice, F.J. (1968) Distribution of iron cations in the crocidolite structure. Amer. Mineral. 53, 770-776.

_____ and Law, A.D. (1970) Hydroxyl stretching frequencies in the infrared spectra of anthophyllites and gedrites. Nature 226, 73-75.

_____ and Greaves, C.J. (1971) Correlation of infrared and Mössbauer site population measurements of actinolites. Amer. Mineral. 56, 2010-2033.

Caruba, R. and Iacconi, P. (1983) Les zircons des pegmatites de Narssârssuk (Groënland)- l'eau et les groupements OH dans les zircons métamictes. Chemical Geol. 38, 75-92.

_____, Baumer, A., Ganteaume, M., and Iacconi, P. (1985) An experimental study of hydroxyl groups and water in synthetic and natural zircons: a model of the metamict state. Amer. Mineral. 70, 1224-1231

Chakraborty, D. and Lehmann, G. (1976) Distribution of OH in synthetic and natural quartz crystals. J. Solid State Chem. 17, 305-311.

Farmer, V.C. (1974) Layer silicates. In: Infrared Spectra of Minerals, V.C. Farmer, ed. Mineral. Soc., London.

Farrell, E.F. and Newnham, R.E. (1967) Electronic and vibrational absorption spectra in cordierite. Amer. Mineral. 52, 380-388.

Flörke, D.W., Köhler-Herbertz, B., Langer, K. and Törges, I. (1982) Water in microcrystalline quartz of volcanic origin: agates. Contrib. Mineral. Petrog. 80, 329-333.

Frondel, C. (1982) Structural hydroxyl in chalcedony (type B quartz). Amer. Mineral. 67, 1248-1257.

Freund, F. and Oberheuser, G. (1986) Water dissolved in olivine: a single-crystal infrared study. J. Geophys. Res. 91, 745-761.

Gilkes, R.J., Young, R.C., and Quirk, J.P. (1972) The oxidation of octahedral iron in biotite. Clays & Clay Minerals 20, 303-315.

Goldman, D.S, Rossman, G.R. and Dollase, W.A. (1977) Channel constituents in cordierite. Amer. Mineral. 62, 1144-1157.

_____, Rossman, G.R. and Parkin, K.M. (1978) Channel constituents in beryl. Phys. Chem. Minerals 3, 225-235.

Graetsch, H., Flörke, O.W. and Miehe, G. (1985) The nature of water in chalcedony and opal-c from Brazilian agate geodes. Phys. Chem. Minerals 12, 300-306.

Hanisch, K. (1966) Messung des Ultrarot-Pleochroismus von Mineralen. VI. Der Pleochroismus der OH-Streckfrequenz in Riebeckit. N. Jahrb. Mineral. Mh. 1966, 109-112.

Hawthorne, F.C. (1981) Amphibole spectroscopy. In: D.R. Veblen, ed., Amphiboles and Other Hydrous Pyriboles- Mineralogy. Rev. Mineral. 9A, 103-139.

_____ (1983) Characterization of the average structure of natural and synthetic amphiboles. Period. Mineral. Roma 52, 543-581.

Hofmeister, A.M. and G.R. Rossman (1985a) A spectroscopic study of amazonite: irradiative coloration of structurally hydrous Pb-bearing feldspar. Amer. Mineral. 70, 794-804.

_____ and G.R. Rossman (1985b) A model for the irradiative coloration of smoky feldspar and the inhibiting influence of water. Phys. Chem. Minerals 12, 324-332.

_____ and G.R. Rossman (1986) A spectroscopic study of blue radiation coloring in plagioclase. Amer. Mineral. 71, 95-98.

Kats, A. (1962) Hydrogen in alpha quartz. Philips Research Reports 17, 133-195; 201-279.

Kekulawala, K.R.S.S., Paterson, M.S., and Boland, J.N. (1981) An experimental study of the role of water in quartz deformation. In: Mechanical Behavior of Crustal Rocks, Geophysical Monograph 24, Amer. Geophys. Union., 49-60.

Kitamura, M., Kondoh, S., Morimoto, N., Miller, G.H., Rossman, G.R. and Putnis, A. (1987) Planar OH-bearing defects in mantle olivine. Nature 328, 143-145.

Kronenberg, A.K., Kirby, S.H., Aines, R.D., and Rossman, G.R. (1986) Solubility and diffusional uptake of hydrogen in quartz at high water pressures: implications for hydrolytic weakening. J. Geophys. Res. 91, 12723-12744.

Labotka, T.C. and Rossman, G.R. (1974) The infrared pleochroism of lawsonite: The orientation of the water and hydroxide groups. Amer. Mineral. 59, 799-806.

Langer, K. and Flörke, O.W. (1974) Near infrared absorption spectra (4000-9000 cm^{-1}) of opals and the role of "water" in alpha-quartz. Fortschr. Mineral. 52, 17-51.

Langer, K. and Lattard, D. (1980) Identification of a low-energy OH-valence vibration in zoisite. Amer. Mineral. 65, 779-783.

Law, A.D. (1973) A model for the investigation of hydroxyl spectra of amphiboles. In: Physics and Chemistry of Minerals and Rocks, R.G.J. Strens, ed. NATO Adv. Study Inst. "Petrophysics". Wiley, New York.

Mackwell, S.J., Kohlstedt, D.L. and Paterson, M.S. (1985) The role of water in the deformation of olivine single crystals. J. Geophys. Res. 90B, 11319-11333.

Maresch, W.V. and Langer, K. (1976) Synthesis, lattice constants and OH-valence vibrations of an orthorhombic amphibole with excess OH in the system $Li_2O-MgO-SiO_2-H_2O$. Contrib. Mineral. Petrol. 56, 27-34.

McMillan, P., Jakobsson, S., Holloway, J., and Silver, L.A. (1983) A note on the Raman spectra of water-bearing albite glasses. Geochim. Cosmochim. Acta 47, 1937-1943.

_____, Peraudeau, G., Holloway, J., and Coutures, J.P. (1986) Water solubility in a calcium aluminosilicate melt. Contrib. Mineral. Petrol. 94, 178-182.

Miller, G.H., Rossman, G.R., and Harlow, G.E. (1987) The natural occurrence of hydroxide in olivine. Phys. Chem. Minerals 14, 461-472.

Mottana, A. and Griffen, W.L. (1986) Crystal chemistry of two coexisting K-richterites from St. Marcel (Val d'Aosta, Italy). Amer. Mineral. 71, 1426-1433.

Nakamoto, K., Margoshes, M. and Rundle, R.E. (1955) Stretching frequencies as a function of distances in hydrogen bonds. J. Amer. Chem. Soc. 77, 6480-6488.

_____ (1978) Infrared and Raman Spectra of Inorganic and Coordination Compounds. J. Wiley & Sons, New York.

Newman, S., Stolper, S.M., and Epstein, S. (1986) Measurement of water in rhyolitic glasses: calibration of an infrared spectroscopic technique. Amer. Mineral. 71, 1527-1541.

Nikitina, L.P., Petkevich, E.Z., Sverdlova, O.V., and Khristoforov, K.K. (1973) Determination of occupancy of octahedral positions in the structures of amphiboles $(Ca,Na,K)_{2-3}$ $(Fe,Mg,Al)_5[Si,Al]_4O_{11}(OH)_2$ from vibrations of OH^-. Geochem. Int'l 11, 1233-1239. Translated from Geokhimiya, 1661-1668

Novak, A. (1974) Hydrogen bonding in solids. Correlation of spectroscopic and crystallographic data. Structure and Bonding 18, 177-216.

Paterson, M. (1982) The determination of hydroxyl by infrared absorption in quartz, silicate glasses, and similar materials. Bull. Minéral. 105, 20-29.

Rossman, G.R. (1975) Joaquinite: the nature of its water content and the question of four-coordinated ferrous iron. Amer. Mineral. 60, 435-440.

_____ (1984) S.W. Bailey, ed., Micas. Rev. Mineral. 13, 145-181

_____ (1986) The hydrous component in garnets. Abstracts with Program, The 14th general meeting of the Int'l Mineral. Assoc., Stanford, Calif., p. 216

_____ and Aines, R.D. (1986) Spectroscopy of a birefringent grossular from Asbestos, Quebec, Canada. Amer. Mineral. 71, 770-780.

_____, Rauch, F., Livi, R., Tombrello, T.A., Shi, C.R., and Zhou, Z.Y. (1988) Nuclear reaction analysis of hydrogen in almandine, pyrope, and spessartite garnets. N. Jahrb. Mineral. Mh. (in press).

Rouxhet, P.G. (1970) Hydroxyl stretching bands in micas: A quantitative interpretation. Clay Minerals 8, 375-388.

Rowbotham, G. and Farmer, V.C. (1973) The effect of "A" site occupancy upon the hydroxyl stretching frequency in clinoamphiboles. Contrib. Mineral. Petrol. 38, 147-149.

Sahl, K. (1966) Messung des Ultrarot-Pleochroismus von Mineralen. VI. Der Pleochroismus der OH-Streckfrequenz in Datolith. N. Jahrb. Mineral. Mh. 1966, 42-57.

Sanz, J. and Stone, W.E.E. (1983) NMR study of minerals. III. The distribution of Mg^{2+} and Fe^{2+} around the OH groups in micas. J. Phys. C16, 1271-1281.

Scholze, H. (1959) Der Einbau des Wassers in Gläsern II. Glastechn. Ber. 32, 142-152.

Schreyer, W. (1986) The mineral cordierite: structure and reactions in the presence of fluid phases. Ber. Bunsenges. Phys. Chem. 90, 748-755.

Strens, R.G.J. (1966) Infrared study of cation ordering and clustering in some (Fe,Mg) amphibole solid solutions. Chem. Comm. 15, 519-520.

_____ (1974) The common chain, ribbon, and ring silicates. In: The infrared structure of minerals, V.C. Farmer, ed., 305-330. Mineralogical Society, London

Thompson, W.K. (1965) Infrared spectroscopic studies of aqueous systems, I. Trans. Faraday Soc., 61, 1635-1640.

Velde, B. (1983) Infrared OH-stretch bands in potassic micas, talcs and saponites; influence of electronic configuration and site of charge compensation. Amer. Mineral. 68, 1169-1173.

Wilkins, R.W.T. and Sabine, W. (1973) Water content of some nominally anhydrous silicates. Amer. Mineral. 58, 508-516.

Wood, D.L. and Nassau, K. (1967) Infrared spectrum of foreign molecules in beryl. J. Chem. Phys. 47, 2220-2228.

OPTICAL SPECTROSCOPY

INTRODUCTION

Optical spectra are concerned with the quantitative measurement of the absorption, reflection and emission of light in the near-ultraviolet (~250 nm) through the mid-infrared (~3000 nm) portions of the spectrum. The reflection spectroscopy of minerals has been motivated largely by interest in remote sensing. Studies of metal ion site occupancy, oxidation states and concentrations have generally been done with absorption spectroscopy. This chapter will concentrate on single crystal absorption spectroscopy.

RESOURCES

A variety of books and articles review the spectroscopy of minerals. These are listed in Table 1. The book by Burns (1970) remains the most comprehensive general introduction to the field of optical spectroscopy of minerals, with many examples of mineral spectra. Chapters by Burns (1985a,b) serve as useful abbreviated introductions to the field. The book by Marfunin (1979) is a wide ranging introduction which goes more deeply into the theoretical aspects of mineral spectroscopy, but which does not have many examples of actual data. Both Burns (1970) and Marfunin (1979) provide extensive lists of references to mineral spectra. The language of spectroscopy makes use of the terminology of group theory to describe individual orbitals, symmetry properties of molecular units, and individual electronic transitions. A variety of books from the chemical sciences discuss this nomenclature and the theory behind it, although none of these books is directed at mineralogical applications; Cotton (1971) is one of the most useful introductions to group theory.

GENERAL CONCEPTS

A number of textbooks present discussions of the theories which govern electronic transitions. Marfunin (1979), in particular, develops the theory with many mineralogical examples. A number of the important concepts are briefly outlined below.

Four type of processes generally contribute to the optical absorption spectra of minerals:

1) electronic transitions involving electrons in the d-orbitals of ions of the first row transition elements such as Cr^{3+}, Mn^{2+}, Mn^{3+}, Fe^{2+} and Fe^{3+}. These transitions involve rearrangement of the valence electrons, and give rise to absorption in the visible and near-infrared region. The spectra they produce are often called crystal-field spectra after the theory used to describe them. They are a major cause of color in many minerals.

2) electronic transitions which involve displacement of charge density from one ion to another. These charge transfer processes are of two general types.

The first typically involves charge transfer between an anion and a cation. The one most commonly encountered in mineral spectroscopy is the transfer of electron density from a filled oxygen orbital to a partially occupied Fe^{3+} orbital. These transitions require higher energies

Table 1.   Selected references to mineral spectroscopy.

Books:

Burns, R.G. (1970)   Mineralogical Applications of Crystal Field Theory.   Cambridge University Press,   Cambridge. An extensive review of the spectra of minerals.

Cotton, F.A. (1971)   Chemical Applications of Group Theory. 2d edition, J. Wiley, New York.   An excellent text on group theory.   Does not deal with mineralogical examples,   but covers the principles used to interpret their spectra.

Karr, C., editor (1975)   Infrared and Raman Spectroscopy of Lunar and Terrestrial Minerals.   Academic Press,   New York.   Contains chapters about the optical spectra of minerals.

Platonov, A.N. (1976)   The Nature of the Color of Minerals. Scientific Publishers, Kiev.   A comprehensive review of mineral spectra primarily from the author's works and other Russian language papers (in Russian).

Strens, R.J.G., editor (1976) The Physics and Chemistry of Minerals and Rocks.   J. Wiley and Sons,   New York. Includes chapters on optical spectra of minerals.

Egan, W.G. and Hilgeman, T.W. (1979)   Optical Properties of Inhomogeneous Materials.   Academic Press,   New York. Concerned with the theory of reflectance spectroscopy of solids with several mineralogical examples.

Marfunin, A.S. (1979)   Spectroscopy,   Luminescence and Radiation Centers in Minerals.   Springer Verlag,   New York.   A broad review of the principles of mineral spectroscopy.

Nassau, K. (1983) The Physics and Chemistry of Color: the Fifteen Causes of Color.   John Wiley and Sons, New York. A wide ranging introduction to color in solids including many examples of color and spectroscopy in minerals and gemstones.

Berry, F.J. and Vaughan, D.J. (1985)   Chemical Bonding and Spectroscopy in Mineral Chemistry.   Chapman and Hall, London.   Several chapters on the optical spectra of minerals.

Reviews in Mineralogy Series.   Mineralogical Society of America.   Chapters on the spectroscopy of minerals are contained in volumes 2 (Feldspars, 2d edition, 1983), 7 (Pyroxenes, 1980), 9a (Amphiboles, 1981), 13 (Micas, 1984), and 14 (Microscopic to Macroscopic, 1985).

Articles:

Loeffler, B.M. and Burns, R.G. (1976)   American Scientist, 64, 636-647.   A general introduction to color and spectroscopy in minerals.

Lehmann, Gerhard (1978)   Farben von Mineralen und ihre Ursachen.   Fortschr. Mineral. 56, 172-252.   A review on the color and spectroscopy of minerals.   (in German)

---

than crystal field transitions and produce absorption bands which are centered in the ultraviolet region. In the case of ions in higher oxidation states such as Fe^{3+} and Cr^{6+}, the wing of the absorption band will extend into the visible part of the spectrum, causing absorption which is strongest in the violet and extends towards the red. The orange-red color of the CrO_4^{2-} ion and the yellow-brown color of some Fe^{3+} minerals is a result of this wing of absorption. These transitions may be described using molecular orbital theory. There has been very little experimental work on this type of charge transfer in minerals due to the difficulty of preparing samples thin enough to keep these high intensity absorptions "on scale" (Fig. 1).

The second type of charge transfer transition is <u>intervalence charge transfer</u> (IVCT), also called metal-metal charge transfer. It involves movement of electron density between metal ions in different oxidation states. The pairs or clusters of cations typically share edges of coordination polyhedra. Relatively low concentrations of these pairs can produce appreciable absorption. The deep blue of sapphire is a familiar example of color caused by this type of transition. IVCT between Fe^{2+} and Fe^{3+} is frequently encountered in minerals (Fig. 2), and Fe^{2+}/Ti^{4+} interactions are also common (Fig. 3). In meteoritic minerals, the Ti^{3+}/Ti^{4+} interaction is also prominent.

3) Absorption edges result from electronic transitions between the top of a valence band and the bottom of the conduction band. Any photon with energy greater than this band gap will be absorbed. These type of absorption bands are usually encountered in sulfides (Fig. 4). The red color of cinnabar is the result of a band gap which allows light with wavelength longer than 600 nm to pass while absorbing shorter wavelengths. The band-gap of silicate minerals typically is located far into the ultraviolet.

4) Overtones of vibrational transitions. The most commonly encountered bands in the near-infrared are the overtones of OH and H_2O groups. The spectrum of beryl (Fig. 5) contains both absorption from Fe^{2+} and the vibrational modes of H_2O molecules. Vibrational overtones are readily recognized because they have much smaller widths than electronic transitions which can occur in the same spectral region. Vibrational overtones are treated separately in Chapter 3.

Additional processes are important in the case of a few minerals. Noteworthy are transitions involving f-orbitals of uranium and the rare earth elements. The absorption bands from the trivalent rare-earths tend to be much sharper than most of the bands from the third-row transition metals (Fig. 6). Also important in many mineral spectra are the various absorptions associated with electron-hole centers and molecular ions produced by ionizing radiation (Fig. 7). The spectra of these centers can often be quite difficult to interpret. Smoky quartz, blue feldspar, green diamonds and blue calcite are other examples of this process.

UNITS

The basic units of optical spectra are discussed in chapter 1. Interconversions among the units commonly used for mineral spectra are:

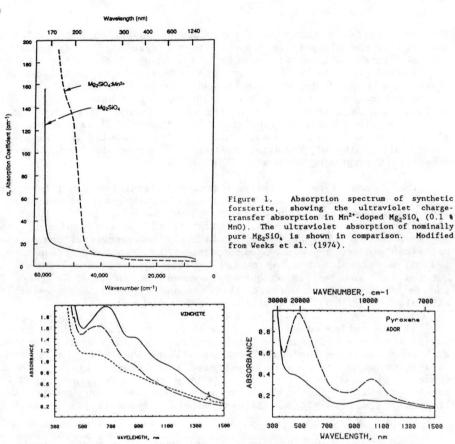

Figure 1.   Absorption spectrum of synthetic forsterite, showing the ultraviolet charge-transfer absorption in Mn^{2+}-doped Mg_2SiO_4 (0.1 % MnO). The ultraviolet absorption of nominally pure Mg_2SiO_4 is shown in comparison. Modified from Weeks et al. (1974).

Figure 2 (left).   Absorption spectrum of the sodic-calcic amphibole, winchite, from Ottawa Valley, Canada, which is dominated by the Fe^{2+}/Fe^{3+} intervalence charge-transfer band near 700 nm.   Solid line γ-spectrum; dashed line β-spectrum; short dashed line: α-spectrum. Fe^{2+} features are present near 900 and 1150 nm, and an Fe^{3+} feature occurs near 440 nm.

Figure 3 (right).   Absorption spectrum of fassaite from the Angra dos Reis meteorite, showing a prominent Fe^{2+}/Ti^{4+} intervalence charge-transfer band at 490 nm. Fe^{2+} features at the M1 and M2 sites occur in the 900 to 1300 nm region.   Dashed line: β-spectrum; solid line: approximately γ orientation; crystal 35 μm thick.

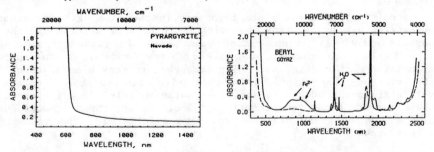

Figure 4 (left).   Absorption spectrum of pyrargyrite, Ag_3SbS_3, from Tohopah, Nevada, showing the abrupt onset of intense absorption at the absorption edge near 630 nm.   Omega (E⊥c) spectrum of a 0.12 mm thick crystal. Data from D. Beaty and G.R. Rossman.

Figure 5.   Absorption spectrum of a golden-yellow beryl crystal from Goyaz, Brazil, showing a pair of overlapping Fe^{2+} absorption bands at 820 and 970 nm and a series of sharp bands near 970, 1150, 1400, and 1900 nm which represent overtones and combination vibrations of molecular H_2O in beryl.   Solid line E∥c; dashed line E⊥c;   Crystal thickness 5.0 mm.   From Goldman et al. (1978)

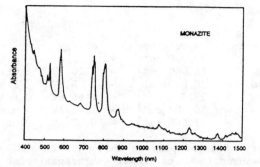

Figure 6. Absorption spectrum of monazite from Montgomery County, North Carolina, showing the comparitively sharp absorption bands of the rare-earth elements Nd^{3+} (all features below 900 nm) and Sm^{3+} (three features near 1075, 1225 and 1375 nm). From Bernstein (1982). Sample thickness: 180 μm.

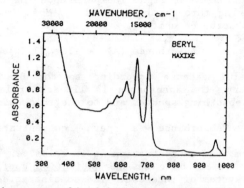

Figure 7. Absorption spectrum of beryl from Minas Gerais, Brazil, showing features near 600 nm from a color center identified as CO_3^- by Edgar and Vance (1977). This variety is known as Maxixe beryl. The sharp features are vibrational features superimposed on the electronic transition.

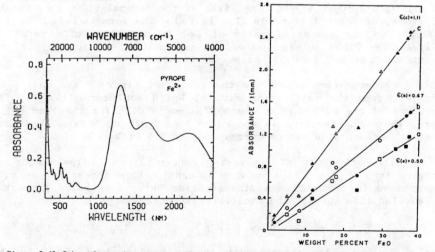

Figure 8 (left). Absorption spectrum of a pyrope-almandine garnet from Tanzania with three components in the near-infrared at about 1250, 1600 and 2200 nm due to Fe^{2+} at the distorted eight-coordinated cubic site. Weaker, sharper features between 350 and 800 nm are spin-forbidden bands. Plotted for 1.0 mm thickness.

Figure 9 (right). Correlation of the absorption intensity of Fe^{2+} at the eight-coordinated site of pyrope-almandine garnets with Fe^{2+} concentration. From White and Moore (1972b). The linear dependence of absorbance on Fe^{2+} concentration in this plot indicates that absolute Fe^{2+} concentrations can be readily obtained spectroscopically.

wavelength		wavenumber		energy	color
333 nm	=	30,000 cm^{-1}	=	3.72 eV	ultraviolet
400 nm	=	25,000 cm^{-1}	=	3.10 eV	violet
500 nm	=	20,000 cm^{-1}	=	2.48 eV	green
750 nm	=	13,333 cm^{-1}	=	1.65 eV	red
1000 nm	=	10,000 cm^{-1}	=	1.24 eV	near infrared
2000 nm	=	5,000 cm^{-1}	=	0.62 eV	near infrared

The intensity of light passing through a crystal at a given wavelength can be measured in units of percent transmission

$$\%T = 100 \cdot I/I_0, \tag{1}$$

where I_0 is the intensity of light incident upon the crystal, and I is the intensity passing through the crystal. Most instruments present the intensity in absorbance units:

$$\text{Absorbance (A)} = -\log_{10}(I/I_0). \tag{2}$$

This unit is useful because according to the Beer-Lambert law of solution colorimetry, the absorbance is linearly related to the concentration of the absorbing species and to the thickness of the sample:

$$\text{Absorbance} = \epsilon \cdot \text{path} \cdot \text{concentration}, \tag{3}$$

where path is in cm, and concentration is in moles per liter. The molar absorptivity, also referred to as the extinction coefficient, ϵ, is a constant for each system of interest. The concentration in a "liter" of crystal can also be calculated as follows:

Consider the concentration of Fe^{2+} in a grossular garnet of density 3.61 and with 1.10 wt % FeO. One liter of the crystal (1000 cm^3) weighs 3610 grams, of which 1.1% or 39.71 g is FeO. The formula weight of FeO is 71.85, so the number of moles of FeO in one liter of crystal is 39.71/71.85 = 0.553 moles. There is one mole of Fe in FeO, so the concentration of Fe^{2+} is 0.553 moles per liter.

The concentration of Fe^{2+} in the garnet of Figure 8 can easily be determined from the calibration data of White and Moore (1972). The absorbance of the Fe^{2+} band at about 1250 nm is 0.66 for the 1 mm thick sample, where the baseline is chosen to be at 920 nm. From Figure 9, the absorbance of 0.66 per mm corresponds to 9.5 wt % FeO.

An alternative way of expressing concentration is the number of absorbing ions in 1 cm^3 of crystal. Absorption band intensity can also be expressed in terms of the dimensionless unit, oscillator strength. For Gaussian line shapes, the oscillator strength is:

$$f = 9.20 \cdot 10^{-9} \cdot \epsilon_0 \cdot \delta, \tag{4}$$

where ϵ_0 is the molar absorptivity at the band maximum and δ is the half-width (in wavenumbers, cm^{-1}) at half-height.

THE EXPERIMENT

Qualitative spectra which tell where the absorption bands are can be obtained by measuring the transmission of light through the sample without regard to sample preparation, as long as the sample is not too thick to allow adequate transmission for the instrument to measure. However, the samples are usually oriented (by morphology, optics or X-ray) so that light propagates along a crystal axis or an axis of the optical indicatrix. Windows are polished on the two sides of the sample to improve the transmission quality and to exactly define the sample thickness. For optimum results, the thickness of the sample should be adjusted (often by trial and error) so that the absorption bands are "on scale". For a darkly colored mineral such as augite, the optimal thickness may be a few tens of micrometers, whereas a pale blue beryl (aquamarine) may need to be a centimeter thick. Because much of the important spectroscopic information is in the infrared region (especially in the case of Fe), it is often difficult to visually estimate the appropriate sample thickness.

Because the sample area required is determined by the transmission quality of the sample and by the spectrophotometer being used, there are no fixed rules on sample size. However, some guidelines can be presented. Microscope spectrophotometers can work with samples a few tens of micrometers in diameter (Langer and Abu-Eid, 1977; Burns, 1970), although necessarily compromising the orientational purity of the incident light because of the convergence of the microscope optics (Goldman and Rossman, 1978b). Conventional spectrometers can work with samples on the order of 100 μm in diameter in the visible spectral region using photomultiplier tubes, but usually require larger diameters in the near-infrared because of the lower sensitivity of the PbS photo-detectors used in this energy region. Fourier transform spectrometers with cryogenic indium antimonide detectors can work in the near-infrared region with samples a few tens of μm in diameter. In general, larger sample diameters give better quality data.

For minerals of symmetry lower than cubic, it is necessary to obtain separate spectra with linearly polarized light vibrating along each axis of the optical indicatrix. In the case of biaxial crystals, they are called the alpha, beta and gamma spectra, where the alpha spectrum is obtained in the X vibration direction in which the alpha refractive index would be measured. Unfortunately, particular care must be used in correlating the morphological directions of an orthorhombic crystal with the optical directions because different authors have used different conventions for naming both the morphologic and crystal-lographic axes. Dowty has discussed some of the special requirements for a full optical description of monoclinic minerals (Dowty, 1978)

For uniaxial crystals, two spectra are required. They are taken with polarized light vibrating both parallel and perpendicular to the c axis. The spectra are usually called E$\parallel$c and E⊥c spectra or the ϵ and ω spectra where ϵ corresponds to the ϵ index of refraction obtained with light polarized E$\parallel$c.

INTENSITIES AND SELECTION RULES

Absorption bands can vary greatly in their intensity. Ruby with 0.1 % Cr^{3+} is deep red when 1 mm thick, whereas orthoclase with 0.1% Fe^{3+} is essentially colorless at the same thickness. A variety of selection rules derived from quantum mechanics governs the intensity of the various types of absorption phenomena.

Laporte selection rule

Transitions between two d-orbitals or two p-orbitals are forbidden, but transitions between s- and p-orbitals or between p- and d-orbitals are allowed (Laporte Selection Rule).

This means that oxygen-to-metal charge-transfer transitions (from an oxygen p-orbital to a metal d-orbital) will occur with high probability. Photons of the appropriate energy will stand a high probability of being absorbed, and the resulting absorption band will be intense. Transitions within the d-orbitals will occur with low probability and will produce correspondingly low intensity absorptions. The electronic configuration of Fe^{2+} in octahedral coordination is shown in Figure 10, together with the configuration after the absorption of a photon (an excited state configuration).

In practice, these "rules" do not rigidly control the intensity of spectra because there are various ways to weaken them. A common factor is mixing of d- and p-orbitals to produce a hybrid state which has character intermediate between the two extremes. Nevertheless, the intensity of crystal-field absorption bands from ions of the common transition metals will be about a factor of 1000 times less than the oxygen-to-metal charge-transfer bands.

Spin-forbidden transitions

An additional selection rule is the spin-multiplicity selection rule. It states that the total number of unpaired electrons on an atom must remain the same before and after an electronic transition occurs. Because all orbitals in Mn^{2+} and Fe^{3+} are half-filled, the only possible electronic transitions involve pairing electrons in an orbital with the necessity of changing the spin of one of the electrons (Fig. 11). Consequently, the Mn^{2+} and Fe^{3+} transitions are spin-forbidden. In fact, they can occur, but with very low probability. Spin-forbidden bands can occur in the spectrum of any ion with two or more valence electrons (Table 2)

The rules can be bent when ions get together

The intensities of metal ion absorption bands do not always follow these guidelines. A number of authors have noted that the intensity of absorption from Fe^{3+} paired with another Fe^{3+} through a shared anion is often much higher than the absorption from an isolated Fe^{3+} (Krebs and Maisch, 1971; Rossman, 1975, 1976a,b; Bakhtin and Vinokurov, 1978). This effect is related to the strength of the magnetic coupling between the pair of cations. Figure 12 compares the spectra of two minerals which contain planar clusters of four octahedrally coordinated Fe^{3+} (Fig. 13). In amarantite, the bridging anion is O^{2-} which is conducive to strong magnetic coupling; in leucophosphite, the bridging anion is OH which leads to weak coupling. The absorption intensity of amarantite is more than an order of magnitude greater than that of leucophosphite. The dark orange to red colors of many ferric iron minerals (such as hematite) are caused by such interactions.

The intensity of Fe^{2+} absorption bands (d-d transitions) can also be greatly enhanced by interactions with neighboring Fe^{3+} ions (Smith,

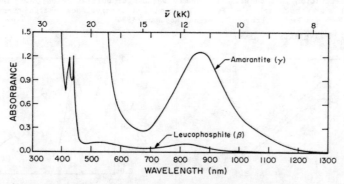

Fe^{2+} d^6 (left)   Fe^{3+} d^5 (right)

Figure 10 (left). Diagram of a spin-allowed electronic transition of Fe^{2+}. When the cation is in octahedral coordination, the 3d orbitals are split into two sets of different energy. The orbitals in the higher energy set are called the e_g orbitals and ones in the lower energy set are called the t_{2g} orbitals. An electronic transition frequently involves excitation of an electron from the t_{2g} set to the e_g set. In the case of Fe^{2+} with six valence electrons, the paired electron in the t_{2g} set can be promoted to the e_g set without changing the total electron spin. Such a transition is spin-allowed.

Figure 11 (right). On the left, the ground state electron configuration for the 3d valence orbital of Fe^{3+}. The ion contains 5 valence electrons which are distributed 1 each into the 5 valence orbitals of the 3d shell. The pair of orbitals at higher energy is the e_g pair; the triply degenerate orbitals at lower energy are the t_{2g} orbitals. On the right, one of several possible excited state electronic configurations which change the total spin of the ion by pairing two electrons in the same orbital. All possible electronic transitions necessarily involve pairing electrons in the excited state. This can only happen if an electron "flips" its spin in the process. Such transitions are spin-forbidden.

Figure 12. Comparison of the absorption spectrum of amarantite, $[Fe_4O_2(SO_4)_4]\cdot14H_2O$, from Sierra Gorda, Chile, and leucophosphite, $K_2[Fe_4(OH)_2(PO_4)_4]\cdot4H_2O$, from the Tip Top pegmatite, South Dakota, both plotted for a 100 μm thick crystal. The absorption intensity of the Fe^{3+} band in the amarantite spectrum near 880 nm is an order of magnitude more intense than for leucophosphite in spite of the similar Fe^{3+} concentrations in both crystals. This is due to magnetic interaction between the cations. From Rossman (1976b).

Table 2. Valence orbital configuration of metal ions encountered in minerals.

			Ti^{3+} d^1	Ti^{4+} d^0					
			V^{3+} d^2	V^{4+} d^1	V^{5+} d^0				
	Cr^{2+} d^4	Cr^{3+} d^3				Cr^{6+} d^0
	Mn^{2+} d^5	Mn^{3+} d^4	Mn^{4+} d^3			
	Fe^{2+} d^6	Fe^{3+} d^5				
	Co^{2+} d^7					
	Ni^{2+} d^8					
Cu^{1+} d^{10}	Cu^{2+} d^9					

d^2 means two electrons in the 3d valence orbital. For example, the total electronic configuration for Ti^{3+} is 1s^2 2s^2 2p^6 3s^2 2p^6 3d^1.

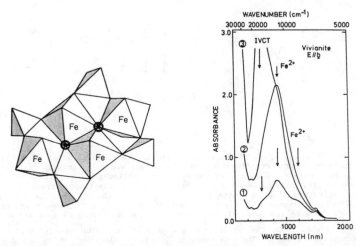

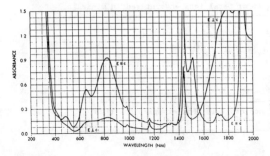

Figure 13(left).   The structure of the tetrameric cluster common to amarantite and leucophosphite.   In amarantite, the four Fe^{3+} are bridged by O^{2-} which is conducive to strong magnetic interaction among the cations.   In leucophosphite, the four Fe^{3+} are bridged by OH^- which is not conducive to strong interactions.   From Rossman (1976b).

Figure 14 (right).   Vivianite absorption spectrum of a single crystal with zones of three different degrees of oxidation.   The colorless zone (1) shows predominantly Fe^{2+} absorption; the light blue zone (2) shows enhanced Fe^{2+} absorption even though the total Fe^{2+} concentration must be slightly reduced because of partial oxidation.   The dark blue zone (3) shows both enhanced Fe^{2+} absorption and intervalence charge transfer (IVCT) between Fe^{2+} and Fe^{3+}.   From Amthauer and Rossman (1984)

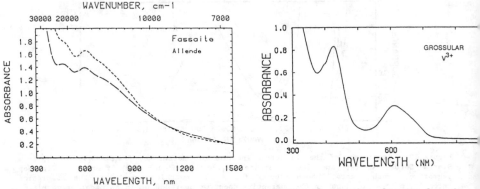

Figure 15 (left).   Absorption spectrum of a thin section of titanian fassaite from the Allende meteorite.   The two features at 500 and 620 nm are tentatively interpreted as Ti^{3+} absorption sitting on top of an intense Ti^{3+}/Ti^{4+} intervalence charge-transfer feature centered near 750 nm.

Figure 16 (right).   Absorption spectrum of V^{3+} in grossular from Kenya. The features near 426 and 607 nm are from the V^{3+}.   Sample 1.00 mm thick.   These garnets with typically 0.5 to 2 wt % V are a brilliant green color.

Figure 17.   Absorption spectrum   of gre apophyllite from Poona, India, showing t features of V^{4+} near 830 nm. H_2O vibration bands occur near 1450 nm and 1900 nm. Sample 2.64 mm thick with 1600 ppm vanadium.   Fr Rossman (1974a)

1978b; Amthauer and Rossman, 1984; Mattson and Rossman, 1987b). This effect is in addition to the intervalence charge-transfer absorption which generates a new absorption feature. Figure 14 shows spectra from a single crystal of vivianite. The structure of vivianite contains pairs of edge-sharing Fe^{2+} octahedra in which the Fe^{2+} cations are subject to partial oxidation. This crystal has three zones of different degrees of oxidation (Fe^{3+} content). The first zone contains predominantly Fe^{2+} and gives a spectrum dominated by the Fe^{2+} features. The second zone contains a small amount of Fe^{3+} which produces enhancement of the Fe^{2+} absorption. The third zone contains a greater amount of Fe^{3+} which produces both the enhanced Fe^{2+} band and the intervalence charge-transfer band. The theoretical details of such enhancements have yet to appear.

Beer's Law plots

To determine quantitative site-occupancies or total cation content, the ϵ value from Beer's law must be known for the system of interest, and it must be reasonably constant in a solid solution series. Very few mineral systems have been examined in enough detail to establish Beer's law plots. When such information is available, quantitative site occupancy data can be obtained. Those which have a nearly linear correlation between absorbance and concentration are Fe^{2+} in garnets (White and Moore, 1972), olivines (Hazen et al., 1977b), orthopyroxenes (Goldman and Rossman, 1978b) and feldspars (Mao and Bell, 1973b; Hofmeister and Rossman, 1984). ϵ values for a number of other ions in minerals are available, but they represent a single measurement and do not explore the variation of ϵ with composition along a solid solution series or with concentration of the absorbing ion.

IDENTIFICATION OF THE OXIDATION STATES OF CATIONS

The spectra of many cations are sufficiently distinctive to allow identification of their oxidation states (e.g. Fe^{2+} vs. Fe^{3+}; Mn^{2+} vs. Mn^{3+}; Cr^{3+} vs. Cr^{6+}). When a metal ion is present in two oxidation states, it is difficult to determine the quantitative ratio because of inadequate calibration standards available for the optical data.

Titanium

Ti^{4+} has no valence d-electrons and as such does not have absorption in the visible region by itself. Ti^{3+} has one valence electron which can be excited to a higher energy by visible light. Consequently, Ti^{3+} minerals will have absorption in the optical region. This oxidation state is formed only under conditions more reducing than those found in the terrestrial environment; Ti^{3+} is found primarily in meteorites, lunar samples and synthetics. Ti^{3+} in octahedral coordination is a comparitively weak absorber and is easily overpowered by other cations which might be present. Two complicating factors must be considered: Ti^{4+} can enter into intervalence charge transfer with Fe^{2+} to produce strong absorption (Manning, 1977; Burns, 1981). This process has been implicated in the color of a variety of minerals such as blue sapphire and orange-brown micas. Many early reports of Ti^{3+} in the spectra of terrestrial minerals are probably due to the Fe^{2+}-Ti^{4+} interaction.

Ti^{4+} can also enter into intervalence charge transfer with Ti^{3+} producing intense absorption in the red to near-infrared. The blue color easily induced in synthetic rutile results from this process. The effect of mixed titanium oxidation states is seen in the spectra of meteoritic samples (Fig. 15).

Vanadium

Trivalent vanadium is occasionally encountered in minerals and is the cause of vivid colors in a number of gemstones. It produces a diversity of colors because the locations of its absorption bands are sensitive to the details of the size and symmetry of the vanadium site (Fig. 16). The spectroscopy of vanadium in minerals has been reviewed by Schemtzer (1982). The positions of the V^{3+} absorption bands are very close those of Cr^{3+} at the same site. Consequently, the color of a particular mineral is frequently similar regardless of whether the chromophore is Cr^{3+} or V^{3+} (e.g., both green grossular and green beryl can be colored by either Cr or V).

Vanadate fluxes are frequently used in mineral synthesis. The blue color of some of the resulting products comes from ~0.1% substitutional vanadium.

Tetravalent vanadium is less commonly encountered and produces a different spectrum which is centered in the red. Few mineralogical examples of this have been presented (Fig. 17).

Chromium

Chromium contributes to the spectra of minerals most commonly in the trivalent state. The familiar red color of rubies and green color of emeralds are both due to Cr^{3+}. Like vanadium, the exact position of the Cr^{3+} absorption bands will depend upon the particulars of the Cr site (Fig. 18). Because the molar absorptivity (ϵ value) of Cr^{3+} is high (typically 40-60) compared to Fe^{2+} (typically 3-6), a minor amount of Cr^{3+} (0.X%) can dominate the spectrum of many common ferro-magnesian silicates. The colors of green micas, pyroxenes, and amphiboles often result from traces of Cr^{3+} as much as from the primary iron component (Fig. 19).

Chromate, CrO_4^{2-}, produces intense colors through an oxygen to Cr^{6+} charge-transfer band centered in the ultraviolet. The brilliant orange and yellow colors of crocoite ($PbCrO_4$) and vanadinite ($Pb_5(VO_4)_3Cl$) are from the chromate ion. In the case of vanadinite, a small amount of CrO_4^{2-} is involved in a solid solution with VO_4^{3-}. Because of the experimental difficulty of obtaining spectra of the intensely absorbing chromate bands, few spectra of chromate in minerals are available.

Manganese

Manganese is widely distributed as Mn^{2+}. Because all electronic transitions of Mn^{2+} are spin-forbidden, the absorption by small amounts of Mn^{2+} in minerals is very weak and usually overpowered by the spectra of almost any other cations that may be present. The spectrum of Mn^{2+} in octahedral coordination consists of two weak bands at longer wavelength and a sharp band near 412 nm (Fig. 20). Only the sharp 412 nm band is seen in the spectrum of many minerals with minor amounts of Mn^{2+} in the presence of greater quantities of Fe^{2+}.

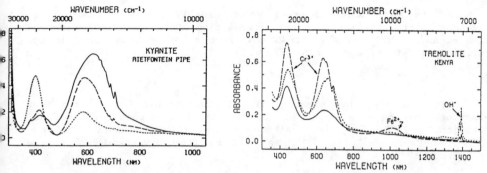

Figure 18. Absorption spectrum of blue kyanite from the Rietfontein pipe, South Africa. The structured absorption feature near 600 nm and the feature near 400 nm are due to Cr^{3+}. Sample plotted as 2.0 mm thick. α = short dash; β = long dash; γ = solid.

Figure 19. Absorption spectrum of tremolite from Kenya, with 0.22% chromium showing the characteristic pair of Cr^{3+} absorption bands near 650 and 440 nm. Small amounts of Cr^{3+} frequently contribute to the green color of calcic amphiboles. Plotted for 1.0 mm thickness.

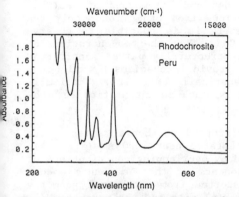

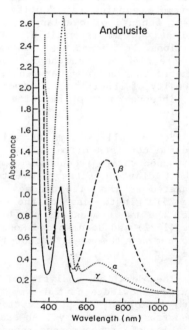

Figure 20. Absorption spectrum of Mn^{2+} in rhodochrosite from Pasto Bueno, Peru, showing the multitude of spin-forbidden Mn^{2+} bands. In octahedral coordination, Mn^{2+} spectra usually show two broader bands at low energy, followed by a sharper band near 412 nm. Plotted for 2.0 mm thick, omega orientation.

Figure 21. Absorption spectrum of andalusite with Mn^{3+} at the octahedral Al site. The two prominant absorption features near 450 and 700 nm are due to Mn^{3+}. The 260 μm thick crystal from Minas Gerais, Brazil, contains 1.09% manganese.

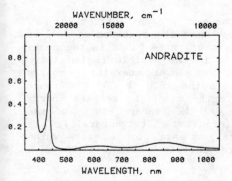

Figure 22. Andradite from Val Malenco, Italy. It shows a typical Fe^{3+} pattern with two broad bands at longer wavelengths (near 860 and 630 nm) followed by a sharp band near 440 nm. Plotted for 0.455 mm thick.

Mn^{3+} is an intense absorber which usually produces red through lavender colors. Common examples include piemontite, red tourmaline and pink micas. Mn^{3+} can also produce a green color such as found in the viridine variety of andalusite (Fig. 21).

Iron

More than any other element, iron is responsible for the color of most rock-forming minerals. The colors arise from a combination of absorption by Fe^{2+}, Fe^{3+}, intervalence charge transfer between Fe^{2+} and Fe^{3+} or between Fe^{2+} and Ti^{4+}, as well as from oxygen-to-Fe^{3+} charge transfer.

All electronic transitions of Fe^{3+}, like Mn^{2+}, are spin forbidden. Consequently, minerals with dilute Fe^{3+} are pale colored due to weak absorption. They generally show a characteristic pattern of two broad, low-energy absorptions, labeled T_{1g} and T_{2g}, and a sharp band near 440 nm (Fig. 22). Because the wavelengths of the two broad bands are sensitive to the exact structural details of the Fe^{3+} site, they are of use in remote sensing to identify iron oxides and polymorphs of FeO(OH). Often, because of overlapping contributions from other ions, only the band at 440 nm is visible. Occasionally it is possible to observe additional Fe^{3+} bands near 376 nm (Fig. 23). When the concentrations of Fe^{3+} are high or when the Fe^{3+} ions are structurally paired, antiferromagnetic interactions between Fe^{3+} ions provide mechanisms by which the intensity of absorption can increase dramatically (Fig. 12). The familiar, intense color of hematite arises from such an intensified Fe^{3+} absorption.

Fe^{2+} usually occurs at slightly distorted octahedral sites. Its spectrum consists of a pair of bands centered near 1000 nm (Fig. 24). These spectra are frequently complicated by overlapping contributions of Fe^{2+} at distinct sites. The two components of the Fe^{2+} bands become separated as the site becomes more distorted from octahedral geometry (Fig 25). In the case of the amphibole M(4) site and the pyroxene M(2) site, the two components can be separated by about 1000 nm (Fig. 26). Faye (1972) and Goldman and Rossman (1977b) have discussed the relationship between the magnitude of the splitting the the nature of the Fe^{2+} site.

The detailed analysis of an optical absorption spectrum first requires knowledge of the symmetry of the cation site. Selection rules which establish the polarization of the various bands are then determined from group theory. An example of such an analysis for Fe^{2+} at the M(2) site of orthopyroxene appears in Appendix A.

Sites of different coordination number present different Fe^{2+} spectra. The most extensively studied has been the spectrum of Fe^{2+} in the garnet 8-coordinated site. It consists of three bands in the near-infrared region plus spin-forbidden bands in the visible region (Fig. 8). Tetrahedral Fe^{2+} transitions occur at lower energies, usually consisting of a pair of bands centered in the 1800 to 2000 nm range (Fig. 27). If the Fe-O bond lengths were the same, the average energy of the tetrahedral absorptions would be 4/9 the average energy of the octahedral absorptions. The absorption intensity of tetrahedral Fe^{2+} is greater than octahedral Fe^{2+}. In general, a metal ion at site which lacks a center of symmetry (non-centrosymmetric site) will have a

Figure 23. Absorption spectrum of a 1.00 mm thick crystal of phosphosiderite, $FePO_4 \cdot 2H_2O$, showing the characteristic features of Fe^{3+} near 750, 540 and 440 nm. The feature near 376 nm is also characteristic of Fe^{3+} but is only ocassionaly seen because of interference from the oxygen to metal charge tranfer bands. Solid line E∥c; dashed line E⊥c. From Rossman (1976)

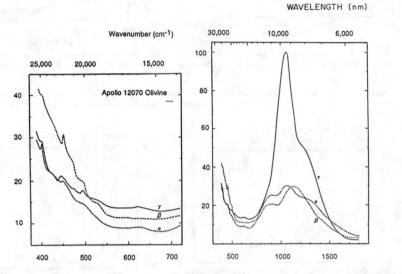

Figure 24. Absorption spectrum of a lunar olivine showing the features of Fe^{2+}. The features between 400 and 700 nm are spin-forbidden bands and the features between 700 and 1700 nm are spin-allowed bands. The two strongest, overlapping absorption bands in the gamma spectrum near 1000 nm are from Fe^{2+} at the M(2) site. The features in the alpha and beta spectra between 800 and 1500 nm are the superposition of M(1) and M(2) absorptions. From Hazen et al. (1977d)

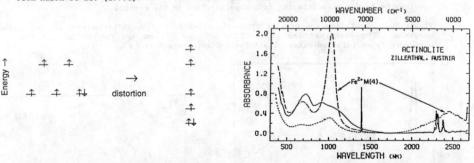

Figure 25. Diagram of splittings induced in the Fe^{2+} d-orbitals by distortion of the iron environment from octahedral symmetry. In octahedral symmetry, the 3d orbitals are split into two sets. In a lower symmetry environment (distortion from octahedral symmetry), all five 3d orbitals will be at different energies. The "second" electron in the lowest energy orbital can now be excited to either of the two higher energy orbitals. The energy separation between them will depend on the magnitude of the distortion.

Figure 26. Absorption spectrum of a 1.0 mm thick actinolite from Zillerthal, Austria, showing prominent features at 1030 and 2470 nm from Fe^{2+} at the large, distorted M(4) site. Other features include a prominant OH overtone near 1400 nm, intervalence charge-transfer absorption near 700 nm, and absorption from Fe^{2+} at the M(1), M(2), and M(3) sites near 1150 nm. Modified from Goldman and Rossman (1977).

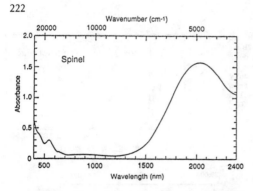

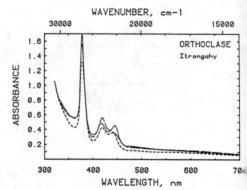

Figure 27.   Absorption spectrum of Fe^{2+} at the tetrahedral site of a lunar spinel dominated by the Fe^{2+} absorption at 2000 nm. Data from P. Bell. See also Mao and Bell (1975)

Figure 28.   Absorption spectrum of Fe^{3+} in the tetrahedral Al site of orthoclase feldspar with 0.42% Fe_2O_3 from Itrongahy, Madagascar. Sample thickness 2.0 cm.   From Hofmeister and Rossman (1983).

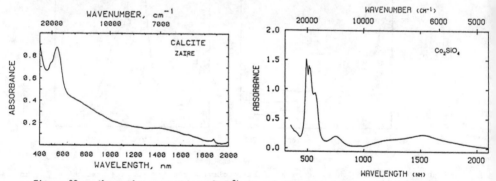

Figure 29.   Absorption spectrum of Co^{2+} in calcite from Zaire.   The Co^{2+} absorption feature near 500 nm is the cause of the pink color of the mineral.   A broad, weak Co^{2+} feature is also present near 1500 nm.   The sharp feature near 1880 nm is an overtone of a carbonate vibration.   Sample 7.0 mm thick, unpolarized spectrum through a cleavage section.   The gradual rise in absorbance level which becomes more intense at shorter wavelengths is due to scattering from imperfections in the crystal .

Figure 30.   Absorption spectrum of synthetic olivine. Co_2SiO_4. Thickness: 20 μm. Alpha spectrum.   The features near 1500, 1200, 800 and 500 nm are due to Co^{2+} at the two octahedrally coordinated sites of olivine.   Olivine synthesized by J. Ito.

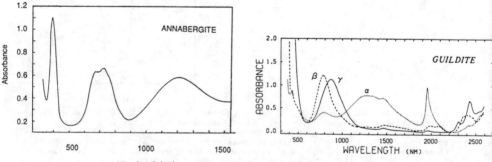

Figure 31.   Unpolarized absorption spectrum of Ni^{2+} in a 0.015 cm thick (010) cleavage slab of annabergite, $Ni_3(AsO_4)_2 \cdot 8H_2O$. From Faye (1971).

Figure 32.   Absorption spectrum of Cu^{2+} and Fe^{3+} in a 150 μm thick crystal of guildite, $CuFe(SO_4)_2(OH) \cdot 4H_2O$. Features of Cu^{2+} occur at 780 nm and 1285 nm. Fe^{3+} features are at 874 nm and 430 nm; H_2O features are at 1950 and 1450 nm.   From Wan et al. (1978).

greater probability of absorbing light then one at a centrosymmetric site.

Tetrahedral Fe^{3+} is encountered in a number of minerals including framework silicates such as feldspars, micas and pyroxenes (Fig. 28). Its absorption intensity is several times greater than that of Fe^{3+} in an octahedral site because of the non-centrosymmetric site.

Cobalt

Cobalt is rarely encountered in minerals. When in octahedral co-ordination, Co^{2+} causes a pink color such as in cobaltian-calcite (Fig. 29). In hosts with multiple sites (such as forsterite), color closer to brown results (Fig. 30). At a tetrahedral site, it is a strong absorber producing a blue color in synthetic spinels and contributes to the blue color of some rare natural crystals.

Nickel

Nickel is also an infrequent contributor to the spectrum of minerals. It is usually present as Ni^{2+} at an octahedral site and gives rise to a green color. (Fig. 31).

Copper

Cu^{2+} in an intense absorber which produces the familiar green and blue colors of the copper minerals (Fig. 32). Although there are a great number of Cu^{2+} minerals, there has been very little work on their spectroscopy. The Cu^+ ion by itself does not contribute to the optical absorption spectrum of minerals. The color of Cu(I) oxides is due to band gap absorptions. Cu(0) was implicated in the color of some unusual green and red plagioclase feldspars from basalt flows; absorption was attributed to small clusters of copper atoms (Hofmeister and Rossman, 1985b).

GUIDE TO MINERAL OPTICAL SPECTRAL DATA

Table 3 is intended to be an extensive list of published mineral optical spectra data. The emphasis is on papers published after 1965 which deal with natural minerals. A number of papers which present data for synthetic minerals are also included. Although the emphasis is on single-crystal transmission data, occasionally, when such data are not available, reflectance data are referenced. The extensive Russian language literature on mineral spectroscopy is under-represented, although the book by Platinov (1976) provides an extensive entry into the pre-1976 literature.

Table 3.

References to Optical Spectroscopic data on Minerals.

Akaganeite	Fe^{3+}	Taylor et al. (1974)
Amarantite	Fe^{3+}	Rossman (1976b)
Amblygonite	V^{3+}	Platonov (1976)
Amphiboles	Review paper	Hawthorne (1981)
actinolite	Fe^{2+}	Goldman and Rossman (1982)
		Goldman and Rossman (1977)
		Burns (1970a)
		White and Keester (1966)
	Fe IVCT	Burns (1970a)
	Cr^{3+}	Hawthorne (1981)
cummingtonite	Fe^{2+}	Burns (1970a)
ferrohastingsite	Fe^{2+}, IVCT	Faye and Nickel (1970)
glaucophane	Fe^{3+}, Fe^{2+}, IVCT	Smith and Strens (1976)
		Burns (1970a)
		Faye and Nickel (1970)
		Bancroft and Burns (1969)
		Chesnokov (1961)
grunerite	Fe^{2+}	Burns (1970a)
hornblende	Fe	Faye and Nickel (1970)
holmquistite	Fe^{3+}, IVCT	Faye and Nickel (1970)
pargasite	Fe^{2+}	Goldman and Rossman (1977)
riebeckite	Fe	Allen (1976)
		Manning and Nickel (1970)
		Faye and Nickel (1970)
		Burns (1970a)
tremolite	Fe^{2+}	Goldman and Rossman (1977)
	Mn^{3+}	Hawthorne (1981)
	Cr^{3+}	Hawthorne (1981)
		Manning and Nickel (1969)
		Faye and Nickel (1970)
		Mao and Seifert (1974)
		Faye and Nickel (1970)
winchite	Mn^{3+}	Ghose et al. (1986)
Andalusite	Mn^{3+}	Smith et al. (1982)
		Abs-Wurmbach et al. (1981)
		Hålenius (1978)
		Abs-Wurmbach et al. (1977)
		Smith and Strens (1976)
	theory	Kai et al. (1980)
	Fe, Ti	Smith (1977)
		Faye and Harris (1969)
blue	Fe IVCT	Langer et al. (1984)
kanonaite	Mn^{3+}	Smith et al. (1982)
		Abs-Wurmbach et al. (1981)
Annabergite	Ni^{2+}	Faye (1971b)
Apatite	Co, Ni, REE	Graisafe and Hummel (1970)
Apophyllite	V^{4+}	Rossman (1974a)
	V^{3+}	Platonov (1976)
Astrophyllite	Ti, Fe	Manning (1969)
Axinite	Fe^{2+}	Faye (1972)
	V^{3+}	Schmetzer (1982)

Babingtonite	Fe, IVCT	Mattson and Rossman (1987a)
		Amthauer and Rossman (1984)
Benitoite		Burns et al. (1964)
Beryl	Fe	Blak et al. (1982)
		Goldman et al. (1978)
		Platonov et al. (1978)
		Price et al. (1976)
		Samoilovich et al. (1971)
		Lakshman and Reddy (1970)
		Wood and Nassau (1968)
maxixe	CO_3^-	Edgar and Vance (1977)
		Nassau et al. (1976)
emerald	Cr^{3+}	Wood and Nassau (1968)
		Wood (1965)
		Schmetzer and Bank (1981)
	Mn^{3+}	Nassau and Wood (1968)
	V^{3+}	Ghera and Lucchesi (1987)
		Wood and Nassau (1968)
Botryogen	Fe^{3+}	Rossman (1975)
Bustamite	Mn^{2+}	Manning (1970)
		Manning (1968b)
Butlerite	Fe^{3+}	Rossman (1976a)
Cassiterite	Sn	Cohen et al. (1985)
	Fe	Calas and Cottrant (1982)
Celesite	radiation	Bernstein (1979)
Chalcanthite	Cu^{2+}	Lakshman and Reddy (1973b)
Chlorite	Fe	Smith (1977)
		Smith and Strens (1976)
		White and Keester (1966)
clinochlore	Cr^{3+}	Calas et al. (1984)
chlorotoid	Fe, IVCT, Mn^{2+}	Hålenius et al. (1981)
	Fe, IVCT	Faye et al. (1968)
Chrysoberyl	Fe^{3+}	Farrell and Newnham (1965)
Coalingite	Fe^{3+}	Mattson and Rossman (1984)
Copiapite	Fe^{3+}	Rossman (1975)
Coquimbite	Fe^{3+}	Rossman (1975)
Cordierite	Fe^{2+}, Fe^{3+}, IVCT	Vance and Price (1984)
		Goldman et al. (1977)
		Smith (1977)
		Smith and Strens (1976)
		Faye (1972)
		Faye et al. (1968)
		Robbins and Strens (1968)
		Farrell and Newnham (1967)
sekaninite	Fe^{2+}	Goldman et al. (1977)
		Smith and Strens (1976)
Corundum		
sapphire	Fe^{2+}, Fe^{3+}, Ti^{4+}	Schmetzer (1987)
		Burns (1985)
		Burns and Burns (1981)
		Schmetzer and Bank (1980)
		Nikolskaya et al. (1978)
		Smith (1977)
		Smith and Strens (1976)
		Smith (1978b)
		Eigenmann and Günthard (1972)
		Eigenmann et al. (1972)

Mineral	Ion	Reference
Parabutlerite	Fe^{3+}	Rossman (1976a)
Periclase		Goto et al. (1980)
		Mao (1976)
		Mao (1973)
Phosphosiderite	Fe^{3+}	Rossman (1976b)
Pyroxenes	review article	Rossman (1980)
bronzite	Fe^{2+}	Goldman and Rossman (1977a)
		Hazen et al. (1977)
		Goldman and Rossman (1976)
		Bell and Mao (1973c)
		Runciman et al. (1973)
		Burns (1970a)
		Bancroft and Burns (1967)
enstatite	Fe^{2+}	Goldman and Rossman (1978b)
		White and Keester (1966)
	Ni^{2+}	Rossman et al. (1981)
orthopyroxenes		Goldman and Rossman (1978b)
		Hazen et al. (1978)
		Burns (1970a)
hypersthene	Fe^{2+}	Goldman and Rossman (1976)
		Cohen (1973)
		Burns (1966)
ferrohypersthene	Fe	Burns (1970a)
orthoferrosilite	Fe	Langer and Abu-Eid (1977)
		Bell and Mao (1971)
		Burns (1970a)
Clinopyroxenes		Knomenko et al. (1982)
		Hazen et al. (1978)
acmite	Fe^{3+}, IVCT	Amthauer and Rossman (1984)
		Langer and Abu-Eid (1977)
aegirine	Fe^{3+},Fe^{2+}, IVCT	Amthauer and Rossman (1984)
augite	Fe^{3+}	Bell and Mao (1972b)
	Fe^{2+}	Burns et al. (1972a,b)
	Fe, Ti	Burns et al. (1976)
		Cohen (1973)
		Bell and Mao (1972a,d)
		Manning and Nickel (1970)
		Burns (1970a)
		Manning and Nickel (1969)
diopside	Fe^{2+}	Cohen (1973)
		White and Keester (1966)
	Cr^{3+}	Khomenko and Platonov (1985)
		Ikeda and Ohashi (1974, 1982)
		Boksha et al. (1974)
		Mao et al. (1972)
		Grun-Grzhimailo et al. (1968)
	Cr^{2+}	Burns (1975)
	V^{3+}	Mao et al. (1972)
	Co^{2+}	Platonov et al. (1984)
hedenbergite	Fe, IVCT	White et al. (1971)
		Amthauer and Rossman (1984)
		Burns and Huggins (1973)
jadeite	Fe^{3+}, IVCT	Rossman (1974b)
	Cr^{3+}	Khomenko and Platonov (1985)
kosmochlor	Cr^{3+}	Khomenko and Platonov (1985)
omphacite	Fe^{2+}, Ti^{4+}	Strens et al. (1985)
	Fe, IVCT	Khomenko and Platonov (1985)
		Abu-Eid (1976)
other	Mn^{3+}	Ghose et al. (1986)
pigeonite	Fe^{2+}	Cohen (1973)
		Bell and Mao (1973c)
		Bell and Mao (1972b,d)
		Burns et al. (1972a,b)
		Bell and Mao (1971)
spodumene	Mn	Schmitz and Lehmann (1975)
		Cohen and Janezic (1983)
	radiation centers	Rossman and Qui (1982)
	Cr^{3+}	Khomenko and Platonov (1985)

Platonov et al. (1971)

Spinel

    Fe — Taran et al. (1987)
Fe_2Cr, V — Schmetzer and Gübelin (1980)
Fe^{2+}, Fe^{3+} — Dickson and Smith (1976)
Mao and Bell (1975)
Shankland et al. (1974)
Gaffney (1973)
Slack et al. (1966)

    Cr^{3+} — Mao and Bell (1975, 1974b)
Sviridov et al. (1973)
Reed (1971)
Wood et al. (1968)

    Co^{2+} — Shigley and Stockton (1984)

  chromite Cr^{3+}, Fe^{2+} — Mao and Bell (1974b)

  spinel polymorph of olivine — Burns and Sung (1978)
Yagi and Mao (1977)
Mao (1976)
Mao and Bell (1972)

Sphalerite Cu^{2+} — Manning (1966)
Fe^{2+} — Manning (1967c)

Staurolite Co^{2+} — Burns (1970a)
Fe^{2+} — Burns (1970a)
Bancroft and Burns (1967)

Stewartite Fe^{3+} — Rossman (1976a)

Stichtite Cr^{3+} — Calas et al. (1984)

Suzukiite V^{4+} — Kato et al. (1974)

Tephroite Mn^{2+} — Takei et al. (1984)
Takei (1976)
Burns (1970a)

Tetrahedrite — Jeanloz and Johnson (1984)

Theory — Sherman (1987a,b)
Zhao et al. (1987)
Runciman (1987)
Zhao et al. (1986)
Zhao et al. (1984)
Strens et al. (1982)
Dowty (1978)
Strens and Freer (1978)
Manning (1973a)
Faye (1972)
Manning and Townsend (1970)
Manning (1970)

Topaz   various — Petrov (1977)
Petrov and Berdesinski (1975)
Cr^{3+} — Petrov et al. (1977)
radiation colors — Aines and Rossman (1986)
Nassau and Prescott (1975a)

Tourmaline

  burgerite Fe^{3+} — Manning (1969c)
  dravite Fe^{3+} — Mattson and Rossman (1984)
review — Smith (1978)
Fe, Ti — Smith (1977)
Faye et al. (1974)
Wilkins et al. (1969)
Manning (1969d)
Cr^{3+} — Manning (1969c)
  elbaite, schorl   Fe — Mattson and Rossman (1987)
Smith and Strens (1976)
Bakhtin et al. (1975)
Faye et al. (1974)

		Townsend (1970)
		Wilkins et al. (1969)
		Robbins and Strens (1968)
		Faye et al. (1968)
	Mn	Schmetzer and Bank (1984)
		Manning (1973b)
		Bershowv et al. (1969)
		Vershov et al. (1967)
		Manning (1969d)
	Fe, Ti	Smith (1977)
	Mn, Ti	Mattson and Rossman (1986)
schorl	Fe	Manning (1969c)
Triphylite	Fe^{2+}	Leckebusch and Recker (1973)
Tugtupite	S	Povarennysh et al. (1971)
Turquoise	Cu^{2+}	Diaz et al (1971)
Ussingite		Povarennykh et al. (1970)
V in minerals	review	Schmetzer (1982)
Varascite	Fe^{3+}	Lehmann (1978)
Vermiculite	Fe	Karickhoff and Bailey (1973)
Vesuvianite	Fe,Ti IVCT	Manning (1977)
	Fe	Manning (1976)
		Manning and Tricker (1975)
		Manning (1975)
	Fe^{3+}	Manning (1968a)
Vivianite	Fe^{2+}, IVCT	Amthauer and Rossman (1984)
		Güdel (1983)
		Mao (1976)
		Bell and Mao (1974)
		Robins and Strens (1972)
		Townsend and Fry (1970)
		Faye et al. (1968)
		Faye (1968)
		Hush (1967)
Voltaite	Fe, IVCT	Beveridge and Day (1979)
Wavellite	V^{3+}	Platonov (1976)
Wulfenite		Bell and Mao (1974)
Xenotime	V^{3+}	Demiray et al. (1970)
Yoderite		Langer et al. (1982)
		Abu-Eid et al. (1978)
Zircon	U^{4+}, U^{5+}	Vance and Mackey (1978)
	U^{5+}	Vance and Mackey (1974)
	U^{4+}	Richman et al. (1967)
	V^{4+}	Demiray et al. (1970)
	Nb^{4+}	Fielding (1970)
Zoisite	V^{3+}	Platonov (1976)
		Faye and Nickel (1971)
		Tsang and Ghose (1971)
	Cr^{3+}	Schmetzer and Berdesinski (1978)

REFERENCES

Abs-Wurmbach, I., K. Langer and Tillmanns, E. (1977) Structure and polarized absorption spectra of Mn^{3+} -substituted andalusites (viridines). Naturwissenschaften 64, 527-528.

_____, K. Langer, Seifert, F. and Tillmanns, E. (1981) The crystal chemistry of $(Mn^{3+}Fe^{3+})$-substituted andalusites (viridines and kanonaite), $(Al_{1-x-y} Mn^{3+}_x Fe^{3+}_y)_2(O|SiO_4)$: crystal structure refinements, Mössbauer, and polarized optical absorption spectra. Z. Kristallogr. 155, 81-113.

Abu-Eid, R.M., Mao, H.K. and Burns, R.G. (1973) Polarized absorption spectra of gillespite at high pressure. Carnegie Instit. Washington Yearbook 72, 564-567.

_____ (1976) Absorption spectra of transition metal-bearing minerals at high pressures. In: The Physics and Chemistry of Minerals and Rocks. R.G.J. Strens, ed., J. Wiley, New

York, p 641-675.

_____, Langer, K., and Siefert, F. (1978) Optical absorption and Mössauer spectra of purple and green yoderite, a kyanite-related mineral. Phys. Chem. Minerals 3, 271-289.

Adams, J.W. (1965) The visible absorption spectra of rare earths in minerals. Amer. Mineral. 50, 356-360.

Aines, R.D. and Rossman, G.R. (1986) Relationships between radiation damage and trace water in zircon, quartz, and topaz. Amer. Mineral. 71, 1186-1193.

Allen, G.C. (1976) Mixed-valence interaction absorption in the electronic absorption spectrum of the iron-containing mineral crocidolite. Transition Metal Chem. 1, 143-146.

Amthauer, G. and Rossman, G.R. (1984) Mixed valence of iron in minerals with cation clusters. Phys. Chem. Minerals 11, 37-51.

Annersten, H. and Hassib, A. (1979) Blue sodalite. Canadian Mineral. 17, 39-46.

Balitsky, V.S. and Balitskay, O.V. (1986) The amethyst-citrine dichromatism in quartz and its origin. Phys. Chem. Minerals 13, 415-421.

Bakhtin, A. I., Minko, O.Y. and Vinokurov, V. M. (1975) Isomorphism and colour of tourmaline. Izv. Akad. Nauk SSSR., Ser. Geol. 6, 73-83.

_____ and Vinokurov, V. M. (1978) Exchange-coupled pairs of transition metal ions and their effect on the optical absorption spectra of rock-forming silicates. Geokhimiya, 1, 87-95. Trans., Geochem. Int'l 1978, 53-60.

Bancroft, G.M. and Burns, R.G. (1967) Interpretation of the electronic spectra of iron in pyroxenes. Amer. Mineral. 52, 1278-1287.

Bell, P.M. and Mao, H.K. (1969) Crystal-field spectra at high pressure. Ann. Rept. Geophys. Lab., Yearbook, 68, 253-258.

_____ and Mao, H.K. (1972a) Crystal-field determinations of Fe^{3+}. Ann. Rept. Geophys. Lab., Yearbook, 68, 531-538.

_____ and Mao, H.K. (1972b) Crystal-field effects of iron and titanium in selected grains of Apollo 12, 14, and 15 rocks, glasses and fine fractions. Proc. Third Lunar Sci. Conf., Suppl. 3, Geochim. Cosmochim. Acta, Vol 1, 545-553.

_____ and Mao, H.K. (1972c) Apparatus for measurement of crystal-field spectra of single crystals. Carnegie Inst. Washington Yearbook 71, 608-611.

_____ and Mao, H.K. (1972d) Crystal-field studies of lunar samples. Carnegie Inst. of Washington Yearbook 71, 480-489.

_____ and Mao, H.K. (1973a) Optical and chemical analysis of iron in Luna 20 plagioclase. Geochim. Cosmochim. Acta 37, 755-759.

_____ and Mao, H.K. (1973b) Measurements of the polarized crystal-field spectra of ferrous and ferric iron in seven terrestrial plagioclases. Carnegie Inst. Washington Yearbook 72, 574-576.

_____ and Mao, H.K. (1973c) Optical absorption studies of the Russian Luna 20 soil. Carnegie Inst. Washington Yearbook 72, 656-665.

_____ and Mao, H.K. (1974) Pressure effect on charge-transfer processes in minerals. Carnegie Inst. Washington Yearbook 73, 507-510.

_____ and Mao, H.K. (1976) Crystal-field spectra of fassaite from the Angra Dos Reis meteorite. Carnegie Inst. Washington Yearbook 75, 701-705.

Belov, V.F., Martirosyan, V.O., Marfunin, A.S., Platonov, A.N., and Tarashchan, A.N. (1969) Color centers in lithium tourmaline (elbaite). Krystallografiya 13, 730-732. Trans. from Soviet Phys. Cryst. 13, 629-630.

Belsky, H.L. and Rossman, G.R. (1984) Crystal structure and optical spectroscopy (300 to 2200 nm) of $CaCrSi_4O_{10}$. Amer. Mineral. 69, 771-776.

Bernstein, L.R. (1979) Coloring mechanisms in celestite. Amer. Mineral. 64, 160-168.

_____ (1982) Monazite from North Carolina having the alexandrite effect. Amer. Mineral. 67, 356-359.

Berry, F.J. and Vaughan, D.J. (1985) Chemical Bonding and Spectroscopy in Mineral. Chemistry. Chapman and Hall, London.

Bershowv, L.V., Martirosyan, V.O., Marfunin, A.S., Platonov, A.N. and Tarashehan, A.N. (1969) Color centers in lithium tourmaline (elbaite). Soviet Phys. - Crystallogr. 13, 629-630.

Bill, H., Sierro, J., and Lacroix, R. (1967) Origin of coloration in some fluorites. Amer. Mineral. 52, 1003-1008.

_____ and Calas, G. (1978) Color centers, associated rare-earth ions and the origin of coloration in natural fluorites. Phys. Chem. Minerals 3, 117-131.

_____ (1982) Origin of color of yellow fluorite: the O_3^- center: structure and dynamical aspects. J. Chem. Phys. 76, 219-224.

Billing, D.E., Hathaway, B.J. and Nicholls, P. (1969) Polarizations of the copper(II) and uranyl electronic spectra of meta-zeunerite single crystals. J. Chem. Soc. (A) 316-319.

Blak, A.R., Isotani, S. and Watanabe, S. (1982) Optical absorption and electron spin resonance in blue and green natural beryl. Phys. Chem. Minerals 8, 161-166.

Boksha, O.N., Varina, T.M. and Kostyukova,I.G. (1974) Structure of the absorption spectrum of chrome-diopsides. Kristallographya 19, 392-394.

Bonnin, D., Calas, G., Suquet, H., and Pezerat, H. (1985) Sites occupancy of Fe^{3+} in Garfield nontronite: a spectroscopic study. Phys. Chem. Minerals 12, 55-14.

Burns, R.G. (1966) Origin of optical pleochroism in orthopyroxenes. Mineral. Mag. 35, 715-719.

_____ (1966) Apparatus for measuring polarized absorption spectra of small crystals. J. Sci. Instrum. 43, 58-60.

_____, Clark, M.G. and Stone, A.J. (1966) Vibronic polarization in the electronic spectra of gillespite, a mineral containing iron(II) in square-planar coordination. Inorg. Chem. 1268-1272.

_____ and Strens R.G.J. (1967) Structural interpretation of polarized absorption spectra of the Al-Fe-Mn-Cr epidotes. Mineral. Mag. 36, 204-226.

_____ (1970a) Mineralogical Applications of Crystal Field Theory. Cambridge University Press, Cambridge.

_____ (1970b) The crystal field spectra and evidence of cation ordering in olivine minerals. Amer. Mineral. 55, 1608-1632.

_____, Abu-Eid, R.M, and Huggins, F.E. (1972a) Crystal field spectra of lunar pyroxenes. Proc. third Lunar Sci. Conf., Suppl. 3, Geochim. Cosmochim Acta. Vol 1, 533-543. The MIT Press.

_____, Huggins, F.E. and Abu-Eid, R.M (1972b) Polarized absorption spectra of single crystals of lunar pyroxenes and olivines. The Moon 4, 93-102.

_____, and Huggins, F.E. (1973) Visible-region absorption spectra of a Ti^{3+} fassaite from the Allende meteorite: A discussion. Amer. Mineral. 58, 955-961.

_____, Vaughan, D.J., Abu-Eid, R.M. and Witner, M. (1973) Spectral evidence for Cr^{3+}, Ti^{3+}, and Fe^{2+} rather than Cr^{2+} and Fe^{3+} in lunar ferromagnesian silicates. Proc. 4th Lunar Science Conf. (Suppl. 4, Geochim. Cosmochim. Acta) 1, 983-994.

_____ (1974) The polarized spectra of iron in silicates: Olivine. A discussion of neglected contributions from Fe^{2+} ions in M(1) sites. Amer. Mineral. 59, 625-629.

_____ (1974) On the occurrence and stability of divalent chromium in olivines included in diamonds. Contrib. Mineral. Petrol. 51, 213-221.

_____, Parkin, K.M., Loeffler, B.M., Leung, I.S. and Abu-Eid, R.M. (1976) Further characterization of spectral features attibutable to titanium on the moon. Proc. Lunar Sci. Conf. 7th , 2561-2578.

_____ and Sung, CM (1978) The effect of crystal field stabilization on the olivine - spinel transition in the system Mg_2SiO_4 - Fe_2SiO_4. Phys. Chem. Minerals 2, 349-364.

_____, Nolet, D.A., Parkin, K.M., McCammon, C.A. and Schwartz, K.B. (1979) Mixed valence minerals of iron and titanium: Correlations of structural, Mössbauer and electronic spectral data . In: Mixed-Valence Compounds: Theory and Applications in Chemistry, Physics, Biology, D.B. Brown, ed. D. Reidel Pub. Co., Boston. p. 295-336.

_____ (1981) Intervalence transitions in mixed-valence minerals of iron and titanium. Ann. Rev. Earth Planet. Sci. 9 345-383.

_____ (1982) Electronic spectra of minerals at high pressures: how the Mantle excites electrons. in: High-Pressure Researches in Geoscience. W. Schreyer, ed., E. Schwiezerbart'sche Verlagsbuchhandlung, Stuttgart, p. 223-246.

_____ (1985) Thermodynamic data from crystal field spectra. In: SW Kieffer and A Navrotsky, eds., Microscopic to Macroscopic, Rev. Mineral. 14, 277-316.

Calas, G. (1972) On the blue colour of natural banded fluorites. Mineral. Mag. 38, 977-979.

_____ and Cottrant, J.F. (1982) Cristallochime du fer dans les cassitérites bretonnes. Bull. Minéral. 105, 598-605.

_____, Manceau, A., Novikoff, A. and Boukili, H. (1984) Comportement du chrome dans les minéraux d'altération du gisement de Campo Formoso (Bahia, Brésil). Bull. Minéral. 107, 755-766.

Cemic, L., Grammenopoulou-Bilal, S. and Langer, K. (1986) A microscope-spectrometric method for determining small Fe^{3+} concentrations due to Fe^{3+}-bearing defects in fayalite. Ber. Bunsenges. Phys. Chem. 90, 654-661.

Chesnokov, B.V. (1961) Spectral absorption curves of glaucophane from eclogite of the southern Urals. Zap. Vses. Mineral. Obshch. 90, 700-703.

Clark, C.D. and Walker, J. (1973) The neutral vacancy in diamond. Proc. Royal Soc. London A. 334, 241-257.

_____, Mitchell, E.W.J. and Parsons, B.J. (1982) Colour centers and optical properties. In: J.E. Filed, ed., The Properties of Diamond. Academic Press, London, pp. 23-77.

Clark, M.G. and Burns, R.G. (1967) Electronic spectra of Cu^{2+} and Fe^{2+} square planar coordinated by oxygen in $BaXSi_4O_{10}$. J. Chem. Soc. A, 1034-1038.

Cohen, A.J. (1973) Anisotropy of absorption bands in some lunar, meteorite and terrestrial pyroxenes. The Moon 7, 307-321.

_____ and Hassan, F. (1974) Ferrous and ferric ions in synthetic Á-quartz and natural amethyst. Amer. Mineral. 59, 719-728.

_____ and Janezic G.G. (1983) The crystal-field spectra of the $3d^3$ ions, Cr^{3+} and Mn^{4+} in green spodumenes. In: The Significance of Trace Elements in Solving Petrogenetic Problems & Controversies. Theophrastus Pubs. S.A., Athens.

_____ and Makar, L.N. (1984) Differing effects of ionizing radiation in massive and single crystal rose quartz. N. Jb. Mineral. Mh. 11, 513-521.

_____ and Makar, L.N. (1985) Dynamic biaxial absorption spectra of Ti^{3+} and Fe^{2+} in a natural rose quartz crystal. Mineral. Mag. 49, 709-715.

_____, Adekeye, J.I.D., Hapke, B. and Partlow, D. (1985) Intersitial Sn^{2+} in synthetic and natural cassiterite crystals. Phys. Chem. Minerals 12, 363-369.

_____ (1985) Amethyst color in quartz, the result of radiation protection involving iron. Amer. Mineral 70, 1180-1185.

Collins, A.T. (1982) Colour centres in diamonds. J. Gemmology, 18, 37-75.

Cox, R.T. (1977) Optical absorption of the d4 ion Fe^{4+} in pleochroic amethyst quartz. J. Phys, C: Solid State Phys. 10, 4631-4643.

Davies, G. (1977a) The optical properties of diamond. Chem. Phys. of Carbon 31, 1-143.

_____ (1977b) The H3 centre. Diamond Research 1977, 15-24.

_____ and Thomaz, M.F. (1979) The N3 centre. Diamond Research 1979, 13-24.

_____ and Foy, C. (1980) Jahn-Teller coupling at the neutral vacancy in diamond. J. Phys. C: Solid St. Phys. 13, 2203-2213.

Demiray, T., Nath, D.K. and Hummel, F.A. (1970) Zircon-vanadium blue pigment. J. Amer. Ceramic

Soc. 53, 1-4.

Diaz, J., Garach, H.A. and Poole, C.P. (1971)   An electron spin resonance and optical study of turquoise. Amer. Mineral. 56, 773-781.

Dickson, B.L. and Smith, G. (1976) Low temperature optical absorption and Mössbauer spectra of staurolite and spinel. Canadian Mineral. 14, 206-215.

Dowty, E. (1971)   Crystal chemistry of titanian and zirconian garnet: I. Review and Spectral studies. Amer. Mineral. 56, 1983-2009.

_____ and Clark, J.R. (1973a) Crystal structure and optical properties of a Ti^{3+} fassaite from the Allende meteorite. Amer. Mineral. 58, 230-242.

_____ and Clark, J.R. (1973b) Crystal structure and optical properties of a Ti^{3+} fassaite from the Allende meteorite: Reply. Amer. Mineral. 58, 962-964.

_____   (1978)   Absorption optics of low-symmetry crystals - applications to titanian clinopyroxene spectroscopy. Phys. Chem. Minerals 3, 173-181.

Edgar A. and Vance E.R. (1977) Electron Paramagnetic resonance, optical absorption, and magnetic circular dichroism studies of the CO_3^- molecular-ion in irradiated natural beryl.   Phys. Chem. Minerals 1, 165-178.

Egan, WG and Hilgeman, TW (1979)   Optical Properties of Inhomogeneous Materials. Academic Press, New York.

Eigenmann, K. and Günthard, H. (1972) Valence states, redox reactions and biparticle formation of Fe and Ti doped sapphire. Chem. Phys. Letters 13, 58-61.

_____, Kurtz, K. and Günthard, H.H. (1972) The optical spectrum of $Á-Al2O3:Fe^{3+}$.   Chem. Phys. Letters 13, 58-61.

Farmer, G.L. and Boettcher, A.L. (1981) Petrologic and crystal-chemical significance of some deep-seated phlogopites. Amer. Mineral. 66, 1154-1163.

Farrell, E.F. and Newnham, R.E. (1965)   Crystal-field spectra of chrysoberyl, alexandrite, peridot, and sinhalite. Amer. Mineral. 50, 1972-1981.

Faye, G.H. (1968a) The optical absorption spectra of iron in six-coordinate sites in chlorite, biotite, phlogopite and vivianite. Some aspects of pleochroism in the sheet silicates. Canadian Mineral. 9, 403-425.

_____   (1968b)   The optical absorption spectra of certain transition metal ions in muscovite, lepidolite, and fuchite. Canadian J. of Earth Sci. 5, 31-38.

_____ and Nickel, E.H. (1968) The origin of pleochroism in erythrite. Canadian Mineral. 9, 493-504.

_____, Manning, P.G. and Nickel, E.H. (1968)   The polarized optica absorption spectra of tourmaline, cordierite, chloritoid and vivanite: ferrous-ferric electronic interaction as a source of pleochroism. Amer. Mineral. 53, 1174-1201.

_____   (1969)   The optical absorption spectrum of tetrahedrally-bonded Fe^{3+} in orthoclase. Canadian Mineral. 10, 112-117.

_____ and Harris, D.C. (1969)   On the colour and pleochroism in andalusite from Brazil. Canadian Mineral. 10, 47-56.

_____ and Hogarth, D. D. (1969)   On the origin of 'reverse pleochroism' of a phlogoite. Canadian Mineral. 10, 25-34.

_____ and Nickel, E.H. (1969) On the origin of colour and pleochroism of kyanite. Canadian Mineral. 10, 35-46.

_____ and Nickel, E.H. (1970)   The effect of charge-transfer processes on the colour and pleochroism of amphiboles. Canadian Mineral. 10, 616-635.

_____   (1971a)   On the optical spectra of di-and trivalent iron in corundum: a discussion. Amer. Mineral. 56, 344-348.

_____   (1971b) A semi-quantitative microscope technique for measuring the optical absorption spectra of mineral and other powders. Canadian Mineral. 10, 889-895.

_____ and Nickel, E.H. (1971) Pleochroism of zoisite from Tanzania. Canadian Mineral 10, 812-821.

_____   (1972)   Relationship between crystal-field splitting parameter "$\tilde{S}_{yI}$" and M_{host}-O bond distance as an aid in the interpretation of absorption spectra of Fe^{2+}-bearing materials. Canadian Mineral. 11, 473-487.

_____, Manning, P.G., Gosselin, J.R. and Tremblay, R.J. (1974) The optical absorption spectra of tourmaline: importance of charge-transfer processes. Canadian Mineral. 12, 370-380.

_____   (1974)   Optical absorption spectrum of Ni^{2+} in garnierite: A discussion. Canadian Mineral. 12, 389-393.

_____   (1975) Spectra of shock-affected rhodonite: a discussion. Amer. Mineral. 60, 939-941.

Ferguson, J. and Fielding, P.E. (1971) The origins of the colours of yellow, green and blue sapphires. Chem. Phys. Letters 10, 262-265.

_____ and _____   (1972) The orgins of the colours of natural yellow, blue, and green sapphires. Australian J. Chem. 25, 1371-1385.

Fielding, P.E. (1970) The distribution of uranium, rare earths, and color centers in a crystal of natural zircon. Amer. Mineral. 55, 428-440.

Finch, J., Gainsford, A.R., and Tennant, W.C. (1982)   Polarized optical absorption and 57Fe Mössbauer study of pegmatitic muscovite. Amer. Mineral. 67, 59-68.

Gaffey, S.J. (1986) Spectral reflectance of carbonate minerals in the visible and near infrared (0.35-2.55 microns): calcite, aragonite, and dolomite. Amer. Mineral. 71, 151-162.

Gaffney, E.S. (1973) Spectra of tetrahedral Fe^{2+} in MgAl2O4. Phys. Rev. B8, 3484-3486.

Ghera, A., Graziani, G. and Lucchesi, S. (1986) Uneven distribution of blue colour in kyanite. N. Jb. Mineral. Abh. 155, 109-127.

_____ and Lucchesi, S. (1987) An unusual vanadium-beryl from Kenya. N. Jb. Mineral. Mh. 1987, 263-274.

Ghose, S., Kersten, M., Langer, K., Rossi, G. and Ungaretti, L. (1986) Crystal field spectra and

Jahn Teller effect of Mn^{3+} in clinopyroxenes and clinoamphioles from India. Phys. Chem. Minerals 13, 291-305.

Goldman, D.S. and Rossman, G.R. (1976) Identification of a mid-infrared electronic band of Fe^{2+} in the distorted M(2) site of orthopyroxene $(Mg,Fe)SiO_3$. Chem. Phys. Lett. 41, 474-475.

_____ and _____ (1977a) The spectra of iron in orthopyroxene revisited: the splitting of the ground state. Amer. Mineral. 62, 151-157.

_____ and _____ (1977b) The identification of Fe^{2+} in the M(4) site of calcic amphiboles. Amer. Mineral. 62, 205-216.

_____, _____ and Dollase, W.A. (1977) Channel constituents in cordierite. Amer. Mineral. 62, 1144-1157.

_____ and _____ (1978a) The site distribution of iron and anomalous biaxiality in osumilite. Amer. Mineral. 63, 490-498.

_____ and _____ (1978b) Determination of quantitative cation distribution in orthopyroxenes from electronic absorption spectra. Phys. Chem. Minerals 4, 43-55.

_____, _____ and Parkin, K.M. (1978) Channel constituents in beryl. Phys. Chem. Minerals 3, 225-235.

_____, _____ (1982) The identification of Fe^{2+} in the M(4) site of calcic amphiboles: Reply. Amer. Mineral. 67, 340-342.

Gonschorek, W. (1986) Electron density and polarized absorption spectra of fayalite. Phys. Chem. Minerals 13, 337-339.

Goto, T, Ahrens, T.J., and Rossman, G.R. (1979) Absorption spectra of Cr^{3+} in Al2O3 under shock compression. Phys. Chem. Minerals 4, 253-263.

_____, Ahrens, T.J., Rossman, G.R. and Syono, Y. (1980) Absorption spectrum of shock-compressed Fe^{2+}-bearing MgO and the radiative conductivity of the lower mantle. Phys. Earth Planet. Interiors. 22, 277-288.

Graisafe, D.A. and Hummel, F.A. (1970) Crystal chemistry and color of apatites containing Co, Ni, and rare-earth ions. Amer. Mineral 55, 1131-1145.

Grun-Grzhimailo, S.V., Boksha, O.N. and Varina, T.M. (1968) The absorption spectrum of chrome-diopside. Soviet Phys. Crystallogr. 12, 935.

Gübelin, E. and Weibel, M. (1975) Vanadium-Grossular von Lualenyi bei Voi, Kenja. N. Jb. Mineral. Abh. 123, 191-197.

Güdel, H.U. (1983) Pair excitations in vivianite, $Fe_3(PO_4)_2 \cdot 8H_2O$. Inorg. Chem. 22, 3812-3815.

Hålenius, U. (1978) A spectroscopic investigation of manganian andalusite. Canadian Mineral. 16, 567-575.

_____ (1979) State and location of iron in sillimanite. N. Jb. Mineral. Mh. 1979, 165-174

_____, Annersten, H. and Langer, K. (1981) Spectral studies on natural chloritoids. Phys. Chem. Minerals 7, 117-123.

Hassan, F. and Cohen, A.J. (1974) Biaxial color centers in amethyst quartz. Amer. Mineral. 59, 709-718.

Hawthorne, F.C. (1981) Amphibole spectroscopy. In: D.R. Veblen, ed., Amphiboles and Other Hydrous Pyriboles - Mineralogy, Rev. Mineral. 9A, 103-139.

Hazen, R.M. Mao, H.K. and Bell, P.M. (1977) Effects of compositional varation on absorption spectra of lunar olivines. Proc. 8th Lunar Sci. Conf., Pergamon Press, New York, 1081-1090.

_____, Bell, P.M. and Mao, H.K. (1977a) Effects of compositional variation on absorption spectra of lunar pyroxenes. Proc. 9th Lunar Planet. Sci. Conf. 2919-2934.

_____ and _____ (1977b) Comparison of absorption spectra of lunar and terrestrial olivines. Carnegie Inst. Washington Yearbook 76, 508-512.

_____ and _____ (1977c) Polarized absorption spectra of the Angra dos Reis fassaite to 52 Kbar. Carnegie Inst. Washington Yearbook 76, 515-516.

_____ Bell, P.M. and Mao, H.K. (1978) Systematic variations of pyroxene absorption spectra with composition. Carnegie Inst. Washington Yearbook 77, 853-855.

Hofer, S.C. (1985) Pink diamonds from Australia. Gems & Gemology 21, 147-155.

Hofmeister, A.M. and Rossman, G.R. (1983) Color in feldspars. In: P.H. Ribbe, ed., Feldspars. Rev. Mineral. 2, 2d ed., 271-280.

_____ and _____ (1984) Determination of Fe^{2+} and Fe^{3+} concentrations in feldspar by optical and EPR spectroscopy. Phys. Chem. Minerals 11, 213-224.

_____ and _____ (1985a) A spectroscopic study of amazonite: irradiative coloration of structurally hydrous Pb-bearing feldspar. Amer. Mineral. 70, 794-804.

_____ and _____ (1985b) Exsolution of metallic copper from Lake county labradorite. Geology 13, 644-647.

_____ and _____ (1985c) A model for the irradiative coloring of smoky feldspar and the inhibiting influence of water. Phys. Chem. Minerals 12, 324-332.

_____ and _____ (1986) A spectroscopic study of blue radiation coloring in plagioclase. Amer. Mineral. 71, 95-98.

Hunt G.R. and Salisbury, J.W. (1970) Visible and near infrared spectra of minerals and rocks. I. Silicate minerals. Modern Geol. 1, 283-300.

_____ and _____ (1971) Visible and near infrared spectra of minerals and rocks. II. Carbonates. Modern Geol. 2, 23-30.

_____, _____, and Lenhoff, C.J. (1971a) Visible and near infrared spectra of minerals and rocks. III. Oxides and hydroxides. Modern Geol. 2, 195-205.

_____, _____ and _____ (1971b) Visible and near infrared spectra of minerals and rocks. IV. Sulphides and sulfates. Modern Geol. 3, 1-14.

_____, _____, and _____ (1972) Visible and near infrared spectra of minerals and rocks. V. Halides, phosphates, arseniates, vanadates and borates. Modern Geol. 3, 121-132.

Hush, N.S. (1967) Intervalence-transfer absorption. In: F.A. Cotton, ed., Progress in Inorganic

238

Chemistry, Vol. 8, Interscience Publishers, New York. p.391-444.

Ikeda, K. and Ohashi, H. (1974) Crystal field spectra of diopside-kosmochlor solid solutions formed at 15 kb pressure. J. Japan Assoc. Mineral. Petrol Econ. Geol. 69, 103-109

Ikeda, K and Yagi, K. (1982) Crystal-field spectra for blue and green diopsides synthesized in the join $CaMgSi_2O_6$-$CaCrAlSiO_6$. Contrib. Mineral. Petrol. 81, 113-118.

Ihinger, P.D. and Stolper, E.M. (1986) The color of meteoritic hibonite: an indicator of oxygen fugacity. Earth Planetary Sci. Letters 78, 67-79.

Jeanloz, R. and Johnson, M.L. (1984) A note on the bonding, optical spectrum and composition of tetrahedrite. Phys. Chem. Minerals 11, 52-54.

Kai, A.T., Larsson, S., and Hålenius, U. (1980) The electronic structure and absorption spectrum of MnO_6^{9-} octahedra in manganian andalusite. Phys. Chem. Minerals 6, 77-84.

Kane, R.E. (1983) The Ramaura synthetic ruby. Gems and Gemology 1983, 130-148.

Karickhoff, S.W. and Bailey, G.W. (1973) Optical absorption spectra of clay minerals. Clays Clay Minerals 21, 59-70.

Karr, C. (1975) Infrared and Raman Spectroscopy of Lunar and Terrestrial Minerals. Academic Press, New York.

Kato, K, Sugitani, Y. and Nagashima, K. (1974) Absorption spectrum of a new barium vanadyl silicate $BaVOSi_2O_6$. Mineral. J. 7, 421-430.

Khomenko, V.M., Platonov, A.N., Matsyuk, S.S. and Kharkiv, A.D. (1982) Colouring and pleochroism of clinopyroxenes from deep inclusions in the "Mir" kimberlite pipe. Mineral Zh. 4, 41-51. (in Russian)

_____ and Platonov, A.N. (1985) Electronic absorption spectra of Cr^{3+} ions in natural clinopyroxenes. Phys. Chem. Minerals 11, 261-265.

Kleim, W. and Lehmann, G. (1979) A reassignment of the optical absorption bands in biotites. Phys. Chem. Minerals 4, 65-75.

Köhler P. and Amthauer, G. (1979) The ligand field spectrum of Fe^{3+} in garnets. J. Solid State Chem. 28, 329-343.

Krebs J.J. and Maisch W.G. (1971) Exchange effects in the optical-absorption spectrum of Fe^{3+} in Al2O3. Phys. Review B, 757-769.

Lakshman, S.V.J. and Reddy, B.J. (1970) Optical absorption spectrum of iron in beryl. Spectrochim. Acta 26A, 2230-2234.

_____ and Reddy, B.J. (1973a) Optical absorption spectrum of Ni^{2+} in garnierite. Proc. Indian Acad. Sci. 77A, 269-279.

_____ and Reddy, B.J. (1973b) Optical absorption spectrum of Cu^{2+} in chalcanthite and malachite. Canadian Mineral. 12, 207-210.

_____ and Reddy, B.J. (1973c) Optical absorption spectrum of Mn^{2+} in rhodonite. Physica 66, 601-610.

Langer, K. and Abu-Eid, R.M. (1977) Measurement of the polarized absorption spectra of synthetic transition metal-bearing silicate microcrystals in the spectral range 44,000-4,000 cm-1. Phys. Chem. Minerals 1, 273-299.

_____, Smith, G. and Hålenius, U. (1982) Reassignment of the absorption spectrum of purple yoderite. Phys. Chem. Minerals 8, 143-145.

_____ and Lattard, D. (1984) Mn^{3+} in garnets II. Optical absorption spectra of blythite-bearing, synthetic calderites, $Mn_3^{2+}(Fe^{3+},Mn^{3+})_2[SiO_4]_3$. N. Jb. Mineral. Abh. 149, 129-141.

_____, Hålenius, U. and Fransolet, A.-M. (1984) Blue andalusite from Ottré, Venn-Stavelot Massif, Belgium: A new example of intervalence charge-transfer in the aluminum silicate polymorphs. Bull. Minéral. 107, 587-596.

Leckebusch, R. and Recker, K. (1973) Die Absorptionsspektren von Triphylin und Lithiophilit. N. Jb. Mineral. Mh. 1973, 70-75.

Lehmann, G. (1969) Zur Farbe von Rosenquarz. N. Jb. Mineral. Mon. 1969, 222-225.

_____ and Harder, H. (1970) Optical spectra of di- and trivalent iron in couundum. Amer. Mineral. 55. 98-105.

_____ (1971) On the optical spectra of di- and trivalent iron in corundum: A reply. Amer. Mineral. 56. 349-350.

_____ and Bambauer, H.U. (1973) Quartz crystals and their colors. Angewandte Chem. Int'l. 12, 283-291.

_____ (1978) Farben von Mineralen und ihre Ursachen. Fortschr. Mineral. 56, 172-252.

Lin, C (1981) Optical absorption spectra of Fe^{2+} and Fe^{3+} in garnets. Bull. Minéral. 104, 218-222.

Loeffler, B.M., Burns, R.G. and Tossell, J.A. (1975) Metal-metal charge transfer transitions: interpretation of visible-region spectra of the moon and lunar materials. Proc. Lunar Sci. Conf. 6th, 2663-2672.

_____ and Burns, RG (1976) Shedding light on the color of gems and minerals. Amer. Scientist, 64, 636-647.

MacKenzie, K.J.D. (1972) The possible role of sulphur in the coloration of Blue John fluorite. Mineral. Mag. 38, 979-981.

Manning, P.G. (1966) Cu(II) in octahedral sites in sphalerite. Canadian Mineral. 8, 567-571.

_____ (1967a) The optical absorption spectra of the garnets almandine-pyrope, pyrope, and spessartine and some structural interpretations of mineralogical significance. Canadian Mineral. 9, 237-251.

_____ (1967b) The optical absorption spectra of some andradites and the identification of the $^6A_1$ - $^4A_1{}^4E(G)$ transition in octahedrally bonded Fe^{3+}. Canadian J. Earth Sci. 4, 1039-1047.

_____ (1967c) Absorption spectra of Fe(III) in octahedral sites in sphalerite. Canadian Mineral. 9, 57-64.

_____ (1968a) Optical absorption spectra of octahedrally bonded Fe^{3+} in vesuvianite. Canadian J. Earth Sci. 5, 89-92.

_____ (1968b) Absorption spectra of the manganese-bearing chain silicates pyroxmanganite, rhodonite, bustamite and serandite. Canadian Mineral. 9, 348-357.
_____ (1969a) On the origin of colour and pleochroism of astrophyllite and brown clintonite. Canadian Mineral. 9, 663-677.
_____ (1969b) Optical absorption studies of grossular, andradite (var. colophonite) and uvarovite. Canadian Mineral. 9, 723-729.
_____ (1969c) Optical absorption spectra of chromium-bearing tourmaline, black tourmaline and buergerite. Canadian Mineral. 9, 57-70.
_____ (1969d) An optical absorption study of the origin of colour and pleochroism in pink and brown tourmalines. Canadian Mineral. 9, 678-690.
_____ (1970) Racah parameters and the relationship to lengths and covalence of Mn^{2+}- and Fe^{3+}-oxygen bonds in silicates. Canadian Mineral. 10, 677-688.
_____ and Townsend, MG. (1970) Effect of next-nearest neighbour interaction on oscillator strengths in garnets. J. Phys. Chem. 3, L14-15.
_____ and Harris, D.C. (1970) Optical-absorption and electron-microprobe studies of some high-Ti andradites. Canadian Mineral. 10, 260-271.
_____ and Nickel E.H. (1970) A spectral study of the origin of colour and pleochroism of a titanaugite from Kaiserstuhl and of a riebeckite from St. Peter's Dome, Colorado. Canadian Mineral. 10, 71-83.
_____ (1972) Optical absorption spectra of Fe^{3+} in octahedral and tetrahedral sites in natural garnets. Canadian Mineral. 11, 826-839.
_____ (1973a) Extinction coefficients of Fe^{3+} spectral bands in oxides and silicates as indicators of local crystal composition. Canadian Mineral. 12, 120-123.
_____ (1973b) Effect of second-nearest-neighbour interaction on Mn^{3+} absorption in pink and black tourmalines. Canadian Mineral. 11, 971-977
_____ and Tricker, M.J. (1975) Optical absorption and Mössbauer spectral studies of iron and titanium site-populations in vesuvianites. Canadian Mineral. 13, 259-265.
_____ (1975) Charge-transfer processes and the origin of colour and of pleochroism of some Ti-rich vesuvianites. Canadian Mineral. 13, 110-116.
_____ (1976) Ferrous-ferric interaction on adjacent face-sharing antiprismatic sites in vesuvianites: evidence for ferric ion in eight coordination. Canadian Mineral. 14, 216-220.
_____ (1977) Charge-transfer interactions and the origin of color in brown vesuvianite. Canadian Mineral. 15, 508-511
Mao, H.K. and Bell, P.M. (1971) Crystal field spectra. Carnegie Inst. Washington Yearbook 70, 207-215.
_____ and Bell, P.M. (1972a) Interpretation of the pressure effect on the optical absorption bands of natural fayalite to 20 kb. Crystal field stabilization of the olivine-spinel transition. Carnegie Inst. Washington Yearbook 71, 524-527.
_____ and Bell, P.M. (1972b) Electrical conductivity and red shift of absorption in olivine and spinel at high pressure. Science, 176, 403-406.
_____ and Bell, P.M. (1972c) Optical and electrical behavior of olivine and spinel (Fe2SiO4) at high pressure. Carnegie Inst. Washington Yearbook 71, 520-524.
_____, Bell, P.M. and Dickey, J.S. (1972) Comparison of the crystal-field spectra of natural and synthetic chrome diopside. Carnegie Inst. Washington Yearbook 71, 538-541.
_____ (1973) Observations of optical absorption and electrical conductivitty in mangesiowustite at high pressures. Carnegie Inst. Washington Yearbook, 72, 554-557.
_____ and Bell, P.M. (1973a) Polarized crystal-field spectra of microparticles of the moon. In : Analytical Methods Developed for applicaton to Lunar Sample Analyses. ASTM STP 539, 100-119. Amer. Soc. for Testing and Materials.
_____ and Bell, P.M. (1973b) Luna 20 plagioclase: Crystal-field effects and chemical analysis of iron. Carnegie Inst. Washington Yearbook 72, 662-665.
_____ and Bell, P.M. (1974a) Crystal-field effects of ferric iron in goethite and lepidocrocite: band assignment and geochemical applications at high pressure. Carnegie Inst. Washington Yearbook 73, 502-507
_____ and Bell, P.M. (1974b) Crystal-field effects in spinel: oxidation states of iron and chromium. Carnegie Inst. Washington Yearbook, 73, 332-341.
_____ and Bell, P.M. (1974c) Crystal field effects of trivalent titanium in fassaite from the Pueble de Allende meteorite. Carnegie Inst. Washington Yearbook 73, 488-492.
_____ and Seifert, F. (1974) A study of the crystal-field effects of iron in the amphiboles anthophyllite and gedrite. Carnegie Inst. Washington Yearbook 73, 500-502.
_____ and Bell, P.M. (1975) Crystal-field effects in spinel: oxidation states of iron and chromium. Geochim. Cosmochim. Acta 39, 865-874.
_____ (1976) Charge-transfer processes at high pressure. In: The Physics and Chemistry of Minerals and Rocks. R.G.J. Strens, ed., J. Wiley, New York, p. 573-581
_____, Bell, P.M. and Virgo, D. (1977) Crystal-field spectra of fassaite from the Angra Dos Reis meteorite. Earth Planet. Sci. Let. 35, 352-356.
Maschmeyer, D. and Lehmann, G. (1983) A trapped-hole center causing rose coloration of natural quartz. Zeit. Kristallogr. 163, 181-196.
Matsyuk, S.S., Platonov, A.N. and Belichenko, V.P. (1978) On the dependence spectroscopic parameters of Cr^{3+} ions on garnet composition. Konst. Svoystv. Mineral. 12, 112-115. (in Russian)
Mattson S.M. and Rossman, G.R. (1984) Ferric iron in tourmaline. Phys. Chem. Minerals 14, 225-234.
_____ and _____ (1986) Yellow, manganese-rich elbaite with manganese-titanium intervalence charge transfer. Amer. Mineral. 71, 599-602.
_____ and _____ (1987a) Identifying characteristics of charge transfer transitions

in minerals. Phys. Chem. Minerals 14, 94-99.

_____ and _____ (1987b) $Fe^{2+}-Fe^{3+}$ interactions in tourmaline. Phys. Chem. Minerals, 14, 163-171.

McClure, D.S. (1962) Optical spectra of transition metal ions in corundum. J. Chem. Phys. 36, 2757-2779.

_____ (1963) Comparison of the crystal fields and optical spectra of Ce_2O_3 and ruby. J. Chem. Phys. 38, 2757-2779.

Mitra, S. (1981) Nature and genesis of colour centers in yellow and colorless fluorite from Amba Dongar, Gujrat, India. N. Jb. Mineral. 141, 290-308.

Moore, R.K. and White, W.B. (1971) Intervalence electron transfer effects in the spectra of the melanite garnets. Amer. Mineral. 56, 826-840.

_____ and _____ (1972) Electronic spectra of transition metal ions in silicate garnets. Canadian Mineral. 11, 791-811.

Nassau, K. and Wood, D.L. (1968) An examination of red beryl from Utah. Amer. Mineral. 53, 801-806.

_____ and Prescott, B.E. (1975a) Blue and brown topaz produced by gamma irradiation. Amer. Mineral. 60, 705-709.

_____ and Prescott, B.E. (1975b) A reinterpretation of smoky quartz. Phys. stat. sol. 29, 659-663.

_____, Prescott, B.E., and Wood, D.L (1976) The deep blue Maxixe-type color center in beryl. Amer. Mineral. 61, 100-107.

_____ and Prescott, B.E. (1977) Smoky, blue, greenish-yellow, and other irradiation-related colors in quartz. Mineral. Mag. 41, 301-312.

_____ (1983) The Physics and Chemistry of Color: the Fifteen Causes of Color. J. Wiley & Sons, New York.

Neuhaus, A. and Richartz, W. (1958) Absorptionsspektrum und Koordination allochromatisch durch Cr^{3+} gefärbter natürlicher und synthetischer Einkristalle unk Kristallpulver. Angew. Chem. 70, 430-434.

Newnham, R.E. and Santoro, R.P. (1967) Magnetic and optical properties of dioptase. Phys. Stat. Sol. 19, K87-K90.

Nikolskaya, L.V., Terekhova, V.M. and Samoilovich, M.I. (1978) On the origin of natural sapphire color. Phys. Chem. Minerals 3, 213-224.

Noack, Y., Decarreau, A. and Manceau, A. (1986) Spectroscopic and oxygen isotope evidence for low and high temperature origin of talc. Bull. Mineral. 109, 253-263.

Parkin, K.M and Burns, R.G. (1980) High temperature crystal field spectra of transition metal-bearing minerals: relevance to remote-sensed spectra of planetary surfaces. Proc. 11th Lunar Planet. Sci Conf., Suppl. 12, Geochim. Cosmochim. Acta 1, 731-755.

Partlow, E.P. and Cohen, A.J. (1986) Optical studies of biaxial Al-related color centers in smoky quartz. Amer. Mineral. 71, 589-598.

Petrov, I. and Berdesinski, W. (1975) Untersuchung künstlich farbveränderter blauer Topase. Z. Dt. Gemmol. Ges. 24, 16-19.

_____ (1977) Farbeuntersuchungen an Topas. Neus Jb. Mineral. Abh. 130, 288-302.

_____, Schmetzer, K. and Eysel, H.H. (1977) Absorptionsspektren von Chrom in Topas. N. Jb. Mineral. Monat. 365-372.

_____ (1978) Farbe, Farbursachen and Garbveränderung bei Topasen. Z. Dt. Gemmol. Ges. 27, 1-11.

Pizani, P.S., Terrile, M.C., Farach, H.A. and Poole, C.P. (1985) Color centers in sodalite. Amer. Mineral. 70, 1186-1192.

Platonov, A.N., Tarashchan, A.N., Belichenko, V.P. and Povarennykh, A.S. (1971) Spectroscopic study of sulphide sulphur in some framework aluminosilicates. Konst. Svoistva. Mineralov 5, 61-72. (in Russian)

_____, Taran, M.N., Minko, O.E., and Polshyn, E.V. (1978) Optical absorption spectra and nature of color of iron-containing beryls (A) Phys. Chem. Minerals 4, 87-88.

_____, A.N., Khomenko, V.M. and Belichenko, V.P. (1984) Optical absorption spectra of vanadium diopside; lavrovite Zapiski Vsesoyuznogo Mineralogicheskogo Obshchestva 1984, 724-727. (in Russian)

_____, Tarashchan, A.N. and Taran, M.N. (1984) Color centers in amazonites. Mineralogicheskiy Zhurnal 6,: 6, 3-16. (in Russian)

_____, Taran, M.N., Sobolev, N.V., Matsyuk, S.S. and Spetsius, Z.V. (1985) Optical absorption spectra and the color of natural specimens of chromium-containing kyanite. Mineralogicheskiy Zhurnal 7: 3, 22-30. (in Russian)

Povarennykh, A.S., Platonov, A.N. and Belichenko, V.P. (1970) On the colour of ussingite from the Ilimaussaq (South Greenland) and Lovozero (Kola Peninsula) alkaline intrusions. Bull. Geol. Soc. Denmark 20, 20-26.

_____, Platonov, A.N., Tarashchan, A.N. and Belichenko. V.P. (1971) The color and luminescence of tugtupite (beryllosodalite) from Ilimaussaq, South Greenland. Meddelelsev om Gronland 181, 1-12.

Price, D.C., Vance, E.R., Smith, G., Edgar, A., and Dickson, B.L. (1976) Mössbauer effect studies of beryl. J. Physique 12, C6-811 - C6-817.

Rager, H. and Weiser, G. (1984) Polarized absorption spectra of trivalent chromium in forsterite, Mg_2SiO_4. Bull. Minéral. 104, 603-609.

Recker, K., Neuhaus, A., and Leckebusch, R. (1968) Vergleichende Untersuchungen der Garb- und Lumineszenzeigenschaften natürlicher und gezüchteter definiert dotierter Fluorite. Proc. Internat. Mineral. Assoc., Cambridge, 145-152.

Reddy, B.J. and Lakshman, S.V.J. (1975) Optical absorption spectra of Ni^{2+} in garnierite: a reply. Canadian Mineral. 13, 300-301.

Reed, J.S. (1971) Optical absorption spectra of Cr^{3+} in $MgO-Al_2O_3$ - $MgO-3.5Al_2O_3$ spinels. J. Amer. Ceram. Soc. 54, 202-204.

Richardson, S.M. (1973) A pink muscovite with reverse pleochroism from Archer's Post, Kenya. Amer. Mineral. 60, 73-78.

_____ (1976) Ion distribution in pink muscovite: A reply. Amer. Mineral. 61, 1051-1052.

Richman, I., Kisliuk, P. and Wong, E.Y. (1967) Absorption spectrum of U^{4+} in zircon ($ZrSiO_4$). Phys. Rev. 155, 262-267.

Robbins, D.W. and Strens, R.G.J. (1968) Polarization-dependence and oscillator streingths of metal-metal charge-transfer bands in iron (II,III) silicate minerals. Chem. Comm. 1968, 508-509.

_____ and Strens, R.G.J. (1972) Charge-transfer ferromagnesian silicates: The polarized electronic spectra of trioctahedral micas. Mineral. Mag. 38, 551-563.

Rossmam, G.R. (1974a) Optical spectroscopy of green vanadium apophyllite from Poona, India. Amer. Mineral. 59, 621-622.

_____ (1974b) Lavender jade. The optical spectrum of Fe^{3+} and $Fe^{2+}-Fe^{3+}$ intervalence charge transfer in jadeite from Burma. Amer. Mineral. 59, 868-870

_____ (1975) Spectroscopic and magnetic studies of ferric iron hydroxy sulfates: Intensification of color in ferric iron clusters bridged by a single hydroxide ion. Amer. Mineral. 60, 698-704

_____ (1976a) Spectroscopic and magnetic studies of ferric iron hydroxy sulfates: The series $Fe(OH)SO_4 \cdot nH_2O$ and jarosite. Amer. Mineral. 61, 398-404

_____ (1976b) The optical spectroscopic comparison of the ferric iron tetrameric clusters in amarantite and leucophosphite. Amer. Mineral. 61, 933-938.

_____ (1980) Pyroxene spectroscopy. In: C.T. Prewet, ed., Pyroxenes. Rev. Mineral. 7, 93-116

_____, Shannon, R.D. and Waring, R.K. (1981) Origin of the yellow color of complex nickel oxides. J. Solid State Chem. 39, 277-287.

_____, Grew, E.S. and Dollase, W.A. (1982) The colors of sillimanite. Amer. Mineral. 67, 749-761.

_____ (1984) Spectroscopy of Micas. In: S.W. Bailey, ed., Micas. Rev. Mineral. 13, 145-181.

Rumyantsev, V.N., Grum-Grzhimailo, S.V. and Boksha, O.N. (1971) Optical absorption spectra of hydrothermal corundum crystals containing impurities. Soviet Phys. Crystallogr 16, 373-374. Translated from Kristallogr. 16, 445-447.

Runciman, W.A., Sengupta, D. and Marshall, M. (1973) The polarized spectra of iron in silicates. I. Enstatite. Amer. Mineral. 58, 444-450.

_____, Sengupta, D. and Gourley, J.T. (1973) The polarized spectra of iron in silicates: II. Olivine. Amer. Mineral. 58, 451-456.

_____, Sengupta, D. and Gourley, J.T. (1974) The polarized spectra of iron in silicates: II. Olivine: a reply. Amer. Mineral. 59, 630-631.

_____ and Sengupta, D. (1974) The spectrum of Fe^{2+} ions in silicate garnets. Amer. Mineral. 59, 563-566.

Runciman, W.A. (1987) Comment on the spin-forbidden absorption spectrum of Fe^{2+} in orthopyroxene by Zhao, SB, Wang, HS, Zhou, KW and Xiao, TB. Phys. Chem. Minerals 14, 482.

Samoilovich, M.I., Novozhilov, A.I., Radyanskii, V.M., Davydchenko, A.G. and Smirnova, C.A. (1973) On the nature of the blue color in lazurite. Izv. A. N. SSSR, Ser. Geol. 1937-7, 95-102. (in Russian)

Scarratt, K.V.G. (1982) The identification of artificial coloration in diamond. Gems & Gemology 18, 72-78.

Scheetz, B.E. and White, W.B. (1972) Synthesis and optical absorption spectra of Cr^{2+} containing silicates. Contrib. Mineral. Petrol. 37, 221-227.

Schirmer, O.F. (1976) Smoky coloration of quartz caused by bound small polaron optical absorption. Solid State Commun. 18, 1349-1353.

Schmetzer, K. and Berdesinski, W. (1978) Das Absorptionsspektrum von Cr^{3+} in Zoisit. N. Jb. Mineral. Mh., 1978, 197-202.

_____ and Ottemann, J. (1979) Kristallchemie und Farbe Vanadium-haltiger Granate. N. Jb. Mineral. Abh., 136, 146-168.

_____ and Bank, H. (1980) Explanations of the absorption spectra of natural and synthetic Fe- and Ti-containing corundums. N. Jb. Miner. Abh. 139, 216-225.

_____ and Gübelin, E. (1980) Alexandrite-like spinel from Sri Lanka. N. Jb. Mineral. Mh. 428-432.

_____ and Bank, H. (1981) An unusual pleochroism in zambian emeralds. J. Gemmology 17, 443-446.

_____ (1982) Absorption spectroscopy and color of vanadium(3+)-bearing natural oxides and silicates - a contribution to the crystal chemistry of vanadium. N. Jb. Mineral, Abh. 144, 73-106.

_____ (1982b) An unusual garnet from Umba Valley, Tanzania. J. Gemmology 18, 194-200.

_____, Bosshart, G., and Hanni, H.A. (1983) Naturally-coloured and treated yellow and orange-brown sapphires. J. Gemmology 18, 607-622.

_____ and Bank, H. (1984) Crystal chemistry of tsilaisite (manganese tourmaline) from Zambia. N. Jb. Mineral. Mh. 1984, 61-69.

_____ (1987) Zur Deutung der Farbursache blauer Saphire - eine Diskussion. N. Jb. Mineral.

Mh. 1987, 337-343.

Schmitz, B. and Lehmann, B. (1975) Color centers of manganese in natural spodumene, $LiAlSi_2O_6$. Ber. Bunsenges. Phys. Chem. 79, 1044-1049.

Seal, M., Vance, E.R. and Demago, B. (1981) Optical spectra of giant radiohaloes in Madagascan biotite. Amer. Mineral. 66, 358-361.

Shankland, T.J., Duba, A.G. and Woronow, A. (1974) Pressure shifts of optical absorption bands in iron-bearing garnet, spinel, olivine, pyroxene and periclase. J. Geophys. Res. 79, 3273-3282.

_____, Nitsan, U. and Duba, A.G. (1979) Optical absorption and radiative heat transport in olivine at high temperature. J. Geophys. Res. 84, 1603-1610.

Sherman, D.M. and Waite, T.D. (1985) Electronic spectra of Fe^{3+} oxides and oxide hydroxides in the near IR to near UV. Amer. Mineral. 70, 1262-1269.

_____ (1987a) Molecular orbital (SCF-Xα-SW) theory of metal-metal charge transfer processes in mineral. I. Applications to Fe^{2+}-Fe^{3+} charge transfer and "electron delocalization" in mixed-valence iron oxides and silicates. Phys. Chem. Minerals 14, 355-363.

_____ (1987b) Molecular orbital (SCF-Xα-SW) theory of metal-metal charge transfer processes in mineral. II. Applications to Fe^{2+}-Ti^{4+} charge transfer transitions in oxides and silicates. Phys. Chem. Minerals 14, 364-367.

Shigley, J.E. and Stockton, C.M. (1984) Cobalt-blue gem spinels. Gems & Gemology 20, 34-41.

_____, and Foord, E.E. (1985) Gem-quality red beryl from the Wah Wah Mountains, Utah. Gems & Gemmology 20, 208-221.

_____, Kane, R.E. and Manson, D.V. (1986) A notable Mn-rich gem elbaite tourmaline and its relationship to "tsilaisite". Amer. Mineral. 71, 1214-1216.

_____, Fritsch, E., Stockton, C.M., Koivula, J.I, Fryer, C.W. and Kane, R.E. (1986) The gemological properties of the Sumitomo gem-quality synthetic yellow diamonds. Gems & Gemmology 22, 192-208.

Singer, R.B. (1982) Spectral evidence for the mineralogy of high-albedo soils and dust on Mars. J. Geophys. Res. B87, 10159-10168.

Slack, G.A. and Chrenko, R.M. (1971) Optical absorption of natural garnets from 1000 to 30000 wavenumbers. J. Optical Soc. Amer. 61, 1325-1329.

Smith, G. and Strens, R.G.J. (1976) Intervalence transfer absorption in some silicate, oxide and phosphate minerals. In: Strens, R.G.J., ed, The Physics of Minerals and Rocks. Wiley, New York. p. 583-612.

_____ (1977) Low-temperature optical studies of metal-metal charge-transfer transitions in various minerals. Canadian Mineral. 15, 500-507.

_____ (1978a) A reassessment of the role of iron in the 5,000-30,000 cm-1 region of the electronic absorption spectra of tourmaline. Phys. Chem. Minerals 3, 343-373.

_____ (1978b) Evidence for absorption by exchange-coupled Fe^{2+}-Fe^{3+} pairs in the near infra-red spectra of minerals. Phys. Chem. Minerals 3, 375-383.

_____, Vance, E.R., Hasan, Z., Edgar, A., and Runciman, W.A. (1978) A charge transfer mechanism for the colour of rose quartz. Phys. Status Solidi (a) 46, K135-K140.

_____ (1980) Evidence for optical absorption by Fe^{2+}-Fe^{3+} interactions in MgO:Fe. Phys. Status Solidi A61, K191.

_____, Howes, B., and Hasan, Z. (1980) Mössbauer and optical spectra of biotite: a case for Fe^{2+}-Fe^{3+} interactions. Phys. Status Solidi (a), 57, K187-K192.

_____ and Langer, K. (1982) Single crystal spectra of olivines in the range 40,000 - 50,000 cm-1 at pressures up to 200 kbars. Amer. Mineral. 67, 343-348.

_____, Hlenius, U. and Langer, K. (1982) Low temperature spectral studies of Mn^{3+}-bearing andalusite and epidote type minerals in the range 30000-5000 cm-1. Phys. Chem. Minerals 8, 136-142.

_____ and Langer, K. (1983) High pressure spectra up to 120 Kbars of the synthetic garnet end members spessartine and almandine. N. Jb. Mineral. Mn. 1983, 541-555.

_____, Hlenius, U., Annersten, H., and Ackermann, L. (1983) Optical and Mössbauer spectra of manganese-bearing phlogopites: $Fe^{3+}(IV)$-$Mn^{2+}(VI)$ pair absorption as the origin of reverse pleochroism. Amer. Mineral., 68, 759-768.

Speit, B. and Lehmann, G. (1976) Hole centers in the feldspar sanidine. Phys. Status Solidi A36, 471-481.

Stolper, E.M., Paque, J. and Rossman, G.R. (1982) The influence of oxygen fugacity and cooling rate on the crystallization of Ca-Al-Rich inclusions from Allende. Lunar Planet. Sci. 14, 772-773.

Strens, R.G.J. and Freer, R. (1978) The physical basis of mineral optics: I. Classical theory. Mineral. Mag. 42, 19-30.

Strens, R.G.J., Mao, H.K. and Bell, P.M. (1982) Quantitative spectra and optics of some meteoritic and terrestrial titanian clinopyroxenes. In: Advances in Physical Geochemistry, Vol. 2, S.K. Saxena, ed. Springer Verlag, New York. pp 327-346.

Sung, C.M., Abu-Eid, R.M. and Burns, R.G. (1974) Ti^{3+}/Ti^{4+} ratios in lunar pyroxenes: implications to depth of origin of mare basalt magma. Proc. 5th Lunar Sci. Conf., Suppl. 5, Geochim. Cosmochim Acta, Vol 1, 717-726.

_____, Singer, R.B., Parkin, K.M. and Burns, R.G. (1977) Temperature dependence of Fe^{2+} crystal-field spectra: implications to mineralogical mapping of planetary surfaces. Proc. 8th Lunar Sci. Conf., Pergamon Press, New York, p. 1063-1079.

Sviridov, D.T., Sevastyanow, B.K., Orekhova, V.P., Sviridova, R.K. and Veremeichik, T.F. (1973) Optical absorption spectra of excited Cr^{3+} ions in magnesium spinel at room and liquid nitrogen temperatures. Opt. Spect. 35, 59-61.

Takei, H. (1976) Czochralski growth of Mn_2SiO_4 (tephroite) single crystal and its properties. J. Crystal Growth 34, 125-131.

_____, Hosoya, S. and Ozima, M. (1984) Synthesis of large single crystals of silicates and titanates. In: Materials Science of the Earth's Interior, I. Sunagawa, ed. pp 107-130. Terra Scientific Publishing Co., Tokyo.

Taran, M.N., Platonov, A.N., Pol'shin, E.V. and Matsyuk, S.S. (1987) Optical spectra and coloration of natural spinels of the $(Mg,Zn,Fe)(Al,Fe)_2O_4$ composition. Mineralogicheskiy Zhurnal 9, 3-15. (in Russian)

Taylor, L.A., Mao, H.K. and Bell, P.M. (1974) Identification of the hydrated iron ixide mineral akaganeite in sample 66095,85. Carnegie Inst. Washington Yearbook 73, 477-480.

Tippens, H.H. (1970) Charge-transfer spectra of transition-metal ions in corundum. Phys. Rev. B., B 1, 126-135.

Townsend, M.G. (1968) Visible charge-transfer band in blue sapphire. Solid State Commun. 6, 81-83.

_____ (1970) On the dichroism of tourmaline. J. Phys. Chem. Solids 31, 2481-2488.

_____ and Faye, G.H. (1970) Polarized electronic absorption spectrum of vivianite. Phys. Status Solidi 38, K57-K60.

Tsang, T. and Ghose, S. (1971) Electron paramagnetic resonance of V^{2+}, Mn^{2+}, Fe^{3+} and optical spectra of V^{3+} in blue zoisite, $Ca_2Al_3Si_3O_{12}(OH)$. J. Chem. Phys. 54, 856-862.

Vance, E.R. (1974) The anomalous optical absorption spectrum of low zircon. Mineral. Mag. 39, 709-714.

_____ and Mackey, D.J. (1974) Optical study or U^{5+} in zircon. J. Phys. C: Solid State Phys. 7, 1898-1908.

_____ and _____ (1978) Optical spectra of U^{4+} and U^{5+} in zircon, hafnon, and thorite. Phys. Rev. B 18, 185-189.

_____ and Price, D.C. (1984) Heating and radiation effects on optical and Mössbauer spectra of Fe-bearing cordierites. Phys. Chem. Minerals 10, 200-208.

Veremechik, T.F., Grechusnikov, B.N., Varina, T.M., Sviridov, D.T and Kalinkina, I.N. (1975) Absorption spectra and calculation of energy-level diagrams of Fe^{3+} and Mn^{2+} ions in single crystals of yttrium aluminum garnet, orthoclase, and manganese silicate. Kristallografiya 19, 1194-1199. (Trans: Sov. Phys. Crystallogr. 19, 742-744.)

Weeks, R.A. (1970) Paramagnetic resonance and optical absorption in gamma-ray irradiated alpha quartz: The "Al" center. J. Amer. Ceramic Soc. 53, 176-179.

_____, Pigg, J.C. and Finch, C.B. (1974) Charge-transfer spectra of Fe^{3+} and Mn^{2+} in synthetic forsterite. Amer. Mineral. 59, 1259-1266.

White, E.W. and White, W.B. (1967) Electron probe and optical study of kyanite. Science 158, 915.

White, W.B. and Keester, K.L. (1966) Optical absorption spectra of iron in the rock-forming silicates. Amer. Mineral. 51, 774-791.

_____, Roy, R. and Crichton, J.M. (1967) The "Alexandrite" effect: and optical study. Amer. Mineral. 51, 774-791.

_____, McCarthy, G.J. and Scheetz, B.E. (1971) Optical spectra of chromium, nickel and cobalt-containing pyroxenes. Amer. Mineral. 56, 72-89.

_____ and Moore, R.K. (1972) Intrepetation of the spin-allowed bands of Fe^{2+} in silicate garents. Amer. Mineral. 57, 1692-1710.

_____, Matsumura, M., Linnehan, D.G., Furukawa, T. and Chandrasekhar, B.K. (1986) Absorption and luminescence of Fe^{3+} in single-crystal orthoclase. Amer. Mineral. 71, 1415-1419.

Wilkins, R.W.T., Farrell, E.F. and Naiman, C.S. (1969) The crystal field spectra and dichroism of tourmaline. J. Phys. Chem. Solids 30, 43-56.

Wood, B.J. (1974) Crystal field spectrum of Ni^{2+} in olivine. Amer. Mineral. 59, 244-248.

Wood, D.L. and Nassau, K. (1968) The characterization of beryl and emerald by visible and infrared absorption spectroscopy. Amer. Mineral. 53, 777-800.

_____, Imbusch, G.F., MacFarlane, R.M., Kisliuk, P. and Larkin, P.M (1968) Optical spectrum of Cr^{3+} in spinels. J. Chem. Phys. 48, 5255-5263.

Yagi, T. and Mao, H.K. (1977) Crystal-field spectra of the spinel polymorph of Ni_2SiO_4 at high pressure. Carnegie Inst. Washington Yearbook 17, 505-508.

Zhao, S.B., Wang, H.S. and Zhou, K.W. (1986) A simplified strong-field scheme and the absorption spectrum of Mn^{2+} in rhodonite. J. Phys. C: Solid State Phys. 19, 2729-2740.

_____, _____, _____ and Xiao, T.B. (1986) The spin-forbidden absorption of Fe^{2+} in orthopyroxene. Phys. Chem. Minerals 13, 96-101

_____, _____, and _____ (1987) On the approach to the assignment of spin-forbidden bands of complex ions: A reply. Phys. Chem. Minerals 14, 483-484.

Zhau, K.W. and Zhao, S.B. (1984) The spin-forbidden spectrum of Fe^{2+} in silicate garnets. J. Phys. C: Solid State Phys. 17, 4625-4632.

APPENDIX

Detailed Example: Fe^{2+} at the M(2) Site in Orthopyroxene.

Many mineral spectroscopic studies have been concerned with the orthopyroxenes. The pyroxenes are of interest because of their impor- tance in a variety of rocks and because of the crystal chemical problem of partitioning of cations between the M(1) and M(2) sites. The interpretation of the spectrum of Fe^{2+} at the distorted, six-coordinated M(2) site has become a testing ground for the methods of interpreting optical spectra of minerals. In the extended example which follows, the analysis of the spectrum of Fe^{2+} at the M(2) site will be worked out in detail. The example follows the analysis of Goldman and Rossman (1977b).

THE SYMMETRY OF THE CATION POLYHEDRA IN ORTHOPYROXENES

To interpret a spectroscopic experiment in detail, the point group symmetry of the cation site must first be determined. In general, this information comes from an analysis of the parameters obtained from the single-crystal X-ray structural determination. The structural details of the orthopyroxene M(1) and M(2) sites will be used to discuss the relative distortions of the sites in conjunction with spectroscopic and site-occupancy data.

It is also necessary to establish the orientation of the individual cation sites within the external morphology of an orthopyroxene crystal. This information is used to interpret the pleochroism (spectroscopic anisotropy).

Part I. Site Projections From X-Ray Structural Refinement Data.

The structural parameters for the orthopyroxene hypersthene are ob- tained from Ghose (1965); Table A1 lists selected interatomic distances.

Projections of the cation sites can be prepared following the method outlined in Megaw (1973). The projections of the M(2) coordination polyhedra on both (100) and (010) are presented in Fig. A1. Comparison of these projections with the projection of the M(1) site (Fig. A2) shows that the M(2) site is more distorted then the M(1) site. Volume 1 of **International Tables for X-Ray Crystallography** (1969) provides additional data about the pyroxene space group (Pbca) and atomic coordinates which are useful in preparing the projections.

Part II. Correlation of Crystallographic, Morphological and Optical Properties.

Next to the projections in Figure A1, the *a*, *b*, and *c* crystallographic axes are indicated. They are related to the principal vibration directions (orientation of the optical indicatrix) within the crystal, as listed in Table A2.

One must be careful not to change conventions in going from the unit cell axes to the morphological axes to the optical directions. Not all authors use the same choice of labels for the crystallographic and morphological axes. This problem is particularly prevalent in the

Table A1: Atomic coordinates for orthopyroxene from Ghose (1965)

M(1)-O1	2.087 Å	M(2)-O2	2.066 Å
-O1	2.166	-O4	2.175
-O2	2.038	-O3	2.519
-O4	2.152	-O1	2.119
-O4	2.036	-O5	2.037
-O5	2.075	-O6	2.405
mean	2.092	mean	2.220

Table A2. Correlation of optical, morphological and crystallographic directions in orthopyroxene.

optical direction	unit cell axis	unit cell length	morphological axis
α	b	8.927 Å	b
β	a	18.310 Å	a
γ	c	5.226 Å	c

Table A3: Character table for C_{2v}

C_{2v}	E	C_2	$\sigma_v(xz)$	$\sigma_v'(yz)$		
A_1	1	1	1	1	z	$x^2y^2z^2$
A_2	1	1	-1	-1	xy	R_z
B_1	1	-1	1	-1	x	xz	R_y
B_2	1	-1	-1	1	y	yz	R_x

Table A4. Subset of the octahedral site character table preserved in C_{2v}.

O_h	E	$6C_2$	$3\sigma_h$	$6\sigma_d$
T_{2g}	3	1	-1	1
E_g	2	0	2	0

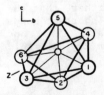 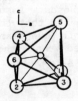 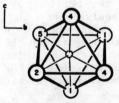

ORTHOPYROXENE M(2) SITE

ORTHOPYROXENE M(1) SITE

Figure A1 (left). Projection of the M(2) octahedron of orthopyroxene. From Goldman and Rossman (1977b).

Figure A2 (right). Projection of the M(1) octahderon of orthopyroxene.

orthorhombic system.

Part III. Determination of the Point Group for the M(2) Site.

There is a body of data which indicates that Fe^{2+} at the M(2) site is responsible for most of the absorption features in the orthopyroxene spectrum. For example, Evans et al. (1967) used Mössbauer spectra to show that most of the Fe is at the M(2) site in low-iron orthopyroxenes. Therefore, we will concentrate on the analysis of the spectroscopy of Fe^{2+} at the M(2) site.

In the analysis of the orthopyroxene spectrum, the first step it to determine the point group for the M(2) site, using the exact metal-oxygen bond distances and bond angles in Table A1. To do this, it is necessary to have some understanding of point group symmetry and the nomenclature of point groups. The book by Cotton (1971) is one of the more useful texts for this material.

Because all the Fe-O bond distances are different, it follows immediately that there are no symmetry elements at all in the M(2) polyhedron. Therefore, the symbol for the M(2) site point group is C_1.

In C_1 symmetry, all optical absorptions are allowed in each polarization. However, the experimental optical data show that the absorption bands are strongly polarized, indicating that the Fe^{2+} experiences a higher electrostatic pseudo-symmetry. It is therefore useful to model the M(2) site with a higher effective symmetry.

Part IV. Symmetry Modeling of the M(2) Site

The answer from part III, although correct, is not of much use in the detailed analysis of the pyroxene spectra. The problem is that we need symmetry to aid us in the interpretation of all sorts of useful things. Therefore, we need to approximate the symmetry of the M(2) site with a point group of higher symmetry. The reasons for doing this will be obvious only after we work through the entire analysis and find that we come up with something useful when we are done.

Some simplifying assumptions must be made. Note that there are three pairs of Fe-O bonds of approximately the same length. Therefore, assume that the two long bonds (2.4 and 2.5 Å) are close enough to be considered equivalent. The two shortest bonds are assumed to be the same length as are the two intermediate bonds. Assume further, that the four longest bonds are in the same plane and that the two shortest bonds are exactly normal to this plane. Both Runciman et al. (1973) and Goldman and Rossman (1977b) recognized that the symmetry of this approximation to the M(2) site is C_{2v} (Fig. A3).

Part V. Character Tables.

Chapter 4 in Cotton (1971) develops the concepts of the character table and the matrix representation of groups. An understanding of these concepts is necessary for the analysis which follows. We will be concerned with the character table for C_{2v} in our analysis of the M(2) site of orthopyroxenes (Table A3).

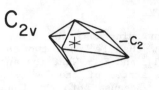

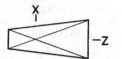

Figure A3. The C_{2v} model for the M(2) site. The two-fold rotation axis labeled C_2 corresponds to the molecular z-axis. From Goldman and Rossman (1977b).

^5E_g —

—— state 2

—— state 1

↑ energy

O_h C_{2v}

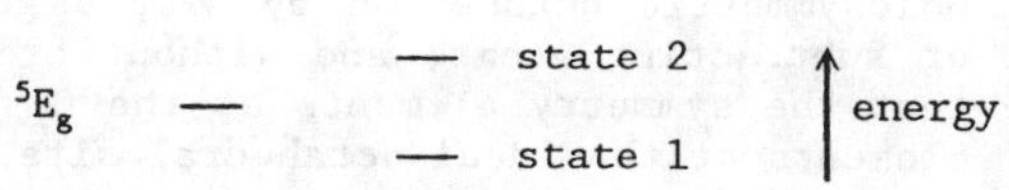

Figure A4. Splitting of the octahedral E_g state brought about by distortion of the octahedron to C_{2v} symmetry.

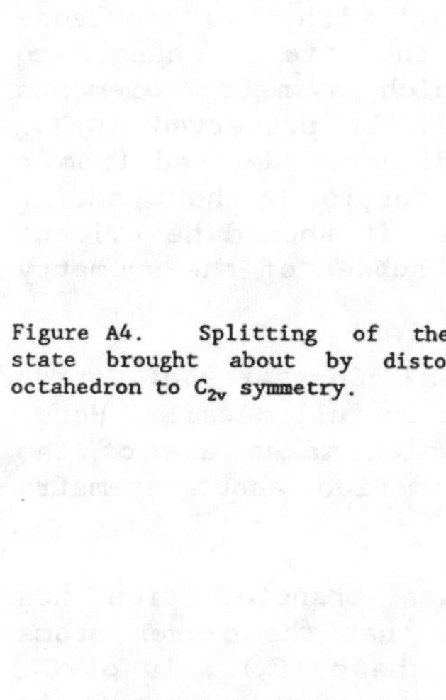

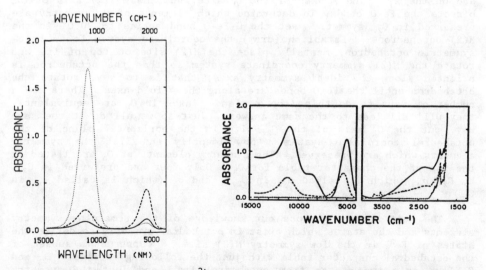

Figure A5 (left). Spectrum of Fe^{2+} at the M(2) site of the orthopyroxene, bronzite. From Goldman and Rossman (1977b). Polarizations: a = dots; b = dash; c = solid.

Figure A6 (right). Spectrum of three orthopyroxenes of different iron contents in gamma polarization showing the midinfrared absorption near 2350 cm^{-1}. Modified from Goldman and Rossman (1977b).

Part VI. Energy States of Fe^{2+} in C_{2v} Symmetry.

The energy level diagrams (Tanabe-Sugano diagrams) in references such as Cotton (1971), or Burns (1970) show that Fe^{2+} has two energy states in an octahedral environment, namely T_{2g} and E_g. What the energy level diagrams do not show is what are the states of the Fe^{2+} system when it is in a lower symmetry such as C_{2v}. Because both the T and E states are degenerate, we should suspect that these degeneracies might be lifted in the lower symmetry. We will have to work this out in what will be a moderately long, but not particularly complicated process.

The energy states of Fe^{2+} at the M(2) site would be known if the site were an octahedron. The approach is to start with this knowledge and to see how the octahedral states split when the site is changed to lower symmetry. First we must determine which symmetry elements originally present in octahedral symmetry are still preserved in C_{2v} symmetry. Because the M(2) polyhedron can be directly derived from a holosymmetric octahedron by simple geometric distortion without adding or subtracting atoms, and without breaking bonds, it should be evident that the symmetry elements at the C_{2v} site are a subset of the symmetry elements at the ideal octahedral site.

But how to determine which ones are preserved? Chapter 9 of Cotton (1971) explains a general method to work this out in full detail. Here, we will proceed along a less detailed route which makes use of the symmetry properties of the site and the information about symmetry encoded in the character tables.

Our primary problem is that a coordinate transformation has occurred in going from O_h to C_{2v} symmetry. Note that the oxygen atoms are on the x, y and z axes of the O_h site, but the z (C_2) axis of C_{2v} bisects the Fe-O bonds. To determine which elements of O_h symmetry are preserved in C_{2v} symmetry, set the O-Fe-O bond angles to 90.0° in the M(2) octahedron.   Establish (draw) the coordinate system for a holo-symmetric octahedron, mentally place the M(2) site on top of it, and rotate the M(2) symmetry coordinate system so that the octahedron is oriented along the ideal symmetry axes; that is to say, rotate the octahedron until the Fe-O bonds are along the 4-fold axes. There are a number of ways to do this (the x,y and z axes in O_h are equivalent); they will all lead to the same answer. Just so we all do it the same way, put the C_2 axis of the C_{2v} site in the horizontal plane of the octahedral coordinate system. Now identify and list the symmetry elements which are preserved. The symmetry elements of O_h are listed at the top of the character table for O_h. Only four are preserved in C_{2v}: E, C_2, σ_d (which is called σ_{yz} in C_{2v}), and σ_h (which is called σ_{xz} in C_{2v}).

The next step is to use our knowledge of the remaining symmetry elements and the states which exist in octahedral symmetry to derive the states of Fe^{2+} in the low symmetry M(2) site.   Construct a subset of the octahedral character table with just the following entries: T_{2g} and E_g (the only two states from octahedral Fe^{2+}) and just the symmetry elements which are preserved in C_{2v} symmetry (Table A4, above). These are the only symmetry elements and entries in the character table with which we must be concerned. If we can solve the problem for this subset, we will have solved it for C_{2v}.

The octahedral states are degenerate. The fact that the E operator has a character of 2 for the E_g symmetry indicates that the E_g state is doubly degenerate. Our task is to determine which two components of C_{2v} add together to give the entry for E_g in our subset of the O_h character table.

This can be done by trial and error (inspection): Specifically, determine which two rows of the C_{2v} table add together to form the E_g row of the subset of the O_h character table. The answer to this provides the symmetry designations of the two states which are derived from a C_{2v} distortion of the E_g state of Fe^{2+} in an octahedral site (Fig. A4). Of course, this answer could have been read directly from the descent of symmetry tables in Cotton (1971).

The T_{2g} case could also be determined by trial and error. However, there are many situations where trial and error is not a practical way to do this. In these cases, the general formula (formula 4.3-11 of Cotton, 1971) must be used; this determines the number of times the a_ith irreducible representation occurs in a reducible representation. By either method, it can be determined that the $^5A_1$, $^5A_2$, and $^5B_2$ states form in C_{2v} from the splitting of the octahedral $^5T_{2g}$ state.

We now know how the states of Fe^{2+} split when the symmetry is reduced from holosymmetric octahedral to that of the pyroxene M(2) site. That is to say, we know how many non-degenerate states exist, and we know their symmetry designations. Still to be determined are the relative energies of these states. We must go to experimental data (polarized optical spectra) to address these aspects, making use of the symmetry designations.

Part VII. Experimentally Determined Energy States of Fe^{2+} in C_{2v} Symmetry.

Figures A5 and A6 present the orthopyroxene spectra. Figure A4 shows the α (dots), β (dash) and γ (solid line) spectra of a 100 μm thick bronzite (Fs_{14}) in the visible and near-infrared range. Figure A5 shows only the gamma spectrum of two crystals in the visible through the mid-infrared range. The solid line is for hypersthene ($Fs_{39.5}$), and the dotted line is for synthetic enstatite ($Fs_{0.00}$). Bands are present at 10,900 cm^{-1} polarized dominantly along the b-axis, at 5400 cm^{-1} polarized along the a-axis, and at 2350 cm^{-1} polarized along the c-axis.

Part VIII. Origin of Polarizations and Pleochroism.

The observed polarization properties of the spectrum of Fe^{2+} at the pyroxene M(2) site are used to establish the order of the C_{2v} states. To do this, we must first establish some predictions about the polarization dependence of each possible electronic transition. We will then compare our predictions (models) to the actual data and hope that afterwards, only a single model will remain viable.

Electronic transitions are generally polarized. They correspond to a redistribution of electronic charge in three-dimensional space. These redistributions of electrons have vector properties and must match the electric dipole of the exciting radiation if there is to be effective coupling of energy between the incident radiation and the electron cloud.

We want to know the polarizations of transitions such as:

Ground State $\Rightarrow$ Excited State

That is to say, we want to know along which C_{2v} molecular axis (or axes) a particular electronic transition (such as $A_1 \Rightarrow B_1$) will occur. Once we know this, we need only to align a linear polarizer in this direction to make the transition occur (this assumes that we know the orientation of the molecular axes with respect to the morphological axes of the crystal).

Quantum mechanics tells us that the intensity of an electronic transition will be proportional to an integral of the form:

$$I = k \int (\text{excited state wavefunction}) \cdot \mu \cdot (\text{ground state wavefunction}) \, d\tau \quad (A1)$$

or
$$I = k \cdot \int \Psi_{\text{excited state}} \, \mu \, \Psi_{\text{ground state}} \, d\tau \, , \quad (A2)$$

where μ is an operator which corresponds to the components of an electric dipole having the form:

$$\mu = (\, k_1 \cdot X + k_2 \cdot Y + k_3 \cdot Z) \, , \quad (A3)$$

where X, Y, and Z are the molecular axes.

We want to evaluate integrals of the form:

$$\int \Psi_{ex} \, (X + Y + Z) \, \Psi_{gs} \, d\tau \quad (A4)$$

or by components:

$$I_x = k \int \Psi_{ex} \, X \, \Psi_{gs} \, d\tau \, , \quad I_x = k \int \Psi_{ex} \, X \, \Psi_{gs} \, d\tau \quad \text{and} \quad I_x = k \int \Psi_{ex} \, X \, \Psi_{gs} \, d\tau. \quad (A5)$$

From purely symmetry considerations, we can tell if the integrals are non-zero. This is analogous to evaluation of integrals of the odd (sine) and even (cosine) functions.

Assume for a moment that the A_1 state is the ground state (in fact, ground states are often A_1 states). We need to evaluate a whole series of integrals of the type:

$$\int \Psi_{A2} \, X \, \Psi_{A1} \, d\tau \qquad \int \Psi_{A2} \, Y \, \Psi_{A1} \, d\tau \qquad \int \Psi_{A2} \, Z \, \Psi_{A1} \, d\tau \quad (A6)$$

$$\int \Psi_{B1} \, X \, \Psi_{A1} \, d\tau \qquad \int \Psi_{B1} \, Y \, \Psi_{A1} \, d\tau \qquad \int \Psi_{B1} \, Z \, \Psi_{A1} \, d\tau \quad (A7)$$

$$\int \Psi_{B2} \, X \, \Psi_{A1} \, d\tau \qquad \int \Psi_{B2} \, Y \, \Psi_{A1} \, d\tau \qquad \int \Psi_{B2} \, Z \, \Psi_{A1} \, d\tau \, , \quad (A8)$$

where A1, B1, A2 and B2 are the designations for all the possible states.

Because A1, B1, A2 and B2 are not only designations for the

possible states but are also the designation for their symmetry, we can forget the details of the (complicated) mathematical description of the wavefunctions, and can instead concentrate ONLY on their symmetry properties.

Thus we need to evaluate integrals of the type:

$$\int A2 \ Z \ A1 \ d\tau \ . \tag{A9}$$

From the character table, we learn that the molecular z axis in C_{2v} has symmetry A1. (This can be worked out by going through all the symmetry operations with the axis [+ and - lobes, don't forget] or can be read directly from the character table; the letter z in the A1 row indicates that anything which is mathematically described by a function first order in just z [such as the z axis, which is just +z] has A1 symmetry.)

Thus, we need to evaluate integral (A9) which has components with the symmetry property:

$$A2 \cdot A1 \cdot A1 \ . \tag{A10}$$

Is this integral non-zero? Here is how we proceed:

We need to work out a multiplication table for C_{2v} wherein we multiply all possible symmetries by each other. Take the case of expression (A10). We can do this with pairs of functions: $(A2 \cdot A1) \cdot A1$

First, to evaluate $(A2 \cdot A1)$.

from the character table, we write the appropriate rows. The character for $A1 \cdot A2$ is just the product of the two component rows.

	E	C2	$\sigma(xz)$	$\sigma(yz)$
A1	1	1	1	1
A2	1	1	-1	-1
A1·A2	1	1	-1	-1

This answer is equivalent to A2. In fact, it can be shown that A1 times any other row is the other row.

Our problem of evaluating $(A2 \cdot A1) \cdot A1$ has now been reduced to one of evaluating $(A2) \cdot A1$. We already know that the answer to this is A2.

In order for the integral to be non-zero, the integrand must be invariant under the operation of the symmetry elements of the group. This requires that the totally symmetric representation (A1) must be a component of the integrand. (This comes back to the even, odd function problem; we now are concerned with it in three dimensions).

Thus, we can now conclude that the transition from an A1 state to an A2 state will not occur if it is stimulated with linearly polarized light vibrating in the direction of the molecular z axis.

Once we work out the answer for the X and Y components of this transition, we will conclude that this transition is not allowed for any vibration direction of linearly polarized light. <u>It is forbidden</u>.

Next we need to work out the answer for all three components of a possible transition from:

a) an A1 state to an A2 state.
b) an A1 state to a B1 state.
c) an A1 state to a B2 state.

This work will show that only one polarization component is allowed for each of these possible transitions.

2) It is necessary to establish the multiplication table for all possible pairs of symmetry elements:

```
C_2v│ A1  A2  B1  B2
-------------------
A1 │ A1  A2  B1  B2
A2 │ A2  A1  B2  B1
B1 │ B1  B2  A1  A2
B2 │ B2  B1  A2  A1
-------------------
```

This table will allow us to fill in the remaining entries in the following table which states the polarization directions for all possible transitions in C_{2v}.

```
              excited
               state
        C_2v│ A1  A2  B1  B2
        -------------------
        A1 │ Z  na   X   Y
ground  A2 │ na  Z   Y   X
state   B1 │ X   Y   Z  na
        B2 │ Y   X  na   Z
        -------------------
        na = not allowed
```

Our next step will be interpret the experimental data on the basis of our theoretical modeling.

Part IX. Comparison of Models to Experimental Data

1) The highest energy absorption band is polarized along the X vibration direction (the α-spectrum). From the projections of the crystal structure in part I, we can establish that molecular z-axis of the C_{2v} M(2) site is oriented close to the crystallographic b axis which is the direction of the α-spectrum.

2) We can use the table of selection rules (polarizations) from part VIII to list all possible transitions which are polarized along the

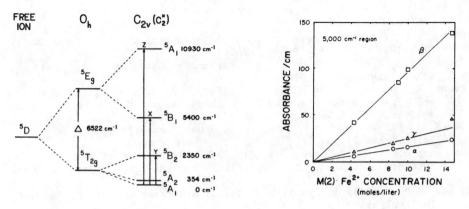

Figure A7 (left). Energy level diagram for Fe^{2+} at the M(2) site of orthopyroxene. From Goldman and Rossman (1977b).

Figure A8 (right). Beer's law plot relating absorbance in the 5000 cm^{-1} region to Fe^{2-} concentration at the M(2) site of orthopyroxene. From Goldman and Rossman (1979).

molecular axis (z) closest to the α indicatrix direction. These are the candidates for the transition: **ground state** $\Rightarrow$ **the highest excited state**. In the case of Fe^{2+} at the M(2) site, there is only one possible transition polarized along the molecular z-axis. It is the $^5A_1 \Rightarrow {}^5A_1$ transition.

3) Once the states involved in the highest energy transition are identified, we can establish that the ground state of Fe^{2+} at this site must be A1. From the polarization data, we are able to assign the remaining experimentally observed transitions to particular excited states. At that point, an energy level diagram for Fe^{2+} at the M(2) site can be constructed; the states are identified and the polarization rules are indicated for each state (Fig. A7).

4) We still have one state, $^5A_2$, left over and unaccounted for. This remaining state has been theoretically predicted to be at 354 cm^{-1} and is included on Figure A7. It has not yet been experimentally detected. The details of the calculation of the energy of this state can be found in Goldman and Rossman (1977b).

A FINAL COMMENT

Both Mössbauer and optical spectra can determine the proportion of Fe^{2+} at the M(2) site. The $^5A_1 \Rightarrow {}^5B_1$ transition near 5000 cm^{-1} follows Beer's law over a wide range of concentrations (Fig. A8). Because this spectroscopic region is free of interferences, it is possible to make sensitive quantitative determinations of the M(2) Fe^{2+} contents of pyroxene crystals. The major disadvantage of this method is the requirement that the crystal be oriented so that the interrogating light will vibrate along the crystallographic axes. The major advantages are that textural relationships in the original rock can be preserved during the measurement, and that low concentrations can be determined, limited only by the thickness of the sample available.

REFERENCES - APPENDIX

Burns, R.G. (1970) "Mineralogical Applications of Crystal Field Theory" Cambridge University Press, Cambridge.

Cotton, F.A. (1971) "Chemical Applications of Group Theory", 2nd ed., Wiley-Interscience, New York.

Evans, H., Ghose, S. and Hafner, S. (1967) Hyperfine splitting of ^{57}Fe and Mg-Fe order-disorder in orthopyroxene ($MgSiO_3$-$FeSiO_3$ solid solutions). J. Geology 75, 306-322.

Ghose, S. (1965) Mg^{2+}-Fe^{2+} order in an orthopyroxene, $Mg_{0.93}Fe_{1.07}Si_2O_6$ Zeit. Kristallogr. 122, 81-94.

Goldman, D.S. and Rossman, G.R (1977b) The spectra of iron in ortho-pyroxene revisted: the splitting of the ground state. Amer. Mineral. 62, 151-157.

Megaw, H.D. (1973), "Crystal Structures: A Working Approach", Saunders, Philadelphia.

Runciman, W.A., Sengupta, D., and Marshall, M. (1973) The polarized spectra of iron in silicates. I. Enstatite. Amer. Mineral. 58, 444-450.

Chapter 8   Frank C. Hawthorne
MÖSSBAUER SPECTROSCOPY

INTRODUCTION

The Mössbauer effect (Mössbauer, 1958) is the recoil-free emission and resonant absorption of γ-rays by specific atomic nuclei in solids. In Mössbauer spectroscopy, γ-rays are used as a probe of nuclear energy levels which, in turn, are sensitive to the details of both the local electron configuration and the electric and magnetic fields of the solid. Of mineralogical interest are the abilities to differentiate between various oxidation states (local electron configuration), spin states (local electron configuration, magnetic interactions) and structural environments (electric/crystal field effects). Hence oxidation ratios and site-occupancies of sensitive elements (isotopes) can often be derived.

Soon after the discovery of the Mössbauer effect, work on minerals began appearing, but it was the systematic work of Bancroft, Burns and co-workers (Bancroft, 1967; Bancroft and Burns, 1967; Bancroft et al., 1966,1967a,b,c,d) that showed the strength of Mössbauer spectroscopy in the determination of Fe valence states and site-occupancies in minerals. More recently, there has been an upsurge of interest in the magnetic properties of minerals, and as Mössbauer spectroscopy is also sensitive to the magnetic environment of the nucleus, it has played an important role in these studies. Although mineralogical applications have been dominated by ^{57}Fe spectroscopy, there has been some recent interest in the use of other nuclides, particularly ^{119}Sn, ^{121}Sb and ^{197}Au. There is no shortage of texts (Wertheim, 1964; Greenwood and Gibb, 1971; Bancroft, 1973; Gibb, 1977; Cranshaw et al., 1985; note that Bancroft (1973) contains many mineralogical applications), review volumes (Herber, 1984; Dickson and Berry, 1986) or review articles (Amthauer, 1980,1982; Van Rossum, 1982; Maddock, 1985). This chapter is intended to provide an introduction to the method, together with some mineralogical examples.

We begin with a review of some of the properties of the nucleus that are important in understanding the origin of the Mössbauer effect.

RADIOACTIVE DECAY

Mössbauer spectroscopy involves the emission and absorption of γ-radiation by specific atomic nuclei, and thus we need to start with a brief discussion of the nucleus, nuclear energy levels and decay (emission) processes.

In a nucleus, the number of protons is Z (the atomic number) and the number of neutrons is N: the mass number is thus Z+N. A nuclide (nuclear species) is conventionally written as $^{N+Z}X_Z$ where X is the chemical symbol; thus $^{57}Fe_{26}$ denotes iron with a mass number of 57, having 26 protons and 31 neutrons. Nuclides having the same atomic number but different mass numbers are called isotopes. Some isotopes are stable (in a ground state), some are radioactive (in an excited state) and spontaneously decay by emission of radiation to form other (stable or radioactive) isotopes. Most common decay processes are α, β and γ emission: (i) $\underline{\alpha\text{-emission}}$ (particles) are helium nuclei ($^4He_2$); (ii) $\underline{\beta\text{-emission}}$ is any process in which Z changes while N+Z is constant; of particular importance in Mössbauer spectroscopy is electron capture (denoted by $^0\beta_{-1}$), in which a neutron-deficient nuclide "captures" an inner valence electron, changing a proton to a neutron; (iii) $\underline{\gamma\text{-emission}}$ involves high-energy electromagnetic radiation, and is of central importance in Mössbauer spectroscopy. Transition from an excited state to the ground state is normally a multiple process, often involving α and/or β emission, together with several γ-ray emissions.

As an example, consider the decay scheme of ^{57}Co, a radioactive isotope that is used as a source of γ-radiation in ^{57}Fe Mössbauer spectroscopy. As shown in Figure 1, ^{57}Fe is formed by electron capture from ^{57}Co:

$$^{57}Co_{27} + {}^0\beta_{-1} \Rightarrow {}^{57}Fe_{26} \qquad (1)$$

The ^{57}Fe thus formed is in an excited state, and emits three γ-rays of energies 14.4, 123 and 137keV on decaying to the stable ^{57}Fe ground state. This decay scheme is the source of the 14.4 keV γ-radiation used in ^{57}Fe Mössbauer spectroscopy.

Nuclear spin

A nucleus contains an assemblage of nucleons (protons and neutrons), each of which has angular momentum. In a given nuclear state, the vector sum of these moments gives the net observable angular momentum hn/2, where n is an integer; the net angular momentum is thus quantized, and n/2 is the nuclear spin quantum number I. Thus if a nucleus has spin I, it can have discrete states $m_I = I$, $I-1...-I$ (e.g., I=1/2, $m_I=1/2,-1/2$; I=3/2, $m_I=3/2,1/2,-1/2,-3/2$). In some cases, the states characterized by different m_I are degenerate (they have the same energy), but this degeneracy can be lifted by the imposition of extranuclear fields.

The net angular momentum of the nucleus generates a nuclear magnetic moment μ_N; if I>1/2, the nucleus may also have a colinear quadrupole moment. In a magnetic field H, the interaction between the field and the nuclear magnetic moment causes the energy of the nucleus to be dependent on its orientation w.r.t. (with respect to) the magnetic field. The magnetic field defines a symmetry axis about which the angular momentum must be quantized; the magnetic moment may only assume a certain set of orientations w.r.t. the symmetry axis, those for which the components of the angular momentum along the field direction are Ih,(I-1)h...-Ih.

EMISSION, ABSORPTION AND MOMENTUM TRANSFER

When a nucleus emits a quantum of radiation of energy E_g, that quantum of energy has momentum E_g/c; because of the conservation of momentum, the nucleus must recoil with equal and opposite momentum $-E_g/c$. The energy E_g of a quantum of γ-radiation is so large that the recoil velocity of the nucleus is significant. The energy of this total process of emission and recoil comes from the decay of the nucleus from its excited state to its ground state; hence

$$E_g \equiv E_\gamma = E_t - E_R \qquad \text{(for emission)} \quad , \qquad (2)$$

where E_t is the nuclear transition energy,
     E_R is the recoil energy,
     E_g is the emitted γ-radiation energy.

When the nucleus absorbs a quantum of γ-radiation, the γ-radiation has to supply the energy (E_t) to excite the nucleus and the recoil energy (E_R) of the nucleus; hence

$$E_g = E_t + E_R \qquad \text{(for absorption)} \quad . \qquad (3)$$

Figure 2 is a cartoon of these relationships, indicating that nuclei cannot in general re-absorb γ-radiation that they emit because the required γ-ray energies differ by the sum of the recoil energies of the two events.

This argument considers just a free nucleus, whereas we are concerned with solids, in which the nucleus is part of an atom held in place by a network of chemical bonds. For the γ-ray transitions that we are considering, the recoil energy is far too small to break the bonds holding the atom in the structure. Consequently, the recoil energy becomes the property of the whole structure, and can only be "radiated" by transferral to the phonon spectrum of the solid. As phonons are quantized, the momentum is transferred in integral amounts with a finite probabilty that in some cases there will be no momentum transfer. For this 'zero phonon' event, the recoil energy E_R of Figure 2 is zero and $E_t = E_g$ for both emission and absorption.

Although there is no momentum transfer in the zero phonon event, the conservation of momentum is still macroscopically satisfied, as when the average is taken over many emission processes, the energy transferred per event is equal to the free-atom recoil energy E_R. Now let f be the probability of a zero-phonon event, the recoil-free fraction. Multi-phonon processes are rare, and neglecting these, we can write the mean recoil energy E_R of single-phonon processes as

$$E_R = (1 - f)\hbar\omega \quad , \qquad (4)$$

where ω is the phonon frequency.

Going back to our free atom (nucleus) model, the emitter must recoil with momentum p_e given by

$$p_e = Mv_R = \frac{-E_\gamma}{c} = \hbar k \quad , \qquad (5)$$

where M is the mass of the nucleus,
     v_R is the recoil velocity,
     k is the wave-vector of the γ-ray.

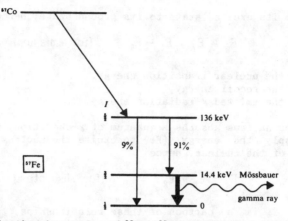

Figure 1. Radioactive decay scheme of $^{57}Co \rightarrow \,^{57}Fe$ with the emission of various γ-rays; the $3/2 \rightarrow 1/2$ transition emits the 14.4 keV γ-ray that is used in ^{57}Fe Mössbauer spectroscopy.

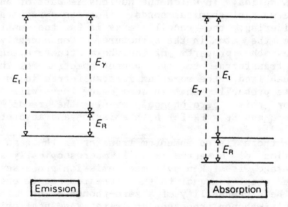

Figure 2. Energy level diagrams for emission and absorption of γ-rays: E_t is the transition energy and E_R is the recoil energy, and both of these are the same in each process; E_{gamma} is the energy of the emitted or absorbed γ-ray. As is apparent, the energy of the γ-ray associated with each process (in the same nuclide) is different.

The recoil energy of the emitter, E_R, is given by

$$E_R = \frac{Mv_R^2}{2} = \frac{p_e^2}{2M} = \frac{E_\gamma^2}{2Mc^2} .$$ (6)

Substituting into equation (4) and rearranging gives

$$f = 1 - \frac{E_n}{\hbar\omega} = 1 - \frac{\hbar^2 k^2}{\hbar\omega 2M} .$$ (7)

In the simple Einstein model, the vibrational characteristics of the structure are described by the vibrational oscillations of the atoms with angular frequency ω. For harmonic forces, the total energy

E$_{TOT}$ can be written as

$$E_{TOT} = M\omega^2\langle ax^2\rangle \quad , \tag{8}$$

where $\langle x^2\rangle$ is the mean square vibrational amplitude in the x direction. As the energy of the oscillators is quantized,

$$E_{TOT} = h\omega\left(n + \frac{1}{2}\right) \quad , \tag{9}$$

where n is the quantum number.

For the lowest Einstein oscillator level, n=0 and hence

$$\langle x^2\rangle = \frac{h}{2M\omega} \quad . \tag{10}$$

Returning now to equation (7) with x as the propagation direction of the γ-ray,

$$f = 1 - k^2\langle x^2\rangle \quad . \tag{11}$$

A more rigorous treatment gives

$$f = \exp\left[-k^2\langle x^2\rangle\right] \quad , \tag{12}$$

such that equation (11) $\approx$ equation (12) with $k^2\langle x^2\rangle \ll 1$.

As the temperature of a solid increases, $\langle x^2\rangle$, the mean square vibrational amplitude of the atom (nucleus) increases, leading to a decrease in the recoil-free fraction. This is also to be expected as at higher temperatures, increasing numbers of higher quantum states are excited, leading to a decrease in the proportion of zero-phonon events. Thus the recoil-free fraction is some inverse function of temperature.

Second-order Döppler shift

In the harmonic approximation, we can write an equation of motion for the Mössbauer nucleus:

$$m \frac{d^2x}{dt^2} = -Ax \quad , \tag{13}$$

where A is the force constant.

According to general relativity theory, this acceleration can be considered through the equivalence principle as arising from a mean (gravitional) potential $\langle V\rangle$; there will thus be a relativistic shift in the γ-ray energy given by

$$\Delta E = 3m_\gamma\langle v\rangle = 3m_\gamma\langle \int_o^x - \frac{A}{M} xdx\rangle \quad , \tag{14}$$

where $m_\gamma = \hbar\omega/c^2$ and ΔE is the energy shift.

The fractional energy change is thus

$$\frac{\Delta E}{E} = \frac{3m_\gamma <v>}{m_\gamma c^2} = -\frac{3A}{2mc^2}\langle x^2\rangle \quad . \tag{15}$$

This is called the <u>second order Döppler shift</u> (<u>SOD</u>); SOD is related to the recoil-free fraction by the equation

$$SOD = \frac{3\hbar}{2mE_\gamma^2} A \ln f \quad . \tag{16}$$

SOD thus has the same temperature dependence as $\ln f$, and slightly shifts the resonance line, adding to the normal isomer shift to produce the observed shift. Some authors (e.g., Bancroft, 1973) distinguish between the isomer shift (IS) and the chemical shift (CS, or centre shift), the latter being the sum of the isomer shift and the second order Döppler shift (CS = IS + SOD). However, this distinction is not commonly made in the literature; the SOD is similar in most solids, and as we normally measure the chemical shift relative to some standard materials, the SOD quantities (nearly) cancel, leaving us with a relative isomer shift. Consequently, we will use the term isomer shift here.

Resonant Mössbauer absorption

If we take ^{57}Co diffused into stainless steel as a source of γ-radiation (Fig. 1) and irradiate a sample of stainless steel, then the γ-rays will be absorbed by the ^{57}Fe in the stainless steel, exciting the ^{57}Fe nuclei into an excited state (cf., Fig. 2); this process is known as resonant absorption. If we take the same source of γ-radiation and irradiate a sample of hercynite ($Fe^{2+}Al_2O_4$), then the γ-rays will not be absorbed by the ^{57}Fe in the sample; the change in environment of the nucleus is sufficient to alter the nuclear energy levels of ^{57}Fe in the spinel and bring the system out of resonance. In this case, the transition energy for absorption by ^{57}Fe in the spinel is changed by the different environment, and thus the γ-rays no longer have the correct energy to excite the transition.

However, we can modulate the energy of the γ-rays by vibrating the source, which applies a continuously varying Döppler shift to the emitted radiation. If we use this vibrating source and irradiate the stainless steel sample, then we get the absorption spectrum shown in Figure 3a, a single absorption peak at a velocity (relative energy scale) of zero, corresponding to the identical transition energies in the source and in the absorber. If we used the vibrating source to irradiate the hercynite sample, then we get the absorption spectrum shown in Figure 3b, a single absorption peak at a non-zero velocity corresponding to the energy difference between the transition energies in the source and the absorber.

Thus one can use a constant (modulated) source and monitor the relative changes in the nuclear levels of the same isotope in a variety

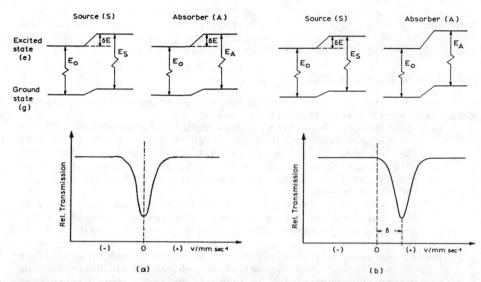

Figure 3. Resonant Mössbauer absorption. In (a), the nuclide has the same ground state and excited state energies in the source and in the absorber; hence resonant absorption can occur without the need to modulate the energy of the γ-rays emitted by the source; the resulting spectrum (taken with modulation of the source γ-ray energy) shows an absorption at 0.0 mm/s. In (b), the nuclide has different ground and excited state energies in the source and in the absorber; hence the transition energy in the absorber, E_A, is different from that in the source E_S; thus resonant absorption cannot occur without modulation of the energy of the γ-rays emitted by the source; the resulting spectrum shows an absorption shifted from zero by the difference in transition energies, $E_A - E_S$ (= δ), of the source and absorber.

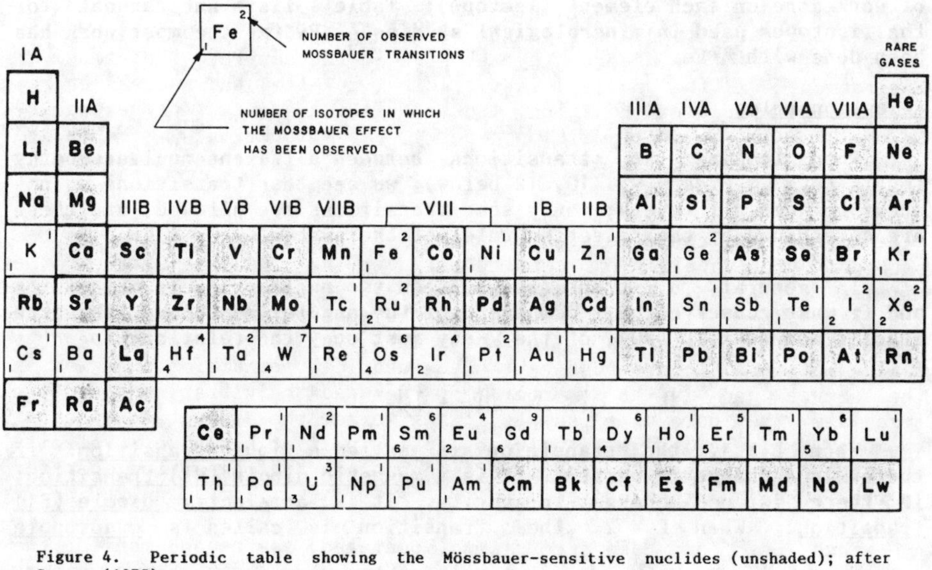

Figure 4. Periodic table showing the Mössbauer-sensitive nuclides (unshaded); after Gonser (1975).

of solid state environments. As the nuclear energy levels are a function of the extranuclear environment, one can use the relative changes in these absorption spectra as a probe of the factors that cause these changes in nuclear energy levels.

Which isotopes are useful?

Not all nuclei are suitable for Mössbauer spectroscopy; certain conditions must be met before it is feasible:

(a) the nuclear transition energy must be large enough to give useful γ-radiation, but not large enough to produce a recoil the energy of which is significantly greater than the vibrational quanta $(10 < E^- < 150eV)$; an excessively large energy will tear the atom loose from the structure and cause radiation damage.

(b) a significant fraction of the excited nuclei must decay by the emission of γ-radiation; other decay processes (e.g., emission of conversion electrons) must obviously not be too dominant.

(c) the excited state half-life must be short enough to give lines of sufficient width, but long enough to avoid introducing excessive uncertainty in the transition energy E_t through the uncertainty principle; $1 < \tau < 100$ ns are suitable.

(d) the excited state of the emitter must have a precursor that has a fairly long half-life and is reasonably easy to handle (i.e., is not excessively radioactive).

(e) the ground state isotope should be stable and have a fairly high natural abundance; the latter restriction can be relaxed for synthetic minerals and glasses if the enriched isotope is reasonably easy to obtain.

(f) the cross-section for absorption should be high.

Figure 4 shows those elements for which suitable isotopes exist for Mössbauer spectroscopy, together with some indication of the amount of work done on each element (isotope); Table 1 lists nuclear data for the isotopes used in mineralogical studies. By far the most work has been done with ^{57}Fe.

Selection rules

When looking at transitions between different nuclear energy levels (e.g., Figs. 9, 10, 12 below), we see that transitions do not occur between all levels; only some transitions are allowed, and these are determined by the selection rules of the system.

In general, a γ-transition between two nuclear levels of spin I_1 and I_2 must conserve the z-component of the angular momentum; thus the angular momentum, L $(\neq 0)$, of the γ-ray must obey the relationship

$$|I_1 - I_2| \leq L \leq |I_1 + I_2| \quad . \tag{17}$$

When $L = 1$, the transition is called a dipole transition: if there is a change in parity, it is a magnetic dipole (M1) transition; if there is no change in parity, it is an electric dipole (E1) transition. When $L = 2$, the transition is called a quadrupole

Table 1. Nuclear data for Mössbauer isotopes of mineralogical interest.

Isotope	Gamma energy(keV)	#Half-life of source	Half width(mms^{-1})	Spin	σ_0 (x10^{-19} cm^2)	Natural abundance(%)	δR R (x10^4)
^{57}Fe	14.4	270d	0.19	$^1/_2$-->$^3/_2$	25.7	2.17	-18 ±4
^{119}Sn	23.9	245d	0.62	$^1/_2$-->$^3/_2$	13.8	8.58	+3.3±1
^{121}Sb	37.2	76y	2.10	$^5/_2$-->$^7/_2$	2.04	57.25	-8.5±3
^{153}Eu	83.4			$^5/_2$-->$^7/_2$	0.72	52.2	5.7
^{197}Au	77.3	65h	1.85	$^3/_2$-->$^1/_2$	0.44	100	+

#d = days; y = years; h = hours.

Table 2. Selection rules for nuclear transitions.

L	Parity change	Transition
1	No	Electric dipole (E1)
1	Yes	Magnetic dipole (M1)
2	No	Electric quadrupole (E2)
2	Yes	Magnetic dipole (M2)

transition, with both **magnetic (M2)** and **electric (E2) quadrupole transitions**. These relationships are summarized in Table 2. For a given L, transitions are restricted by the relationship

$$|\Delta m| \leq L \quad . \tag{18}$$

Line shape

The γ-radiation is the product of nuclear decay and obeys the usual exponential decay law:

$$N_t = N_o \, e^{-\lambda t} = N_o \, e^{\frac{-t}{\tau}} \quad , \tag{19}$$

where N_t is the number of quanta emitted at time t,
  λ is the decay constant (=1/τ),
  τ is the mean lifetime of the excited state.

The energy distribution of the emitted γ-radiation is given by its Fourier transform

$$N(E)de \; \alpha \; \frac{\hbar}{2\pi\tau} \; \frac{1}{(E-E_\gamma)^2 + (\frac{\hbar}{2\tau})^2} \; dE \quad , \tag{20}$$

where E_g is the mean γ-radiation energy.

The cross-section for recoilless decay is given by

$$\sigma_e(E) = f_e \, N(E)_{NAT} \quad , \tag{21}$$

where f is the recoil-free fraction of the source nuclei. When γ-radiation is absorbed, the cross-section for resonant nuclear absorption is given by the <u>Breit-Wigner relation</u>

$$\sigma_a(E) = \sigma_o \frac{\left(\frac{\hbar}{2\tau}\right)^2}{(E-E_\gamma)^2 + \left(\frac{\hbar}{2\tau}\right)^2} \cdot f_a \quad , \tag{22}$$

where

$$\sigma_o = \frac{1}{2\pi} \left(\frac{hc}{E_\gamma}\right)^2 \frac{2I_e + 1}{2I_g + 1} \frac{1}{1+\alpha} \quad ,$$

in which I_g and I_e are the ground and excited state nuclear spins, and α is the internal conversion coefficient of the transition (a measure of how many γ-rays are emitted as compared with how many conversion electron processes occur, see section on CEMS).

The experimental shape of the resonance line results from the convolution of $\sigma_e(E)$ with $\sigma_a(E)$, which we can write in terms of a normalized resonance intensity as

$$I(E) = \frac{h}{\pi\tau} \frac{1}{(E-E_\gamma)^2 + \left(\frac{\hbar}{\tau}\right)^2} \quad . \tag{23}$$

Writing the excited state mean lifetime in terms of the line width Γ_E (see next section), $\Gamma_o = 2\Gamma_E = 2h/\tau$, our resonance line shape becomes

$$I(E) = \frac{\hbar_o}{2\pi} \frac{1}{(E-E_\gamma)^2 + \left(\frac{\hbar_o}{2}\right)^2} \quad , \tag{24}$$

a <u>Lorentzian</u> curve with Γ_o as the resonant line width. Although this is only ideally for zero source and absorber thickness, it seems to be a good approximation for reasonably thin materials.

Line width

The intrinsic line-width of the zero-phonon event is limited by Heisenberg's uncertainty principle as applied to the γ-ray transition, as no energy is lost to the crystal for this particular event. Thus

$$(\Delta E) \cdot (\Delta t) \geq \hbar \quad , \tag{25}$$

where ΔE, Δt are the uncertainties in the energy and time respectively. The excited state has a mean lifetime τ (a half-life $t_{\frac{1}{2}} = \tau \, \ell n \, 2$), and the ground state is stable (that is, it has a long lifetime with a well-defined energy level). For this case, the uncertainty relation can be written as

$$\Delta E = \Gamma_E = \frac{\hbar}{\tau} = \frac{\hbar}{t_{\frac{1}{2}} \ln 2} \quad . \tag{26}$$

The uncertainty in the energy is actually the (natural) line width of the emission or absorption line, Γ_E, defined as the <u>full width at half maximum intensity</u>.

For ^{57}Fe, the first excited state (14·4 keV) has $t_{\frac{1}{2}} \approx 10^{-7}$ sec, and thus the natural line width $\Gamma \approx 5 \times 10^{-9}$ eV; in terms of the usual Mössbauer energy scale, this corresponds to (a half width of) 0.097 mm/s. Table 1 contains similar information for isotopes of geological interest.

MÖSSBAUER PARAMETERS

Mössbauer spectroscopy is concerned with interactions between the nucleus and extra-nuclear electric and magnetic fields - these are called <u>hyperfine interactions</u>. In principle, we should consider the multipolar interaction of the nuclear charge with the enveloping electric and magnetic fields (Maddock, 1985): monopole, dipole, quadrupole, octupole, etc. interactions. Schematic illustrations of the pertinent multipoles are shown in Figure 5. No nucleus has a static dipole moment (note however that it can have a magnetic dipole moment): the charge density at any point is given by $\phi^*\phi$, where $\phi(\phi^*)$ is the (complex conjugate of the) total normalized wave-function for charge-bearing nuclear particles; the charge density is thus an even function of the co-ordinates, and must be invariant w.r.t. inversion; consequently there can be no static dipole moment. Similarly, a nucleus cannot have an octupole moment. Interactions only occur between moments of equal rank; these decrease strongly with increasing rank, and hence high moment interactions are negligible. Consequently, we are only concerned with <u>electric monopole</u> and <u>quadrupole interactions</u> and <u>magnetic dipole interactions</u>.

Chemical (isomer) shift

This is the <u>electric monopole interaction</u>, arising from the interaction between the positive nuclear charge (which occupies a small but finite volume of the nucleus) and the electric field of the surrounding electrons. This interaction leads to a shift of the nuclear energy levels compared to those for a bare nucleus; the magnitude of this shift is a function of the electric field at the nucleus, specifically the charge density distribution that is sensitive to the details of the local electronic structure. Thus similar nuclei in different environments will have energy levels that are shifted by different amounts according to the specific details of the nuclear environments. A nucleus is generally of different sizes in its ground and excited states; consequently the monopole interaction causes both a change in the absolute energies of the ground and excited states <u>plus</u> a relative change in the energy levels, a change that stems from this size difference (Fig. 6). This relative change in the energy levels may be expressed as:

$$v = \frac{4\pi c}{5E_\gamma} Ze^2R^2 \left(\frac{\delta R}{R}\right) |\psi(0)|^2 \quad , \tag{27}$$

where E_g is the energy difference between ground and excited states in the bare nucleus,
 Z is the nuclear charge,

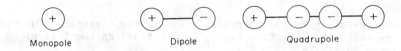

Monopole   Dipole   Quadrupole

Figure 5. Diagrammatic representation of a monopole, a dipole and a quadrupole.

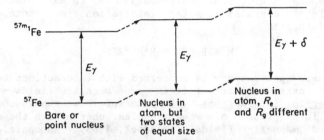

(Level displacements greatly exaggerated)

Figure 6. Energy levels of a nuclide, showing diagrammatically the effects of different nuclide radii in the ground and excited states. The monopole interaction causes a change in the absolute energies of the ground and excited states, plus a relative change, δ, in the energy levels that stems from this size difference; after Maddock (1985).

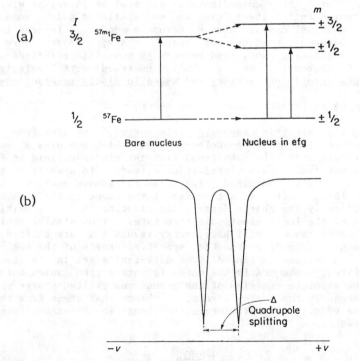

Figure 7. (a) The electric quadrupole interaction caused by an EFG acting on the nucleus; this splits the energy levels of a nucleus with a nuclear spin quantum number I > 1, and hence the excited state (I=3/2) of the ^{57}Fe nucleus is split; thus there can be two nuclear transitions (indicated by arrows) rather than just one (as occurs when there is no non-cubic component of the electric field gradient); (b) the Mössbauer spectrum resulting from the transitions indicated in Figure 7(a); this is a quadrupole-split doublet; modified after Maddock (1985).

δR is the difference between the radii of the ground and excited states,

R is the mean radius of the ground and excited states,

$|\psi(0)|^2$ is the electron density at the nucleus, often called the contact density,

v is the velocity for the equivalent Döppler change of E_g.

In Mössbauer resonance, one is comparing the relative energy levels between nuclei in a source (emitter) and a sample (absorber), that is the difference in relative changes of the ground and excited state levels in the source and sample. This energy difference is called the chemical or isomer shift, δ, and can thus be written as

$$\delta = \frac{4\pi}{5} Ze^2R^2 \left(\frac{\delta R}{R}\right) \left[|\psi(0)|^2_{ABS} - |\psi(0)|^2_{SOURCE}\right] \quad . \tag{28}$$

The prefactors are constant for a given nucleus, and the contact density of the source is constant for a given source. Consequently when a standard source is used, δ is actually a linear measure of the contact density at the absorber nucleus. Ideally, absorber contact density consists of the s-electron density at the nucleus, as all the other orbitals have zero probability at the nucleus. However, these other orbitals have a significant perturbing effect on the outer occupied s orbitals and consequently affect the isomer shift, a circumstance that is extremely useful in the determination of valence states.

In the Mössbauer resonance experiment, one is comparing the relative energy levels between a γ-ray source (the emitter) and a sample (the absorber). Thus in terms of the recorded spectrum (Figs. 3 and 12), the isomer shift is the velocity at which maximum absorption occurs, the energy shift caused by the Döppler effect of this particular velocity being a measure of the difference in relative energy levels between source and sample. In fact, it is customary to refer all isomer shifts to some standard material such as iron foil or sodium nitroprusside (isomer shift of sodium nitroprusside w.r.t. iron foil = 0.257 mm/sec).

Quadrupole splitting

This is the electric quadrupole interaction, arising from the interaction between the nuclear quadrupole moment and the electric field in which it occurs. When the nuclear spin quantum number I>1, the nucleus has a quadrupole moment which may interact with any noncubic component of the electric field; this leads to a splitting of the nuclear energy levels, called the quadrupole splitting (Fig. 7).

The magnitude of the quadrupole splitting (QS) is a function of the electric field gradient (EFG), that property of the external field that is capable of interaction with the nuclear quadrupole moment. The electric field (E) at the nucleus is the negative gradient of the potential V : E = - ∇V where ∇ is the grad operator. The EFG is the gradient of E. Writing

$$V_{ij} = \frac{\delta^2 V}{\delta_i \delta_j} \quad ,$$

we may represent the electric field gradient as

$$EFG = - \begin{bmatrix} V_{xx} & V_{xy} & V_{xz} \\ V_{yx} & V_{yy} & V_{yz} \\ V_{zx} & V_{zy} & V_{zz} \end{bmatrix} \quad . \tag{29}$$

The simplest model considers a point charge $-e$ with polar coordinates (r, θ, ϕ) and the nucleus as origin. We may write the components of the EFG in terms of spherical harmonics (Table 3).

The EFG tensor is symmetrical, and can always be diagonalized by an appropriate choice of (electric field) axes, whereupon all off-diagonal elements are zero:

$$EFG = - \begin{bmatrix} V_{xx} & 0 & 0 \\ 0 & V_{yy} & 0 \\ 0 & 0 & V_{zz} \end{bmatrix} \quad . \tag{30}$$

Such a diagonalization is equivalent to orienting the axes such that the charge lies on the z axis; this can easily be shown using the spherical harmonic representation of Table 3, and is left as an exercise for the reader.

Furthermore, for an electric field, the trace of the matrix is zero, and hence

$$V_{xx} + V_{yy} + V_{zz} = 0 \quad . \tag{31}$$

Conventionally, the axes of the electric field are labelled such that $|V_{zz}| > |V_{yy}| > |V_{xx}|$; note that these axes do not bear any obvious relationship to the crystallographic axes. From equation (31), it can be seen that there are only two independent parameters of the EFG. Conventionally these are chosen as V_{zz} and $(V_{xx} - V_{yy})/V_{zz} = \eta$; because $|V_{zz}| > |V_{yy}| > |V_{xx}|$, $0 < \eta < 1$.

The magnitude of the quadrupole splitting is proportional to V_{zz}; for $I = 3/2$ (^{57}Fe),

$$Q.S. = \frac{-e}{2} V_{zz} Q \sqrt{\left(1 + \frac{\eta^2}{3}\right)} = \frac{e^2}{2} qQ \sqrt{\left(1 + \frac{\eta^2}{3}\right)} \quad , \tag{32}$$

where Q is the quadrupole moment of the nucleus, $eq = V_{zz}$ = the $-z$ component of the EFG, and e is the proton charge. The selection rule for transitions between such states is $\Delta m = 0, \pm 1$, and the possible transitions are shown in Figure 7, together with the resulting Mössbauer spectrum. In the spectrum, there are two peaks, corresponding to the two transitions, and the energy separation between these two peaks, the quadrupole splitting, is a measure of the energy separation between the two excited states of the nucleus.

The transition probabilities are dependent on the angle between the direction of the EFG at the nucleus and the direction of propagation of the γ-radiation. The angular dependence of the allowed

Table 3. The electric field gradient tensor for a charge of Z electronic units.

$V_{xx} = Zer^{-3} (3 \sin^2\theta \cos^2\phi - 1)$	$V_{xy} = V_{yx} = Zer^{-3} (3 \sin^2\theta \sin\phi \cos\phi)$
$V_{yy} = Zer^{-3} (3 \sin^2\theta \sin^2\phi - 1)$	$V_{xz} = V_{zx} = Zer^{-3} (3 \sin\theta \cos\theta \cos\phi)$
$V_{zz} = Zer^{-3} (3 \cos^2\theta - 1)$	$V_{yz} = V_{zy} = Zer^{-3} (3 \sin\theta \cos\theta \sin\phi)$

ϕ = latitudinal angle; θ = longitudinal angle

Table 4. Angular dependence of nuclear transitions.

Electric Quadrupole (E2) Transitions					
Transitions	*Δm	Angular dependence	**$\theta=0°$	$\theta=90°$	$\langle\theta\rangle$
$\pm^3/_2 \longrightarrow \pm^1/_2$	-	$^3/_2 (1 + \cos^2\theta)$	3	$^3/_2$	1
$\pm^1/_2 \longrightarrow \pm^1/_2$	-	$1 + ^3/_2 \sin^2\theta$	1	$^5/_2$	1

Magnetic Dipole (M1) Transitions					
Transitions	Δm	Angular dependence	$\theta=0°$	$\theta=90°$	$\langle\theta\rangle$
$^3/_2 \longrightarrow ^1/_2$ $-^3/_2 \longrightarrow -^1/_2$	+1 -1	$^9/_4 (1 + \cos^2\theta)$	$^9/_2$	$^9/_4$	3
$^1/_2 \longrightarrow ^1/_2$ $-^1/_2 \longrightarrow -^1/_2$	0 0	$3 \sin^2\theta$	0	3	2
$-^1/_2 \longrightarrow ^1/_2$ $^1/_2 \longrightarrow -^1/_2$	+1 -1	$^3/_4 (1 + \cos^2\theta)$	$^3/_2$	$^3/_4$	1

* see equation (18); **θ = angle between EFG direction at nucleus and γ-ray propagation direction; $\theta = 0$, 90° denote relative line intensities at these values of θ; $\langle\theta\rangle$ denotes relative line intensities averaged over all orientations.

transitions is shown in Table 4. Assuming isotropic atomic vibrations, one can calculate the relative line intensities for a quadrupole split doublet:

$$\frac{I\left[\left|\frac{3}{2}\right| \Rightarrow \left|\frac{1}{2}\right| \right]}{I\left[\left|\frac{1}{2}\right| \Rightarrow \left|\frac{1}{2}\right| \right]} = \frac{\frac{3}{2}(1 + \cos^2\theta)}{(1 + \frac{3}{2}\sin^2\theta)} = R; \quad R(\theta=0) = \frac{3}{1}; \quad R(\theta=90) = \frac{3}{5} \quad . \quad (33)$$

Averaged over all orientations (i.e., for a randomly oriented powder),

$$\langle R \rangle = \frac{\int_0^\pi \frac{3}{2}(1 + \cos^2\theta) \sin\theta \, d\theta}{\int_0^{2\pi} (1 + \frac{3}{2}\sin^2\theta) \sin\theta \, d\theta} = 1 \quad , \quad (34)$$

and thus quadropole split doublets of randomly oriented absorbers have (ideally) equal intensities.

Inspection of equation (32) shows that the quadrupole splitting is controlled by the magnitude and the sign of V_{zz} and the magnitude of η (when the $\pm(3/2)$ state is the excited state, then the sign of the quadrupole splitting is positive). It is also obvious from equation

(32) that we cannot determine that magnitude and sign of V_{zz} and the magnitude of η just from a single powder spectrum, and it is perhaps fortunate that for many mineralogical uses we do not need to do so.

To measure the sign of the quadrupole splitting (and hence of V_{zz}), it is necessary to exploit the angular dependence of the transition probabilities (Table 4). When the angle between the z-axis of the EFG and the direction of propagation of the γ-radiation is zero ($\theta = 0$), the $I_{3/2}/I_{1/2}$ intensity ratio approaches 3:1 (it is 3:5 for $\theta=90°$) as shown in equation (33). If the most intense line, which corresponds to the 3/2 line, is at positive velocities, then the sign of the QS is positive. For ^{57}Fe, Q, the nuclear quadrupole moment, is positive and hence a positive QS denotes a positive value of V_{zz}. For ^{119}Sn, Q is negative and hence a positive QS denotes a negative value of V_{zz}.

Alternately the sign of the QS can (often) be determined from a powder sample in an intense magnetic field. Figure 8a,b show the room-temperature powder spectra for almandine and andradite respectively, taken in an external magnetic field of 35kG. The spectra may be calculated by diagonalization of the Hamiltonian (see Chapter 2) for a mixed electrostatic and magnetic interaction (Amthauer et al., 1976), assuming positive and negative signs for the quadrupole interaction; details of the calculation may be found in Gabriel (1965). The calculated spectra are shown below the experimental spectra in Figure 8. For almandine, visual comparison of the shape of the three spectra shows Fe^{2+} at the {X} site in almandine to have a negative quadrupole interaction. A similar comparison for andradite shows Fe^{3+} at the [Y] site in andradite to have a positive quadrupole interaction.

For nuclei with spin states I >(3/2), the splitting is more complex as there are more non-degenerate levels. Figure 9 shows the case for ^{121}Sb with I = (5/2) and (7/2) in the ground and excited states respectively. There are three non-degenerate levels in the ground state and four non-degenerate levels in the first excited state, with the possibility of eight separate transitions. Also shown in Figure 9 is the ideal "bar spectrum" for this pure quadrupole interaction, together with an experimental spectrum; the quadrupole split lines are only partly resolved, as is often the case in ^{121}Sb spectra.

Magnetic splitting

Also called nuclear Zeeman splitting, this arises from the interaction between the magnetic dipole moment, μ_N, of the nucleus and a magnetic field H at the nucleus. The magnetic field removes the spin degeneracy to form 2I+1 energy levels given by

$$E_m = - g_n \mu_N m_I H = - \frac{\mu_N m_I H}{I} \quad , \tag{35}$$

in which the energy separation of each level is $g_n \mu H$, where g_N is the nuclear Landé g factor, β_N is the Bohr magneton (= eh/2Mc); the nuclear magnetic moment is related to the nuclear Bohr magneton by the nuclear Landé splitting factor g_N; $\mu_N = g_N \beta_N I$. For ^{57}Fe, the ^{57mI}Fe

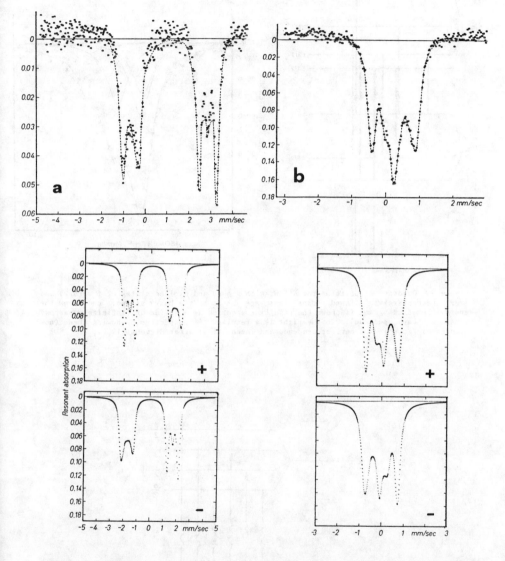

Figure 8. Mössbauer spectra of ^{57}Fe in garnets, taken at room temperature with the powder samples in an intense (35 kG) magnetic field: (a) [Fe^{2+} in] almandine; (b) [Fe^{3+} in] andradite. Shown below are the calculated spectra for different signs of the quadrupole-splitting; simple comparison of the spectrum shapes allows determination of the sign; after Amthauer et al. (1976).

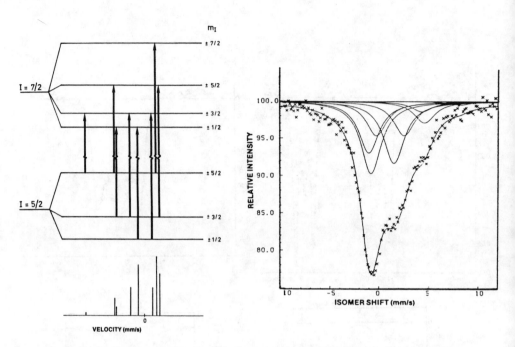

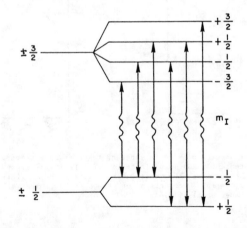

Figure 9. Quadrupole splitting for ^{121}Sb, with ground and excited states of $I = (5/2)$ and (7/2) respectively; allowed transitions are shown by the heavy arrows, and below the energy level diagram is shown the resulting spectrum as a series of (infinitely narrow) lines (a "bar" spectrum); to the right is a resolved ^{121}Sb spectrum, showing these (now broadened) lines in the correct sequence and intensities; after Stevens (1984).

Figure 10. Nuclear energy level diagram showing magnetic (nuclear Zeeman) splitting for ^{57}Fe; the allowed transitions are shown by heavy arrows.

excited state gives two levels (= $\frac{1}{2}$, $-\frac{1}{2}$); this is shown in Figure 10. Because μ_N is different for the ground state and the excited state, the energy separations are different in the ground and excited states, and because μ_N has a different sign in the ground and excited states (being positive in the ground state and negative in the excited state), the sequence of magnetic quantum numbers (m_I) is different. The selection rule governing these transitions is $\Delta m_I = 0, \pm 1$, and the allowed transitions are shown in Figure 10; the corresponding Mössbauer spectrum is shown in Figure 11a. Note that the lines in the Mössbauer spectrum are not of equal intensity. This occurs because the transition probabilities are dependent on the angle between the direction of the magnetic field at the nucleus and the propagation direction of the γ-radiation. The angular dependence of the allowed transitions is given in Table 4. Assuming isotropic atomic vibrations, one can calculate the relative line intensities of a pure Zeeman hyperfine pattern: for $\theta=0°$, they are 3:0:1:1:0:3; for $\theta=90°$ they are 3:4:1:1:4:3. Integrating over all directions, one gets the intensities for a randomly oriented magnetic material.

$$\left(\frac{3}{2} \Rightarrow \frac{1}{2}\right), \left(-\frac{3}{2} \Rightarrow \frac{1}{2}\right): \frac{1}{2} \int_0^\pi \frac{9}{4}(1 + \cos^2\theta) \sin\theta \ d\theta = 3 \quad , \tag{36}$$

$$\left(\frac{1}{2} \Rightarrow \frac{1}{2}\right), \left(-\frac{1}{2} \Rightarrow -\frac{1}{2}\right): \frac{1}{2} \int_0^\pi 3\sin^2\theta.\sin\theta \ d\theta = 2 \quad , \tag{37}$$

$$\left(-\frac{1}{2} \Rightarrow \frac{1}{2}\right), \left(\frac{1}{2} \Rightarrow -\frac{1}{2}\right): \frac{1}{2} \int_0^\pi \frac{3}{4}(1 + \cos^2\theta\sin\theta \ d\theta = 1 \quad . \tag{38}$$

Thus the intensity ratios are 3:2:1:1:2:3. The corresponding Mössbauer spectra for α-Fe are shown in Figure 11.

The magnetic field at the nucleus can be imposed through an externally applied magnetic field, or can be intrinsic and due to unpaired orbital electrons. The magnetic field necessary to produce such splittings is extremely large. The field produced by unpaired electrons of the same atom are sufficient to cause such splittings, but it is necessary that the field exist in a specific direction longer than the life-time of the excited state (^{57mI}Fe in the case of ^{57}Fe).

For ferromagnetic and antiferromagnetic minerals (see Figure 39), co-operative interactions keep the fields aligned, and below the Curie and Néel temperatures respectively, magnetically split spectra occur. For paramagnetic minerals, the magnetic field is normally changing due to spin-spin and spin-lattice relaxation effects; as these changes are fast compared to the lifetime of the ^{57mI}Fe excited state ($t_\frac{1}{2}$ = 97.8 nsecs), the average magnetic field at the nucleus is zero and no magnetic splitting is observed. At low temperatures (usually <30° K for ^{57}Fe minerals), these relaxation effects slow down such that their half-life becomes longer than that of the ^{57mI}Fe excited state, and magnetic spectra can then be observed.

Goldanskii-Karyagin effect

Vibrational anisotropy of the Mössbauer nucleus will give rise to

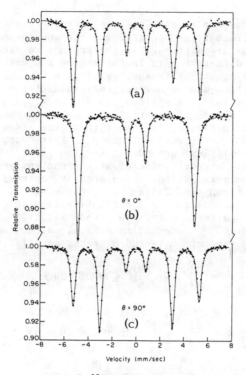

Figure 11.   Mössbauer spectrum of ^{57}Fe showing magnetic splitting: (a) is a powder
spectrum;   (b) is with the angle between the direction of the magnetic field at the
nucleus and   the   direction of propagation of the γ-radiation, $\theta = 0°$;   (c) is with the
angle θ = 90°; after Gonser (1975).

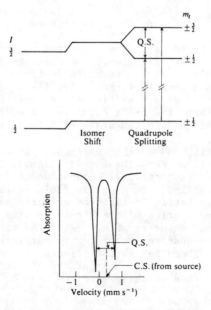

Figure 12.   Nuclear energy level diagram and spectrum showing the combined effects of
electric monopole and electric quadrupole interactions;   allowed transitions are shown by
single-ended arrows; after Bancroft (1973).

directional anisotropy in the recoil-free fraction. This in turn can give rise to relative intensities of absorption lines which are different from those expected for a randomly oriented polycrystalline sample. This result arises essentially because the various nuclear transition probabilities have an angular dependence on the relationship between the propagation direction of the γ-radiation and the symmetry axis of the field. The angular dependences of the transition probabilities for electric quadrupole and magnetic dipole transitions were discussed in the last two sections and are shown in Table 4.

Consider the simple case when the absorbing nucleus has axial symmetry; the recoil-free fraction can then be written as

$$f(\theta) = \exp\left[-k^2 \langle x_\perp^2 \rangle - k^2 \left(\langle x_\parallel^2 \rangle - \langle x_\perp^2 \rangle\right)\cos^2\theta\right] \quad , \qquad (39)$$

where $\langle x_\perp^2 \rangle$ and $\langle x_\parallel^2 \rangle$ are the mean square vibrational amplitudes perpendicular and parallel to the symmetry axis,
θ is the angle between the symmetry axis and the direction of propagation of the γ-radiation.

The relative intensity of the quadrupole split doublet for ^{57}Fe in a randomly oriented sample is then given by the ratio of the transition probabilities averaged over all possible orientations:

$$\frac{I\left(\left|\frac{3}{2}\right| \Rightarrow \left|\frac{1}{2}\right|\right)}{I\left(\left|\frac{1}{2}\right| \Rightarrow \left|\frac{1}{2}\right|\right)} = \frac{\int_0^\pi f(\theta)\ 2(1 + \cos^2\theta)\ \sin\theta\ d\theta}{\int_0^\pi f(\theta)\left(\frac{2}{3} + \sin^2\theta\right)\ \sin\theta\ d\theta} \quad . \qquad (40)$$

If $f(\theta)$ is isotropic, this ratio reduces to unity (equation 34). Most minerals show some vibrational anisotropy (as judged by anisotropic temperature factors derived from crystal structure refinements), suggesting that this effect should be present; however, as preferred orientation will also cause such asymmetric line intensity variations, it is difficult to separate the two effects (although of course it should be noted that they both have the same root cause: vibrational anisotropy in the structure).

Combined hyperfine interactions

Thus far, we have been considering a single hyperfine interaction at a time, together with the resulting Mössbauer spectrum. In practice, most materials we look at have more than one hyperfine interaction operative, and thus the nuclear energy levels and the resulting spectrum are correspondingly more complicated. We may recognize several different combinations of effects; these are illustrated below using the example of ^{57}Fe:

Electric monopole and quadrupole interactions: the nuclear energy levels and the corresponding spectrum are shown in Figure 12. The ground and excited state energy levels are shifted relative to their free nucleus values, and then the shifted excited state is split into two levels, with the possible transitions indicated. The spectrum shows a quadrupole split doublet, and the position of the centroid of that doublet defines the isomer shift (relative to some external standard). This is the most common type of (single site) ^{57}Fe spectrum in a paramagnetic mineral.

Electric monopole and magnetic dipole interactions: the nuclear energy levels and the corresponding spectrum are shown in Figure 13. Again the ground and excited states are shifted relative to the free nucleus, and then the spin degeneracy is removed for both states. The spectrum shows a magnetic spectrum whose centroid position defines the isomer shift (relative to some external standard).

Electric quadrupole and magnetic dipole interactions: this case is much more complicated than the previous two cases because the electric field and the magnetic field will not usually have the same orientation. Normally, one interaction is small compared with the other and can be treated as a perturbation: thus the magnetic hyperfine interaction is perturbed by the quadrupole interaction, or vice versa. Such interactions are susceptible to (non-trivial) calculation, whereby spectra can be simulated using variable interaction parameters, but a unique solution needs external information on structural and magnetic properties.

In "high symmetry" cases, there can be a simple solution. For example, Figure 14 shows the case where there is an axially symmetric EFG tensor with its symmetry axis parallel to that of the magnetic field, H, and the magnetic field is dominant. All four of the magnetic sublevels are displaced the same amount by the quadrupolar interaction.

INSTRUMENTATION

A Mössbauer spectrometer is simple and comparatively cheap; an adequate room-temperature spectrometer can be obtained commercially for less than $20,000. Figure 15 shows a schematic diagram of a spectrometer. The radioactive γ-ray source is attached to a vibration mechanism which is attached to a drive. The vibrating mechanism, often as simple as a spring, imparts a Döppler shift to the γ-ray energy of the source. The modulated γ-ray passes through the sample, undergoing absorption, and then on to a dectector. The resulting signals are accumulated (as a function of source velocity) in a multichannel analyzer, giving us the raw spectrum.

The main stability problem in the spectrometer is caused by the recoil of the drive system. The force that drives the source also causes a reaction in the supporting parts of the system, potentially producing motion that can adversely affect the behaviour of the source. Hence it is important to make the mass of the moving parts as small as possible, and make the mount of the drive system as rigid and "acoustically dead" as possible.

Drive mechanism

It is desirable that the drive transmit either a constant velocity or a constant acceleration to the source. In the constant velocity mode, a range of fixed velocities are imparted (one by one) to the source, and counts are recorded for a specified length of time at each velocity. Under constant acceleration (the more common mode of

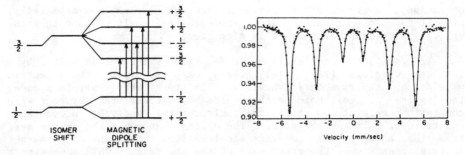

Figure 13. Nuclear energy level diagram and spectrum showing the combined effects of electric monopole and magnetic dipole interactions; allowed transitions are shown by arrows; after Wertheim (1964).

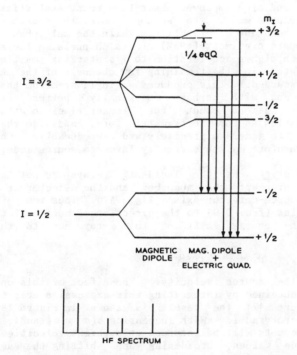

Figure 14. Nuclear energy level diagrams and (bar) spectrum showing the combined effects of magnetic dipole and electric quadrupole interactions; allowed transitions are shown by arrows; after Wertheim (1964).

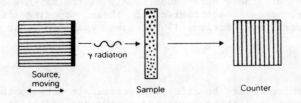

Figure 15. Schematic diagram of an experimental setup for Mössbauer spectroscopy.

operation), a range of velocities is scanned linearly and repeatedly, with the counts recorded in a multichannel analyzer such that the velocity increment per channel is a constant.

In constant acceleration mode, the vibrator is driven by a sawtooth waveform (Fig. 16); there are two types, the symmetric waveform and the asymmetric waveform. In the symmetric waveform mode, the spring or coil holding the source moves toward the absorber with constant acceleration while the multichannel analyzer accumulates counts over half the channels. The spring then moves the source away from the absorber at constant acceleration, while the multichannel analyzer counts over the other half of the channels. This generates a mirror-image spectrum of the absorber, with the mirror plane at the middle channel. Usually during data processing, the spectrum is folded back upon itself. This is a good test of the linearity of the velocity scan, and also improves absorption count statistics. In the asymmetric waveform mode, the spring moves the source towards the absorber with constant acceleration, while the multichannel analyzer accumulates counts over most (>95%) of its channels. The spring then returns at a much higher acceleration to its starting position, and the counts accumulated in the remaining few channels of the multichannel analyzer are discarded. This produces a single spectrum that does not need folding, and with perhaps marginally better resolution. Conversely, one has to count for longer times to attain as good counting statistics as for the symmetric form, and also the motion is more difficult to generate precisely and reproducably. Consequently the symmetric waveform is the currently favoured source mode.

Cosine smearing: as the incident γ-rays do not have a point source, they intercept the absorber and the detector at a range of angles to the source-detector axis (Fig. 17). Thus the γ-rays are at a range of angles (from 0-θ) to the direction of motion of the source; consequently the energy shift of the γ-ray due to the Döppler motion is

$$\Delta E = \frac{v_o}{c} E_\gamma \cos\theta_i \quad , \tag{41}$$

where v_o is the source velocity. The effect of this on the total γ-ray beam is obtained by integrating this expression over the conical volume in Figure 17; the result is shown in Figure 18 for equal source and detector radii. With increasing α($\alpha = \frac{1}{2}\tan\theta$), a greater proportion of counts will be recorded at apparent velocities different from their true values, broadening and shifting the energy of the absorption lines.

This effect can be minimized by increasing the source-detector distance ($\alpha < 0.05$ is desirable) and/or by collimating the beam; however, both of these are done only at the expense of decreased beam-intensity, and a compromise of these opposing factors is necessary. Bara and Bogacz (1980) have examined this problem in detail.

Detectors

The detector must be able to discriminate between the Mössbauer γ-ray and other radiation (unwanted γ-rays and radiation produced by the absorber). For ^{57}Fe, good resolution of the 14.4 keV γ-ray can be obtained with either scintillation or proportional counters; because

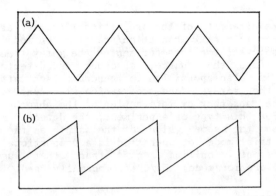

Figure 16. Symmetric (a) and asymmetric (b) saw-tooth wave-forms.

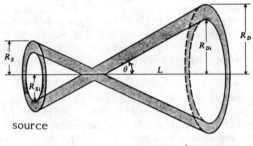

source

detector

Figure 17. The cosine smearing effect: a schematic of the geometry involved; after Bancroft (1973).

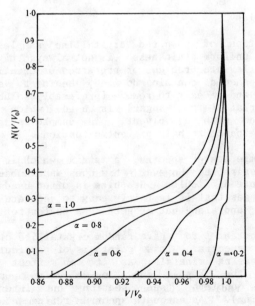

Figure 18. The effect of cosine smearing, shown as a function of the angle subtended at the absorber by the source, for equal source and detector radii; after Bancroft (1973).

the voltage requirements of the proportional counter are less, it is generally used. In normal absorption/transmission mode, the equipment is arranged so that the detector senses the γ-ray transmitted by the absorber; therefore the source, absorber and detector are linearly aligned. It is also possible to generate backscattered Mössbauer spectra from the x-rays and conversion electrons generated at the surface of the absorber by interaction of the absorber ^{57}Fe with the ^{57}Co γ-ray. For this type of experiment, the detector is located near the source (on the same side of the sample as the source). The detector for this mode of operation is a proportional counter, but with a different ionizing gas than that used for the transmission/absorption mode; He/CH$_4$ is normally used.

Heaters and Cryostats

Good vacuum sample heaters are commercially available, and they are also relatively easy and cheap to build. They can usually be heated over a temperature range of 300-1000 K and are composed of a single-piece cylinder of metal with transparent windows (usually beryllium or some other non-iron-bearing material). This contains a sample holder/heater, usually graphite. The entire ensemble is evacuated by a roughing pump which must be vibrationally isolated from the Mössbauer spectrometer.

A good liquid nitrogen or helium cryostat is probably the single most expensive piece of equipment for Mössbauer instrumentation (including the spectrometer itself!). However, is is possible to get a good variable-temperature cryostat for the temperature range 12-300 K; these use no liquid helium, relying on a closed cycle refrigerator. A cheaper alternative is to cool sample temperatures to approximately 150 K using a copper cold finger suspended in a tank of liquid nitrogen.

Absorber preparation

The preparation of powdered crystalline samples of random orientation and uniform thickness is not always trivial. An inert matrix can help alleviate problems of preferred orientation. An ideal matrix material should contain no iron, be inert, softer than the mineral studied and be easy to remove (preferably soluble). Usually an organic material (e.g., sugar) is used, which can then be completely removed with a solvent. The sample is ground, usually under acetone or ethanol to help prevent oxidation.

For high-temperature spectra, a matrix which is stable at high temperatures is desirable. However, this does make subsequent recovery of the sample more difficult. Graphite is often used for such work, having the advantages that is is fairly easy to separate out subsequent to the experiment, and also that it acts as a buffer for Fe^{2+}.

It is important to have random crystallite orientation when recording powder spectra. This point is of particular interest in studies of Fe-bearing minerals, as it can effect the relative intensities of the high- and low-velocity components of quadrupole-split doublets (see section on interpretation of quadrupole-splitting). As adequate spectrum refinement often requires the use of linear constraints between doublet components (see section on spectrum refinement), the attainment of random absorber orientation if often a key part of the experimental procedure.

Most of the theoretical arguments concerning Mössbauer spectroscopy are developed for infinitely thin absorbers. Excessive γ-ray absorption leads to peak-shape degradation, a non-linear baseline, peak broadening, and saturation effects. A general "rule-of-thumb" is that for Fe-bearing oxide/oxysalt minerals, an Fe-content of $5mg/cm^2$ is a good compromise between a "thin absorber" and good counting statistics. A more detailed consideration of absorption is given by Long et al. (1983). The percentage absorption should not significantly exceed 10%, and most careful work usually shows absorption in the range 2-6%, although smaller absorptions may be unavoidable in Fe-poor samples.

Calibration

In Mössbauer spectroscopy, peak positions are not measured in terms of their true energies but in relation to a zero energy point and an energy scale derived from a standard absorber spectrum. There are several different types of absorbers used for this purpose (^{57}Fe: iron foil, stainless steel, sodium nitroprusside; ^{119}Sn: SnO_2; ^{121}Sb: InSb). For ^{57}Fe, there is no concensus on what to use. The correct material (i.e., we use it) is iron-foil; to convert other standardized values to this datum, the following conversion factors obtain: stainless steel = 0.10 mm/s; sodium nitroprusside = 0.257 mm/s; ^{57}Co in Pd: -0.185 mm/s).

The spectrum of natural iron foil has six lines, the positions of which are very accurately known. Usually the inner four lines are used for calibration for paramagnetic substances. One plots the known values for the peak positions versus the observed positions. The slope of the resulting regression line gives the velocity increment per channel, and the zero velocity position is taken as the centroid of the four (or six) peaks.

Spectrum quality

The variance of the counts in a specific channel is equal to the number of counts (Chapter 3, equation 4), and hence the "quality" of the spectrum increases with increased counts, assuming only random error. Thus one can improve the quality of a spectrum by counting for a longer time. Unfortunately, this is a process of diminishing returns, as the improvement gradually tails off with increasing time until only trivial improvements result from great increases in counting time. Generally, most spectra are accumulated for times such that the baseline (off-resonance) count is of the order of $1-5 \times 10^6$ counts/channel, usually a matter of 24-48 hours for ^{57}Fe. The effect of counting time (as measured by the off-resonance channel count) on standard deviation of the channel count is shown in Figure 19. For off-resonance counts of $\approx 10^5$, the standard deviation does not fall below $\approx$ 8% (relative) for reasonable (< 10%) absorptions; an order-of-magnitude increase in the off-resonance count greatly improves this situation, with standard deviations of < 5% for reasonable (3-10%) absorptions. A further order-of-magnitude increase in the off-resonance count only extends the standard deviations of < 5% to the

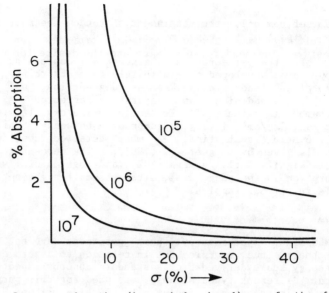

Figure 19. Percentage absorption (in a single channel) as a function of absorption standard deviation for different values (10^5, 10^6, 10^7 counts/channel) of the off-resonant count.

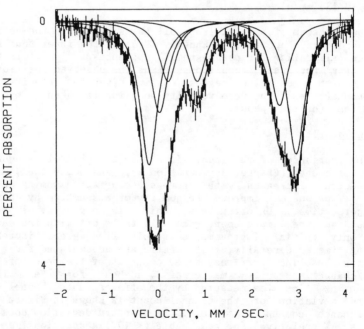

Figure 20. An experimental Mössbauer spectrum (of ^{57}Fe in ferro-tschermakite); the ordinate shows the energy scale, expressed as mm/s (the instantaneous modulating velocity applied to the source); the abcissa shows the γ-ray absorption by the sample (usually expressed as a %). In this spectrum, the experimental data are represented by vertical dashes, the magnitude of which represents 2σ based on counting statistics (see Chapter 3). The count rate at zero absorption is called the off-resonance count. From Hawthorne (1973).

1-10% absorption range, and most experimenters feel that this range (10^6-10^7 off-resonance counts/channel) provides an acceptable balance between the need for precision and the desire for experimental efficiency.

SPECTRUM REFINEMENT

An experimental Mössbauer spectrum is shown in Figure 20. Each datum represents the number of counts recorded over that particular source velocity interval, and the length of the vertical dash represents two standard deviations based on counting statistics. The counts at the edges of the spectrum represent the background counts, that is the intensity of the γ-radiation where no absorption occurs; the background should be approximately constant across the spectrum, and the average value is called the off-resonance count. Towards the center of the spectrum, the counts decrease due to the resonant absorption of γ-rays by the sample. As we have seen, the ideal line shape is Lorentzian (see Chapter 3), and an observed spectrum ideally consists of a series of Lorentzian lines, the number and characteristics of which are a function of the Mössbauer nucleus and the crystal structure of the sample. The relative magnitudes of the half-width, isomer shift and quadrupole splitting often lead to complex overlap of lines, and the derivation of quantitative information from such spectra (peak positions, half-widths and areas) requires numerical spectrum fitting. General aspects of spectrum fitting are covered in Chapter 3; details specific to Mössbauer spectroscopy are discussed here.

Mathematical description of the spectrum

The intensity of the γ-radiation transmitted by the sample as a function of its energy x can be written as

$$y = b - \sum_{i=1}^{l} \frac{2A_i/\pi\Gamma_i}{1+4[x-x_i]^2/\Gamma_i^2} \quad , \tag{42}$$

where l is the number of (Lorentzian) lines,
  b is the background intensity,
  A_i is the area (intensity) of the ith line,
  Γ_i is the half-width of the ith line,
  x_i is the peak-position of the ith line,
  y is the channel count,
  x is the channel number.

Slight sinusoidal and linear deviations can occur in the background intensity due to source movement and instrumental drift. Consequently the background intensity b can be written as a function of channel number, with various refinable parameters; thus for example

$$b = b_o + b_1 x + b_2 \sin [\pi x/w] \quad , \tag{43}$$

where b_j (j = 0 --> 2) are variable parameters,
  w is the total width of spectrum (in channel numbers).

The term b_0 accounts for the bulk of the background intensity; the term involving b_1 accounts for linear drift with channel number; the term involving b_2 accounts for source movement. Of course, more complex background functions can be designed.

Although the ideal lineshape is Lorentzian, there are a number of factors that can result in a Gaussian component in the lineshape. This may be incorporated into the fitting procedure by introducing a Gaussian component into the lineshape equation. The amount of this Gaussian component can either be fixed (by prior modelling on simple spectra) or it can be incorporated as a variable in the spectrum-fitting procedure. Omitting the more complex background function for clarity, the resulting spectrum function can be written as

$$y = b - \sum_{i=1}^{l} \left[\frac{(1-\alpha)2A_i/\pi\Gamma_i}{1+4(x-x_i)^2/\Gamma_i^2} + \frac{\alpha 2A_i}{\Gamma_i}\sqrt{\frac{\ln 2}{\Gamma}} \exp\left\{ -4\ln 2\left(\frac{x-x_i}{\Gamma_i}\right)^2 \right\} \right] \quad , \quad (44)$$

where α is the fractional Gaussian character of the line; note that we have assumed that the half-widths of the Lorentzian and Gaussian components are the same.

Least-squares refinement

This is usually done by the Newton-Gauss method. The equation of the spectrum given above is obviously not linear in all variables, and consequently we have to linearize this equation by application of some expansion approximation (see Chapter 3). This means that the least-squares equations with which we calculate our parameter shifts during the minimization process are only approximations, and so we have to iterate the fitting procedure, gradually approaching the optimum values.

Where there is little or no line overlap in the spectrum, the refinement procedure is usually straightforward. One needs a set of starting parameters for the refinement; most of these may be derived by inspection of the spectrum. The background intensity may be estimated quite accurately from the off-resonance count; at this stage, drift and sinusoidal corrections are not yet used. Peak positions may be estimated to the nearest channel from the position of the maxima in the spectra; similarly half-widths may be measured directly from the spectrum (or estimated from refinements of similar spectra). Relative peak areas can be estimated by eye, but the actual numerical values cannot be so easily derived, and a judicious guess may be necessary. This being the case, the numerical values of the peak areas are likely to be the most inaccurate of the starting parameters; the quickest way of deriving better values is to hold all other variables constant and just refine the peak areas. Once this is done, full-matrix least-squares refinement of all variables should prove no problem.

When there is significant peak overlap in the spectrum, the refinement procedure is more difficult as the variables can now interact with one another in the refinement procedure, and the spectrum inherently contains less information about the mineral than it would if there were no overlapping peaks. The background may be estimated as before. However, deriving the peak positions is now more difficult because each peak position will not necessarily coincide with a maximum in the spectrum. Normally, peaks are assigned to the positions of the spectrum maxima, and then the sides of the spectrum envelope are inspected for "shoulders" (discontinuties) that indicate the presence of "hidden" peaks; such positions are also used as starting parameters, but they will normally not be as good as in the case where there is no overlap. Half-widths are difficult to estimate, and use of values from previously recorded spectra (or simpler spectra) is best. Again, only relative areas can be estimated, and not as well as in the non-overlapped case. Refinement now has to proceed much more carefully, as many of the variables can interact strongly in the refinement process, even to the extent of precluding refinement at all. Again it is best to initially refine just the peak areas, essentially to ensure that the problem is scaled correctly. This is (usually) best followed by release of the peak positions, and then the peak widths. After convergence at each stage, it is best to inspect the observed and calculated spectra and the "residuals" to ensure that no obvious changes (e.g., addition or subtraction of a peak) are necessary. This can be followed by releasing the background variables. All variables must be varied simultaneously in the final cycles of refinement. For very badly overlapped spectra, an even more conservative approach may be necessary, with successive sets of refinement cycles not varying highly correlated variables simultaneously (blocked-matrix method). However, all variables must be refined simultaneously in the final cycle(s).

Use of constraints

Consider two peaks with the same position and half-width. If we attempt to refine the areas, we see that there is no distinct solution (ignoring for the moment the fact that the refinement will fail). As one area increases, the other decreases by the same amount; the two variables show a correlation of -1, and are indeterminate. If we have external knowledge of the sample, e.g., the area of one peak is half the area of the other, we can refine the area of one peak and can constrain the area of the other to be half that of the first:

Area of peak 1 = A_1 Area of peak 2 = $\frac{1}{2}A_1$.

We now only have one variable in the least-squares process, and we get a unique solution for the areas of the two peaks. What we are doing here is taking a piece of information external to the spectrum, and incorporating it with the spectrum information to get a distinct result.

Consider two overlapping peaks separated by a fixed number of channels, where the area of peak A is approximately ten times the area of peak B. If the half-width of the larger peak decreases, it overlaps less with the smaller peak, the half-width and area of which increase to compensate. Thus the areas and half-widths of the two peaks are strongly correlated because of the interaction of all four variables. If we have external information to say (for example) that the half-widths of the peaks are equal, then we can input this information into the fitting process to suppress the high variable correlation and give precise results.

Two types of constraints are normally used in ^{57}Fe Mössbauer spectroscopy: (a) half-width constraints, and (b) area constraints.

Half-width constraints are often used in multisite Fe-bearing minerals; indeed, in many cases, spectrum refinement will fail without such constraints. Their origin (e.g., Bancroft, 1973) stems from the observation that the half-widths of component peaks of Fe^{2+} quadrupole -split doublets tend to be equal and to fall in the region 0.28-0.35 mm/s. Thus

(a) the half-width of component quadrupole doublets are usually constrained to be equal (both Fe^{2+} and Fe^{3+});
(b) the half-width of all Fe^{2+} peaks are constrained to be equal; this constraint is also applied to Fe^{3+} peaks.

Constraint (a) is extremely common, and constraint (b) is used for most complex spectra. Deviations from such relationships have been used to suggest the presence of additional peaks in a spectrum.

A spectrum contains only a certain amount of information, and in the case of badly overlapping peaks, this information may not be sufficient to satisfactorily refine the spectrum, that is to extract sufficient information to solve the mineralogical problem. When we add constraints, we are adding information to the fitting process, and this information combines with that in the spectrum to give a solution to our problem. However, the accuracy (as distinct from the precision) of the final result depends on the correctness of the constraints. If the constraints are not appropriate, then the information added to the fitting process is wrong, and the final result will be wrong. Thus it is necessary to be a lot more careful about half-width constraints than has often been the case in the past, as it is possible that next-nearest-neighbour effects can cause variable peak broadening in complex multisite minerals.

Area constraints usually are restricted to quadrupole (or magnetic hyperfine) split component peaks, where it is generally assumed that the areas of the high- and low-velocity components are equal. This is a reasonably good constraint for randomly oriented absorbers, but the problem of residual orientation (and the Goldandskii-Karyagin effect) can cause asymmetry. If this is the case, it is possible in complex spectra to constrain the ratio of the areas of quadrupole-split doublet components to be equal, with the ratio as a refinable parameter; although an approximation, it is a better one than equal areas for quadrupole-split doublets.

The same caveat applies to area constraints as to width constraints; if they are not correct, the final answer will be wrong.

Goodness-of-fit criteria

The most common criterion for the "best-fit to the data" for a specific model is the minimization of the weighted sum of the squares of the deviations between observed and "calculated" data (see Chapter 3):

$$\chi_o^2 = \sum_{i=1}^{N} \frac{1}{\sigma_i^2} \left[y_i - f(x_i) \right]^2 \quad , \tag{45}$$

where y_i is the total observed counts in the ith channel,
  σ_i is the weight assigned to the ith observation,
 $f(x_i)$ is the calculated counts for the ith channel,
  N   is the number of channels.

We may define an ideal residual χ_I^2 as

$$\chi_I^2 = \sum_{i=1}^{n} \frac{1}{\sigma_i^2} \left[y_i - f_I(x_i) \right] \quad , \tag{46}$$

where $f_I(x_i)$ are the calculated counts for the "correct function f_I". χ_I^2 follows the chi-squared distribution, and if the parameters of the fitted function $f(x)$ are a valid approximation to the true function $f_I(x)$, then χ_o^2 is a value from this distribution (Law, 1973). Note that here we are using the method of maximum likelihood (see Chapter 3, equations (12) - (15)) to overcome the fact that we cannot know $f_I(x)$. It is unfortunate that the symbol χ_o^2 of equation (45) is used, as it is χ_I^2 that follows the chi-squared distribution, not χ_o^2 (Law, 1973). To test the hypothesis that $f(x)$ is a valid approximation to $f_I(x)$, one uses the distribution to assess the probability that $R_o < R_i$ at a certain percentage confidence limit. At the 1% point on the chi-squared distribution, the probability is 1% that χ_i^2 will exceed that value. Thus if $\chi_o^2 > 1\%$ point (= N-n + 2.2-3.3$\sqrt{(N-n)}$, where N is the number of data points and n is the number of fitted parameters, i.e., N-n is the number of degrees of freedom), then $f(x)$ is generally accepted as a good "fit".

It should be realized (Law, 1973) that <u>within</u> the 1% and 99% points of the chi-squared distribution, there is no statistical justification for preferring a fit with a lower χ_o^2 value, as the probability that χ_I^2 will exceed the value of χ_o^2 is quite high.

Another version of the residual is the reduced χ_R^2 defined as

$$\chi_R^2 = \frac{\chi_o^2}{N-n} \quad . \tag{47}$$

This has the advantage that the values are close to 1.0, rather than being large numbers that are dependent on the number of channels used to collect the data.

As discussed by Ruby (1973) and Waychunas (1986), χ_o^2 has some problems as a general goodness-of-fit parameter for Mössbauer spectra. In particular, it is affected by the magnitude of the background counts, and can be reduced by using poorer quality data. Ruby (1973) has introduced a new goodness-of-fit parameter called MISFIT, designed to overcome these difficulties. Details of the development of MISFIT (M) are given in Chapter 3. It is defined in the following way:

$$M = \frac{D}{S} \quad , \tag{48}$$

where

$$D = \sum_{i=1}^{N} \left\{ \frac{1}{\sigma_i^2} [y_i - f(x_i)]^2 - 1 \right\} \tag{49}$$

and

$$S = \sum_{i=1}^{N} \left\{ \frac{1}{\sigma_i^2} \left[y_o - f(x_i) \right]^2 - 1 \right\} \quad . \tag{50}$$

The uncertainty in MISFIT is denoted as ΔMISFIT and is given by

$$\Delta M = \frac{1}{S} \sqrt{[N(1+M^2) + 4D(1+M)]} \quad . \tag{51}$$

D is known as the _distance_ and S is the _signal_. The numerical value of M gives the fraction of the experimental signal that remains unfitted, and together with ΔM, is useful in testing fit models. Waychunas (1986) has examined the performance of χ_R^2, M and ΔM on a series of numerically simulated spectra, and concludes that
(1) χ_o^2 is strongly dependent on data quality with an improper fit model;
(2) M is less sensitive to data quality than χ_o^2, but can show anomalous behaviour for low-quality data;
(3) M and M are of high value in testing fit models.

Standard deviations and error-of-propagation analysis

As discussed in Chapter 3, the reason why we use least-squares methods is because we get our results, together with a standard deviation that we can use to test the significance of the result using linear hypothesis tests. However, it is important that the standard deviations be assigned correctly; if this is not done, one cannot assess the significance of the result, which is thus useless.

When encountering problems of variable correlation in complex overlapped spectra, it is common practice to refine highly correlated variables in separate cycles, to damp variable interaction and possible oscillation; this is known as blocked-matrix least-squares. However, it should be realized that this is merely a technique to aid convergence. When at final convergence, it is _essential_ to run a series of _full-matrix_ least-squares cycles, so that all covariances are taken into account in the calculation of the parameter standard deviations. If this is not done, the standard deviations of some of the results will be wrong (Hawthorne and Grundy, 1977), and it will be impossible to confidently assess the significance of the results.

When deriving site-occupancies from peak areas, it is necessary to ensure that the "standard error" is propagated through the complete calculation to obtain correct standard deviations for the site-occupancies. Some least-squares programs already handle this, giving the peak areas as a fraction of the total area. When this is not done, the total area, A, must be calculated by summing the individual areas:

$$A = \sum_{i=1}^{n} a_i \quad , \tag{52}$$

where n is the number of lines and a_i is the area under the ith peak. Then the fractional area is calculated from the sum of the doublet component peaks $F_j = (a_j + a_j')/A$, where F_j is the fractional area of the jth doublet, and a_j, a_j' are its component peaks.

Alternatively, the site-occupancies may be calculated from individual peak area ratios, as outlined in the section on site-occupancies. The standard error of the individual peak areas <u>must</u> be propagated thought these calculations (as discussed in Chapter 3); if this is not done, the whole reason for doing the least-squares refinement is negated and the results are of little use. Unfortunately, this has often been the case in many studies on minerals.

For most problems in mineralogy, Mössbauer spectra consist of a complex overlap of peaks that reflect differential occupancy of more than one crystallographic site in the mineral structure. In such cases, the precision with which the parameters of interest can be determined is strongly a function of the degree of overlap of the constituent peaks. This has been investigated by Dollase (1975) for the case of two overlapped peaks using simulated spectra. Several interesting features emerged from this work, First, for equal area and equal width Lorentzian peaks, separate maxima in the envelope occur for peak separations greater than $1/\sqrt{3}(\approx 0.6)$ peak width. Second, the uncertainty associated with the derived peak area is a function of the peak separation and the peak-to-background ratio. Because Fe^{3+} shows much less variation than Fe^{2+} in quadrupole splitting values for most minerals, complex Fe^{3+} spectra generally show far greater overlap than comparably complex Fe^{2+} spectra. Consequently, Fe^{3+} site-occupancies are usually far less-precisely determined than Fe^{2+} site-occupancies.

Standard deviations derived from non-linear least-squares refinement must be treated with caution. If the estimated standard deviations are small enough that the functions are linear over a range of several standard deviations, then the normal methods of testing linear hypotheses can be applied; this is hopefully the general case in most uses of least-squares refinement. However, if there are departures from linearity in this range, as is to be expected if high correlations give rise to large standard deviations, then variance-ratio tests are no longer exact (Hawthorne, 1983a). This is presumably the origin of the "imprecisely determined" standard deviations in the study of Dollase (1975), and this effect should be borne in mind when assessing the significance of site-populations derived from spectra in which high correlations are encountered.

The correct answer

Statistical tests are used as a measure of comparison between alternate models, with the residual as a measure of the agreement between the observed data and a corresponding set of calculated values derived from the least-squares solution. Statistical acceptability and a low residual are no guarantee that the derived model is <u>correct</u>; they merely indicate that the derived model adequately (but not exclusively) explains the observed data. The most obvious examples of this are the Mössbauer spectra of the amphiboles of the magnesio-cummingtonite--grunerite series (Bancroft et al., 1967a; Hafner and Ghose, 1971). The C2/m amphibole structure indicates that the correct (ideal) spectrum model should consist of four quadrupole-split Fe^{2+} doublets. However, statistically acceptable fits are obtained for two quadrupole-split Fe^{2+} doublets. Doublet intensity (Bancroft et al., 1967a; Hafner and Ghose, 1971; Ghose and Weidner, 1972) and local stereochemistry (Hawthorne, 1983b) show that three of the four doublets overlap almost perfectly in the spectra and have very similar IS and QS values. The correct (ideal) four-doublet model cannot be resolved, as the spectrum does not contain sufficient information for this to be done. This is not to say that useful information cannot be derived from the two-peak solution in this particular case; however, it does illustrate that statistical acceptability and correctness are not necessarily concomitant. Thus one should judge whether or not the refined model is correct on the basis of whether or not it is physically and/or chemically reasonable.

INTERPRETATION OF ISOMER SHIFT

As we saw above, the isomer shift, δ, is proportional to the difference in contact density between the sample and a standard material:

$$\delta = C \left[|\psi(0)|^2_{SAM} - |\psi(0)|^2_{STAN} \right] \quad , \tag{53}$$

where

$$C = \frac{4\Pi}{5} Ze^2R^2 \left(\frac{\delta R}{R} \right) \tag{54}$$

(equation 28) and depends primarily on δR, the difference in nuclear radius between the ground and excited states. Thus the isomer shift is affected by a nuclear property, δR, and an extranuclear property, the contact density. The contact density at the nucleus affects the magnitude of the isomer shift, and δR affects the sign of the isomer shift. When δR is positive (as for ^{119}Sn), an increase in contact density results in a more positive isomer shift. When δ is negative (as for ^{57}Fe), an increase in contact density results in a more negative isomer shift. Values of $\delta R/R$ are given in Table 1.

Only s-electrons have a finite probability of overlapping with the nuclear charge density, and thus of affecting the isomer shift. However, p- and d-electrons can indirectly affect the contact density by shielding effects, and increase in the number of p- and/or d-electrons will decrease the contact density. Hence the isomer shift is sensitive to any factor that affects the number and/or distribution of valence-shell electrons, and thus is a probe of oxidation state, co-ordination and covalency.

Oxidation state

The isomer shift is sensitive to variation in absorber oxidation state. This may occur directly by variation in the number of valence s-electrons (as is the case for ^{119}Sn and ^{121}Sb) or through shielding effects (as is the case for ^{57}Fe, in which variation in the number of 3d-electrons occurs). In the former cases (^{119}Sn and ^{121}Sb), the isomer shift differences (Table 5) are large, and the ranges for the different valence states do not overlap (Fig. 21). In the latter (^{57}Fe), the isomer shift differences are much smaller (Table 5, Fig. 22); however, for most mineralogical applications, we are dealing with high spin Fe^{2+} and Fe^{3+}, and distinguishing these is usually not a problem.

Let us see if we can rationalize the shifts we see in Figures 21 and 22 by equation (53) and Table 1. For ^{119}Sn, going from Sn^{2+}-->Sn^{4+} involves removing s-electrons, thus decreasing the contact density; as $\delta R/R$ is positive, the isomer shift should decrease, as is the case. For ^{121}Sb, again Sb^{3+}-->Sb^{5+} involves the removal of s-electrons, but $\delta R/R$ is negative, and so the isomer shift should increase. For ^{57}Fe, Fe^{2+}-->Fe^{3+}-->Fe^{4+}-->Fe^{6+} involves the progressive removal of 3d-electrons, which reduces the shielding and progressively increases the contact density; as $\delta R/R$ is negative, the isomer shift should decrease, and this is the case (Fig. 22).

Ligancy

Variation in oxidation state changes the number of (valence) electrons, affecting the contact density very directly. Changes in the character of the bonded atoms (ligands) affect the partial electron energy density of states of the (Mössbauer) atom, and hence affect the contact density which, in turn, affects the isomer shift. Figure 23 shows the experimental ^{119}Sn isomer shifts for a series of Sn compounds versus the valence-electron contribution to the contact density (Svane and Antoncik, 1987), calculated via tight-binding theory. Essentially we are seeing an increase in isomer shift with increase in covalency of the Sn-X bond. Such calculations are involved and not inexpensive, and a more empirical treatment is also desirable.

It is common to use the electronegativity difference between the atoms of a bond as a measure of the degree of covalency, and thus isomer shifts are commonly systematized in this fashion. Figure 24 sumarizes some of these trends for various simple inorganic solids. The general idea is that the more electronegative ligand withdraws electron density from the "Mössbauer" atom to decrease the contact density and decrease the isomer shift for positive $\delta R/R$ nuclei (Fig. 24); however, because s-electrons and p- and d-electrons act in an opposite fashion on the contact density, such relationships may be perturbed by different degrees of orbital hybridization. Such differences may account for the irregularities in the relationship for the Sn^{2+} halides (Fig. 24).

Coordination number

Here again we are looking at a change in ligancy, but involving the number of ligands rather than the type. However, we still are involved with changes in the degree of covalency, and so for positive $\delta R/R$, we expect a negative correlation of co-ordination number with isomer shift. In minerals, we are primarily concerned with ^{57}Fe, and here we see a distinct trend (Fig. 25); there is some overlap, but by and large the populations are sufficiently distinct for isomer shift values to play a useful role in site-occupancy studies.

Partial isomer shifts

As we have seen (Fig. 24), the isomer shift is often a linear function of the electronegativity of the ligands. This infers that the isomer shift can be expressed as an additive function of the ligands (Parish, 1985), and leads directly to the idea of partial chemical shifts whereby a (fictive) value of isomer shift, $p\delta_e$, can be assigned

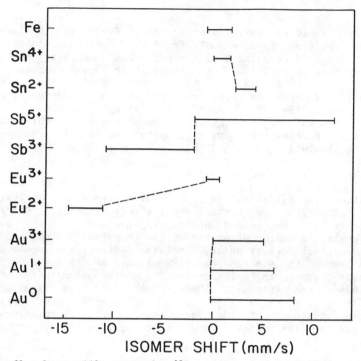

Figure 22. Isomer shift ranges for ^{57}Fe, given relative to Fe-foil; note the much narrower range (for all valence states), typical of species for which valence changes involve p-, d- or f-electrons, and which affect the contact density via shielding effects only.

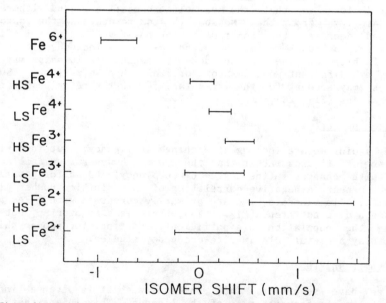

Figure 21. Isomer shift ranges for ^{57}Fe, ^{119}Sn, ^{121}Sb, ^{151}Eu and ^{197}Au in various valence states; ^{119}Sn shifts are relative to SnO$_2$; ^{121}Sb shifts are relative to InSb; ^{151}Eu shifts are relative to Eu$_2$O$_3$; ^{197}Au shifts are relative to pure ^{197}Au; note the very wide range for the species for which the valence changes directly involve s-electrons, with the exception of ^{197}Au.

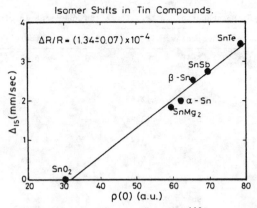

Figure 23. Isomer shift values for a series of ^{119}Sn compounds versus the valence-electron contribution to the contact density, calculated using tight-binding theory; after Svane and Antoncik (1987).

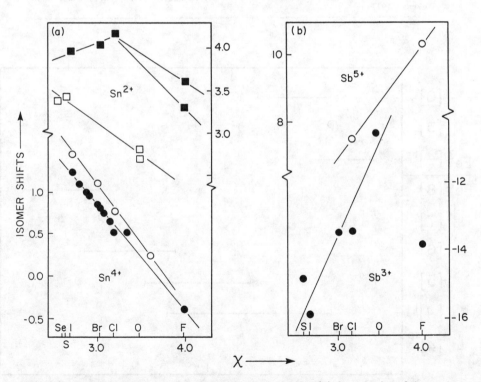

Figure 24. Isomer shift versus ligand electronegativity (X) for a series of inorganic ^{119}Sn and ^{121}Sb compounds: (a): O = Sn^{4+}X$_4$Y$^{2-}_2$; O = Sn^{4+}X$_4$; □ = Sn^{2+} chalcogenides; ■ = Sn^{2+} halides; (b): O = Sb^{3+} halides and chalcogenides; O = Sb^{5+} halides; values for ^{119}Sn relative to SnO$_2$; values for ^{121}Sb relative to InSb; note the negative correlations for the ^{119}Sn compounds and the positive correlations for the Sb compounds, indicative of the sign difference in δR/R; also note the large separation between valence states.

Table 5. Mössbauer parameters for ^{119}Sn and ^{121}Sb in minerals.

Minerals	Species	IS(mm/s)	QS(mm/s)	Γ(mm/s)	T(°K)	Ref.
Sn-rich andradite*	Sn^{4+}	-0.14	0.42	0.85	295	(1)
Pb-rich franckeite**	Sn^{4+}	1.25	–	1.02	78	(2)
"	Sn^{2+}	(2.69)	–	0.70	78	(2)
"	Sn^{2+}	(3.98)	–	1.04	78	(2)
Sn-rich incaite**	Sn^{4+}	1.21	–	1.00	78	(2)
"	Sn^{2+}	2.85	–	0.96	78	(2)
"	Sn^{2+}	3.95	–	0.75	78	(2)
Ordonezite#	Sb^{5+}	9.08	5.8	3.4	78	(3)
Pb-free franckeite##	Sb^{3+}	-5.56	10.43	3.08	4.2	(2)

* ^{119}Sn quoted relative to SnO2; ** ^{119}Sn quoted relative to BaSnO3; values for Sn^{2+} in parentheses indicate peak positions in an unfitted spectrum; # ^{121}Sb quoted relative to InSb; ## ^{121}Sb quoted relative to BaSnO3.

References: (1) Amthauer et al. (1979); (2) Amthauer (1986); (3) Baker and Stevens (1977)

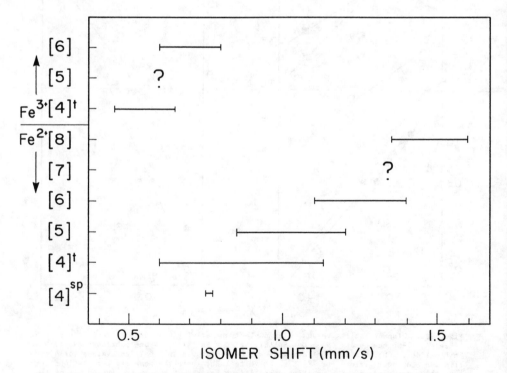

Figure 25.   Isomer shift for ^{57}Fe (in minerals) as a function of co-ordination number and valence state; values relative to iron-foil.

to a specific ligand, and the actual isomer shift observed in a specific compound, δ, can be expressed as the sum of the partial isomer shifts of the constituent ligands:

$$\delta = \sum_{CN} p\delta_e \quad . \tag{55}$$

This approach seems to be successful when the range of isomer shifts is fairly small, working reasonably well for transition metals (in which the changes affect δ indirectly through shielding effects, and consequently the range of δ is not large) but not for main group elements (in which the changes affect δ directly through occupancy of the s-orbitals). It has be used extensively in organometallics (see reviews by Bancroft (1973) and Parish (1985)), and also in organics.

This use of partial isomer shifts implies that the isomer shift is insensitive to small geometrical and topological changes in structure. For many mineralogical uses, a rather more subtle use of this idea is of interest. In many minerals, the ligands are always O^{2-}, and local changes occur only in the next-nearest-neighbour "co-ordination" shell. This being the case, it is (potentially) possible to assign partial isomer shifts to next-nearest-neighbour configurations, and to measure isomer shifts as a probe of local structure.

INTERPRETATION OF QUADRUPOLE SPLITTING

When considering the structural (crystal and electronic) factors that affect the QS, it is convenient to assume that $\eta = 0$ and to partition the EFG into three additive functions:

$$V_{zz} = V_{zz}(ve) + V_{zz}(mo) + V_{zz}(str)$$

$$= (1-\gamma_{\infty})V_{zz}(str) + (1-R)[V_{zz}(ve) + V_{zz}(mo)] \quad , \tag{56}$$

where $V_{zz}(ve)$ is the part of the EFG due to the (formal) valence electrons;
$V_{zz}(mo)$ is the part of the EFG due to covalent transfer of electron density;
$V_{zz}(str)$ is the part of the EFG due to the rest of the crystal;
γ and R are the <u>Sternheimer antishielding factors</u> (see following discussion).

The contribution from the local valence electron density $[V_{zz}(ve) + V_{zz}(mo)]$ is given by the following summation over the valence electrons:

$$-\sum_{\substack{valence \\ electrons}} k \ (3\cos^2\theta - 1)\langle r^{-3}\rangle \quad , \tag{57}$$

where k is a set of numerical constants that are orbital dependent θ is the polar co-ordinate angle;
$\langle r^{-3}\rangle$ is the expectation value for the third inverse power of the radial polar co-ordinate.

As the inner filled shells have (ideal) spherical or cubic symmetries, they should not contribute to the EFG. However, both $V_{zz}(str)$ and $[V_{zz}(ve) + V_{zz}(mo)]$ will polarize and distort the inner orbitals, affecting $V_{zz}(str)$ and $[V_{zz}(ve) + V_{zz}(mo)]$. The Sternheimer factors take this into account: γ_∞ is negative, and of the order of -10, whereas R is positive and about 0.2 (for ^{57}Fe and ^{119}Sn). Consequently the Sternheimer antishielding factor considerably enhances the V(str) component, which tends to be the weaker component because of the much larger values of r involved. If the local symmetry is cubic, both $V_{zz}(str)$ and $[V_{zz}(ve) + V_{zz}(mo)]$ are zero and the EFG is zero. With deviation from cubic symmetry, the various terms of V_{zz} become non-zero and a QS results. This QS is sensitive to the various electronic and structural factors affecting V_{zz}, and can be considered as a local probe of these features in a mineral.

Oxidation state

For transition metals with non-spherical/non-cubic symmetry (high-spin Fe^{2+}, low-spin Fe^{3+}), $V_{zz}(ve) \gg V_{zz}(mo)$; for transition metals having spherical or cubic symmetry (Sn^{4+}, low-spin Fe^{2+}), $V_{zz}(ve) \ll V_{zz}(mo)$.

Consider first the case of non-spherical/non-cubic symmetry, using the example of high-spin $^{VI}Fe^{2+}$ (Fig. 26). If the structural environment has octahedral symmetry, then ideally the extra electron equally populates the three t_{2g} orbitals and $[V_{zz}(ev) + V_{zz}(mo)] = 0$; thus $N(d_{xy}) = N(d_{xz}) = N(d_{yz})$. However, this configuration is electronically degenerate and can gain stabilization energy by spontaneously distorting such that one of the t_{2g} orbitals lowers its energy relative to the other two, and is preferentially occupied by the extra electron; this is known as a Jahn-Teller distortion.

The relative contributions of an electron in d_{xy}, d_{xz} and d_{yz} to V_{zz} are -2:1:1 (controlled by the expectation values of the angular wave functions). Consequently a small Jahn-Teller distortion from cubic symmetry localizes the extra electron in d_{xy}; this causes a large $V_{zz}(ve)$ contribution without materially affecting $V_{zz}(mo)$ or $V_{zz}(str)$, and leads to a large QS. Increasing distortion increases $V_{zz}(ev)$ but also increases $V_{zz}(mo)$ and $V_{zz}(str)$; as $V_{zz}(mo)$ and $V_{zz}(str)$ usually act in the opposite direction to $V_z(ev)$, then the increasing $V_{zz}(mo)$ and $V_z(str)$ with distortion serve to detract from the effect of increasing $V_{zz}(ev)$, and lead to a decrease in the QS. Thus one expects the relationship between distortion and QS shown in Figure 27. By and large, this is what is observed in minerals.

The stabilization energy induced by Jahn-Teller distortion is usually fairly small, and there is thermal excitation of electrons among the originally degenerate orbitals. Such orbital occupancies are temperature dependent (being subject to Boltzmann statistics), and hence the resulting QS has a considerable temperature dependence.

It is a little difficult to quantify the term "distortion", as in this context, it refers to the total electric field felt by the Mössbauer nucleus, a quantity for which we have no easy structural expression. It is most often represented by some index that purports to represent the distortion of the co-ordination polyhedron in which the atom resides; however, such polyhedra show both distance and

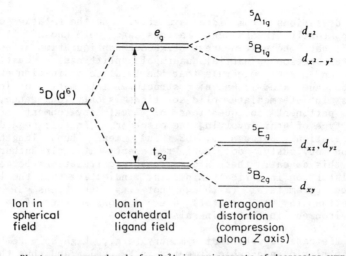

Figure 26. Electronic energy levels for Fe^{2+} in environments of decreasing symmetry; for a d^6 configuration (high spin Fe^{2+}), four electrons occupy the t_{2g} orbitals, the "extra" electron equally populating the d_{xy}, d_{xz} and d_{yz} orbitals in an octahedral field; distortion (tetragonal) splits these orbitals such that the d_{xy} orbital energy is lowered and the d_{xy} and d_{yz} orbital energies are raised; the extra electron can then occupy the d_{xy} orbital and gain stabilization energy; modified from Maddock (1985).

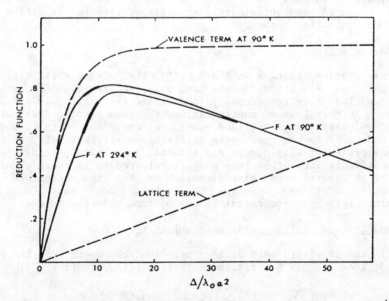

Figure 27. Relation between the reduction function, F, and distortion ($\Delta/\lambda_o\alpha^2$) parameter; the quadrupole splitting is directly related to the reduction function, F, which has two components, a valence term (essentially due to the effect of the sixth d-electron being localized in the lowest-energy orbital) and a "lattice" term (due to deviations from cubic symmetry of the crystal field). These two terms have the functional dependence shown, but in octahedral geometry have the opposite sign, and hence combine to give the relation for F shown as the solid line. Hence QS rises rapidly with small distortions from cubic symmetry, and then falls more gradually after reaching a maximum value; from Dowty and Lindsley (1973), after Ingalls (1964).

angular deviations from ideal symmetry, and the relative effects of these apparently independent factors are not known. It is also probable that the next-nearest-neighbour configuration affects the QS (see later section on partial quadrupole splittings). Lastly, it is only in end-member minerals that the local co-ordination environment of the Mössbauer atom is known; structural information on interatomic distances in intermediate solid solutions is only long-range, and not directly pertinent to questions of local environment around any specific type of atom involving the disorder. In this regard, many of the correlations between δ, QS and mean bond length or local distortion obviously do not represent what their authors intend. Despite this caveat, there are reasonable correlations between $^{57}Fe^{2+}$ QS and distortion in silicate minerals, indicating that the long-range distortion parameters (although not representations of any real configuration in the crystal) are a gross measure of the changing local environment in these minerals.

For the case of spherical symmetry (e.g., high-spin Fe^{3+}), there is no intrinsic non-cubic symmetry in $V_{zz}(ve)$, and thus the QS will tend to increase with gradually increasing local distortion. However, because all of the V_{zz} terms are now of similar magnitude, the QS values will be much less than can be the case for high-spin Fe^{2+}. In addition, as the terms $V_{zz}(mo)$ and $V_{zz}(str)$ are of opposite sign from $V_{zz}(ve)$, there is no guarantee of a regular relationship between QS and any structural distortion index. QS ranges for various co-ordination numbers do overlap for both Fe^{2+} and Fe^{3+}, and thus QS cannot be used indiscriminately as a fingerprint for oxidation state and/or co-ordination number.

Lone pair electrons

For species such as Sn^{2+} and Sb^{3+}, the valence shell contains a lone pair of electrons. This lone pair can have high 5s character, whereupon the electron density is disposed very uniformly around the nucleus. In this type of configuration, the lone pair is designated as stereochemically inactive, and does not contribute significantly to V_{zz}, and hence to the quadrupole splitting. Alternatively, this lone pair may preferentially occupy a p orbital, such that $5p_z >> 5p_x \approx 5p_y$. In this configuration, the lone pair is referred to as stereochemically active; it contributes significantly to V_{zz} and hence to the quadrupole splitting values. QS values can thus be used as an indicator of the stereochemical nature of lone pair species.

Coordination and partial quadrupole splittings

If the $V_{zz}(str)$ term of the principal component of the EFG is ignored, then V_{zz} may be written as follows (cf., equation 57):

$$V_{zz} = -k \sum_{\substack{valence \\ electrons}} \langle 3\cos^2\theta - 1\rangle\langle r^{-3}\rangle \quad . \tag{58}$$

Let us consider the ligands as point charges, and assign to each different ligand a partial quadrupole splitting value, pQS, such that these values are additive over the local coordination. Consequently, we can write

$$QS = \sum_{\text{ligands}} pQS_i(3\cos^2\theta_i - 1) \quad , \tag{59}$$

where $pQS_i = Kr^{-3}$. The angular terms take account of the distortional contribution to the QS. Thus for ideal octahedral geometry with identical ligands and the Z axis along an L-M-L direction (M = metal; L = ligand),

$$QS = pQS\left[(3\cos^2 0° - 1) + 4(3\cos^2 90° -1) + (3\cos^2 180° - 1)\right] \quad . \tag{60}$$

Note that this expression makes no allowance for variations in the M-L bond length and is not appropriate for extremely (distance) distorted co-ordinations. In addition, as the orbital hybridisation is different for different coordination numbers, then different pQS values must be used for different coordination numbers. This technique has been used extensively for Sn^{4+}, Au^{1+} and low-spin Fe^{2+} with a wide variety of ligands (halides to complex organic species) and co-ordination numbers (Bancroft, 1973; Parish, 1985). However, in cases when there is a strong V_{zz}(ve) component (high-spin $^{VI}Fe^{2+}$, Sn^{2+}, Sb^{3+}), the QS is affected by local distortion in a non-linear fashion (Fig. 27). This additive model of partial quadrupole splittings can be used provided the range of local distortions lies to one side or the other of the maximum QS value in Figure 27. If this is the case, then the relationship between QS and distortions is approximately linear, with a positive correlation at small distortions and a negative correlation at larger distortions. In cases where the distortion for various ligands spans both regions of Figure 27, the additive model is obviously no longer valid. In addition, the wide range of ligands considered in chemical applications are not encountered in minerals, limiting the applicability of the ideas as formulated above. However, this idea has proved extremely useful as a means of looking at next-nearest-neighbor (NNN) effects in minerals.

In most solid solutions, the anions remain the same whereas the cations change with varying composition. Consequently, in equation (59), the pQS values will remain the same and the θ values will not be known (note that they cannot be derived from crystal structure data, as this is long-range information, whereas equation (59) needs short-range information). V_{zz} will be affected through V_{zz}(str) whereby the NNN configuration changes; of course there will also be an inductive effect that affects V_{zz}(mo) as well. Thus we expect different NNN configurations to cause different EFGs at the Mössbauer nucleus, and hence different QS values. This has not been explored extensively in minerals, but NNN effects have been proposed in pyroxenes (Dowty and Lindsley, 1973; Aldridge et al., 1978; Seifert, 1983; Dollase and Gustafson, 1982) and examined in some detail in spinels (Bancroft et al., 1983).

The idea behind the quantitative application of partial quadrupole splittings to NNN configurations is that a change in a NNN(s) around one ligand will produce a QS change; this constitutes an additive pQS whereby the same NNN change around two ligands would produce twice the QS. Thus the QS may be expressed in the following manner, in which the sum is over the NNN configurations:

$$QS = \sum_{\substack{NNN \\ \text{configuration}}} pQS_{NNN} \quad . \tag{61}$$

As emphasized earlier, this model requires linear additivity of configurations, and must lie on either branch of the QS versus distortion relationship, and not span the two branches.

DETERMINATION OF OXIDATION STATES

Mössbauer spectroscopy is a powerful method for the determination of oxidation states, and has been used extensively for this purpose in crystal-chemical studies of Fe-bearing rock-forming minerals. More recently, applications of other Mössbauer nuclei (^{119}Sn, ^{121}Sb, ^{151}Eu, ^{197}Au) to such problems have begun appearing. Both isomer shift and quadrupole splitting are sensitive to oxidation state, However, isomer shift values are intrinsically more informative, and we will look at a series of examples in which the oxidation state has been determined by isomer shift values.

As we have seen in equation (28), isomer shift is sensitive to the s-electron density at the nucleus. It is affected directly by the s-electron configuration, and indirectly by p- and d-electrons through shielding effects. Although most of the work done on oxidation states in minerals has concerned ^{57}Fe, there are other isotopes for which such work is conceptually more straightforward. Useful mineralogical examples are ^{119}Sn and ^{121}Sb, elements in which the different oxidation states directly involve the s-electron configuration.

^{119}Sn

Sn may be divalent ($5s^2 5p^o$) or quadrivalent ($5s^o 5p^o$) in minerals, the difference in valence state directly involving the $5s^2$ electrons. As we saw in the section on the interpretation of isomer shift, $\delta R/R$ is positive (see equation 54), and as $Sn^{2+} \rightarrow Sn^{4+}$ involves removing s-electrons and decreasing the contact density, the isomer shift should decrease with increase in oxidation state. Amthauer et al. (1979), Baker and Stevens (1977) and Amthauer (1986) have characterized Sn oxidation states in both silicates and sulphide/sulphosalt minerals.

^{121}Sb

Sb may be trivalent ($5s^2 5p^o$) or pentavalent ($5s^o 5p^o$) in minerals, and as with Sn, the difference in valence state directly involves the $5s^2$ electrons. However, for Sb, $\delta R/R$ is negative, and hence the removal of s-electrons results in an increase in isomer shift. The differences are large, and the valence states are easily differentiated (Baker and Stevens, 1977; Amthauer, 1986).

^{57}Fe

The situation for valence state determination of Fe is a little more involved, as the s-electron configurations for Fe^{2+} ($4s^0 3d^6$) and Fe^{3+} ($4s^0 3d^5$) are the same. However, the differing number of 3d-electrons causes differential shielding of the s-electrons at the nucleus, and indirectly affects the contact density and thus the isomer shift. Consequently, the difference in isomer shift between the two valence states is generally much less than is the case for valence state differences resulting from direct s-shell effects, as in the case of ^{119}Sn or ^{121}Sb. Fortunately the natural line-width for ^{57}Fe is somewhat smaller than is the case for these other isotopes, and so such smaller differences can still be resolved fairly easily. Consequently, Mössbauer spectroscopy is a very useful probe for oxidation state in Fe-bearing minerals (see next section) and glasses (Mysen and Virgo, 1985; Dyer et al., 1987).

^{151}Eu

Eu is usually trivalent, but the geochemical Eu anomaly common in many rocks has been explained by the substitution of Eu^{2+} for Ca in plagioclase feldspars. However, there has been no previous determination of the valence state of Eu in plagioclase. Aslani-Samin et al. (1987) have looked at this problem using ^{151}Eu Mössbauer spectroscopy. The spectrum of a Eu_2O_3-containing anorthite glass shows a singlet assigned to Eu^{3+}. Synthesis of crystalline feldspar gave spectra showing a sharp Eu^{3+} line (characteristic of residual glass) and a Eu^{2+} line (due to Eu^{2+} in the feldspar); the authors used the Eu^{3+}/Eu^{2+} ratio as a measure of the residual glass.

DETERMINATION OF SITE-OCCUPANCIES

By far the most common application of Mössbauer spectroscopy to mineralogy involves the determination of ^{57}Fe site-occupancies. All of the Fe-bearing rock-forming minerals have been extensively investigated in this regard, together with the more common of the minor and trace minerals; usually this work has also involved the simultaneous determination of the Fe^{3+}/Fe^{2+} ratio. Bancroft (1969) examined the relevant equations for a mineral containing Fe at two sites. Here, this is generalized to a mineral containing Fe at n sites, assuming all absorptions are quadrupole-split doublets.

Let A_i and A_j be the (average) areas of the ith and j^{th} quadrupole-split doublets respectively, If there are n distinct sites in the mineral, we may write $\Sigma(n-1)$ equations of the form

$$\frac{A_i}{A_j} = \frac{\Gamma_i \ G(\chi_i) \ f_i \ N_i \ r_i}{\Gamma_j \ G(\chi_j) \ f_j \ N_j \ r_j} \quad , \tag{62}$$

where Γ_i is the half-width of the ith peak,
$\quad G(\chi_i)$ is the saturation correction for the ith peak,
$\quad f_i$ is the recoil-free fraction at the ith site,
$\quad N_i$ is the Fe-occupancy of the ith site,
$\quad r_i$ is the equipoint rank of the ith site.

In equation (62), we measure the individual peak areas, A_i, by Mössbauer spectroscopy and the total amount of Fe present, $\Sigma N_i r_i$, by chemical analysis. Our goal is to derive N_i, and this can only be done by making some simplifying assumptions (Bancroft, 1969):

$$[1] \quad \Gamma_i \approx \Gamma_j \; .$$
$$[2] \quad f_i \approx f_j \; .$$
$$[3] \quad G(\chi_i) \approx G(\chi_j) \; .$$

If this is the case, then equation (62) simplifies as follows

$$\frac{A_i}{A_j} = \frac{N_i \; r_i}{N_j \; r_j} \qquad (63)$$

Writing $\Sigma N_i r_i = T$, the amount of Fe in the unit cell,

$$\frac{N_j}{T} = \frac{N_j}{\sum_i N_i \; r_i} = \frac{1}{\sum_i \left(\frac{N_i \; r_i}{N_j \; r_j} \right)} \qquad (64)$$

Substituting from equation (63) and rearranging gives

$$N_j = \frac{T}{\sum_i \left(\frac{A_i}{A_j} \right)} \qquad (65)$$

Thus we may derive Fe site-occupancies from our peak area ratios.

However, the results are only as good as the assumptions (1)-(3) used to derive equation (63):

(1) $\Gamma_i \approx \Gamma_j$. In end members, this usually seems to be a reasonable assumption. However, in solid solutions, variations in short-range order can cause peak broadening to different extents at various sites, and hence this approximation can potentially break down in this (perhaps not uncommon) situation.

(2) $f_i \approx f_j$. This assumption has produced some contention when the sites involved have different co-ordination numbers (e.g. M(4) and M(1,2,3) in monoclinic amphiboles; the dodecahedral, octahedral and tetrahedral co-ordinate sites in garnet). This may be resolved to some extent by examination of synthetic Fe end-members, but normally such materials are not available and often resist synthesis. However, where available, known f_i/f_j ratios may be substituted into a suitably modified equation (63). Details of specific examples of different f_i and f_j values are given in the sections on garnets and orthopyroxenes.

(3) $G(X_i) \approx G(X_j)$. These saturation corrections approach unity for thin absorbers (Bancroft, 1969), but significant differences could exist if the peaks were to differ markedly in intensity.

Of course, equation (62) is not quite as restrictive as these conditions indicate, as the simplification requires only that

$$\Gamma_i \; G(\chi_i) \; f_i \approx \Gamma_j \; G(\chi_j) \; f_j \quad , \tag{66}$$

and hence compensating deviations from conditions (1)-(3) do not necessarily invalidate the approximation.

SITE-OCCUPANCIES IN MINERALS

In this section, I will briefly survey some of the work done on rock-forming silicates. Obviously, this cannot be a detailed survey, and here is not the place to dissect the niceties of conflicting spectral assignments. Consequently, I will emphasise the success of the method in obtaining useful results, and only occasionally mention problems of interpretation when they strongly affect the application of such results to current geological problems.

Garnets

Most of the common garnets are quite straightforward from a spectroscopic point of view. However, the Ti-rich garnets have much more complex spectra that are not amenable to a simple interpretation. Consequently, we will consider just the non-Ti-rich garnets.

Selected spectra are shown in Figure 28; Figures 28a-c show Fe^{2+} at the {X} (dodecahedral) site, and Fe^{3+} at the [Y] (octahedral) and (Z) (tetrahedral) sites respectively. Table 6 shows the ranges of isomer shift and quadrupole splitting for Fe at these sites; Amthauer et al. (1976) examined a range of garnets and suggested that part of the range exibited in Table 6 is due to experimental error, as their results on a wide compositional range of garnets showed much more restricted variations. This suggests that these peaks should be resolvable in more complex garnets, and Figure 28d shows this to be the case. This melanite (TiO_2 = 2.30 wt %) has all three doublets present, and as can be seen (Fig. 33d below), they are easily resolved.

In the derivation of site-occupancies, a matter of concern in the general procedure described above relates to the usual assumption that the recoil-free fractions in the sample are the same at each non-equivalent site. The recoil-free fraction is related to the mean squared vibrational displacement of the Mössbauer nucleus (equation 12). It is well-known from crystal structure refinements of ordered compounds that the mean-squared vibrational displacement of atoms is related to their co-ordination number and mean bond-valence. Consequently it would not be surprising to encounter significant differences in recoil-free fraction at the {X}, [Y] and (Z) sites in garnet. We will consider this in detail here as the garnets are a good example of this problem, and how to deal with it.

In spectra of a grossular recorded at 295 and 4.53 K, the area of the doublet due to Fe^{2+} at the {X} site is much larger at low temperature, indicating a strong differential temperature dependence of the recoil-free fractions at these two sites in the garnet structure. Obviously this must be taken into account in the derivation of site-occupances, as f_x cannot be equal to f_z except perhaps at one particular temperature. We will look at this example in detail, as it has been well-characterized by Amthauer et al. (1976) and this point

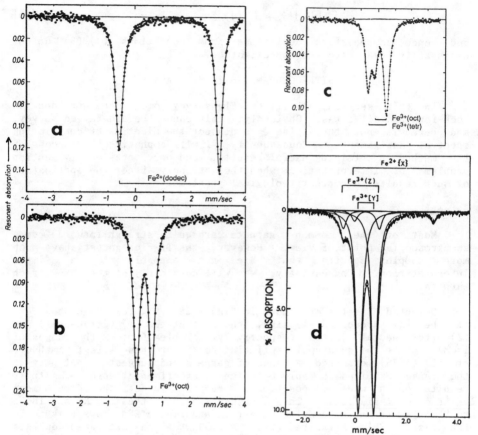

Figure 28. Mössbauer spectra of ^{57}Fe in garnets: (a) Fe^{2+} at {X} in pyrope; (b) Fe^{3+} at [Y] in andradite; (c) Fe^{3+} at [Y] and (Z) in a synthetic Ti-andradite; (d) Fe^{2+} at {X}, Fe^{3+} at [Y] and (Z) in a melanite. Temperatures: (a), (b), 77° K; (c) 15° K; (d) 295° K. Spectra are modified from: (a), (b), (c) Amthauer et al. (1976); (d) Schwartz et al. (1980).

is particularly important with regard to the derivation of accurate site-occupancies.

Table 7 shows the area ratios of Fe at the different sites in garnet, as a function of absorber temperature. The area ratio, r_{xy}, of the doublet ratios $A_x^{t(1)}/A_y^{t(1)}$ is given by

$$r_{xy} = \left[\frac{A_x^{t(1)}}{A_y^{t(1)}} \right] \bigg/ \left[\frac{A_x^{t(2)}}{A_y^{t(2)}} \right] , \qquad (67)$$

in which t(i) denotes temperature, and the subscripts X, Y and Z denote the various sites. The similar values for r_{xy} at 77 and 295 K in each garnet suggest that the r_{xy} values are relatively insensitive to composition. However, the strong variation in r_{xy} with temperature shows that the relative magnitudes of the corresponding recoil-free

Table 6. Mössbauer parameters for ^{57}Fe in garnets; values for garnet show
the ranges from a literature survey; values for named minerals
correspond to Figure 33.

Minerals	Species	IS(mm/s)	QS(mm/s)	Γ(mm/s)	T($^\circ$K)	Ref.
Garnets	[X]Fe^{2+}	1.20-1.39	3.47-3.70	-	298	(1)
"	"	1.33-1.44	3.55-3.73	-	77	(1)
"	(Z)Fe^{2+}	0.68-0.79	1.53-1.99	-	298	(1)
"	"	0.63	1.62	-	77	(1)
"	[Y]Fe^{3+}	0.35-0.45	0.29-0.75	-	298	(1)
"	"	0.42-0.52	0.26-0.64	-	77	(1)
"	(Z)Fe^{3+}	0.04-0.20	1.05-1.28	-	298	(1)
"	"	0.14-0.29	1.05-1.15	-	77	(1)
Almandine	[X]Fe^{2+}	1.43	3.66	0.30	77	(1)
Andradite	[Y]Fe^{3+}	0.50	0.55	-	77	(1)
Ti-andradite	[Y]Fe^{3+}	0.50	0.75	0.37	77	(1)
"	(Z)Fe^{3+}	0.30	1.15	0.32	77	"
Melanite	[X]Fe^{2+}	1.47	3.09	0.28	298	(2)
"	[Y]Fe^{3+}	0.42	0.59	0.29	298	"
"	(Z)Fe^{3+}	0.19	1.41	0.27	298	"

* values quoted relative to Fe-foil.

References: (1) Amthauer et al. (1976); (2) Schwartz et al. (1980)

Table 7. Area ratios in garnets as a function of temperature for
^{57}Fe at the [X], [Y] and (Z) sites.

Garnet	T($^\circ$K)	Area ratio			
		[X]Fe^{2+}	[Y]Fe^{3+}	(Z)Fe^{3+}	*r
Pyrope	4.5	0.871	0.129	-	-
"	77	0.862	0.138	-	0.93(5)
"	295	0.840	0.161	-	0.77(5)
Grossular	4.5	0.212	0.788	-	-
"	77	0.196	0.805	-	0.91(5)
"	295	0.168	0.832	-	0.75(5)
Ti-andradite	25	-	0.431	0.569	-
"	77	-	0.429	0.571	1.01(4)
"	295	-	0.431	0.569	1.00(4)

After Amthauer et al. (1976); * see text for definition of r.

fractions f_x and f_y are strongly temperature dependent, and that the approximation $f \approx f_y$ will lead to errors of at least 20% in site-occupancies derived from room-temperature spectra. Use of area ratios derived from liquid N_2 spectra (77 K) will reduce these errors to around 5%.

On the other hand, the results for Ti andradite (Table 7) indicate that f_y and f_z show the same temperature dependence, and Amthauer et al. (1976) suggest that errors will be of the order of 3-5% if site-occupancies are determined from room-temperature spectra without the appropriate corrections. Thus variation in the recoil-free fraction of Fe at the cation sites in silicate garnets poses a considerable difficulty to rapid characterization of site-occupancies, even though the spectra themselves are fairly well-resolved.

Olivines

There has been less site-occupancy work done on the ferromagnesian olivines than on most of the other rock-forming ferromagnesian silicate minerals, primarily because they show little or no Mg/Fe ordering. Very small amounts of ordering have been detected, and there seems to be some correlation with cooling history, but the marginal nature of the effect prevents it from being a useful petrologic indicator (i.e., usually a better indicator is available). On the other hand, the olivine-group minerals with additional cations (Mn^{2+}, Ca, Ni, etc.) have more recently been examined in some detail, as they show very interesting ordering behaviour (Annersten et al., 1982, 1984: Nord et al., 1982).

The crystal structure of olivine has two octahedrally co-ordinated sites [M(1) and M(2)] containing (almost) equal amounts of Fe^{2+}. Thus the spectrum should consist of two quadrupole-split Fe^{2+} doublets. In both forsterite and fayalite spectra, these doublets are not visually apparent (Stanek et al., 1986; Schaefer, 1985; Shinno, 1981).

The resolution can be increased by raising the absorber temperature. There is a strong decrease in the quadrupole splitting with increasing temperature, but the effect differs at the two sites, with a concomitant splitting of the upper velocity peak (Eibschütz and Ganiel, 1967; Stanek at al., 1986). Thus the ease and precision of site-occupancy determinations can be significantly improved by recording the spectra at high temperatures, provided allowance is made for possible differences in the recoil-free fraction at each site. Shinno (1974) has determined the $f_{M(1)}/f_{M(2)}$ ratio in fayalite (at room temperature) to be 1.058(28); Annersten et al. (1982) give $f_{M(1)}/f_{M(2)}$ for fayalite (at 673 K) as 0.946(-).

Ferromagnesian olivine can also incorporate significant amounts of Fe^{3+} as fine-scale intergrowths of laihunite, $Fe^{2+}Fe_2^{3+}(SiO_4)_2$ (Schaefer, 1985).

Pyroxenes

In terms of site-occupancies, the pyroxenes have probably been the most extensively and successfully examined group of minerals. In most

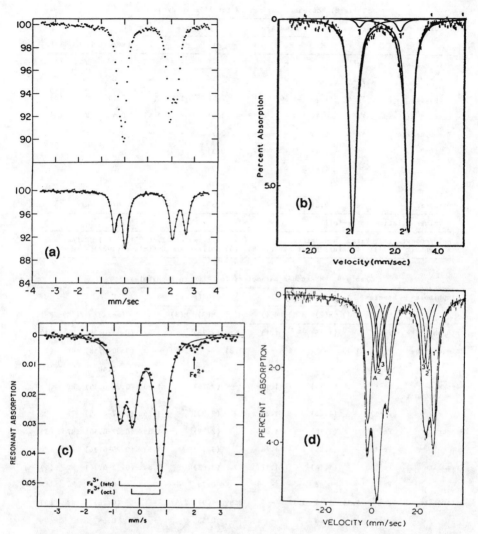

Figure 29. Mössbauer spectrum of ^{57}Fe in pyroxenes: (a) orthopyroxene (upper spectrum at 293, lower spectrum at 77° K); (b) diopside; note the weak Fe^{2+} doublet assigned to Fe^{2+} at M(2); (c) ferrian diopside, with significant tetrahedrally co-ordinated Fe^{3+}; (d) titanium ferro-omphacite; (b), (c), (d) at room temperature; spectra after Virgo and Hafner (1971), Bancroft et al. (1971), Hafner and Huckenholz (1971) and Aldridge et al. al. (1978).

pyroxenes, there are generally two octahedrally co-ordinated sites, and the principal chemical variables at these sites are the occupancies of Fe^{2+} and Mg. Furthermore, the degree of order often shows significant variation with conditions of equilibration, and a detailed knowledge of site occupancy behavior is necessary for an adequate thermodynamic description of these minerals. Typical spectra (Fig. 29) and Mössbauer parameters (Table 8) are given.

　　　　Orthopyroxenes. Typical orthopyroxene spectra are characterized by two Fe^{2+} doublets; the outer doublet can be assigned to Fe^{2+} at

Table 8. Mössbauer parameters for ^{57}Fe in pyroxenes.

Pyroxene	Species	*IS(mm/s)			QS(mm/s)			Γ		
		M1	M2	T	M1	M2	T	(mm/s)	T(°K)	Ref.
Orthopyroxene	Fe^{2+}	-	-	-	-	-	-	-	RT	(1)
"	Fe^{2+}	-	-	-	-	-	-	-	78	"
Diopside	Fe^{2+}	1.50	≈1.50	-	2.32	2.11	-	0.35	RT	(2)
Ferrian diopside	Fe^{3+}	0.42	-	0.18	1.01	-	1.49	0.43	RT	(3)
Ti-omphacite	Fe^{2+}	1.45	-	-	2.96	-	-	0.31	RT	(4)
"	"	1.42	-	-	2.31	-	-	"	"	"
"	"	1.41	-	-	1.94	-	-	"	"	"
"	Fe^{3+}	0.86	-	-	0.30	-	-	"	"	"
Di80Ac20	Fe^{3+}	0.21	-	-	0.59	-	-	0.49	RT	(5)

* values relative to Fe-foil

References: (1) Virgo and Hafner (1970); (2) Williams et al. (1971); (3) Hafner and Huckenholz (1971); (4) Aldridge et al. (1978); (5) Dollase and Gustafson (1982)

Table 9. Mössbauer parameters for ^{57}Fe in amphiboles.

Amphibole	Species	*IS(mm/s)				QS(mm/s)				Γ		
		M(1)	M(2)	M(3)	M(4)	M(1)	M(2)	M(3)	M(4)	(mm/s)	T(°K)	Ref.
Anthophyllite	Fe^{2+}	(1.15)		1.13		(2.78)		1.82		(0.25-0.39)	RT	(1)
Holmquistite	Fe^{2+}	1.13	-	1.11	-	2.81	-	2.03	-	(0.30-0.38)	RT	(2)
"	Fe^{3+}	-	0.49	-	-	-	0.33	-	-	0.40	"	"
Magnesio-cummingtonite	Fe^{2+}	(1.15)		1.12		(2.74)		1.80		(0.26-0.33)	RT	(3)
"	Fe^{2+}	(1.26)		1.24		(2.99)		1.85		(0.27-0.38)	77	"
Grunerite	Fe^{2+}	(1.16)		1.07		(2.79)		1.55		(0.27-0.30)	RT	(3)
"	Fe^{2+}	(1.28)		1.18		(3.10)		1.54		(0.29-0.36)	77	"
Tremolite	Fe^{2+}	(1.14)		1.17		(2.82)		1.84		(0.32-0.34)	RT	(4)
"	Fe^{3+}	-	0.48	-	-	-	0.74	-	-	0.53	"	"
Actinolite	Fe^{2+}	1.14	1.14	1.12	-	2.89	1.91	2.57	-	(0.26)	RT	(5)
"	Fe^{3+}	-	0.28	-	-	-	0.53	-	-	0.68	"	"
Pargasite	Fe^{2+}	1.15	1.14	1.15	1.10	2.59	2.26	2.59	1.88	(0.31)	RT	(6)
"	Fe^{3+}	-	0.56	-	-	-	0.61	-	-	0.49	"	"
Magnesio-hornblende	Fe^{2+}	1.39	1.26	1.38	-	2.76	2.02	2.35	-	(0.32)	295	(7)
"	Fe^{3+}	-	0.75	-	-	-	0.53	-	-	0.38	"	"
Crossite	Fe^{2+}	1.13	-	1.11	-	2.85	-	2.36	-	(0.31-0.41)	RT	(8)
"	Fe^{3+}	-	0.38	-	-	-	0.47	-	-	0.34	"	"
Riebeckite	Fe^{2+}	1.13	-	1.10	-	2.83	-	2.32	-	(0.29)	RT	(8)
"	Fe^{3+}	-	0.37	-	-	-	0.43	-	-	(0.29)	"	"

* values quoted relative to Fe-foil

References: (1) Seifert (1977); (2) Law (1973); (3) Hafner and Ghose (1971); (4) Goldman and Rossman (1977); (5) Burns and Greaves (1971); (6) Goldman (1979); (7) Bancroft and Brown (1975); (8) Bancroft and Burns (1969)

M(1) and the inner doublet to Fe^{2+} at M(2). A notable feature of these spectra is the strong differential temperature dependence of the QS of the two doublets; this can be exploited to increase the resolution of the doublets by recording the spectra at liquid N temperatures.

With increasing substitution of Al and Fe^{3+}, the M(2) doublet shows increasing half-width, together with increasingly poor fitting statistics. For this and other reasons, Seifert(1983) fit such spectra to three Fe^{2+} doublets, corresponding to Fe^{2+} at M(1) and two doublets to Fe^{2+} at M(2) with different next-nearest-neighbour environments. These spectra are discussed further in the section on next-nearest-neighbour effects.

Clinopyroxenes. Pyroxenes close to (and on) the diopside-hedenbergite join can consist of either one (Dollase and Gustafson, 1982) or two (Bancroft et al., 1971) quadrupole split Fe^{2+} doublets. In the second case (Fig. 29b), the outer intense doublet represents Fe^{2+} at M(1) and the inner weak doublet is due to Fe^{2+} at M(2). In this regard, recent X-ray structure refinements (Rossi et al., 1987) have also shown the presence of Fe^{2+} at the M(2) site in some calcic and subcalcic clinopyroxenes.

In a series extending from hedenbergite ($CaFe^{2+}Si_2O_6$) to ferrosilite ($Fe_2^{2+}Si_2O_6$), Dowty and Lindsley (1973) showed that fitting two doublets (assigned to Fe^{2+} at M(1) and M(2) respectively for the larger and smaller QS values) led to M(2) site-occupancies that were physically impossible; this area ratio anomaly had been noted by many previous workers (Williams et al., 1971), and has been ascribed to next-nearest-neighbour effects.

Figure 29c shows the spectrum of a synthetic ferrian diopside $Ca(Mg_{0.74}Fe^{2+}_{0.26})(Si_{1.74}Fe^{3+}_{0.26})O_6$. There are two prominent doublets, the isomer shift values of which clearly identify them as due to Fe^{3+} in both octahedral and tetrahedral coordinations; a minor absorption at about 2 mm/s can be assigned to the upper velocity component of a weak Fe^{2+} doublet. With this spectrum as a model, tetrahedrally co-ordinated Fe^{3+} has been identified in many clinopyroxenes.

Perhaps the most complex pyroxene spectra are for the P2/n omphacites; Figure 29dc shows an omphacite spectrum for material the structure of which has also been refined (Curtis et al., 1975). The spectrum is fit to one octahedrally co-ordinated Fe^{3+} doublet (assigned to the M1 + M1(1) sites) and three octahedrally coordinated Fe^{2+} doublets. The Fe^{2+} doublets were all assigned to Fe^{2+} at the M1 + M1(1) sites with different next-nearest-neighbour arrangements of Ca and Na at the M2 and M2(1) sites. Amthauer (1982) has reported thermally-activated electron hopping in aegirine.

General considerations. Clearly the chemically complex pyroxenes give rise to more doublets than there are formal (Fe) sites in the structures. The two suggested explanations both involve short-range order, but the details of each model are different, and they have not yet been tested experimentally.

Applications. Site occupancies derived by Mössbauer spectroscopy have been used extensively in the characterization of orthopyroxenes. Virgo and Hafner (1969, 1970) showed that site-occupancies in orthopyroxenes were sensitive to equilibration temperature, and that the site-occupancies in natural orthopyroxenes reflect the rate of cooling. This site-occupancy data has been used extensively in the thermodynamic analysis of pyroxenes (Saxena and Ghose, 1971; Saxena, 1973; Seifert, 1987), and in cation ordering rate studies (Besancon, 1981).

Amphiboles

The first use of Mössbauer spectroscopy in amphibole studies was that of Gibb and Greenwood (1965) on oxidation of fibrous amphiboles. Following the initial work of Bancroft et al. (1966,1967a,b), there has been a significant amount of effort in deriving Fe^{2+} and Fe^{3+} site-occupancies in all varieties of amphiboles. The general appearance of amphibole Mössbauer spectra varies considerably with amphibole type (Fig. 30; Table 9), and the peak assignment in these spectra is frequently not straightforward.

Fe-Mg-Mn amphiboles. The anthophyllite spectrum consists of two doublets: the outer doublet is attributable to Fe^{2+} at the M1, M2 and M3 sites, with the inner doublet due to Fe^{2+} at the M4 site. This contrasts with the gedrite spectrum, in which the resolution decreases (Seifert, 1977) and only a single wide doublet occurs.

The Mössbauer spectrum of grunerite is shown in Figure 30a. It is similar to the anthophyllite spectra, with the outer and inner doublets assigned to Fe^{2+} at the $M(1,2,3)$ ($\equiv M(1) + M(2) + M(3)$) and $M(4)$ sites, respectively. Perhaps the most notable feature of the Fe-Mg-Mn amphibole spectra is the complete overlap of doublets due to Fe^{2+} at the $M(1)$, $M(2)$ and $M(3)$ sites. Although complete site-occupancies cannot be derived, the relative order of Fe^{2+} at $M(4)$ vs. $M(1,2,3)$ can be characterized.

Calcic amphiboles. The interpretation of the Mössbauer spectrum of calcic amphiboles is not straightforward, and peak assignments differ from study to study; a detailed review of the assignments is given by Hawthorne (1981,1983b). Fe^{2+} may occupy the $M(1)$, $M(2)$, $M(3)$ and $M(4)$ sites; Fe^{3+} may be disordered over the $M(1)$, $M(2)$ and $M(3)$ sites, but the differences in environment are not sufficient to resolve different Fe^{3+} doublets, and a single Fe^{3+} doublet is seen. Thus there is the possibility for four Fe^{2+} doublets and one Fe^{3+} doublet. However, none of the observed spectra show four such doublets.

Goldman and Rossman (1977) resolved the spectrum of tremolite (Fig. 30b) into two doublets due to Fe^{2+} at $M(4)$ - inner doublet - and $M(1,2,3)$ - outer doublet - respectively, and presented supporting electronic absorption spectral data; thus the doublets from Fe^{2+} at the $M(1)$, $M(2)$ and $M(3)$ sites (the assumption is made that these are all occupied by Fe^{2+}) overlap, as in the Fe-Mg-Mn amphiboles. Burns and Greaves (1971) resolved the spectrum of an actinolite into one Fe^{3+} and three Fe^{2+} doublets, the three Fe^{2+} doublets assigned as Fe^{2+} at $M(1)$, $M(2)$ and $M(3)$ respectively. Bancroft and Brown (1975) similarly resolved and assigned the spectrum of a magnesio-hornblende (Fig. 30c). Goldman (1979) resolved the spectrum of a pargasite in a similar

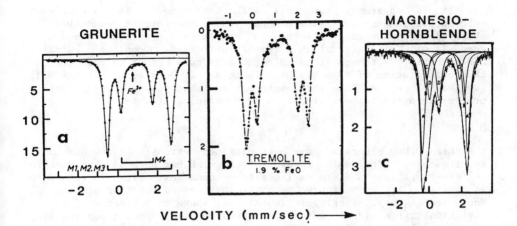

Figure 30. Mössbauer spectra of ^{57}Fe in amphiboles: (a) grunerite (at 78° K), the inner doublet is due to Fe^{2+} at M(4), the outer doublet is due to Fe^{2+} at M(1,2,3); (b) tremolite, the inner doublet is assigned to Fe^{2+} at M(4) and the outer doublet is assigned to Fe^{2+} at M(1,2,3); (c) magnesio-hornblende, AA' = Fe^{2+} at M(1), BB' = Fe^{2+} at M(2), CC' = Fe^{2+} at M(3), DD' = Fe^{3+}; (b) and (c) are at room temperature, after Hafner and Ghose (1971), Goldman and Rossman (1977), and Bancroft and Brown 1975).

fashion to Bancroft and Brown (1975), but assigned the doublets to Fe^{3+}, and Fe^{2+} at M(1,3), M(2) and M(4) respectively. As is apparent, the assignments of the doublets in amphiboles is somewhat uncertain: in addition, if pyroxenes show NNN effects, one would expect the analogous amphiboles to do so as well. It is possible that some amphibole compositions give spectra that are too complicated to resolve solely from the Mössbauer data alone. Additional complications are suggested by the hastingsite spectra of Thomas (1982).

Alkali amphiboles. Mössbauer studies of alkali amphiboles have concentrated on the glaucophane - ferro-glaucophane - magnesio-riebeckite - riebeckite series. In these amphiboles, the M(2) site is generally filled with trivalent cations, and the Mössbauer spectra consist of three quadrupole-split doublets due to Fe^{2+} at M(1) and M(3), and Fe^{3+} predominantly at M(2).

General considerations. As is apparent from the above discussion, the refinement and assignment of amphibole spectra are not always straightforward; however, some of the problems can be circumvented by use of a combination of experimental techniques.

Applications. Several authors have shown that the distribution of Fe^{2+} over the M(4) and $\Sigma(M(1) + M(2) + M(3))$ sites is strongly temperature dependent. Ghose and Weidner (1972) characterized site-occupancies as a function of annealing temperature in cummingtonite-grunerite amphiboles. Similar work on actinolites has been done by Skogby and Annersten (1985) and Skogby (1987). However, they interpreted their equilibrium distribution temperature in terms of kinetic effects, assuming that their equilibrium distribution reflected the cooling history of the sample.

It is apparent from this work that there is significant information in amphibole site-occupancies. However, it seems that the observed occupancies represent a complete interaction between bulk composition, crystallization temperature and cooling rate. Consequently, although it is going to take more effort to characterize the quantitative effects of these factors, it does mean that one will eventually get a lot more information (than just a crystallization or equilibration temperature) from such measurements.

Micas

Mica group minerals have been examined extensively by Mössbauer spectroscopy. The main thrust of the work has been to characterize Fe^{3+}/Fe^{2+} ratios and to derive site-occupancies, often in regard to mechanisms of oxidation/dehydroxylation as a function of heating. Much recent crystallographic work has shown the presence of subtle deviations from ideal (usually centrosymmetric) space group symmetry in micas (Bailey, 1984a,b; Guggenheim, 1984), usually connected with cation ordering of the convergent type. Such details are not (as yet) visible in Mössbauer spectra, and spectra are normally modelled in terms of the highest space-group symmetry consonant with the geometry of the structure. Here we use the site-nomenclature of Bailey (1984a), whereby M(1) is co-ordinated by four oxygens and two hydroxyls in the trans arrangement, and M(2) is co-ordinated by four oxygens and two hydroxyls in the cis arrangement. There has been some confusion in the literature with regard to this nomenclature (e.g., Goodman, 1976), and so one should be careful to check the nomenclature used in any specific paper; it should be noted that many Mössbauer papers use the wrong nomenclature. Typical spectra (Fig. 31) and Mössbauer parameters (Table 10) are given here; systematics of Mössbauer parameters are examined by Dyar (1987).

Dioctahedral micas. Typical spectra are dominated by a single quadrupole-split doublet of Fe^{3+} in octahedral co-ordination. Structure refinement studies show Fe^{3+} to be ordered at the M(2) site (Bailey, 1984b). However, even in well-ordered ferrian muscovites, the half-width is very large; in the absence of any obvious mechanism, this broadening must be ascribed to next-nearest-neighbour effects involving Al/Si and possibly Mg/[] disorder. Most dioctahedral micas also show significant Fe^{2+} (Fig. 31a). Although assignments have varied, the doublet with the larger QS is now assigned to Fe^{2+} at M(2) and the doublet with the smaller QS to Fe^{2+} at M(1); note that the multiplicity of M(2) is twice that of M(1). The half width of the Fe^{3+} doublet broadens significantly with increasing trioctahedral component (Table 10), with maximum values (≈ 1.2 mm/s) almost twice those in (almost) ideal dioctahedral micas (≈ 0.60 mm/s); again this suggests the presence of significant next-nearest-neighbour effects.

Trioctahedral micas. Fe plays a much more important role in the chemistry of the trioctahedral micas, and consequently there has been much more work in this area. Figure 31b shows the spectrum of a synthetic fluorannite ($KFe^{2+}_3AlSi_3O_{10}F_2$), fit to two doublets that were assigned to Fe^{2+} at M(2) -outer doublet- and M(1) -inner doublet-respectively; the area ratio closely approximates its ideal value of 2:1 in this end-member mica, and provides definite confirmation of this particular assignment. Natural biotites are usually more complex than these synthetic micas, and can contain both octahedrally

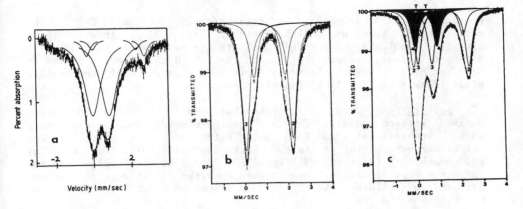

Figure 31. Mössbauer spectra of ^{57}Fe in micas: (a) muscovite containing significant Fe^{3+} and Fe^{2+}; (b) synthetic fluor-annite [(1,1) and (2,2) respectively]; (c) Cape Ann annite with Fe^{2+} doublets from M(1) and M(2) [(1,1) and (2,2) respectively], and Fe^{3+} doublets from M(1), M(2) and T sites [(4,4) shaded, (3,3) and (T,T) respectively]; spectra from Finch et al. (1982), and Dyar and Burns (1986).

Table 10. Mössbauer parameters for ^{57}Fe in micas.

Mica	Species	*IS(mm/s)			QS(mm/s)			Γ		
		M1	M2	T	M1	M2	T	(mm/s)	T(°K)	Ref.
Muscovite	Fe^{3+}	(0.361)			(0.724)			(0.63)	RT	(1)
Muscovite	Fe^{2+}	1.13	1.19		2.19	2.96	-	(0.48)	RT	(2)
"	Fe^{3+}	(0.38)			(0.81)			"	"	"
Muscovite	Fe^{2+}	1.16	1.20	-	2.09	2.96	-	0.44	RT	(2)
"	Fe^{3+}	(0.39)			(0.82)			0.96	"	"
(Ge,Si) Mica	Fe^{3+}	0.40	0.40	-	1.50	0.98	-	0.43	RT	(3)
Tetraferriphologopite	Fe^{3+}	-	-	0.17	-	-	0.50	0.45	RT	(4)
Biotite	Fe^{2+}	1.10	1.13	-	1.12	2.57	-	0.39	RT	(5)
"	Fe^{3+}	0.39	0.44	-	1.16	0.47	-	"	"	"
Ferriannite	Fe^{2+}	1.16	1.10	-	2.16	2.55	-	0.28	RT	(5)
"	Fe^{3+}	(0.42)		0.20	(1.01)		0.45	"	"	"
Annite	Fe^{2+}	1.15	1.12	-	1.90	2.50	-	0.40	RT	(5)
"	Fe^{3+}	0.42	0.37	0.20	0.98	0.61	0.37	"	"	"
Zinnwaldite	Fe^{2+}	1.15	1.14	-	2.93	2.66	-	0.31	RT	(6)
"	Fe^{3+}	0.42	0.38	-	0.88	0.54	-	0.34	"	"
Clintonite	Fe^{2+}	(1.06)		-	(2.34)		-	0.23	295	(7)
"	Fe^{3+}	-	-	0.28	-	-	0.66	0.33	"	"

* relative to Fe-foil

References: (1) Goodman (1976); (2) Finch et al. (1982); (3) Ferrow (1987a); (4) Annersten et al. (1971); (5) Dyar and Burns (1986); (6) Levillain et al. (1981); (7) Joswig et al. (1986)

coordinated Fe^{2+} and Fe^{3+}, and tetrahedrally coordinated Fe^{3+}. The most complex spectrum is that of annite (Fig. 31c), in which there are five doublets: three Fe^{3+} doublets are assigned to Fe^{3+} at M(1), M(2) and T sites, respectively, and the Fe^{2+} doublets are assigned to M(1) and M(2). Work on Li micas (Levillain et al., 1981) and brittle micas (Joswig et al., 1984) has also been done.

Applications. As yet, there has been little use of Mössbauer-derived site-occupancies for any geothermometry and geobarometry work. However, it has been used extensively for characterizing oxidation/dehydroxylation (e.g. Hogg and Meads,1975; Ferrow,1987b) and weathering (Rice and Williams,1969; Bowen et al.,1968) mechanisms in micas, particularly when combined with infrared spectroscopy (Vedder and Wilkins,1969). This is discussed further in the section on oxidation studies.

NEXT-NEAREST-NEIGHBOR EFFECTS

Short-range ordering is difficult to characterize by traditional diffraction techniques, and it is only relatively recently that we have been able to gather reliable data on this phenomenon. Mössbauer spectroscopy is a probe of local structure, and can provide information on such ordering. Short-range order can give rise to different NNN arrangements, and as discussed previously in the sections on the interpretation of isomer shift and quadrupole splitting, such local differences can give rise to significantly different absorptions in the Mössbauer spectrum. Such effects must be quite common in minerals, and yet have been recognized only in the less complex structures. What we will do here is examine a series of examples in which NNN effects are well documented; as we will see, with increasing structral complexity, such effects become less obvious in the Mössbauer spectrum, suggesting that they may well have been overlooked in more complex minerals.

Wüstite

Wüstite (Fe_xO, x > 0.95) has the periclase (halite) structure, with the additional complication that the Fe:O ratio is generally less than unity; the resulting vacancies are charge-compensated by the presence of Fe^{3+}. These defects tend to cluster and produce short-range order; this has been studied extensively by X-ray and electron diffraction, and by Mössbauer spectroscopy.

The spectrum of stoichiometric FeO should consist of an Fe^{2+} singlet. Any decrease in the amount of Fe should be accompanied by oxidation to maintain overall electroneutrality, and hence an Fe^{3+} absorption is to be expected, gradually increasing in intensity with decreasing overall Fe content. As shown in Figure 32, the spectra are far more complex than this, and seem to increase in complexity with decreasing Fe content. The spectra are dominated by Fe^{2+} quadrupole-split doublets in all cases, and it is only the sample with x = 0.981 that shows a significant Fe^{2+} singlet. The quadrupole-split Fe^{2+} doublets show the presence of local non-cubic environments around Fe^{2+} atoms, and a weak and extremely broad Fe^{3+} absorption shows that presence of a number of different Fe^{3+} environments. The resolved Fe^{2+} doublets are also broad, indicating a range of non-cubic environments for each Fe^{2+} doublet. Thus NNN effects dominate the

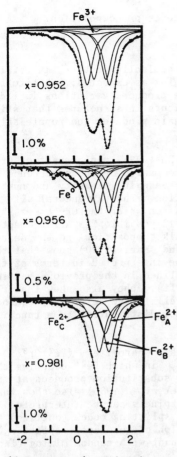

Fe³⁺

x=0.952

1.0%

Fe°

x=0.956

0.5%

Fe²⁺_C   Fe²⁺_A

Fe²⁺_B

x=0.981

1.0%

-2   -1   0   1   2

Velocity → (mm/sec)

Figure 32 (left). Mössbauer spectra of ^{57}Fe in wüstite, showing the fits for three different compositions; for x = 0.981, there are two Fe^{2+} doublets (Fe^{2+}_B and Fe^{2+}_C) and an Fe^{2+} singlet (Fe^{2+}_n); for x = 0.956, the Fe^{2+} singlet has disappeared, the Fe^{2+}_C doublet has increased in intensity relative to the Fe^{2+}_B doublet, and note the broad weak Fe^{3+} singlet and the doublet due to metallic Fe; for x = 0.952, the fit is similar, but the Fe^{2+}_C doublet is even stronger; after McCammon and Price (1985).

Figure 33 (below). Mössbauer spectra of ^{57}Fe in synthetic spinels: (a) $Fe(Cr_{1.90}Al_{0.10})O_4$; (b) $Fe(Cr_{1.75}Al_{0.25})O_4$; $Fe(Cr_{1.90}Al_{0.10})O_4$ is fit to one Fe^{2+} singlet, one strong Fe^{2+} doublet and one weak Fe^{2+} doublet; after Bancroft et al. al. (1983).

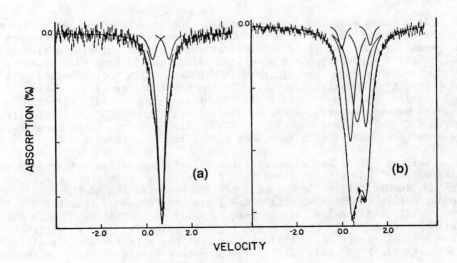

ABSORPTION (%)

0.0

(a)

-2.0   0.0   2.0

0.0

(b)

-2.0   0.0   2.0

VELOCITY

spectrum of Fe$_x$O, and show a strong dependence on composition.

Spinels and thiospinels

Again we are dealing here with nominally cubic materials. Ordered Fe^{2+} end-menbers show an Fe^{2+} singlet as expected; substitution of different cations into the structure produces variations in local NNN configurations, and non-cubic environments in structures that still have long-range cubic symmetry. This again produces quadrupole-split Fe^{2+} doublets.

In the thiospinel series Fe^{2+}(Cr^{3+}$_{2-x}$Rh^{3+}$_x$)S$_4$, replacement of Cr^{3+} by Rh^{3+} at the octahedral sites is accompanied by the occurrence of Fe^{2+} quadrupole-split doublets that increase in intensity and number with progressive increase in the substitution. The tetrahedral site in spinel has 12 NNN octahedral sites on which the Cr^{3+} <--> Rh^{3+} substitution takes place. Thus there are 12 possible NNN arrangements, the frequency of which depends on the bulk composition <u>and</u> the degree of local short-range order. Riedel and Karl (1980) show that the isomer shifts of the doublets can be interpreted in terms of the additive partial isomer shift model (outlined in the previous section on the interpretation of isomer shift); they also show that the intensities of the doublets correspond well to the relative frequency of NNN configurations calculated for random distribution as a function of bulk composition.

A similar situation exists for the chromite (Fe^{2+}Cr^{3+}$_2$O$_4$) -hercynite (Fe^{2+}Al$_2$O$_4$) series (Fig. 33), in which Fe^{2+} occupies the tetrahedral site and the Cr^{3+} <--> Al substitution proceeds at the neighbouring octahedral site. Bancroft et al. (1983) also show that the quadrupole splittings for the oxide spinels can be interpreted in terms of the additive partial quadrupole splitting model (outlined in the previous section on the interpretation of quadrupole splitting). Again, the relative area ratios are consonant with random mixing (lack of short-range order).

Pyroxenes

In the two examples outlined above, the occurrence of NNN effects in the Mössbauer spectra is accentuated by the fact that in the ordered end-member structure, the Mössbauer-sensitive cation occupies a site of cubic point symmetry; hence any local deviation from this symmetry produces one or more quadrupole-split doublets. In most minerals, the symmetry of the site(s) occupied by the Mössbauer-sensitive cation is not cubic, and quadrupole-split doublets are already present in ordered end-member structures. Consequently, it will be intrinsically more difficult to recognize NNN effects in such spectra. Nevertheless, NNN effects have been identified in pyroxenes.

Seifert (1983) has shown that (Al + Fe^{3+})-containing orthopyroxene spectra may be fit to three Fe^{2+} doublets; he interprets the three Fe^{2+} doublet fit in terms of NNN behavior around the M2 site. Octahedrally coordinated (Al + Fe^{3+}) are ordered at M1 (Hawthorne and Ito, 1977), and the additional peak is assigned to Fe^{2+} at M2 with NNN cations of both R^{2+}(Mg + Fe^{2+}) <u>and</u> R^{3+}(Al + Fe^{3+}). An extremely interesting result from this work is that the relative peak areas for Fe^{2+} at M2(3R^{2+}) and M2(R^{2+}, R^{3+}) systematically differ from those expected for a random distribution. If this fitting model is correct, this shows the presence of significant short-range order in R^{3+}-containing orthopyroxenes.

NNN effects have also been identified in hedenbergite ($CaFe^{2+}Si_2O_6$)-ferrosilite ($Fe^{2+}Fe^{2+}Si_2O_6$) solid solutions (Dowty and Lindsley, 1973), natural C2/c and P2/n omphacites (Aldridge et al., 1978), and synthetic Ca-Na pyroxene solid solutions (Dollase and Gustafson, 1982).

Whereas the NNN models for wüstite, the spinels and the thiospinels are convincing, the situation with the pyroxenes is far more uncertain. Although NNN effects do seem to be present, it is difficult to come up with a consistent model. Thus, for example, the doublets assigned to NNN configurations in pyroxenes all show significant variation of Mössbauer parameters with composition, whereas this is not a feature of the spinels and thiospinels, in which the Mössbauer parameters for specific NNN configurations are constant and (additive) partial isomer shift and partial quadrupole splitting models can be applied.

Other minerals

The results for wüstite and the spinels show that NNN effects can easily be seen in Mössbauer spectroscopy. On the other hand, the results for the pyroxenes indicate that although NNN effects may be present and observable in Mössbauer spectra, the fitting and interpretation of such spectra is far from straightforward. This suggests that in more complex minerals such as the amphiboles and the micas, we may get NNN effects contributing to the Mössbauer spectra, but because the spectra are so complex, their contribution is not sufficiently resolved to be recognisable. Unfortunately, such a case would give rise to significant systematic error in site-occupancy studies of such minerals.

INTERVALENCE CHARGE-TRANSFER IN MINERALS

When a mineral contains a transition element in more than one oxidation state, there is the possibility that electrons will transfer between atoms with differing oxidation states. The probability of such a transfer varies greatly, depending on the activation energy; this can be supplied by photons, or the process can be thermally activated. Let us start by looking at the energetics of a simple $Fe^{2+}Fe^{3+}$ dimer in an oxide environment (Fig. 34a). The electron is initially localized on the Fe^{2+}, which consequently has a much larger ionic radius than the neighbouring Fe^{3+}. The local relaxation around the Fe^{2+} due to its "extra" electron may be thought of as a localized state that traps the electron at that site; this is called a small polaron. If the electron moves to the adjacent Fe^{3+}, the local distortion will follow it, as diagrammatically expressed in Figure 34a. However, in this static model, there is no impetus for electron exchange to occur. Figure 34b shows the potential curves for the two states shown as a function of (i.e., incorporating) vibrational effects; this is known as a configuration co-ordinate model, and is drawn for crystallographically equivalent sites. Where the potential curves cross, the electronic interaction modifies each curve as shown. If the electronic interaction is large, the central point is lowest in

318

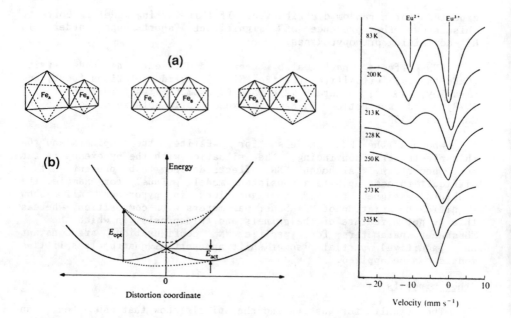

(a)

(b)

Energy

E_{opt}

E_{act}

0

Distortion coordinate

Velocity (mm s^{-1})

Figure 34 (left). (a) Diagrammatic representation of electron exchange in an $(Fe^{2+}Fe^{3+}O_{10})$ dimer: the initial configuration is $Fe^{2+}Fe^{3+}(Fe_AFe_B)$; an electron exchanges to the configuration $Fe^{3+}Fe^{2+}(Fe_AFe_B)$; the central configuration represents the intermediate situation; (b) Configurational co-ordinate representation of the potentials for this dimer; ...... represents a strong electronic interaction, ------ represents a weak electronic interaction; after Cox (1987).

Figure 35 (right). Mössbauer spectra of ^{151}Eu in Eu_3S_4, taken at different temperatures; note the existence of distinct Eu^{2+} and Eu^{3+} peaks at low temperatures, and their coalescence to a single peak at high temperatures, the isomer shift of which indicates a "valence" of 2.67, the weighted mean of the valences observed at low temperatures; after Cox (1987).

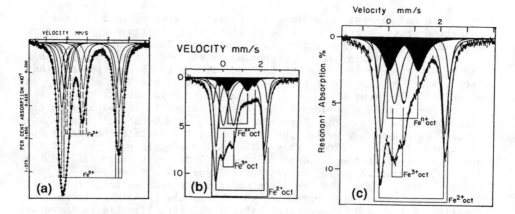

Figure 36. Mössbauer spectra of ^{57}Fe in deerite at different temperatures; (a) 72 K; (b) 295 K; (c) 335 K; note the absence of "intermediate valence" bands in (a), and their gradual increase in intensity relative (shown in black) with increasing temperature; after Amthauer et al. (1980).

energy (dotted curve) and the electron is delocalized. If the electronic interaction is small, the central energy is lowered, but the lowest energy states are still localized on either atom. To move from one atom to another, the electron must cross the barrier between the two sites, and hence there is an activation energy associated with movement or hopping of the electron. It is this activated mobility of the electrons which is most characteristic of small polaron trapping (Cox, 1987).

In terms of what we can observe in a Mössbauer spectrum, we may recognize three distinct situations (Maddock, 1985):
(1) the interaction is sufficiently slow (with regard to the lifetime of the excited state of the Mössbauer atom) that Mössbauer spectroscopy sees Fe^{2+} and Fe^{3+} at specific sites. The specific case of no exchange is in this class, with an infinite lifetime of Fe^{2+} and Fe^{3+} at their constituent sites. In this case, the Mössbauer spectrum is sharp with "normal" Mössbauer parameters.
(2) the interaction is sufficiently fast that the lifetimes of $Fe2+$ and Fe^{3+} at their specific sites are very short when compared to the timescale of the Mössbauer transition. In this case, the Mössbauer spectrum shows narrow peaks, but the Mössbauer parameters are anomalous.
(3) the life-times of the two interactions are comparable, such that the Mössbauer spectrum records a partly averaged situation. In this case, it may be possible to run through the sequence (1)-->(3)-->(2) such that a normal (low-temperature) spectrum broadens and becomes ill-defined with increasing temperature, gradually sharpening up again at high-temperature (when a weighted average valence is observed).

According to the Franck-Condon principle (see Chapter 2), a spectroscopic electron transition occurs vertically between the potential energy curves as the atoms do not have time to relax appreciably during such a (fast) transition. This is shown in Figure 34b as the E_{opt} transition, and corresponds to an excitation of the electron from its trapped state on one atom to an adjacent atom not in the correct ground-state geometry to receive it. Such transitions are excited by photons. The curves of Figure 34b have the form of a parabola, and it is easy to show (and is left as an exercise for the reader) that

$$E_{opt} = 4E_{act}$$ [68]

when the electronic interaction is (infinitely) small. Behavior with increasing interaction is described by Sherman (1987).

As suggested by the nomenclature of equation (68), optically-activated electron transfer is observed in optical (electronic) absorption spectroscopy (see Chapter 7). Electron delocalization and thermally-activated electron transfer are visible in Mössbauer spectroscopy. It is of great interest to distinguish between these three mechanisms (optically-activated electron transfer, electron delocalization, and thermally-activated electron transfer), and Mössbauer spectroscopy has played a major role in this work.

Eu_3S_4

Before we look at the complexities of natural minerals, we will first consider the simple example of Eu_3S_4. The crystal structure shows all the Eu atoms to be equivalent; the Mössbauer spectra (Fig. 35) show a slightly different story. At low temperatures, there are two distinct peaks due to Eu^{2+} and Eu^{3+} respectively, with an area ratio of $\approx$ 2:1. With increasing temperature, the two peaks gradually coalesce, until at high temperatures, there is a single peak with an isomer shift suggesting a valence of 2.67 (as required for the (time- and space-averaged) crystal structure). This is due to thermally-activated electron hopping (polaron hopping), as the temperature behavior testifies.

Minerals with finite Fe-containing octahedral clusters

For many years, it has been well-known that InterValence Charge Transfer (IVCT) is an important process in minerals showing very strong pleochroism. However, the details of the charge-transfer process(es) and their relation to the nature of the polyhedral linkage in the structures were not well-understood. An important step towards a better understanding of these problems was the work of Amthauer and Rossman (1984) on a series of (Fe^{2+}, Fe^{3+})-rich minerals with different polymerizations of $(Fe\emptyset_6)$ octahedra. They showed that in minerals with finite octahedral clusters containing both Fe^{2+} and Fe^{3+}, there is optically-activated IVCT but no thermally-activated electron exchange.

Minerals with chains of Fe octahedra

Minerals containing infinite edge-sharing chains of Fe octahedra have very different temperature-dependent behaviour of their Mössbauer spectra, when compared with minerals based on finite Fe octahedral clusters (Amthauer, 1982).

Deerite. Deerite is an hydrous iron silicate with the (ideal) formula $Fe^{2+}_6Fe^{3+}_3O_3(Si_6O_{17})(OH)_5$; the structure (Fleet, 1977) consists of an $[Si_6O_{17}]$ tetrahedral chain, linked to an infinite strip of edge-sharing Fe octahedra. There are 9 unique Fe positions, and consequently one does not expect to completely resolve the Mössbauer spectrum. On the other hand, these 9 sites can be divided into 3 groups of similar constituent polyhedra.

The Mössbauer spectra are shown in Figure 36. The low-temperature spectrum can be resolved into three Fe^{2+} doublets and three Fe^{3+} doublets, corresponding to Fe^{2+} and Fe^{3+} occupancy of the three distinct groups of octahedral sites. Deerites from different localities can be fit in the same way. The room-temperature and high-temperature spectra are dramatically different from the low-temperature spectrum. In particular, they show significant absorption in the 1-1.5 mm/s region, between the low- and high-velocity components of the absorption envelope. There is also less resolution in both the Fe^{2+} and Fe^{3+} absorptions, preventing fitting of three peaks in each case. Not much significance should be attached to the differing number of doublets (denoted by Fe^{n+}oct in Fig. 36) in the "mixed-valence" region; a more realistic model has since been developed by Litterst and Amthauer (1984). The important point is that the presence of the Fe^{n+}

absorption, with its rapid increase in intensity with increasing temperatures, is characteristic of thermally-activated charge-transfer (electron exchange); we are observing the spectrum at stage (2) of the time-variable exchange process, as discussed above.

Ilvaite. Ilvaite is an iron silicate with the ideal formula $Ca(Fe^{2+} Fe^{3+})Fe^{2+}Si_2O_8(OH)$. It has edge-sharing double-chains of Fe octahedra that are // c-axis, and cross-linked by $[Si_2O_7]$ groups and [7]-co-ordinate Ca cations. There has been a lot of work on the crystallography, spectroscopy, magnetic and electrical properties of ilvaite; this is reviewed by Ghose (1988).

The Mössbauer spectra of ilvaite at a variety of temperatures and pressures are given in Figure 37. At room temperature, there is a significant doublet with an upper velocity peak at about 1.5 mm/s (see arrow in Fig. 37); this is characteristic of "mixed-valent" Fe. With increasing temperature (Fig. 37a), this doublet becomes better resolved and rapidly increases in relative intensity; the same features is displayed with increasing pressure (Fig. 37b). This behavior is characteristic of thermally-activated electron exchange. Litterst and Amthauer (1984) have successfully fit the ilvaite spectra (as a function of temperature) using just doublets due to Fe2+ and Fe^{3+} at the 8d site and Fe^{2+} at 4c, together with a relaxation model for electron exchange between Fe^{2+} and Fe^{3+} in the 8d (edge-sharing) octahedra.

Summary. In these minerals with infinite octahedral chains containing both Fe^{2+} and Fe^{3+}, there is thermally-activated electron hopping that can be observed in their Mössbauer spectra.

Some general considerations

The results of this section indicate that thermally-activated electron exchange only occurs in minerals with infinite chains of (edge- and face-sharing) Fe^{2+}, Fe^{3+} octahedra; where polymerized (Fe^{2+}, Fe^{3+}) octahedra occur only as finite clusters, such behaviour is not found. Thus, rather than the dimeric model suggested by Figure 34, an extended configuration coordinate diagram would seem more appropriate. The various models associated with this behaviour (small polaron transport, intermediate polarons, (solitonic) band conduction) are a subject of great current interest; details can be found in Amthauer (1982), Sherman (1987), Ghosh et al. (1987), Güttler and Salje (1987) and Ghose (1988).

MAGNETIC PROPERTIES OF MINERALS

Mössbauer spectroscopy has contributed significantly to our understanding of the magnetic properties of minerals. Much of the early work was directed towards the characterization of the properties of magnetite, and the literature on this mineral is enormous. However, in the last ten years, there has been a major surge of interest in the properties of oxysalt minerals, particularly silicates. Although Mössbauer spectroscopy has played a major role in this work, magnetic neutron diffraction, magnetization and susceptibility studies are also of considerable importance.

322

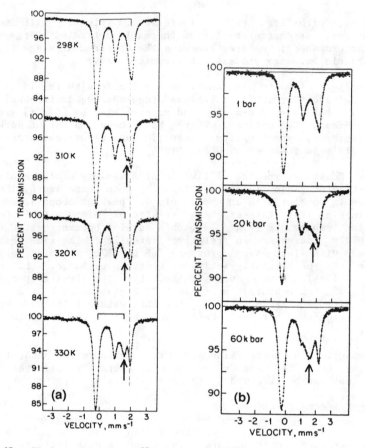

Figure 37. Mössbauer spectra of ^{57}Fe in ilvaite: (a) with increasing temperature; (b) with increasing pressure; note the growth of the "intermediate valence" band in each case; after Evans and Amthauer (1980).

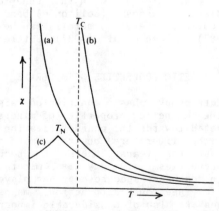

Figure 38. Magnetic susceptibility, X, as a function of temperature, showing the behaviour of (a) paramagnetic; (b) ferromagnetic; (c) antiferromagnetic materials; T_C is the Curie temperature, T_N is the Néel temperature; after Cox (1987).

As the magnetic properties of solids may be a little unfamiliar to many of us, we will briefly review the major ideas and define some of the principal terms used in this section.

Origin of magnetism

The orbital motion of an electron is analogous to an electric current flowing in a coil, and hence the electron has an associated magnetic moment, M given by

$$M = -\left(\frac{e}{2m_e}\right) L \quad , \tag{69}$$

where L is the angular momentum of the electon; the constant $e/2m_e$ is called the gyromagnetic ratio (see Chapter 12) of the electron. To completely describe an electron, four quantum numbers are necessary (see Chapter 2): n (the principal quantum number), ℓ (the angular quantum number), m_e (the magnetic quantum number) and m_s (the spin angular quantum number); the possible values of m_s are $\pm\frac{1}{2}$.

For any complete shell of electrons, the total orbital angular momentum is zero, hence the orbital contributions to the magnetic moment of an atom are only non-zero in the case of incomplete shells of electrons. Only unpaired electrons can contribute to the total spin angular momentum of an atom or ion, and hence to the total spin magnetic moment of the atom.

Magnetic properties of solids

With regard to magnetic properties, we can think of solids as being divided into two classes: materials which have a magnetic moment only in the presence of an external magnetic field (diamagnetic and paramagnetic), and materials which have an intrinsic magnetic moment in the absence of an external magnetic field (ferromagnetic, antiferromagnetic and ferrimagnetic).

Diamagnetism. When subject to an external magnetic field, a diamagnetic material develops a magnetic moment antiparallel to the direction of the applied magnetic field. This occurs because all the electrons precess in the same direction about the principal axis of the applied magnetic field; hence contributions to the magnetic moment from electrons with opposite values of m_e or m_s do not cancel. Note that all the electrons in an atom contribute to the diamagnetic susceptibility, not just those electrons in incomplete shells. Despite this, diamagnetic effects are very small when compared with other types of magnetism, and can only be observed in the absence of any other kind of magnetism (except in superconductors).

Paramagnetism. In a paramagnetic material, the atoms have permanent magnetic moments resulting from the spins of unpaired electrons and/or the orbital motions of the electrons. In the absence of a magnetic field, the magnetic moments of the individual atoms are randomly oriented. In an external magnetic field, the individual magnetic moment of each atom aligns itself with respect to the principal axis of the applied magnetic field; the atoms adopt orientations such that their components parallel to the axis are governed by values of their magnetic quantum numbers. The paramagnetic susceptibility, X_p, obeys Curie's law:

$$\chi_p = \frac{C}{T} \quad , \tag{70}$$

where

$$C = \frac{\mu_o M_s^2}{3nk} \left[\frac{J+1}{J}\right] \quad , \tag{71}$$

and M_s is the saturization magnetism, when all individual atoms have the largest possible value for their magnetic quantum number ($m_j = +J$).

Ideally, paramagnetism corresponds to the case when there is no magnetic interaction between ions. With falling temperature, the magnetic susceptibility rises, as shown in Figure 38.

Ferromagnetism. In a ferromagnetic material, all the magnetic moments of the atoms are aligned parallel to some specific direction. Consequently, the material has a significant permanent magnetic moment in the absence of an external magnetic field. However, the magnetic moments of the individual atoms are not perfectly aligned; at non-zero temperatures, thermal fluctuations perturb the perfect orientation of spins, resulting in an average magnetic moment that is less than the ideal for perfect spin alignment. With increasing temperature, the magnetic moment decreases until, at the Curie temperature, the magnetic susceptibility is given by the Curie-Weiss law:

$$\chi_p = \frac{C}{T - \theta} \quad , \tag{72}$$

where θ is the Weiss constant (positive in the case of a ferromagnetic material). The behaviour of X_p as a function of temperature is shown in Figure 38. The susceptibility rises faster with falling temperature for (a paramagnetic phase of) a ferromagnetic material than for a paramagnetic material because of interaction between magnetic atoms in the former case.

Antiferromagnetism. In an antiferromagnetic material, the magnetic moments are equally distributed over two similar sublattices. On each lattice, the magnetic moments of the atoms are parallel, but the net magnetic moments of each lattice are antiparallel, and the material has no net magnetic moment. At high temperatures, antiferromagnetic materials become paramagnetic. With falling temperature, the magnetic susceptibility increases until, at the Néel temperature, the spins align and the material becomes antiferromagnetic, with a decrease in the susceptibility (Fig. 38).

Characterization of the detailed configurations of unpaired electron spins by neutron diffraction has shown that more complex types of antiferromagnetic ordering are possible; thus in addition to the antiparallel arrangement described above, canted (Fig. 39), helical, spiral and modulated antiferromagnetic structures have been observed.

Ferrimagnetism. In a ferrimagnetic material, the magnetic moments are unequally distributed over two or more sublattices; although the component magnetic moments are antiparallel (or otherwise partly

Ferromagnetic	Antiferromagnetic	Ferrimagnetic	Canted antiferromagnetic (weak ferro-magnetic)
↑↑ ↑↑	↑↑ ↓↓	↑↑ ↓↓	↗↗ ↘↘
No splitting of the spectral lines in an applied field	Splitting of the spectral lines in an applied field directed along the antiferro-magnetic axis. Sharp spin reorientation	Distinct magnetically split spectra with different hyperfine fields	Splitting of the spectral lines in an applied field directed along the antiferromagnetic axis. Continuous spin reorientation

Figure 39. The common magnetic structures and their identification from magnetically-split Mössbauer spectra; after Thomas and Johnson (1985).

opposed), their magnitude is not the same and there is a net spontaneous magnetic moment. As the magnetic moments of the various sublattices are dependent on temperature, the total resultant magnetization may be zero at a specific temperature, the compensation temperature; this is observed in some of the ferrimagnetic rare-earth garnets.

Mössbauer spectroscopy and magnetic structure

As we have seen above, the interaction of the nuclear magnetic dipole moment with the magnetic field at the nucleus removes the spin degeneracy to produce nuclear Zeeman splitting (Figs. 10 and 11). The magnetic field at the nucleus can be imposed through an externally applied magnetic field, or it can be due to the unpaired orbital electrons of the Mössbauer-sensitive atom.

For paramagnetic minerals, the intrinsic magnetic field at the nucleus is changing due to spin-spin and spin-lattice relaxation effects; as these changes are fast compared to the timescale of the Mössbauer transition, the average magnetic field at the nucleus of the Mössbauer atom is zero, and no magnetic splitting is observed. For ferromagnetic, antiferromagnetic and ferrimagnetic materials, cooperative interactions keep the individual magnetic moments aligned, and below the Curie and Néel temperatures, magnetically-split spectra are observed.

Mössbauer spectroscopy can also distinguish between the common magnetic structures. A ferrimagnetic material has (at least) two different magnetic spectra in the powder (or single-crystal)spectrum recorded below the ordering temperature, each with a different value of the hyperfine field (the magnetic field due to the unpaired electrons); the other common magnetic materials give just a single magnetic spectrum under these conditions. Ferromagnetic and antiferromagnetic materials can be distinguished from the Mössbauer spectra of single-crystals in an applied magnetic field, which adds to the hyperfine field. In a ferromagnetic material, the fields are parallel to give a single value of the effective hyperfine field; thus there is no further splitting of the spectrum. In an antiferromagnetic material, the lines of the magnetic spectrum are split into two by the component of the applied magnetic field along the antiferromagnetic axis. Ferromagnetic and antiferromagnetic materials can also be distinguished by their powder spectra in an applied field. In a ferromagnetic material, the applied field can (usually) align the ferromagnetic spin system parallel to itself, and the spectrum is sharp. In an antiferromagnetic material, the applied field is much less effective in aligning the spins; thus the internal hyperfine field is randomly oriented with regard to the applied field, and the lines of the spectrum are broadened. This information is summarized in Figure 39.

Applications to minerals

Mössbauer spectroscopy is often used, together with magnetic susceptibility and magnetization measurements and neutron diffraction, to define the magnetic structure of minerals.

Garnets. In their paramagnetic state, the spectra of andradite ($Ca_3Fe^{3+}_2Si_3O_{12}$) and almandine ($Fe^{2+}_3Al_2Si_3O_{12}$) each consist of a quadrupole-split doublet. However, at low temperatures, both show magnetic ordering (Fig. 40) and both show two distinct hyperfine patterns. This is just visible for the 4.2 K spectrum of andradite, in which an asymmetry of the central peaks in particular is apparent; this spectrum was fitted to two sextets with an intensity ratio of 0.3/07 and QS values of 0.32 and -0.16 mm/s respectively. The magnitudes of the hyperfine fields were the same for both patterns (519 kOe), and the angles between the fields and the EFG were 31.8(8) and 68.3(4)° respectively. The two resolved hyperfine patterns for almandine are shown in the spectrum (Fig. 40b). Murad and Wagner (1987) suggest the spins to be oriented at (two different) angles of 73 and 90° to the EFG (//[110]). The complete interpretation of these spectra are not clear. Certainly, both garnets order antiferromagnetically at low temperatures, but the presence of two hyperfine patterns suggests either some sort of canted spin structure or even L-type ferrimagnetism if the garnets are non-cubic.

Fayalite. Fayalite is paramagnetic at high temperatures and undergoes a phase transition at the Néel point (65 K) to become collinear (antiparallel) antiferromagnetic. At 23 K, it has a second transition to a canted antiferromagnetic state (Santoro et al., 1966), and should saturate to a collinear antiferromagnetic state at very low temperatures, although this was not observed. The pressure variation of the Néel temperature (T_N) was measured by high-pressure Mössbauer spectroscopy (Hayashi et al., 1987); T_N increases linearly with increasing pressure (dT_N/dP = 2.2(2) K/GPa).

Chain silicates. Magnetic hyperfine splitting in amphiboles was first observed in cummingtonite. The magnetic ions (Fe^{2+}) in cummingtonite are separated by diamagnetic cations (Mg), and this is actually a spin glass. Here, the different (local) distances between magnetic ions leads to different magnetic interactions between neighbouring magnetic ions. Thus the resulting magnetic state embodies both a random character and spin-frustration (indicating that adjacent magnetic moments cannot set at optimum orientations with regard to all their magnetic neighbours).

Grunerite (Eisenstein et al., 1975; Linares et al., 1983: Ghose et al., 1987) magnetic transitions at 47 and 7 K, and determined orders (collinear) antiferromagnetically at 47 K, and undergoes a spin-canting transition at 7 K. Within the range 47-8 K, there is ferromagnetic coupling parallel to b within the octahedral strip, and each strip couples antiferromagnetically to adjacent strips.

The sheet silicates have been examined extensively by Coey et al. al. (1981) and Ballet and Coey (1982). The magnetically ordered spectra consists of superimposed components due to Fe^{2+} and Fe^{3+} at different structural sites. Within the octahedral sheets, there is Fe^{2+}-Fe^{3+} ferromagnetic coupling and the magnetic moments tend to lie in the plane of the sheets; thus there is a strong 2-dimensional magnetic character. Coupling between the sheets is antiferromagnetic.

Summary. Mössbauer spectroscopy is very useful in determining the coocurrence of magnetic ordering transitions at low temperature, and can also show the type of magnetic structure present, although susceptibility, magnetization and neutron scattering techniques are normally used in conjunction with it to thoroughly characterize the magnetic structure.

OXIDATION STUDIES

Mössbauer spectroscopy has been used extensively in the study of oxidation processes in minerals. This is particularly the case in OH- and H_2O-bearing minerals, in which oxidation often proceeds by an electron-exchange reaction such that the oxidation process is coupled with dehydrogenation. Typically, such work is combined with infrared studies, focusing on the OH region.

Trioctahedral micas

A series of Mössbauer spectra of a biotite, heated at gradually increasing temperatures, is shown in Figure 41a->f. The relative areas of the Fe^{2+} and Fe^{3+} absorptions allow us to follow the progressive oxidation process. The 200°C spectrum (Fig. 41a) shows little difference from the unheated spectrum (not shown). It consists of two quadrupole-split Fe^{2+} doublets with an intensity ratio of about 1:2, assigned to Fe^{2+} at the trans M(1) and cis M(2) sites respectively; in addition, there is a broad quadrupole-split Fe^{3+} doublet consisting of unresolved contributions from Fe^{3+} at M(1) and M(2). With progressive heating, the intensity of the broad Fe^{3+} doublet gradually increases in intensity at the expense of the the two Fe^{2+} doublets, paralleling the well-characterized changes in the OH-stretching spectra (e.g., Vedder and Wilkins, 1969). Comparison of Figures 41a and b shows that this oxidation occurs preferentially at the M(2) site, as the relative intensity of this doublet, when compared to that of Fe^{2+} at M(1), is much decreased in the higher-temperature spectra. Accompanying this is a broadening of the Fe^{3+} doublet, suggesting increasing resolution of the Fe^{3+} doublets. At 350°C (Fig. 41c), only one broad Fe^{2+} doublet can be distinguished, but two Fe^{3+} doublets are now apparent, the more intense Fe^{2+} doublet having the larger quadrupole splitting. By 500°C, all the Fe^{2+} is now oxidized.

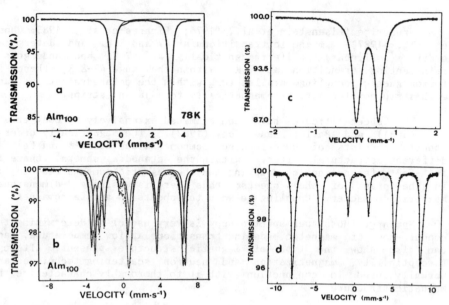

Figure 40. Mössbauer spectra of ^{57}Fe in (a) andradite and (b) almandine garnets, above and below their magnetic transition temperatures; after Murad (1984) and Murad and Wagner (1987).

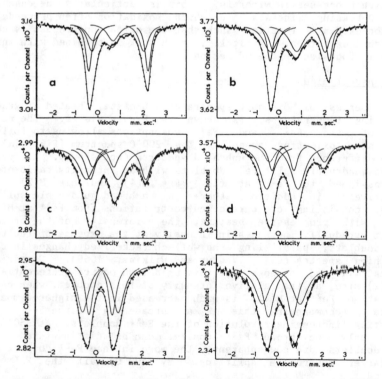

Figure 41. Mössbauer spectra of ^{57}Fe in biotite after heating in air: (a) 200; (b) 300; (c) 350; (d) 400; (e) 500; (f) 800°C; from Hogg and Meads (1975).

In combination with infrared studies, the major features of biotite oxidation/dehydroxylation are now well-known. Similar work has been done on the weathering of micas (Bowen et al., 1968; Rice and Williams, 1969); however, further work on the kinetics and surface effects is still needed, and it is here that Conversion Electron Mössbauer Spectroscopy (CEMS) will play a role.

^{197}Au MÖSSBAUER SPECTROSCOPY

We have seen applications of ^{57}Fe, ^{119}Sn, ^{121}Sb and ^{153}Eu Mössbauer spectroscopy in the section on oxidation states and their determination; no mention was made of ^{197}Au, despite the fact that there has been some Mössbauer spectroscopy done on this isotope in minerals. This is due to the fact that it is very difficult to interpret the Mössbauer parameters of ^{197}Au in order to reliably distinguish between the valence states 0 (metallic gold), +1 and +3 (see Fig. 21). Despite these problems, ^{197}Au Mössbauer spectroscopy has been used effectively to characterize gold ores, and to follow the progress of gold extraction in the smelting and roasting processes used for gold recovery.

Figure 42 shows the spectra of two gold ores before and after roasting (Wagner et al., 1986,1988; Marion et al., 1986). Ore #2 consisted of 80% arsenopyrite and 20% pyrite; it contained 300 ppm gold, none of which was visible. The asymmetric absorption can be fit to two broad lines, and this spectrum is characteristic of "invisible gold" in arsenopyrite. Upon partial roasting, about two-thirds of the gold is liberated (i.e. converted to metallic gold, with an isomer shift of -1.29 mm/s), but this process is very selective. The more intense lower velocity peak in the untreated ore has entirely disappeared in the heated ore, whereas the higher velocity peak seems entirely unaffected; thus we see a very selective liberation of Au by this specific treatment. Ore #5 consists of 50% arsenopyrite, 30% pyrite, 20% unspecified, with 180 ppm Au, most of which is visible in the microscope. The untreated ore gives a broadened singlet at -0.4 mm./s, together with a weak singlet at 3.2 mm/s. The low-velocity singlet can be assigned to an $Au_{0.6}Ag_{0.4}$ alloy, whereas the origin of the upper velocity doublet is unknown. Roasting converts all gold to the metallic state, giving a broadened singlet with an isomer shift of -0.97 mm/s; this indicates partial exsolution/segregation of the Ag component, but the process is not complete.

Thus although the exact details of ^{197}Au Mössbauer spectra of gold minerals are not well characterized as yet, very useful work can still be done using this technique.

CONVERSION ELECTRON MÖSSBAUER SPECTROSCOPY

Basic principles

When the Mössbauer sensitive nuclide in the sample absorbs a γ-ray from the incident beam, it is promoted to an excited state. The half-life of this excited state is short, and the nuclide decays back to its ground state. It does this by a variety of processes that include emission of γ-rays and additional processes involving electrons and X-rays. These additional processes are called <u>internal conversion processes</u>, and their importance in the decay process is

330

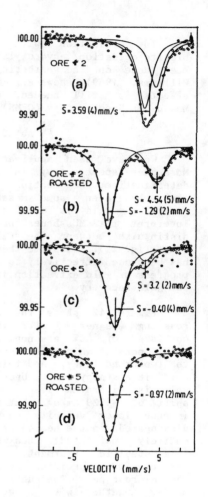

Figure 42 (right). Mössbauer spectra of ^{197}Au in gold ores: (a) 80% arsenopyrite, 20% pyrite, with 300 ppm "invisible" gold; (b) partially roasted (a), showing the preferential release of chemically bound gold to form metallic gold; (c) 50% arsenopyrite, 30% pyrite, 20% unspecified, with 180 ppm visible gold; (d) roasted (c) showing the release of the chemically bound gold, and the shift of the "metallic" gold peak to more negative isomer shifts. Isomer shifts relative to the Pt-metal source; spectra after Wagner et al. (1988).

Figure 43 (below left). Conversion electron Mössbauer spectra of ^{57}Fe in biotite: (a) before and (b) after heat treatment at 550 K for 100 hours; note the increase in Fe^{3+} relative to Fe^{2+} in the spectrum of the heat-treated material; after Tricker (1981).

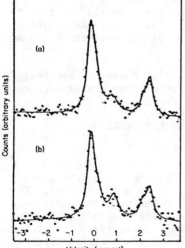

Table 11. Summary of major events during the decay of $I=3/2$ excited-spin states of ^{57}Fe and ^{119}Sn.

	Energy (keV)	Number/100 Absorption Events
^{57}Fe		
γ-photons	14.4	9
K-X-rays	6.3	27
K-conversion electrons	7.3	81
L-conversion electrons	13.6	9
M-conversion electrons	14.3	1
KLL-Auger electrons	5.4	63
LMM-Auger electrons	0.53	
^{119}Sn		
γ-photons	23.8	17
X-rays	3.6	9
L-conversion electrons	19.6	83
LMM-Auger electrons	2.8	74

after Tricker (1981)

represented by the <u>internal conversion coefficient</u>, α, which is a measure of the number of decay processes which proceed via internal conversion. The decay of the $I = {}^3/_2$ excited spin state for ^{57}Fe is given in Table 11. Only about 10% of the decay events involve the emission of a 14.4 keV γ-ray; the rest involve internal conversion processes. Of most importance is the conversion event that results in the ejection of a 7.3 keV K-conversion electron, together with subsequent X-ray photons and Auger electrons. It should be possible to detect these ejected conversion electrons, and gain the same sort of information that is measured in the recording of the excitation (absorption) rather than the emission event.

The next question is "Why go to the trouble?" if the information from the two processes is the same. The answer to this lies in the different absorption characteristics of (γ-ray and X-ray) photons and electrons. X-rays and γ-rays have much lower absorption coefficients than electrons in the sample material. Consequently, detected X-rays and γ-rays will carry information characteristic of the bulk sample, whereas detected conversion electrons can only carry information concerning the characteristics of the first few surface layers, as conversion electrons emanating from any deeper in the sample will be totally absorbed. In Fe-foil, $^2/_3$ of the detected conversion electrons come from the top 550 Å. Thus it is not a true surface technique (in that it does not examine just the top layer), but can be of great use in the characterization of thin films; furthermore, experimental refinements allow depth profiling of surface and near-surface layers.

Instrumentation

As the conversion electrons cannot be transmitted through the sample, the detector must be on the same side of the sample as the source, in order to collect the back-scattered electrons. The electrons are usually recorded using a He/CH$_4$ flow-proportional counter (that is insensitive to the 6.3 keV X-rays and 14.4 keV γ-rays of ^{57}Fe) that registers all of the back-scattered conversion electrons. Alternatively, the electrons can be energy-analyzed, and spectra collected with different electron energies; the individual spectra are then weighted towards a particular depth in the sample, providing a depth-profile through the surface "film". Further details are given in Tricker (1981).

Application to minerals

Only a few applications to minerals have been made so far. However, with the current interest in surface processes, Conversion Electron Mössbauer Spectroscopy (CEMS) can contribute significantly in this area. CEM spectra of biotite before and after heat-treatment are given in Figure 43. Although no change was apparent in the normal transmission Mössbauer spectrum at this stage of the heat-treatment, the CEM spectra definitely show that Fe^{3+} has increased (compare peaks at about 0.8 mm/s) in the surface layers. Additional work on vivianite (Tricker, 1981), ion-implanted enstatite (Stanek, 1986), goethite (Zahn et al., 1986) and surface tarnishing of sulphides Richardson and Vaughan, 1987a,b) has been done. Conversion electron Mössbauer spectroscopy holds considerable promise for the study of surface processes in minerals, and the use of this technique should increase in the future.

SUMMARY

It is now thirty years since the Mössbauer effect was discovered
(Mössbauer, 1958) and just over twenty years since it began to be used
extensively in mineralogy (Bancroft et al., 1966, 1967a,b,c,d); it must
now be considered as a mature technique in the characterization of
geological materials. The strength of Mössbauer spectroscopy is that
it is element-specific; it doesn't matter how complex the material
examined, one only gets a response from the Mössbauer-sensitive
nucleus. Consequently, one can often extract the information one
wants, be it the determination of valence state, coordination number
or site-occupancy, without having to be concerned about the total
complexity of the material. On the other hand, this element
specificity does mean that Mössbauer spectroscopy is generally a
complementary technique, to be used in conjunction with some other
diffraction or spectroscopic technique to more fully characterize
materials.

We may summarize the principal uses of Mössbauer spectroscopy of
geological materials as follows:

(1) <u>Determination of valence states</u>. The first use of Mössbauer
spectroscopy of minerals involved the determination of valence states
of Fe, and this must still be considered as one of its most important
uses. Over the past twenty years, it has been used extensively to
characterize Fe^{2+}/Fe^{3+} ratios, not only in minerals but also in
glasses and poorly crystalline materials such as muds and soils. ^{57}Fe
is not the only Mössbauer-sensitive nucleus of geological interest, and
more recent work has involved valence-state characterization of ^{119}Sn,
^{121}Sb and ^{151}Eu in minerals and glasses. Current interest in surfaces
and the mechanisms of alteration/oxidation has seen the start of
Conversion Electron Mössbauer Spectroscopy as a characterization tool
for valence-state determination on or near surfaces. The use of
Mössbauer spectroscopy for valence-state determination seems likely
to continue more strongly into the future.

(2) <u>Determination of site-occupancies</u>. It was this use of
Mössbauer spectrosocopy that had such an impact on mineralogy in the
late 1960's, focusing on Mg/Fe^{2+} order/disorder in (usually
rock-forming) silicates. Hindsite shows that the technique was rather
"over-sold" in this particular application. Although successful for
certain of the simpler rock-forming silicates (e.g., orthopyroxenes),
problems of resolution and assignment have plagued work on the more
complex silicates (e.g., clino-amphiboles, sub-calcic clinopyroxenes).
More recent work has shown the effects of next-nearest-neighbour
variations and thermally-activated charge-transfer in such minerals;
consequently, the use of Mössbauer spectroscopy in site-occupancy
determination is strongly limited by the existence of these other
phemonena that can complicate spectra to the point of preventing
adequate resolution. Nevertheless, the use of Mössbauer spectroscopy
for Fe^{2+} and Fe^{3+} site-occupancy determinations in less-complex
minerals will continue to be an important aspect of the technique.

(3) <u>Determination of local order</u>. Mössbauer spectroscopy is a
probe of local structure, and as such, can provide us with information
on short-range-order. The most successful work has been done on
high-symmetry (cubic) minerals, in which local structure (variation in

NNN arrangements) has given rise to local non-cubic symmetry, resulting in quadrupole-splitting of an ideal Fe singlet. When the long-range symmetry is non-cubic, the spectrum already consists of quadrupole-split doublets, and short-range effects are far more difficult to see; nevertheless, such NNN effects have been identified in pyroxenes. Such information on short-range-order is difficult to obtain experimentally, and this is one area of Mössbauer spectroscopy that could be profitably explored in future years.

(4) Thermally-activated charge-transfer. Mössbauer spectroscopy has played an important role in the characterization of Inter-Valence Charge-Transfer in minerals. A combination of optical absorption spectroscopy and Mössbauer spectroscopy has allowed the identification of the important structural characteristics associated with optically-activated and thermally-activated electron-exchange. There is considerable potential here for further work in identifying and characterizing thermally-activated charge-transfer in minerals.

(5) Magnetic structure of minerals. Mössbauer spectroscopy is only one of a number of techniques used for characterizing the magnetic structure of minerals. There has been significant interest in this area in the past ten years, particularly with regard to the relationship between crystal and magnetic structure, whereby the chain- and sheet-silicate minerals respectively show predominantly one-and two-dimensional magnetic behaviour. The magnetic structures of many Fe-rich minerals are not known, and there is still considerable work to be done.

Thus Mössbauer spectroscopy is of wide applicablility in the characterization of geological materials. Over the past twenty years, the diversity of use has gradually been increasing and should continue into the future.

ACKNOWLEDGMENTS

This was written while on sabbatical leave at the CNR Centro di Studio per la Cristallografia Strutturale, and the Dipartimento di Mineralogia, Università di Pavia, Italy, and the Department of Earth Sciences, Cambridge University, England. I thank these institutions for their facilities during this time. Support was provided by the CNR Centro di Studio per la Cristallogafia Strutturale, and the Natural Science and Engineering Research Council of Canada. Most of all, I thank my wife, Robin, for word-processing this chapter in less than ideal conditions, while moving around Europe. Margery Osborne (University of Manitoba) and Catherine McCammon (University of British Columbia) provided information concerning experimental aspects of Mössbauer spectroscopy; I thank them for their help and co-operation.

REFERENCES

Aldridge, L.P., Bancroft, G.M., Fleet, M.E. and Herzberg, C.T. (1978) Omphacite studies, II. Mössbauer spectra of C2/c and P2/n omphacites. Amer. Mineral. 63, 1107-1115.
Amthauer, G. (1980) High pressure ^{57}Fe Mössbauer studies on minerals. In W. Schreyer, ed., High-Pressure Researches in Geoscience. E. Geoscience. E. Schweizerbart'sche Verlagsbuchhandlung, Stuttgart.

334

───── (1982) Gemischte Valenzzustäande des Eisens in Mineralen. Fortschr. Miner. 60, 119-152.

───── (1986) Crystal chemistry and valences of iron, antimony, and tin in franckeites. N. Jahrb. Mineral. Abh. 153, 272-278.

───── and Rossman, G.R. (1984) Mixed valence of iron in minerals with cation clusters. Phys. Chem. Minerals 11, 37-51.

─────, Annersten, H. and Hafner, S.S. (1976) The Mössbauer spectrum of ^{57}Fe in silicate garnets. Z. Kristallogr. 143, 14-55.

─────, McIver, J.R. and Viljoen, E.A. (1979) ^{57}Fe and ^{119}Sn Mössbauer studies of natural tin-bearing garnets. Phys. Chem. Minerals 4, 235-244.

─────, Langer, K. and Schliestedt, M. (1980) Thermally activated electron delocalization in deerite. Phys. Chem. Minerals 6, 19-30.

Annersten, H., Devanaryanan, S., Häggström, and Wäppling, R. (1971) Mössbauer study of synthetic ferriphogopite. Phys. Stat. Sol. 48B, 137-138.

─────, Ericsson, T. and Filippidis, A. (1982) Cation ordering in Ni-Fe olivines. Amer. Mineral. 67, 1212-1217.

─────, Adetunji, J. and Filippidis, A. (1984) Cation ordering in Fe-Mn silicate olivines. Amer. Mineral. 69, 1110-1115.

Aslani-Samin, S., Binczydka, H., Hafner, S.S., Pentinghaus, H. and Schürmann, K. (1987) Crystal chemistry of europium in feldspars. Acta Crystallogr. A43 (Supplement), C155.

Bailey, S.W. (1984a) Crystal chemistry of the true micas. Reviews in Mineralogy 13, 13-60.

───── (1984b) Review of cation ordering in micas. Clays and Clay Minerals 32, 81-92.

Baker, R.J. and Stevens, J.G. (1977) ^{121}Sb Mössbauer spectroscopy. Part II. Comparison of structure and bonding in Sb(III) and Sb(V) minerals. Rev. Chim. Mineral. 14, 339-346.

Ballet, O. and Coey, J.M.D. (1982) Magnetic properties of sheet silicates; 2:1 layer minerals. Phys. Chem. Minerals 8, 218-229.

Bancroft, G.M. (1967) Quantitative estimates of site populations in an amphibole by the Mössbauer effect. Phys. Lett. 26A, 17-18.

───── (1969) Quantitative site population in silicate minerals by the Mössbauer effect. Chem. Geol. 5, 255-258.

───── (1973) Mössbauer Spectroscopy. An Introduction for Inorganic Chemists and Geochemists. McGraw Hill, New York.

───── and Brown, J.R. (1975) A Mössbauer study of coexisting hornblendes and biotites: quantitative Fe^{3+}/Fe^{2+} ratios. Amer. Mineral. 60, 265-272.

───── and Burns, R.G. (1967) Distribution of iron cations in a volcanic pigeonite by Mössbauer spectroscopy. Earth Planet. Sci. Lett. 3, 125-127.

───── and ───── (1969) Mössbauer and absorption spectral study of alkali amphiboles. Mineral. Soc. Amer. Spec. Pap. 2, 137-148.

─────, ───── and Maddock, A.G. (1967a) Determination of the cation distribution in the cummingtonite-grunerite series by the Mössbauer spectroscopy. Amer. Mineral. 52, 1009-1026.

─────, Maddock, A.G. and Burns, R.G. (1967b) Applications of the Mössbauer effect to silicate mineralogy. I. Iron silicates of known crystal structure. Geochim. Cosmochim. Acta 31, 2219-2242.

─────, Burns, R.G. and Stone, A.J. (1967c) Applications of the Mössbauer effect to silicate mineralogy. II. Iron silicates of unknown and complex crystal structures. Geochim. Cosmochim. Acta 32, 547-559.

-----, ----- and Howie, R.A. (1967c) Determination of cation distribution in orthopyroxenes by the Mössbauer effect. Nature 213, 1221-1223.

-----, Maddock, A.G., Burns, R.G. and Stone, A.J. (1966) Cation distribution in anthophyllite from Mössbauer and infrared spectroscopy. Nature 212, 913-915.

-----, Williams, P.G.L. and Burns, R.G. (1971) Mössbauer spectra of minerals along the diopside-hedenbergite tie line. Amer. Mineral. 56, 1617-1625.

-----, Osborne, M.D. and Fleet, M.E. (1983) Next-nearest-neighbour effects in the Mössbauer spectra of Cr-spinels: an application of partial quadrupole splittings. Sol. State Comm. 47, 623-625.

Bara, J.J. and Bogacz, B.F. (1980) Geometric effects in Mössbauer transmission experiments. Möss. Effect Ref. Data J. 3, 154-163.

Besancon, J.R. (1981) Rate of cation ordering in orthopyroxenes. Amer. Mineral. 66, 965-973.

Borg, R.J. and Borg, I.Y. (1980) Mössbauer study of behaviour of oriented single crystals of riebeckite at low temperatures and their magnetic properties. Phys. Chem. Minerals 5, 219-234.

Bowen, L.H., Weed, S.B. and Stevens, J.G. (1968) Mössbauer study of micas and their potassium-depleted products. Amer. Mineral. 54, 72-84.

Buckley, A.N. and Wilkins, R.W.T. (1971) Mössbauer and infrared study of a volcanic amphibole. Amer. Mineral. 56, 90-100.

Burns, R.G. and Greaves, C.J. (1971) Correlation of infrared and Mössbauer site population measurements of actinolites. Amer. Mineral. 56, 2010-2033.

Coey, J.M.D., Ballet, O.. Moukarika, A. and Soubeyroux, J.L. (1981) Magnetic properties of sheet silicates; 1:1 layer minerals. Phys. Chem. Minerals 7, 141-148.

Cox, P.A. (1987) The Electronic Structure and Chemistry of Solids. Oxford Science Publications, Oxford.

Cranshaw, T.E., Dale, B.W., Longworth, G.O. and Johnson. C.E. (1986) Mössbauer Spectroscopy and its Applications. Cambridge University Press, Cambridge.

Curtis, L., Gittins, J., Kocman, V., Rucklidge, J.C., Hawthorne, F.C. and Ferguson, R.B. (1975) Two crystal structure refinements of a P2/n titanium ferro-omphacite. Can. Mineral. 13, 62-67.

Dickson, D.P. and Berry, F.J. eds., (1986) Mössbauer Spectroscopy. Cambridge University Press, Cambridge.

Dollase, W.A. (1975) Statistical limitations of Mössbauer spectral fitting. Amer. Mineral. 60, 257-264.

----- and Gustafson, W.I. (1982) ^{57}Fe Mössbauer spectral analysis of the sodic pyroxenes. Amer. Mineral. 67, 311-327.

Dowty, E. and Lindsley, D.H. (1973) Mössbauer spectra of synthetic hedenbergite-ferrosilite pyroxenes. Amer. Mineral. 58, 850-868.

Dyar, M.D. (1987) A review of Mössbauer data on trioctahedral micas: evidence for tetrahedral Fe^{3+} and cation ordering. Amer. Mineral. 72, 102-112.

----- and Burns, R.G. (1986) Mössbauer spectral study of ferruginous one-layer trioctahedral micas. Amer. Mineral. 71, 955-965.

-----, Naney, M.T. and Swanson, S.E. (1987) Effects of quench methods on Fe^{3+}/Fe^{2+} ratios: a Mössbauer and wet-chemical study. Amer. Mineral. 72, 792-800.

Eibschütz, M. and Ganiel, V. (1967) Mössbauer studies of Fe^{2+} in paramagnetic fayalite (Fe_2SiO_4). Sol. State Comm. 5, 267-270.

Eisenstein, J.C., Taragin, M.F. and Thorton, D.D. (1975) Anti-ferromagnetic order in amphibole asbestos. Proc. 20th AIP Conf. (1974): Magnetism and Magnetic Materials 24, 357-358.

Evans, B.J. and Amthauer, G. (1980) The electronic structure of ilvaite and the pressure and temperature dependence of its ^{57}Fe Mössbauer spectrum. J. Phys. Chem. Solids 41, 985-1001.

Ferrow, E. (1987a) Mössbauer effect and X-ray diffraction studies of synthetic iron bearing trioctahedral micas. Phys. Chem. Minerals 14, 276-280.

----- (1987b) Mössbauer and X-ray studies on the oxidation of annite and ferriannite. Phys. Chem. Minerals 14, 270-275.

Finch, J., Gainsford, A.R. and Tennant, W.C. (1982) Polarized optical absorption and ^{57}Fe Mössbauer study of pegmatitic muscovite. Amer. Mineral. 67, 59-68.

Fleet, M.E. (1977) The crystal structure of deerite. Amer. Mineral. 62, 990-998.

Gabriel, J.R. (1965) Computation of Mössbauer spectra. Mössbauer Effect Methodology 1, 121-132.

Ghose, S. (1988) Electron ordering and associated crystallographic and magnetic phase transitions in ilvaite, a mixed-valence iron silicate. In S. Ghose, J.M.D. Coey and E. Salje, eds., Structural and Magnetic Phase Transitions in Minerals. Springer-Verlag, New York.

----- and Weidner, J.R. (1972) $Mg^{2+}-Fe^{2+}$ order-disorder in cummingtonite, $(Mg,Fe)_7Si_8O_{22}(OH)_2$: a new geothermometer. Earth Planet. Sci. Lett. 16, 346-354.

-----, Cox, D.E. and Van Dang, N. (1987) Magnetic order in grunerite, $Fe_7Si_8O_{22}(OH)_2$ - a quasi-one dimensional antiferromagnet with a spin canting transition. Phys. Chem. Minerals 14, 36-44.

Ghosh, D., Kundu, T., Dasgupta, S. and Ghose, S. (1987) Electron delocalization and magnetic behaviour in a single crystal of ilvaite, a mixed valence iron silicate. Phys. Chem. Minerals 14, 151-155.

Gibb, T.C. (1977) Principles of Mössbauer Sepctroscopy. Chapman and Hall, London.

----- and Greenwood, N.N. (1965) Chemical applications of the Mössbauer effect. Part 2. Oxidation state of iron in crocidolite and amosite. Trans. Farad. Soc. 61, 1317-1323.

Goldman, D.S. (1979) A reevaluation of the Mössbauer spectroscopy of calcic amphiboles. Amer. Mineral. 64, 109-118.

----- and Rossman, G.R. (1977) The identification of Fe^{2+} in the M(4) site of calcic amphiboles. Amer. Mineral. 62, 205-216.

Gonser, V. (1975) From a strange effect to Mössbauer spectroscopy. Topics in Applied Physics 5, 1-51.

Goodman, B.A. (1976) The Mössbauer spectrum of a ferrian muscovite and its implications in the assignment of sites in dioctahedral micas. Mineral. Mag. 40, 513-517.

Greenwood, N.N. and Gibb, T.D. (1971) Mössbauer Spectroscopy. Chapman and Hall, London.

Guggenheim, S. (1984) The brittle micas. Reviews in Mineralogy 13, 61-104.

----- and Bailey, S.W. (1977) the refinement of zinnwaldite-1M in subgroup symmetry. Amer. Mineral. 62, 1158-1167.

Güttler, B. and Salje, E. (1987) Temperature dependence of optical absorption and charge transport in ilvaite. Z. Kristallogr. 178, 77-78.

Hafner, S.S. and Ghose, S. (1971) Iron and magnesium distribution in cummingtonites $(Fe,Mg)_7Si_8O_{22}(OH)_2$. Z. Kristallogr. 133, 301-326.

----- and Huckenholz, H.G. (1971) Mössbauer spectrum of synthetic ferrodiopside. Nature (Phys. Sci.) 233, 9-11.

Hawthorne, F.C. (1973) the Crystal Chemistry of the Amphiboles. Ph. D. Thesis, McMaster University, Hamilton.

----- (1981) Amphibole spectroscopy. Reviews in Mineralogy 9A, 103-139.

----- (1983a) Quantitative characterization of site-occupancies in minerals. Amer. Mineral. 68, 287-306.

----- (1983b) The crystal chemistry of the amphiboles. Can. Mineral. 21, 173-480.

----- and Grundy, H.D. (1977) The crystal structure and and site-chemistry of a zincian tirodite by least-squares refinement of X-ray and Mössbauer data. Can. Mineral. 15, 309-320.

----- and Ito, J. (1977) synthesis and crystal structure refinement of transition-metal orthopyroxenes. I: Orthoenstatite and (Mg,Mn,Co) orthopyroxene. Can. Mineral. 15, 321-338.

Hayashi, M., Tamura, I., Shimomura, O., Sawamoto, H. and Kawamura, H. (1987) Antiferromagnetic transition in fayalite under high pressure studied by Mössbauer spectroscopy. Phys. Chem. Minerals 14, 341-344.

Herber, R.H. (1984) ed., Chemical Mössbauer Spectroscopy. Plenum Press, New York.

Hogg, C.S. and Meads, R.E. (1975) A Mössbauer study of thermal decomposition of biotites. Mineral. Mag. 40, 79-88.

Ingalls, R. (1964) Electric-field gradient tensor in ferrous compounds. Phys. Rev. 133A, 787-795.

Joswig, W., Amthauer, G. and Takéucki, Y. (1986) Neutron-diffraction and Mössbauer spectroscopic study of clintonite (xanthophyllite). Amer. Mineral. 71, 1194-1197.

Law, A.D. (1973) Critical evalutation of "statistical best fits" to Mössbauer spectra. Amer. Mineral. 58, 128-131.

Levillain, C., Maurel, P. and Menil, F. (1981) Mössbauer studies of synthetic and natural micas on the polylithionite-siderophyllite join. Phys. Chem. Minerals 7, 71-76.

Linares, J., Regnard, J.-R. and Van Dang, N. (1983) Magnetic behaviour of grunerite from Mössbauer spectroscopy. J. Mag. Magn. Mater. 31-34, 715-716.

Litterst, F.J. and Amthauer, G. (1984) Electron delocalization in ilvaite, a reinterpretation of its ^{57}Fe Mössbauer spectrum. Phys. Chem. Minerals 10, 250-255.

Long, G.T., Cranshaw, T.E. and Longworth, G. (1983) The ideal Mössbauer effect absorber thickness. Möss. Effect Ref. Data J. 6, 42-49.

Maddock, A.G. (1985) Mössbauer spectroscopy in mineral chemistry. In F.J. Berry and D.J. Vaughan eds., Chemical Bonding and Spectroscopy in Mineral Chemistry. Chapmann and Hall, London.

Marion, P., Regnard, J.-R. and Wagner, F.E. (1986) Étude de l'état chimique de l'or dans des sulfures aurifères par spectroscopie Mössbauer de ^{197}Au: premiers résultats. C.R. Acad. Sci. Paris 302, 571-574.

McCammon, C.A. and Burns, R.G. (1980) The oxidation mechanism of vivianite as studied by Mössbauer spectroscopy. Amer. Mineral. 65, 361-366.

----- and Price, D.C. (1985) Mössbauer spectra of Fe_xO (x > 0.95). Phys. Chem. Minerals 11, 250-254.

338

Mössbauer, R.L. (1958) Kernresonanzfluoresent von Gammastrahlung in Ir[191]. Z. Phys. 151, 124-143.

Moukarika, A., Coey, J.M.D. and Van Dang, N. (1983) Magnetic order in crocidolite asbestos. Phys. Chem. Minerals 9, 269-275.

Murad, E. (1984) Magnetic ordering in andradite. Amer. Mineral. 69, 722-724.

----- and Wagner, F.E. (1987) The Mössbauer spectrum of almandine. Phys. Chem. Minerals 14, 264-269.

Mysen, B.O., Virgo, D., Neumann, E.R. and Seifert, F.A. (1985) Redox equilibria and the structural states of ferric and ferrous iron in melts in the system $CaO-MgO-Al_2O_3-SiO_2-Fe-O$: relationships between redox equilibria, melt structure and liquidus phase equilibria. Amer. Mineral. 70, 317-330.

Nord, A.G., Annersten, H. and Filippidis, A. (1982) The cation distribution in synthetic Mg-Fe-Ni olivines. Amer. Mineral. 67, 1206-1211.

Parish, R.V. (1984) Mössbauer spectroscopy and the chemical bond. In G.J. Long, ed., Mössbauer Spectroscopy. Applications in Inorganic Chemistry. Plenum Press, New York, 527-575.

Rice, C.M. and Williams, J.M. (1969) A Mössbauer study of biotite weathering. Mineral. Mag. 37, 210-215.

Richardson, S. and Vaughan, D.J. (1987a) Surface alterations of pentlandite and spectroscopic evidence for secondary violarite formation. Abstracts, Mineral. Soc. Winter Meet. 1987, 35-36.

----- and ----- (1987b) Asenopyrite: a spectroscopic examination of oxidized surfaces. Abstracts, Mineral. Soc. Winter Meet. 1987, 36-37.

Riedel, E. and Karl, R. (1980) Mössbauer studies of thiospinels. I. The system $FeCr_2S_4-FeRh_2S_4$. J. Sol. State Chem. 35, 77-82.

Rossi, G., Oberti, R., Dal Negro, A., Molin, G.M. and Mellini,M. (1987) Residual electron density at the M2 site in C2/c clinopyroxenes: relationships with bulk chemistry and sub-solidus evolution. Phys. Chem. Minerals 14, 514-520.

Ruby, S.L. (1973) Why MISFIT when you already have X^2? Mössbauer Effect Methodology 8, 263-276.

Santoro, R.P., Newnham, R.E. and Nomura, S. (1966) Magnetic properties of Mn_2SiO_4 and Fe_2SiO_4. J. Phys. Chem. Solids 27, 655-666.

Saxena, S.K. (1973) Thermodynamics of Rock-forming Crystalline Solutions. Vol. 8, Rocks, Minerals and Inorganic Materials. Springer-Verlag, New York.

----- and Ghose, S. (1971) $Mg^{2+}-Fe^{2+}$ order-disorder and the thermodynamics of the orthopyroxene crystalline solution. Amer. Mineral. 56, 532-559.

Schaefer, M.W. (1985) Site occupancy and two-phase character of "ferrifayalite". Amer. Mineral. 70, 729-736.

Schwartz, K.B., Nolet, D.A. and Burns, R.G. (1980) Mössbauer spectroscopy and crystal chemistry of natural Fe-Ti garnets. Amer. Mineral. 65, 142-153.

Seifert, F. (1978) Equilibrium $Mg-Fe^{2+}$ cation distribution in anthophyllite. Amer. J. Sci. 278, 1323-1333.

----- (1983) Mössbauer line-broadening in aluminous orthopyroxenes: evidence for next-nearest-neighbour interactions and short-range order. N. Jahrb. Mineral. Abh. 148, 141-162.

----- (1987) Intracrystalline distributions in minerals and the problem of solid solutions. In G. Calas ed., Méthodes Spectroscopiques Applquées au Minéraux. Soc. Franc. Minéral. Crist., 505-526.

Sherman, D.M. (1987) Molecular orbital (SCF-Xα-SW) theory of metal-metal charge transfer processes in minerals. I. Application to $Fe^{2+}-->Fe^{3+}$ charge transfer and "electron delocalization" in mixed-valence iron oxides and silicates. Phys. Chem. Minerals 14, 355-363.

Shinno, I. (1974) Mössbauer studies of olivines. The relation between Fe^{2+} site occupancy number T_{Mi} and interplanar distance d_{130}. Mem. Geol. Soc. Jap. 11, 11-17.

----- (1981) A Mössbauer study of ferric iron in olivine. Phys. Chem. Minerals 7, 91-95.

Skogby, H. (1987) Kinetics of intracrystalline order-disorder relations in tremolite. Phys. Chem. Minerals 14, 521-526.

----- and Annersten, H. (1985) Temperature dependent Mg-Fe cation distribution in actinolite-tremolite. N. Jahrb. Mineral. Mn. 1985, 193-203.

Stanek, J. (1986) CEMS study of enstatite single crystal implanted with Fe ions. Hyperfine Interactions 29, 1245-1248.

-----, Hafner, S.S. and Sawicki, J.A. (1986) Local states of Fe^{2+} and Mg^{2+} in magnesium rich-olivines. Amer. Mineral. 71, 127-135.

Stevens, J.G. (1984) Mössbauer spectroscopy of antimony compounds. In Chemical Mössbauer Spectroscopy: Proc. Sym. 25 Yrs. Chem. Möss. Spec., 319-342.

Svane, A. and Antoncik, E. (1987) Theoretical investigation of the isomer shifts of the ^{119}Sn Mössbauer isotope. Phys. Rev. B 35, 4611-4624.

Thomas, W.M. (1982) ^{57}Fe Mössbauer spectra of natural and synthetic hastingsites, and implications for peak assignments in calcic amphiboles. Amer. Mineral. 67, 558-567.

Tricker, M.J. (1981) Conversion electron Mössbauer spectroscopy and its recent development. Adv. Chem. Ser. 1981, 63-100.

Van Rossum, M. (1982) Characterization of solids by Mössbauer spectroscopy. Prog. Cryst. Growth 5, 1-45.

Vedder, W. and Wilkins, R.W.T. (1969) Dehydroxylation and rehydroxylation, oxidation and reduction of micas. Amer. Mineral. 54, 482-509.

Virgo, D. (1972) Preliminary fitting of ^{57}Fe Mössbauer spectra of synthetic Mg-Fe richterites. Carnegie Inst. Wash. Year Book 71, 534-538.

----- and Hafner, S.S. (1969) Fe^{2+},Mg order-disorder in heated orthopyroxenes. Mineral. Soc. Amer. Spec. Pap. 2, 67-81.

----- and ----- (1970) Fe^{2+},Mg order-disorder in natural orthopyroxenes. Amer. Mineral. 55, 201-223.

Wagner, F.E., Marion, Ph. and Regnard, J.-R. (1980) Mössbauer study of the chemical state of gold in gold ores. Gold 100: Proc. Int. Proc. Int. Conf. Gold, Vol. 2: Extractive Metallurgy of Gold. Johannesburg, SAIMM, 435-443.

-----, ----- and ----- (1988) ^{197}Au Mössbauer study of gold ores, mattes, roaster products, and gold minerals. Hyperfine Interactions (to be published).

Waychunas, G.A. (1986) Performance and use of Mössbauer goodnes-of-fit parameters: response to spectra of varying signal/noise ratio and possible misinterpretations. Amer. Mineral. 71, 1261-1265.

Wertheim, G.K. (1964) The Mössbauer Effect. Principles and Applications. Academic Press, New York.

Williams, P.G.L., Bancroft, G.M., Bown, M.G. and Turnock. A.C. (1971) Anomalous Mössbauer spectra of C2/c clinopyroxenes. Nature (Phys. Sci.) 230, 149-151.

340

Zhou, X.Z., Pollard, R.J. and Morrish, A.H. (1986) Conversion-electron and transmission Mössbauer study of natural microcrystalline goethite. Hyperfine Interactions 28, 863-866.

Chapter 9   R. James Kirkpatrick

MAS NMR Spectroscopy of Minerals and Glasses

INTRODUCTION

Nuclear magnetic resonance spectroscopy (NMR) is a powerful probe of the static structure and dynamic behavior of condensed phases, which, because it directly examines the properties of a specific element, often offers significant advantages over diffraction methods and vibrational spectroscopy. ^{1}H and ^{13}C NMR have been standard tools for determining the structure of organic molecules in solution for many years. Only with the advent of high-field superconducting magnets, magic-angle spinning (MAS), and the ability to examine nuclides with a quadrupole moment, has it been possible to routinely work with a wide range of inorganic solids. Since about 1980, when the first systematic study of the ^{29}Si NMR behavior of inorganic solids was published (Lippmaa et al., 1980), the application of NMR spectroscopic methods to solids has grown explosively, and it is fast becoming an important tool in mineralogy, catalyst and polymer chemistry, and ceramics.

Specific classes of problems concerning inorganic solids to which NMR has been applied include the structures of amorphous and poorly crystalline materials and crystalline phases too fine grained to examine by single crystal diffraction (e.g., glasses, gels, zeolite catalysts, clays), order/disorder phenomena in crystals (especially Al/Si), the structure and dynamic behavior of melts, the detection of small amounts of phases or molecules in mixtures, and the investigation of reaction mechanisms. This chapter will briefly summarize some aspects of the theory and experimental practice of NMR spectroscopy, but will focus primarily on the structural interpretation of NMR spectra and the application of NMR methods to mineralogical and geochemical problems involving inorganic solids. The following chapter by Stebbins discusses NMR relaxation and the investigation of the static structure and dynamic behavior of melts in more detail.

THEORY

The NMR phenomenon can be described by complementary classical and quantum mechanical theories, both of which are useful in experimental design and data interpretation. This chapter presents without support only those theoretical points needed to understand the behavior of nuclear spin systems, the operation of NMR spectrometers, and the interpretation of experimental results. More complete developments are available in numerous texts at many levels. Akitt (1983) is a good introductory text. Wilson (1987) also presents a useful elementary introduction and an extensive discussion of applications to organic geochemistry. Standard intermediate texts, each with a slightly different viewpoint, are Farrar and Becker (1971), Becker (1980), Fukushima and Roeder (1981), Harris (1983), Fyfe (1983), Gerstein and Dybowski (1985), and Sanders and Hunter (1987). The standard advanced texts are Abragam (1961) and Schlicter (1978). The background needed to understand NMR at the level of the intermediate texts is quantum mechanics of the type taught in junior-year chemistry-major's physical chemistry courses in U.S. universities.

Quantum mechanical description

Many atomic nuclei have a quantized property called <u>spin</u>, which can use-fully be thought of as being caused by the physical spinning of the nucleus. The angular momentum, J, of such a nucleus is given by

$$J=\hbar[I(I+1)]^{\frac{1}{2}} , \tag{1}$$

where $\hbar$ is Planck's constant over 2π and I is the spin quantum number, which can be either integer or half-integer. Nuclides with even mass number and even charge (^{12}C, ^{16}O) have zero spin and are of no interest to NMR spectroscopy. Nuclides with odd mass numbers (^{29}Si, ^{27}Al) have half-integer spins and are of the most interest for solid-state NMR. Nuclides with even mass numbers and odd charge (^{2}H, ^{14}N) have integer spins and are often difficult to examine in solids. Most nuclides have spins between 0 and 7/2. The <u>magnetic moment</u>, $\bar{\mu}$, of a nucleus is a fundamental property and is given by

$$\bar{\mu}=\gamma\bar{J} , \tag{2}$$

where γ is the <u>gyromagnetic ratio</u>, which is defined by equation (2) and is a con-stant for each nuclide.

Each nuclide has $2I+1$ energy levels, described by the quantum number m, which takes on the values I, I-1, I-2, ..., -I. In the absence of a magnetic field, these energy levels are degenerate (have the same energy), but when a magnetic field is present, interaction between them and the field (the Zeeman interaction) lifts this degeneracy (Fig. 1). The energy difference between these states, ΔE, is given by

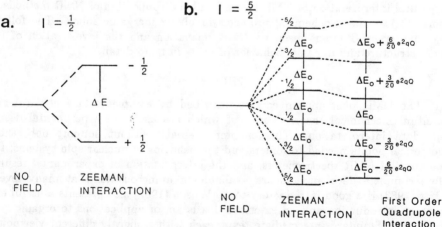

Figure 1. (a) Nuclear spin energy level diagram for a spin $I=1/2$ nuclide. When the sample is placed in a magnetic field, the Zeeman interaction lifts the degeneracy of the energy levels to a value, ΔE, proportional to the magnetic field at the nucleus. The NMR frequency is directly proportional to ΔE. (b) Nuclear spin energy level diagram for a spin $I=5/2$ nuclide, showing the 6 ($=2I+1$) energy levels split by the Zeeman interaction and further split by the interaction of the nuclear quadrupole moment with the electric field gradient at the nucleus. This quadrupole interaction causes the five allowed transitions to have different energies. Usually only the central $(1/2,-1/2)$ transition is observed, although recent work with the spinning sidebands due to the satellite $\pm(1/2,3/2)$ transi-tions has shown that higher resolution can sometimes be obtained with them (Samoson et al., 1986).

$$\Delta E = | \gamma \hbar H | \quad , \tag{3}$$

where H is the magnetic field at the nucleus. The relative populations of higher and lower energy levels, N_a and N_b, are given by the Boltzmann distribution,

$$\frac{N_a}{N_b} = \exp(-\Delta E / kT) \quad . \tag{4}$$

Because the energy differences are small, the populations in each level are nearly the same, causing the signal intensity to be low relative to spectroscopies involving electronic energy states.

A nucleus can increase or decrease its energy only by absorbing or emitting a photon with a frequency, ν, given by

$$\nu = \frac{\gamma}{2\pi} H \quad . \tag{5}$$

The operative selection rules allow transitions only between adjacent energy levels. It is the frequency of this radiation, which is in the radio-frequency (r.f.) range, that is measured in an NMR experiment.

The NMR phenomenon is chemically and structurally useful, in large part, because the electrons in the vicinity of the atom shield the nucleus a tiny (ppm) amount from the applied magnetic field, H_0. Nuclei in different structural environments thus see slightly different magnetic fields and consequently absorb and emit photons of slightly different frequencies. This shielding is described by the shielding tensor, σ_{ij}, with the isotropic shielding, σ_i, defined as $1/3$ Tr σ_{ij}. In general, the shielding is anisotropic, with the chemical shift anisotropy (CSA, $\Delta\sigma$) defined by $\Delta\sigma = \sigma_{33} - \frac{1}{2}(\sigma_{11} + \sigma_{22})$. Due to this shielding,

$$H = H_0(1-\sigma) \quad , \tag{6}$$

and

$$\nu = \gamma \frac{H_0}{2\pi}(1-\sigma) \quad . \tag{7}$$

Because the absolute values of H_0 are difficult to measure, sufficiently accurate absolute values of NMR frequencies are impossible to obtain, and the resonance frequencies are normally reported as chemical shifts, δ, relative to an experimentally useful standard.

$$\delta = \left(\frac{\nu_{sample} - \nu_{standard}}{\nu_{standard}}\right) \times 10^6 \quad . \tag{8}$$

The chemical shift corresponding to the isotropic shielding is called the isotropic chemical shift, δ_i. More negative or less positive chemical shifts correspond to larger shieldings. For determining structure via NMR the chemical shift is normally the most useful parameter.

Classical description

To understand many of the experimental aspects of NMR spectroscopy, it is often useful to consider the NMR phenomenon in strictly classical terms, without direct reference to the quantum description, although the two can be integrated. In this description (Fig. 2) the spin of the nucleus causes the magnetic moment to precess about the orientation of the magnetic field (taken to be the z-axis) in the same way a spinning gyroscope precesses about the orientation of a gravitational field. The frequency of this precession, the Larmor frequency, is exactly equal to the resonance frequency of the quantum description.

The magnetic moments of all the nuclei of a particular nuclide in a sample, called the spin system, can be viewed as forming a cone about the z-axis , yielding a net magnetization, M, parallel to H_0. When viewed in a reference frame rotating about the z-axis at the Larmor frequency with respect to the laboratory reference frame (the so-called rotating frame), the spin system appears to be stationary.

An NMR experiment is done by irradiating the sample with plane-polarized radio-frequency radiation at the Larmor frequency such that a small magnetic field (H_1) is generated parallel to the x-axis in Figure 2 (see below for a brief description of a pulse-Fourier-transform NMR spectrometer). This magnetic field causes the spin system to tip with respect to H_0. In the simplest experiment, this tip angle is 90° (corresponding to equal populations of the two nuclear energy levels involved), and the net magnetization is now parallel to the y'-axis.

At this point in a simple 1-pulse experiment, the r.f. field is turned off, and two things happen to the spin system. It dephases in the x-y plane (called T_2 relaxation), and it returns to the z-axis (called T_1 relaxation); T_1 and T_2 are the time constants which describe these two relaxation processes. For solids, T_2 is usually shorter than T_1. The radiation emitted by the sample during dephasing in the x-y plane is detected by an antenna (usually the same as the transmitter antenna). The signal is recorded as a function of time (in the so-called time domain) to give a free induction decay (FID, Fig. 3). In most experiments the sample is irradiated with many pulses and the signal due to each pulse added together. Thus, the signal/noise (S/N), which is proportional to (number of pulses)$^{1/2}$, is greatly enhanced relative to traditional continuous-wave NMR methods. The r.f. pulses are typically spaced at about $5T_1$. The highest S/N in the least time is obtained by using an H_1 pulse that tips the spin system an angle, α, less than 90.KS°, α is given by

$$\cos\alpha = \exp[-t/T_1] \quad , \tag{9}$$

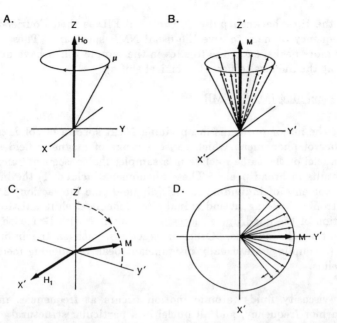

Figure 2. (a) Diagram showing the precession of the magnetic moment of a nucleus about the principle axis of the applied magnetic field, H_0. (b) Diagram showing the magnetic moments of many individual nuclei (the spin system) fixed with respect to the rotating frame (the primed axies) and resulting in a net magnetization, M, parallel to the orientation of H_0 in a static magnetic field. (c) Diagram showing the tipping of the net magnetization, M, of a spin system due to a small magnetic field, H_1, perpendicular to H_0 in the rotating frame. In a simple 1-pulse experiment, H_1 is applied long enough to orient M perpendicular to H_0. (d) Diagram showing the dephasing of a spin system in the x'-y' plane of the rotating frame. During this time signal is emitted from the sample and detected by the spectrometer.

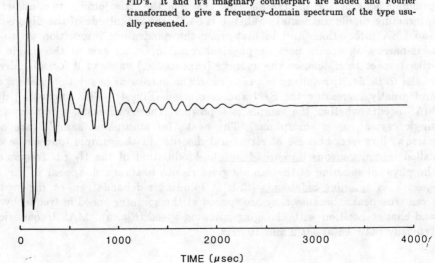

Figure 3. Real part of the free induction decay (FID) of ^{13}C in calcite. This time-domain spectrum is the sum of 186 individual FID's. It and it's imaginary counterpart are added and Fourier transformed to give a frequency-domain spectrum of the type usually presented.

where t is the time between pulses. The total FID is then Fourier-transformed into the frequency domain to give the usual NMR spectrum. These spectra contain one or more peaks, each due to sites in the sample which have different magnetic fields at the nucleus (different chemical shifts).

Magic-angle spinning (MAS) NMR

One of the major problems in obtaining NMR spectra of solids is that there are a number of phenomena which cause a range of magnetic field strengths at individual nuclei of the same nuclide in a sample; the consequent range of chemical shifts results in broad peaks. These phenomena include: 1) the interaction of the dipole moments of individual nuclei (dipole-dipole interaction) 2) anisotropy of the electronic shielding at individual sites (chemical shift anistropy), and 3) the interaction of the quadrupole moment of a nuclide with $I \geq 1$ with the electric field gradient at the nucleus. Other interactions, such as the indirect nuclear interaction (J-coupling) are usually insignificant peak broadening mechanisms for inorganic solids.

In low-viscosity fluids, atomic motion occurs at frequencies much higher than the Larmor frequency, and all nuclei in a particular structural environment see the same average magnetic field. The resulting peaks are often extremely narrow. In solids such motional narrowing normally does not occur, and the peaks for powdered samples (which contain a large number of crystallographic orientations) are broad and often uninterpretable.

This peak broadening is now usually overcome by the use of an experimental technique called Magic-Angle Spinning (MAS). I know of no simple description of how MAS works. It is best understood in terms of the time-dependent Hamiltonians describing the various interactions, which can be treated as perturbations on the Zeeman interaction. These interactions contain terms involving $(3\cos^2\theta - 1)$, where θ is the angle between H_0 (the applied magnetic field) and the principle axis of the interaction (for example the axis joining two nuclei with interacting dipole moments). When θ is 54.7°, the magnitude of the dipole-dipole and CSA interactions and to first-order the quadrupole interaction go to zero. Line-narrowing occurs because physical rotation on an axis at the magic angle with respect to H_0 causes the average (expectation) value of θ for all individual nuclei to be 54.7°, resulting in peaks nearly as narrow as in solution state spectra, and greatly increasing the S/N and resolution (Fig. 4). This narrowing due to MAS occurs whether the sample is a powder (the usual case), a few crystals, a single crystal, or is amorphous. The peaks for amorphous samples are not as narrow, however, because of structural disorder in the sample (one cause of so-called inhomogeneous line-broadening). Modulation of the H_1 r.f. frequency by the physical spinning of the sample gives rise to beats which appear in the NMR spectra as spinning sidebands (SSB). Spinning sidebands can be distinguished from true peaks, because they are spaced at the spinning speed in frequency units and change position with changing spinning speed (Fig. 5). MAS frequencies are typically between about 2 and 10 kHz.

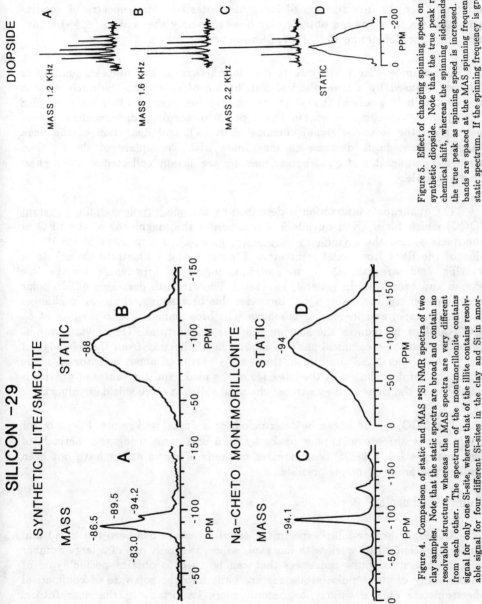

SILICON –29

SYNTHETIC ILLITE/SMECTITE

MASS   STATIC

A   B

–83.0 –86.5 –89.5 –94.2   –88

Na–CHETO MONTMORILLONITE

MASS   STATIC

C   D

–94.1   –94

Figure 4. Comparison of static and MAS ²⁹Si NMR spectra of two clay samples. Note that the static spectra are broad and contain no resolvable structure, whereas the MAS spectra are very different from each other. The spectrum of the montmorillonite contains signal for only one Si-site, whereas that of the illite contains resolvable signal for four different Si-sites in the clay and Si in amorphous silica (at -110 ppm). The static and MAS spectra of each sample are on the same scale. Spectra courtesy of C.A. Weiss, Jr.

DIOPSIDE

A   MASS 1.2 KHz

B   MASS 1.6 KHz

C   MASS 2.2 KHz

D   STATIC

Figure 5. Effect of changing spinning speed on the ²⁹Si spectrum of synthetic diopside. Note that the true peak remains at the same chemical shift, whereas the spinning sidebands move farther from the true peak as spinning speed is increased. The spinning sidebands are spaced at the MAS spinning frequency and map out the static spectrum. If the spinning frequency is greater than the static peak width, no sidebands will be present. Spectra courtesy of K.A. Smith.

Quadrupole interaction

Although the first-order term in the Hamiltonian describing the quadrupole perturbation to the Zeeman interaction contains $(3\cos^2\theta-1)$ terms, the second-order term does not. Thus, if the quadrupole interaction for a nuclide with $I \geq 1$ is significant, as it is for all but the most symmetrical and ionic solids, these second-order effects must be considered. Because 81 out of the approximately 110 NMR active nuclides have a quadrupole moment, such effects are extremely important for the investigation of inorganic materials. Most spectra of quadrupole nuclides in solids are obtained by observing only the central ($\frac{1}{2}$, -$\frac{1}{2}$) transition (Fig. 1b; Cohen and Reif, 1957; Meadows et al., 1982).

Quadrupole effects are due to the interaction of the nuclear quadrupole moment (caused by a non-spherical distribution of charge on the nucleus) with the electric field gradient (EFG) at the nucleus (see Harris, 1983, for a detailed description). Quadrupole effects cause peak broadening, displacement of the peak from the isotropic (true) chemical shift (δ_i), and distortion of the peak shape. These effects decrease in magnitude with the square of the H_0 field strength, and spectra of quadrupolar nuclides are usually collected at the highest field available.

The quadrupole interaction is described by the quadrupole coupling constant (QCC) which for a given nuclide is a measure of the magnitude of the EFG at the nucleus, and the asymmetry parameter, η, which is a measure of the deviation of the EFG from axial symmetry. Figures 6 and 7 illustrate the effects of varying field strength, QCC, η, and inhomogeneous broadening on the line shapes and breadths. In general, increasing field strength decreases quadrupolar peak broadening, increasing QCC increases this broadening, changing η changes the peak shape, and increasing inhomogenous broadening (due to a range of δ_i, QCC, and/or η) reduces the sharpness of the singularities. Under static conditions, the isotropic chemical shift is about 1/3 of the way from the left edge of the peak. Under MAS conditions, the peak is narrowed about a factor of 3, the isotropic chemical shift is at the left edge of the peak, and the center of gravity is displaced to the right of the isotropic chemical shift (to more shielded values).

The QCC, η, and δ_i can be determined for a single, well resolved peak or for up to about three overlapping peaks by trial and error using the methods of Ganapathy et al. (1982). Least-squares methods of curve fitting have not been applied but are in principle possible.

Other experiments

Although the one-pulse experiment described above has been the bread and butter of recent NMR work with inorganic solids, it is only one of a large number of well-understood pulse sequences that can be used to obtain specific types of data. Many of these pulse sequences are built into the software of commercial spectrometers and are likely to become more important in the near future. Fukushima and Roeder (1981), Akitt (1983), and Gerstein and Dybowski (1985)

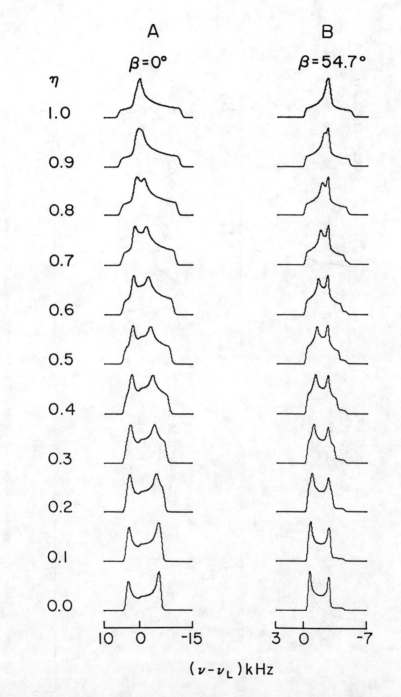

Figure 6. Calculated peak shapes for a spin I=3/2 nuclide on a site with a QCC of 3 MHz with different asymmetry parameters under static and MAS conditions. The isotropic chemical shift is 0 ppm. Note the different scales for the static and MAS spectra. Courtesy of S. Schramm.

350

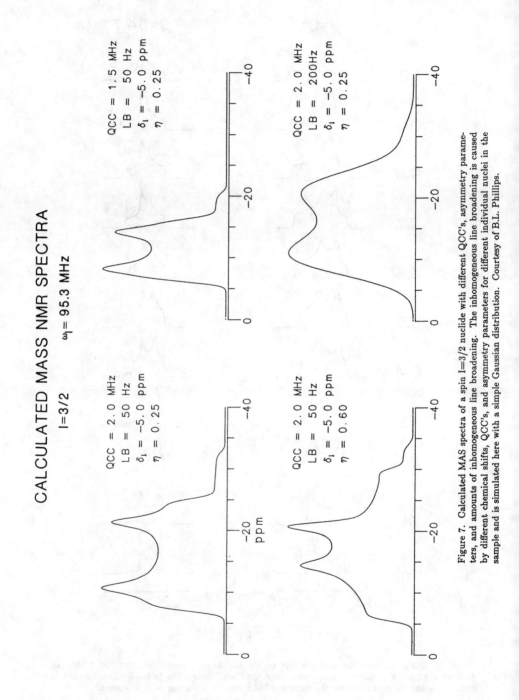

CALCULATED MASS NMR SPECTRA

I=3/2 ω_I = 95.3 MHz

QCC = 2.0 MHz
LB = 50 Hz
δ_I = −5.0 ppm
η = 0.25

QCC = 1.5 MHz
LB = 50 Hz
δ_I = −5.0 ppm
η = 0.25

QCC = 2.0 MHz
LB = 50 Hz
δ_I = −5.0 ppm
η = 0.60

QCC = 2.0 MHz
LB = 200Hz
δ_I = −5.0 ppm
η = 0.25

Figure 7. Calculated MAS spectra of a spin I=3/2 nuclide with different QCC's, asymmetry parameters, and amounts of inhomogeneous line broadening. The inhomogeneous line broadening is caused by different chemical shifts, QCC's, and asymmetry parameters for different individual nuclei in the sample and is simulated here with a simple Gaussian distribution. Courtesy of B.L. Phillips.

describe the theory and application of some of these pulse sequences, and this paper will only discuss a few of the major types.

One of the most important groups of experiments involves <u>double</u> <u>resonance</u>, that is, exciting two nuclides in the same experiment. Of these, <u>decoupling</u> and <u>cross-polarization</u> are the most commonly used.

Decoupling experiments. These experiments remove the dipole-dipole peak broadening, which is sometimes too large to remove by MAS. Such peak-broadening may be due to either the interaction of the dipole moments of nuclei of the same nuclide (homonuclear broadening) or between nuclei of the nuclide of interest and a second nuclide (heteronuclear broadening). Most commonly the second (decoupled) nuclide is ^{1}H. In this experiment the protons are continuously irradiated at their Larmor frequency, putting the ^{1}H spin system along the y'-axis, effectively removing its dipole moment from consideration. The usual NMR spectrum of the nuclide of interest is collected during this decoupling period and is, thus, free of ^{1}H dipolar broadening.

Cross polarization (CP). This technique, again often used with ^{1}H, is an extension of decoupling and provides increased S/N and additional structural information by transferring spin from the ^{1}H spin system to the spin system of interest (Yannoni, 1982). In this experiment, both nuclides are irradiated at the correct frequencies to fulfill the Hartmann-Hahn condition:

$$\gamma_H H_H = \gamma_i H_i \ , \tag{10}$$

where the H's are the H_1 fields of protons (H_H) and the observed nuclide (H_i).

Under these conditions in the rotating frame both spin systems precess about H_1, but not H_0, at the same frequency, causing magnetization transfer from the high energy ^{1}H spin system to the other, lower energy spin system. An FID of the nuclide of interest is then collected at the higher magnetization level while ^{1}H decoupling continues. Thus, the T_1 of the second nuclide is the same low value as that of the protons, often allowing much more rapid data collection. Because of the higher magnetization and larger number of scans that can be collected in a given time, the S/N is often greatly enhanced. In addition and perhaps more importantly, signal from nuclei closer to the protons is preferentially enhanced relative to that of nuclei further from them (Fig. 8), providing additional structural information and eliminating from the spectrum signal from anhydrous phases. Cross polarization under MAS conditions carries the acronym CPMAS.

Spin echo experiments. In these experiments (described well by Harris, 1983) the spin system is given a pulse, allowed to partially decay, and then given another pulse or pulse sequence which causes the dephasing spin system to rephase (refocus). An FID is then collected and Fourier-transformed. Such experiments are often used to measure relaxation times and collect spectra with

HYDROTHERMALLY REACTED ALBITE
250°C, pH 1.0–1.9, 28 DAYS

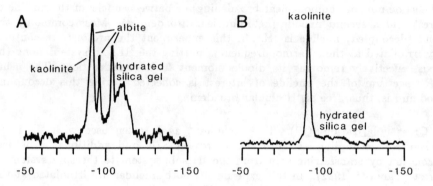

MASS ^{1}H CP/MASS

SILICON–29

A   albite   kaolinite   hydrated silica gel

B   kaolinite   hydrated silica gel

PPM (FROM TMS)

ALUMINUM–27

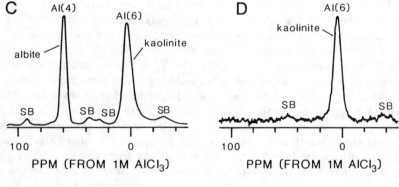

C   Al(4)   Al(6)   albite   kaolinite   SB

D   Al(6)   kaolinite   SB

PPM (FROM 1M AlCl$_3$)

SB: SPINNING SIDEBAND

Figure 8. Comparison of ^{27}Al and ^{29}Si MAS and CPMAS spectra of hydrothermally altered albite. The MAS spectra show signals from both the unhydrated albite and the hydrous alteration products, kaolinite and silica gel. The CPMAS spectra show signals only from the hydrous phases due to magnetization transfer from the ^{1}H spin system to the ^{27}Al or ^{29}Si spin system. In these spectra, signal from anhydrous phases is totally supressed. Spectra courtesy of W.H. Yang.

broad peaks free of the complications that can arise due to the 7 - 50 μsec instrumental dead time (pulse breakthrough period) at the beginning of the FID obtained from a one-pulse experiment.

2-D NMR. Another class of experiments involves Fourier-transformation over two time variables and is called 2-D NMR (Bax, 1984; Bax and Lerner, 1986; Ernst et al., 1986; Sanders and Hunter, 1987). These experiments are not yet entirely routine, even for solutions, but can provide information about site to site connectivities, relative QCC's, and other variables. In general, an experiment is done by collecting a series of spectra, each with a different value of some time variable, such as the time between pulses in a multiple-pulse experiment (evolution period). The spectra are Fourier-transformed in the usual way and then over the second time variable. The data are usually displayed as a contour plot, often displaying the chemical shift along one axis and the effects related to the evolution period along the other. Correlated spectroscopy (COSY) is an especially useful technique, because it provides information about which sites are directly connected. Spectra obtained using COSY contain the usual 1-D spectrum along the diagonal and cross peaks off the diagonal indicating which sites are connected (Fig. 9).

PRACTICAL MATTERS

NMR spectrometer operation and data collection are somewhat more difficult than, for example, operation of an electron microprobe, although the newer, automated instruments are easier to use. This section provides a brief introduction to NMR spectrometers to bring the ideas just discussed into more concrete terms. Gerstein and Dybowski (1985) provide a detailed, practical description of modern NMR instrumentation, and the Radio Amateurs Handbook is a good introduction to r.f. theory and practice.

NMR frequencies are all in the radio frequency (r.f.) range, and NMR spectrometers are, thus, computer operated radio transmitting and receiving systems. A receiver and transmitter antenna surrounds the sample, which is located in the bore of a superconducting magnet. Software and hardware are now available to fully automate both data collection and analysis.

The computers on which new spectrometers are based are usually PC size, often with an array processor board, although some systems now operate off mini-VAX-size computers. Data are usually stored on hard disks, and all systems have some sort of plotter for producing hard copies of the spectra.

The radio transmitter system consists of a tuneable r.f. generator, a pulse programmer and pulse gating system that is controlled by the computer, an amplifier, and band-pass filters. The generator produces continuous wave r.f. signal at the desired frequency. This signal then passes through the pulse programmer and gating box, where it is converted to pulses of the appropriate length which appear in the system at the appropriate time. The signal then enters the amplifier, where it is boosted to a large enough power (often hundreds of watts)

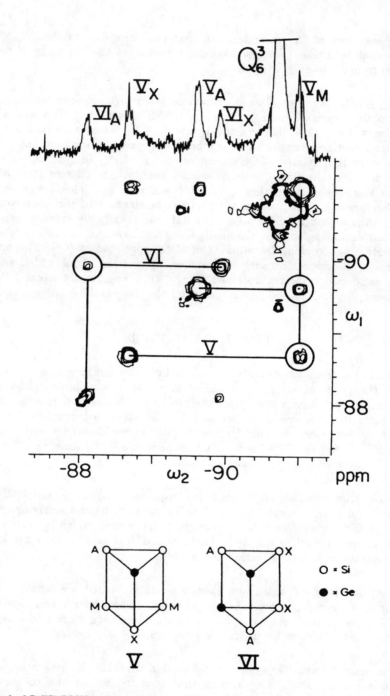

Figure 9. 2-D ^{29}Si COSY spectrum of the two silicate anions shown in aqueous solution. The cross peaks arise from spin-spin (J) coupling of the ^{29}Si nuclei in this isotopically enriched sample. The peaks in the ordinary 1-D spectrum shown at the top are keyed to the labels of the atoms in the structure diagrams. Spectrum from Knight et al. (1987).

to excite the nuclear spin system of interest. The band-pass filters reduce noise.

The output from the transmitter system then enters the sample probe, which contains tuneable capacators, the transmitting/receiving antenna, a mechanical assembly for spinning the sample (the stator), and the sample in a rotor in which it is spun. The capacitors are used to fine-tune the circuits to produce maximum S/N. The antenna is usually a coil 1-2 cm long and about 1 cm in diameter. The stator is an air-bearing emitting high pressure gas which forces the rotor, which always has some grooves or flutes on it, to spin. The rotor is usually between 0.5 and 1 cm in diameter and can be cylindrical or mushroom-shaped. The spinning speed attainable depends on the gas pressure and the uniformity of the sample packing in the rotor, and increases with decreasing rotor diameter. S/N increases with sample volume, however, and the largest possible rotors consistent with completely separating the sidebands from the real peaks are usually used. Sample sizes are typically from 100 to 500 mg and are usually powders, but can be large pieces, single crystals, or even slurries of liquid and solid.

The receiving system is simply a very high quality radio receiver, again with appropriate band-pass filters, which sends the detected signal on to the computer, where it is stored for future processing.

In a simple 90° one-pulse experiment, the transmitter emits r.f. at the resonance frequency of the nuclide being observed in a pulse long enough to equalize the populations of the higher and lower nuclear energy levels. In the classical description this is equivalent to a 90° pulse which puts the spin system along the y'-axis. Typical pulses are 1 to 15 µs long. After this pulse, the sample then begins to emit r.f. at the resonance frequency as the spin system dephases in the x-y plane (T_2 relaxation). After an instrumental deadtime period of from 7 to 50 µs, the antenna picks up this signal and sends it on to the reciever. The spin system then relaxes back to the z-axis, equivalent to a Boltzmann distribution of spin state occupancies, without emitting additional signal. After some time period which is typically 0.1 to 5 times the T_1 relaxation time (depending on the flip angle, equation 9) the process is repeated and the signal added in the computer to the signal from previous pulses. T_1 can be milliseconds to many minutes or even hours or days. This process is repeated as many times as needed to produce the desired S/N, often hundreds or thousands of times. S/N increases only as the square root of the number of pulses (see Chapter 3), and few data-collection runs last longer than a day, although runs of 4 - 12 hours are common.

The magnets now used in NMR spectrometers are mostly superconducting solenoids. These magnets contain toroid-shaped windings of a superconducting alloy immersed in a Dewar of liquid He, which is surrounded by a Dewar of liquid N_2 to reduce He-loss. The magnets contain a room-temperature cylindrical bore down the center into which the sample probe is inserted. The field across the sample is made uniform by room-temperature gradient magnets (shims).

Magnetic field strengths commonly range from 1 Tesla (T) to 11.7T, although 14T magnets are now being manufactured. Field strengths are often described in terms of the ^{1}H resonance frequency at that field (4.7T = 200 MHz, 8.45T = 360 MHz, 9.4T = 400 MHz, 11.7T = 500 MHz).

One difficulty with using NMR spectroscopy, especially for many natural samples, is that paramagnetic components or impurities (most commonly Fe or Mn) cause extensive peak broadening (Oldfield et al., 1983; Grimmer et al., 1983; Sheriff and Hartman, 1985). In the worst case the peak can be so broadened that it is not detectable above noise. This broadening is due to a range of chemical shifts caused by magnetic field inhomogenity due to the interaction of unpaired d- or f-electrons with the applied magnetic field.

Despite this problem being well known, there has been no systematic study of it. In general, it appears that phases with Fe or Mn as major components are not normally observable. 1-2 oxide wt % Fe often does not totally destroy the spectrum, but at least for ^{29}Si, broadens the peaks sufficiently to cause poor resolution of closely spaced peaks (e.g., Altaner et al., in press). For quadrupolar nuclides with intrinsically broad peaks, paramagnetic effects may be less important. In principle, Fe^{+2} should have a larger peak broadening effect than Fe^{+3}, because Fe^{+2} has more unpaired d-electrons.

STRUCTURAL CONTROLS OF CHEMICAL SHIFTS AND QUADRUPOLE PARAMETERS

The objective in obtaining NMR spectra for a sample is often to use the values of the chemical shifts and quadrupole parameters to infer structural information about that sample. Unfortunately, it is not now possible to accurately calculate NMR chemical shifts for a specific structural environment or to uniquely invert an observed spectrum to a structure. Thus, a large part of the recent experimental NMR work with inorganic solids has been to obtain spectra of phases with well-known structures to empirically determine how the various parameters are related to structural environment. These correlations must be obtained for each nuclide of interest, clearly a daunting task if the entire periodic table is to be usefully available (Table 1).

There is, however, a relatively well developed theory describing the electronic shielding of nuclei that works quite well for small molecules, and recent quantum chemical calculations for small molecules containing Si, Al, P, O, F, and H have yielded variations in chemical shifts and QCC's with composition and structure which qualitatively parallel the observed variations. Thus, there is some theoretical basis for understanding the structural controls of NMR parameters and for predicting how they should vary for a poorly understood nuclide. This section first presents the known empirical relationships between various structural and bonding parameters and NMR chemical shifts and QCC's, and then discusses the theoretical understanding of these relationships.

Table1 I. Some nuclides of Potential Use in NMR Studies of Solids

Nucleus	Readily observed	Spin	Quadrupole Moment (10^{-24} cm^2)	Natural Abundance	Frequency MHz (11.7 T)
H-1	yes	1/2	-	99.985	500
H-2	yes	1	0.0028	0.015	76.8
Li-7	yes	3/2	-0.03	92.58	194.3
Be-9	yes	3/2	0.0512	100	70.3
B-10	yes	3	0.074	19.58	53.7
B-11	yes	3/2	0.0355	80.42	160.4
C-13	yes	1/2	-	1.1	125.7
N-14	yes	1	0.016	99.6	36.1
N-15	yes	1/2	-	0.37	50.7
O-17	yes	5/2	-0.026	0.037	67.8
F-19	yes	1/2	-	100	470.4
Na-23	yes	3/2	0.14	100	132.3
Mg-25	yes	5/2	N.D.	10.1	30.6
Al-27	yes	5/2	0.149	100	130.3
Si-29	yes	1/2	-	4.7	99.3
P-31	yes	1/2	-	100	202.4
S-33	no	3/2	-0.064	0.76	38.4
Cl-35	yes	3/2	-0.0789	75.5	49.0
K-39	yes	3/2	0.11	93.1	23.3
Sc-45	yes	7/2	-0.22	100	121.5
Ti-49	no	7/2	N.D.	5.5	28.2
V-51	yes	7/2	-0.04	99.76	131.4
Cu-63	yes	3/2	0.16	69.1	132.5
Zn-67	yes	5/2	0.15	4.1	31.3
Ga-71	yes	3/2	0.112	39.6	152.5
Ge-73	yes	9/2	-0.2	7.8	17.4
Se-77	no	1/2	-	7.6	95.3
Br-79	yes	3/2	0.33	50.5	125.3
Rb-85	yes	5/2	0.27	71.25	48.3
Sr-87	no	9/2	0.2	7.0	21.7
Y-89	yes	1/2	-	100	24.5
Zr-91	no	5/2	N.D.	11.2	46.7
Nb-93	yes	9/2	-0.2	100	122.2
Mo-95	yes	5/2	0.12	15.7	32.6
Ag-109	yes	1/2	-	48.18	23.3
Cd-113	yes	1/2	-	12.26	110.9
In-115	yes	9/2	1.14	95.72	109.6
Sn-119	yes	1/2	-	8.58	186.4
Te-125	no	1/2	-	6.99	158.0
Cs-133	yes	7/2	-0.003	100	65.6
Ba-137	no	3/2	0.2	11.3	55.6
La-139	yes	7/2	0.21	99.9	70.6
Yb-171	no	1/2	-	14.3	88.1
W-183	yes	1/2	-	14.4	20.8
Pt-195	no	1/2	-	33.8	107.5
Hg-199	yes	1/2	-	16.8	89.1
Tl-205	yes	1/2	-	70.5	288.5
Pb-207	yes	1/2	-	22.6	104.6

One should always remember that the existence of an empirical relationship between an NMR parameter and a structural or bonding parameter does not necessarily indicate a cause and effect relationship. Because the NMR parameters for a given nuclide are controlled by the behavior of the electrons in the vicinity of an atom, an understanding of causality must be sought from quantum chemical considerations. It is more useful to think of the structural parameters and the NMR parameters as separate, co-varying effects that are both the result of how the atoms are bonded together.

Empirical relationships

^{29}Si. ^{29}Si is the most thoroughly investigated and widely applied nuclide of mineralogical interest; dozens of papers using ^{29}Si have now been published. ^{29}Si has spin $I=1/2$ and, thus, does not suffer from quadrupolar peak broadening and distortion. The spectral resolution is, therefore, relatively high. The natural abundance of ^{29}Si is only 4.7%, and spectra are often time-consuming to collect (4-24 hours). The low natural abundance greatly reduces the homonuclear dipole interaction. Figure 10 presents a selection of ^{29}Si MASS NMR spectra of crystalline silicate phases which illustrate the range of data available and the spectral resolution possible.

Clearly, the largest effect on ^{29}Si chemical shift, and in fact the chemical shifts of all nuclides, is due to the nearest-neighbor structural environment. Octahedrally coordinated Si (Si^{VI}) has chemical shifts between -180 and -221 ppm (Grimmer et al., 1986), whereas tetrahedrally coordinated Si (Si^{IV}) has chemical shifts between about -60 and -126 ppm (Kirkpatrick et al., 1985). Because few phases that are stable at atmospheric pressure contain octahedral Si (silico-phosphates are an exception), the full range of chemical shifts for this coordination is probably not known. The behavior of tetrahedral Si, however, has been thoroughly studied.

The most useful correlations of ^{29}Si chemical shift are with tetrahedral polymerization, number of Al^{IV} next-nearest neighbors (NNN), mean Si-O-T bond angle per tetrahedron ($<$Si-O-T$>$; T=Si, Al), and group electronegativity (reviewed by Mägi et al., 1984; Oldfield and Kirkpatrick, 1985, and Kirkpatrick et al., 1985). ^{29}Si becomes less shielded with decreasing polymerization (Lippmaa et al., 1980), with increasing number of Al^{IV} NNN for framework and sheet silicates (Lippmaa et al., 1981; Kinsey et al., 1985) , and with decreasing $<$Si-O-T$>$ for framework aluminosilicates (Smith and Blackwell, 1983; Ramdas and Klinowski, 1984; Engelhardt and Radeglia, 1984). Figures 11 and 12 illustrate these relationships. The group electronegativity approach includes all types of structures and considers the total electronegativity of the oxygens of a Si tetrahedron and the other cations to which it is coordinated (Janes and Oldfield, 1985). The ^{29}Si chemical shift also correlates with the sum of the Brown and Shannon (1973) bond strengths to the four tetrahedral oxygens (Smith et al., 1983), the mean Si-Si distance (Ramdas and Klinowski, 1984), and less well with the mean Si-O distance (Higgins and Woessner, 1982). The individual components of the shielding tensor appear to correlate well with individual Si-O bond distances when their

SILICON-29

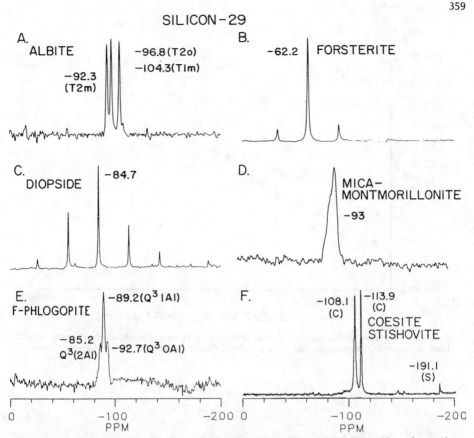

A. ALBITE
−92.3 (T2m)
−96.8 (T2o)
−104.3 (T1m)

B. FORSTERITE
−62.2

C. DIOPSIDE
−84.7

D. MICA-MONTMORILLONITE
−93

E. F-PHLOGOPITE
−89.2 (Q³1Al)
−85.2 Q³(2Al)
−92.7 (Q³0Al)

F. COESITE STISHOVITE
−108.1 (C)
−113.9 (C)
−191.1 (S)

0   −100   −200
PPM

Figure 10. ²⁹Si MAS NMR spectra of the minerals indicated. Forsterite and diopside each contain only one Si-site in the average structure and one local Si-environment, giving rise to one NMR peak. Albite contains 3 Si-sites and coesite 2 Si-sites in their average structures and the same number of local Si-environments and NMR peaks. Fluoro-phlogopite contains only 1 tetrahedral site in the average structure but 3 chemically (magnetically) different local Si-environments due to Si with from 0 to 2 Al NNN. The peaks in the clay sample are broadened due to paramagnetic effects probably associated with Fe in the sample.

SILICON − 29

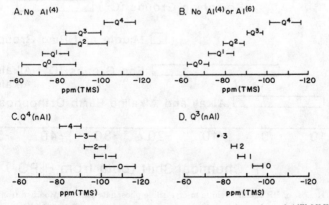

A. No Al⁽⁴⁾
B. No Al⁽⁴⁾ or Al⁽⁶⁾
C. Q⁴(nAl)
D. Q³(nAl)

Figure 11. Variation of ²⁹Si chemical shifts with polymerization and number of Al^{IV} NNN in crystalline silicate and aluminosilicate phases. After Kirkpatrick et al. (1985).

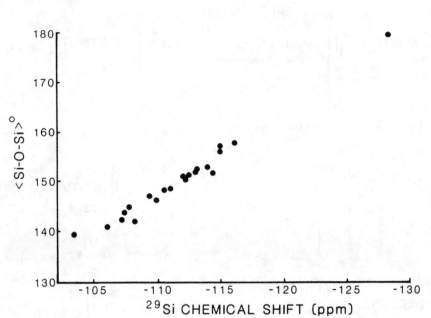

Figure 12. Variation of ^{29}Si chemical shift of Q^4(0Al) sites in crystalline silicates with mean Si-O-Si bond angle per SiO$_4$ tetrahedron (<Si-O-Si>). From Oestrike et al. (1987) and references therein.

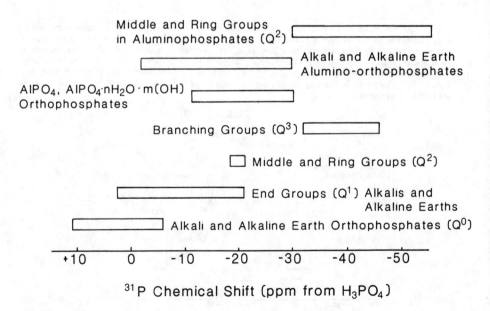

Figure 13. Ranges of ^{31}P chemical shifts in crystalline phosphate phases. Modified from Turner et al. (1986b).

directions coincide, with shorter bonds correlated with greater shielding (Grimmer, 1985). For ortho-, soro-, and chain-silicates the nature of the large cation has more effect than for more polymerized structures. For example, substitution of one Al^{VI} for a Mg^{VI} in orthosilicates causes about a 20 ppm shielding, whereas for sheet-silicates it causes only about a 2 ppm shielding (Kirkpatrick et al., 1985; Weiss et al, 1987).

Based on these correlations, it is possible to use ^{29}Si chemical shifts to infer quite accurate values of structural and bonding parameters, especially bond angles. This approach has been used extensively for glasses (discussed more fully below), but is limited for many glasses of petrological interest because of the overlapping effects of increasing Al^{IV} NNN and depolymerization.

These correlations are also useful in assigning peaks in ^{29}Si spectra of phases for which some compositional and structural information is available. Mineralogical applications are described in a later section.

^{31}P. ^{31}P also has spin $I=1/2$ and, because it is 100% abundant and resonates at a high frequency (Table 1); it is very easy to observe. The variation of its chemical shift with structure and composition is similar to that of ^{29}Si (Fig. 13), but because most natural phosphates are orthophosphates it will probably be of more use in materials science than mineralogy. For orthophosphates without paramagnetic cations, the ^{31}P chemical shift becomes more shielded with increasing electronegativity of the large cation (Turner et al., 1986b, and references therein).

^{27}Al. Next to ^{29}Si, ^{27}Al is the most thoroughly studied nuclide of mineralogical interest. It has spin $I=5/2$, is 100% abundant, and is easy to observe. Figure 14 presents representative spectra. In addition to the recent work using MAS NMR methods, there was considerable single-crystal NMR work on aluminosilicate minerals in the 1960's and 1970's (reviewed by Ghose and Tsang, 1973). Single-crystal experiments can yield the orientation of the quadrupolar coupling tensor with respect to the crystallagraphic axis and more accurate values for the QCC and asymmetry parameter than can be obtained using MAS methods, but cannot give chemical shifts and require large single crystals.

The ^{27}Al chemical shift ranges for Al^{IV} and Al^{VI} are well separated, with Al^{IV} between about 50 and 80 ppm and Al^{VI} between about -10 and 15 ppm in aluminosilicates and aluminates, paralleling the variation for ^{29}Si (reviewed by Kirkpatrick et al., 1985). The chemical shifts of five-coordinate aluminum (Al^{V}) are in the range +35 to +40 ppm (Alemany and Kirker, 1986; Phillips et al., 1987). For Al^{IV} in the same polymerization the chemical shift generally increases with decreasing $Si/Si+Al^{IV}$, with the shielding less in sheet silicates than in framework silicates (Kinsey et al, 1985; Kirkpatrick et al., 1985). Al^{IV} and A^{VI} sites in aluminophosphates are more shielded, +35 - +45 ppm and -21 to -8 ppm respectively (Müller et al., 1984; Blackwell and Patton, 1984). In framework aluminosilicates the ^{27}Al chemical shift becomes more shielded with increasing

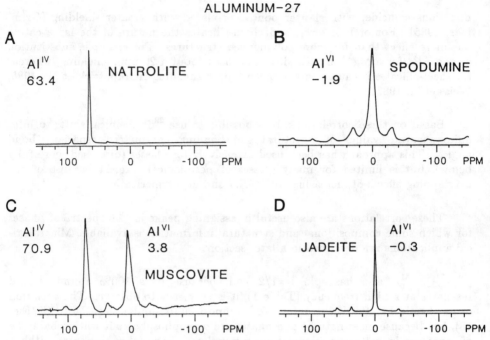

Figure 14. ^{27}Al MAS NMR spectra of aluminosilicate crystals, showing the excellent resolution of signals for Al^{IV} and Al^{VI}. Al^{IV} falls in the range +50 to +80 ppm, and Al^{VI} in the range -10 to +20 ppm at H$_0$ = 11.7T. Al^{V} falls in the range of about +30 to +40 ppm.

mean Al-O-Si bond angle per tetrahedron(<Al-O-Si>), paralleling the behavior of ^{29}Si (Lippmaa et al., 1986).

^{27}Al NMR is useful for detecting Al in different coordinations in both crystalline and amorphous phases, but because of 2nd-order quadrupole broadening does not have the resolution of ^{29}Si. Recently developed methods of observing the spinning side bands of the satellite ±(1/2, 3/2) transitions (Samoson, 1985; Phillips et al., 1987; Phillips et al., submitted) may help overcome this limitation in some cases (Fig. 15).

^{11}B. Boron is a common element that is difficult to analyze for chemically but easy to examine with ^{11}B NMR methods. We have been able to detect and characterize it at concentrations as low as 10 ppm. In oxide phases boron occurs in three-coordinated planar triangles (B^{III}) and tetrahedrally-coordinated (B^{IV}) sites. These sites are easy to distinguish by MAS NMR, although the range of chemical shifts is relatively small. B^{III} in borate minerals has isotropic chemical shifts between 12 and 19 ppm and QCC's between 2.3 and 2.5 MHz. B^{IV} has isotropic chemical shifts between 2 and -4 ppm and QCC's between 0 and 0.5 MHz (Turner et al., 1986a). Figure 16 shows that peaks for these sites are readily resolved in spectra taken at 11.7T for both crystals and glasses. One distinction that can be made is between B^{III} sites with three identical NNN (symmetrical B^{III} sites) versus those with two similar and one different NNN (asymmetrical B^{III} sites). Symmetrical sites have asymmetry parameters of about 0, whereas asymmetrical sites have asymmetry parameters of about 0.6 (G. Turner personal communication). One problem with ^{11}B NMR is that because B^{III} has a relatively large QCC, it's peaks become broader more rapidly with decreasing field strength than those of B^{IV}, and at field strengths less than 11.7T, the peaks overlap and resolution is poor. ^{11}B MAS NMR methods are proving to be very useful for analysis of glass structure.

^{17}O. Oxygen is the most abundant anion in the earth, and investigation of its NMR behavior should provide important structural information complimentary to the data for the more extensively investigated cations. Unfortunately, the only useful nuclide is ^{17}O, which has a 0.037% natural abundance and cannot be observed in unenriched solids such as natural minerals. ^{17}O-enriched phases are usually made from enriched oxide starting materials made by reaction of the appropriate chlorides with commercially available ^{17}O-enriched water. Because of these difficulties, there are fewer data for ^{17}O than its geochemical significance warrants. There is, however, a general understanding of the relationships of its chemical shifts and QCC's to nearest neighbor (NN) structural environment based on the work of Schramm et al. (1983), Schramm and Oldfield (1984), Turner et al. (1985), Timken et al. (1986a,b, 1987), Janes and Oldfield (1986), and Walter et al. (submitted).

Isotropic chemical shifts of ^{17}O for bridging oxygens (those linking two Si, Al^{IV}, or P sites) do not vary much. For zeolites, cristobalite, and AlPO$_4$ phases, all of which have fully polymerized framework structures, oxygens in Si-O-Si sites fall in the range 44-52 ppm, those in Si-O-Al sites in the range 31-40 ppm, and those in P-O-Al sites in the range 61-63 ppm. For alkaline earth metasilicates (chain and ring structures) the chemical shifts of the bridging oxygens fall in the range 62-87 ppm, with the shielding decreasing from Mg to Ba (increasing cation size and decreasing cation electronegativity).

Isotropic chemical shifts of ^{17}O for non-bridging oxygens (those linking Si, Al, or P to one or more large cations) are less well understood, but are more sensitive to the nature of the large cation than those of bridging oxygens. For alkaline earth metasilicates, their chemical shifts fall in the range 42-169 ppm, with the shielding decreasing from Mg to Ba (Fig. 17).

^{17}O QCC's also seem to vary systematically and may be more sensitive indicators of chemical environment than the chemical shifts (Table 2). In general, the QCC decreases with increasing average ionic character of the bonds to the oxygen, which is proportional to the cation electronegativity. Thus, QCC's for oxygens in Si-O-Si sites are in the range 3.7 - 5.3 MHz, those for Si-O-Al sites in the range 3.1 - 3.4 MHz, and those for non-bridging oxygens in the range of 1.0 - 3.2 MHz.

Because ^{17}O cannot be observed at natural abundance in solids, its major applications will probably be in understanding bonding in oxide and oxy-salt phases (e.g., Janes and Oldfield, 1986) and in examining the structure of amorphous phases and the products of hydrothermal reaction (e.g., Bunker et al., in press).

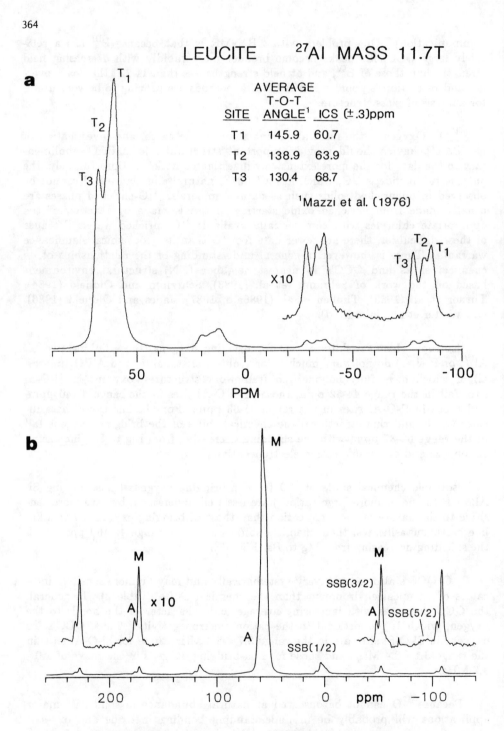

Figure 15. (a) ^{27}Al MAS NMR spectrum of leucite, showing the increased resolution for the three Al-sites in the spinning sidebands of the satellite $\pm(1/2,3/2)$ transitions. (b) ^{27}Al MAS NMR spectrum of a microcline perthite, showing the increased resolution for Al in the albite and microcline in the spinning sidebands of the satellite $\pm(1/2,3/2)$ transitions. Spectra courtesy of B.L. Phillips.

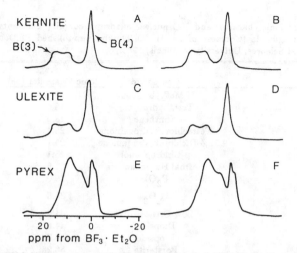

Figure 16. ^{11}B MAS NMR spectra of the borate minerals kernite and ulexite and synthetic pyrex (borosilicate) glass obtained at 11.7T. Note the excellent resolution of signals for B^{III} and B^{IV} and for the glass two B^{IV} sites. After Turner et al. (1986a).

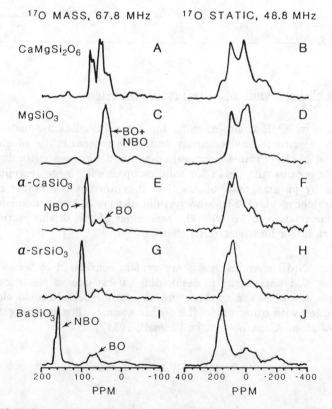

Figure 17. ^{17}O MAS and static NMR spectra of alkaline earth metasilicates, showing the progressively decreased shielding for the non-bridging oxygen sites (NBO's) with increasing size of the large cation. In the MAS spectra the signal for the bridging oxygens (BO's) and NBO's overlap for $MgSiO_3$, but for the other phases the signal is well resolved. After Timken et al. (1987).

Table 2   Percent ionic character and ^{17}O nuclear quadruple coupling constants for a series of A-O-B fragments (Courtesy of Dr. H. K. Timkin, unpublished Ph.D. thesis, School of Chemical Sciences, University of Illinois, 1986.)

A-O-B Fragment	Compound	Average Ionic Character (%)	QCC (MHz)
N-O-C	N-Methylsydnone	11.3	12
C-O-C	Tetrahydropyran	18.6	11.2
C-O-C	Xanthene	18.6	9.868
C-O-H	Tetrachlorohydroquinone	25.1	9.175
C-O-H	2,5-Dichlorohydroquinone	25.1	8.909
C-O-H	p-Chlorophenol	25.1	8.617
H-O-H	Normal Hexagonal Ice	31.6	6.525
B-O-B	B_2O_3	39.0	5.75
B-O-B	B_2O_3	39.0	4.69
Si-O-Si	Low Cristobalite	44.6	5.3
Si-O-Si	Diopside	44.6	4.4
Si-O-Ca,Mg	Diopside	59.0	2.7
Si-O-Mg	Forsterite	56.25	2.35
Si-O-Mg	Forsterite	56.25	2.70
Al-O-Al	α-Al_2O_3	56.2	2.2
Zn-O-Zn	ZnO	54.6	0.13
W-O-K	K_2WO_4	53.8	0.0
Mg-O-Mg	MgO	67.9	0.0

QCC (MHz) = -0.222 I(%) + 14.33
Correlation Coefficient = 0.965

Origin of the chemical shift and quadrupole parameters

Variations in NMR chemical shifts for a given nuclide are caused by variations in local electronic environments and bulk susceptibility effects (reviewed well by Harris, 1983). From a practical standpoint, bulk susceptibility effects are usually small, because the peaks for solid samples with large susceptibilities are so broadened by paramagnetic effects that they are not observable above background. Grimmer et al. (1983), however, did observe a 2 ppm deshielding of ^{29}Si in Fo95 olivine relative to Fo100. The remainder of this of this section will consider the much larger local electronic effects.

For solids, NMR chemical shifts are usually considered in terms of diamagnetic shielding and paramagnetic deshielding effects due to electrons associated primarily with the observed nucleus and the same effects due to electrons primarily associated with other nuclei (i.e., other atoms). The shielding constant for the nucleus of atom A can be written (Tossell, 1984) ,

$$\sigma_A = \sigma_A^{dia} + \sigma_A^{para} + \sum_{B \neq A} \sigma_{AB} \ , \qquad (11)$$

where σ_A^{dia} is the diamagnetic shielding term due to electrons in atom A, σ_A^{para} is the paramagnetic deshielding term due to electrons in atom A, and σ_{AB} is the effect due to the electrons in atom B. σ_{AB} contains both diamagnetic and paramagnetic contributions.

The diamagnetic term can be viewed as arising from the motion of the electrons (i.e., an electrical current) in the applied magnetic field, H_0. Classically, the induced magnetic field along the z-axis, H_{iz}, is given by

$$H_{iz} = \frac{\mu_0 e^2 H_0}{8\pi m_e} \int_v \left(\frac{\rho_e}{r}\right)\sin^2\theta dV \quad , \tag{12}$$

where μ_0 is the permeability constant, e is the charge on an electron, m_e is the mass of an electron, and θ is the angle between H_0 and the line connecting the nucleus and a volume element dV, ρ_e is the electron density in that volume element, and r is its distance from the nucleus. The minus sign indicates that the orientation of H_{iz} is opposite that of H_0 (by the right hand rule), and thus shields the nucleus. σ_A^{dia} is given by

$$\sigma_A^{dia} = \frac{\mu_0 e^2}{8\pi m_e} \int_v \left(\frac{\rho_e}{r}\right)\sin^2\theta dV \quad . \tag{13}$$

A similar expression can be derived from quantum mechanics (see Tossell, 1984). Note that this term decreases only as $1/r$, and, thus, that diamagnetic contributions to σ_{AB} from electrons not associated primarily with the observed nucleus can be significant.

Consideration of this term alone indicates that a bare nucleus, if it existed, would be the least shielded and have the most positive chemical shift, and an isolated atom with its full compliment of electrons would be the most shielded and have the most negative chemical shift. The magnitude of σ_A^{dia} increases by orders of magnitude with increasing atomic number, from 17.8 for H to 10,060.9 for Pb (Harris, 1983).

The paramagnetic term, σ_A^{para}, is difficult to analyze classically, because it is due to occupation of excited electronic states, which can be thought of as being caused by mixing of ground and excited states by H_0. The largest of these effects is simply the Zeeman splitting of the electronic orbital magnetic moments, which allows EPR spectroscopy (see Chapter 9). Because all the excited states must be considered, this term is very difficult to calculate (Tossell and Lazzeretti, 1986) but may be approximated by considering an "average excitation energy" for the ground state to excited state transitions. The paramagnetic term is positive, i.e., it reinforces H_0, and is proportional to $1/\{r\}^3$, where $\{r\}$ is the expectation value of the distance of an electron from the nucleus. Because of this $1/\{r\}^3$ dependence, the paramagnetic effects due to electrons not associated primarily with the observed nucleus decrease rapidly with distance. An important additional feature is that only electrons with non-spherical orbitals contribute to the paramagnetic effects. In particular, s-electrons do not contribute. In general, for a given nuclide, the paramagnetic term decreases as the bonding environment becomes more ionic, (i.e., as the electron distribution becomes more spherical).

The total shielding at a nucleus due to electrons associated primarily with it, σ_A^{loc}, is the sum of the diamagnetic and paramagnetic terms. Both these terms may be large, 10's to 1000's of ppm, and with absolute values of the same order of magnitude. Because they are of opposite sign, accurate calculations of σ_A^{loc} is often difficult for the same reason as for thermochemical calculations - subtraction of two large numbers to give a small number. Because we are usually interested in the variation of the net shielding to interpret structure, and because these terms are very difficult to calculate accurately, this difficulty is often very important in interpreting the results of calculations.

The conventional chemical wisdom concerning NMR shielding constants is that for a given NN structural environment (say SiO_4^{-4}), variation in the paramagnetic term is much larger than variation in the diamagnetic term. For example, Tossell (1984) indicates that σ_A^{dia} for tetrahedral Si should vary only about 1 ppm from orthosilicates to framework silicates, far less than the observed 60 ppm range. If this is true, for many applications we must focus primarily on σ_A^{para}, which by Murphy's law is the most difficult to calculate. Some support for this idea comes from the correlation between less shielding at Si in forsterite relative to the silica polymorphs with a lower difference in the energy between the ground and excited electronic states for forsterite (Tossell, 1984).

Based on these considerations, it appears that the contribution to the shielding from electrons not primarily associated with the observed nucleus, $\sum \sigma_{AB}$, should be small. The paramagnetic term goes as $1/\{r\}^3$, and electrons far from the nucleus simply do not have a big effect. In addition, the long-range contributions to both the paramagnetic term and especially the diamagnetic term from electrons on opposite sides of the observed nucleus tend to cancel. Some support for these ideas comes from the data for small molecules, in which non-local effects are greatest for light elements and negligible for elements heavier than F (Harris, 1983). Even for small molecules containing only H, C, O, and N, non-local contributions are one and a half to three orders of magnitude smaller than the local contributions (Harris, 1983). Although there have been no attempts to calculate the magnitude of these non-local effects in crystals or to evaluate the effects of the repetitive structure of crystals and crystallographic anisotropy, the idea that chemical shifts are insensitive to the effects electrons not associated primarily with the observed nucleus has some basis in experiment and theory and makes a useful working hypothesis.

If this idea is correct, we can think of the effects of NNN substitution of, say, Al^{IV} for Si^{IV} or Mg^{VI} for Al^{VI} on ^{29}Si chemical shifts as being due not to differences in electron distribution in the NNN atom, but to the effects that the substitution produces on the electron distribution in the Si atom and in the Si-O bonds. For instance, NNN substitution of Al^{IV} for Si^{IV}, might increase the electron density in the Si-O bond, thus increasing the paramagnetic deshielding and causing the observed change to less negative chimical shifts.

The calculation of accurate NMR chemical shifts from first principles requires ab initio molecular orbital methods using large basis sets. The diamagnetic shielding and the quadrupolar parameters can be calculated quite easily, because they depend only on the ground states. Calculation of the paramagnetic deshielding is more demanding, because all the excited electronic states must also be included. Methods for doing such calculations are well developed, and quite accurate values can be obtained for small molecules containing light elements (i.e., molecules containing few electrons). The complexity of the calculations and the computer time needed increases as a large power of the number of electrons in the molecular fragment. Precise chemical shifts have not yet been calculated for large fragments that might be sufficient to quantitatively reproduce the values for minerals or to predict accurate chemical shifts for an assumed environment. For instance, to investigate ^{29}Si, such fragments might be be based on a central silica tetrahedron surrounded by four other silica tetrahedron with hydrogen atoms charge balancing the outer oxygens ($Si_5O_{16}H_{12}$). Investigation of the effects of structural variation could be done by changing bond angles and distances and by substitution of Al and large cations plus the appropriate number of oxygens for the outer Si's.

The beginnings of such work and important qualitative insight into the origin of variation in the NMR parameters of inorganic solids have been presented in a series of recent papers by Tossell and Lazzeretti (1986a,b,c, 1987, submitted a, b). The calculated chemical shifts are not quantitatively correct, in part at least because the size of the molecular fragments has been significantly smaller than suggested above, but the following conclusions can be drawn. 1) The observed increased shielding with increased coordination number for cations is due to a combination of increased diamagnetic shielding and decreased paramagnetic deshielding. 2) Varying cation-oxygen bond distance by itself has only a small effect on both oxygen and cation chemical shifts. 3) Changes in the components of ^{29}Si shielding tensors, are apparently related directly to different Si-O bond distances within one tetrahedron, with both the diamagnetic and paramagnetic contributions different for different bonds. 4) The sagging, U-shaped, chemical shift patterns often observed for a series of compounds related by atom for atom substitution is due, at least for SiF_xH_{4-x}, to a combination of an approximately linear change in diamagnetic shielding and a non-linear change in paramagnetic deshielding. 5) The calculations reproduce the observed trend of increased shielding at ^{29}Si with increased Si-O-Si bond angle and indicate that it is due primarily to a decrease in the paramagnetic deshielding term, consistent with an increased ionic character of the Si-O bonds. 6) The calculations also reproduce the observed decrease in ^{17}O QCC of non-bridging oxygens relative to bridging oxygens.

Although these calculations have not yet been able to quantitatively reproduce NMR parameters, the increasing availability of supercomputers may soon allow routine calculations for fragments large enough to be reasonable models of crystalline structures. The ability to predict the NMR behavior of known or postulated structures will greatly increase the range of useful applications of NMR spectroscopic methods.

APPLICATION TO CRYSTALLINE PHASES

The power and the limitation of NMR spectroscopy as a probe of structure is that the chemical shift is sensitive only to the local (primarily NN and NNN) environment. Thus, especially for phases disorderd in some way it can often provide site-specific information not normally obtainable by diffraction methods, which give average structures. NMR does not provide sufficient long-range information to solve or refine crystal structures, although information about the abundances of different types of local environments and their distribution can sometimes be obtained by modeling NMR spectra. The most complete picture of the structure and behavior of a phase is usually obtained by a combination of several complimentary methods.

The crystal chemical problems to which MAS NMR has been most successfully applied include the occupancy of tetrahedral sites, especially Al,Si order/disorder and the occurance or lack of Al-O-Al linkages, the structure of alkali sites in mixed alkali phases, and the direct detection of structural environments of a particular element not detected by diffraction methods. Single crystal and wide-line methods have also been used to investigate structural phase transitions, although not for many phases of direct mineralogical or geochemical interest (Blinc, 1981; Rigamonti, 1984).

This section discusses the application of NMR methods to questions about the structures of crystalline phases, with emphasis on mineralogically important materials. It will not be a comprehensive review of all the work that has been done, but will focus on illustrating the kinds of information that NMR yields about crystal structures and how this information is different from that obtained by more traditional diffraction methods.

Aluminum-silicon order/disorder

When an element occupies more than one crystallographically unique site in a crystal, the nuclei of the atoms on each site will in general have different chemical shifts. Thus Si on the three Si-sites of ordered albite have different chemical shifts (Fig. 10a). In addition, Si's in the same sites in the average structure but with different numbers of Al^{IV} and Si NNN will have different ^{29}Si chemical shifts. For instance, for synthetic phlogopite (Fig. 10e) there is only one tetrahedral site in the average structure, with an occupancy of 1/4 Al and 3/4 Si. The ^{29}Si MASS NMR spectrum, however, contains three peaks due to Si on sites with from 0 to 2 Al NNN. If ^{29}Si spectra for phases with different local Si-environments are collected under conditions of complete T_1 relaxation, the relative areas of the peaks are equal to the relative abundances of the different Si sites.

This quantitative detection of relative site abundances has been used extensively to investigate Si,Al distributions in framework and sheet silicates. A major objective of this work has been to determine if Al-O-Al linkages occur, to determine Si/Al ratios, and to determine if the Al,Si distribution is more ordered than just the avoidance of Al-O-Al linkages.

If the aluminum avoidance rule (the lack of Al-O-Al linkages; Loewenstein, 1954) is obeyed , the Si/Al ratio of the site determined by ^{29}Si NMR, $(Si/Al)_{NMR}$, is given by

$$(Si/Al)_{NMR} = \frac{\sum_{0}^{n=4} I(Si(nAl))}{0.25 \sum_{0}^{n=4} nI(Si(nAl))} , \qquad (14)$$

where n is the number of Al NNN to a Si-site, and $I(Si(nAl))$ is the relative intensity of the ^{29}Si NMR peak with nAl NNN (e.g. Fyfe et al., 1983b). If the value calculated in this way agrees with the value from chemical analysis or stoichiometry (assuming no significant impurities), there are no detectable Al-O-Al linkages. This result is true even if the Al,Si distribution is more ordered than one which just avoids Al-O-Al linkages.

For synthetic zeolite molecular sieves and catalysts, which are the group of phases most extensively investigated by ^{29}Si NMR, there were a few false starts (clarified by Smith and Pluth, 1981, and briefly summarized by Seff and Mellum, 1984), but there is now consensus that the aluminum avoidance rule does hold for most of them. These phases are usually synthesized from solution at relatively low temperatures, and complete order is to be expected. For the faujasite group, the Al,Si distribution over the one T-site does not seem to be completely random. Rather, the relative ^{29}Si peak intensities are consistent with distributions which minimize the number of Al-O-Si-O-Al linkages (Klinowski et al., 1982; Peters et al., 1982). The details of these ordering patterns change with changing Si/Al ratio.

Similar conclusions have been reached for sheet silicates (micas and clays) in which there is one tetrahedral site occupied by both Si and Al (Lipsicas et al., 1984; Barron et al., 1985; Herrero et al., 1985a,b). For all such samples which have been examined, aluminum avoidance seems to hold, but the relative intensities of the peaks are those expected for a distribution more ordered than just aluminum avoidance. This more ordered distribution can be considered either in terms of having a uniform distribution of negative charge over the tetrahedral sheet due to Al for Si substitution (Herrero et al., 1985a,b) or in terms of the minimization of Al-O-Si-O-Al linkages (Barron et al., 1985).

Cordierite

In contrast to the results for phases formed at relatively low temperatures, phases crystallized directly from melts or glasses at high temperature seem to have Al,Si distributions which contain Al-O-Al linkages, with the number of such linkages decreasing as the sample is annealed at high temperature. The best studied example of this behavior is synthetic Mg-cordierite ($Mg_2Al_4Si_5O_{18}$)

crystallized from cordierite-composition glass (Fyfe et al., 1983, 1985; Putnis et al., 1985; Putnis and Angel, 1985). The ^{29}Si MAS NMR spectrum of substantially ordered orthorhombic cordierite (Fig. 18c) consists of predominantly two peaks, corresponding to Q^4(4Al) Si on the T_1-sites and Q^4(3Al) Si on the T_2-sites. The T_1-sites are commonly called "chain" sites, and the T_2-sites "ring" sites, although both are fully polymerized, Q^4. The spectra of cordierite samples annealed for shorter times consist of many more peaks (Figs. 18a,b), indicating substantial Al,Si disorder over both sites.

Quantitatively, these spectra show that aluminum avoidance does not hold for these samples, because the Si/Al ratio calculated from equation (14) does not equal the actual value of 1.25. The number of Al-O-Al linkages per Al-atom, N_{Al-Al}, can be calculated from the completely general relationship (Fyfe et al., 1985):

$$N_{Al-Al} = 2 - \frac{2(Si/Al)_{actual}}{(Si/Al)_{NMR}} , \tag{15}$$

where $(Si/Al)_{actual}$ is determined by stoichiometry or chemical analysis, and $(Si/Al)_{NMR}$ is obtained from equation (14). The number of these linkages decreases linearly with log time (Fig. 19), even though during this time the symmetry of the cordierite changes from hexagonal to orthorhombic, and domains with short-range order become visible with TEM. From these data and the heat of solution data of Carpenter et al. (1983), Putnis and Angel (1985) calculate an enthalpy for the reaction (Al-O-Al) + (Si-O-Si) = 2(Si-O-Al) of -8.1 kcal/mole. They also determine that the NMR data are not consistent with a random distribution of Al,Si but rather with a particular local ordering scheme developed by McConnell (1985).

This kind of local structural information is not obtainable with any method other than NMR, and the most complete picture of the structural changes taking place in complicated phases transitions is best obtained with a combination of methods. In this example, optical and TEM observations, electron and x-ray diffraction, calorometric measurements, and ^{29}Si NMR data combined to give a story than none by themselves could even hint at. Similar combined studies are also certain to be effective with a wide variety of other phases.

Plagioclase feldspars

The picture of the Al,Si distribution in relatively well-ordered anorthite-rich plagioclase feldspars developed from ^{29}Si NMR is different from that of the synthetic zeolites and cordierite (Kirkpatrick et al., 1986). In this case, the Si in excess of that needed for Si/Al = 1 of pure anorthite enters specific sites which are occupied by Al in pure anorthite, and the Si sites of An100 remain Si. Figure 20 shows the ^{29}Si MAS NMR spectrum of Val Pasmeda An100 (the purest and most ordered anorthite known) the spectrum of an An89 metamorphic anorthite (Sittampundi), computer fitted spectra of each, and a schematic diagram of the linkages of Si and Al tetrahedra in plagiocalse. The An100 yields

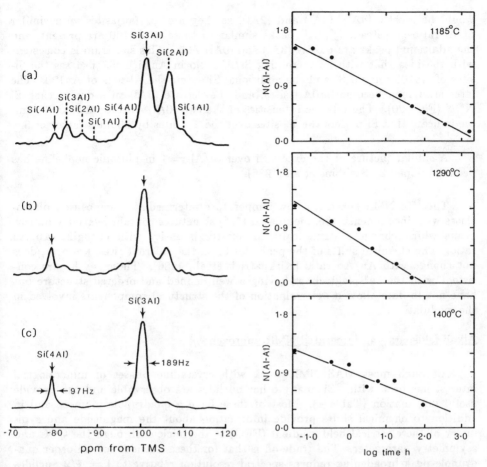

Figure 18(left). ^{29}Si MAS NMR spectra of synthetic cordierite annealed at 1185°C for (a) 2 min, (b) 24 hr, and (c) 2000 hr. After Putnis et al. (1985).

Figure 19(right). Time dependence of the number of Al-O-Al linkages in cordierite annealed at the temperatures indicated. After Putnis et al. (1987).

six readily visible peaks or shoulders, one of which is of greater intensity than the others. These peaks occur in three groups with an intensity ratio of 3:3:2. Anorthite (space group P$\bar{1}$) contains eight equally abundant Q^4(4Al) sites (see Ribbe, 1983, for a review of feldspar structures); the fitted spectrum contains six peaks of equal intensity and one of double intensity, representing Si on the eight sites. The reason for the poor resolution of the three central peaks and the apparently larger breadth of one of them is not known.

The spectrum of the An89 sample is less well resolved (although substantially better than those of many feldspars) but contains additional peaks for Si-sites with fewer than 4Al NNN. The Si/Al ratio for this sample is 1.12. If the Si for Al distribution obeyed aluminum avoidance but were otherwise random, with the Si and Al rearranged over all the T-sites, the ^{29}Si peaks for this sample would be mostly (4Al), (3Al) and (2Al), as they are for faujasites with similar Si/Al ratios. Instead, Q^4(4Al) peaks similar to those of An100 are present, and the additional peaks are mostly (3Al) and (0Al) sites. This spectrum is consistent with the idea that with an increasing Si/Al ratio in An-rich plagioclases the Si-sites of An100 remain Si and the additional Si enters the Al-sites of An100. The later sites then become Si(0Al) sites, and they create (3Al) sites out of each Si NNN (Fig. 20). The relative intensities of the two (0Al) sites for the An89 sample indicate that Si prefers the T_2-sites over the T_1- sites by about a factor of 5.

A similar picture of the excess Si over Si/Al = 1 in plutonic nepheline has been developed by Stebbins et al. (1986).

The ^{29}Si NMR spectra of low-temperature intermediate composition plagioclases with incommensurate, modulated ("e") structures contain relatively narrow peaks which change systematically in relative intensity with changing Ab/An ratio. The chemical shifts of the peaks for Si on the different sites, however, does not change with An/Ab ratio (Kirkpatrick et al., 1986). These results are consistent with the "e"-plagioclases having a well defined and ordered structure but have not to date allowed determination of the structural components involved in the modulation.

Alkali feldspars - an integrated NMR approach

Although most MAS NMR work with crystalline phases of mineralogical interest has been with ^{29}Si, many other nuclides are observable and can provide useful information (Table 1). Most of these have a quadrupole moment and in addition to chemical shifts provide information about the magnitude and symmetry of the electrical field gradient (EFG) at the nucleus through the QCC and asymmetry parameter. The trade-off is that for these nuclides second-order quadrupole peak-broadening reduces spectral resolution relative to I = 1/2 nuclides such as ^{29}Si.

This section describes the results of a number of NMR investigations of alkali feldspars using ^{23}Na, ^{27}Al, and ^{29}Si and shows how data for several nuclides can be used to develop a more complete understanding of structural changes within a solid solution series. It also shows how the quadrupole behavior of nuclides like ^{23}Na and ^{27}Al can be overcome and used to advantage (J. V. Smith et al., 1983; Kirkpatrick et al., 1985; Sheriff and Hartman, 1985; Yang et al., 1986; Phillips et al., submitted). The Al,Si ordered low-temperature Na-K feldspars are of particular interest, because the average structures of the end-members are well known, because no samples with substantially intermediate compositions have been refined by diffraction methods, and because little is known about the distribution of Na and K on the single alkali site of the average structure.

The ^{29}Si spectra of Al,Si ordered alkali feldspars (Fig. 21) consist of three peaks which correspond to the three crystallographically distinct Si sites in these phases. The fourth tetrahedral site is completely occupied by Al, and the ^{27}Al

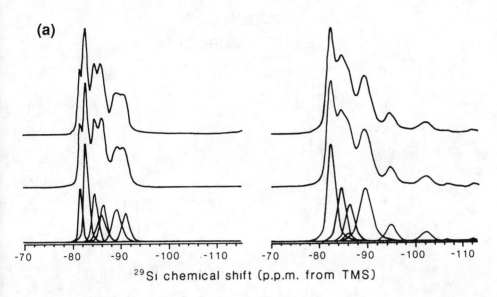

(a)

^{29}Si chemical shift (p.p.m. from TMS)

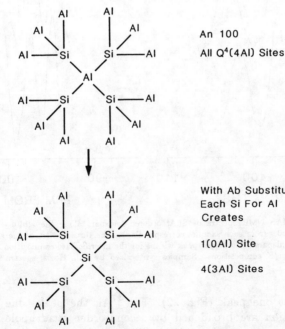

(b)

An 100

All $Q^4(4Al)$ Sites

With Ab Substitution
Each Si For Al
Creates

1(0Al) Site

4(3Al) Sites

Figure 20. (a) ^{29}Si MAS NMR spectra of Val Pasmeda An100 and Sittampundi An89 plagioclase feldspars. After Kirkpatrick et al. (1987). (b) Schematic diagram of the effect of an Si for Al substitution in the An-Ab solid solution series. Each Si on a site occupied by Al in An100 becomes a $Q^4(0Al)$ site and creates $Q^4(3Al)$ sites from its NNN Si's.

SILICON - 29
ALBITE - MICROCLINE

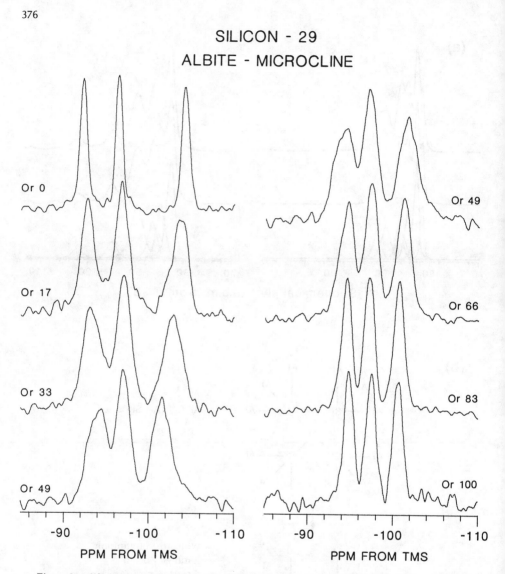

Figure 21. ²⁹Si MAS NMR spectra of Si,Al ordered feldspars along the albite - microcline join. There are three peaks from each sample, corresponding to the three Si-sites of the average structure of ordered alkali feldspars. The larger peak widths for the intermediate-composition samples indicate a range of local Na/K compositions. Samples synthesized by G.L. Hovis; spectra courtesy of B.L. Phillips.

spectra show just one peak (Fig. 22). For ^{27}Al, the peaks due to the central $(1/2,-1/2)$ transition are broadened by second-order quadrupole effects, potentially obscuring important information. The spinning sidebands due to the satellite $\pm(1/2, 3/2)$ transitions, however, are quite narrow, showing that there is only one Al-site, and providing chemical shifts accurate to about ±0.3 ppm. This result clearly demonstrates the usefulness of the satellite sideband technique, even for phases with only one Al-site. The ^{23}Na spectra obtained at 8.45T can also be fitted with one site (Fig. 23), but with inhomogeneous broadening causing less

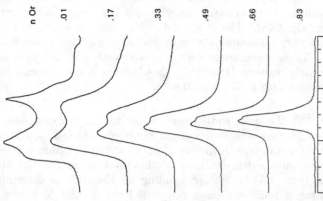

MICROCLINE - LOW ALBITE SERIES

²³Na MASS 8.45T

n Or

.01

.17

.33

.49

.66

.83

PPM

Figure 23. ²³Na MAS NMR spectra of Si,Al ordered feldspars along the join albite - microcline obtained at 8.45T. Note the well developed second-order quadrupole patterns for all the samples at this relatively low field strength and the greater inhomogeneous broadening for intermediate compositions. Same samples as for Figure 21. After Phillips et al. (submitted).

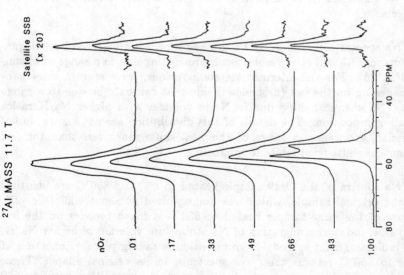

²⁷Al MASS 11.7 T

Satellite SSB
(x 20)

nOr

.01

.17

.33

.49

.66

.83

1.00

PPM

Figure 22. ²⁷Al MAS NMR spectra of Si,Al ordered feldspars along the albite-microcline join obtained at 11.7T. Note the well resolved second-order quadrupole patterns of the albite and microcline samples and the poorer resolution for the intermediate compositions. The spinning sidebands of the satellite $\pm(1/2,3/2)$ transitions, however, contain only one peak, indicating only one Al-site. Same samples as for Figure 21. After Phillips et al. (submitted).

well resolved peaks for the intermediate compositions. The satellite sideband technique is not as useful for ^{23}Na, because it has spin I=3/2 (Samoson et al., 1985).

J.V. Smith et al. (1983) showed that the chemical shifts of ^{29}Si on the three Si-sites in these samples vary systematically across the series and assigned the peaks on this basis (Fig. 24b). The ^{27}Al and ^{23}Na isotropic chemical shifts and the ^{23}Na QCC also vary monotonically with Na/K (Figs. 24a, c and d; Phillips et al., submitted). These parameters correlate well with unit cell edge lengths, angles, and particularly volume (Fig. 25), showing the same decreased rate of change for compositions with K/K+Na>0.6.

For ^{27}Al and ^{29}Si, the peak widths are greater for the intermediate compositions than for the end-members, indicating a range of <Al-O-Si> and <Si-O-T> (T = Si,Al) with a standard deviation of up to about 1° within a given sample. This range is a substantial fraction of the total range for the albite to microcline series of from 1.5° to 5.5° (depending on the site, as determined by diffraction), indicating a relatively large range of local Al and Si environments due to different local Na/K compositions. Simulation of the peaks for the ^{27}Al central transition indicates that not all of this range of bond angles is due to bulk Na,K inhomogeneity in the samples.

The increased average shielding at Na and decreased ^{23}Na QCC with increasing K/(K+Na) clearly indicate that the addition of K increases the coordination number and decreases the anisotropy at the alkali sites occupied not just by K but also those occupied by Na, with most of the effect having occurred by K/(K+Na) = 0.6. Clearly, the nature of such sites is not controlled just by NN bonding effects.

The ^{23}Na spectra collected at 11.7T are less broadened by second-order quadrupole effects and, thus, better resolve peak broadening due to a range of chemical shifts (Fig. 26). For the intermediate compositions, these spectra show substantial broadening on the left (deshielded) side that can only be due to a range of isotropic ^{23}Na chemical shifts due to Na in volumes with higher Na/K ratios than the bulk composition. The details of this distribution are not known, but it is not Gaussian and causes a range of chemical shifts much less than the full range of chemical shifts from albite to microcline.

The ^{23}Na spectra of the Or49 sample heated to 600 and 850°C are identical to that of the original sample, which was homogenized at about 900° (Fig. 27). The spectrum of the same sample heated to 530°C is much broader on the left (deshielded) side, indicating migration of Na atoms into volumes of higher Na/K, apparently indicating that spinodal decomposition is taking place. Reheating of this sample to 850°C returns the ^{23}Na spectrum to its original shape. These results seem to bracket the coherent spinodal for this composition between 530 and 600°C, and apparently show that there is an equilibrium Na,K distribution at temperatures above the coherent spinodal. Thus, a completely homogeneous alkali feldspar maybe impossible to produce. They also show that NMR is likely

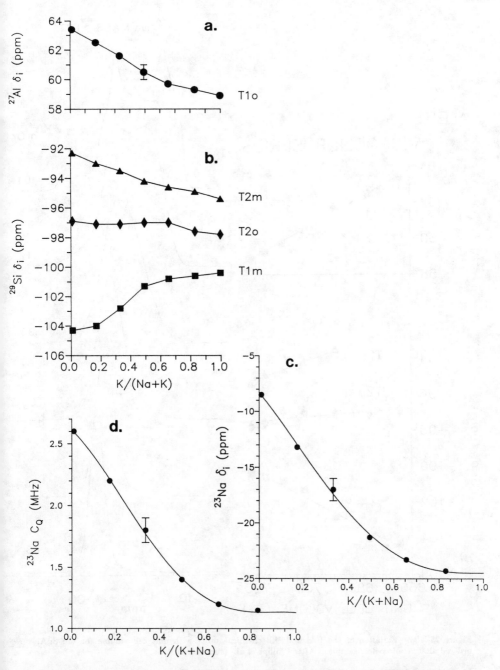

Figure 24. Variation of (a) ^{27}Al; (b) ^{29}Si; (c) ^{23}Na chemical shifts; (d) ^{23}Na QCC with K/Na+K for the Si/Al ordered albite - microcline samples described in Figures 21 - 23. For ^{23}Na and ^{27}Al the estimated precision and accuracy (including inhomogeneous broadening) is given by the error bars. For ^{29}Si the precision and accuracy is about the size of the symbols. ^{29}Si chemical shifts from Smith et al. (1984). After Phillips et al. (submitted).

380

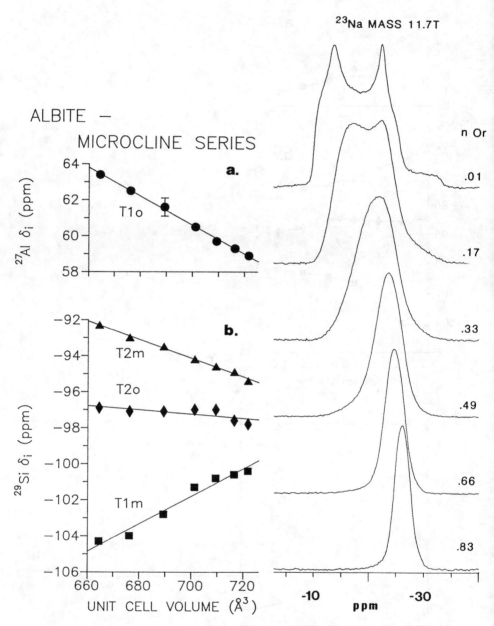

Figure 25(left). Variation of the ^{27}Al and ^{29}Si chemical shifts with unit cell volume for the Si,Al ordered albite - microcline samples. After Phillips et al. (submitted).

Figure 26(right). ^{23}Na MAS NMR spectra of the Si,Al ordered albite - microcline samples obtained at 11.7T. Note the reduced second-order quadrupole broadening relative to the 8.45T spectra (Fig. 23) at this higher field strength and the broadening of the peaks for the intermediate samples to the left (deshielded) side. Second-order quadrupole broadening only causes skewness to the right (shielded) side (see Fig. 6), and the observed broadening must be due to a range of isotropic chemical shifts. Skewness to the deshielded side must be due to Na in local environments more Na-rich than the bulk composition.. After Phillips et al. (submitted).

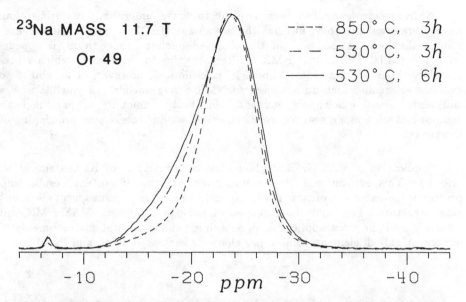

Figure 27. Effect of heating on the ^{23}Na MAS NMR spectrum of the Or49 albite - microcline sample. Heating at 600 and 850°C causes no change from the sample originally annealed at about 900°C. Heating to 530°C causes increased broadening to the left (deshielded) side, indicating the migration of Na atoms into more Na-rich environments, thus probably bracketing the coherent spinodal for this sample between 530 and 600°C. After Phillips et al. (submitted).

to be an effective way of determining the position of coherent spinodals in this system and many other systems of mineralogical importance.

Taken together, these data provide a picture of the structure of the intermediate composition low-temperature alkali feldspars in which there is essentially complete Al,Si order, substantial but not totally random K/Na disorder, a range of bond angles for each of the crystallographically different tetrahedral sites due to the Na,K disorder, and a local Na-site which becomes progressively larger and more symmetrical as K-content increases. Diffraction methods detect only the Al,Si order and indicate that the average alkali site is expanding.

APPLICATIONS TO AMORPHOUS PHASES

Glasses and other amorphous materials by definition lack the long-range order of crystals, and thus their structure cannot be as effectively examined by diffraction methods. Spectroscopic methods, primarily vibrational spectroscopic methods (Raman and IR) and EXAFS, have proven quite effective in characterizing their local (NN and NNN) structural environments. Two of the major conclusions derived from this work are that the NN and NNN environments of many elements are not very different in glasses than they are in crystals but that atomic sites in glasses typically contain a wider distribution of local environments than do individual crystalline phases.

NMR spectroscopy has been applied to fewer amorphous materials than have vibrational methods, but has the advantage that it is unambiguously examining the local environment of the observed nuclide. In addition, it appears from the available data that NMR is often sensitive to elements to which vibrational methods are not (e.g., B and F). It is limited, however, in its ability to examine elements which do not have an NMR active nuclide or a nuclide with a sufficiently small quadrupole moment. Substantial amounts of paramagnetic components such as Fe also reduce its effectiveness due to extreme broadening of the peaks.

Application of MAS NMR to glasses has been reviewed by Kirkpatrick et al. (1986b). This section will only summarize the most important results and present some examples of how NMR can help resolve important questions about glass structure. For amorphous phases of geological interest, MAS NMR will probably find the most application in examining the structural environnments of oxygen, Al, alkali elements, and minor elements such as B, F, C, and P.

^{17}O

As discussed above, ^{17}O is one of the few direct probes of oxygen environments in aluminosilicate materials, and for crystalline phases can often distinguish bridging from non-bridging oxygens (NBO's) and bridging oxygens (BO's) in Si-O-Si, Si-O-Al, and Al-O-Al linkages.

The results of preliminary work with alkaline-earth silicate glasses indicate that it can also resolve BO's from NBO's in these phases if the large cation is not Li, Na, or Mg. The only difference between the spectra of glasses and crystals is that the peaks are broad and featureless for the glasses but have well defined second-order quadrupole patterns for the crystals. Figure 28 shows that BO's and NBO's are well resolved for $CaSiO_3$ glass but not for $MgSiO_3$ glass (cf., Fig. 17, which shows the data for the crystals). For the intermediate $(Ca,Mg)SiO_3$ glasses, the intensity of the peak due to NBO's coordinated to Ca's decreases approximately linearly, as expected for essentially random mixing of Ca and Mg (Kirkpatrick et al., 1986). Alkali silicate glasses show virtually the same pattern (our unpublished data). There are no ^{17}O NMR data for aluminosilicate glasses to date, but it seems likely that this method in conjunction with the published work on crystalline phases will be a very powerful tool for investigating the types of oxygen linkages present in a wide variety of amorphous phases.

These spectra also provide an example of the general result that NMR peaks are broader and less well resolved for glasses than for crystals. This difference is clearly the result of a combination of disorder in bond angle, bond distance, NNN, and perhaps NN environment (Dupree and Pettifer, 1984; Murdoch, 1985; deJong et al., 1983; Kirkpatrick et al., 1986a,b; Oestrike et al., 1987).

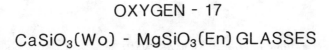

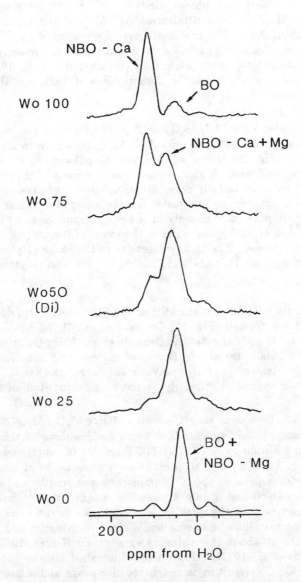

Figure 28. ¹⁷O MAS NMR spectra of glasses along the join CaSiO₃ (Wo) - MgSiO₃ (En); Di = CaMgSi₂O₆. Note that the spectra of the two end-members are very similar to those of the crystal-line phases (Fig. 17) except that the singularities due to second-order quadrupole effects are not present. The NBO's coordinated to Ca's are well separated from the BO's, but the NBO's coordinated to Mg's are not. The signal for NBO's coordinated to only Ca at about 120 ppm decreases in intensity and signal in the range of 70 ppm for NBO's coordinated to both Mg and Ca increases in intensity with increasing Mg/Ca ratio. These results imply extensive local Mg,Ca mixing. After Kirkpatrick et al. (1986a).

^{27}Al

As for crystalline phases, ^{27}Al MAS NMR is a powerful probe of the Al environments in amorphous phases, and easily resolves tetrahedral Al (Al^{IV}), five-coordinate Al (Al^{V}), and octahedral Al (Al^{VI}) if the H$_0$ field strength is greater than about 8.5T. For the most part, the results of the ^{27}Al NMR work are consistent with conclusions based on vibrational spectroscopy, x-ray radial distribution analysis, and calorimetric data (see Oestrike et al., 1987, for a short review), although for peraluminous compositions it indicates that Al^{V} is an important environment.

For all samples with MO/Al$_2$O$_3$ or M$_2$O/Al$_2$O$_3$ $\geq$ 1 (M = a divalent or monovalent cation), NMR detects only Al^{IV}, in agreement with all other spectrographic methods (Fig. 29; Hallas et al., 1983; Engelhardt et al., 1985; Oestrike and Kirkpatrick, in press). It also seems that all or most of the Al in these compositions is concentrated in fully polymerized Q^4 sites (Oestrike and Kirkpatrick, in press). These conclusions are based on the observation that at large field strengths there are peak maxima only in the range from about 50 to 60 ppm (the range of Q^4 Al^{IV} in crystals), and none in the range of 0 ppm (Al^{VI} in crystals) or 30 ppm (Al^{V} in crystals). The lack of intensity in the range of about 70 ppm (Q^3 Al^{IV} in crystals) suggests that there is little Al in sites with polymerizations of Q^3 or less.

^{27}Al in glasses becomes less shielded as the Si/(Si+Al) ratio decreases, paralleling the trend for crystals (Fig. 30; Oestrike et al., 1987), but do not seem to depend greatly on the M/Al ratio (Engelhardt et al., 1985; Oestrike and Kirkpatrick, in press). This latter observation is consistent with the idea that most of the Al is concentrated in fully polymerized sites and that their average Si/(Si+Al) ratio is controlled primarily by the bulk composition of the glass.

Less work has been done on glasses with MO(or M$_2$O)/Al$_2$O$_3$ < 1 (peraluminous compositions), but it appears that some compositions of this type contain Al^{V} and Al^{VI} in addition to Al^{IV}. The 11.7T spectra of rapidly quenched Al$_2$O$_3$-SiO$_2$ glasses (Fig. 31; Risbud et al., 1987) clearly show peaks at about 60, 30 and 0 ppm, along with a peak at about 12 ppm corresponding to undissolved corundum. The peaks at 60 and 0 ppm are readily assigned to Al^{IV} and Al^{VI} respectively. The assignment of the peak at about 30 ppm is the more controversial, but Al^{V} in such crystalline phases as andalusite, vesuvianite, and Ba-aluminum glycolate resonate at about this value (Alemany and Kirker, 1986; Lippmaa et al., 1986; Phillips et al., 1987). Al in distorted tetrahedral sites would most likely have very large QCC's, resulting in much broader peaks and a larger dependence of the chemical shift on the H$_0$ field than is observed. The H$_0$ field dependence of the 30 ppm peak maximum is relatively small and seems to indicate a QCC of about 3-4 MHz for the sites controlling the peak maximum. The most reasonable conclusion is that this peak is due to Al^{V}. Dupree et al. (1985) have also assigned a similar peak in ^{27}Al spectra of amorphous alumina films to Al^{V}.

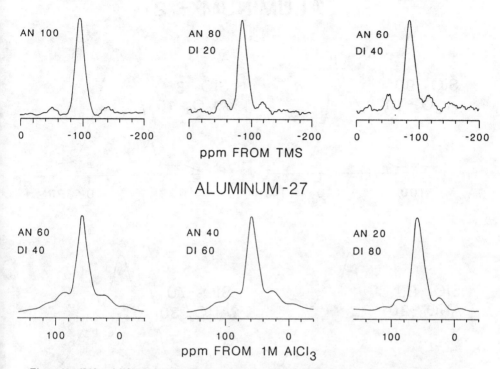

SILICON-29

AN 100

AN 80
DI 20

AN 60
DI 40

ppm FROM TMS

ALUMINUM-27

AN 60
DI 40

AN 40
DI 60

AN 20
DI 80

ppm FROM 1M AlCl$_3$

Figure 29. ^{27}Al and ^{29}Si MAS NMR spectra of glasses along the join diopside - anorthite (basalt ana-log compositions). Note the broad, structureless peaks. For ^{27}Al, the lack of intensity in the range of 0 ppm indicates a lack of Al^{VI}, and the intensity in the range of 58 - 62 ppm, indicates that most of the Al^{IV} is in fully polymerized, Q^4, sites. After Oestrike and Kirkpatrick (in press).

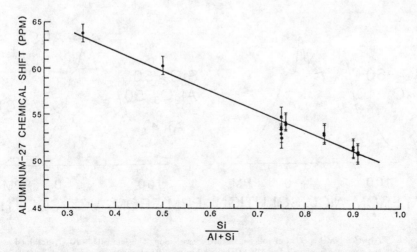

Figure 30. Variation of ^{27}Al chemical shifts (peak maxima at 11.7T) with Si/(Si+Al) for framework-structure glasses (those with just enough alkali or alkaline earth to charge balance all the Al in tetrahedral coordination). The decreased shielding with decreasing Si/(Si+Al) parallels the variation for crystalline aluminosilicates with framework structures. After Oestrike et al. (1987).

ALUMINUM – 27

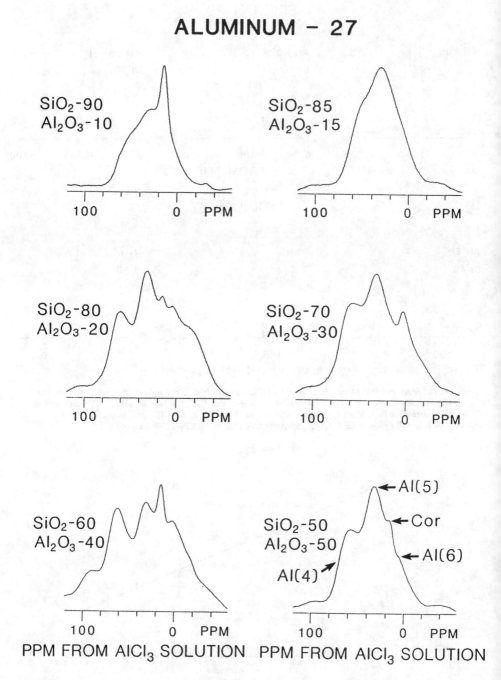

Figure 31. ^{27}Al MAS NMR spectra of roller-quenched glasses along the join SiO$_2$ - Al$_2$O$_3$ obtained at 11.7T. Note the presence of signals for Al^{IV} at about 60 ppm, Al^{VI} at about 0 ppm, and Al^V at about 30 ppm. The narrower peak at about 12 ppm is due to undissolved corundum. After Risbud et al. (1987).

To my knowledge the only other high-field ^{27}Al NMR data available for amorphous aluminosilicate phases with peraluminous composition are for gels along the join $Al_2Si_4O_{11}$-$KAl_3Si_3O_{11}$ (Weiss and Kirkpatrick, unpublished data). Spectra of these samples contain peaks at about $55(Al^{IV})$, $30(Al^V)$, and $0(Al^{VI})$ ppm, and they are much better resolved than for the Al_2O_3-SiO_2 glasses (Fig. 32). The intensities of the peaks due to Al^V and Al^{VI} decrease as the K/Al ratio increases, consistent with the idea that the K is charge balancing Al in tetrahedral coordination, as it does in peralkaline compositions.

Structural models of peraluminous compositions, which are relevant to understanding the behavior of peraluminous magmas, need to explicitly consider the role of Al^V.

Al^V also seems to be a component of Pb-borate calorometric fluxes in which corundum been dissolved (Oestrike et al, 1987b) and in aluminoborate glasses of a wide range of compositions (our unpublished data).

^{29}Si

For glasses, ^{29}Si provides the most information about SiO_2 and alkali silicate glasses and, due to lack of resolution, less information about aluminosilicate compositions. Even for aluminosilicates, however, there are systematic changes in chemical shift and peak width that provide information about the glass structure.

For SiO_2 glass (Fig. 33), the peak maximum is at about -111.5 to -112 ppm and the full width of the peak is from about -100 to -120 ppm (Dupree and Pettifer, 1984; Murdoch et al., 1985; Kirkpatrick, et al., 1986b; Oestrike et al., 1987). Based on the correlation of ^{29}Si chemical shift with $<$Si-O-Si$>$ for $Q^4(OAl)$ sites in crystalline phases for (Fig. 12), these values indicate a mean $<$Si-O-Si$>$ bond angle of about 151° and a range of from about 130° to 170° (see Henderson et al., 1984, and Oestrike et al., 1987, for more thorough reviews).

For alkali silicate glasses, the ^{29}Si spectra often contain multiple peaks or well resolved shoulders due to Si on sites with different degrees of polymerization (Fig. 34; Schramm et al., 1984; Dupree et al., 1984, 1985, 1986; Selvary et al., 1985). It is clear from these data that the average polymerization decreases with increasing M/Si ratio, but many of the spectra published to date have poor signal/noise, and the samples have not been chemically analysed. Thus, it is not possible to effectively address such questions as whether or not special compositions such as M_2SiO_3 (metasilicate) and $M_2Si_2O_5$ (disilicate) contain a range of Si-polymerizations and how the distribution of Si-polymerizations changes with changing large cation at constant M/Si ratio.

For metasilicate (pyroxene stoichiometry) glasses, the ^{29}Si peak width increases with increasing charge/cation radius (Z/r) for the large cation, consistent with the idea that low field-strength cations such as K and Ba cause a

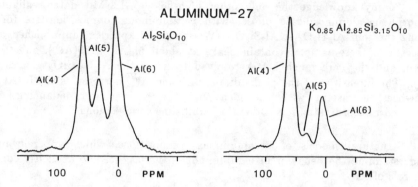

Figure 32. ^{27}Al MAS NMR spectra of dehydrated peraluminous gels obtained at 11.7T. Note the presence of signals for Al^{IV}, Al^{V}, and Al^{VI}. Spectra courtesy of C.A. Weiss, Jr.

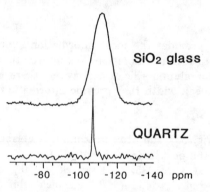

Figure 33. ^{29}Si MAS NMR spectra of SiO$_2$ glass and quartz. The larger peak width for the glass is related to a wider range of <Si-O-Si> bond angles than in crystalline phases. The peak maximum of -112 ppm for the glass indicates a mean <Si-O-Si> of about 152°. After Oestrike et al. (1987).

wider range of polymerizations than do high-field strength cations such as Na and Mg (Murdoch et al., 1985).

For aluminosilicate glasses, peaks for Si on different sites cannot be resolved because both depolymerization and increased Al NNN at constant polymerization cause deshielding at Si (e.g., deJong et al., 1983; Engelhardt et al., 1985; Kirkpatrick et al., 1986a,b; Oestrike and Kirkpatrick, in press).

For framework glasses (those with MO/SiO$_2$ or M$_2$O/SiO$_2$ = 1), the ^{29}Si chemical shift at peak maximum becomes more shielded with increasing Si/(Si+Al), paralleling the variation for crystalline framework aluminosilicates (Fig. 35; Oestrike et al., 1987). There is a linear decrease in shielding from SiO$_2$

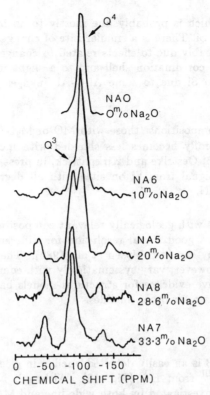

Figure 34. ²⁹Si MAS NMR spectra of glasses in the Na_2O - SiO_2 system. Note the presence of peaks for Si-sites with Q^4 (-111 ppm) and Q^3 (-95 ppm) polymerizations. After Dupree et al. (1984).

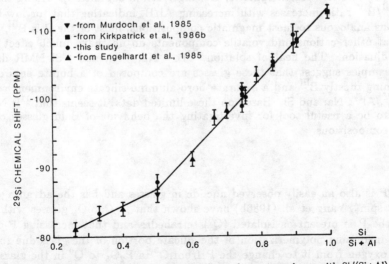

Figure 35. Variation of the ²⁹Si chemical shift of framework-structure glasses with Si/(Si+Al). The deshielding with decreasing Si/(Si+Al) parallels the variation for crystalline aluminosilicates. Note the small effect of the large cation, which includes alkalis, alkaline earths, and Pb. After Oestrike et al. (1987a).

to $Si/(Si+Al) = 0.5$ which is probably due mostly to an increase in the average number of Al NNN to Si. There is a smaller rate of change for compositions with $Si/(Si+Al) < 0.5$, probably due to effects related to changes in the Si/Al ratio in the second tetrahedral corrdination shell and to a slight increase in the average number of Al NNN to Si due to some Al-O-Al linkages in samples with bulk $Si/(Si+Al) \leq 0.5$.

For peralkaline compositions (those with MO or M_2O/Al_2O_3 greater than 1) ^{29}Si, unlike ^{27}Al, generally becomes less shielded with increasing $NBO/(Si+Al)$ (Engelhardt et al., 1985; Oestrike and Kirkpatrick, in press). The peaks for these compositions contain signal from Si on sites with all degrees of polymerization and numbers of Al NNN.

Overall, for glasses with geologically relevant compositions, it seems that ^{29}Si NMR cannot provide as good signal resolution for different types of Si-sites as can Raman spectroscopy can (e.g. Mysen et al., 1982). The NMR chemical shifts and peak widths do, however, vary systematically with composition and can provide at least coroborative evidence for structural models based on other types of data, including NMR data for other nuclides.

^{11}B

As for crystals, ^{11}B is an easily detected nuclide in glasses, and gives spectra that readily resolve B^{III} from B^{IV} (Fig. 16). Although Na-borosilicate glasses have been extensively investigated by both wide-line and MAS NMR (e.g., Dell et al., 1983; Bunker et al., in press), the only glass compositions of direct geochemical interest that have been investigated using ^{11}B are along the join $NaAlSi_3O_8$ - $NaBSi_3O_8$ (albite - reedmergnerite, Geisinger et al., submitted). Along this join, the B^{III}/B^{IV} ratio increases with increasing Al/B, indicating that for low-B compositions analogous to most magmatic compositions, most of the B is B^{III}, assuming that other cations and volatile components do not have much effect on the B-coordination. The heat of solution, ^{11}B, ^{23}Na, ^{27}Al, and ^{29}Si NMR data for these samples suggest that these glasses are composed of a borate environment containing mostly B^{III} and a separate boro-alumino-silicate environment containing B^{IV}, Al^{IV}, Na, and Si. Based on these limited data, it seems that ^{11}B NMR is likely to be a useful tool for investigating the behavior of B in glasses of magmatic compositions.

^{31}P

^{31}P is also an easily observed nuclide in glasses and has the advantage of an $I=1/2$ spin. Yang et al. (1986b) have shown that in $CaSiO_3$ glasses with added P_2O_5 the P is present as isolated (Q^0) tetrahedra and that increasing P-content causes increasing polymerization of the silicate portion of the glass due to removal of oxygen from it to change the P from Q^3 in P_2O_5 to Q^0 in the glass. This interpretation is based on a constant ^{31}P chemical shift of $+2.6$ ppm (the same as Ca-orthophosphate) and an increasingly shielded ^{29}Si chemical shift as the P_2O_5

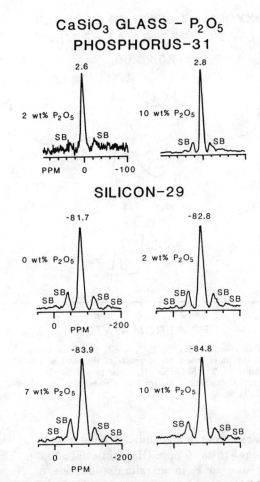

Figure 36. ^{31}P and ^{29}Si MAS NMR spectra of CaSiO$_3$ glasses with added P$_2$O$_5$. The ^{31}P chemical shift of about +2.7 ppm implies an orthophosphate (Q$^\circ$) environment. The progressive shielding at ^{29}Si with increasing P-content indicates progressive polymerization due to removal of O-atoms from the silicate environments in the glass to depolymerize the P from Q^3 in P$_2$O$_5$ to Q^0 in the glass. After Yang et al. (1986b).

content increases from 0.5% to 10% (Fig. 36). In Ca-aluminosilicate glasses more analogous to magmatic compositions, it seems that most of the P co-polymerizes with Al (Yang et al., unpublished data).

^{19}F

^{19}F is a difficult nuclide to observe with MAS NMR, because it often suffers from a large homonuclear dipolar broadening that makes resolution of multiple peaks difficult. Nonetheless, some data are available for silicate and aluminosilicate glasses which indicate that F can occur in several different environments. In SiO$_2$ glasses containing small amounts of F, the ^{19}F chemical shift is about +21 ppm (Duncan et al, 1986). Fluorinated silica gels also give this value (Hayashi et al., 1987), and it seems that this chemical shift is indicative of F in Si-F linkages.

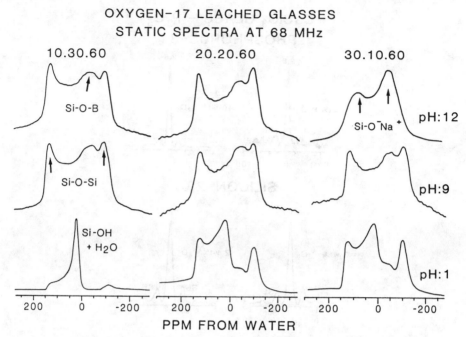

OXYGEN-17 LEACHED GLASSES
STATIC SPECTRA AT 68 MHz

Figure 37. ^{17}O static NMR spectra of Na$_2$O - B$_2$O$_3$ - SiO$_2$ glasses leached in ^{17}O enriched water at 70°C. The oxygen from the water enters the structure of the residual amorphous material and is present in the linkages indicated. The Si-OH and H$_2$O sites present at pH=1 are removed by heating at 115°C, leaving silica gel. Compositions are for base glasses in (in order) mole % Na$_2$O · B$_2$O$_3$ · SiO$_2$. After Bunker et al. (in press).

SiO$_2$-NaAlO$_2$ glasses containing added F give a peak at this position and also a peak in the range -16 to -6 ppm (Hayashi et al., 1987). This latter peak is interpreted as being due to F in an alumino-fluoride or Na-alumino-fluoride environment. A phase-separated peralkaline Na-aluminosilicate glass containing added F gives these two peaks and a peak at -65 ppm, identical to crystalline NaF. This later peak is interpreted to be due to phase separated NaF. These results need to be integrated into a complete structural model for F in alumino-silicate glasses, but it is clear that ^{19}F MAS NMR has the potential to provide useful information about the behavior of F in magmas and lavas and how it influences their physical and chemical behavior.

INVESTIGATION OF ROCK-WATER INTERACTION

Study of the solid phases produced during the reaction of minerals and glasses with aqueous fluids is often hindered by the fine grain-size and amorphous structure of the products. Because NMR investigates only local structure, in many cases it offers information about those phases that other methods cannot. There have been two investigations of such reactions: Bunker et al. (in press), who investigated the dissolution of Na-borosilicate glasses in ^{17}O-enriched, pH-buffered solutions at 70° and Yang (1987), who investigated the reaction of albite, a glass of nearly albite composition, and a rhyolitic glass with pH-buffered solutions at 250°C under autoclave conditions.

For the Na-borosilicate glasses, ^{17}O from the water enters the structure of the residual amorphous phase and occurs in Si-O-Si, Si-O-B, and Si-O-Na linkages, depending on the pH (Fig. 37). In combination with TEM observations, these data clearly indicate a solution-reprecipitation mechanism for the reaction. The reaction products consist of 10-100Å spherules which coarsen with time. As these particles coarsen, they begin to allow bulk movement of water around them, explaining the change in the leaching kinetics from a $t^{\frac{1}{2}}$ to a t^1 time dependence. The molecular-level mechanism of attack by water on the glass is also a function of pH.

For albite at 250°C, the NMR results show that the reaction is surface controlled and that if a layer of hydrated albite occurs on the surface, it is probably less than 100 Å thick. At pH = 1 to 2, the reaction products are kaolinite and silica gel, which is detected only by ^{29}Si NMR (Fig. 8). At pH 9 the reaction products are kaolinite and an amorphous aluminosilicate with the local structure of a smectite. This latter phase is detected only by NMR, and gives signal in the ^{27}Al, ^{29}Si, ^{27}Al CPMAS, and ^{29}Si CPMAS spectra (Fig. 38).

For the albite and rhyolite glasses, on the other hand, there seems to be extensive hydration and H $\Leftrightarrow$ Na exchange in the interior of the 10-40 micron starting fragments, as well as solution/reprecipitation at the glass/water interface. The glass fragments seem to be fully hydrated (about 10 wt % H_2O) in about 20 days, but the Al/Si ratio does not change. The NMR, IR, and electron microprobe data suggest that about 20% of the protons are present as OH^- groups and the rest as water molecules hydrating the resuidual +1 and +2 cations. These hydration shells apparently weaken the interaction between the large cations and the framework, because the $<Si-O-T>$ bond angles increase about 5° upon hydration, as indicated by an increased shielding at ^{29}Si (Fig. 39). As for albite, the product phases are sheet silicates at low pH's, but the glass eventually crystallizes to a mixture of analcime and a phillipsite-like zeolite at high pH's.

ADSORBED AND EXCHANGABLE IONS

The structural environment of surface adsorped and exchanged ions in, e.g. zeolites and clays, is often difficult or impossible to investigate directly. Recent results, however, have shown that NMR methods are very useful for investigating the environments of such ions. The limitations appear to be sensitivity (i.e., is it possible to collect enough signal in a reasonable amount of time?) and spectral resolution (i.e., are the peaks narrow enough that we can learn something?).

^{133}Cs has been an especially useful nuclide for the initial work in this area, because it is readily adsorbed and exchanged, has a large chemical shift range, is 100% abundant, has a relatively high Larmor frequency, and a small quadrupole moment (so that the second-order quadrupole broadening is small, the peak overlap is minimized, and the spectral resolution enhanced relative to nuclides with

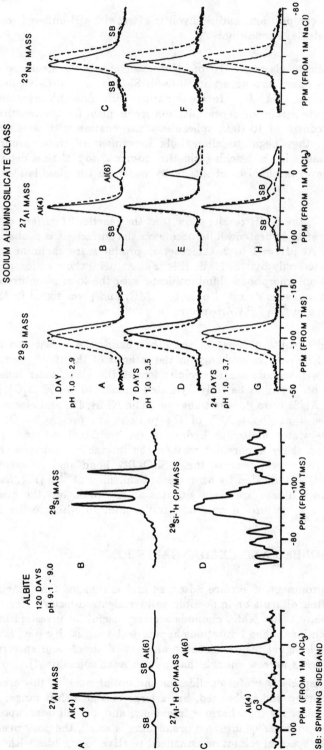

Figure 39. ^{23}Na, ^{27}Al, and ^{29}Si MAS NMR spectra of a Na-aluminosilicate glass close to albite in composition (dashed lines) and the same glass reacted with buffered pH 1 - 3 water at 250°C under autoclave conditions (solid lines). Note the increased shielding at ^{29}Si due to increased $<Si-O-T>$ (T=Si,Al), the formation of kaolinite (indicated by the Al^{IV} and ^{29}Si shoulder at -92 ppm), and the narrowing of the ^{23}Na peak due to formation of a shell of hydration water around Na-atoms in the glass. Spectra courtesy of W.-H. Yang.

Figure 38. ^{27}Al and ^{29}Si MAS and CPMAS spectra of albite reacted with buffered pH 9 water at 250°C. In the MAS spectra most of the signal is due to unreacted albite, but in the CPMAS spectra, all the signal is from the hydrous product phase. This phase is amorphous to x-rays, but contains a small amount of Al^{IV}, Al^{VI}, and a broad range of Si-sites consistent with it having a local structure similar to a sheet silicate. Spectra courtesy of W.-H. Yang.

SB: SPINNING SIDEBAND

large quadrupole moments).

The results of Weiss and Turner (work in progress) on Cs exchanged onto montmorillonite with a high cation exchange capacity illustrate the kinds of data that can be obtained (Fig. 40). For a slurry of clay in 0.1M CsCl solution both MAS and static spectra show a narrow peak at about 0 ppm for Cs in the solution and a broader peak at about -10 ppm for Cs in the clays (Figs. 40a,b). Resolution is much better in the MAS spectrum. When the same sample is dried, washed with deionized water, and kept at 100% relative humidity, the solution peak disappears and only the peak at about -10 ppm remains (Fig. 40c). When the same sample is kept at room humidity (~30%), the main peak moves to about -20 ppm, and a small peak at about 38 ppm appears (Fig. 40d). The same sample dried at 100°C gives a spectrum with greatly increased intensity of the peak at about 38 ppm, reduced the intensity of the peak near -20 ppm, and a much more shielded peak at -106 ppm (Fig. 40e). The peak at +38 ppm is relatively broad. It also has many spinning sidebands, probably due to a relatively large CSA. Further drying of the smple at 400°C eliminates the peak at about -20 ppm (Fig. 40f).

Because of it's progressive disappearance with dehydration, apparently small CSA and narrow width, the peak at -10 to -20 ppm is probably due to interlayer Cs with water molecules hydrating it. Because of it's appearance with dehydration, the peak at +38 ppm is probably due to interlayer Cs without hydration. The larger peak width may indicate a range of local environments, and the larger CSA probably indicates that these environments are relatively anistropic. The peak at -106 ppm is most likely due to Cs with a much higher coordination, similar to the increased shielding at Na in the alkali feldspars as the average coordination increases. The narrow peak width and lack of spinning sidebands indicate that the site giving rise to this peak is well defined and quite symmetrical. Most likely this peak is due to Cs in 12-coordinate sites with 6NN from the plane of basal oxygens in the two NNN tetrahedral sheets. The relatively low intensity of this peak is probably due to the poor relative orientation of the TOT layers in smectite clays. A similar increased shielding at Cs due to increased coordination number occurs with dehydration of mordenite (Chu et al. 1987).

ACKNOWLEDGEMENTS

This work was supported by NSF Grants EAR 84-08421 and EAR 87-06929. Christopher Knight, Richard Oestrike, Brian Phillips, Suzanne Schramm, Hye-Kyung Timken, Charles A. Weiss, Jr., and Wang-Hong Yang kindly provided some of the spectra presented. Brian Phillips, Charles A. Weiss, Jr., and Frank Hawthorne provided helpful reviews. Jean Daly and Tom Dirks worked far beyond the call of duty to produce the manuscript.

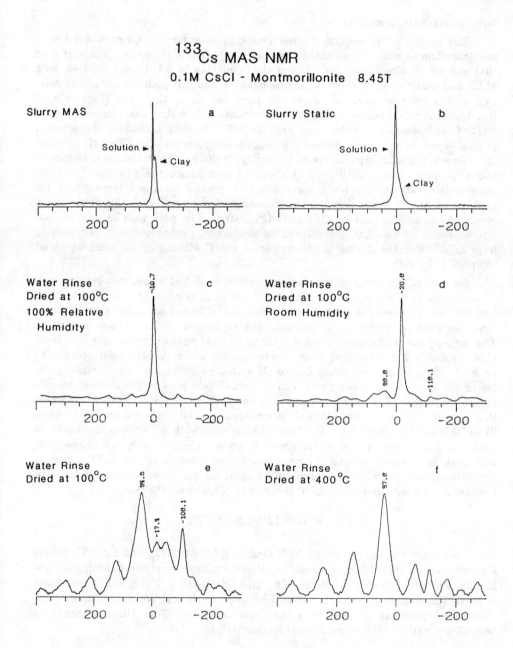

^{133}Cs MAS NMR
0.1M CsCl - Montmorillonite   8.45T

Slurry MAS   a

Solution ►
◄ Clay

200   0   -200

Slurry Static   b

Solution ►
◄ Clay

200   0   -200

Water Rinse   c
Dried at 100°C
100% Relative
Humidity

-19.7

200   0   -200

Water Rinse   d
Dried at 100°C
Room Humidity

-20.8

96.8   -118.1

200   0   -200

Water Rinse   e
Dried at 100°C

94.5   -17.4   -108.1

200   0   -200

Water Rinse   f
Dried at 400°C

97.8

200   0   -200

Figure 40. ^{133}Cs NMR spectra of Cs-exchanged montmorillonite (smectite) clay run as a slurry, and then progressively dehydrated under the conditions indicated. C.A. Weiss, Jr. and G. Turner, unpublished data.

REFERENCES

Abragam, A. (1961) The Principles of Nuclear Magnetism, Oxford. Clarendon, 599 pp.

Akitt, J.W. (1983) NMR and Chemistry, An Introduction to the Fourier Transform-Multinuclear Era. London, Chapman and Hall, 263 pp., 2nd ed.

Alemany, C.B. and Kirker, G.W. (1986) First observation of 5-coordinate aluminum by MAS ^{27}Al NMR in well characterized solids. J. Am. Chem. Soc. 108, 6158-6162.

Altaner, S.P., Weiss, C.A. Jr., and Kirkpatrick, R.J. (in press) Evaluation of structural models of mixed-layer illite/smectite: ^{29}Si NMR evidence. Nature.

Andrew, E.R. (1971) The narrowing of NMR spectra of solids by high-speed specimen rotation and the resolution of chemical shift and spin multiplet structures for solids. Progress in NMR Spectroscopy. 8, 1-39.

Andrew, E.R. (1981) Magic angle spinning. Int. Rev. Phys. Chem. 1, 195-224.

Barron, P.F. , Slade, P. and Frost, R.L. (1985) Ordering of aluminum in tetrahedral sites in mixed-layer 2:1 phyllosilicates by solid-state high-resolution NMR. J. Phys. Chem. 89, 3880-3885.

Bax, A. (1984) in Topics in ^{13}C NMR spectroscopy, vol. 4., Levy, G. C., ed., p. 197-238, New York, John Wiley & Sons.

Bax, A. and Lerner, L. (1986) Two-dimensional nuclear magnetic resonance spectroscopy. Nature 232, 960-967.

Becker, E.D. (1980), High Resolution NMR, Theory and Application. New York, Academic Press, 354 pp., 2nd ed.

Blackwell, C.S. and Patton, R.L. (1984) Aluminum-27 and phosphorous-31 nuclear magnetic resonance studies of aluminophosphate molecular sieves. J. Phys. Chem. 88, 6135-6139.

Blinc, R. (1981) Magnetic resonance and relaxation in structurally incommensurate systems. Phys. Repts. 79, 331-398.

Bunker, B.C., Tallant, D.R., Headley, T.J., Ashley, C.A., Turner, G.L. and Kirkpatrick, R.J. (1987) The structure of leached sodium borosilicate glass. J. Am. Ceram. Soc., in press.

Carpenter, M.A., Putnis, A., Navrotsky, A. and McConnell, J.D.C. (1983) Enthalpy effects associated with Al, Si ordering in anhydrous Mg cordierite. Geochim. Cosmochim. Acta 47, 899-906.

Chu, P.J., Gerstein, B.C., Nunan, J. and Klier (1987) A study by solid- state NMR of ^{133}Cs and ^{1}H of a hydrated and dehydrated cesium mordenite. J. Phys. Chem. 91, 3588-3592.

Cohen, M.H. and Reif, F. (1957) Quadrupole effects in nuclear magnetic resonance studies of solids. Solid State Physics 5, 321-438.

Cruickshank, M.C., Dent Glasser, L.S., Barni, S.A.I. and Poplett, I.J.F. (1986) Penta-coordinated aluminum: a solid-state ^{27}AL NMR study. J. Chem. Soc., Chem. Comm., 23-24.

Davis, J.C. (1965) Advanced Physical Chemistry. New York, Ronald, 632 pp.

deJong, B.H.W.S., Schramm, C.M. and Parziale, V.E. (1983) Polymerization of silicate and aluminate tetrahedra in glasses, melts, and aqueous solutions-IV, Aluminum coordination in glasses and aqueous solutions and comments on the aluminum avoidance principle. Geochim. Cosmochim. Acta 47, 1223-1236.

deJong, B.H.W.S., Schramm, C. M. and Parziale, V.E. (1984) Polymerization of silicate and aluminate tetrahedra in glasses, melts, and aqueous solution-V. The polymeric structure of silica in albite and anorthite composition glass and the devitrification of amorphous anorthite. Geochim. Cosmochim. Acta 48, 2619-2629.

Dell, W.J. , Bray, P.J. and Xiao, S.Z. (1983) ^{11}B NMR studies and structural modeling of Na_2O-B_2O_3-SiO_2 glasses of high soda content. J. Non-Cryst. Solids 58, 1-16.

Duncan, T.M. and Douglass, D.C. (1984) On the ^{31}P chemical shift anisotropy in condensed phosphates. Chem. Phys. 87, 339-349.

Dupree, R., Farnan, I., Forty, A.J., El-Mashri, S. and Bottyan, L. (1985) A MAS NMR study of the structure of amorphous aluminous flims. J. Physique 46, C8, 113-117.

Dupree, R., Holland, D., McMillan, P.W. and Pettifer, R.F. (1984) The structure of soda-silicate glasses: A MAS NMR study. J. Non-Cryst. Solids 68, 399-410.

Dupree, R., Holland, D. and Williams, D.S. (1985) A magic angle spinning study of the effect of modifer and intermediate oxides on the local structure in vitreous silicate networks. J. Physique 46, C8, 119-123.

Dupree, R., Holland, J.D. and Williams, D.S. (1986) The structure of binary alkali silicate glasses. J. Non-Cryst. Solids 81, 185-200.

Dupree, E. and Pettifer, R.F. (1984) Determination of the Si-O-Si bond angle distribution in vitreous silica by magic angle spinning NMR. Nature. 308, 523-525

Engelhardt, G. and Radeglia R. (1984) A semi-empirical quantum-chemical rationalization of the correlation between SiOSi angles and ^{29}Si NMR chemical shifts of silica polymorphs and framework aluminosilicates (zeolites). Chem. Phys. Lett. 108, 271-274.

Engelhardt, G., Nofz, M., Forkel, K., Wihsmann, F.G., Mägi, M., Samoson, A. and Lippmaa, E. (1985) Structural studies of calcium aluminosilicate glasses by high resolution solid state ^{29}Si and ^{27}Al magic angle spinning nuclear magnetic resonance. Phys. Chem. Glasses. 26, 157-165.

Ernst, R. R., Bodenhausen, G. and Wokaun, A. (1986) Principles of nuclear magnetic resonance in one and two dimensions, Oxford, Clarendon Press.

Farrar, T.C. and Becker, E.D. (1971) Pulse and Fourier transform NMR: Introduction to theory and methods. New York, Academic Press.

Fujiu, T. and Ogino, M. (1984) ^{29}Si NMR study on the structure of lead-silicate glasses. J. Non-Cryst. Solids 64, 287-290.

Fukushima, E. and Roeder, S.B. (1981) Experimental Pulse NMR, a Nuts and Bolts Approach, Addison-Wesley, Reading, MA, 519 p.

Fyfe, C.A. (1984) Solid State NMR for Chemists. C.R.C. Press, Guelph, Ontario, Canada.

Fyfe, C.A., Gobbi, G.C., Hartman, J.S., Klinowski, J. and Thomas, J.M. (1982) Solid-state magic-angle spinning aluminum-27 nuclear magnetic resonance studies of zeolites using a 400-MHz high-resolution spectrometer. J. Phys. Chem. 86, 1247-1250.

Fyfe, C.A., Gobbi, G.C., Klinowski, J., Putnis, A. and Thomas, J.M. (1983a) Characterization of local atomic environments and quantitative determination of changes in site occupancies during the formation of ordered synthetic cordierite by Si and Al magic-angle spinning NMR spectroscopy. Chem. Commun., 556.

Fyfe, C.A., Gobbi, G.C., Murphy, W.J., Ozubko, R.S. and Slack, D.A. (1984) Investigation of the contributions to the ^{29}Si MAS NMR line widths of zeolites and the detection of crystallographically inequivalent sites by the study of highly siliceous zeolites. J. Am. Chem. Soc. 106, 4435-4438.

Fyfe, C.A., Gobbi, G.C. and Putnis,A. (1986) Elucidation of the mechanism and kinetics of the Si, Al ordeing process in synthetic magnesium cordierite by Si-29 magic angle spinning NMR spectroscopy. J. Am. Chem. Soc. 108, 3218-3223.

Fyfe, C.A., O'Brien, J.H. and Strobl, H. (1987) Ultra-high resolution ^{29}Si MAS NMR spectra of highly siliceous zeolites. Nature 326, 281-283.

Fyfe, C.A. Thomas, J.M., Klinowski, J. and Gobbi, G.C. (1983b) Magic angle spinning NMR spectroscopy and the structure of zeolites. Angew Chem. 22, 259-336.

Ganapathy, S., Schramm, S. and Oldfield, E. (1982) Variable-angle sample spinning high resolution NMR of solids. J. Chem. Phys. 77, 4360-4365.

Gerstein, B.C. and Dybowski, C.R. (1985) Transient techniques in NMR of solids. Academic Press, New York, 295 p.

Geisinger, K., Oestrike, R., Navrotsky, A., Turner, G. and Kirkpatrick, R.J. (submitted) Thermochemistry and structure of glasses along the join $NaAlSi_3O_8$ - $NaBSi_3O_8$. Geochim. Cosmoochim. Acta.

Ghose, S. and Tsang, T. (1973) Structural dependence of quadrupole coupling constant e^2qQ/h for ^{27}Al and crystal field parameter D for Fe^{3+} in aluminosilicates. Am. Mineral. 58, 748-755.

Gladden, L.F., Carpenter, T.A. and Elliott, S.R. (1986) ^{29}Si MAS NMR studies of the spin-lattice relaxation time and bond-angle distribution in vitreous silica. Philos. Mag. B53, L81-87.

Gladden, L.F., Carpenter, T.A., Klinowski, J. and Elliott, S.R. (1986) Quantitative interpretation of exponentially broadened solid-state NMR signals. J. Mag. Res. 66, 93-104.

Grimmer, A.R. (1985) Correlation between individual Si-O bond lengths and the principal values of the ^{29}Si chemical shift tensor in solid silicates. Chem. Phys. Lett. 119, 416-420.

Grimmer, A.R., Mägi, M., Hähnert, M., Stade, H., Samoson, A., Weiker, W. and Lippmaa, E. (1984) High-resolution solid-state ^{29}Si nuclear magnetic resonance spectroscopic studies of binary, alkali silicate glasses. Phys. Chem. Glasses 25, 105-109.

Grimmer, A.R., Peter, R., Fechner, E. and Molgedey, G. (1981) High-resolution ^{29}Si NMR in solid silicates. correlations between shielding tensor and Si-O bond length. Chem. Phys. Lett. 77, 331-335.

Grimmer, A.D., von Lampe, F., Mägi, M. and Lippmaa, E. (1983) Hochauflösende ^{29}Si-NMR and festen silicaten; einfluss von Fe^{2+} in olivinen. Zeit. Chemie 23, 343-344.

Grimmer, A.D., von Lampe, F. and Mägi M. (1986) Solid-state high-resolution ^{29}Si MAS NMR of silicates with sixfold coordinated silicon. Chem. Phys. Lett. 132, 549-553.

Hallas, E. Haubenreisser, U., Hänert M. and Müller, D. (1983) NMR-untersuchungen an Na_2O-Al_2O_3-SiO_2 - gläsern mit hilfe der chemischen verschiebung, von ^{27}Al-kernen. Glastech. Berichte 56, 63-70.

Harris, R.K. (1983) Nuclear magnetic resonance spectroscopy. London, Pitman Books, Ltd.

Hayashi, S., Kirkpatrick, R.J. and Dingwell, D. (1987) MASS NMR study of F-containing Na_2O-Al_2O_3-SiO_2 glasses (Abs). Trans. Am. Geophys. Union (EOS) 68, 430.

Henderson, G.S., Fleet, M.E. and Bancroft, G.M. (1984) An X-ray scattering study of vitreous $KFeSi_3O_8$ and $NaFeSi_3O_8$ and reinvestigation of vitreous SiO_2 using quasi-crystalline modelling. J. Non-Crystal. Solids 68, 333-349.

Herrero, C.P., Sanz, J. and Serratosa, J.M. (1985a) Si, Al distribution in micas: analysis by high-resolution ^{29}Si NMR spectroscoy. J. Phys. C.: Solid State Physics 18, 13-22.

Herrero, C.P., Sanz, J. and Serratosa, J.M. (1985b) Tetrahedral cation ordering in layer silicates by ^{29}Si NMR spectroscopy. Solid State Comm. 53, 151-154.

Higgins, J.B. and Woessner, D.E. (1982) ^{29}Si, ^{27}Al, and ^{23}Na NMR spectra of framework silicates (abstr.). Trans. Am. Geophys. Union (EOS) 63, 1139.

Janes, N. and Oldfield, E. (1985) Prediction of silicon-29 nuclear magnetic resonance chemical shifts using a group electronegativity approach: applications to silicate and aluminosilicate structures. J. Am. Chem. Soc. 107, 6769-6775.

Janes, N. and Oldfield, E. (1986) Oxygen-17 NMR study of bonding in silicates: the d-orbital controversy. J. Am. Chem. Soc. 108, 5743-5753.

Kinsey, R.A., Kirkpatrick, R.J., Hower, J., Smith, K.A. and Oldfield, E. (1985) High resolution aluminum-27 and silicon-29 nuclear magnetic resonance spectroscopic study of layer silicates, including clay minerals. Am. Mineral. 70, 537-548.

Kirkpatrick, R.J., Carpenter, M.A., Yang, W.-H. and Montez, B. (1987) ^{29}Si magic-angle NMR spectroscopy of low-temperature ordered plagioclase feldspar. Nature 325, 236-238.

Kirkpatrick, R.J., Dunn, T., Schramm, S., Smith, K.A., Oestrike, R. and Turner, F. (1986b) Magic-angle sample-spinning nuclear magnetic resonance spectroscopy of silicate glasses: a review. In Structure and Bonding in Noncrystalline Solids (eds. G.E. Walrafen and A.G. Revesz). pp. 303-327. New York, Plenum Press.

Kirkpatrick, R.J., Oestrike, R., Weiss, C.A. Jr., Smith, K.A. and Oldfield, E. (1986a) High-resolution ^{27}Al and ^{29}Si NMR spectroscopy of glasses and crystals along the join $CaMgSi_2O_6$-$CaAl_2SiO_6$. Am. Mineral. 71, 705-711.

Kirkpatrick, R.J., Smith, K.A., Schramm, S., Turner, G. and Yang, W.-H. (1985) Solid-state nuclear magnetic resonance spectroscopy of minerals. Ann. Rev. Earth Planet. Sci. 13,

29-47.

Klinowski, J., Ramdas, S., Thomas, J.M., Fyfe, C.A. and Hartman, J.S. (1982) A re-examination of Si, Al ordering in zeolites NaX and NaY. J. Chem. Soc., Faraday Trans. 278, 1025-1050.

Knight, C.T.C., Kirkpatrick, R.J. and Oldfield, E. (1987) Silicon-29 2D evidence of four novel doubly germainian substituted silicate cages in a tetramethylammonian germanosilicate solution. J. Amer. Chem. Soc. 109, 1632-1635.

Komarneni, S., Roy, R., Fyfe, C.A., Kennedy, G.J. and Strobl, H. (1986) Solid state ^{27}Al and ^{29}Si magic-angle spinning NMR of aluminosilicate gels. J. Am. Ceram. Soc. 69, C42-C44.

Lippmaa, E., Mägi, M. Samoson, A., Tarmak, M. and Engelhardt, G. (1980) Structural studies of silicates by solid-state high-resolution ^{29}Si NMR. J. Amer. Chem. Soc. 102, 4889-4893.

Lippmaa, E., Mägi, M. Samoson, A., Tarmak, M. and Engelhardt, G. (1981) Investigation of the structure of zeolites by solid-state high-resolution ^{29}Si NMR spectroscopy. J. Amer. Chem. Soc. 103, 4992-4996.

Lippmaa, E. Samoson, A. and Mägi, M. (1986) High-resolution ^{27}Al NMR of aluminosilicates. J. Amer. Chem. Soc. 108, 1730-1735.

Lipsicas, M., Raythatha, R.H., Pinnavaia, T.J., Johnson, I.D., Giese, R.F., Costanzo, P.M. and Robert, J.L. (1984) Silicon and aluminum site distributions in 2:1 layered silicate clays. Nature 309, 604-609.

Loewenstein, W. (1954) The distribution of aluminum in the tetrahedra of silicates and aluminates. Amer. Mineral. 39, 92-96.

Mägi, M., Lippmaa, E., Samoson, A., Engelhardt, G. and Grimmer, A.R. (1984) Solid-state high-resolution silicon-29 chemical shifts in silicates. J. Phys. Chem. 88, 1518-1522.

McConnell, J.D.C. (1985) Symmetry aspects of order-disorder and the application of Landau theory, in Revs. in Mineral. 14, Ribbe, P.H., ed, M.S.A., Washington, D.C.

Meadows, M. P., Smith, K. A., Kinsey, R. A., Rothgeb, T. M., Skarjune, R. P. and Oldfield, E. (1982) High-resolution solid-state NMR of quadrupole nuclei. Proc. Nat. Acad. Sci. USA, 79, 1351-1355.

Müller, D., Jahn, E., Ladwig, G. and Haubenreisser, U. (1984) High-resolution solid-state ^{27}Al and ^{31}P NMR: correlation between chemical shift and mean Al-O-P angle in AlPO polymorphs. Chem. Phys. Lett. 109, 332-336.

Murdoch, J.B,, Stebbins, J.F. and Carmichael, I.S.E. (1985) High-resolution ^{29}Si NMR study of silicate and aluminosilicate glasses. The effect of network-modifying cations. Am. Mineral. 70, 332-343.

Mysen, B.O., Virgo, D. and Seifert, F.A. (1982) The structure of silicate melts: implications for chemical and physical properties of natural magma. Rev. Geophys. Space Phys. 20, 353-383.

Oestrike, R. and Kirkpatrick, R.J. (in press) ^{27}Al and ^{29}Si MASS NMR spectroscopy of glasses in the system anorthite-diopside-forsterite. Am. Mineral.

Oestrike, R., Yang, W.-H., Kirkpatrick, R.J., Hervig, R.L., Navrotsky, A. and Montez, B. (1987a) High-resolution ^{23}Na, ^{27}Al and ^{29}Si NMR spectroscopy of framework aluminosilicate glasses. Geochim. Cosmochim. Acta 51, 2199-2209.

Oestrike, R., Navrotsky, A., Turner, G., Montez, B. and Kirkpatrick, R.J. (1987b) Structural environment of Al dissolved in $2Pb \cdot B_2O_3$ glasses used for solution calorimetry: an ^{27}Al NMR study. Am. Mineral. 72, 788-791.

Oldfield, E., Kinsey, R.A., Smith, K.A., Nichols, J.A. and Kirkpatrick, R.J. (1983) High-resolution NMR of inorganic solids. Influence of magnetic centers on magic-angle sample-spinning lineshapes in some natural alumino- silicates. J. Magn. Reson. 51, 325-327.

Oldfield, E. and Kirkpatrick, R.J. (1985) High-resolution nuclear magnetic resonance of inorganic solids. Science 227, 1537-1544.

Peters, A.W. (1982) Random siting of aluminum in faujasite. J. Phys. Chem. 86, 3489-3491.

Phillips, B.L., Allen, F.M. and Kirkpatrick, R.J. (1987) High-resolution solid-state ^{27}Al NMR of Mg-rich vesuvanite. Am. Mineral. 72, 1190-1194.

Phillips, B.L., Kirkpatrick, R.J. and Hovis, G.L. (submitted) ^{23}Na, ^{27}Al, and ^{29}Si MAS NMR study of an Al/Si ordered alkali feldspar solid solution series, Phys. Chem. Mineral.

Putnis, A. and Angel, R.J. (1985) Al, Si ordering in cordierite using magic angle spinning NMR II: models of Al, Si order from NMR data. Phys. Chem. Mineral. 12, 217-222.

Putnis, A., Fyfe, C.A. and Gobbi, G.C. (1985) Al, Si ordering in cordierite using magic angle spinning NMR. I: Si-29 spectra of synthetic cordierites. Phys. Chem. Mineral. 12, 211-216.

Putnis, A., Salje, E., Redfern, S.A.T., Fyfe, C.A. and Strobl, H. (1987) Structural states of Mg-cordierite I: order parameters from synchrotron X-ray and NMR data. Phys. Chem. Mineral. 14, 446-454.

Ramdas, S. and Klinowski, J. (1984) A simple correlation between isotropic ^{29}Si-NMR chemical shifts and T-O-T angles in zeolite frameworks. Nature 308, 521-523.

Ribbe, P., ed. (1983) Feldspar Mineralogy, 2nd ed., Reviews in Mineral., v.2, Min. Soc. Am., Washington, D.C., 362 p.

Rigamonti, A. (1984) NMR-NQR studies of structural phase transitions. Adv. in Physics 33, 115-191.

Risbud, S.H., Kirkpatrick, R.J., Taglialavore, A.P. and Montez, B. (1987) Solid-state NMR evidence of 4-, 5-, and 6-fold aluminum sites in roller-quenched SiO_2-Al_2O_3 glasses. J. Am. Cer. Soc. 70, C10-C12.

Samoson, A. (1985) Satellite transition high-resolution NMR of quadrupolar nuclei in powders. Chem. Phys. Lett. 119, 29-32.

Samoson, A., Kundla, E. and Lippmaa, E. (1982) High resolution MAS-NMR of quadrupolar nuclei in powders. J. Magn. Reson. 49, 350-357.

Sanders, J.K.M. and Hunter, B. K. (1987) Modern NMR spectroscopy, Oxford, Oxford Univ. Press., 308 p.

Schlicter, C.P. (1978) Principles of magnetic resonance. 2nd Ed, Berlin, Springer-Verlag.

Schramm, C.M., deJong, B.H.W.S. and Parziale, V.E. (1984) ^{29}Si magic angle spinning NMR study on local silicon environments in amorphous and crystalline lithium silicates. J. Amer. Chem. Soc. 106, 4396-4402.

Schramm, S.E., Kirkpatrick, R.J. and Oldfield, E. (1983) Observation of high-resolution oxygen-17 NMR spectra of inorganic solids. J. Am. Chem. Soc. 105, 2483-2485.

Schramm, S.E. and Oldfield, E. (1984) High-resolution oxygen-17 NMR of solids. J. Am. Chem. Soc. 106, 2502-2506.

Self, K. and Mellum, M.D. (1984) The silicon/aluminum ratio and ordering in zeolite A. J. Phys. Chem. 88, 3560-3563.

Selvary, U., Rao, K.J., Rao, C.N.R., Klinowski, J. and Thomas, J.M. (1985) MAS NMR as a probe for investigating the distribution of Si-O-Si angles in lithium silicate glass. Chem. Phys. Lett. 114, 24-27.

Sherriff, B.L. and Hartman, J.S. (1985) Solid-state high-resolution ^{29}Si NMR of feldpars: Al-Si disorder and the effects of paramagnetic centers. Can. Mineral. 23, 205-212.

Sindorf, D.W. and Macil, G.E. (1983) ^{29}Si NMR study of dehydrated/rehydrated silca gel using cross polarization and magic-angle spinning. J. Am. Chem. Soc. 105, 1487-1493.

Smith, J.V. and Blackwell, C.S. (1983) Nuclear magnetic resonance of silica polymorphs. Nature 303, 223-225.

Smith, J.V., Blackwell, C.S. and Hovis, G.L. (1984) NMR of albite-microcline series. Nature 309, 14-42.

Smith, J.V. and Pluth J.J. (1981) Si,.Al ordering in Linde A zeolite. Nature 291. 265.

Smith, K.A., Kirkpatrick, R.J., Oldfield, E. and Henderson, D.M. (1983) High-resolution silicon-29 nuclear magnetic resonance spectroscopic study of rock-forming silicates. Am. Mineral. 68, 1206-1215.

Stebbins, J.F., Murdoch, J.B., Carmichael, I.S.E. and Pines, A. (1986) Defects and short range order in nepheline group minerals: a silicon-29 nuclear magnetic resonance study. Phys. Chem. Minerals 13, 371-381.

Thomas, J.M., Klinowski, J., Wright, R.A. and Roy, R. (1983) Probing the environment of Al atoms in noncrystalline solids: Al_2O_3-SiO_2 gels, soda glass, and mullite precursors. Angewandte Chemie, Int'l. Ed. (English) 22, 614-616.

Timken, H.K.C., Janes, N., Turner, G., Lambert, S.L., Welch, L.B. and Oldfield, E. (1986) Solid state oxygen-17 nuclear magnetic resonance spectroscopic study of zeolites and related systems. 2. J. Am. Chem. Soc. 108, 7236-7241.

Timken, H.K.C., Schramm, S.E., Kirkpatrick, R.J. and Oldfield, E. (1987) Solid state oxygen-17 nuclear magnetic resonance spectroscopic studies of alkaline earth metasilicates. J. Phys. Chem. 91, 1054-1058.

Timken, H.K.C., Turner, G.L. Gilson, J.-P., Welch, L.B. and Oldfield, E. (1986) Solid-state oxygen-17 nuclear magnetic resonance spectroscopic studies of zeolites and related systems. 1. J. Am. Chem. Soc. 108, 7231-7235.

Tossell, J.A. (1984) Corelation of ^{29}Si nuclear magnetic resonance chemical shifts in silicates with orbital energy differences obtained from X-ray spectra. Phys. Chem. Mineral. 10, 137-141.

Tossell, J.A. and Lazzeretti, P. (1986a) Ab initio calculations of ^{29}Si chemical shifts of some gas phase and solid state silicon fluorides and oxides. J. Chem. Phys. 84, 369-374.

Tossell, J.A. and Lazzeretti, P. (1986b) Ab initio coupled Hartree-Fock calculation of the ^{29}Si NMR shielding constants in SiH_4, Si_2H_6, Si_2H_4 and H_2SiO. Phys. Chem. Lett. 128, 420-424.

Tossell, J.A. and Lazzeretti, P. (1986c) On the sagging pattern of ^{29}Si NMR shieldings in the series SiH_aF_b, a+b=4. Phys. Chem. Lett. 132, 464-466.

Tossell, J.A. and Lazzeretti, P. (1987) Ab initio calculations of oxygen nuclear quadrupole coupling contants and oxygen and silicon NMR shielding constants in molecules containing Si-O bonds. Chem. Phys. 112, 205-212.

Tosell, J.A. and Lazzeretti, P. (in press) Calculation of NMR parameters for bridging oxygens in H_3T-O-$T'H_3$ linkages (T, T'= Al, Si, P, H) and for bridging fluorine in $H_3SiFSiH_3^+$. Phys. Chem. Mineral.

Turner, D.L. (1985) Basic two-dimensional NMR. Prog. in NMR Spect. 17, 281-358.

Turner, G.L., Chung, S.E. and Oldfield, E. (1985) Solid-state oxygen-17 nuclear magnetic resonance spectroscopic study of the group II oxides. J. Mag. Res. 64, 316-324.

Turner, G.L., Smith, K.A., Kirkpatrick, R.J. and Oldfield, E. (1986a) Boron nuclear magnetic resonance spectroscopic study of borate and borosilicate minerals, and a borosilicate glass. J. Mag. Res. 67, 544-550.

Turner, G.L., Smith, K.A., Kirkpatrick, R.J. and Oldfield, E. (1986b) Structure and cation effect on phosphorous-31 NMR chemical shifts and chemical shift anisotropies of model orthophosphates. J. Mag. Res. 70, 408-415.

Walter, T.H., Turner, G.L. and Oldfield, E. (submitted) Oxygen-17 cross-polarization NMR spectroscopy of inorganic solids. J. Mag. Res.

Weiss, C.A. Jr., Altaner, S.P. and Kirkpatrick, R.J. (1987) High-resolution ^{29}Si NMR spectroscopy of layer silicates: correlations among chemical shifts, structural distortions, and chemical variations. Am. Mineral. 72, 935-942.

Welsh, L.B., Gilson, J.-P. and Gattuso, M.J. (1985) High resolution ^{27}Al NMR of amorphous silica-aluminas. Appl. Catalysts 15, 327-331.

Wilson, M.A. (1987) NMR techniques and applications in geochemistry and soil chemistry. Oxford, Pergamon Press.

Yang, W.-H., Kirkpatrick, R.J. and Henderson, D.M. (1986a) High-resolution ^{29}Si, ^{27}Al, and ^{23}Na NMR spectroscopic study of Al-Si disordering in annealed albite and oligoclase. Am. Mineral. 71, 712-726.

Yang, W.-H., Kirkpatrick, R.J. and Turner, G. (1986b) ^{31}P and ^{29}Si magic-angle sample-spinning NMR investigation of the structural environment of phosphorous in alkaline-earth silicate glasses. J. Am. Cer. Soc. 69, C222-C223.

Yang, W.-H. (1987) Hydrothermal reaction of crystalline albite, sodium aluminosilicate glass, and a rhyolite-composition glass with aqueous solutions: a solid-state study. Ph.D. Thesis, Dept. of Geology, Univ. of Illinois at Urbana-Champaign.

Yannoni, C.S. (1982) High-resolution NMR in solids: the CPMAS experiments. Accts. Chem. Res. 15, 201-208.

NMR SPECTROSCOPY AND DYNAMIC PROCESSES IN MINERALOGY AND GEOCHEMISTRY

INTRODUCTION

As reviewed in the previous chapter, the mineralogical community knows NMR spectroscopy primarily as a relatively new tool to help determine the static, local structure of crystalline and amorphous solids. Here, "static" means unchanging at the time scale sampled by the experiment, which for NMR is generally between a few seconds down to the nanosecond range. Atoms in minerals are of course not fixed rigidly in the structure, but at all real temperatures undergo the rapid periodic interatomic vibrations (at the picosecond to femtosecond timescale) studied so extensively by some of the other spectroscopies (IR, Raman) described in this volume. At the NMR timescale, however, it is generally only the time-average of the positions explored during vibration that influences the observable spectrum.

In contrast, to much of the chemistry community, NMR is a well-established, everyday technique for determining the structures of organic molecules in liquids and solutions. In the most common applications, the structure determined by the technique is "static" in the sense that the molecules do not change shape or bonding arrangement during the experiment. Here, however, the situation is strongly influenced by another sort of dynamics: molecules in solution generally rotate or tumble rapidly. For the reasons described below, this motion produces narrow-line spectra that are analogous to those seen during a magic-angle-spinning (MAS) NMR experiment on a solid. As for vibrations, information about the details of these dynamics in low-viscosity liquids is usually lost because the motion is so rapid. As in MAS NMR, this motional averaging often does result in better resolution of structural details than when motion is slow or absent.

In between these realms of "solid state" and "liquid state" NMR is an important region where atomic and molecular motion takes place at the Hz to 100's of MHz frequency scale. In this range, the structural nature, frequency, and energetics of the motion itself can often be studied by NMR spectroscopy. This region is particularly well explored in organic chemistry, where such problems as the dynamics of molecular shape change, of rotation in viscous liquids and molecular solids, and the kinetics of species exchange are familiar topics for many NMR spectroscopists. The motion of molecules associated with surfaces in multi-phase systems is often in this intermediate frequency range, as well. These dynamics and the kinetics of the reactions among surface species, have long been explored by chemists interested in catalysis. High surface area minerals, such as zeolites and clays, have been particularly well studied. An excellent introduction to this field is by Resing (1980).

There are many dynamical processes at the NMR timescale which are or will be of importance to mineralogy and geochemistry. These include the hindered motion of molecules (such as water, CO_2, or hydrocarbons) at structural sites in crystal structures, in or on the internal or external surfaces of minerals or dissolved in melts and glasses; diffusion of molecules or ions in solutions, melts and minerals; displacive phase transitions; and viscous flow and creep. Detailed information about how such motion takes place may be essential in understanding the energetics and kinetics of such processes, as well as in predicting bulk thermodynamic and transport properties.

The detection and characterization of atomic motion at this "diffusional" timescale by NMR can be broken down into two closely related experimental approaches. The first involves the relatively straightforward techniques of studying the shapes of NMR spectra and how various styles of motion affect them. This will be described first. A second aspect of NMR may provide even more dynamical information, but is often harder to interpret in detail. When a nuclear spin system is excited to a high energy state by a radio-frequency pulse during an NMR experiment, the rate at which energy is distributed among spins, or between the spins and thermal energy, is strongly influenced by atomic motion. These processes fall under the heading of "nuclear spin relaxation."

Much of the general discussion here is based on standard texts, such as Becker (1980), Akitt (1983), and Harris (1983). The recent book by Fyfe (1983) is a particularly good source of examples and information on NMR techniques applied to solids, and that of Fukushima and Roeder (1981) gives a great deal of practical guidance to NMR experimental techniques and data analysis.

NMR LINESHAPES AND MOTION

Review of influences on static NMR lineshapes

The general effect of atomic or molecular motion at the "NMR timescale" is to cause NMR lines to decrease in width as motion becomes more rapid and line broadening influences are averaged to zero. In order to understand how this can occur, it is first necessary to briefly discuss why NMR lines in most solids and in very viscous liquids are broad in the first place.

Consider first an isotope that is naturally dilute and has a nuclear spin of 1/2, such as ^{29}Si (4.7% natural abundance) or ^{13}C (1.1% natural abundance). Assume that a perfect single crystal of a compound of the nuclide is available, and that there are no interactions between the nuclei of interest and other nuclear spins, unpaired electrons, ferromagnetic inclusions, etc. Finally, let us suppose that only one type of structurally distinct site is present for the nuclei of interest, and that they therefore all experience the same local electronic environment and have the same NMR resonant frequency or chemical shift. If this single crystal is placed in a magnetic field and its NMR spectrum is observed, a single, narrow line will be seen. The width of this line could possibly be limited by the NMR spectrometer itself: if the magnetic field is not perfectly homogeneous, then the resonant frequency at different places in the crystal will vary and the line will be spread out.

In the real world, a more likely cause of broadening of this ideal single line is the presence of imperfections in the crystal structure. As discussed in the previous chapter, variations in local structural parameters such as interatomic distances or angles will change the distribution of electrons around a nucleus and affect its chemical shift. This effect can range from a broadening of a single line to the creation of multiple lines for multiple types of sites.

If the local symmetry about the nucleus of interest is not cubic, then the chemical shift will depend also on the orientation of the crystal in the external magnetic field. As the orientation of a single crystal is changed, the position of the NMR line (or lines) will change in a way related to the site symmetry. Direct measurement of this kind of orientational effect in single crystals has been made on a few silicates (Weiden and Räger, 1985) and can yield important information about local crystal structure and its relation to the NMR spectra.

More commonly, however, randomly oriented, polycrystalline powders are studied. Here, as discussed in the previous chapter, a range of chemical shifts is covered by the presence of crystals in all possible orientations, giving rise to a "chemical shift powder pattern," whose shape is characteristic of the local site symmetry. This "powder" pattern can be described by the three principal values of the chemical shift anisotropy (CSA) tensor. Figure 1 shows the full, static ^{29}Si NMR spectra of two crystalline alkali silicates as examples of line shapes due mostly to chemical shift anisotropy. The influence of site symmetry is apparent. In the disilicate, all silicon atoms have only a single bridg-

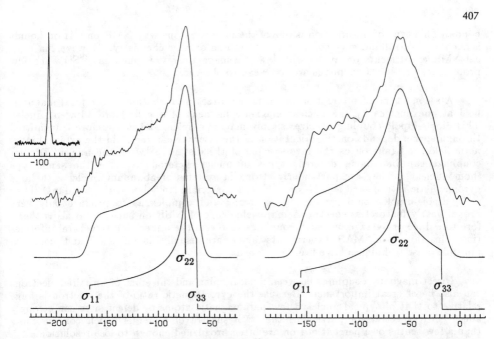

Figure 1. Static (e.g. *non*-magic angle spinning) ^{29}Si NMR spectra of crystalline $Na_2Si_2O_5$ (left) and Na_2SiO_3 (right). Upper curves are data, lower curve are simulations, middle curves are the same simulated spectra convoluted with a Gaussian broadening function to account for a small amount of disorder and dipole-dipole interaction. σ_{11}, σ_{22}, and σ_{33} are the principal components of the chemical shift anistropy (CSA) tensor. The inset at upper left shows the spectrum of the liquid with the same horizontal scale.

ing oxygen, so that the topological symmetry is uniaxial. In the powder pattern, σ_{22} is close to σ_{33}, indicating that the electron distribution around the silicon is also close to uniaxial. For the metasilicate, each silicon has two bridging oxygens, and the lower topological (and crystallographic) symmetry is reflected in the NMR powder pattern, where σ_{22} is far from either σ_{11} or σ_{33}. (The mean of σ_{11}, σ_{22} and σ_{33} is the isotropic chemical shift σ_i that is observed in a liquid or an MAS experiment. The divergence from uniaxial symmetry is often expressed as the asymmetry parameter, $\eta = (\sigma_{22}-\sigma_{11})/(\sigma_{33}-\sigma_i)$). The divergence from spherical symmetry is often expressed as $\Delta\sigma = \sigma_{33} - 1/2(\sigma_{11}+\sigma_{22})$. Also shown in Figure 1 is the very narrow single line that comprises the spectrum for the melt. In typical experiments on solids at the present time, of course, orientational effects are averaged away by magic-angle spinning in order to better resolve slight differences among sites in the isotropic chemical shift. Sometimes, however, static spectra can provide important structural information not obtainable from MAS spectra, as shown for alkali silicate glasses by Stebbins (1987).

This simple picture of spectra dominated by the chemical shift will be perturbed if there are interactions present among nuclei or between nuclei and unpaired electrons. These couplings can be of several sorts. In organic molecules, one of the weaker but most informative type of coupling involves an indirect perturbation of the local magnetic fields of atoms that are bonded together; this coupling is transmitted through the bonding electrons. This is commonly called a "spin-spin" or "scalar" coupling because, unlike the direct dipole-dipole interactions described below, its major component is independent of orientation in the external field. The strength of a spin-spin coupling between two nuclei j and k is usually symbolized by J_{jk}, giving the interaction a third name of "J-coupling." This kind of interaction typically causes discrete splittings of lines that can be of crucial importance in determining molecular structure, because the interactions depend strongly on the connectivity of the structure and the lengths and angles of bonds. This type of information is one of the major reasons why NMR on ^{1}H on liquids is such a powerful and very commonly used tool in organic chemistry. However, in typical NMR experiments on nuclei with low gyromagnetic ratios (such as ^{29}Si) when few protons are present, J couplings are too weak to observe.

A second type of interaction involves the direct perturbation of the local magnetic field at one nucleus by the nuclear magnetic moment of one or more adjacent nuclei. This dipole-dipole coupling involves simply magnetic fields, and is therefore transmitted through space, depends on the orientation of the internuclear vector in the external magnetic field, and falls off as the inverse cube of the distant between the nuclei. Dipolar couplings can be among different types of nuclei (heteronuclear) or the same type (homonuclear). They are particularly strong in systems of abundant nuclei with large gyromagnetic ratios, such as ^{1}H- or ^{19}F- rich materials. In solids with abundant ^{1}H or ^{19}F, dipole-dipole couplings can be so strong and complex as to produce only very broad, nearly featureless spectra. A major challenge of NMR on such materials is therefore to reduce or average out such couplings to reveal the underlying structural information. In this case, MAS experiments alone are usually insufficient, and complex, multiple-pulse techniques have been developed.

Direct magnetic couplings between nuclear spins and the spins of unpaired electrons can also be of great importance. Because the gyromagnetic ratio of the electron is huge relative to that of typical nuclei, a few paramagnetic atoms or defect sites can strongly affect NMR spectra. In particular, the ^{29}Si spectra of minerals which contain more than a few tenths of a percent of iron are often broadened enough to lose resolution.

An even more extreme effect is the influence of the conduction electrons in a metal. These often broaden NMR resonances so severely that only very limited structural information can be obtained, although some information about electronic interactions may be derived.

Another source of broadening involves quadrupolar effects. For a nucleus with spin I, 2I NMR transitions can potentially be observed, if their separation in frequency is not greater than the bandwidth of the spectrometer. (The separations of such higher order transitions *are* often so large for nuclei such as ^{23}Na and ^{27}Al in typical silicates that only the "central," 1/2 to -1/2 transition can be seen.) To first order, then, a series of possibly overlapping NMR lines may be present, and in a random powder, a series of overlapping powder patterns will result. To second order, the transitions interact. For a non-integral spin quadrupole nucleus (such as ^{23}Na or ^{27}Al) therefore, even the central, 1/2 to -1/2 transition will be broadened or split. As described in the previous chapter, the size of such interactions can provides clues as to site geometry for quadrupolar nuclei.

Motion and NMR lineshape

Several of the mechanisms described above broaden lines because they introduce perturbations on the local magnetic field that are orientation-dependent. These include the chemical shift anisotropy, dipole-dipole interactions, and first-order quadrupole interactions. These effects can be described as traceless tensors, and in the case of rapid random, isotropic motion will therefore average to zero. This motion must be rapid with respect to the linewidth in order for complete averaging to take place. In the example in Figure 1, the total range of the CSA is about 150 ppm. Suppose the sample is placed in a magnetic field of 8.5 Tesla. Since γ for ^{29}Si is -5.32×10^7 rad s^{-1}T^{-1}, the Larmor frequency (ν_0) will be $8.5 \times 5.32 \times 10^7 / 2\pi$ or about 72 MHz. The width of the spectrum will then be $150 \times 72 = 10800$ Hz. At half the field strength, the linewidth will still be 150 ppm, but this will correspond to 5400 Hz. If reorientation takes place at a rate similar to this linewidth (expressed as $2\pi \times \nu_0$ in rad s^{-1}), narrowing will begin. If motion is much more rapid than the linewidth (the "fast motion" regime), then the remaining spectral shape will then be dominated by the isotropic, averaged chemical shifts of nuclei at distinct structural sites, by scalar couplings, and by instrumental effects such as external field heterogeneity.

Other broadening mechanisms scale differently with frequency. Dipolar interactions, for example, involve a fixed amount of energy, and are therefore constant in absolute frequency (in Hz), not relative frequency (in ppm). The result is one of the most important motivations to work at higher magnetic field strengths. Suppose the NMR lines for nuclei at two different sites in a molecule are separated by 2 ppm, and some kind of dipolar interaction produces a broadening of 500 Hz. At a low external field strength which gives a Larmor frequency of 100 MHz, the chemical shift separation would be only $2 \times 100 = 200$ Hz and the lines would not be resolved. At a high field strength which produces a Larmor frequency of 500 MHz, the separation of $2 \times 500 = 1000$ Hz would be twice as great as the broadening, and the two sites could be distinguished.

The point of this example is that the mechanism of broadening can often be explored by collection of data at different frequencies. At some temperature, motion may be rapid enough to average the CSA when the observation is made at low field, but too slow to narrow the line at high field. Motional averaging of field-strength independent broadening will, on the other hand, remain constant. The relative rarity of comparative studies at varying external field strengths is the result of one of the few disadvantages of modern high field NMR spectrometers: it is difficult to adjust the field of a standard superconducting magnet.

Random, isotropic motion is generally associated with the tumbling of small molecules in low-viscosity liquids. However, rapid rotations or librations can be quite important in solids, and are particularly well known in molecular solids such as NH_3, CO, and a wide variety of organic compounds. A striking example is that of crystalline white phosphorus (P_4), which transforms from a rigid solid to one in which the constituent tetrahedral molecules rapidly jump from one orientation to another (Spiess et al., 1974) (Fig. 2). At 25 K, the ^{31}P NMR spectrum of the compound shows a simple, uniaxial powder pattern, indicating that the local tetrahedral symmetry of the molecule has been reduced to trigonal by crystallographic constraints. The width of the spectrum is dominated by the chemical shift anisotropy, and is about 405 ppm. At the external magnetic field strength used in these measurements, which corresponds to a ^{31}P resonant frequency of 92 MHZ, this width translates to a frequency range of 37 kHz, or an angular frequency range $\Delta\omega$ of $2\pi \times 37000 \approx 230000$ rad s^{-1}. As the material is heated, the lineshape progressively changes, reaching a relatively narrow final form at 126 K. This narrowing process was accurately modeled by a single type of motion, the rotation of the tetrahedra about any and all of their four three-fold axes. The important qualitative point in this non-mineralogical example is that as the inverse of the time between "jumps" between configurations (τ^{-1}) becomes greater than the linewidth, substantial narrowing begins. Complete narrowing in this example requires motion about an order of magnitude faster than the total angular frequency range of the spectrum. In this study, the motion and structure were further characterized by relaxation time measurements (see below).

These kinds of mobile molecular solids are of course of minimal importance in inorganic geochemistry (perhaps until we extend our reach to the surface of Europa!), but the analogous behavior of mobile molecules on the internal and external surfaces of minerals are of crucial importance in the interactions of fluids with minerals.

NMR has a relatively low sensitivity: roughly 10^{18} to 10^{20} nuclei of interest must be present in the sample to obtain high quality results. Surface studies have therefore

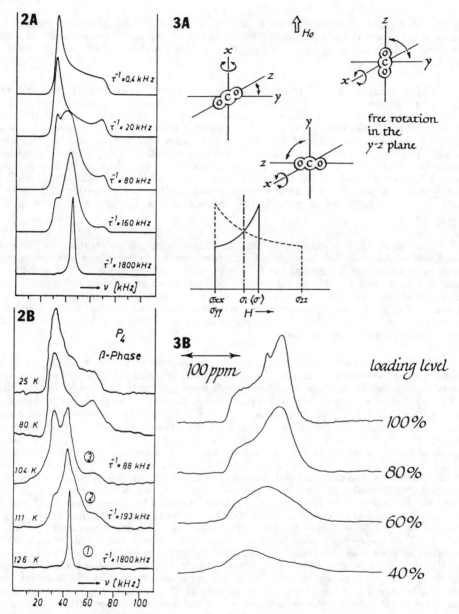

Figure 2. (A) Calculated spectra for P_4 molecule jumping between its four equilibrium orientations at the frequencies shown. (B) ^{31}P NMR spectra of the β phase of solid white phosphorus at various temperatures. From Spiess et al. (1974)

Figure 3. (A) Theoretical ^{13} NMR spectra of sorbed CO_2. Dashed line is for a random array of non-rotating molecules ($\sigma_{XX} = \sigma_{YY}$ because of the symmetry of the molecule); solid line is for molecules rotating anisotropically (in the Y-Z or X-Z plane: σ_{ZZ} and σ_{YY} or σ_{XX} are averaged to $<\sigma>$); single vertical dot-dash line is for isotropic rotation (in all directions: σ_{ZZ}, σ_{YY}, and σ_{XX} are all averaged together). (B) Experimental ^{13}C NMR spectra for CO_2 sorbed in Na-mordenite zeolite, as a function of % of maximum CO_2 content. Note transition from static pattern to partially averaged pattern, then finally, the development of a narrow, fully averaged peak. Both from Sefcik et al. (1976).

been limited to systems with very high surface areas, such as clays and zeolites, which can easily contain this amount of material in 1 g of bulk sample. To date, much of this kind of research has been oriented towards an understanding of catalysis. In very mobile systems, the rates of reactions among organic molecules can be directly studied by what is essentially liquid-state ^{13}C and ^{1}H NMR (e.g. Resing, 1980). In less mobile systems, a great deal can be learned about the details of how molecules interact with the surface.

A good example is CO_2 trapped in the pores of the zeolite mordenite, as studied by Sefcik et al. (1976, 1977). As shown in Figure 3, the ^{13}C NMR spectra can be modeled as a combination of lineshapes. At low CO_2 concentrations, the spectrum is dominated by the uniaxial powder pattern expected for molecules that are rigidly attached to the surface. At higher concentrations, the spectrum is still quite wide, but has taken on a narrower form which can be attributed to molecules that are loosely attached and rotating about an axis perpendicular to the O-C-O axis. At the highest concentration obtained, a small percentage of a very narrow, symmetrical line appeared which must be from molecules that interact only weakly with the surface and undergo isotropic rotation. Because the area of an NMR line is in principle strictly proportional to the number of spins involved, it may be possible to quantitatively simulate such "mixed" spectra and determine how many molecules undergo different types of motion. In this case, the authors were able to suggest the type of sites that were progressively filled during CO_2 loading: smaller, chemically active but motionally hindered cages are apparently filled before larger cages.

In another intriguing example of this sort, Nagy et al. (1983) were able to show at a molecular scale why two different isomers of xylene (a benzene ring with two added methyl groups) show a three orders of magnitude difference in their diffusivities and reaction rates in the industrial zeolite ZSM-5. The ^{13}C NMR lineshape revealed that the slightly "wider" molecule, o-xylene (with two adjacent methyl groups on the ring) was essentially immobile compared to the timescale of its linewidth, with jumps between positions or orientations less often than about once per 10^{-4} s. In contrast, the "narrower" p-xylene molecule (with methyl groups on opposite sides of the ring) diffused locally at least 10 times faster, giving a motionally narrowed spectrum. These differences in both the local motion and bulk diffusion can be well-correlated with the relative sizes of the two molecules and the various available "windows" between the zeolite pores. Other details of these dynamics were determined by ^{2}H NMR lineshape analysis by Eckman and Vega (1983).

In some cases, even more specific information about orientations of molecules and their motions can be determined by NMR. By studying oriented preparations of clay minerals, Resing et al. (1980) were able to show, surprisingly enough, that benzene molecules stand on edge between the platelets in hectorite. At liquid nitrogen temperatures the molecules are static, but room temperature rotate rapidly about their six-fold axes. Information on both the orientation and its dynamics proved to be important in understanding reactions involving the molecule on surfaces.

^{1}H NMR has been used in a number of studies to explore the mobility and orientation of water molecules on the surfaces of clay minerals. It has long been known that the single, motionally completely averaged NMR line of liquid water is partially split as the result of reduced, anisotropic motion near a clay particle surface (e.g. Walmsey and Shporer, 1978). ^{1}H NMR lineshapes have been used to determine the number and orientation of water molecules that hydrate interlayer cations in sheet silicates such as vermiculite, and to show that their diffusion rates were several orders of magnitude lower than those in non-interacting water (Hougardy et al., 1976; Fripiat et al., 1980). Important details of the re-orientation rates of water molecules on clay surfaces, and their relationship to nuclear spin relaxation (see below), were derived from a combined study of ^{1}H and ^{2}H NMR by Woessner (1980). A variety of NMR techniques have been recently applied to study the structure and dynamics of organic molecules on metal surfaces (Wang et al., 1986), and may prove useful in some cases to work on oxide systems.

Geochemically important molecules can be integral parts of solid structures as well as being located on internal and external surfaces. The most common example is that of structural water in minerals. The ^{1}H NMR spectrum of static, isolated water molecules consists of a doublet caused by the splitting of the single 1/2 to -1/2 resonance into two peaks by the dipolar interactions of the two protons. The splitting is given by Harris (1983):

$$\Delta H = \frac{3\mu}{2} \left(\frac{1-3\cos^2\theta}{r^3} \right) , \tag{1}$$

where μ is the nuclear magnetic moment, r is the separation between the pair of protons, and θ is the angle between the proton internuclear vector and the magnetic field. The maximum splitting for a stationary molecule can easily be calculated from its known geometry as 88 kHz at $\theta=0$. When $\cos\theta = (1/3)^{1/2}$ (and $\theta = 54.7°$, or the "magic angle"), the splitting goes to zero.

When water molecules are present in crystalline (or amorphous) solid structures, this simple picture is perturbed in ways that can provide information about both structure and dynamics. In fact, one of the very first NMR studies of any solid was of the ^{1}H spectrum of structural H_2O in natural gypsum by Pake (1948). A more recent and somewhat more complex example is that of H_2O in cordierite $((Mg,Fe)_2Al_4Si_5O_{18} \cdot nH_2O)$ by ^{1}H NMR by Carson et al. (1982). They reported a doubled doublet, with the minor component the result of the paramagnetic shift of water molecules relatively close to iron atoms. The splitting of the main doublet increased from about 60 kHz at room temperature to about 80 kHz at 100 K, indicating that some kind of motional narrowing of the static 88 kHz width was occurring that slowed down at lower temperature. Furthermore, data taken at varying orientations of a single crystal (varying values of θ) showed that the average H-H vector was oriented parallel or nearly parallel to the [001] channels, confirming the earlier NMR results of Tsang and Ghose (1972). The fact that infra-red spectroscopy showed the presence of two different water molecule positions in the structure (Goldman et al., 1977) indicates that the motional averaging of the ^{1}H NMR spectrum is due to the hopping between two orientations at a time scale long compared to that of vibrations, but short compared to the NMR linewidth, perhaps at the microsecond timescale. The NMR data thus show that no static ordering of water molecules into two types of "sites" is required. It should be noted that a slight discrepancy still remains between the IR and NMR results and a determination of the water molecule orientation by neutron diffraction (Hochella et al., 1979).

Molecular water in zeolite cages is even more mobile, as one might expect from their very open structure and multi-directional channels. Typically, two components are seen in the ^{1}H NMR spectrum of zeolites (Kasai and Jones, 1984). The first is a relatively broad resonance, typically several kHz in width, which can be attributed to H_2O molecules at small cage sites. These are clearly free to undergo some kind of rotational motion, as the wide dipolar splitting described above is averaged out. The motion is, however, hindered enough to prevent complete narrowing to a typical "free" water linewidth of a few tenths of a Hz to a few Hz. A second component is much narrower, a few tens or 100's of Hz. Studies of water in a variety of zeolites and of single structures with a variety of exchangeable cations (Kasai and Jones, 1984) indicate that this component is from the water that is present in the larger pores and channels, water that can readily diffuse throughout the structure. The linewidth of this component increases systematically with the size of the exchangeable cation as they begin to hinder diffusional

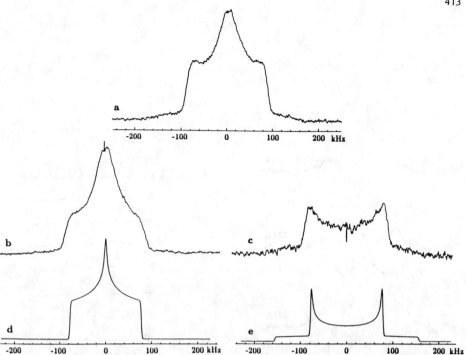

Figure 4. ^{2}H NMR spectra of rhyolitic glass containing 4.8 wt % D_2O, taken at 76.8 MHz, using composite pulse sequence. a) Fully relaxed spectrum taken with relaxation delay of 3 seconds. Contributions from both static OD and mobile D_2O molecules are visible. (b) Spectrum taken with relaxation delay of 0.01 seconds. The fast-relaxing ("flipping") D_2O groups are selectively observed (compare with d). (c) Difference spectrum (spectrum (a) minus scaled spectrum (b)), assigned to rigid OD groups. (d) Simulated spectrum for a rapidly flipping D_2O molecule using a single quadrupole coupling constant of 104 MHz, $\eta=1$, and convolution with a Gaussian function of 4.4 KHz half-height width. (e) Simulated spectrum for a rigid OD group using a single quadrupole coupling constant of 208 kHz, $\eta=0$, and convolution with a Gaussian function of 4.4. kHz half-height width. From Eckert et al. (1987).

motion. Rather surprisingly to those used to the stability of the constituents of mineral structures, experiments on mechanical mixtures of zeolites show that water molecules exchange among particles at a timescale of the order of milliseconds (Kasai and Jones, 1984). The effect of cation size on the ^{29}Si MAS NMR spectra of synthetic zeolite Linde A was used by Fyfe et al. (1986) to show that even at room temperature, Na^+ and Li^+ can diffuse among discrete (but touching) zeolite particles on a time scale of minutes to hours.

Another example that shows the relationship between structure and dynamics of water molecules in solids is the study of D_2O dissolved in rhyolitic glasses by Eckert et al. (1987). In this work, ^{2}H (or "D") was substituted for ^{1}H for several reasons. Because its gyromagnetic ratio is much lower than that of ^{1}H, dipole-dipole couplings do not dominate the spectra, as is often the case in ^{1}H NMR of solids with relatively immobile water. ^{2}H has a nuclear spin of 1, and its spectra are thus determined almost completely by quadrupolar interactions, which reflect the electric field gradients and therefore the bonding environment around the nucleus. ^{2}H NMR lineshapes are therefore well known to be a powerful tool in studying molecular motions in solids (e.g. Fyfe, 1983). One of the key results of this study is illustrated in Figure 4. The ^{2}H spectrum at room temperature can be explained as the overlapping spectra of deuterium in two type of sites. The first of these (c and e) is that expected for an isolated, static spin 1 nucleus. It consists of two overlapping, mirror image quadrupolar powder patterns, one for each of the two transitions (0 to -1 and 0 to +1). As for the CSA powder patterns

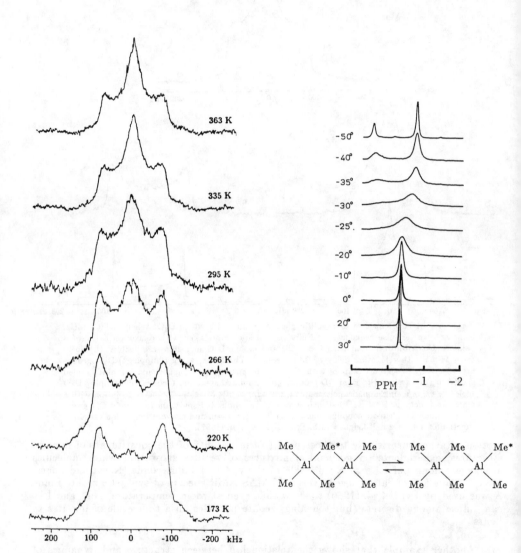

Figure 5 (left). ^{2}H NMR spectra of glass as in Figure 4. Note disappearance of narrow-line component as motion of D_2O molecule slows at low temperature. From Eckert et al. (1987).

Figure 6 (right). ^{1}H NMR spectra of $(Al(CH_3)_3)_2)$ dimer in cyclopentane solution. Exchange of methyl groups ("Me") is illustrated schematically (Al-C-Al bonds link monomer units). Temperatures are in °C. From Ramey et al. (1965).

described above for spin 1/2 nuclei, the shape of these patterns is controlled by the effects of orientation in the external magnetic field. The authors attributed this doublet to rigid OD groups.

The second type of site has a differently shaped spectrum, typical of that for water molecules in many deuterated crystalline hydrates, including minerals such as gypsum (Hutton and Pedersen, 1969). Here, the linewidth is reduced and a narrow central peak begins to appear because of partial motional narrowing caused by $180°$ "flips" of the molecule about its two-fold axis. Again, this motion must occur rapidly when compared with the inverse of the over-all linewidth (on a time scale of the order of 10^{-5} s). Simulation of the spectra using these two sub-spectra gave ratios of OD/D_2O that agreed well with the results of IR spectroscopy.

The temperature dependence of the ^2H lineshape was also studied. The mobility of the D_2O species clearly decreased with cooling, and the narrow spectral component was lost (Fig. 5). It is conceivable that detailed simulations in this or similar situations could provide more details of the temperature dependence, and therefore of the energetics of this motion; this would be significant to the understanding of thermodynamic properties. The authors point out that such refinement is difficult in this case because of the disorder introduced by the amorphous character of the samples and the technical difficulties of quantitatively observing wide NMR lines.

Chemical exchange

The types of motion described above are the rotation or translation of coherent molecules, and thus involve the breaking of only very weak intermolecular bonds; this requires relatively little energy. In many situations, even at complete thermodynamic equilibrium, more drastic types of motion can occur. Many organic molecules undergo conformational changes that require a substantial activation energy. In more extreme rearrangements, strong *intra*-molecular bonds, or covalent or ionic bonds in non-molecular solids or liquids, may continually break and re-form as the exchange of an atom or atoms among various sites takes place. In the NMR literature, this kind of process is usually called "chemical exchange," and is described in detail by Harris (1983), and by Kaplan and Fraenkel (1980). If the exchange takes place at a relatively slow ("diffusive") time scale, it may again be directly observable by NMR.

We turn again to a non-geochemical example as a first simple case, in a study by Ramey et al. (1965). Trimethyl aluminum forms the dimer Al_2Me_6, where "Me" symbolizes a methyl ($-CH_3$) group. As shown in Figure 6, the dimer has two bridging methyl groups and four terminal methyl groups. The protons in the two types of methyl groups are structurally and magnetically inequivalent and should have different chemical shifts. At low temperature, the ^1H NMR spectrum of the compound in cyclopentane solution indeed shows two peaks. (Chemical shift anisotropy and dipole-dipole couplings are averaged out by the fast tumbling of the molecules in solution). As temperature is increased, however, the two peaks coalesce and form a single narrow line at room temperature. The most likely explanation is that the methyls or their protons somehow jump back and forth between the two types of sites at a rate that, on warming to 0°C becomes much faster than the separation in frequency between their NMR lines. A careful study of the temperature dependence of the line shapes resulted in an estimate for the activation energy for the exchange that was similar to that expected for the dissociation of the dimer, suggesting that the exchange is the result of breaking and re-forming of the dimer molecules, not an exchange of protons among molecules.

Aqueous solutions of silicates, especially at high pH, can contain a wide assortment of complex hydrated molecular species. The details of their structures have been revealed by ^{29}Si NMR, in particular in a series of studies by Engelhardt et al. (1975), Harris et al. (1984), and Knight et al. (1987). Chemical exchange among the various

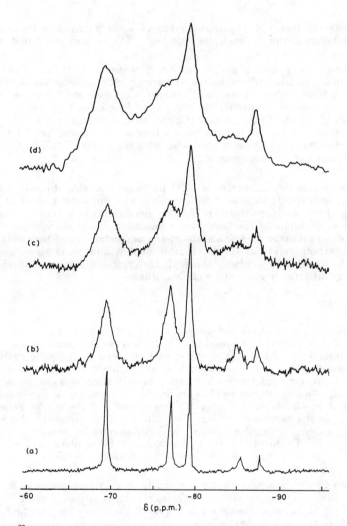

Figure 7. ^{29}Si NMR spectra of a 4M aqueous sodium silicate solution with Na/Si = 2.0. (a) through (d) are at 4, 26, 45, and 60°C, respectively. Lines begin to merge as chemical exchange rates increase with temperature. Broadening is due in part to an atypical decrease in T_2 with temperature. From Engelhardt and Hoebbel (1984).

species may be an important effect. As shown in Figure 7 from Engelhardt and Hoebbel (1984), the spectrum of an aqueous sodium silicate solution at 4°C contains sharp, distinct peaks characteristic of a series of silicate polymer molecules. As the temperature is raised, these peaks begin to merge together, at least in part because of the exchange of silicate groups from one type of molecule to another. This exchange must involve the breaking and re-forming of the strongest, most stable bonds in the system: the Si-O linkages. Griffiths et al. (1986) and others have also pointed out that part of the rather surprising broadening at higher temperature is due to some unusual spin-spin relaxation effects, as introduced below.

Again, these types of motion are not expected to be very common in the rigid structures of most minerals, except when structural or "guest" molecules are present, or when surface species are considered. One possible exception could be during displacive phase transitions.

When minerals are melted, of course, the picture is rather different. By definition, a liquid can flow and atomic mobility plays a fundamental role in the transport properties. Recently, a theory that was developed for glass-forming organic polymers has been applied to silicate melts (Richet, 1984, 1986), and has given a central role to configurational change in the explanation of the thermodynamic properties of these liquids. In terms of bulk properties, there is thus a clear link between the motion at a macroscopic scale that defines viscosity, and the entropy and heat capacity which is due to structural change at the atomic scale. An analogous link is in the properties of water (Eisenberg and Kauzmann, 1969): the heat capacity of the liquid is almost twice that of the crystal, because the liquid has a coherent structure which becomes more and more mobile with rising temperature. (The same transition is seen as water absorbed in the pores of a zeolite becomes mobile with rising temperature, as shown by Vučelic and Vučelic, 1985).

Because silicate melts are usually very viscous, it is expected that the local atomic motions responsible for both flow and configurational change should at some temperature be at the timescale accessible by NMR measurements. Stebbins et al. (1985) and co-workers (Schneider, 1985; Liu et al., 1987,1988) have recently shown that this is indeed the case. Part of their results involve the observation of chemical exchange not too dissimilar to the examples given above. In Figure 8 are shown the ^{29}Si NMR spectra for $Na_2Si_2O_5$ glass, supercooled, and stable liquid, all taken without magic angle spinning. At room temperature, the main part of the spectrum is due to a nearly uniaxial powder pattern of the silicons with three bridging and one non-bridging oxygens (Q^3 sites). This is broadened somewhat by the disorder in the glass structure and possibly by a minor amount of Si-Si and Na-Si dipole-dipole coupling. On top of this component is a relatively narrow, symmetrical peak due to about 8% of silicons with four bridging oxygens (Q^4) (Stebbins, 1987). The broad, low-symmetry pattern of a comparable number of Q^2 sites is hidden in the main Q^3 pattern. The spectrum of the melt at about 100°C above the glass transition is very similar to that of the glass, except for the much poorer signal-to-noise ratio inherent in high temperature NMR measurements. As temperature is increased further, the "powder patterns" coalesce into what eventually becomes a single narrow peak. This change results from the motional averaging of the chemical shift anisotropy as the structure becomes mobile. However, the presence of one line and not two or three requires that an exchange of silicon, oxygen and sodium atoms takes place to give an averaged isotropic chemical shift. This process must again involve the breaking of strong Si-O bonds, and is probably closely related to the atomic-scale mechanism for viscous flow. Extrapolated to liquidus temperatures, these results suggest that the average lifetime of an Si-O bond is very short, in the micro- to nanosecond range.

NUCLEAR SPIN RELAXATION AND MOTION

Simple theory: spin-lattice relaxation

The changes in nuclear magnetism that take place in NMR spectrometers are not instantaneous, and in many situations the rates of these changes can provide important information about styles and rates of atomic and molecular motion.

A material containing nuclei with non-zero spins, when placed in a magnetic field, will be slightly magnetized along the external field direction because a slightly unequal distribution of spin alignments is generated. Because the lowering of energy caused by this alignment is small when compared to the energy of thermal motion, the excess of spins aligned with the field at room temperature is on the order of only 1 in 10^5. Nonetheless, it is the absorption and redistribution of this small amount of energy which constitutes the NMR signal. When a sample is first placed in a magnetic field, energy

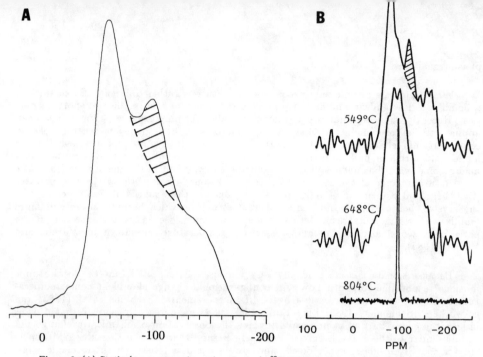

Figure 8. (A) Static (e.g. *non*-magic angle spinning) ^{29}Si NMR spectrum of $Na_2Si_2O_5$ glass at room temperature. Cross-hatched area is due to Q^4 sites. The predominate, nearly uniaxial powder pattern is due to silicon in Q^3 sites with a minor proportion of Q^2 sites. From Stebbins (1987). (B) ^{29}Si NMR spectra of $Na_2Si_2O_5$ melt above the glass transition. Cross-hatched areas as in (A). From Liu et al. (1988).

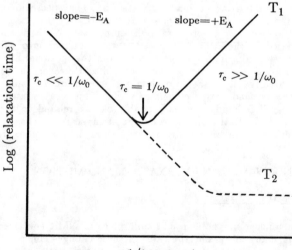

Figure 9. Typical effect of temperature on relaxation times in a liquid where isotropic, random motion prevails. τ_c is the correlation time for the motion that is causing the relaxation; ω_0 is the Larmor frequency.

must initially be transferred into the nuclear magnetization. This transfer takes a certain period of time. Similarly, after a radio frequency (rf) pulse in an NMR spectrometer, it takes a finite amount of time for the full, equilibrium magnetization of the sample to return, because energy must be transferred back out of the spin system. This recovery from a perturbation is called longitudinal or spin-lattice relaxation, because it involves transfer of energy from the component of the nuclear magnetization parallel to the external field to other forms of energy (usually thermal motion) of the surrounding atoms and bonds of the "lattice". This can often be characterized by an exponential decay of this Z component M_z, such that

$$(M_z)_t = (M_z)_\infty(1 - \exp(-t/T_1)) \quad , \tag{2}$$

where $(M_z)_\infty$ is the equilibrium value. A relaxation time T_1 characterizes this process. Simple exponential relaxation may not always be the rule, but the inverse of T_1 can always be thought of as an instantaneous relaxation *rate*.

Spin-lattice relaxation is non-instantaneous (and in fact can be excruciatingly slow) because it requires the stimulated, rather than spontaneous, emission of energy. This is in marked contrast with visible or IR spectroscopy, and is a direct consequence of the very low energies of nuclear spin transitions in available magnetic fields. In order for relaxation to take place, a spin $1/2$ nucleus must experience fluctuations in its local magnetic field at its Larmor (NMR resonance) frequency. A nucleus with spin $> 1/2$ can, in addition, couple to fluctuations in the local electric field gradient. The most common way in which such fluctuations are generated is in relatively slow (10's to 100's of MHz) atomic or molecular motions. Measurements of NMR relaxation times can therefore serve as additional ways in which to probe the mechanisms of molecular rotation, translation and diffusion.

In real systems, any given type of diffusive motion may have a wide spread in frequencies, but can usually be characterized by some kind of average time between fluctuations called the correlation time τ_c. This variable can sometimes be described as the typical time for a molecule to diffuse through its own diameter, or to rotate through one radian (Akitt, 1983), but here must remain more loosely defined. The key phenomenon which affects relaxation is the proportion of the atoms of interest involved in the motion at a given angular frequency ω called the "spectral density" or $J(\omega)$ ($\omega = 2\pi\nu$, where ω is in rad s^{-1}, ν is in Hz). This quantity can be difficult to determine directly, but can often be shown to obey an approximation such that $J(\omega)$ is proportional to a term of the form $\tau_c/(1+\omega^2\tau_c^2)$. In addition, in many simple models of relaxation, $1/T_1$ is directly proportional to J at the angular Larmor frequency ω_0. This yields a common expression for the relaxation rate (Farrar and Becker, 1971):

$$1/T_1 = \text{constant} \times \tau_c/(1+\omega_0^2\tau_c^2) \quad . \tag{3}$$

The form of this relationship is more important here than the details of various relaxation models. At small values of ω, the spectral density is nearly independent of ω. As ω approaches $1/\tau_c$, $J(\omega)$ begins to decrease as $1/\omega^2$, and becomes negligible at frequencies much greater than $1/\tau_c$. This means that if all molecular motion is at frequencies very low when compared to the Larmor frequency, it will not contribute to spin-lattice relaxation. On the other hand, if τ_c is very short, the spectral density will be spread out over a very wide range of frequencies, and again the contribution at the Larmor frequency will be small. This is true for typical inter-atomic vibrations in solids.

Another key general relationship is that for $\omega_0 \gg 1/\tau_c$, $1/T_1$ becomes proportional to $1/\tau_c$. The inverse is true when $\omega_0 \ll 1/\tau_c$, with $1/T_1$ proportional to τ_c. For relaxation dominated by a single mechanism and a single type of motion, the former relationship holds at low temperatures where motion is slow and τ_c is long, whereas the latter is true at high temperatures when motion is rapid and τ_c is short.

In simple cases, particularly in liquids, the motion responsible for relaxation can often be described by an Arrhenius expression, such that

$$\tau_c = \tau_0 \exp(E_A/kT) \quad , \tag{4}$$

where E_A is the activation energy, T is absolute temperature, and k is Boltzmann's constant. In this idealized case, a plot of $\ln T_1$ against the inverse of temperature will have a slope of $-E_A$ at high temperature, a slope of $+E_A$ at low temperature, and a minimum value at the point where $\omega_0 = 1/\tau_c$ (Fig. 9). As will be illustrated below, this simple behavior is of course not always seen, but the qualitative relationships described are fundamental: if a single type of relaxation persists over a wide enough range of temperature, then at low temperature most of the motion will be too slow to cause efficient relaxation. T_1 will therefore be long and will *decrease* with increasing temperature as motional frequencies increase. At some point with rising temperature, a minimum in T_1 will be reached when the match between motional and Larmor frequencies becomes optimum. This can be a very significant point, because it can allow an independent estimation of τ_c at this temperature. At higher temperature, most of the motion causing the relaxation becomes faster than the Larmor frequency, and T_1 begins to *increase* with rising T.

This simple theory is based on the classic work of Bloembergen, Pond and Purcell (1948). BPP theory is generally the starting point for approximating relaxation behavior in many systems, including complex, viscous liquids and even solids, in which it cannot be expected to strictly apply. However, it remains a useful starting point even in these systems.

More specific aspects of relaxation theory can only be touched on here, and are most well-understood for liquids. A general point is that the spectral density gives only the number or rate of fluctuations. The other part of the equation is the strength of the coupling between the nuclear spins of interest and the fluctuations, which determines the constant in equation (3). This can vary enormously. For example, as summarized by Harris (1983), one type of fluctuation is caused by dipole-dipole interactions between nuclei in a molecule in a liquid: the coupling is orientation-dependent and therefore changes as the molecule rotates. For a homonuclear coupling, the spin lattice relaxation time varies as the inverse fourth power of the gyromagnetic ratio and the sixth power of the internuclear distance. The effect on relaxation is thus a very strong function of the molecular structure, and will vary greatly from nucleus to nucleus.

Another type of fluctuation in local magnetic field is that caused by a change in orientation or position that alters the chemical shift because chemical shift anisotropy is present, or because sites with different local structure can be occupied. In typical organic systems, this effect is weak, but is proportional to the square of the external magnetic field strength and can therefore be distinguished from other mechanisms.

Relaxation is often caused by coupling to nuclei that are not an inherent part of the molecule of interest, or to unpaired electrons in paramagnetic impurities such as molecular O_2 or Fe^{3+}. In both cases, the coupling is again a very strong function of the distance, and therefore of the concentration. As mentioned above, the very high gyromagnetic ratio of the electron means that small amounts of paramagnetics often are the primary cause of relaxation.

Nuclei with spin $> 1/2$ can relax through fluctuations in the local electric field gradient, which are often much more effective than magnetic field fluctuations. Thus, it is often the case that quadrupolar nuclei have shorter T_1 values than do dipolar nuclei in the same structural environment.

For both spin-lattice and spin-spin relaxation (see below), it is often the case that several relaxation mechanisms are present simultaneously. As these are generally uncorrelated and take place in parallel, the *rates* add arithmetically:

$$1/T_{observed} = \sum 1/T_i \qquad (5)$$

where T_i is the relaxation time for an individual mechanism or an individual site. Relaxation times generally vary over several orders of magnitude in a typical variable temperature study, so that unless the individual mechanisms give similar relaxation rates and have similar temperature dependencies, one or the other will generally be seen to dominate at a given point. Changes in relaxation mechanism, and therefore in the magnitude or even the sign of the slope in a T_1 versus temperature plot, often occur at phase transitions in solids if these involve some change in the local motion.

Spin-spin relaxation

In a pulse NMR experiment, an rf pulse rotates the magnetization vector away from the Z axis (the axis of the external magnetic field), introducing a component in the XY plane and reducing the Z component. The spin-lattice relaxation time characterizes the time it takes the Z component to return to equilibrium. The spin-spin relaxation time T_2 characterizes the time required for the XY component to return to zero, and can be formulated as a simple exponential decay analogous to the expression above for T_1:

$$(M_{xy})_t = (M_{xy})_0 \exp(-t/T_2) \quad . \qquad (6)$$

In contrast to spin-lattice relaxation, spin-spin (or "transverse") relaxation does not require the transfer of energy to or from the nuclear spin system. Instead, it involves simply the dispersal of spin energy among nuclei with slightly different resonant frequencies due to varying chemical shifts or couplings. As this dispersion takes place after a single rf pulse, the rf signals being emitted by the various nuclei get out of phase, and the signal received by the spectrometer eventually degenerates to noise. This decay is directly observed in the length of the free-induction decay (FID) signal.

The spin-spin relaxation time is directly related to the width of a line when this time-domain FID is Fourier-transformed to the frequency domain spectrum. In the case (common in liquids) of a simple, single exponential decay of the transverse magnetization, the full width of the resulting Lorentzian line at half maximum height (FWHM) is $1/\pi T_2$. The shape of the line in solids is generally not Lorentzian, and the FID is not well represented by the simple exponential of equation 6. However, for a given line shape, simple relationships still hold. For a Gaussian lineshape, for example, the FWHM is approximately $1/1.88 T_2$ (Fukushima and Roeder, 1981). The important generalization is that anything that causes the FID to decay rapidly will result in a wide NMR line, and vice versa. These effects are not limited to NMR spectroscopy, of course: a vibrational state with a short lifetime will result in a broadened infra-red spectral line, just as a rapidly decaying NMR signal will appear as a broad line in the spectrum. As discussed above, various types and rates of motion can strongly effect the lineshape, and these changes are also manifested in the spin-spin relaxation as well.

Finally, the difference in relaxation behavior between liquids and solids needs to be emphasized. In very mobile systems, NMR lines are fully narrowed by motion. Here, in most cases, T_2 becomes equal to T_1. In less mobile systems where linewidths are greater, T_2 can be much shorter than T_1 and provides useful additional information about dynamics.

Techniques for measuring relaxation times and observing wide lines

The simplest type of pulse NMR experiment is done by giving a single rf pulse and recording the FID signal from the sample. The pulse may be repeated many times and the data averaged to obtain a better signal to noise ratio. The one-pulse experiment is useful for obtaining relatively simple, narrow-line spectra, and is commonly used in MAS experiments.

If an NMR line is very wide, T_2 will be short and the time-domain signal that is observed will decay away rapidly. For a variety of instrumental reasons, it is never possible to begin to record the FID until several (often 10-20 or more) microseconds after the end of the rf pulse. (For example, the rf coil used for both excitation and observation "rings" for a certain period of time after the end of the rf pulse, and the tiny NMR signal will be entirely masked until the ringing dies out. The very sensitive receiver used to observe the signal must also generally be turned on after the pulse to prevent damage.) During this period of "dead time", much (or even all) of a wide-line signal may decay away. There are, however, effective multi-pulse techniques for allowing fast-decaying signals to be observed. These rely on the possibility of forming "spin echoes", where the entire FID, including the critical initial part, can be made to reappear well after such instrumental dead times. The terms "spin echo", "Hahn echo" (i.e. Hahn, 1950), "solid echo", and "quadrupole echo" are used for this basic method and its modifications, which are explained in some detail in texts such as Fyfe (1983) and Fukushima and Roeder (1981). Spin echo techniques were used in many of the NMR experiments described above, particularly in the observation of wide ^1H and ^2H static spectra.

Related multi-pulse techniques are used for the measurement of relaxation. For T_1, for example, a typical sequence is termed "inversion-recovery". Here, the magnetization is initially rotated 180° by an rf pulse twice as long as the usual 90° pulse. After a delay time τ, a 90° pulse is given, which rotates any remaining Z-component magnetization into the XY plane and allows its signal to be observed. A series of such pulses with varying values of τ allows the full time history of the spin-lattice relaxation to be determined. Other labels used for this technique are the "180-τ-90" or "π-τ-$\pi/2$" sequence (180° = π radians). Again, there are several variations on this and related techniques that are explained in detail in the texts cited above.

Relaxation and mobility in geochemistry: single phase systems

NMR has been applied a number of times to the diffusion of alkalis in crystalline and glassy solids. A good example of this type of work is the ^7Li NMR study of Li^+ motion in lithium borate, phosphate, and silicate glasses by Göbel et al. (1979). Figure 10 shows a plot of $1/T_1$ versus 1/temperature for these results (and is thus upside down from the usual plots in the NMR literature). Maxima in the relaxation rates (T_1 minima) are clearly seen for most of the compositions shown, illustrating the general principles outlined above. Calculations show that the relaxation process involves the quadrupolar coupling among the lithium nuclei. In detail, a complex picture emerges. The magnitude of the slope, or apparent activation energy, increases progressively with increasing temperature, and is higher above the T_1 minimum than below. This pattern has been seen in a number of similar studies (Hendrickson and Bray, 1974; Balzer et al., 1984; Jain et al., 1985; Liu et al., 1987,1988). Slopes at high temperatures, which at least in the work of Göbel et al. (1979) and Liu et al. (1987,1988) are above the glass transition, are generally close to those for electrical conductivity and cation tracer diffusivity, whereas those near room temperature may be less than half as large.

Interpretation of this pattern has been somewhat controversial. It may be due to a wide distribution of correlation times for the motion responsible for the relaxation (Gobel et al., 1979), but is perhaps more likely the result of a progressive change in the type of motion that dominates the relaxation (Jain et al., 1985, 1983). At low temperature, through-going diffusive motion is at too low a frequency to contribute much, and local "cage rattling" motions may be most effective. This motion has a low activation energy

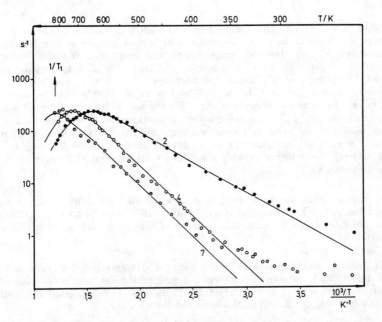

Figure 10. Spin-lattice relaxation rates $(1/T_1)$ versus reciprocal temperature for $Li_2Si_2O_5$ (curve 2), $LiNaSi_2O_5$ (curve 4), and $LiKSi_2O_5$ (curve 7) glasses and supercooled liquids. Note maxima in curves, corresponding to minima in T_i. From Göbel et al. (1979).

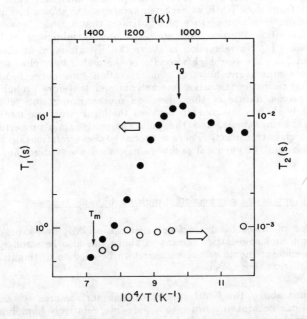

Figure 11. ^{29}Si T_1 (filled circles, left scale) and T_2 (open circles, right scale) for ^{29}Si -enriched $NaAlSi_3O_8$ (albite) glass and liquid. T_m is liquidus temperature, T_g is bulk glass transition temperature. From Liu et al. (1987).

because bonds are stretched and not broken. At high temperatures, this local motion is undoubtedly still present, but may become too high in frequency to be effective in causing relaxation $(\tau_c >> \omega_0)$. Instead, longer range, less frequent and higher energy jumps from interstitial site to site come to dominate. These type of data may be important examples of "coupling theory", which has been used to quantify a variety of transport processes in glasses (Ngai and Jains, 1986). Liu et al. (1987) also reported a small inflection in the T_1 versus temperature curve for sodium in $NaAlSi_3O_8$ melt at the glass transition temperature (T_g), suggesting some cooperative behavior between the motions of network modifier and network former cations. When T_g is above the T_1 minimum (as in this composition), interactions with network oxygen atoms apparently slow the motion of some of the sodiums, making relaxation more efficient and lowering T_1 slightly at temperatures above T_g.

Much related work has been done in liquids, which have some structure but are of much lower viscosity. For example, Harold-Smith (1974) and Filho et al. (1982) characterized the rotational and translational motions of ions in molten alkali nitrates at high temperatures.

NMR has had numerous other applications in determining the types and rates of motion of defects and vacancies in amorphous and crystalline materials. Kanert (1985), for example, reviews studies in ionic crystals. Fuda et al. (1985) present interesting ^{17}O NMR results on vacancy motion in the ionic conductor CeO_2.

The anionic frameworks of silicate solids are probably not mobile, except perhaps at phase transitions. In silicate melts, the situation is again very different. When a glass transforms to a melt (a liquid in thermodynamic metastable equilibrium) at the glass transition, the structure must begin to become mobile and must begin to change with increasing temperature. Liu et al. (1987,1988) present intriguing clues as to the nature of this mobility from ^{29}Si NMR at high temperature. As shown in Figure 11, the slope of the T_1 versus inverse temperature plot indicates that at least between about $1200°C$ and the glass transition, temperatures are below any T_1 minimum, and motion is therefore relatively slow (^{23}Na relaxation is above the T_1 minimum at these temperatures). This is expected from the very high viscosity of the melt. Very close to the macroscopic glass transition temperature, however, the relaxation time curve bends over abruptly, indicating a change in the relaxation mechanism and therefore probably in the type of motion. The precise nature of the relaxation mechanism is not yet known, but this result does clearly show a relationship between the local structural motion of the silicons and their surrounding atoms, and the bulk thermodynamic properties which change abruptly at the glass transition. There is probably also a relationship between the spin-lattice relaxation and the chemical exchange discussed above, but the details are rather speculative.

Relaxation and mobility in geochemistry: multiple phase systems

The motion of molecules on surfaces affects the NMR lineshapes in ways outlined above. Spin-spin and spin-lattice relaxation times can also be strongly influenced, and relaxation time measurements can often therefore be used to distinguish among molecular species at different types of sites.

As described above, the NMR spectra of high surface area systems with adsorbed molecules often can be broken down into a wide-line, relatively immobile component and a narrow-line, mobile component. From the linewidth alone, it is clear that the former will have a relatively short T_2, and the latter a relatively long T_2. The effect on T_1 will depend on how tightly bound the molecules are. In typical highly fluid systems, even the adsorbed species have motional frequencies that place their relaxation times above the T_1 minimum. Therefore, if motion is slowed (and therefore has a greater spectral density at the Larmor frequency), T_1 as well as T_2 becomes shorter. These relationships are described in detail in several of the papers cited above in the section entitled "Motion and NMR lineshapes."

If a single molecular species with a single, exponential spin-lattice relaxation mechanism is present, then a plot of the measured Z component of the magnetization against time can be fitted with a single value of T_1. If there is more than one species, with more than one type of motion, the curve may need to be fitted by the sum of exponentials, and more than one value of T_1 can be derived (Fig. 12). In many cases, the relative abundances of the multiple species can be quantified.

An example of the usefulness of these relationships is the study of water in porous rocks by Schmidt et al. (1987). In this work, ^1H T_1 measurements over wide ranges of time scales defined two or more components to the spin-lattice relaxation. This allowed the relative proportions of surface-bound and free pore water to be measured, and the change in this distribution as a function of total water content to be determined.

A final example of the types of structural and kinetic data that can be derived from relaxation time data also is based on a silicate-water system of a more reactive nature. Portland cement in unreacted form is a complex, anhydrous mixture of calcium silicates and aluminates. When mixed with water, a complicated set of hydration reactions occurs with rates stretching from minutes to months. During setting of the cement, a series of amorphous phases form which are very difficult to characterize by techniques such as x-ray diffraction or microscopy. As shown in Figure 13, several of these phases can be distinguished, and the amount of water in each determined by ^1H T_1 measurements (Schreiner et al., 1985). During time period I, the water is present in fluid gels coating the grains, and exchanges freely enough so that only a single T_1 is observed. As time progresses, three distinct phases can be distinguished. In relatively immobile gel layers, the shortest T_1 values are observed. A smaller percentage of the protons are present in a more mobile liquid-like phase in small pores, which has a longer T_1. Some of the protons also are incorporated into solid calcium hydroxide. Here the T_1 is very long, because of the lack of any significant proton motion and the relative isolation of the protons from each other and from paramagnetic Fe^{3+} ions. These results are complemented by a number of ^{29}Si MAS NMR studies which provide details of the change in static silicate structure during the hydrate formation (Clayden et al., 1984). Analogous studies of metamorphic or hydrothermal reactions while they take place are an intriguing future possibility.

CONCLUSIONS

NMR spectroscopy has proven to be an important tool in determining the relatively slow dynamics of molecular and atomic diffusion and rotation, in addition to its better known role in studies of structure. This information is potentially useful in understanding transport properties, transitions, and structures involving mobile species in both organic and inorganic chemistry. Classic mineralogy deals almost exclusively with static structures. However, the interaction of those structures with the surrounding environment, whether aqueous, organic, or molten, involves highly mobile atoms and molecules. These interactions are at the heart of many of the most interesting problems in geochemistry and petrology. It is hoped that the basic concepts and examples given here will help to inspire the future exploration of the mechanisms of these dynamics by geochemists. NMR is one technique which may help along the way.

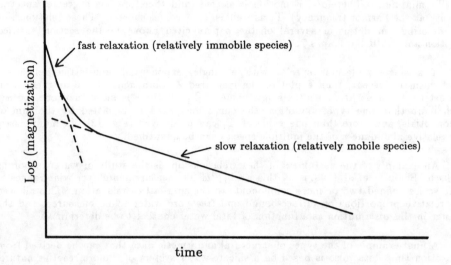

Figure 12. Decay of the Z component of magnetization in a system with two spin-lattice relaxation times. The slope of the lines are proportional to $1/T_1$.

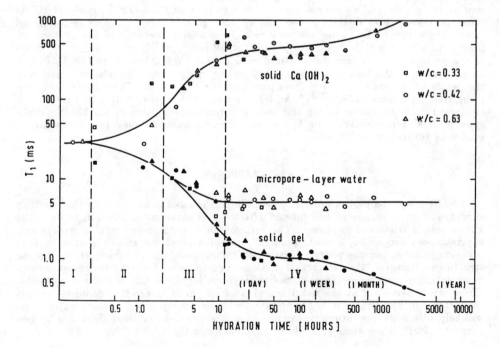

Figure 13. Time evolution of T_1 components in curing Portland cement. Various samples with different water to cement ratios are shown. From Schreiner et al. (1985).

ACKNOWLEDGMENTS

This work was supported by the National Science Foundation, NSF EAR-8507925 and EAR-8707175. The static ^{29}Si spectra were collected at the Stanford Magnetic Resonance Laboratory, supported by the National Institutes of Health, NIH RR00711. I would particularly like to thank Dr. Shang-Bin Liu, who made many valuable comments on an early version of this chapter.

REFERENCES

Akitt, J.W. (1983) NMR and Chemistry. Chapman and Hall, New York, NY.

Balzer, G., Jain, H. and Kanert, O. (1984) Dynamical properties of borate glasses by nuclear magnetic resonance. In: Proc. XXII Congress Ampere on Magnetic Resonance and Related Phenomena, K.A. Müller, R. Kind and J. Roos, eds. University of Zürich, 95-96.

Becker, E.D. (1980) High Resolution NMR. Academic Press, New York, NY.

Bloembergen, N., Purcell, E.M. and Pound, R.V. (1948) Relaxation effects in nuclear magnetic absorption. Phys. Rev. 73, 679-712.

Carson, D.G., Rossman, G.R. and Vaughn, R.W. (1982) Orientation and motion of water molecules in cordierite: a proton nuclear magnetic resonance study. Phys. Chem. Minerals 8, 14-19.

Clayden, N.J., Dobson, C.M., Hayes, C.J. and Rodger, S.A. (1984) Hydration of tricalcium silicate followed by solid-state ^{29}Si N.M.R. spectroscopy. J. Chem. Soc. Commun., 1984, 1396-1397.

Eckert, H., Yesinowski, J.P., Stolper, E.M., Stanton, T.R. and Holloway, J. (1987) The state of water in rhyolitic glasses: a deuterium NMR study. J. Non-Cryst. Solids 93, 93-114.

Eckman, R. and Vega, A.J. (1983) Deuterium NMR study of organic molecules absorbed by zeolites. J. Am. Chem. Soc. 105, 4841-4842.

Eisenberg, D. and Kauzmann, W. (1969) The Structure and Properties of Water. Oxford University Press, Oxford, England.

Engelhardt, G., Zeigan, D., Jancke, H., Hoebbel, D. and Wieker W. (1975) Zur Abhängigkeit der Struktur der Silicatanionen in wassrigen Natriumsilicatlosungen vom Na:Si Verhältnis. Z. Anorg. Allg. Chem. 418, 17-28.

Engelhardt, G. and Hoebbel, D. (1984) ^{29}Si N.M.R. spectroscopy reveals dynamic $SiO_4{}^{4-}$ group exchange between silicate anions in aqueous alkaline silicate solutions. J. Chem. Soc. 1984, 514-516.

Farrar, T.C. and Becker, E.D. (1971) Pulse and Fourier Transform NMR. Academic Press, New York, NY.

Filho, W.W., Havill, R.L. and Titman, J.M.(1982) Nuclear spin relaxation in molten alkali nitrates. J. Magnet. Res. 49, 296-303.

Fripiat, J.J., Kadi-Hanifi, M., Conrad, J. and Stone, W.E.E. (1980) NMR study of adsorbed water-III molecular orientation and protonic motions in the one-layer of a Li hectorite. In: Magnetic Resonance in Colloid and Interface Science, J.P. Fraissard and H.A. Resing, eds., D. Reidel Publishing Co., New York, NY, 529-535.

Fuda, K., Kishio, K., Yamauchi, S. and Fueki, K. (1985) Study on vacancy motion in Y_2O_3-doped CeO_2 by ^{17}O NMR technique. J. Phys. Chem. Solids 46, 1141-1146.

Fukushima, E. and Roeder, S.B.W. (1981) Experimental Pulse NMR. Addison-Wesley, New York, NY.

Fyfe, C.A. (1983) Solid State NMR for Chemists. CFC Press, Guelph, Ontario, Canada.

Fyfe, C.A., Kokotailo, G.T., Graham, J.D., Browning, C., Gobbi, G.C., Hyland, M., Kennedy, G.J. and DeShutter, C.T. (1986) Demonstration of contact induced ion exchange in zeolites. J. Am. Chem. Soc. 108, 522-523.

Göbel, E., Müller-Warmuth, W. and Olyschlager, H. (1979) ^{7}Li NMR spectra , nuclear relaxation, and lithium ion motion in alkali silicate, borate, and phosphate glasses. J. Magnet. Res. 36, 371-387.

Goldman, D.S., Rossman, G.R. and Dollase, W.A. (1977) Channel constituents in cordierite. Am. Mineral. 62, 1144-1157.

Griffiths, L., Cundy, C.S. and Plaisted, R.J. (1986) The temperature dependence of ^{29}Si nuclear magnetic resonance linewidths in aqueous silicate solutions and their effect on rate determinations. J. Chem. Soc. Dalton Trans. 1986, 2265-2269.

Hahn, E.L. (1950) Spin echoes. Phys. Rev. 80, 580-594.

Harris, R.K. (1983) Nuclear Magnetic Resonance Spectroscopy. Pitman Books, Ltd., London.

Harold-Smith, D. (1974) Nuclear magnetic relaxation in molten salts. II. Spin-lattice relaxation time of ^7Li in lithium nitrate and its mixtures with potassium nitrate. J. Chem. Phys. 60, 1405-1409.

Harris, R.K., O'Conner, M.J., Curzon, E.H. and Howarth, O.W. (1984) Two-dimensional silicon-29 NMR studies of aqueous silicate solutions. J. Magnet. Res. 57, 115-122.

Hendrickson J.R. and Bray P.J. (1974) Nuclear magnetic resonance studies of ^7Li ionic motion in alkali silicate and borate glasses. J. Chem. Phys. 61, 2754-2764.

Hochella, M.F. Jr., Brown, G.E. Jr., Ross, F.K. and Gibbs G.V. (1979) High temperature crystal chemistry of hydrous Mg- and Fe-cordierites. Am. Mineral. 64, 337-351.

Hougardy, J., Stone, W.E.E. and Fripiat J.J. (1976) NMR study of adsorbed water. I. Molecular orientation and protonic motions in the two-layer hydrate of a Na vermiculite. J. Chem. Phys. 64, 3840-3851.

Hutton, G. and Pedersen, B. (1969) Proton and deuteron magnetic resonance in partly deuterated crystals-III. gypsum. J. Phys. Chem. Solids 30, 235-242.

Jain H., Balzer-Jollenbeck G. and Kanert O. (1985) ^7Li nuclear magnetic resonance in (^7Li, ^6Li) and (Li,Na) triborate glasses. J. Am. Ceram. Soc. 68, C24 - C26.

Jain H., Peterson N.L. and Dowling H.L. (1983) Tracer diffusion and electrical conductivity in sodium-cesium silicate glasses. J. Non-Cryst. Solids 55, 283-300.

Kanert, O. (1985) Dynamical properties of defects in insulating solids by nuclear magnetic resonance. Cryst. Latt. Defects Amorph. Mat. 12, 41-57.

Kaplan, J.I. and Fraenkel, G. (1980) NMR of Chemically Exchanging Systems. Academic Press, New York, NY.

Kasai, P.H. and Jones, P.M. (1984) MAS-NMR spectroscopic study of water in zeolites. J. Molec. Catal. 27, 81-93.

Knight, C.T.G., Kirkpatrick, R.J. and Oldfield, E. (1987) Silicon-29 2D NMR evidence of four novel doubly germanium substituted silicate cages in a tetramethylammonium germanosilicate solution. J. Am. Chem. Soc. 109, 1632.

Liu S.-B., Pines A., Brandriss M. and Stebbins J.F. (1987) Relaxation mechanisms and effects of motion in albite ($NaAlSi_3O_8$) liquid and glass: A high temperature NMR study. Phys. Chem. Minerals 15, 155-162.

Liu, S. -B., Stebbins, J.F., Schneider, E. and Pines, A. (1988) Diffusive motion in alkali silicate melts: an NMR study at high temperature. Geochem. Cosmochim. Acta, in press.

Nagy, J.B., Derouane, E.G., Resing, H.A. and Miller, G.R. (1983) Motions of o- and p-xylenes in ZSM-5 catalyst. Carbon-13 nuclear magnetic resonance. J. Phys. Chem. 87, 833-837.

Ngai K.L. and Jain H. (1986) Conductivity relaxation and spin lattice relaxation in lithium and mixed alkali borate glasses: activation enthalpies, anomalous isotope- mass effect and mixed alkali effect. Solid State Ionics 18&19, 362-367.

Pake, G.E. (1948) Nuclear resonance absorption in hydrated crystals: fine structure of the proton line. J. Chem. Phys. 16, 327-336.

Ramey, K., O'Brian, J., Hasegawa, I. and Borchert A.E. (1965) Nuclear magnetic resonance study of aluminum alkyls. J. Phys. Chem. 69, 3418-3423.

Resing, H.R. (1980) NMR techniques for chemical kinetics in adsorption systems. In: Magnetic Resonance in Colloid and Interface Science, J.P. Fraissard and H.A. Resing, eds., D. Reidel Publishing Co., New York, NY, 219-238.

Resing, H.A., Slotfeldt-Ellingsen, D., Garroway, A.N., Weber, D.C., Pinnavaia, T.J. and Unger K. (1980) ^{13}C chemical shifts in adsorption systems: molecular motions, molecular orientations, qualitative and quantitative analysis. In: Magnetic Resonance in Colloid and Interface Science, J.P. Fraissard and H.A. Resing, eds., D. Reidel Publishing Co., New York, NY, 239-258.

Richet P. (1984) Viscosity and configurational entropy of silicate melts. Geochim. Cosmochim. Acta 48, 471-483.

Richet P., Robie R.A. and Hemingway B.S. (1986) Low temperature heat capacity of diopside glass ($CaMgSi_2O_6$): A calorimetric test of the configurational entropy theory applied to the viscosity of liquid silicates. Geochim. Cosmochim. Acta 50, 1521-1535.

Schneider E. (1985) Applications of High Resolution NMR to Geochemistry: Crystalline, Glassy, and Molten Silicates. Ph.D. dissertation, Univ. of California, Berkeley, and Lawrence Berkeley Laboratory report LBL-20936.

Schreiner, L.J., MacTavish, J.C., Miljkovic, L., Pintar, M.M., Blinc, R., Lahajnar, G., Lasic, D. and Reeves, L.W. (1985) NMR line shape-spin-lattice relaxation correlation study of Portland cement hydration. J. Am. Ceram. Soc. 68, 10-16.

Schmidt, E.J., Velasco, K.K. and Nur, A.M. (1986) Quantification of solid-fluid interfacial phenomena in porous rocks with proton nuclear magnetic resonance. J. Appl. Phys. 59, 2788-2797.

Sefcik, M.D., Schaefer, J. and Stejskal, E.O. (1976) Characterization of the small-port mordenite adsorption sites by carbon-13 NMR. In: Magnetic Resonance in Colloid and Interface Science, H.A. Resing and C.G. Wade, eds., American Chemical Society, Washington, DC, 109-122.

Sefcik, M.D., Schaefer, J. and Stejskal, E.O. (1977) Characterization of the mordenite sorption sites by carbon-13 NMR. In: Molecular Sieves-II, J.R. Katzer, ed., American Chemical Society, Washington, DC, 344-356.

Spiess, H.W., Grosescu, R. and Haeberlen, U. (1974) Molecular motion studied by NMR powder spectra. II. Experimental results for solid P_4 and solid $Fe(CO)_5$. Chem. Phys. 6, 226-234.

Stebbins, J.F. (1987) Identification of multiple structural species in silicate glasses by ^{29}Si NMR. Nature 330, 465-467.

Stebbins, J.F., Murdoch, J.B., Schneider, E., Carmichael, I.S.E. and Pines, A. (1985) A high temperature nuclear magnetic resonance study of ^{23}Na, ^{27}Al, and ^{29}Si in molten silicates. Nature 314, 250-252.

Tsang, ,T. and Ghose, S. (1972) Nuclear magnetic resonance of H and Al and Al-Si order in low cordierite, $Mg_2Al_4Si_5O_{18} \cdot nH_2O$. J. Chem. Phys. 56, 3329-3332.

Vučelic, V. and Vučelic, D. (1985) Heat capacities of water on zeolites. In: Zeolites, B. Držaj, S. Hočevar and S. Pejovnik, eds., 475-480. Elsevier, Amsterdam.

Walmsley, R.H. and Shporer, M. (1978) Surface-induced NMR line splittings and augmented relaxation rates in water. J. Chem. Phys. 68, 2584-2590.

Wang, Po-Kang, Ansermet, J.-P., Rudaz, S., Wang, Z., Shore, S., Slichter, C.P. and Sinfelt, J.H. (1986) NMR studies of simple molecules on metal surfaces. Science 234, 35-41.

Weiden N. and Räger H. (1985) The chemical shift of the ^{29}Si nuclear magnetic resonance in a synthetic single crystal of Mg_2SiO_4. Z. Naturforsch. A, 40A, 126-130.

Woessner, D.E. (1980) An NMR investigation into the range of the surface effect on the rotation of water molecules. J. Magnet. Res. 39, 297-308.

Chapter 11

G. E. Brown, Jr., G. Calas,
G. A. Waychunas and **J. Petiau**

X-Ray Absorption Spectroscopy and its Applications in Mineralogy and Geochemistry

I. INTRODUCTION AND OVERVIEW

This chapter reviews the principles of x-ray absorption spectroscopy (XAS) and some of its applications to problems in mineralogy and geochemistry. These applications are relatively new because XAS has been developed as a quantitative structural method only during the past 10 to 15 years. XAS is still undergoing improvements in theory, particularly that dealing with near-edge spectra, and in experimental practice. Many of the advances in the capabilities of XAS are tied to development of synchrotron radiation sources, which provide very intense x-rays tunable over a broad spectral range [infrared (E < 1 eV, λ > 12,000 Å) to hard x-ray (E $\leq$ 100 keV, $\lambda \geq$ 0.12 Å)]. XAS has become a powerful element-specific probe which can be used to determine the local structure (bond distance, number and type of near neighbors) around a specific absorbing element, even when the element is at low concentration levels ($\geq$ 1000 ppm). Furthermore, XAS can be used to probe most elements of the periodic table in most types of phases (crystalline or amorphous solids, liquids, gases) and at structural sites ranging from those in crystals and glasses to those at interfaces, such as a mineral/water interface. Like other spectroscopic methods, XAS is not without limitations; however, it is a remarkably versatile method capable of providing unique information on structure and bonding in earth materials that complements information from other spectroscopic and scattering methods.

The chapter is divided into six parts, including: (I) introduction to and general overview of XAS; (II) experimental methods of XAS, including discussion of synchrotron radiation sources; (III) overview of the theoretical framework and data analysis of <u>Extended X-ray Absorption Fine Structure</u> (EXAFS) spectra; (IV) discussion of the <u>X-ray Absorption Near-Edge Structure</u> (XANES); (V) examination of the information content of XAS spectra of common earth materials; and (VI) selective review of recent applications of EXAFS and XANES spectroscopies to problems in mineralogy, geochemistry, and earth materials research. Another function of this chapter is to serve as a guide to the rapidly expanding literature on XAS applications in a variety of disciplines (earth sciences, condensed matter physics, materials science, chemistry, catalysis, surface science). Finally, some of the future applications of XAS are discussed in the context of the next generation of synchrotron radiation sources.

The absorption of electromagnetic radiation by matter is a relatively old concept which has been used to study the discrete (quantized) energy levels of electrons in atoms, molecules, and condensed matter (e.g., Glenn and Dodd, 1968). X-ray absorption was revitalized in the early 1970s as a molecular-level structural probe. This change was stimulated by advances in theory (Sayers et al., 1970; Sayers et al., 1971; Stern, 1974; Ashley and Doniach, 1975; Lee and Pendry, 1975) and by the availability of intense sources of continuous x-radiation referred to as <u>synchrotron radiation</u> (Winick and Doniach, 1980). The x-ray absorption process was first explained by Kossel (1920) who suggested that the sharp increase in absorption of x-rays by matter over a narrow range of incident x-ray energies (the <u>absorption edge</u>) is caused by the excitation of an electron from a deep core state of an atom to an empty bound state or to a continuum state. Such a transition will occur when the energy of the incident x-ray photons equals the energy required for excitation. The production of <u>photoelectrons</u> by this process is the primary cause of x-ray attenuation by matter. Other processes also occur when x-rays interact with matter, including x-ray scattering (both elastic and inelastic), electron-positron pair production, production of optical photons, and production of phonons (Fig. 1-1). However, in the x-ray energy range (0.5 keV to 100 keV), photoelectric absorption is the dominant process. The excited atom returns to its ground state through secondary processes (see Fig. 2-9 below) such as x-ray emission (fluorescence) (see Chapter 14) and Auger electron production (see Chapter 13), both of which provide valuable information on bonding and composition.

Table 1. Electron binding energies, in eV, for the elements in their natural forms.
(From Williams, in Vaughan, 1986)

Element	K 1s	L_I 2s	L_{II} $2p_{1/2}$	L_{III} $2p_{3/2}$	M_I 3s	M_{II} $3p_{1/2}$	M_{III} $3p_{3/2}$	M_{IV} $3d_{3/2}$	M_V $3d_{5/2}$	N_I 4s	N_{II} $4p_{1/2}$	N_{III} $4p_{3/2}$
1 H	16*											
2 He	24.6*											
3 Li	54.7*											
4 Be	111.5*											
5 B	188*											
6 C	284.2*											
7 N	409.9*	37.3*										
8 O	543.1*	41.6*										
9 F	696.7*											
10 Ne	870.2*	48.5*	21.7*	21.6*								
11 Na	1070.8†	63.5†	30.4†	30.5*								
12 Mg	1303.0†	88.6*	49.6†	49.2†								
13 Al	1558.98*	117.8*	72.9*	72.5*								
14 Si	1839	149.7*b	99.8*	99.2*								
15 P	2149	189*	136*	135*								
16 S	2472	2309*b	163.6*	162.5*								
17 Cl	2833	270*	202*	200*								
18 Ar	3205.9*	326.3*	250.6*	248.4*	29.3*	15.9*	15.7*					
19 K	3608.4*	378.6*	297.3*	294.6*	34.8*	18.3*	18.3*					
20 Ca	4038.5*	438.4†	349.7†	346.2†	44.3†	25.4†	25.4†					
21 Sc	4492	498.0*	403.6*	398.7*	51.1*	28.3*	28.3*					
22 Ti	4966	560.9†	461.2†	453.8†	58.7†	32.6†	32.6†					
23 V	5465	626.7†	519.8†	512.1†	66.3†	37.2†	37.2†					
24 Cr	5989	695.7†	583.8†	574.1†	74.1†	42.2†	42.2†					
25 Mn	6539	769.1†	649.9†	638.7†	82.3†	47.2†	47.2†					
26 Fe	7112	844.6†	719.9†	706.8†	91.3†	52.7†	52.7†					
27 Co	7709	925.1†	793.3†	778.1†	101.0†	58.9†	58.9†					
28 Ni	8333	1008.6†	870.0†	852.7†	110.8†	68.0†	66.2†					
29 Cu	8979	1096.7†	952.3†	932.5†	122.5†	77.3†	75.1†					
30 Zn	9659	1196.2*	1044.9*	1021.8*	139.8*	91.4*	88.6*	10.2*	10.1*			
31 Ga	10367	1299.0*b	1143.2†	1116.4†	159.5†	103.5†	103.5†	18.7†	18.7†			
32 Ge	11103	1414.6*b	1248.1*b	1217.0*b	180.1*	124.9*	120.8*	29.0*	29.0*			
33 As	11867	1527.0*b	1359.1*b	1323.6*b	204.7*	146.2*	141.2*	41.7*	41.7*			
34 Se	12658	1652.0*b	1474.3*b	1433.9*b	229.6*	166.5*	160.7*	55.5*	54.6*			
35 Br	13474	1782*	1596*	1550*	257*	189*	182*	70*	69*			
36 Kr	14326	1921	1730.9*	1678.4*	292.8*	222.2*	214.4	95.0*	93.8*	27.5*	14.1*	14.1*
37 Rb	15200	2065	1864	1804	326.7*	248.7*	239.1*	113.0*	112*	30.5*	16.3*	15.3*
38 Sr	16105	2216	2007	1940	358.7†	280.3†	270.0†	136.0†	134.2†	38.9†	20.3†	20.3†
39 Y	17038	2373	2156	2080	392.0*b	310.6*	298.8*	157.7†	155.8†	43.8*	24.4*	23.1*
40 Zr	17998	2532	2307	2223	430.3†	343.5†	329.8†	181.1†	178.8†	50.6†	28.5†	27.7†
41 Nb	18986	2698	2465	2371	466.6†	376.1†	360.6†	205.0†	202.3†	56.4†	32.6†	30.8†
42 Mo	20000	2866	2625	2520	506.3†	410.6†	394.0†	231.1†	227.9†	63.2†	37.6†	35.5†
43 Tc	21044	3043	2793	2677	544*	447.6*	417.7*	257.6*	253.9*	69.5*	42.3*	39.9*
44 Ru	22117	3224	2967	2838	586.2†	483.3†	461.5†	284.2†	280.0†	75.0†	46.5†	43.2†
45 Rh	23220	3412	3146	3004	628.1†	521.3†	496.5†	311.9†	307.2†	81.4*b	50.5†	47.3†
46 Pd	24350	3604	3330	3173	671.6†	559.9†	532.3†	340.5†	335.2†	87.1*b	55.7†a	50.9†a
47 Ag	25514	3806	3524	3351	719.0†	603.8†	573.0†	374.0†	368.0†	97.0†	63.7†	58.3†

Element	K 1s	L_I 2s	L_{II} $2p_{1/2}$	L_{III} $2p_{3/2}$	M_I 3s	M_{II} $3p_{1/2}$	M_{III} $3p_{3/2}$	M_{IV} $3d_{3/2}$	M_V $3d_{5/2}$	N_I 4s	N_{II} $4p_{1/2}$	N_{III} $4p_{3/2}$
48 Cd	26711	4018	3727	3538	772.0†	652.6†	618.4†	411.9†	405.2†	109.8†	63.9†a	63.9†a
49 In	27940	4238	3938	3730	827.2†	703.2†	665.3†	451.4†	443.9†	122.7†	73.5†a	73.5†a
50 Sn	29200	4465	4156	3929	884.7†	756.5†	714.6†	493.2†	484.9†	137.1†	83.6†a	83.6†a
51 Sb	30491	4698	4380	4132	946†	812.7†	766.4†	537.5†	528.2†	153.2†	95.6†a	95.6†a
52 Te	31814	4939	4612	4341	1006†	870.8†	820.8†	583.4†	573.0†	169.4†	103.3†a	103.3†a
53 I	33169	5188	4852	4557	1072*	931*	875*	631*	620*	186*	123*	123*
54 Xe	34561	5453	5104	4782	1148.7*	1002.1*	940.6*	689.0*	676.4*	213.2*	146.7	145.5*
55 Cs	35985	5714	5359	5012	1211*b	1071*	1003*	740.5*	726.6*	232.3*	172.4*	161.3*
56 Ba	37441	5989	5624	5247	1293*b	1137*b	1063*b	795.7*	780.5*	253.5†	192	178.6†
57 La	38925	6266	5891	5483	1362*b	1209*b	1128*b	853*	836*	247.7*	205.8	196.0*
58 Ce	40443	6548	6164	5723	1436*b	1274*b	1187*b	902.4*	883.8*	291.0*	223.2	206.5*
59 Pr	41991	6835	6440	5964	1511	1337	1242	948.3*	928.8*	304.5	236.3	217.6
60 Nd	43569	7126	6722	6208	1575	1403	1297	1003.3*	980.4*	319.2*	243.3	224.6
61 Pm	45184	7428	7013	6459	—	1403	1357	1052	1027	—	242	242
62 Sm	46834	7737	7312	6716	1723	1541	1419.8	1110.9*	1083.4*	347.2*	265.6	247.4
63 Eu	48519	8052	7617	6977	1800	1614	1481	1158.6*	1127.5*	360	284	257
64 Gd	50239	8376	7930	7243	1881	1688	1544	1221.9*	1189.6*	378.6*	286	271
65 Tb	51996	8708	8252	7514	1968	1768	1611	1276.9*	1241.1*	396.0*	322.4*	284.1*
66 Dy	53789	9046	8581	7790	2047	1842	1676	1333	1292*	414.2*	333.5*	293.2*
67 Ho	55618	9394	8918	8071	2128	1923	1741	1392	1351	432.4*	343.5	308.2*
68 Er	57486	9751	9264	8358	2206	2006	1812	1453	1409	449.8*	366.2	320.2*
69 Tm	59390	10116	9617	8648	2307	2090	1885	1515	1468	470.9*	385.9*	332.6*
70 Yb	61332	10486	9978	8944	2398	2173	1950	1576	1528	480.5*	388.7*	339.7*

Table 1 (continued from previous page). 433

Element	K 1s	L_I 2s	L_{II} $2p_{1/2}$	L_{III} $2p_{3/2}$	M_I 3s	M_{II} $3p_{1/2}$	M_{III} $3p_{3/2}$	M_{IV} $3d_{3/2}$	M_V $3d_{5/2}$	N_I 4s	N_{II} $4p_{1/2}$	N_{III} $4p_{3/2}$
71 Lu	63314	10870	10349	9244	2491	2264	2024	1639	1589	506.8*	412.4*	359.2*
72 Hf	65351	11271	10739	9561	2601	2365	2107	1716	1662	538*	438.2†	380.7†
73 Ta	67416	11682	11136	9881	2708	2469	2194	1793	1735	563.4†	463.4†	400.9†
74 W	69525	12100	11544	10207	2820	2575	2281	1949	1809	594.1†	490.4†	423.6†
75 Re	71676	12527	11959	10535	2932	2682	2367	1949	1883	625.4†	518.7†	446.8†
76 Os	73871	12968	12385	10871	3049	2792	2457	2031	1960	658.2†	549.1†	470.7†
77 Ir	76111	13419	12824	11215	3174	2909	2551	2116	2040	691.1†	577.8†	495.8†
78 Pt	78395	13880	13273	11564	3296	3027	2645	2202	2122	725.4†	609.1†	519.4†
79 Au	80725	14353	13734	11919	3425	3148	2743	2291	2206	762.1†	642.7†	546.3†
80 Hg	83102	14839	14209	12284	3562	3279	2847	2385	2295	802.2†	680.2†	576.6†
81 Tl	85530	15347	14698	12658	3704	3416	2957	2485	2389	846.2†	720.5†	609.5†
82 Pb	88005	15861	15200	13055	3851	3554	—	3066	2586	891.8†	761.9†	643.5†
83 Bi	90526	16388	15711	13419	3999	3696	3177	2688	2580	939†	805.2†	678.8†
84 Po	93105	16939	16244	13814	4149	3854	3302	2798	2683	995*	851*	705*
85 At	95730	17493	16785	14214	4317	4008	3426	2909	2787	1042*	886*	740*
86 Rn	98404	18049	17337	14619	4482	4159	3538	3022	2892	1097*	929*	768*
87 Fr	101137	18639	17907	15031	4652	4327	3663	3136	3000	1153*	980*	810*
88 Ra	103922	19237	18484	15444	4822	4490	3792	3248	3105	1208*	1958*	879*
89 Ac	106755	19840	19083	15871	5002	4656	3909	3370	3219	1269*	1080*	890*
90 Th	109651	20472	19693	16300	5182	4830	4046	3491	3332	1330*	1168*	966.4†
91 Pa	112601	21105	20314	16733	5367	5001	4174	3611	3442	1387*	1224*	1007*
92 U	115606	21757	20948	17166	5548	5182	4303	3728	3552	1439*b	1271*b	1043†

Element	N_{IV} $4d_{3/2}$	N_V $4d_{5/2}$	N_{VI} $4f_{5/2}$	N_{VII} $4f_{7/2}$	O_I 5s	O_{II} $5p_{1/2}$	O_{III} $5p_{3/2}$	O_{IV} $5d_{3/2}$	O_V $5d_{5/2}$	P_I 6s	P_{II} $6p_{1/2}$	P_{III} $6p_{3/2}$
71 Lu	206.1*	196.3*	8.9*	7.5*	57.3*	33.6*	26.7*					
72 Hf	220.0†	211.5†	15.9†	14.2†	64.2†	38*	29.9†					
73 Ta	237.9†	226.4†	23.5†	21.6†	69.7†	42.2*	32.7†					
74 W	255.9†	243.5†	33.6*	31.4†	75.6†	45.3*b	36.8†					
75 Re	273.9†	260.5†	42.9*	40.5*	83†	45.6†	34.6*b					
76 Os	293.1†	278.5†	52.4†	50.7†	83†	58*	44.5†					
77 Ir	311.9†	296.3†	63.8†	60.8†	95.2*b	63.0*b	48.0†					
78 Pt	331.6†	314.6†	74.5†	71.2†	101.7*b	65.3*b	51.7†					
79 Au	353.2†	335.1†	87.6†	83.9†	107.2*b	74.2†	57.2†					
80 Hg	378.2†	358.8†	104.0†	99.9†	127†	83.1†	64.5†	9.6*	7.8†			
81 Tl	405.7†	385.0†	122.2†	117.8†	136.*b	94.6†	73.5†	14.7†	12.5†			
82 Pb	434.3†	412.2†	141.7†	136.9†	147*b	106.4†	83.3†	20.7†	18.1†			
83 Bi	464.0†	440.1†	162.3†	157.0†	159.3*b	119.0†	92.6†	26.9†	23.8†			
84 Po	500*	473*	184*	184*	177*	132*	104*	31*	31*			
85 At	533*	407*	210*	210*	195*	148*	115*	40*	40*			
86 Rn	567*	541*	238*	238*	214*	164*	127*	48*	48*	26		
87 Fr	603*	577*	268*	268*	234*	182*	140*	58*	58*	34	15	15
88 Ra	636*	603*	299*	299*	254*	200*	153*	68*	68*	44	19	19
89 Ac	675*	639*	319*	319*	272*	215*	167*	80*	80*	—	—	—
90 Th	712.1†	675.2†	342.4†	333.1†	290*a	229*a	182*a	92.5†	85.4†	41.4†	24.5†	16.6†
91 Pa	743*	708*	371*	360*	310*	232*	232*	94*	94*	—	—	—
92 U	778.3†	736.2†	388.2*	377.4†	321*ab	257*ab	192*ab	102.8†	94.2†	43.9†	26.8†	16.8†

Element	N_{IV} $4d_{3/2}$	N_V $4d_{5/2}$	N_{VI} $4f_{5/2}$	N_{VII} $4f_{7/2}$	O_I 5s	O_{II} $5p_{1/2}$	O_{III} $5p_{3/2}$	O_{IV} $5d_{3/2}$	O_V $5d_{5/2}$	P_I 6s	P_{II} $6p_{1/2}$	P_{III} $6p_{3/2}$
48 Cd	11.7†	10.7†										
49 In	17.7†	16.4†										
50 Sn	24.9†	23.9†										
51 Sb	33.3†	32.1†										
52 Te	41.9†	40.4†										
53 I	50*	50*										
54 Xe	69.5*	67.5*	—	—	23.3*	13.4*	12.1*					
55 Cs	79.8*	77.5*	—	—	22.7	14.2*	12.1*					
56 Ba	92.6†	89.9†	—	—	30.3†	17.0†	14.8†					
57 La	105.3*	102.5*	—	—	34.3*	19.3*	16.8*					
58 Ce	109*	—	0.1	0.1	37.8	19.8*	17.0*					
59 Pr	115.1*	115.1*	2.0	2.0	37.4	22.3	22.3					
60 Nd	120.5*	120.5*	1.5	1.5	37.5	21.1	21.1					
61 Pm	120	120	—	—	—	—	—					
62 Sm	129	129	5.2	5.2	37.4	21.3	21.3					
63 Eu	133	133	0	0	32	22	22					
64 Gd	—	127.7*	8.6*	8.6*	36	20	20					
65 Tb	150.5*	150.5*	7.7*	2.4*	45.6*	28.7*	22.6*					
66 Dy	153.6*	153.6*	8.0*	4.3*	49.9*	26.3	26.3					
67 Ho	160*	160*	8.6*	5.2*	49.3*	30.8*	24.1*					
68 Er	167.6*	167.6*	—	4.7*	50.6*	31.4*	24.7*					
69 Tm	175.5*	175.5*	—	4.6	54.7*	31.8*	25.0*					
70 Yb	191.2*	182.4*	2.5*	1.3*	52.0*	30.3*	24.1*					

434

The attenuation of x-rays by atoms varies smoothly with increasing photon energy except in the vicinity of an element's absorption edge. Figure 1-2a shows x-ray absorption by copper metal over the photon energy range 0 to 12,000 eV. Absorption increases abruptly above the smoothly increasing background at incident photon energies of ≈ 8,980 eV and in the range 930 to 1,100 eV for copper metal. The higher energy absorption edge, referred to as the K edge, is caused by excitation of K-level (1s) electrons to the lowest energy, unoccupied or partially occupied level in copper metal (3d level); the lower energy edge, referred to as the L edge, is made up of three distinct features or edges caused by excitation of L_I (binding energy = 1,097 eV), L_{II} (952 eV), and L_{III} (932 eV) electrons to unoccupied levels in copper metal. Table 1 lists electronic binding energies for the elements, which range from ≈ 55 eV for Li to over 115,000 eV for U for the K electronic level. The energies of absorption edges for elements in various compounds will vary relative to these values by as much as ±10 eV because of differences in electronic energy levels caused by different nearest-neighbor bonding and geometry and different absorber oxidation states. Historically, <u>the absorption edge energy of an element (also referred to as the absorption or energy threshold) is defined as the inflection point on the low energy side of the edge.</u> However, as will be discussed later (§3.3.1), the energy threshold is not so easily defined in practice.

Figure 1-1 shows the type of experiment required to measure x-ray absorption by a sample of thickness x (cm) and density ρ (gcm^{-3}). The incident x-ray beam has energy E = $h\upsilon = hc/\lambda = 12.39/\lambda$ (E is in keV, h is Planck's constant, υ is the frequency, c is the velocity of light, and λ is wavelength in Å) and intensity I_0. Attenuation of the incident beam intensity I_0 is

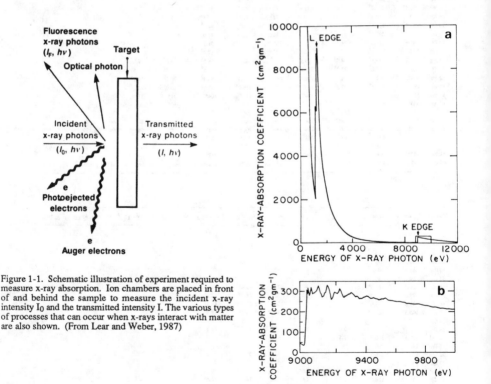

Figure 1-1. Schematic illustration of experiment required to measure x-ray absorption. Ion chambers are placed in front of and behind the sample to measure the incident x-ray intensity I_0 and the transmitted intensity I. The various types of processes that can occur when x-rays interact with matter are also shown. (From Lear and Weber, 1987)

Figure 1-2. (a) X-ray absorption spectrum of copper metal over the energy range 0 to 12 keV; (b) X-ray absorption spectrum of copper metal over the energy range 8.8 to 10 keV, showing the Cu K edge and EXAFS oscillations. (From Stern and Heald, 1983)

described by an equation of the form

$$I = I_0 \exp(-\mu_m \rho\, x) \quad , \tag{1-1}$$

where μ_m is the mass absorption coefficient ($cm^2 g^{-1}$), which is independent of the physical and chemical state of the absorber but dependent on incident x-ray energy. In x-ray absorption spectroscopy, we are most interested in a different form of the absorption coefficient which is dependent on the physical and chemical state of the absorber and on x-ray energy. This form is the linear absorption coefficient, defined as $\mu = \mu_m \rho$ (cm^{-1}), and equation (1-1) becomes:

$$I = I_0 \exp(-\mu\, x) \quad . \tag{1-2}$$

Tabulated values for μ can be found in many standard references, including the tables of McMaster et al. (1969). μ_m can be calculated for a material of known composition and density using a summation of the form:

$$\mu_m \text{(compound)} = \sum_i f_i\, \mu_i/\rho \quad , \tag{1-3}$$

where f_i is the weight percent of each element i present in the compound, with linear absorption coefficient μ_i for each element specified for a particular x-ray energy. A detector is placed in front of the sample to detect the incident intensity I_0 and after the sample to detect the transmitted intensity I.

The plot in Figure 1-2b -- a blow-up of the K-edge region of copper metal shown in Figure 1-2a -- is correctly viewed as variation of the linear absorption coefficient μ [$= (x^{-1}) ln\, I_0/I$] with incident photon energy. The absorption spectrum of an element in the vicinity of an absorption edge can be divided into two major regions (XANES and EXAFS) (Fig. 1-3), depending on the values of incident photon energy, E, and binding energy of a core-level electron, E_b, in the absorber. The XANES region is further divided into two regions (pre-edge and XANES proper). The regions are defined as follows:

(1) If $E < E_b$, electronic transitions have low probabilities, and no significant absorption occurs, except for that caused by localized electronic transitions to unfilled or partially-filled atomic levels or that due to processes other than photoelectron production (the smoothly increasing type).

(2) If $E \approx E_b$, electronic transitions occur with high probability from a core level, defined by E_b, to unoccupied bound states or continuum states .

(3) If $E > E_b$, electronic transitions occur with lower probability than at the edge energy to continuum states, and the excited photoelectrons remain in the sample for a short time with an excess kinetic energy $E_k = E - E_b$.

Region (1) is referred to as the pre-edge when E is $\approx$ 2-10 eV below the main absorption edge . The pre-edge region may show features (Fig. 1-3b) due to electronic transitions from a core level (1s, 2s, etc.) to the lowest energy, unoccupied or partially occupied level. These transitions are considered to be localized, involving atomic or molecular orbital energy levels. The transition probabilities and intensity of pre-edge features are determined in part by the symmetry of the ligands surrounding the absorber. The quantum mechanical selection rules controlling these transition probabilities are the same as those for optical absorption spectra (see Chapter 7). The energy and intensity of pre-edge features provide information on the absorber's oxidation state, the site geometry of the absorber, and absorber-ligand bonding. In the case of the K edge of Mn in MnO_2 (Fig. 1-3b), the vertical arrow indicates the pre-edge feature, which is due to a 1s→3d transition.

Region (2), which extends from a few eV above the pre-edge to $\approx$ 50 eV above the edge, is referred to as the X-ray Absorption Near Edge Structure (XANES) or sometimes as the Near Edge X-ray Absorption Fine Structure (NEXAFS). This spectral region is often characterized

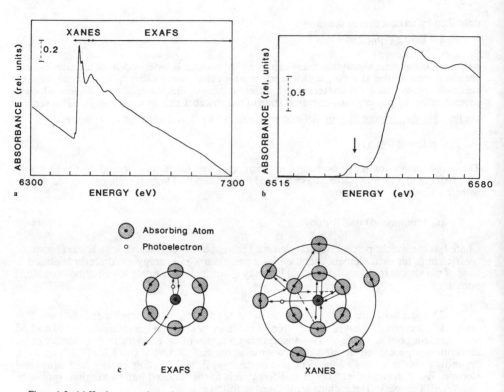

Figure 1-3. (a) K-edge x-ray absorption spectrum of Mn in MnO_2 showing the XANES and EXAFS regions; (b) K-edge XANES spectrum of Mn in MnO_2, showing the pre-edge region and the near-edge region; (c) Schematic illustration of the single-scattering process in the EXAFS energy region and the multiple-scattering process in the XANES energy region. (From Calas et al., 1987)

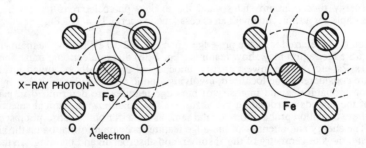

Figure 1-4. Schematic representation of the backscattering of iron photoelectron waves by oxygen nearest neighbors and the resulting interference with outgoing photoelectron waves. The left-hand figure shows backscattered waves at a position and energy (E_1) where constructive interference will occur, resulting in an amplification of EXAFS oscillations, whereas the right-hand figure shows the backscattered waves at a different position and greater energy (E_2) where destructive interference will occur, resulting in a reduction in the amplitude of EXAFS oscillations.

by intense, sharp features (Fig. 1-3a), relative to the EXAFS oscillations in region (3). These features arise from strong multiple scattering of photoelectrons with moderate kinetic energy from atoms surrounding the absorber (Fig. 1-3c). This scattering process involves multiple-scattering centers that often extend beyond the first nearest-neighbor shell of ligands. Therefore, the transition is not localized; instead, the electron is excited to a continuum level. Because of this origin, XANES is particularly sensitive to the geometrical arrangement of first and more distant neighbors around the central absorber, and can be used to obtain information on site geometry and bond angles. Recent advances in the interpretation of XANES spectra make it possible to obtain distance information from XANES.

Region (3), which extends from ≈ 50 eV above the edge to as much as 1000 eV above the edge, is referred to as the <u>Extended X-ray Absorption Fine Structure (EXAFS)</u>. The EXAFS region is characterized by weak oscillations of low frequency relative to the XANES region (Fig. 1-3a). These oscillations arise from weak backscattering of photoelectrons of high kinetic energy in a single-scattering process (Fig. 1-3c). This fine structure above the absorption edge was recognized in the early 1920s (Fricke, 1920; Hertz, 1920; Kronig, 1921; Kronig, 1931) and was called "Kronig oscillations"; however, its origin was not well understood until the early 1970s (Sayers et al., 1970; 1971). Lytle et al. (1982) give an excellent historical account of these developments. It is the constructive and destructive interference between the outgoing and backscattered photoelectron waves that produce the EXAFS oscillations (Fig. 1-4). EXAFS can be analyzed to give information about the distance from the absorber to near neighbors, in some cases extending out to several shells of ligands (< 5 Å), and about the number and type of backscatterers. Figure 1-5 shows schematically that the frequency of EXAFS oscillations is inversely related to average absorber-backscatterer distance and that the amplitude of these oscillations is directly related to the number of backscatterers. Isolated atoms do not exhibit EXAFS (Fig. 3-2). Only when the absorber is surrounded by other atoms are the outgoing photoelectron waves backscattered.

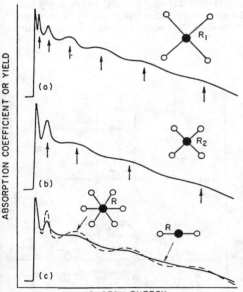

Figure 1-5. Changes in EXAFS frequency and amplitude caused by different absorber-backscatterer distances [(a,b)] and by differences in number of nearest-neighbors around the absorber [(c)]. Notice that long absorber-backscatterer distance results in high frequency oscillations (a), whereas a short distance results in low frequency oscillations (b). In (c), both molecular groups have the same distance R; only coordination number changes. (From Stohr, 1984)

Absorption spectra cannot be analyzed in the XANES region with the widely-used single-scattering, plane-wave approximation of the EXAFS process because of the multiple-scattering origin of XANES. However, definition of the energy where EXAFS begins and XANES ends is somewhat arbitrary. One way of thinking about the difference between these two scattering regimes is in terms of the wavelength of the photoelectron relative to the absorber-backscatterer distance or the photoelectron wave vector k (defined by equation (3-3) in §3.1) (Bianconi, 1984). For simple absorber-backscatterer complexes, in which the backscatterers are at a single distance d from the absorber, it is possible to define a minimum energy, E_c, for the EXAFS region. The lower limit is defined when the photoelectron wavelength is smaller than d or where its wave vector $k > 2\pi/d$ (Fig. 1-6). Therefore, for situations where the photoelectron wavelength is greater than d (i.e., below E_c), the spectral features should be due to XANES resonances, in this simple picture. E_0 is the energy above which the ejected electron becomes delocalized and enters the XANES multiple-scattering energy regime. Another way of distinguishing between EXAFS and XANES is to think of the former as a type of high energy electron diffraction, with the absorber as the electron source, whereas the latter can be thought of as a type of low energy electron diffraction.

II. EXPERIMENTAL METHODS

X-ray absorption spectroscopy (XAS) requires a continuous and intense x-ray source covering a broad range of energies, a tunable monochromator with which to select the incident photon energy, and appropriate x-ray detectors, typically gas-filled ionization chambers. Each part of a typical x-ray absorption spectroscopy system will be examined separately.

2.1 Synchrotron light sources

The two sources of continuous x-ray radiation commonly used for x-ray absorption studies are the bremsstrahlung (continuous radiation) output from a rotating anode x-ray tube and the synchrotron radiation produced by an electron (or positron) storage ring. Conventional x-ray emitting tubes are of limited use in XAS because of their relatively low x-ray intensities

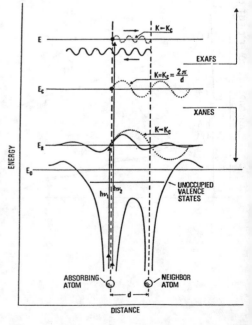

Figure 1-6. Schematic representation of the final state wavefunctions produced by excitation of a core-level electron in a simple molecule with a single distance d between absorber and backscatterer. Final state wave functions (dotted lines) are shown for photoelectrons of high kinetic energy (E) in the EXAFS region, for photoelectrons of low kinetic energy (E_R) in the XANES region, and for photoelectrons of kinetic energy (E_c) defined as separating the two regions. The energy level at which the photoelectron becomes delocalized is represented by the horizontal line at energy E_0. Atomic potentials are shown for the absorber and nearest-neighbor atoms. Two electronic transitions are shown with energies $h\nu_1$ and $h\nu_2$, representing transitions with energies in the XANES and EXAFS regions, respectively. Also shown is the energy of an unoccupied valence level. An electronic transition from a deep core level to the valence level would produce a spectral feature in the pre-edge region of an x-ray absorption spectrum. (After Bianconi, 1984)

and the long times required for collecting a single spectrum of adequate signal-to-noise level, even when the element of interest is at relatively high concentration levels. Instead, much more intense synchrotron x-ray sources are now commonly used for XAS, and the increasing availability of these sources has led to rapid development of this spectroscopic method. Here we will present only the main characteristics of synchrotron radiation. The interested reader is referred to several comprehensive reviews of synchrotron radiation and its applications in fields outside of the earth sciences (Kunz, 1979; Winick and Doniach, 1980; Koch, 1983, Winick, 1987). Synchrotron radiation is being used increasingly in the earth sciences, and reviews of some of these applications are available (Will, 1981; Calas et al., 1984; Waychunas and Brown, 1984; Fischer, 1984; Prewitt et al., 1987).

2.1.1 Storage rings and synchrotron radiation from bending magnets. Elementary particles, such as electrons or positrons, traveling with a velocity close to that of light along a curved trajectory emit electromagnetic radiation referred to as synchrotron radiation or synchrotron light. In practice, synchrotron radiation is produced using high energy storage rings under high vacuum (typically 10^{-8} torr), where quadrupole and sextupole magnets maintain high energy electrons (or positrons) in a planar orbit. The charged particles are injected into the ring from a linear accelerator (linac) or other particle accelerator at energies as high as several GeV, and these particles are further accelerated in the ring to energies as high as 15 GeV (Table 2). Such rings are not simple circles. Instead, they consist of curved sections of a small diameter (5-10 cm), non-magnetic beam tube which are joined together by straight sections; the curved sections are placed inside magnetic fields produced by "bending magnets" (Fig. 2-1). Synchrotron radiation is emitted tangential to these curved sections. Energy loss in these storage rings is inevitable because of particle-particle and particle-gas molecule interactions and production of synchrotron radiation. A radio frequency (rf) cavity, operating in

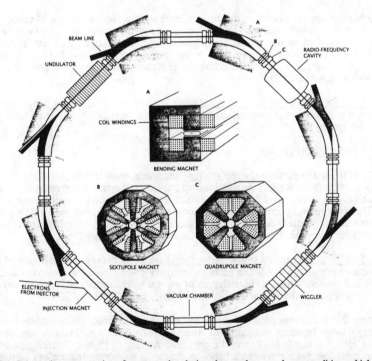

Figure 2-1. Schematic representation of a storage ring designed to produce synchrotron radition, which emerges from the ring into beam lines tangential to the ring. Insertion devices, such as wiggler and undulator magnets are placed into straight section of the ring to produce synchrotron light of greatly enhanced brilliance. (From Winick, 1987)

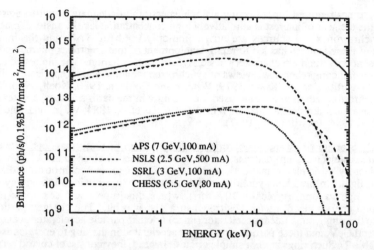

Figure 2-2. Intensity, in units of brilliance, versus energy (keV) curves for synchrotron radiation from several existing (CHESS, SSRL, NSLS) and one proposed (APS) synchrotron radiation sources. See Table 2 for details. (From Shenoy et al., 1988)

the 50-500 MHz range, is placed in the ring to increase the kinetic energy of electrons in the initial fill from the particle injector and to occasionally boost the energy of (or "kick") the particles by means of an oscillating electromagnetic field. Even so, a particular fill of electrons (or positrons) has a finite lifetime of 4-24 hours, during which a usable flux of radiation is produced. The maximum current carried by the present generation of storage rings is on the order of 100-300 mA (Table 2).

2.1.2 Light emitted by an accelerated particle. The energy lost by a particle undergoing circular motion is given, in the ultra-relativistic approximation (velocity of the particle similar to the velocity of light c) by

$$\Delta E = [4\pi e2/3R] \, [E/mc2]4 \quad , \tag{2-1}$$

where e, E and m are the particle charge, energy, and mass, respectively, and R is the curvature radius of the particle trajectory. This relationship implies that synchrotron radiation is produced more effectively by particles of low mass such as electrons or positrons than by heavier particles. The total intensity emitted is proportional to ΔE defined above for a single particle and to the current of the particle beam inside the storage ring. The radiated power is thus in the kilowatt range for a storage ring operating at 1 GeV and a few hundred mA.

2.1.3 Characteristics of synchrotron radiation. The main characteristics of synchrotron radiation important in XAS are its high intensity over a broad and continuous energy range, its pulsed time structure, its high degree of natural collimation, and its polarized character (Winick, 1980). Each of these characteristics will be discussed briefly.

Figure 2-2 compares plots of intensity, measured in units of brilliance (see below), versus energy from bending magnets for three existing U.S. synchrotron sources (SSRL, NSLS, and CHESS -- see Table 2) and the proposed Advanced Photon Source (see §6.7). The synchrotron radiation spectrum from a bending magnet is a smooth, featureless continuum. Its intensity versus energy distribution depends on the energy of the accelerated particles and on the curvature of their trajectory at the emitting point. In order to describe the spectral distribution, one must consider the critical energy ε_c, which divides the spectrum in two parts of equal emitted power. The parameter ε_c is defined (in keV) as

$$\varepsilon_c = 2.2\ E_e^3/R \quad , \tag{2-2}$$

where E_e (in GeV) is the stored electron energy and R (in m) is the bending radius at the emitting point. This spectral distribution is asymmetrical, as a result of two distinct limiting parameters: for photon energy $E > \varepsilon_c$, the intensity decreases exponentially as $\exp(-E/\varepsilon_c)$; for $E < \varepsilon_c$, it decreases smoothly as $(E/\varepsilon_c)^{2/3}$. The low energy limit is imposed, in part, by the

Table 2. A listing of synchrotron radiation sources, including normal operating conditions.
(From Wong, 1986)

Machine	Location	Energy (GeV)	Current (mA)	Bending radius (m)	Critical energy (KeV)	Remarks
PETRA	Hamburg, Germany	15	50	192	39.0	Possible future use for synchrotron radiation research
PEP	Stanford, CA, U.S.A.	15	50	165.5	45.2	Synchrotron radiation facility planned
		12	45	(23.6)	(163)	(From 17 kG wiggler)
CESR (Cornell)	Ithaca, NY, U.S.A.	8	50	32.5	35.0	Used parasitically
VEPP-4	Novosibirsk, U.S.S.R.	7	10	16.5	46.1	Initial operation at 4.5 GeV
		4.5		(18.6)	(10.9)	(From 8 kG wiggler)
DORIS	Hamburg, F.R.G.	5	50	12.1	22.9	Partly dedicated
		2.5	300		2.9	
SPEAR	Stanford, CA, U.S.A.	4.0	50	12.7	11.1	50% dedicated
		3.0	100		4.7	
		3.0		(5.5)	(10.8)	(From 18 kG wiggler)
SRS	Daresbury, Gt. Britain	2.0	500	5.55	3.2	Dedicated
				(1.33)	(13.3)	(For 50 kG wiggler)
VEPP-3	Novosibirsk, U.S.S.R.	2.25	100	6.15	4.2	Partly dedicated
				(2.14)	(11.8)	(From 35 kG wiggler)
DCI	Orsay, France	1.8	500	4.0	3.63	Partly dedicated
ADONE	Frascati, Italy	1.5	60	5.0	1.5	Partly dedicated
				(2.8)	(2.7)	(From 18 kG wiggler)
VEPP-2M	Novosibirsk, U.S.S.R.	0.67	100	1.22	0.54	Partly dedicated
ACO	Orsay, France	0.54	100	1.1	0.32	Dedicated
SOR Ring	Tokyo, Japan	0.40	250	1.1	0.13	Dedicated
SURF II	Washington, DC, U.S.A.	0.25	25	0.84	0.041	Dedicated
TANTALUS I	Stoughton, WI, U.S.A.	0.24	200	0.64	0.048	Dedicated
PTB	Braunschweig, F.R.G.	0.14	150	0.46	0.013	Dedicated
N-100	Kharkov, U.S.S.R.	0.10	25	0.50	0.004	
Photon factory	Tsukuba, Japan	2.5	500	8.33	4.16	Dedicated
				(1.67)	(20.5)	(For 50 kG wiggler)
NSLS	Brookhaven National Laboratory, NY, U.S.A.	2.5	500	6.88	5.01	Dedicated
				(1.67)	(20.5)	(For 50 kG wiggler)
BESSY	West Berlin, F.R.G.	0.80	500	1.83	0.62	Dedicated; industrial use planned
NSLS	Brookhaven National Laboratory, NY, U.S.A.	0.70	500	1.90	0.40	Dedicated
ETL	Electrotechnical Laboratory, Tsukuba, Japan	0.66	100	2	0.32	Dedicated
UVSOR	Institute of Molecular Science, Okatabi, Japan	0.60	500	2.2	0.22	Dedicated
MAX	Lund, Sweden	0.50	100	1.2	0.23	Dedicated
KURCHATOV	Moscow, U.S.S.R.	0.45		1.0	0.21	Dedicated

absorption by beryllium windows (typically 100-250 μm thick), which isolate the high vacuum ring from the experimental beam line.

Most experiments involving synchrotron radiation are concerned only with a narrow range of energy (or bandwidth); therefore, the <u>intensity</u> or <u>flux</u> of radiation is defined as the number of photons emitted per second within a given bandwidth ($\Delta E/E$ is typically 0.1%). Both high flux and small geometrical divergence at the source (or low electron beam emittance) result in the exceptional <u>brilliance</u> (flux emitted per unit of source area and per unit of solid angle) of synchrotron x-ray beams. The term brightness is also used on occasion as a measure of spectral intensity and is defined as flux emitted per unit of solid angle. A high brilliance or low emittance source would be desirable for spectroscopy or scattering studies requiring high spatial resolution (e.g., x-ray microprobe analysis, EXAFS studies of very small samples or of cations on surfaces, x-ray diffraction studies of very small single crystals or of extremely small sample volumes as in ultra-high pressure diamond anvil cell experiments), whereas a high emittance or low brilliance source would be most useful when a large sample must be irradiated (e.g., x-ray angiography studies of human hearts, x-ray lithography). In general, synchrotron radiation from a bending magnet is several orders of magnitude more brilliant than the bremsstrahlung emission from a modern, in-laboratory rotating anode x-ray generator. When multi-pole wiggler or undulator magnets are used in place of bending magnets (see §2.1.4), brilliances as high as 10^{17} to 10^{18} can be obtained (see Fig. 2-3). This enormous gain in brilliance over a large spectral range, relative to laboratory x-ray sources, has led to a number of new applications of synchrotron radiation, including EXAFS spectroscopy.

The operating parameters of some storage rings are reported in Table 2. A distinction is generally made between rings that produce "hard" x-rays (2,000-50,000 eV) and those that produce "soft" x-rays and vacuum ultraviolet (VUV) radiation (1-2000 eV). Generally speaking, those rings listed in Table 2 that operate at energies less than 1 GeV are used almost exclusively for soft x-ray -VUV studies.

The high natural collimation of synchrotron light (Fig. 2-4) arises from relativistic effects, resulting in a vertical beam divergence proportional to mc^2/E_e. The observed divergence values are usually small, on the order of a few tenths of a milliradian, and can be varied by adjusting the magnetic lattice of the ring. This high natural collimation and the high x-ray flux make it possible to study samples as small as 1mm², even when the experimental station is located at distances greater than 10 m from the storage ring. Horizontal divergence is related to the size of the window, as the emission occurs tangential to the orbit. Because of this feature, a synchrotron beam line can support several experimental stations simultaneously by dividing the beam into several portions using special mirrors. A beam about 30 mm wide is typically delivered to an individual experimental station at the Stanford Synchrotron Radiation Laboratory (SSRL) and at other synchrotron facilities. With such high natural collimation and small source size, it is possible to achieve spectral resolution of 1-2 eV in the 4-8 KeV range and even higher resolution at lower x-ray energies. Such resolution levels are essential for XANES spectroscopy, as will be discussed below in §2.2.2 and Part IV.

Two other properties of synchrotron light have importance in the context of XAS. Electrons in synchrotrons and storage rings are not accelerated individually but are grouped into bunches or buckets whose length is determined by the rf system. For example, the Stanford storage ring SPEAR operates at an rf frequency of 358 MHz which results in a pulse length of 0.2 to 0.4 nsec. This value has been lowered to as small as 50 psec on SPEAR under special conditions. Thus there is great potential for studies of transient phenomena because of the pulsed time structure of the ring electrons. Finally, the radiation is horizontally polarized in the orbital plane and elliptically polarized outside this plane. This source polarization allows polarization-dependent study of oriented single crystals. Such studies can provide important information on directional bonding (from polarized XANES spectroscopy) and on the distribution of interatomic distances in complex structures (from polarized EXAFS spectroscopy); these methods are just beginning to be used in the study of earth materials (Manceau et al., 1988).

2.1.4 Insertion devices. Modifications to existing storage rings and the design of new rings are driven in large part by the desire to increase the brightness of synchrotron light (see Fig. 2-4). This can be accomplished using "insertion devices" such as multi-pole wiggler or undulator magnets, which are inserted into the straight sections of storage rings between bending magnets. These devices employ two rows of magnets that create fields of alternating polarity perpendicular to the electron beam, which cause the electrons to have a sinusoidal trajectory. In the case of a multi-pole wiggler magnet, these additional bends of the electron beam produce additional flux, increasing the brightness of the emitted synchrotron light by a factor roughly equal to the number of magnetic poles, relative to a bending magnet of the same field strength. A change in the field strength of the wiggler or undulator magnet causes a change in the critical energy of the synchrotron light spectrum. The key difference between the now commonly used wiggler magnet and the less commonly used undulator magnet is the

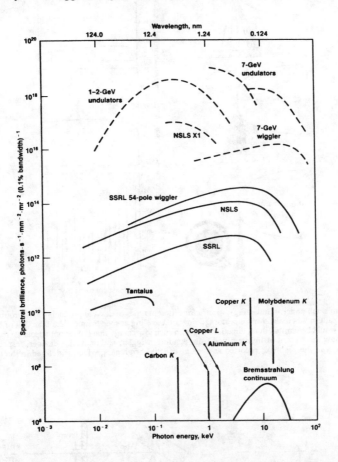

Figure 2-3. Comparison of spectral brilliances for a variety of synchrotron x-ray sources and laboratory x-ray sources. The range of brilliance for the laboratory sources reflects ranges possible for stationary-anode tubes and for rotating-anode tubes, with and without microfocussing. Tantalus is the synchrotron ring at the University of Wisconsin, SSRL is the Stanford Synchrotron Radiation Laboratory, and NSLS is the National Synchrotron Light Source at Brookhaven National Laboratory. The dashed lines represent calculated spectral brilliances for the proposed 7-GeV Advanced Photon Source at Argonne National Laboratory (7-GeV wiggler and 7-GeV undulators - which shows the envelope of the possible output spectrum), the proposed 1-2 GeV soft x-ray, vacuum ultraviolet Advanced Light Source at Lawrence Berkeley Laboratory (1-2-GeV undulators - which shows the envelope of the output spectrum), and proposed beam line X1 at Brookhaven National Laboratory. (From Lear and Weber, 1987)

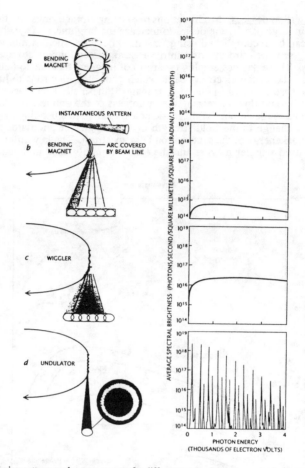

Figure 2-4. Emission patterns and energy spectra for different types of devices producing synchrotron radiation. (a) shows the pattern when electrons in a storage ring are traveling at speeds much less than that of light. In this case radiation would be emitted in all directions but with such low energy that it does not appear on the graph. (b), (c), and (d) show the pattern of light emitted from bending, wiggler, and undulator devices, respectively. The intensity of light in the graphs is reported as average spectral brightness, which is defined in the text. (From Winick, 1987)

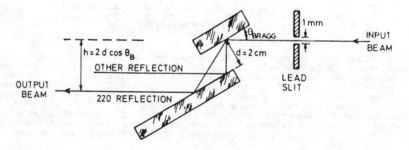

Figure 2-5. Principle of a channel-cut x-ray monochromator. (From Lagarde, 1985)

larger angular deflection of the electron beam produced by wigglers. The smaller deflection in undulators results in constructive and destructive interference between certain wavelengths causing certain energies to be enhanced greatly in brightness (see Fig. 2-4). The increase in brightness for those wavelengths that are enhanced, relative to a bending magnet of the same field strength, is approximately equal to the square of the number of oscillations. The drawback of undulators for XAS studies is the non-uniform brightness over the broad spectral range.

2.2 Experimental spectral resolution

Monochromatization of the polychromatic incident x-rays is essential for XAS measurements, and the monochromator must be tunable to allow step scanning in energy through the XANES and EXAFS region of an element's x-ray absorption spectrum. The energy resolution level required in XAS differs in the two main regions of the spectrum. The XANES region requires relatively high resolution because of the need to observe pre-edge and main-edge features separated by as little as 1 eV. In contrast, the EXAFS region does not require such high energy resolution, because EXAFS oscillations are usually well separated (by 4-5 eV at low energy and increasing to about 10 eV at higher energies). Factors that affect the resolution levels in the near- edge energy region are discussed in §2.2.2.

2.2.1 Monochromatization.
Whereas the divergent character of x-ray beams from conventional x-ray tubes makes it necessary to use curved crystal monochromators, the quasi-parallel geometry of synchrotron radiation is well-suited for the use of flat cut, single- crystal monochromators. The basic monochromator system for synchrotron XAS studies uses two parallel crystals, each cut parallel to the same (hkl) plane. The first crystal is used to monochromatize the incident beam (with typical bandwidth ($\Delta E/E$) on the order of 10^{-4}), and the second crystal is used to keep the outgoing beam parallel to the incident one. The energy scale is thus determined by the crystal rotation. Two types of monochromators are commonly used. In the classical "channel-cut" type (Fig. 2-5), the monochromator is cut from a monolithic single crystal along low-order (hkl) planes (111, 220, 331, 400) of silicon or germanium. The outgoing monochromatized beam is displaced from the incident beam by a distance that is angle-dependent. Thus the sample must be moved vertically in a fashion synchronous with the beam displacement, in order to keep the emerging beam on the sample. A much more elaborate monochromator movement mechanism is required to maintain the emerging beam at a constant height. The main disadvantage of channel-cut monochromators is that the higher energy harmonics present in the synchrotron beam are transmitted through the monochromator because of their diffraction as higher order reflections. These beam harmonics will strongly affect the measurement of absorption spectra because the higher energy photons will be transmitted without significant absorption by the sample. This problem results in an effective decrease in the amplitude of EXAFS oscillations. For that reason, two-crystal monochromators are being increasingly used which consist of separate crystals cut parallel to the same diffracting planes and mounted separately. With these two-crystal monochromators, harmonics in the transmitted beam can be eliminated by slightly "detuning" the second crystal. Such detuning takes advantage of the narrower Darwin reflection profiles of the high energy harmonic reflections relative to the fundamental reflection. Another solution for the harmonic problem consists of inserting a mirror before the monochromator. At grazing incidence, the x-ray beam will be totally reflected by the mirror (the refractive index of x-rays is slightly smaller than 1.0), and the angle at which total reflection occurs depends on the photon wavelength. The mirror is adjusted so that only the highest wavelength is reflected and used in the XAS experiment.

2.2.2 Limitations on experimental resolution.
Several major factors affect the ability to separate or resolve closely spaced features in an x-ray absorption edge spectrum, including source size, beam collimation, choice of monochromator crystal reflecting plane, width of the rocking curve of the monochromator crystal, and finite core-hole lifetime of the absorbing element. The first four effects are experimentally controllable, in part, but the fifth effect is an intrinsic property of the absorbing element and cannot be controlled. The resulting resolution level of the experiment is a convolution of these factors. As seen in §2.2.1, mono-chromatization of the x-ray beam is achieved by double reflection from two parallel single crystals. When monochromator crystals cut parallel to high-order (hkl) planes (400, 511, etc.) are used at large Bragg angles, a smaller line width results. However, the choice of

monochromator crystal is dictated by the energy range of the experiment, by the flux distribution as a function of energy, by the availability of single crystals of adequate size, perfection, and narrow rocking curve, and by their ablility to withstand a high x-ray flux without significant structural damage. Such damage can significantly degrade spectral resolution. Source size is the dominant experimental parameter in controlling resolution, once proper collimation is achieved. A typical beam divergence of 10^{-4} radians may be easily obtained using synchrotron radiation. This value of divergence results in a 1 eV width at 10 keV (K absorption edge of gallium), which is relatively large when other contributions are also considered, and must be lowered as much as possible by reducing divergence. Finally, intrinsic spectral resolution, ΔE, is defined by the width of the core hole of the absorbing element, which is related to the lifetime of the transition Δt by the Heisenberg relation, $\Delta E \Delta t \approx h$, h being Planck's constant. The core-hole lifetime increases regularly and rapidly with increasing atomic number Z for a given electronic level (according to a Z^4 law for the K-levels, Kostroun et al., 1971) (Fig. 2-6). This fact hinders all K-edge studies at energies higher than 15-18 keV. Elements with K-edge energies higher than these values must be studied using lower energy, outer-shell edges, such as one of the L edges. An example of this situation is the element silver with a K-edge at 25,512 eV and a core-hole width of 6.4 eV (Fig. 2-6). The Ag L_{III} level (3351 eV) has a core-hole width of only 2.2 eV. At the Fe K-edge (7,112 eV), the core-hole width (1.15 eV) limits the effective resolution to about 1.3 eV using Si (400) monochromator crystals. The stability of synchrotron radiation results in energy shifts of less than 0.1-0.2 eV due to ring variations with time. Thus careful choice of monochromator crystal and absorption edge to be studied and care in reducing beam divergence as much as possible can provide adequate resolution levels for many elements needed to obtain the bonding and structure information contained in XANES spectra. To a good approximation, the energy resolution possible with a monochromator crystal of spacing d is given by

$$\Delta E/E \;=\; 2.12 \; r_0 \{2d^2 /2\pi(1+n)^2\} \qquad , \tag{2-3}$$

where r_0 is the electron radius and n is the order of the harmonic present in the beam. It is clear from this relationship that low d-spacing planes (i.e., high-order hkl planes) are required for high energy resolution.

2.3 Detection schemes and experimental methods

Several different experimental methods have been developed for XAS studies, depending on the nature of the absorber and matrix. The two most commonly used methods involve transmission and fluorescence measurements. The transmission method directly measures absorption by the sample, and it is used typically for samples with high concentrations (> 2 wt %) of the absorber. The fluorescence method indirectly measures absorption by detecting the fluorescent x-ray yield from the front of the sample and is used for samples with lower absorber concentrations (< 2 wt %) and/or high matrix absorption. This method is also used for very small samples. The most commonly used detectors for either method are gas-filled ion chambers, with the choice of gas or gas mixture dictated by the energy of the beam and the amount of absorption of the incident beam desired. These simple detectors are well-suited for high signal level experiments (the typical flux delivered by storage rings is 10^{10} photons/sec/eV). Several other direct or indirect methods will also be discussed below.

2.4 Direct measurements of absorption

2.4.1 The transmission method. As with optical absorption or infrared spectroscopy performed in the transmission mode, it is common in transmission XAS to measure the intensity of the incident (or reference) beam (I_o) and that of the transmitted beam (I). A typical experimental arrangement is shown in Figure 2-7a. The first ion chamber is filled with an appropriate gas mixture to absorb about 20% of the incident beam. A second ion chamber is placed after the sample to detect the transmitted beam, and its gas mixture is chosen to absorb all of the beam transmitted by the sample. N_2, Ne, Ar (in increasing order of absorbance), and mixtures of these gases with He are commonly used in these ion chambers. The absorption by a sample of thickness x and absorption coefficient μ is related to the ratio of I_o and I as

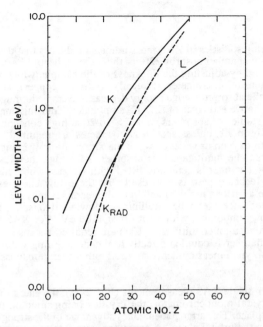

Figure 2-6. K- and L-level core-hole width (ΔE) as a function of atomic number. The dashed line shows the contribution of luminescence to the K width. (From Brown, 1980)

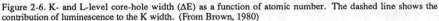

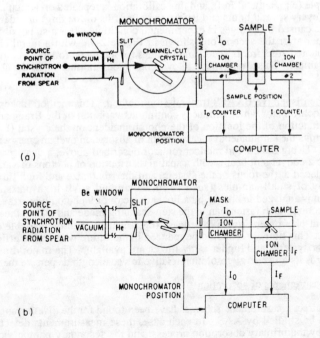

(a)

(b)

Figure 2-7. Experimental arrangement for XAS measurements. (a) Transmission; (b) Fluorescence detection. (From Wong, 1987)

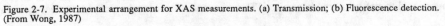

$$\mu x = ln \, (I_0/I) \quad . \tag{2-4}$$

Consideration of counting statistics in XAS measurements indicates that the optimum value of μx is 2.6 when the I_0 ion chamber absorbs 22% of the incident beam (Stern and Heald, 1983). This μx value is achieved by adjusting the thickness of the sample, which is generally on the order a few to several tens of micrometers (typically 10 μm for copper). The optimum sample thickness decreases as the absorption edge energy decreases. Sample homogeneity is important in order to obtain quantitative estimates of EXAFS amplitudes. Transmission measurements are well-adapted for XAS studies of absorbers with $Z > 16\text{-}20$ at high concentrations (> 2 wt %). The sample is generally in air, unless special atmospheres are required, such as when the sample might react with oxygen or water or when the x-rays are low enough in energy for air absorption of the x-rays to be significant. For absorbers with edge energies in the range 2000 to 4000 eV, the sample chamber is generally filled with He gas, which absorbs little in this energy range, and a He beam path is also used. At these low energies, matrix and self absorption can become a serious problem and very thin samples may be required. At lower energies (< 2 keV), absorbance due to the beryllium window, used to isolate the beam line from the storage ring, becomes important, with a cut-off at 1100 eV. For XAS studies at energies lower than 2 keV, high vacuum conditions ($> 10^{-8}$) are required because of air absorption; in this case the sample chamber is coupled directly to the storage ring without a window, and fluorescence or electron yield measurements are used rather than transmission measurements (see §2.5).

For transmission mode XAS measurements (Fig. 2-7a), the energy is varied step-wise by rotating the two-crystal monochromator, the crystals being oriented to give two Bragg reflections in a vertical plane. This arrangement takes advantage of the strong polarization of the synchrotron radiation in the orbital plane. A computer is used to control stepping motors that drive the monochromator crystals and, if necessary, the height of the table on which the sample sits. The voltages from the ion chambers are converted to frequencies at each monochromator step and counted. These I_0 and I values are stored on the computer disk, along with the monochromator and table positions. The energy scale is calibrated using the absorption edge of a reference sample (e.g., a metal foil), and this calibration is repeated on a regular basis during data collection (every several hours and before and after fills of the ring) in order to correct for any energy shifts caused by changes in experimental conditions. It is also possible to arrange a third ion chamber after the second one shown in Figure 2-7a, with a metal calibration foil permanently mounted at its front. The I_0 and I values for the metal foil are recorded along with those from the sample. This arrangement allows continuous monitoring of the absolute energy scale.

2.4.2 The dispersive EXAFS transmission method.

This method makes use of a curved crystal monochromator which produces continuous variation of the Bragg angle (and thus of energy) as a function of the location of the photon incident on the crystal (Fig. 2-8). This arrangement permits the simultaneous measurement of absorption over an energy range of some hundreds of eV. Using the curved monochromator described above, a quasi-parallel beam is transformed into a convergent beam, with a spatial distribution of photons of various energies. The sample is placed at the focus of the system, and the small size and high flux of the beam permit the study of small samples (such as in a diamond anvil high pressure cell). The transmitted beam is analyzed in energy by a linear detector or a photodiode array. This method makes data collection much faster than in the step-scanning mode and allows the possibility of studying transient phenomena down to at least 10^{-2} s using currently available detectors and brilliance levels. This time limit should be reduced in the future when more efficient position sensitive detector systems and higher x-ray fluxes are available. The major drawback of this method currently is its low energy resolution relative to wavelength dispersive methods.

2.5 Indirect measurements of absorption

Many indirect measurement schemes have been found for the diverse range of samples and elements to be studied by XAS. In each case, these measurements detect a secondary process caused by the primary absorption process, and the secondary photon yield is directly proportional to the primary absorption. These secondary processes include fluorescent x-ray

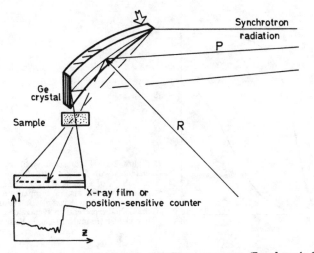

Figure 2-8. Experimental arrangement for dispersive EXAFS measurements. (From Lagarde, 1985)

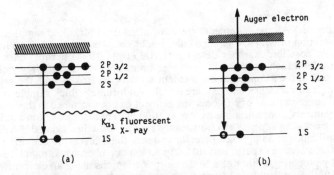

Figure 2-9. Illustration of two possible processes for de-excitation of of the 1s core hole produced by x-ray production of a photoelectron. (a) shows emission of a $K_{\alpha 1}$ fluorescent x-ray. (b) shows emission of an Auger electron. (From Stern and Heald, 1983)

yield, Auger electron yield, total electron yield, and optical emission. A schematic illustration of fluorescent and Auger electron emission processes is shown in Figure 2-9. Variations in experimental arrangements using these secondary yield processes can lead to unique methods for the study of surfaces.

2.5.1 The fluorescence method. Fluorescence x-ray emission arises from de-excitation of the 1s core hole generated by the absorption of an incident x-ray photon (Fig. 2-9). In the case of K-shell absorption, this fluorescence process is dominated by the production of K_α radiation whose energy is characteristic of the absorbing element and is less than that of the exciting x-ray photon. This yield, I_f, is directly proportional to the number of absorption events for dilute samples, and I_f/I_0 is proportional to μ. For samples in which the element of interest is concentrated, this proportionality is not the case, and transmission methods should be used instead, if possible. Fluorescent yield is not a very efficient process (Fig. 2-10; e.g., 0.34 efficiency for the Fe K level; Krause, 1979). Therefore, large solid angle detectors are needed and background must be reduced to give reasonable signal-to-noise levels. The major contributors to background in fluorescent yield measurements are Compton and elastic scattering, which are often much more intense than the signal from fluorescence emission. Fortunately, the energies of these scattered photons are higher than that of the fluorescence and can be conveniently discriminated by appropriate filters, by monochromators, or by direct energy discrimination using a solid state detector. When the element of interest has a K-edge energy below 2000 eV, the fluorescent yield is very small (Fig. 2-10 and Table 1), and electron detection techniques become more feasible than fluorescence techniques.

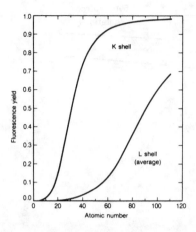

Figure 2-10. Fluorescent yield for K and L levels of elements with $5 \leq Z \leq 110$. These yields represent the probability of a core hole in the K or L shell being filled by a radiative process, in competition with non-radiative processes (such as Auger electron emission). The curve for the L shell represents an average of L_I, L_{II}, and L_{III} effective yields. (From Kortright, in Vaughan, 1986)

Fluorescence XAS measurements are most often used for dilute samples, samples with high matrix absorption or with low K-edge energies, or samples that are very small. A relatively efficient and low cost ion chamber detector (Lytle et al., 1984) permits easy collection of fluorescence data from such samples. In these measurements, fluorescent x-ray yield is measured from the same side of the sample that the incident x-ray beam strikes, with the sample mounted such that the angle of incidence and exit are equal. An infinitely thick sample is assumed, and even very small concentrations of a fluorescing element can be examined in this manner. For first row transition elements, concentrations as low as 1000 ppm of that element in a silicate matrix or at the 0.001 M level in an aqueous electrolyte solution can be studied using this XAS mode. With higher source brilliance and improved detector systems, these levels are certain to be reduced in the future. In fluorescence experiments utilizing an ion chamber, background discrimination is accomplished by preparing a filter of 3 to 6 absorption lengths using an element with a Z value just below that of the element being studied. A schematic drawing of the experimental arrangement for fluorescence detection is shown in Figure 2-7b.

A new technique related to fluorescence XAS is X-ray Excited Optical Luminescence (XEOL), where the detected signal comes from the rearrangement of outer electronic shells in fluorescing species such as in the rare earths (see Goulon et al., 1984, and references therein). Using XEOL, it is possible to separate contributions from absorbers in different sites in a structure, because the fluorescence optical emission from an element in different sites will occur at distinct wavelengths.

2.5.2 Surface EXAFS (SEXAFS) and the electron detection method.

Electrons are emitted by an element during the absorption process (photoelectrons or primary electrons), or during the relaxation process (secondary or Auger electrons) (see Fig. 2-9). It is possible, under high vacuum conditions, to detect the electrons emitted near the surface of a sample (from 20-50 Å depth). Appropriate detectors can be used to count the total number of electrons yielded (using a channeltron or spiraltron detector) or only those electrons in a certain energy range (cylindrical mirror analyzer (CMA) detector). In either case, this method requires the use of a high vacuum, low energy beam line, and thus its use is normally restricted to elements with absorption edges less than about 3000 eV. This method is well-suited for surface studies (Stohr, 1984; Citrin, 1986), and it is comparable in some respects to other electron techniques such as XPS. However, in contrast to x-ray photoelectron spectroscopy (XPS) or Auger electron spectroscopy (AES), SEXAFS spectroscopy yields the same type of direct bonding, interatomic distance, and coordination number information that XAS does. In this respect, XPS and AES are indirect methods.

2.5.3 Reflexion EXAFS (ReflEXAFS) method. When an x-ray beam impinges on a flat surface with a glancing angle smaller than the critical angle for total reflection (< 10 mrad for the K-edges of the 3d elements), the reflected beam intensity is related to the sample absorbance at the same energy (Kramers-Kronig relations). This method permits XAS study of the surface of a sample (the penetration at this angle is about 30 Å) without the experimental constraints (such as high vacuum levels) of the usual surface techniques. ReflEXAFS can yield the same type of information as XAS (Bosio et al., 1984; Heald et al., 1986), and it holds promise for the study of mineral and melt surfaces.

III. EXTENDED X-RAY ABSORPTION FINE STRUCTURE (EXAFS)

3.1 Introduction

Beyond the absorption edge, the absorbance decreases in the next few hundred eV and shows a modulation of the absorption coefficient (μx) referred to as Extended X-ray Absorption Fine Structure (EXAFS). The K-edge absorption spectrum of nickel in natural Ni-kerolite (= nepouite: $Ni_6Si_4O_{10}(OH)_8$) exhibits such a modulation. (Fig. 3-1). These oscillations have been known for more than a half-century and are called "Kronig oscillations". Only recently, however, has EXAFS been related by Stern et al. (1975) to the presence of well-defined atomic shells around the x-ray absorbing atom, which produce backscattering of the ejected photoelectrons and interference between ejected and backscattered photoelectrons. The general aspects of EXAFS spectroscopy have been presented in a number of review papers and books, including Teo and Joy (1980), Brown and Doniach (1980), Lee et al. (1981), Hayes and Boyce (1982), Stern and Heald (1983), Wong (1986), Petiau (1986), Teo (1986), and Koningsberger and Prins (1987). Applications of EXAFS spectroscopy to mineralogical and geochemical problems have been reviewed by Waychunas and Brown (1984), Brown et al. (1986b), and Calas et al. (1987). Only the basic elements of EXAFS spectroscopy will be presented here so that its uses and limitations can be understood.

As shown in Figure 3-2, the modulation is present only for compounds where local structural order exists. For a noble gas like neon, only monoatomic species are present and Ne gas does not exhibit an EXAFS spectrum. In contrast, the element Ne in its crystalline form exhibits a well-defined EXAFS spectrum. The absorption modulation beyond the edge has a period which increases with the photoelectron kinetic energy E, defined as

$$E = h\upsilon - E_0 \quad , \tag{3-1}$$

where $h\upsilon$ is the photon energy (h is Planck's constant and υ is frequency) and E_0 corresponds to the energy of the continuum. The ejected photoelectron has a large energy compared to the interatomic potential energies and, therefore, can be considered a free electron. The EXAFS oscillations are due to interferences between the ejected photoelectron and the backscattered photoelectron, which modulates the absorption coefficient because of the increase (constructive interference) or decrease (destructive interference) in the photoelectron's wave function in the region of the absorbing atomic orbital (initial state). EXAFS can be defined as the function $\chi(E)$ by

$$\chi(E) = \frac{\mu(E) - \mu_o(E)}{\mu_o(E)} \quad , \tag{3-2}$$

where $\mu(E)$ is the experimental absorption coefficient and $\mu_0(E)$ is the atomic contribution to the absorption coefficient. The difference $[\mu(E) - \mu_0(E)]$ represents the EXAFS oscillations subtracted from the background. Division by $\mu_0(E)$, which is proportional to the number of atoms per unit volume, normalizes the EXAFS data to a per atom basis.

In order to relate $\chi(E)$ to structural parameters, it is necessary to convert the energy E into the modulus of the wave vector k of the photoelectron using Equation (3-3):

$$k = \sqrt{\frac{2m}{h^2}(E-E_o)} \quad = \quad \left\{ 0.262\,(E-E_o) \right\}^{1/2} \quad , \tag{3-3}$$

in which m is the mass of the electron, h is Planck's constant, E is the kinetic energy of the photoelectron, and E_0 is the energy of the photoelectron at $k = 0$.

We will discuss briefly an outline of EXAFS theory, before presenting analysis methods and limitations of the technique. More detailed discussions of the theoretical background of EXAFS can be found in the excellent reviews (Hayes and Boyce, 1982; Stern and Heald, 1983).

3.2 Outline of EXAFS theory

A theoretical treatment of EXAFS oscillations relies on several simplifying assumptions:

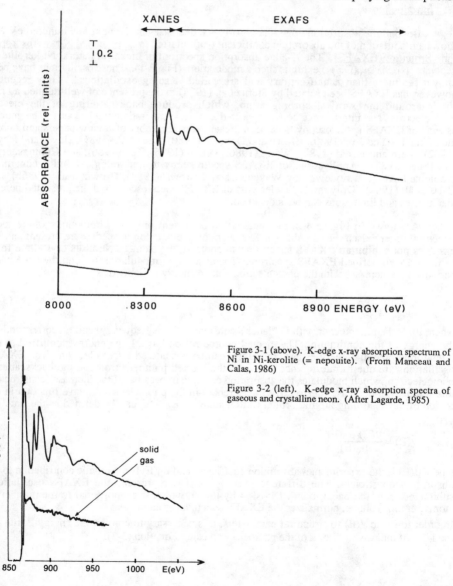

Figure 3-1 (above). K-edge x-ray absorption spectrum of Ni in Ni-kerolite (= nepouite). (From Manceau and Calas, 1986)

Figure 3-2 (left). K-edge x-ray absorption spectra of gaseous and crystalline neon. (After Lagarde, 1985)

(1) as outlined above, the photoelectron has a large enough kinetic energy such that it is assumed to be a free electron in the interatomic potential;

(2) the process producing the oscillations is assumed to be a single-electron process;

(3) only simple scattering is considered. The photoelectron is ejected by the central atom, which is modeled using a simplified plane wave approximation, and backscattered at π (i.e., 180°) by a neighboring atom. Multiple scattering is of low probability in this energy domain and is predominant only in the low-energy domain corresponding to the XANES region ($k < 3.0$).

3.2.1 Simplified formulation of EXAFS. The general expression of the photon absorption by an atom is given by Fermi's golden rule (Brown et al., 1977):

$$\mu \approx \sum_f |<i| \ \mathbf{H} \ |f>|^2 \ \delta(E_i - E_f - h\omega) \quad , \tag{3-4}$$

where $<i|$ is the initial state of energy E_i (e.g., the 1s level for a K edge), and $|f>$ is the final state of energy E_f (a p level for a K edge). $\mathbf{H}$ is the Hamiltonian which describes the electric dipolar interactions. The dipolar approximation for this transition is justified by the localized radial extent of the 1s orbital which is small compared to the wavelength of the radiation (1 Å or less).

In Equation (3-4), the modulation of μ is due only to the matrix element $|f>$. $<i|$ is a core level localized around the nucleus of the absorbing or central atom, which is independent of the energy of the incident photon. The density of states of the photoelectron is a monotonic function of the energy. A consequence of the localized character of the initial state $<i|$ is that it is possible to calculate the absorption coefficient of the central atom, μ_0, as

$$\mu_0 \approx |<i| \ r \ |f_0>|^2 \ \delta(E_{fo} - E_i - h\omega) \quad , \tag{3-5}$$

where $|f_0>$ is the atomic part of the final state, and r is a scaler potential approximating a uniform electromagnetic field; r is proportional to the distance z when the x-ray polarization is in the z direction. If there is an atom at distance R from the central atom, the final state will be perturbed by $|\delta f>$, producing the EXAFS. One of the main approximations used to derive a simplified formulation for EXAFS is that the outgoing photoelectron wave is considered to be a plane wave at the backscattering atom. The wavefunction has the form given by

$$(i \ e^{i \ [kR + \delta(k)]}) \ / \ 2kR \quad , \tag{3-6}$$

The central atom potential creates a phase shift $\delta(k)$ of the outgoing and the backscattered waves (i.e., $2\delta_a$). It depends on the nature of the atom as well as the value of k. The backscattering atom potential also produces a phase shift, $\delta_b(k)$, which is a part of the atomic diffusion function, $f(\pi,k)e^{\delta(k)}$, where $f(\pi,k)$ is a complex number with amplitude $|f(\pi,k)|$ and argument ϕ. The total phase shift then is given by

$$\delta(k) = 2\delta_a + \delta_b \quad , \tag{3-7}$$

where δ_a and δ_b are the phase shift contributions from the absorbing and backscattering atoms, respectively. If $\delta(k)$ were not present in the phtoelectron scattering process in condensed materials, the frequency of the EXAFS oscillation with respect to k would be a direct measure of R. The k dependency of $\delta(k)$ also contributes to the frequency. When $\delta(k)$ is expanded in powers of k [i.e., $\delta(k) = \delta_0 + \delta_1(k)$], the frequency is proportional to $R + \delta_1$, and the phase shift must be determined in order to obtain the correct value of R. The backscattered wave will have the following amplitude at the central atom:

$$(e^{i \, [kR + \delta(k)]}) \, / \, R \quad , \tag{3-8}$$

and the wavefunction of the final state at the central atom is then expressed as

$$|f\rangle = |f_0\rangle \left\{ 1 + \frac{i \, f(\pi,k)}{2kR^2} \, e^{2i \, (kR + \delta(k))} \right\} \quad , \tag{3-9}$$

If we take into account only the first order terms, the absorption coefficient is of the same form, and EXAFS is expressed as an interference phenomenon described by

$$\chi(k) = -\frac{|f(\pi,k)|}{kR^2} \, \sin\{2kR + \phi(k)\} \quad , \tag{3-10}$$

where $\phi(k) = 2\delta_a(k) + \delta_b(k)$.

The complete formulation of an EXAFS theory must take into account the damping of the photoelectron wave and disorder effects.

3.2.2 Damping processes and disorder effects.

These two effects cause EXAFS to have a limited energy and spatial range, and they constitute a limitation of the EXAFS method for structural studies. Inelastic scattering effects cause a damping of the photoelectron wave, resulting in a relatively short mean free path (typically < 6 Å) of the photoelectron, $\lambda(k)$, which is a function of the electron's energy. This damping is modeled by the factor $\exp[-2R/\lambda(k)]$. In the energy range of EXAFS, the variation of λ with energy is nearly linear. Thus, EXAFS is intrinsically a local method. The k dependence of $\lambda(k)$ is often expressed as $\lambda(k) = k/\Gamma$, where Γ depends only on the element studied and is of the order of 2Å^{-2}. The damping factor then becomes $\exp[-2\Gamma R/k]$.

EXAFS provides an almost instantaneous snap shot of the local structure, as the lifetime of an x-ray transition is of the order of 10^{-16} s. This value is small compared to molecular vibrations, $\approx 10^{-12}$ s. The measurement thus constitutes a spatial integration of the instantaneous positions of the absorber and backscattering atoms. This integration includes all factors that affect interatomic distance distributions, and these can be discussed in terms of static and thermal (or vibrational) disorder. Static disorder refers to the distribution of interatomic distances within a given shell, e.g., first nearest neighbors. Static disorder is considered large if these distances are spread over a set of values comparable to EXAFS resolution (≈ 0.15 Å); thus many minerals have cation sites that exhibit a high degree of static disorder for EXAFS analysis. This concept is important since each interatomic distance will give rise to an EXAFS frequency. If the interatomic distances and hence frequencies are over a particular range, they will tend to cancel and remove amplitude from the EXAFS oscillations. If the range of interatomic distances for a cation is small compared to the EXAFS resolution, there is mainly addition of the frequency components, and the average frequency is a reliable indicator of the average interatomic distance. If the range is large compared to the EXAFS resolution, then separate frequencies can be resolved and separate interatomic distances determined from EXAFS analysis.

Within the resolution range where static disorder is a problem, the exact details of the interatomic distance distribution are critical. If this distribution is approximately Gaussian (i.e., binomial distribution), then the average frequency is still a good estimate of the average interatomic distance. If the distribution varies significantly from Gaussian, however, information will be lost from the EXAFS spectrum. Thermal or vibrational disorder is effectively the same as static disorder for a rapid sampling spectroscopy like EXAFS. In the case of thermal disorder, in contrast, the Gaussian approximation for a given bond is probably good except at high enough temperatures where anharmonicity may begin to occur. Thus for EXAFS analysis, the ideal distribution in a given atomic shell of interatomic distances or of vibrations is a narrow Gaussian-like distribution of Gaussian-broadened bond distances. This convolutes to a Gaussian distribution function.

The usual term in the EXAFS amplitude which models the effect of disorder is the Debye-Waller factor derived in the harmonic approximation. This form assumes harmonic vibrations and a Gaussian distribution of interatomic distances. The term is similar to that used in x-ray diffraction analysis, but differs in detail because the relevant disorder is not in terms of the deviations of the atoms from their mean positions. Instead, the EXAFS Debye-Waller factor measures the mean square deviation of the absorber-scatter interatomic distance from its average length (i.e., correlated atomic motion with displacements parallel to the interatomic bond).

The Debye-Waller factor in EXAFS analysis, which is used to simulate disorder effects as defined above, is made up of static and thermal (or vibrational) contributions:

$$\sigma^2 = \sigma^2_{static} + \sigma^2_{vibrational} \quad . \tag{3-11}$$

In principle, it is possible to determine the magnitude of these contributions by temperature dependent EXAFS studies. σ_{static} is related to the symmetric Gaussian pair distribution function:

$$g(r) = \sigma^{-1} (2\pi)^{-1/2} \exp[-(r - r_0)^2 / 2\sigma^2] \quad , \tag{3-12}$$

where $(r - r_0)$ is the deviation from the mean distance r_0. For individual bonds, σ_{static} is related to the root-mean-square deviation given by

$$\sigma_{static} \approx \left\{ \sum_{j=1}^{n} (r_j^2 - r_0)^2 / n \right\}^{1/2} \quad , \tag{3-13}$$

In the harmonic approximation, the vibrational contribution to the Debye-Waller factor is given by

$$\sigma^2_{vibrational} = [h / 8\pi^2 m_r \upsilon] \; Coth \, [h\upsilon / 2kT] \quad , \tag{3-14}$$

where m_r is the reduced mass, T is temperature in K, k is the Boltzman constant, and υ is the vibrational frequency. When carrying out EXAFS studies of atoms with weak bonds (i.e., large υ), it may be necessary to take data at low temperatures in order to reduce $\sigma_{vibratonal}$. For systems with small disorder ($\sigma \leq 0.1$ Å), it is possible to represent disorder in the EXAFS expression by $\exp(-2\sigma^2 k^2)$, where the pair distribution function $g(r)$ is assumed to be a symmetric Gaussian as in Equation (3-12). When static disorder is large ($\sigma \geq 0.1$ Å) due to an asymmetric pair distribution function or to an anharmonic vibrational potential, this harmonic approximation is not valid (Eisenberger and Brown, 1979), and an asymmetric pair distribution function must be used. Detailed discussion of ways to correct for these effects, including the use of a simple Einstein model of lattice vibrations, can be found in Stern and Heald (1983). Large disorder effects can lead to a reduction of EXAFS amplitude and of derived coordination numbers (Eisenberger and Lengeler, 1980). These effects can also cause an apparent contraction in the nearest-neighbor distances as large as 0.15 Å (Eisenberger and Brown, 1979; Crozier and Seary, 1980, de Crescenzi et al., 1983).

An example of the effects of thermal or vibrational disorder on EXAFS spectra is shown in Figure 3-3 for crystalline fayalite (Fe_2SiO_4) at temperatures of 90 K and 300 K. The amplitude of EXAFS oscillations is significantly reduced over the entire k-range with increasing temperature. At higher k-values ($k > 9$ Å^{-1}), significant loss of real EXAFS features occurs at high temperature, and this loss affects the level of resolution and accuracy possible in interatomic distance and coordination number determination. At temperatures higher than 300 K, loss of amplitude and phase information at high k-values becomes a serious problem and second and higher neighbor shells of atoms around an absorber may not be observable. However, in cases where relatively strongly-bound absorbers with coordination numbers of six or less (e.g., Fe in magnesio-wüstite, $Mg_{0.9}Fe_{0.1}O$) reliable first-neighbor distances can be obtained even at temperatures of 1000 K or more (Waychunas et al., 1986b, 1988). An example of the effect of large static disorder is also shown in Figure 3-3, using the EXAFS spectrum of fayalite glass at 300 K (plotted on a different k scale than the crystalline fayalite spectra; also, the χ function of the glass is multiplied by k rather than k^3). In this case, a single, damped, sinusoidal wave is observed with a small, additional peak at $k = 5.5$ Å^{-1} (perhaps due to second neighbor contributions). In contrast, the EXAFS spectra for crystalline fayalite show the contributions from second and higher neighbors. The severe loss of EXAFS information in fayalite glass at $k > 7$ Å^{-1} is caused by the large degree of positional or static disorder in second and higher shells around Fe. This static disorder effect limits the information content of EXAFS spectra of glasses. In spite of this difficulty, distance and coordination information can be obtained for a wide range of elements in silicate glasses and melts (Brown et al., 1986b; Calas et al., 1987; Waychunas et al., 1986b, 1988).

3.2.3 Formulation of the EXAFS modulations. Based on the assumptions made in §3.2, the general formulation of EXAFS is given by:

$$\chi(k) = -1/k \sum_j A_j(k) \sin [2kR_j + \phi_j(k)] \quad . \tag{3-15}$$

The summation is made over the j shells of backscatterers surrounding the central absorber atom. The argument of the sine gives the phase difference between outgoing and backscattered portions of the photoelectron wave. Thus the EXAFS function is a superposition of the photoelectron scattering contributions by atoms in a number of coordination shells, where j refers to the jth shell. R_j is the average distance from the absorbing atom to the backscatterering atom(s) in the jth coordination shell. ϕ_j (k) is the phase shift discussed in §3.2.1. A_j is the amplitude function for the jth shell and is defined as

$$A_j(k) = (N_j/R_j^2) f_j(\pi,k) S_0^2(k) \exp(-2R_j/\lambda) \exp(-2\sigma_j^2 k^2) \quad . \tag{3-16}$$

where N_j is the average number of backscattering atoms in the jth shell, and $f_j(\pi,k)$ is the backscattering amplitude characteristic of a particular type of coordinating atom and dependent on k. The factor R_j^{-2} reflects the product of the amplitudes of the outgoing and backscattered waves, both of which drop off as R_j^{-1} because of their spherical nature. $S_0^2(k)$ is an amplitude reduction factor due to many-body relaxation effects of the absorbing atom and multielectron excitations such as shake-up and shake-off processes. $S_0^2(k)$ ranges in value between 0.7 and 0.8 in typical cases (Carlson, 1975). The term $\exp(-2R_j/\lambda)$ accounts for inelastic losses in the scattering process, with λ being the photoelectron mean free path. The last term containing the Debye-Waller factor σ^2 was discussed in §3.2.2. It should be emphasized that the amplitude function, $A_j(k)$, depends primarily on the type of backscattering atoms, whereas the phase shift function, $\phi_j(k)$, contains contributions from both absorber and backscatterers:

$$\phi_j(k) = 2\delta(k)_{absorber} + \delta(k)_{backscatterer} - a\pi \quad , \tag{3-17}$$

where $a = 1$ for K and L_I absorption edges and $a = 2$ or 0 for L_{II} and L_{III} edges.

When only a single shell of ligands surrounds an absorber at one average distance without much static disorder in the first shell, the EXAFS spectrum is approximated by a damped sinusoid. This is the typical situation encountered around an element in a glass or in an aqueous solution (Fig. 3-3c). However, in high symmetry crystals, the EXAFS spectrum often has a much more complex shape due to the undamped contributions of many distances (Fig. 3-3a,b). Equation (3-15) provides an adequate model for EXAFS in most cases. Using this equation, it is possible to derive structural parameters from an unknown substance once the other parameters are known. Among these parameters, the total phase shift, $\phi(k)$, and amplitude, $A_j(k)$, functions

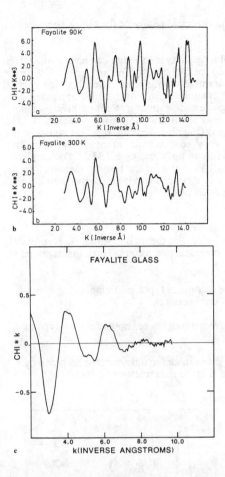

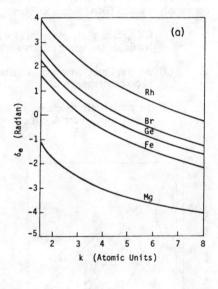

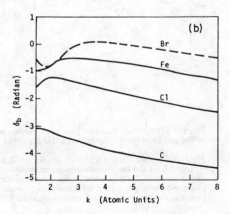

Figure 3-3 (above). Comparison of background-subtracted Fe K-EXAFS spectra of (a) crystalline fayalite (Fe_2SiO_4) at 90 K, (b) fayalite at 300 K (From Waychunas et al., 1986b), and (c) fayalite composition glass at 300 K (From Calas and Petiau, 1983a). The chi function for crystalline fayalite has been weighted by k^3, whereas that for fayalite composition glass has been weighted by k.

Figure 3-4 (left). Calculated phase shifts for selected elements. (a) shows phase shifts introduced by the central absorber atom; (b) shows phase shifts for the neighboring backscattering atom. (From Stern and Heald, 1983)

are the most subject to variations depending on the chemical nature of the atoms (both central absorber and backscatterer) and the local structure.

Appropriate values for phase shift and amplitude parameters either can be extracted from known reference or model compounds or calculated for free atoms (Teo and Lee, 1979) using some approximations. The variation of these parameters with atomic number, for several elements, is shown in Figures 3-4 and 3-5. The magnitude of phase shifts increases with increasing Z, and this fact makes it possible to recognize backscattering atoms of different atomic numbers (e.g., oxygen vs. chlorine; Fig. 3-6) in known and unknown structures. The amplitude function exhibits more complex variations (Fig. 3-5). For the light elements, it is a monotonously decreasing function. For the very low atomic number elements, the backscattering amplitude is so low that these atoms cannot be "seen" by EXAFS. The amplitude functions are larger in magnitude and more complex in shape for the heavier elements. In the first row transition elements, these functions have a pronounced maximum at medium k-range, but several maxima can be seen in the case of the heaviest elements.

3.3 Analysis of the EXAFS spectrum

As the EXAFS signal represents only a few percent of the total absorbance, careful data analysis is necessary to derive reliable structural or chemical parameters. Three assumptions are commonly made in EXAFS data analysis:

(1) EXAFS is a simple sum of waves due to various types of neighboring atoms (multiple-scattering effects are relatively unimportant);

(2) the amplitude function is transferable from model to unknown for each type of backscattering atom;

(3) the phase shift function is transferable from model to unknown for each pair of atoms A-B, where A is the absorber and B is the backscatterer.

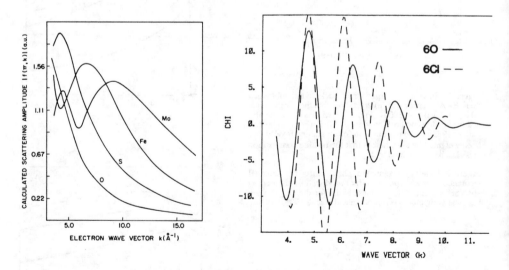

Figure 3-5 (left). Calculated backscattered amplitudes for selected elements using theoretical amplitude functions of Teo and Lee (1979). (From Doniach et al., 1980)

Figure 3-6 (right). Theoretical chi functions for Zr surrounded by six nearest-neighbor oxygens (d[Zr-O] = 2.10 Å) (solid line) and six nearest-neighbor chlorines (d[Zr-Cl] = 2.52 Å) (dashed line). The EXAFS spectra were calculated using the theoretical phase and amplitude functions of Teo and Lee (1979). The large differences in phase and amplitude permit differentiation between oxygen and chlorine ligands during fitting of an analytical expression of the chi function to the experimental EXAFS spectrum.

Numerous published EXAFS analyses for a wide range of materials indicate that these assumptions are generally valid.

The steps in EXAFS data analysis are relatively standardized; some of them are shown schematically in Figure 3-7. They consist of (1) correcting for spectrometer shifts (energy scale calibration); (2) deglitching; (3) pre-edge and post-edge background removal; (4) edge normalization; (5) extraction of the EXAFS signal $\chi(k)$; (6) establishing a value for E_0; (7) Fourier transformation of $\chi(k)$, which produces a radial distribution function (rdf); (8) inverse Fourier

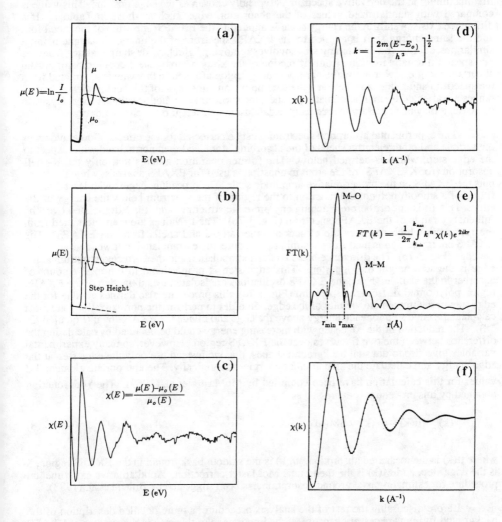

Figure 3-7. Steps in the analysis of an EXAFS spectrum: (a) raw spectrum showing the smooth background absorption; (b) fit of backgrounds to pre-edge and EXAFS regions and definition of step height or edge step; (c) the background-subtracted EXAFS plotted as a function of energy; (d) the background-subtracted EXAFS plotted as a function of wave vector k; (e) Fourier transform of (d) resulting in a radial distribution function made up of pair correlations involving the absorber. A Fourier filter window is defined by r_{min} and r_{max}. The two major peaks in the rdf correspond to absorber (M) first-neighbor oxygen distance and M-M distance in Å; (f) back Fourier transform of the M-O peak in (e) over the filter range r_{min} to r_{max}, resulting in a chi function (solid line). The dots represent a least-squares fit of an analytical chi function to the back-transformed chi function from the experimental data.

transformation of a portion of the rdf to isolate the EXAFS contribution from a selected region in distance space; and (9) least-squares fitting of an analytical EXAFS expression to the EXAFS function produced in step (7), which yields average distance, coordination number, and Debye-Waller factor for the absorber. Each of these steps is discussed below. Additional discussion of Fourier analysis of EXAFS spectra can be found in the Chapter 3 (this volume).

3.3.1 Extraction of the EXAFS modulation.

The first step in EXAFS data analysis is calibration of the energy scale of the spectrum. This is normally accomplished by taking the first derivative of the EXAFS spectrum of a metal "calibration" foil of the absorber of interest. The first maximum in the derivative spectrum is typically chosen as the edge position. This value is compared with standardized values of the absorption edge, such as those in Table 1. If a difference exists, a shift in the energy scale is applied to the EXAFS spectrum of the metal foil, and the same energy shift is applied to the EXAFS spectrum of the unknown sample using a similar procedure. The deglitching step involves removing "glitches" or sharp features from the raw spectrum usually resulting from diffraction of the x-ray beam by the monochromator crystal at certain energies. Normally, the narrow energy region affected in this way is subtracted from the spectrum and a line is fit between the data points on either side of this region. When large k ranges are affected by monochromator glitches, the spectrum may not be suitable for analysis. In this case, it may be necessary to recollect the data using a different monochromator crystal.

The experimental absorption spectrum must be corrected for the effects of other atoms in the sample and other electron shells of the absorbing atom, and then normalized with respect to the edge step, which is defined below. The former operation assures that only the K-shell absorption (for K-EXAFS) of the atom in question is used for EXAFS analysis, while the latter puts the EXAFS on the proper relative amplitude scale. The procedure begins with least-squares fitting of a smooth polynomial function to the region of the spectrum below the energy of the edge. This fit is cut off before reaching any pre-edge structure, and it is extrapolated over the full energy range of the EXAFS spectrum (Fig. 3-7b). The fit values are then subtracted from the raw data, which removes the effects of other atoms and shells from the EXAFS. The EXAFS energy region is next fit by a similar procedure and extrapolated to lower energies over the edge (Fig. 3-7b). The difference between this extrapolation and the corrected spectrum at the edge is the step height or edge step. This edge step is proportional to the total amount of absorber in the sample. Next, the EXAFS oscillations are isolated by subtraction of the EXAFS region polynomial fit from the spectrum (Fig. 3-7c); this procedure also defines a value for the edge step in the immediate vicinity of the edge. In order to perform the normalization, one must calculate the edge step for each energy above the edge and divide the EXAFS by it (Equation (3-18)). The reduction of the edge step with increasing energy could be obtained by calculating the difference between the two fit curves over the EXAFS region. However, so many experimental variables give rise to unusual background shapes, that we instead use only the edge step at the edge energy and calculate the step at other energies theoretically. The appropriate elemental μ values for this calculation have been compiled by McMaster et al. (1969). The $\chi(k)$ function obtained by this procedure is expressed as

$$\chi(k) = [\mu(k) - \mu_0(k)] / SM(k) \quad , \tag{3-18}$$

where $\mu(k)$ is the measured absorption, $\mu_0(k)$ is the smooth background in the EXAFS region, S is the edge step, and M(k) is the theoretical McMaster correction. An alternative experimental procedure for estimating background absorption has been suggested by Boland et al. (1983).

Before discussing the rest of the analysis procedure, a more detailed description of the background fitting is necessary because of the importance of this step in the success of EXAFS analysis. The fitting is done with a polynomial spline function, usually divided into three or four energy regions. Each region generates a separate polynomial curve in any desired order, and the "tie" points at the region boundaries are constrained to match and have continuous first derivatives by the use of Lagrange multipliers in the error equations. Typically three to five such spline functions are adequate for fitting the EXAFS background for data extending 1000 eV above the absorption edge. When the number of fit segments is too small, the background is not well defined; when the number is too large, the background may fit the EXAFS oscillations, especially at low energies, which incorrectly reduces EXAFS amplitude. The total calculated background may be a very complex function, but the non-linear least-squares proceeds in typical fashion via full matrix inversion. Cook and Sayers (1981) have suggested criteria for proper removal of background absorption from EXAFS spectra.

The selection of fitting regions, order of polynomials, and choice of weights is a complex procedure, usually done iteratively. For each good fit, the EXAFS analysis is carried through to the radial distribution function (rdf) which is produced by forward Fourier transformation of the $\chi(k)$ function (see §3.3.2). The quality of the rdf can be evaluated by examination of the low R region. Background fit errors are usually manifested by low frequency oscillations either missed or induced by the polynomial fitting. Such errors lead to spurious low R peaks in the rdf, which interfere with analysis of the first neighbor shell.

The last step in data manipulation before Fourier transformation of the $\chi(k)$ function is the setting of the k-range. The energy in eV above the absorption edge threshold (E_0) is related to the scattering vector by Equation (3-3). E_0 is experimentally located by the first maximum in the first derivative spectrum of the absorption edge as described in the energy calibration step above. If E_0 is chosen incorrectly (as is often the case), we have $E_0' = E_0 + \Delta E_0$. The momentum k is then redefined as

$$k' = (k^2 - 0.262\ \Delta E_0)^{1/2} \quad , \tag{3-19}$$

and the phase shift function is modified as

$$\phi'(k') = \phi(k) - 2r\ (k' - k) \approx \phi(k) + 0.262r\ (\Delta E_0)/k \tag{3-20}$$

for $0.262(\Delta E_0) \ll k^2$, r is the distance to the neighbor involved. Equation (3-20) indicates that the difference $\Delta\phi(k) = \phi'(k') - \phi(k)$ decreases with increasing k, indicating that phase shifts are more sensitive to a change in E_0 at small k than at large k. The error in k from incorrectly setting E_0 is given by:

$$\Delta k = (\partial k\ /\ \partial E_0)_E\ \Delta E \quad , \tag{3-21}$$

where the derivative is given by

$$(\partial k\ /\ \partial E_0)_E\ =\ -\ 0.131\ /\ [0.262\ (E - E_0)]^{1/2} \quad , \tag{3-22}$$

A 5.0 eV error in the choice of E_0 results in errors in k ranging from –1.28 (at k = 0.51 Å^{-1}) to –0.230 (at k = 2.80 Å^{-1}) to -0.065 (at k = 10.24 Å^{-1}); a 10 eV error in E_0 results in double these errors in k. This effect is particularly large at low k values, but it is unimportant beyond k ≈ 3. However, obvious phase shifts occur due to this "E_0 problem", which can affect EXAFS analysis. This problem is generally handled by analysis of a well-characterized, crystalline model compound, where E_0 is an adjustable parameter.

Lee and Beni (1977) suggested an alternate procedure for adjusting E_0 to an optimum value. It involves shifting E_0 until the maxima of the imaginary and real (modulus or absolute value) parts of the Fourier transform are coincident. This procedure assures that the absolute phase is correctly given. Shifts in E_0 by as much as 10 to 15 eV are not uncommon using this procedure. It is important to point out that by adjusting E_0, it is not possible to produce an artificially good fit with the wrong distance r, provided that the phase shift is accurate for large k. This is true because changes in E_0 produce a change in phase that decreases as 1/k according to Equation (3-20), which is opposite to a shift in distance which causes a change in ϕ that increases linearly with k. A different approach for estimating E_0 has been suggested by Boland et al. (1983). It relies on the damping of EXAFS amplitude resulting from convolution with Gaussian functions of different widths and can provide a unique estimate of the threshold energy.

3.3.2 Fourier transform analysis. The most accurate method for deriving structural parameters from EXAFS oscillations is Fourier transformation of the experimental spectrum with a function of the type given by:

$$\text{F.T.} \left[k^n \cdot \chi(k)\right] = (2\pi)^{-1/2} \int_{k_{min}}^{k_{max}} k^n \cdot \chi(k) \cdot W(k) \cdot e^{2ikR} \, dk \quad , \tag{3-23}$$

where $W(k)$ is the window function of the Fourier transform. k^n is a weighting function used to compensate for amplitude reduction as a function of k. Values of n of 1, 2, and 3 have been suggested by Teo and Lee (1979) for backscatterers with $Z > 57$ (lathanum), $36 < Z < 57$, and $Z < 36$ (krypton), respectively. Oxygen backscatterers, for example, would require a weight of k^3.

Notice that the Fourier transform omits $\phi_j(k)$, the phase shift, from the exponential term, which is typically done for simplicity. Therefore, Fourier transforms of $f(k)$ functions computed in this fashion are uncorrected for phase shift, and each peak in the transform is shifted by δ_1 (or α) relative to the correct distance. The phase shift correction is commonly made during the fitting procedure described in §3.3.3. One commonly used window function is a Hanning function (Bingham et al., 1967), which is defined as

$$W(k) = 1/2 \left\{ 1 - \cos 2\pi \left[(k - k_{min}) / (k_{max} - k_{min}) \right] \right\} \quad . \tag{3-24}$$

$W(k)$ is equal to 0 at $k = k_{min}$ and $k = k_{max}$ from Equation (3-24). In the data analysis shown in Figure 3-7, this window function was applied to the first and last 10% of the normalized data.

Equation (3-23) transforms from k-space (or frequency space) to real (or distance) space and produces a type of pair correlation function, with peaks corresponding to average absorber-backscatterer distances (Fig. 3-7e). However, each peak in the pair correlation function is shifted from the true distance by an amount α, which is the change in distance due to phase shift (see §3.2.1). The value of α can range from about 0.2 to 0.5 Å, but is relatively constant for a given absorber-backscatterer pair. The parameters that must be specified for this Fourier transform are the shape and dimensions of the window function and the value of n. The widths of the window and of the peaks in the Fourier transform are inversely proportional according to the fundamental equations relating direct and reciprocal spaces. Thus it is best to use as large a window (i.e., k-space range) as possible. However, the lower k-space limit is generally taken above 3 Å^{-1}, because the single-scattering approximation is not valid at low wave vector values, where multiple scattering dominates. EXAFS oscillations may not be visible at values of $k > 10$ to 15 Å^{-1}, depending on the type of material. The window function minimizes parasitic oscillations or ripples of the Fourier transform without eliminating the distance information at low and high k.

The absolute value (modulus or real part) of the Fourier transform should ideally exhibit each atomic shell surrounding the central absorber as a delta function. The actual distance distribution and the limited reciprocal space studied broaden the peaks. The peaks are located at a distance equal to the actual distance modified by the phase shift, as expressed in Equation (3-15). The quasi-linear dependence of phase shifts on k implies that the difference between apparent and real distances is nearly constant for the same backscattering atom for all interatomic distances.

3.3.3 Curve-fitting analysis. The last step in EXAFS data analysis is the filtering of information from the various shells of ligands surrounding a central absorber. This is accomplished by inverse Fourier transformation of the rdf over the R range of interest (referred to as Fourier-filtering) in order to obtain the structure-dependent parameters relative to this shell (Figs. 3-7e and 3-7f). A peak in the Fourier transform is isolated by an appropriate window, and the inverse Fourier transform of this peak gives the contribution of this isolated shell to the EXAFS. Equation (3-15), applied to the jth shell, simulates a partial EXAFS spectrum that would arise from only the jth shell and central absorber. The experimental, filtered spectrum (Fig. 3-7f) is compared to the one theoretically simulated using an analytical expression for EXAFS. Values of the various parameters are adjusted in the theoretically simulated EXAFS spectrum (by least-squares fitting) until an acceptable fit with the experimental spectrum is achieved. This procedure is discussed in more detail below.

Disorder effects can be determined, in a relative fashion, by comparison with reference compounds through Equation (3-25):

$$\ln\left[\frac{\chi_j^A(k)}{\chi_j^B(k)}\right] = \ln\left[\frac{N_j^A}{N_j^B}\left(\frac{R_j^B}{R_j^A}\right)^2\right] + 2\,(\,\sigma_{j,A}^2 - \sigma_{j,B}^2\,)\,k^2$$

The ratio between the partial EXAFS from the reference compound and the unknown must be linear in k^2. The slope gives the relative disorder, and the ordinate intercept at the origin gives the coordination ratio.

Finally, characterization of the amplitude function in compounds where the coordination numbers are known makes it possible to characterize the chemical nature of the atoms in the jth shell. Estimation of phase and amplitude functions is commonly done by means of least-squares fitting of the $\chi(k)$ function of well-characterized model compounds to a theoretical EXAFS function with adjustable phase and amplitude parameters. Once the amplitude and phase for a given absorber-backscatterer pair in the model compound have been calibrated in this way, the unknown can be analyzed. Determination of average interatomic distance, coordination number, and Debye-Waller factor for the jth shell about an absorber in an unknown material are made by least-squares fitting of the Fourier-filtered $\chi(k)$ function of the unknown to a theoretical EXAFS function. These steps are discussed below.

In the least-squares EXAFS fitting procedure, the amplitude function can be parameterized using a simple sum of Lorentzians as in Equation (3-26):

$$A(k) = \sum_i H_i / [1 + W_i^2\,(k - P_i)^2]\quad, \tag{3-26}$$

where H is the peak height, 2/W is the peak width, and P is the peak position in k space. For backscatterers of $Z < 36$, $36 < Z < 57$, and $57 < Z < 86$, one, two, and three Lorentzians are needed for the weighting schemes of $k^3\chi(k)$, $k^2\chi(k)$, and $k\chi(k)$, respectively. Teo et al. (1977) have provided values for H, W, and P from fits of theoretical backscattering amplitudes for a number of elements. The phase shift functions can be parameterized using an equation such as

$$\varphi(k) = p_0 + p_1 k + p_2 k^2 + p_3/k^3\quad. \tag{3-27}$$

In theory, both absorber and backscatterer have separate phase shift functions, with the absorber phase function fit adequately using a quadratic form of Equation (3-27). Lee et al. (1977) reported values for the coefficients of Equation (3-27) that are useful as starting values in fitting phases to experimental EXAFS spectra of model compounds. The backscatterer phase function requires a third order function like (3-27) when $Z < 36$, or a more complicated form when $Z > 36$. However, in practice the combined phase $\phi(k) = \phi^a(k) + \phi^b(k)$ can be parameterized adequately by (3-27) because $\phi^a(k)$ has a stronger k dependence than $\phi^b(k)$.

Using these parameterized functions in an analytical expression for $\chi(k)$, the EXAFS least-squares fitting procedure consists of the following steps:

(1) EXAFS spectra from a well-characterized, crystalline model compound are processed through the Fourier-filtering step described above. Parameters for the amplitude [Equation (3-26)] and phase shift function [Equation (3-27)] and E_0, are fit to Equation (3-15), keeping distance, coordination number, Debye-Waller factors fixed at their known values. The differences between this fitted EXAFS spectrum and the Fourier-filtered EXAFS spectrum are minimized by a least-squares procedure explained in step (3) below. Theoretical starting values for the amplitude and phase shift parameters can be found in Teo and Lee (1979).

(2) The amplitude and phase shift parameters from (1) are used as fixed parameters in the least-squares fitting of the EXAFS model equation to the Fourier-filtered EXAFS spectrum for the unknown. Distance, coordination number, Debye-Waller factor, and E_0 for the absorber-jth backscatterer shell pair are varied to optimize the fit. Allowing E_0 of the unknown to vary relative to the E_0 value of the model compound can help account for any differences in bonding between model and unknown compound. Step (2) assumes that phases and amplitudes are transferable from the model to the unknown compound for the same types of absorber-backscatterer pairs (see §3.3.4). The k-range of the Fourier transformation in step (2) is normally chosen to be the same as in step (1) so that analysis of model compound and unknown are as parallel as possible.

(3) The least-squares fitting procedure in steps (1) and (2) minimizes the variance S expressed as

$$S = \sum_{i}^{n} (\chi_i{}^F - \chi_i)^2 \tag{3-28}$$

where $\chi_i{}^F$ are the Fourier-filtered experimental data (i.e., back Fourier transformed over a restricted R-range) and χ_i is the analytical form of Equation (3-15). Adjustment of the variables in the least-squares procedure is continued until the improvement in fit is less than a desired amount, usually 1%.

3.3.4 Transferability of the phase-shift and amplitude functions and accuracy of distance and coordination number determinations from EXAFS analysis.

The analysis of EXAFS data relies on knowledge of the electronic functions describing phase-shift and amplitude. The basic hypothesis in this analysis is that these parameters can be "transferred" from a model compound to the unknown, i.e., that for the same central atom-backscattering atom pair, the phase shift and backscattered amplitude values are conserved in systems with similar interatomic distances and chemical bonding. The use of theoretical functions, calculated in an atomic approximation, must be made carefully in condensed systems. In this case, the E_0 value may be adjusted by a limited energy shift (< 20 eV). The validity of this procedure has been verified in numerous analyses.

In cases where the unknown compound is compared to a compound of similar composition and structure, the accuracy of the EXAFS-derived interatomic distances is ±0.01Å, the coordination numbers (CN) are accurate to ±0.2-0.5 atoms, and the structural disorder parameter is accurate to about ±0.01Å^2. Accuracy is higher for distance determinations than for CN or disorder determinations, because only the phase-shift functions are needed for distances. In general, the more periods that are visible on the partial EXAFS, the better the precision level. However, the accuracy is limited by the uncertainty in defining E_0 and by the width of the peaks in the forward Fourier transform. Accuracy also can be affected by the Fourier-filtering procedure when two non-resolved peaks occur. In that case, EXAFS can give mean distances which may not be structurally significant. Distance distributions may be studied through the occurrence of beat patterns due to the presence of more than one shell of nearest neighbors at different distances (see Fig. 6-11). In this situation, EXAFS oscillations of two different frequencies add up to produce the total EXAFS spectrum, and interferences or "beat patterns" result. When two such shells are separated by a value ΔR, the first minimum of the beat pattern is given by:

$$2k\Delta R = \pi. \tag{3-29}$$

Thus, because of limited k-space range generally available in EXAFS analysis (e.g., ≈10Å⁻¹), the minimum distance separation detectable between two discrete atomic shells is $\Delta R = 0.15$ Å. Least-squares fitting of spectra, however, can resolve smaller differences.

The accuracy of EXAFS analysis for intermediate-range distance determination (> 5 Å) is intrinsically limited by currently used single-scattering models and by the fact that low k data (k < 3 Å⁻¹) are normally discarded due to the major contributions of multiple scattering in this low k region. As discussed above, resolution of distances in EXAFS analysis can also be reduced by limited high k-range data, as is often the problem encountered in EXAFS spectra of cations in aqueous solutions or in silicate glasses and melts.

IV. X-RAY ABSORPTION NEAR EDGE STRUCTURE (XANES)

4.1 Introduction

Typical XANES spectra of several crystalline insulators [Mn K-XANES spectra of MnO_2; $Ca_2(Al,Fe,Mn^{3+})_3Si_3O_{12}OH$ (piemontite); Mn_2SiO_4 (tephroite)] are shown in Figure 4-1. The XANES spectrum may be separated into two regions:

(1) the low energy part of the absorption edge (<10 eV below the absorption edge, referred to as the pre-edge region), with features caused by electronic transitions to empty bound states – the transition probabilities are controlled primarily by the selection rules for dipolar electronic transitions;

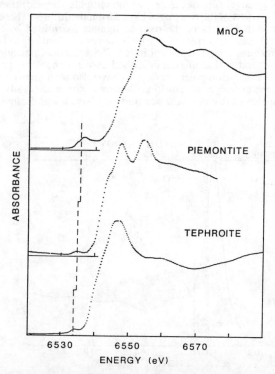

Figure 4-1. Experimental Mn K-edge XANES spectra of $Mn^{4+}O_2$, piemontite $[Ca_2(Al,Fe,Mn^{3+})_3Si_3O_{12}OH]$, and tephroite $[Mn_2^{2+}SiO_4]$. The dotted line shows the location of a pre-edge feature due to a 1s→3d electronic transition. The pre-edge shifts to higher energy with increasing oxidation state of Mn. The significant differences in the XANES region are caused by differences in first, second, and higher neighbor arrangement among these three crystalline phases.

466

(2) the intermediate energy region (a few eV above the pre-edge to ≈50 eV above the absorption edge, referred to as the XANES region), which is dominated by multiple-scattering resonances of the photoelectrons ejected at low kinetic energy.

The high energy part of an XAS spectrum (≈50 to ≈1000 eV above the edge) corresponds to the EXAFS single-scattering regime, which was discussed in Part III. Other effects are also important in the XANES region, such as multi-electronic effects (shake-up and shake-off processes). The two major processes occuring in the XANES energy region – namely bound-state transitions and multiple-scattering processes – are often intimately associated, but can give distinct structural and chemical information concerning the absorber and its local environment. Bound-state transitions are localized, if they are restricted to atomic orbitals of the absorbing atom, or they may be somewhat delocalized, if they involve molecular orbitals (MO's) between the absorber and first-neighbor ligands (Fig. 1-4). Thus XANES features due to bound-state transitions can be used as spectroscopic probes of the local crystal chemistry of the absorber (oxidation state, site symmetry). The higher energy, multiple-scattering features of the XANES spectrum are due to scattering of continuum electrons (i.e., individual electrons that are not bound to an atomic center or to a bound-state MO), and they provide information about the first and more distant coordination shells around an absorber (Fig. 1-3). They are particularly sensitive to the geometrical arrangement of the atoms in these shells.

Selection rules governing these transitions lead to resolution of electronic states of different angular momentum and to information on the bonding characteristics of the different states. Figure 4-2 shows the L_I and L_{III} absorption edges of palladium in PdO (Davoli et al., 1983). Both spectra have been shifted to the same relative energy scale to facilitate comparison. The first final state (4d) is the same for both edges. However, dipole selection rules allow the 2p→4d transition in the L_{III} edge, but not the 2s→4d (quadrupole allowed) transition in the L_I edge. Hence the low energy part of the two edges have distinctly different shapes. A mixing of d states of the absorber with p states of the oxygen ligands (permitted by the point group symmetry of the absorber) leads to a transition which causes the "pre-edge" feature of the L_I edge. Similar pre-edge features are often present in XANES spectra of transition elements. In addition to the absorber-ligand orbital mixing described above, dynamic symmetry reduction (vibronic coupling) and electric quadrupolar interactions have also been invoked to partly explain the observed intensity of the pre-edges of transition elements (Roe et al., 1984). Observation of these pre-edge features indicates that covalent bonding may have a significant influence on the

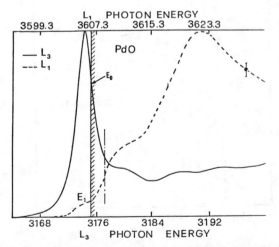

Figure 4-2. Comparison of Pd L_I and L_{III} edges in PdO, an insulator with Pd coordinated by four oxygens in planar configuration. The L_I spectrum has been shifted by -431.3 eV for direct comparison with the L_{III} spectrum. The energy threshold, E_0, was estimated using a method described by Bianconi (1984). The main edge of the L_{III} XANES spectrum is due to a 2p→4d transition, and the weak peak in the L_I spectrum labeled E_1 is the dipole forbidden (quadrupole allowed) 2s→4d transition. (After Davoli et al., 1983)

edge spectra. In contrast to these insulators, in which the near-edge structure can be described using a localized atomic or MO picture, the near-edge structures of semiconductors must be described using partial density of state functions from appropriate band calculations.

In the discussion below, several types of one-electron excitations will be examined, including (1) atomic Rydberg states, which are the easiest type to interpret but are of limited usefulness; (2) bound valence states, which are observed in insulating compounds and molecular groups; (3) local densities of states near the Fermi level, which are present in the spectra of metals and semiconductors; and (4) multiple-scattering resonances at higher kinetic energies.

4.2 The constitution of an elementary absorption edge: XANES of noble-gases

XANES spectra of noble gases are the easiest type to interpret because they should display no features due to transitions to bound-state bonding MO's or due to multiple-scattering resonances. In the case of argon, a Rydberg series is seen in the experimental spectrum (Fig. 4-3). With increasing energy, the absorption lines converge towards a continuum. The energies of these features should correspond to the 1s→np transition energies of neutral potassium because of the influence of the core hole created during the electronic transition, which modifies the final state. This "Z+1 rule" will prove to be useful in the interpretation of 3d-transition element pre-edge features (see §4.3). The high intensity of these transitions arises from the fact that they are allowed by the selection rules. They have a natural (Lorentzian) lineshape (see Chapter 3, this volume) and a width of 0.58 eV. In the limit n→∞ for these transitions to p final states, photoelectron ejection occurs (at the ionization threshold), and this continuum can be approximated with an arctangent (= integrated Lorentzian) lineshape. The various resonances are closely spaced and are resolvable because of the relatively narrow width of the core hole. Atomic features such as these will be only partly conserved in compounds with a defined local structure (molecular gases, liquids or solids), because of the influence of the local bonding environment on the effective final density of states. In practice, generally only the first transitions may be recognized unambiguously, as the edges are dominated by multiple-scattering processes at higher energy. As discussed above, description of the higher energy features (resonances) of XANES spectra can be done using density of states or multiple-scattering terminology, since both refer to the same type of process.

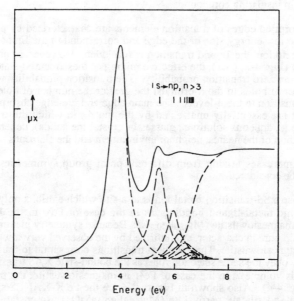

Figure 4-3. X-ray absorption at the K edge of gaseous argon showing the Rydberg series of bound atomic states. (After Paratt, 1959)

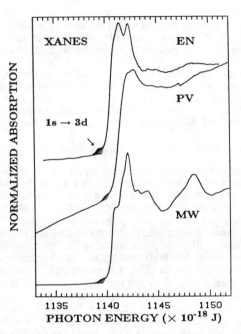

Figure 4-4. Iron (II) K-XANES spectra of magnesiowüstite (MW: $Mg_{0.9}Fe_{0.1}O$), silicate perovskite (PV: $Mg_{0.88}Fe_{0.12}SiO_3$), and enstatite (EN: $Mg_{0.9}Fe_{0.1}SiO_3$). The pre-edge feature (shaded) is very weak in the spectra of MW and PV and about 30% more intense in the spectrum of EN. These observations are consistent with a site of low distortion in MW and PV, and with a site of greater distortion in EN. (From Jackson et al., 1987)

4.3 Bound states in insulating compounds

The K absorption edges of transition elements are characterized by pre-edge features, which occur on the low energy side of the edge and correspond to the first empty states. In a purely atomic description, the 1s→nd transition is forbidden. However, as pointed out for the Pd L_I-state in PdO (Fig. 4-2, §4.1), the effective mixing of these metal d-states with ligand p-states will give a non-zero transition probability. The transition probability will also take into account the number of holes in the d-orbitals: the greater the number of holes, the higher the probability of a transition to the d-levels. The main edge and pre-edge absorption features of a transition element are essentially unaffected by the matrix in which the transition element complex occurs (e.g., aqueous solutions, glasses or crystalline solids), depending only on the geometry and bonding of the nearest neighbor environment and the element's oxidation state.

Two extreme cases arising from different point group symmetries of the absorber environment may be considered:

(1) Suppose a 3d-transition metal ion is at a site which exhibits only slight distortions with relatively ionic metal-ligand bonding, as in the case of Fe^{2+} in the highly-symmetric octahedral site of magnesiowüstite ($Mg_{0.9}Fe_{0.1}O$). Because symmetry prevents d-p mixing, the Fe^{2+} d-states remain pure in character, as confirmed by the observed very low absorption of UV-visible light by magnesiowüstite. These optical transitions correspond to transitions among the 3d-levels, and they obey the same selection rules as the bound-state XAS transitions (see Chapter 7 by Rossman, this volume). In the case of Fe^{2+} in magnesiowüstite, the pre-edge feature is almost absent (Fig. 4-4). Also shown in Figure 4-4 are the Fe K-XANES spectra of enstatite ($Mg_{0.9}Fe_{0.1}SiO_3$) and silicate perovskite ($Mg_{0.88}Fe_{0.12}SiO_3$). In enstatite, most of the Fe occupies the M2 site, which is significantly more distorted than the Fe site in magnesiowüstite;

the pre-edge intensity of the enstatite spectrum is about 30% greater than this feature in the magnesiowüstite or silicate perovskite spectra.

(2) Suppose a 3d-transition metal is at a site, in which there is strong d-p orbital mixing arising from highly covalent M-O bonds and the lack of a center of symmetry. Such is the case for Fe^{3+} in a silicate glass of composition $KFeSi_3O_8$ (Brown et al., 1978). A molecular orbital bonding scheme is thus more applicable to this situation and may be used to explain the high relative intensity of the pre-edge feature (Fig. 4-5). The feature is almost as intense as the main absorption edge of a formally d^0 transition metal complex (e.g., Ti^{4+} in SiO_2 glass; $V^{5+}O_4$ or $Cr^{6+}O_4$ complexes), and considerably more intense than Fe^{2+} in magnesiowüstite (compare the pre-edge intensities with the edge height in each case).

In intermediate cases, corresponding to a distortion of the site and subsequent removal of the inversion center, pre-edge features are clearly visible. Their relative intensity, for a given site, is dependent on the number of electrons present in the d-orbitals. As they become filled, the transition intensity decreases, falling to zero for 3d-ions with filled orbitals, as in the case of Zn^{2+} or Cu^+ compounds. An excellent example of the systematic variation in intensity of the pre-edge feature in XANES spectra was given by Lytle et al. (1988) in their study of the K-XANES of 3d-elements in oxides (Fig. 4-6).

The influence of local geometry around the transition metal ion may be seen in the pre-edge and XANES regions of K-XANES spectra for Fe-containing minerals, with Fe in different coordination environments (Fig. 4-7). In order to quantitatively analyze these spectra, the pre-edge has to be separated from the main edge background by fitting a curve of arctangent shape to describe the boundary between bound (atomic-like) and continuum states (Fig. 4-8) (see discussion on Ar in §4.2). The pre-edge may be fit by one or more Lorentzian components, depending on the number of different sites occupied by the cation. These features correspond to the different final bound states. The natural shape neglects the influence of experimental spectral broadening, which can be approximated by a Gaussian or mixed (Voigt) profile. The Lorentzian approximation is not accurate in many cases. The width of the Lorentzian components is determined mainly by the core-hole lifetime (the width of the final state is negligible), giving intrinsic line widths of about 1 eV for the Fe K-edge. This intrinsic width may be considered constant for a given element, as it is an atomic-like property. In the case of 3d-elements, a simple formalism derived from effects of the crystal-field on 3d-orbital energies and orbital occupation has been used to explain XANES pre-edges (Calas and Petiau, 1983b). Influence of the core hole on the final state is taken into account by considering the electronic levels of the Z+1 element.

In the case of Fe^{3+}, two states are predicted, corresponding to T_2 and E crystal field states. The stronger splitting of an octahedral crystal field, as compared to a tetrahedral crystal field, explains the presence of two pre-edge features in the Fe K-XANES of andradite (Fig. 4-9a), where Fe^{3+} is 6-coordinated. Only one pre-edge feature is apparent in the Fe K-XANES of yttrium-iron garnet (YIG) (Fig. 4-9b), in which Fe^{3+} occurs in both tetrahedral and octahedral coordination. This last example also illustrates the difference in transition probabilities for 3d-transition elements at octahedral versus tetrahedral sites. Although 40% of the Fe^{3+} occupies octahedral sites in YIG, only the spectral signature of the 60% tetrahedral Fe^{3+} is apparent because of the much more intense pre-edge feature of the latter species. The pre-edge feature of Fe^{3+} in a $Na_2Si_2O_5$ glass (Fig. 4-9c) also is relatively intense and narrow, suggesting that ferric iron in a tetrahedral environment is the dominant contributor to this spectrum.

Because they arise from electronic transitions to well-defined bound states, the pre-edge features of transition metal XANES spectra show a dependence on the oxidation state and bonding characteristics of the absorbing atom. In both cases, the observed shift (also called the "chemical shift") derives from the screening of the nuclear charge from the electronic core levels by the valence electrons. Higher oxidation states of metallic ions cause a reduction in screening which results in higher core levels energies and more tightly bound core electrons (Fig. 4-10) (see also Fig. 4-1). A shift of 1.5 to 3 eV is found between the first empty bound states (pre-edge features) of the K-edges for Fe^{2+} and Fe^{3+} in similar environments (Waychunas *et al.*, 1983; Calas and Petiau, 1983a). Opposite effects are observed in the case of more covalent

470

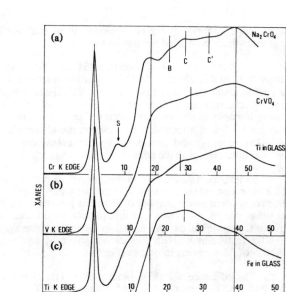

Figure 4-5 (right). K-XANES spectra of Fe^{3+}, Ti^{4+}, V^{5+}, and Cr^{6+} in tetrahedral sites in various compounds. (From Bianconi et al., 1985a)

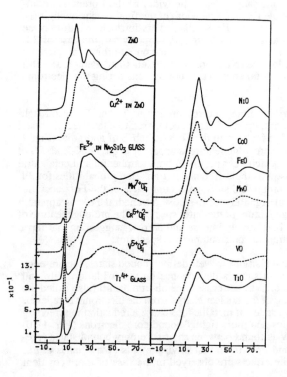

Figure 4-6 (left). K-XANES spectra of 3d-transition metal ions in various compounds. The spectra on the left are for metal ions in 4-coordinated sites, whereas the spectra on the right are for metal ions in 6-coordinated sites. Comparison of the spectra for 4- and 6-coordinated Ti, V, Cr, Mn, and Fe shows that the 4-coordinated metal ions have much more intense pre-edge features than the 6-coordinated metal ions. (From Lytle et al., 1988)

cation-ligand bonding, but in that case the final state rapidly looses its localized character, and an interpretation in terms of bound atomic + molecular final states is no longer valid (as in sulfides). The variation of covalency of metal-ligand bonding also affects the intensity of the pre-edge. For example, tetrahedrally-coordinated Ti^{4+} shows a more intense pre edge in oxides than in chlorides (Dumas and Petiau, 1986). The observed p-character of the final state changes (17% in TiO_4 versus 1.4% in $TiCl_4$), and the change from oxygen to chlorine ligands (L) causes an increase in the Ti-L distance [d(Ti-O) = 1.89 Å and d(Ti-Cl) = 2.19 Å]. This dependence of the pre-edge on the local crystal chemistry around the absorbing atom makes it a sensitive probe for studying minerals (see § 5.5 and 5.6).

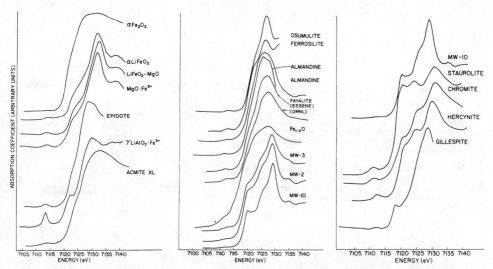

Figure 4-7. Iron K-XANES spectra of various minerals with Fe in different coordination environments. The significant variations in the XANES region among the spectra reflects the differences in second- and higher-neighbor environments among these phases. The spectra on the left are for Fe^{3+} in selected oxides and silicates; the spectra in the middle are for Fe^{2+} in 6-coordinated sites in oxides and silicates; the spectra on the right are for Fe^{2+} in 4-coordinated sites in oxides and silicates, compared with Fe^{2+} in the 6-coordinated of magnesiowustite (MW-10). (From Waychunas et al., 1983)

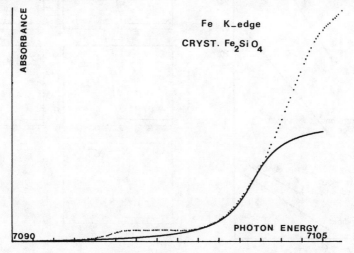

Figure 4-8. Pre-edge region of the Fe K-XANES spectrum of fayalite, showing an arctangent function fit of the background. (From Calas and Petiau, 1983b)

472

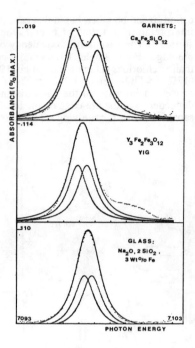

Figure 4-9 (right). Pre-edge spectra of (a) 6-coordinated Fe^{3+} in andradite garnet, (b) 40% 6-coordinated and 60% 4-coordinated Fe^{3+} in yittrium iron garnet (YIG), and (c) 4-coordinated Fe^{3+} in $Na_2Si_2O_5$ glass. The background in the pre-edge regions were fit using arctangent functions, and the pre-edge features were fit using Lorentzians. (From Calas and Petiau, 1983b)

Figure 4-10 (below). Variation of the positions of edge features with metal oxidation state for a series of vanadium compounds. The XANES spectra from which this correlation results are shown on the right. (From Wong et al., 1985)

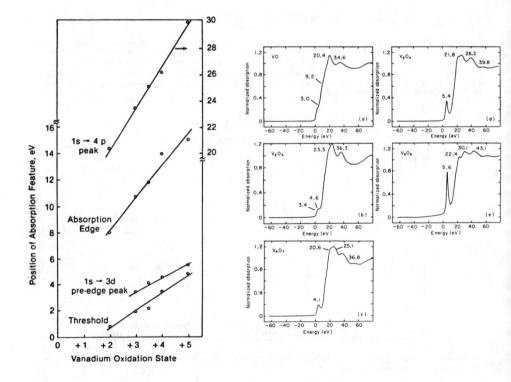

4.4 Electronic structure of semiconducting minerals

The energy range of XANES is narrow enough that transition probabilities should be relatively constant. Consequently, the XANES spectrum will reflect the density of empty states in the conduction band, determined by the transition considered (e.g., 1s $\rightarrow$np). In semiconducting compounds, the core hole created during the absorption process is shielded by the valence electrons, and the absorbing atom can be considered in the ground state. In this case, the density of states is only local because only the electronic structure around the absorbing element is observed. An example of this situation is tetragonal TiO_2 (rutile), which has been studied at various absorption edges, including K- and L_{II-III}-edges of titanium (Grunes, 1983) and the K-edge of oxygen (Grunes, 1983, Brown et al., 1986a). Figure 4-11 shows these spectra aligned at threshold, together with the calculated density-of-states (DOS) and the empirical molecular orbital diagram derived for TiO_6 clusters in TiO_2. Good agreement is found for these three absorption edges, and collectively, they show the contribution of both cation and anion to the final states. In particular, the two first empty states are thought to correspond to the $2t_{2g}$ and e_g anti-bonding molecular orbitals, which are combinations of the Ti 3d orbitals and O 2p orbitals, probed by the Ti-L_{II-III} ($2p\rightarrow3d$ transition) and O-K ($1s\rightarrow3p$) edges, respectively. The corresponding transition at the Ti K-edge is responsible for the pre-edge feature, and it has a non-zero intensity because of the mixed d-p character of these MO's. A feature of low intensity is observed at lower energy, and has been attributed by some authors to a core exciton residing within the band gap. This exciton describes the interaction of the photoelectron with the core hole. It is not predicted by the MO model nor by DOS calculations, which are made in the ground state. Local densities of states in semiconducting minerals have also been studied in sulfides by Sainctavit et al. (1986). The XANES spectra of all the cations (Zn, Fe, Cu) and the sulfur K-XANES have been correlated in the minerals sphalerite (ZnS:Fe) and chalcopyrite ($CuFeS_2$) (Fig. 4-12). In particular, empty d-states have been found at Cu based on the Cu K- and L-XANES in chalcopyrite, which is in apparent contradiction with the formal monovalent state ascribed to this element based on studies of chalcopyrite magnetic properties.

4.5 Multiple-scattering region of XANES spectra

Once the photoelectron is ejected in the continuum, it is subject to the local potential arising from the surrounding atoms. At low kinetic energy, elastic multiple-scattering processes are dominant because inelastic interactions have low probability in this energy regime. With increasing photoelectron energy, we enter the single-scattering regime of the EXAFS oscillations. Description of the photoelectron wavefunction by the single-scattering approximation of EXAFS is not valid when the photoelectron wavelength is larger than the interatomic distances. The expected low energy limit of multiple scattering may be defined as

$$E = \frac{151}{d^2} - V \quad , \tag{4-1}$$

where the interatomic distances d are expressed in Å, and V represents the interatomic potential (V can be taken at the Fermi level in free electron metals) (Natoli, 1984). Equation (4-1) explains why the multiple-scattering energy range is strongly dependent on the type of compound studied. The variation in energy of XANES features with interatomic distance d is shown in Figure 4-13, for some tetrahedrally-coordinated cations.

Recently, Rehr et al. (1986a, b) proposed a unified treatment of both XANES and EXAFS which uses a spherical-wave approximation of the scattering process . The single-scattering formalism of EXAFS is maintained using distance-dependent back-scattering amplitudes which have been calculated by these authors for metallic copper. Using this approach, analysis of EXAFS may be carried out down to energies in the edge region. This approach provides for the use of more experimental data in the analysis of local structures and should lead to increased resolution of distances. It is also particularly useful in increasing the range of k-values in analysis of low-Z elements or of disordered systems, where data are limited at high k-values.

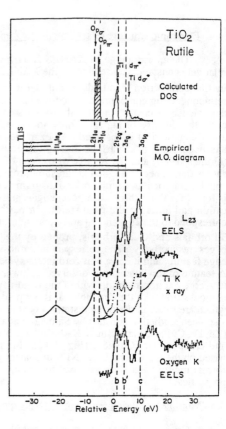

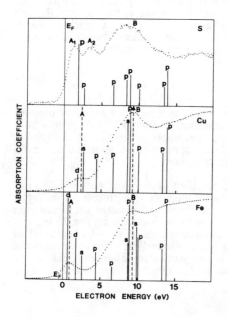

Figure 4-11 (right). Comparison of the near-edge structures of Ti-$L_{2\text{-}3}$, Ti-K and O-K absorption edges in TiO_2 (rutile) and with the calculated density of states (DOS), all shifted to the same relative energy scale. An empirical molecular orbital (MO) diagram is also shown. (From Grunes, 1983)

Figure 4-12 (left). K-XANES of Fe, Cu, and S in chalcopyrite, $CuFeS_2$. The edges have been shifted to the same relative energy scale by aligning them at their first inflection points. Eigenvalues and atomic orbital make-up (s,p,d orbitals) of the eigenvectors are represented by the vertical lines, which are proportional to the calculated eigenvector coefficients. (From Sainctavit et al., 1986)

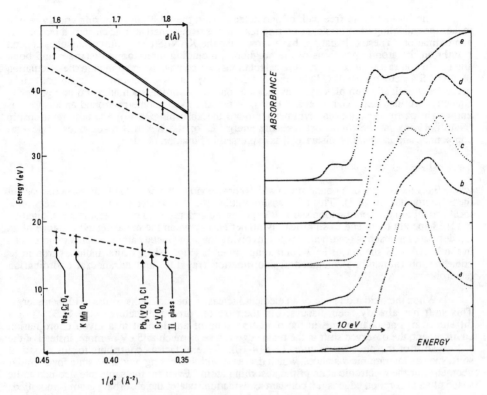

Figure 4-13 (left). Energy positions of the XANES maxima plotted as a function of $1/d^2$ for tetrahedrally coordinated 3d-transition metals in various compounds. (From Bianconi et al., 1985a)

Figure 4-14 (right). Fe K-XANES spectra for (a) almandine (8-coordinated Fe^{2+}), (b) wüstite (6-coordinated Fe^{2+}), (c) hypersthene (6-coordinated Fe^{2+}), (d) staurolite (4-coordinated Fe^{2+}), and (e) pyrite (Fe 6-coordinated by S). (From Calas et al., 1984a)

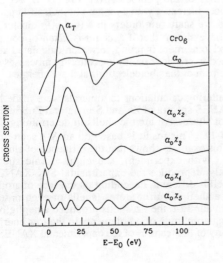

Figure 4-15. Calculated multiple-scattering contributions of various scattering orders (representing different photo-electron path lengths) in the CrO_6 cluster. (From Garcia et al., 1986b)

The limited mean free path of photoelectrons in the multiple-scattering energy region makes the modeling of multiple- scattering resonances using small atomic clusters a reasonable approximation. These calculations have shown that the XANES continuum resonances depend mainly on the atomic positions of the neighbors, including interatomic distances and bond angles. In contrast with band structure calculations, which have been applied to the calculation of XANES spectra of metals (Müller and Wilkins, 1984), multiple-scattering calculations can be made on chemically complex systems lacking long range order, such as metal complexes in aqueous solutions (e.g., Garcia et al., 1986). When the local geometry around an absorber is known (in particular the point symmetry around the absorbing atom), a simple relationship (Natoli, 1984) can be found between the energy E_r of the XANES resonances of several compounds and interatomic distance d, as expressed in Equation (4-2):

$$(E_r - E_b) \, d^2 = \text{const.} \quad , \tag{4-2}$$

where E_b is the energy of a bound state (which correspond to the pre-edge features in the case of the 3d elements, see §4.3). This relationship implies that the scattering of the photoelectron is localized within the coordination shell. It explains the general trend of the features exhibited by XANES spectra to extend to significantly higher energy when the metal-ligand distance of an ion decreases in a series of structure types involving the same metal cation. Such a trend is seen in the K-XANES spectra of Fe^{2+}-containing minerals (Fig. 4-14), which includes iron in 8-coordination (almandine), octahedral coordination (fayalite) and tetrahedral coordination (staurolite).

When the oxidation state of an element increases, the main edge shifts to higher energy. This shift has already been discussed in the case of pre-edge features (see §4.3). For a difference of one charge unit in the oxidation state of an element in a given coordination environment, the observed shift in the main edge can be as much as 5 eV or more, instead of the 1-3 eV shift of the pre-edge features (Fig. 4-10). Several factors contribute to this chemical shift, such as interatomic distances, which decrease with increasing oxidation state for a similar geometry, or the electronic state of the absorbing atom. Even for the same site geometry, the shape of an absorption edge is not constant as oxidation state of the absorbing atom is modified.

Calculations of multiple-scattering resonances have been carried out for various materials, but they generally are very CPU intensive (requiring up to several hours of CPU time for calculating a XANES spectrum of a five to seven atom cluster with a VAX 11/750). The simplest example comes from a metal complex in a dilute aqueous solution, where only the contribution of the first coordination shell to the XAS is expected because of the strong disorder of more distant ligands (Benfatto et al., 1986; Garcia et al., 1986a,b; Bianconi et al., 1985b). Multiple-scattering events occur inside this first coordination shell, and correspond to two, three, and even higher numbers of successive scatterings of the photoelectron (Fig. 1-3). As for EXAFS, the frequency of the observed interferences increases with a longer photoelectron path. The result from such a calculation of the successive scattering orders in a $(CrO_6)^{9-}$ cluster is shown in Figure 4-15. Garcia et al. (1986a,b) have shown that if account is taken of contributions from the various scattering orders, a good agreement is found betweeen calculated and experimental spectra (Fig. 4-16). There is no need to consider multiple-scattering processes of order higher than six because of limited mean free path of the photoelectrons at these energies.

The first applications of full multiple-scattering calculations to minerals were made by Sainctavit et al. (1987) who calculated the K-XANES spectra of Zn and S in natural ZnS in the first 55 eV above the edge. The participation of successive atomic shells in determining the shape of the XANES is represented in Figure 4-17. This result is based on an X-α scattered wave calculation around sulfur atoms in ZnS. This example shows that the influence of cluster size can be important in obtaining a good fit with experiment, and eight atomic shells (representing 99 atoms) were needed to correctly reproduce the experimental S K-XANES spectrum. The effect of adding a new atomic shell is different, depending on the type of atoms that comprise it. For example, the addition of a shell of sulfur atoms (shells of even order) always improved the calculation dramatically. In contrast, the Zn-shells produced a negligible effect on the fit because of the low backscattering amplitude of 3d-elements in the low energy region. The first atomic shells do not reproduce any of the features of the experimental XANES

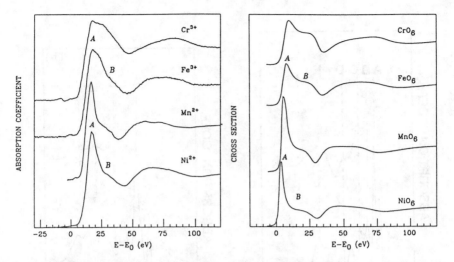

Figure 4-16. Comparison between experimental (left) and calculated (right) K-XANES spectra of 6-coordinated 3d-transition elements in aqueous solutions. (From Garcia et al., 1986b)

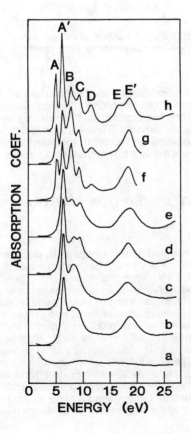

Figure 4-17. Xα multiple scattered wave calculations of the XANES spectrum of a sulfur atom in sphalerite (ZnS), showing the influence of the successive atomic shells around S (a-g). The contribution of Zn shells (a,c,e,g) is relatively small. (After Sainctavit et al., 1987)

478

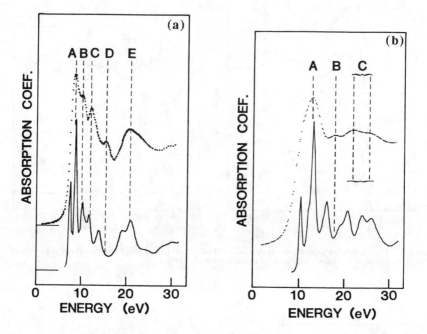

Figure 4-18. Xα multiple scattered wave calculations of K-XANES of (a) S and (b)Zn in ZnS (solid lines) compared with the experimental K-XANES spectra (dotted lines). The effect of the broadening due to the corehole lifetime is shown in (b). (After Sainctavit et al., 1987)

spectrum. Comparison of the final results of these X-α calculations with the experimental K-edge spectra of S and Zn is presented in Figures 4-18a and 4-18b, respectively. In the calculated spectra, lineshape broadening due to the core-hole lifetime and experimental broadening were included. These first results show the power of such calculations. Theoretical modeling of XANES spectra could be used to provide new information on the local geometry around absorbing elements in unknown structures, particularly when only a few shells are required to fit the XANES.

V. INFORMATION CONTENT OF X-RAY ABSORPTION SPECTRA

5.1 Introduction

X-ray absorption spectra can be divided into two main energy regions as discussed earlier: XANES, extending from ≈10 eV below to ≈50 eV above the absorption edge, and EXAFS, extending from ≈50 eV to as much as 1000 eV above the edge. The types of information contained in these spectra arise from two fundamentally different processes: electronic transitions that are localized on the absorbing atom, and electron scattering involving multiple-scattering sites. XANES spectra arise from a combination of these two processes, whereas EXAFS is due to scattering only. This difference in process produces different levels of structural and/or bonding information. Considering XANES spectra first, electronic transitions between a core level and the first empty bound levels of an absorbing atom produce pre-edge features whose intensity, shape, and position provide indirect information about the symmetry of the site occupied by the absorber (including type of coordination environment) and its oxidation state, and direct information about the bonding between absorber and first-neighbor ligands. The pre-edge features contain no direct information about bond lengths or angles. Scattering of electrons at kinetic energies in the vicinity of the edge region gives rise to those features of a XANES spectrum that can be interpreted in terms of multiple scattering from atoms in the first several coordination shells around the absorber. This portion of the spectrum can yield information about interatomic distances and angles. Turning to the EXAFS portion of the spectrum, scattering of electrons at higher kinetic energies causes the interference between outgoing and singly-backscattered photoelectron waves characteristic of an average radial distance between absorber and backscatterer and the number (and type) of backscatterers. Analysis of the EXAFS of an atom also can provide information on local disorder (or order) and on the type(s) of atoms coordinating the absorber (see §3.3.3).

5.2 Local Structure in Minerals

As an example of the type of information EXAFS analysis can provide about cation environments in a compositionally and structurally complex mineral, consider a hypothetical clinopyroxene of composition:

$$(Mg_{0.95}Fe_{0.70}Ca_{0.20}Ti_{0.11}Cr_{0.04})Si_{1.80}Al_{0.20}O_6.$$

The general structural formula for clinopyroxene can be written $^{VI}M1\ ^{VII\text{-}VIII}M2\ ^{IV}T1\ ^{IV}T2\ O_6$. For pyroxene of this composition, it is likely that the M1 site is occupied by all of the Ti^{4+} and Cr^{3+} and some combination of Mg and Fe^{2+} (depending on the thermal history of the pyroxene), and that the M2 site contains all of the Ca and the remainder of the Fe^{2+}; the T1 and T2 sites contain the Si and Al. A conventional x-ray structure refinement will yield the average structure of this phase, including the average geometry of the M1 and M2 sites and the T1 and T2 sites. Thus with x-ray diffraction, it is not possible to determine the true average local structural environment about any of the cations. EXAFS analysis cannot provide as much detailed information about a crystal structure as a careful x-ray structure analysis. However, because of its element specificity, it can provide a more correct picture of the average local structure around the cations that occupy only one type of crystallographic site (Ca, Ti^{4+}, Cr^{3+} in this case). Unfortunately, EXAFS analysis, at current levels of resolution using randomly oriented powdered samples, cannot distinguish between the Mg or Fe^{2+} M1 and M2 environments nor between the Si or Al T1 and T2 environments. For these elements, interatomic distances from an x-ray refinement provide a better indicator of site ordering than EXAFS analysis does. However, in certain situations, EXAFS analysis can provide unique information about the identity of second-nearest neighbors surrounding the absorber, and this information can aid in identifying the type of site that a particular cation occupies, if the sites in a structure have different second neighbor identities. Moreover, this feature of EXAFS analysis can lead to unique information on short-range ordering that conventional diffraction studies cannot provide, as will be discussed in §6.6.2.

5.3 Oxidation State

When a valence electron is removed from an atom, screening of core electrons provided by the valence electrons is reduced, and the core levels become more tightly bound. This change produces a shift in pre-edge and bound-state edge features, and this "chemical shift" can be correlated with differences in oxidation state for an element. For example, the pre-edge feature for iron in Fe^{3+}-containing minerals is generally 2-3 eV higher in energy than the corresponding feature for Fe^{2+} (Calas and Petiau, 1983a; Waychunas et al., 1983). A larger "chemical shift" is observed for greater differences in oxidation state of an element, as shown for arsenic in cobaltite (CoAsS: As^{3-}) versus erythrite (CoAsO$_4$: As^{5+}), where the difference is 7.2 eV between the main edge positions (Fig. 5-1a).

Differences in final state of the electron also affect the magnitude of the chemical shift. For iron, the 3d final state is mainly non-bonding, whereas for arsenic the final state is mainly an anti-bonding 4p state. This difference would favor a larger chemical shift for arsenic relative to iron because anti-bonding molecular orbitals should be more greatly affected by a change in electron population than non-bonding MO's. In spite of these effects, many minerals have mixtures of different oxidation states of a given element, or have the same element in different coordination geometries (e.g., both tetrahedral and octahedral). In these cases, determination of oxidation state from XANES bound-state features is not straightforward and usually requires complementary data from other methods and very careful analysis of the XANES pre-edge features. The analysis of XANES for oxidation state information is made easier when the

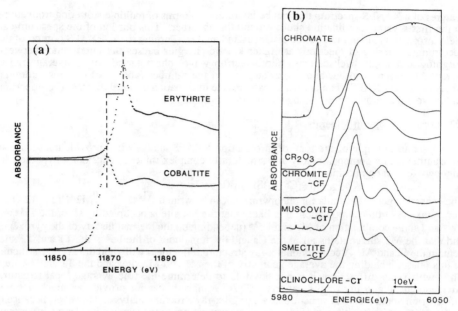

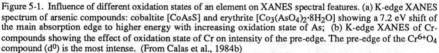

Figure 5-1. Influence of different oxidation states of an element on XANES spectral features. (a) K-edge XANES spectrum of arsenic compounds: cobaltite [CoAsS] and erythrite [$Co_3(AsO_4)_2 \cdot 8H_2O$] showing a 7.2 eV shift of the main absorption edge to higher energy with increasing oxidation state of As; (b) K-edge XANES of Cr-compounds showing the effect of oxidation state of Cr on intensity of the pre-edge. The pre-edge of the $Cr^{6+}O_4$ compound (d^0) is the most intense. (From Calas et al., 1984b)

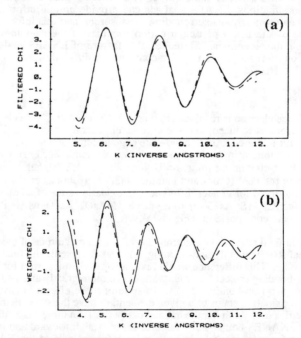

Figure 5-2. Fourier-filtered Fe K-EXAFS spectra corresponding to the first Fe-O contribution in (a) hercynite ($LiAlO_2$:Fe^{3+}) and (b) almandine ($Fe_3Al_2Si_3O_{12}$). The lower frequency in hercynite than in almandine is expected because of the shorter Fe-O distance in hercynite. (After Waychunas et al., 1986a)

element in question occurs in discrete tetrahedral clusters such as in chromates or vanadates (see Fig. 4-5). Tetrahedral geometry favors strong d-p mixing which greatly intensifies the pre-edge and makes the contribution of the high oxidation state element in such clusters easier to measure accurately (see Fig. 4-9 and discussion in §4.3). For example, the absence of a strong pre-edge in Cr-containing clay minerals from the weathering zone of a Brazilian chromite deposit is clear evidence that no CrO_4 clusters are present at levels greater than 2% of the total Cr (Calas et al., 1984). Instead, the Cr K pre-edge and edge are consistent with Cr^{3+} in octahedral coordination (Fig. 5-1b).

Complementary information concerning the oxidation state of an element can be obtained from EXAFS-derived bond lengths. For example, Fe^{2+}-O bond lengths are typically ≈ 0.15 Å longer than Fe^{3+}-O bond lengths for the same coordination number, and this difference is easily detectable. When both Fe^{2+} and Fe^{3+} are present in a mineral, intermediate distances are obtained, and the fitting of two shells can yield approximate Fe^{2+}/Fe^{3+} ratios when EXAFS spectra have good signal-to-noise levels (Petiau et al., 1981).

5.4 Interatomic Distances

5.4.1 Introduction and limitations. Because of the inverse dependence of the energy of scattering features (both single scattering as in EXAFS and multiple scattering as in XANES) on interatomic distance, average distance information can be obtained easily from XAS measurements. As pointed out earlier, the distances obtained always involve the absorbing element. This chemical selectivity makes XAS-derived distances unique relative to other direct structural methods based on scattering, which typically yield distances from all atom pairs. This unique feature of XAS is especially important in multicomponent systems. An important aspect of EXAFS-derived distances is the one dimensional nature of the information, which makes EXAFS analysis particularly well-suited for disordered materials, where the angular dependence of distances reduces the accuracy of symmetry sensitive methods. However, this same disorder can produce inaccuracies in EXAFS distances as was discussed in §3.2.2 and as will be shown below. A major limitation of EXAFS spectroscopy for distance determination arises because of the dominance of multiple-scattering processes at low k values (< 3 Å^{-1}), which cannot be interpreted using the single-scattering, plane wave formalism for EXAFS currently used. Because the low k data are most sensitive to longer distance contributions and to atoms of low backscattering amplitudes, the loss of these data limits the accuracy and sensitivity of EXAFS analysis to second and higher coordination shells. The effect of this limitation is most severe when EXAFS spectra are limited in k-space range (as in a silicate glass) and especially when light atoms (e.g., as in most natural silicates) are present in the first and higher coordination shells around an absorber. An alternative curved-wave formalism was developed by Lee and Pendry (1975) to permit treatment of EXAFS spectra to k values nearer the absorption edge. However, because of the limited use to date of this treatment in studies of minerals and glasses (Greaves et al., 1981; Binsted et al., 1985; Cressey and Steel, 1988), we will not discuss its advantages and problems.

5.4.2 Distances from EXAFS spectra. The effect of differences in interatomic distance on EXAFS phases is clearly seen in the Fe K-EXAFS spectra of almandine [d(Fe-O) = 2.30 Å] versus $LiAlO_2$:Fe^{3+} [d(Fe-O) = 1.87 Å], as shown in Fig.5-2. The frequency of EXAFS oscillations is lower in $LiAlO_2$:Fe^{3+} than in almandine as expected from the inverse dependence of frequency on distance. In practice, derivation of high accuracy distance information from EXAFS spectra requires spectra of high signal-to-noise level and of high k-space range. In addition, accuracy depends on the use of well-characterized, crystalline model compounds with the same absorbing atom as the unknown in crystallographic sites of similar size and topology relative to the unknown compound.

As an example of the accuracy possible for EXAFS-derived distances, consider the Fe-O distances derived for 6-coordinated Fe^{3+} in andradite ($Ca_3Fe_2Si_3O_{12}$) using phase shifts and backscattering amplitudes experimentally determined from EXAFS analysis of magnesiowüstite of composition $Mg_{0.9}Fe_{0.1}O$ (Jackson et al., 1987). The EXAFS-derived Fe-O distance of 2.06 Å for andradite compares favorably with the value of 2.024 Å derived from x-ray refinement (Novak and Gibbs, 1971). This comparison indicates an accuracy on the order of ±0.04 Å.

Other similar analyses in the literature have shown accuracies as good as ±0.01 Å. Precision levels achieved in least-squares fitting of experimental EXAFS spectra to analytical EXAFS functions (see §3.3) are commonly ±0.005 for distances.

The effect of static disorder on the accuracy of distance determination (resulting in shorter distances in all cases) is clearly shown in a study of fayalite (Waychunas et al., 1986a). Using a magnesiowüstite ($Mg_{0.9}Fe_{0.1}O$) as a model compound for phases and amplitudes, an Fe-O distance of 2.11 Å was derived for fayalite using a single-shell fit. The x-ray refined, average Fe-O distance is 2.168 Å. Iron in fayalite resides in two distinct octahedral sites, both with significant distortions from holosymmetry. The range of distances is between 2.07 and 2.29 Å. This static disorder results in loss of information about longer Fe-O distances in fayalite in the analysis, using a single shell fit. However, an improved fit can be obtained using multiple oxygen shells. If thermal disorder is significant in a sample, the quality of distance information from EXAFS analysis can be significantly improved by recording spectra at low temperature (see §3.2.2 and Fig. 3-3).

An important question for EXAFS spectroscopy is its ability to resolve different radial distances involving the same element in a phase. In general, this problem may arise when one element occupies two distinct sites in a structure, such as octahedral and tetrahedral sites or two octahedral sites of different mean sizes. It may also arise when an element occupies one site with a range of distances, resulting in two or more distinct populations of distances. The work of Combes et al. (1986) on hematite and goethite demonstrates the ability to resolve differences in Fe-O distances of 0.17 and 0.14 Å, respectively, with Fe^{3+} occupying one 6-coordinate site with two distinct distances. An EXAFS study of sillimanite (McKeown et al., 1985b) provides an example of the resolution capability of EXAFS in separating Al-O distance populations for 4-coordinate [d(Al-O) = 1.76 Å] and 6-coordinate Al [d(Al-O) = 1.92 Å]. Experience with a variety of compounds with two populations of distances to an absorber shows that resolution of distances for two populations is not generally possible when the difference is less than ≈0.1 Å.

As discussed in part III, the EXAFS probe is a local one due to the limited mean free path of the photoelectron, typically less than 5 Å. This fact limits the range of radial distance information from EXAFS to about 5 Å or less, depending on the types of neighbors and extent of static and thermal disorder. Even in well-ordered crystalline materials such as metallic copper, the structural information from EXAFS is lost beyond about 5.5 Å (Stern and Heald, 1983). In silicates, the range of nearest-neighbor distances sampled depends on the amount of disorder and on the backscattering amplitude of the neighbors of an absorber. For example, analysis of Zr K-EXAFS spectra in crystalline (non-metamict) zircon (with eight oxygens at an average distance of 2.21 Å, and second- and third-nearest neighbor Zr and Si at distances between 3.0 and 5.0 Å) shows distinct distance contributions to above 5 Å. In contrast, similar study of a sodium zirconium silicate glass shows distances only up to 3 Å around Zr, corresponding to six oxygen nearest neighbors and Si in the second coordination shell (Waychunas and Brown, 1984).

5.4.3 Distance information from XANES spectra. Distance information can also be obtained from analysis of XANES spectra for an element, although the procedure is not as straightforward and the theory is not as well developed as for EXAFS analysis. Potentially, the information content of the XANES spectrum exceeds that of the EXAFS spectrum for an element because of the multiple-scattering origin of the former. Because of this origin, XANES spectra can yield information about interatomic angles under favorable circumstances. Currently, extraction of distance information from XANES depends on the theoretical relationship developed by Natoli (1984), which shows that $(E_r - E_b) = constant/d(A-L)^2$, where E_r and E_b are the energies of the resonance feature and the electron's bound state, respectively, and where d(A-L) is the distance from absorber to ligand. This relationship has been successfully applied to determination of the average Fe-S distance in Fe-containing ZnS (sphalerite) (Sainctavit et al., 1986). It has also been used recently to correlate M-O distances in transition metal oxides (Lytle et al., 1988).

5.5 Coordination Number

5.5.1 Coordination geometry from EXAFS spectra. Information on the number of ligands around an atom (the coordination number = CN) can be derived directly from EXAFS analysis and indirectly from XANES spectra. In EXAFS analysis, the amplitude of EXAFS oscillations is related to the number of backscattering ligands. Figure 5-2 shows this effect for almandine (Fe^{2+} surrounded by 8 oxygens) and $LiAlO_2:Fe^{3+}$ (Fe^{3+} surrounded by 4 oxygens). The accurate measurement of amplitude is more difficult than measurement of phase in an EXAFS spectrum, because of several factors, including heterogeneity of sample and structural (static or thermal) and chemical disorder effects. The effect of disorder on amplitude is seen in the study of fayalite by Waychunas et al. (1986a), where an average CN of 4.3 oxygens was found. As discussed in §5.4.2, the distribution of Fe-O distances in the two 6-coordinated iron sites is quite irregular and may account for the low CN when only a single shell of scatterers is considered. As with distance determination, thermal disorder effects gives lower values of CN; however, this problem can be minimized by low-temperature measurements. Generally speaking, the accuracy of CN determined by least-squares fitting of EXAFS data to a model is ±20%; this accuracy level can be improved to ±10%, if well-characterized model compounds are used for calibrating EXAFS phases and amplitudes and if data are of high quality. The precision level of CN information from least-squares fitting is of the order of ±0.2-0.4.

For transmission EXAFS measurements, the problem of sample heterogeneity can be severe, especially for a highly absorbing sample. If the sample thickness is not uniform or if the sample has pinholes, the ratio I_0/I will be lowered and the EXAFS amplitude will show an apparent reduction. This problem will reduce CN but has little effect on distance determination. Another serious problem affecting both transmission and fluorescence measurements, which also reduces EXAFS amplitude, is the high energy harmonic content of the incident synchrotron x-ray beam, which is exasperated by the non-linear response of ion chamber detectors. Such beam harmonics will be diffracted by a conventional single crystal or channel-cut mono-chromator as higher order reflections, which will pass through the sample without significant absorption. These harmonic contributions can be easily eliminated by using a double-crystal monochromator, with one of the two crystals slightly detuned, or by employing a mirror before the incident beam monochromator.

EXAFS-derived distances can help constrain coordination numbers in minerals because of well-established correlations between interatomic distances and CN (e.g., Shannon and Prewitt, 1969). For example, the Fe^{2+}-O distances for 4-, 6-, and 8-coordinate sites are 2.01, 2.16, and 2.25 Å, respectively. Unfortunately, disorder effects tend to reduce both interatomic distance and CN. Therefore, bond distances cannot be used to estimate the magnitude of disorder or its effect on CN in unknown compounds.

5.5.2 Coordination geometry from XANES spectra. XANES spectra are particularly sensitive to the first- and higher-neighbor coordination geometry of the site occupied by an absorber. When comparisons are made between XANES spectra of model compounds with known coordination geometries, and unknown compounds, information on CN may be obtained which is complementary to that from EXAFS. Clear differences in XANES spectra are seen for Fe^{2+} with different CN's (Waychunas et al., 1983; Calas and Petiau, 1984) (see Fig. 4-7). When an element exists with more than one CN in a mineral, however, the distinction between XANES features may not be so clearcut, and determination of the relative percentage of each CN species is hindered. Another example of CN information contained in XANES spectra may be derived from Al K-edge XANES of Al-containing oxides and silicates (Brown et al., 1983) (Fig. 5-3), in which contributions from both 4- and 6-coordinate Al are clearly distinct. In the case of sillimanite (Al_2SiO_5), the presence of both 4- and 6-coordinate Al is revealed by the presence of both a single line, due to Al(IV), and a double line shifted by 2 eV, due to Al(VI).

Another very distinctive indicator of CN is found in the pre-edge feature of XANES spectra when a transition element with unfilled d-orbitals is present in a non-centrosymmetric site. This situation causes a strong intensification of the pre-edge feature due to increased d-p orbital mixing, as is discussed in more detail below. An example of this difference is shown in Figure 5-4 for Fe^{3+} in tetrahedral and octahedral aqueous complexes (Apted et al., 1985). The pre-edge feature in the 1.0 M $FeCl_3$ solution with 15.0 M Cl^- indicates a tetrahedral tetrachloro

484

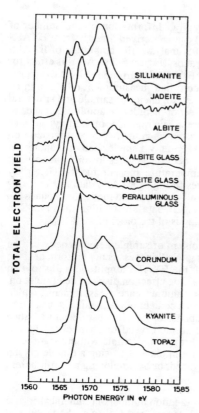

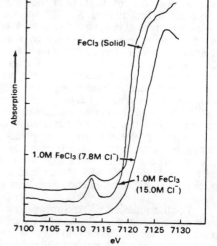

Figure 5-4 (above). Fe K-edge absorption spectra of iron chloride solutions. (From Apted et al., 1985)

Figure 5-3 (above). Al K-XANES spectra of alumino-silicate glasses and crystalline model compounds. (From Brown et al., 1983)

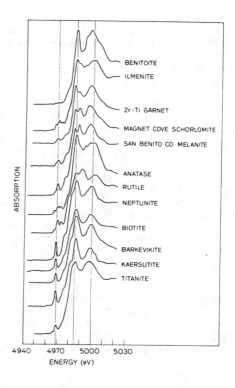

Figure 5-5 (right). Ti K-XANES spectra of various Ti-bearing minerals. (From Waychunas, 1987)

ferric iron complex, whereas the lack of this feature in the 1.0 M FeCl$_3$ solution with 7.8 M Cl$^-$ indicates an octahedral hexa-aquo ferric iron complex. Recent applications of this effect include site determinations for transition elements in spinels (4- versus 6-coordinated) (Lenglet et al., 1986) and in the clay mineral nontronite (Bonin et al., 1986). In favorable cases, this method permits detection of 1% or more of a tetrahedrally coordinated transition element in the presence of majority octahedral species.

5.6 Site Distortion

In this section we define site distortion as departure from a given local point group symmetry, and we defer until a later section (§5.7) consideration of effects of radial and static disorder on XANES and EXAFS spectra. In XANES spectra of transition elements, the presence or absence of a center of symmetry in a coordination site has a major effect on the pre-edge intensity, especially when the d orbitals are empty (see §4.3). This effect is caused by the possibility of d-p mixing when the point group symmetry lacks an inversion center. This fact has been exploited in the study of Ti^{4+} in minerals (Fig. 5-5) (Waychunas, 1987) and silicate glasses (Greegor et al., 1983; Dumas and Petiau, 1986). The plot of Ti site distortion, as estimated from bond angle variance values (σ^2) from x-ray structure refinements, *versus* the relative intensity of the pre-edge feature for a number of Ti^{4+} minerals (Fig. 5-6) shows a good correlation of these variables. Such a correlation can be used to estimate the percentage of tetrahedral versus octahedral Ti^{4+} in minerals and glasses (Greegor et al., 1983). When the titanium resides in both tetrahedral and distorted octahedral sites, as is common in glasses, quantitative estimation of the percentage of each type of species requires care. A different situation arises when a more electronegative element such as chlorine coordinates Ti^{4+}, as in gaseous TiCl$_4$ (Greegor et al., 1983). In this case, the smaller amount of p character of the final state reduces the intensity of the pre-edge. Thus, degree of covalency must be taken into account when using the XANES pre-edge intensity to estimate site distortion.

For 3d-transition elements with a partially filled d-shell, application of this method is not as sensitive to site distortion as for Ti^{4+}-containing solids. This difference is seen in Fig. 4-14 for iron at different coordination geometries. In general, the pre-edge intensity for Fe^{2+} in octahedral sites, relative to the main edge jump, ranges from 0.7% for small deviations from O$_h$ point symmetry to 2.0% for large deviations. This intensity increases to 5 to 7% for Fe^{2+} in tetrahedral coordination and can be up to 15% for Fe^{3+} in 4-fold coordination. These values are significantly lower than those observed for Ti^{4+} in different coordination geometries (Fig. 5-6).

When site distortion is large, as for Fe^{2+}-containing hypersthene and Mn^{3+}-containing epidote, the main absorption edge may be split into two peaks (see Figs. 4-1 and 4-14). This

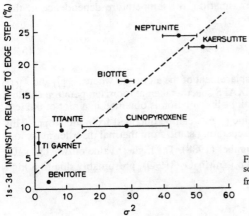

Figure 5-6. Plot of the Ti K-XANES pre-edge intensity of selected Ti-bearing minerals versus site distortion, estimated from bond angle variance (σ^2). (From Waychunas, 1987)

splitting may be explained in terms of differences in multiple-scattering paths caused by extreme site distortion, as is true for the M2 site of orthopyroxene and the M3 site of epidote.

5.7 Disorder

Disorder manifests itself in two fundamental ways in a structure: static and thermal (vibrational). Increasing disorder due to either produces a damping of the EXAFS signal at higher k values, as shown in Figure 3-3 for fayalite at different temperatures. This effect is more important in EXAFS scattering than in Bragg diffraction because the expression for Debye-Waller attenuation in an EXAFS model equation has the form $\exp[-2\sigma^2 k^2]$ instead of $\exp[-\sigma^2 k^2]$ as present in the expression for the x-ray or neutron structure factor. This disorder problem can create a serious limitation in quantitative structural analysis using EXAFS spectroscopy, and emphasizes the short-range character of EXAFS-derived structural information. When a high degree of static disorder is present in a sample (as is the case for glasses or melts with highly coordinated, weakly bonded elements), the bond lengths derived from EXAFS analysis may be too short, if the EXAFS formulation uses a Gaussian distance distribution to model disorder and the distance distribution is non-Gaussian or thermal vibration is anharmonic (for high temperature EXAFS studies). This effect can be corrected for, in part, by an appropriate analytical expression for disorder which accounts for the asymmetric nature of the distribution (see e.g., Teo, 1986). In contrast to EXAFS, relatively high static or thermal disorder is required to significantly broaden multiple-scattering features of the absorption edge. Bound-state features such as the pre-edge observed for certain transition elements in non-centrosymmetric sites will not be affected by disorder.

A direct measure of the local disorder around an absorber can be obtained from the magnitude of the Debye-Waller factor in EXAFS analysis; typical values for iron in crystalline silicates and oxides at room temperature fall in the range 0.01 $Å^2$ (for 4-coordinate Fe) to 0.03 $Å^2$ (for 6- and 8-coordinate Fe) (Waychunas et al., 1986a). In silicate melts at temperatures up to 1200 K, the Debye-Waller factors for Fe are as high as 0.11 $Å^2$ (Waychunas et al., 1988). In general, these Debye-Waller values should be no more accurate than EXAFS-derived coordination numbers since σ^2 is highly correlated with coordination number and is, therefore, affected by inaccuracies in amplitude measurements. Precision levels for EXAFS-derived σ^2 values is of the order of ± 0.01 $Å^2$. A more qualitative measure of local disorder can be obtained by observing the width of a peak corresponding to a particular coordination shell in the Fourier transform; wider peaks indicate a greater range of radial distances and higher static or thermal disorder.

It is not possible to separate static from thermal disorder effects unless temperature-dependent studies are made. However, such studies are feasible using appropriate sample cryostats and heaters. In the harmonic approximation, the temperature dependence of the Debye-Waller factor is given by the expression:

$$\log[\chi_{j(T1)} / \chi_{j(T0)}] = 2k^2 \Delta\sigma^2 \quad , \tag{5-1}$$

where $\Delta\sigma^2$ is the difference in mean-square displacement of the atom and where $\chi_{j(T1)}$ and $\chi_{j(T0)}$ are the experimental, background-subtracted EXAFS spectra taken at room temperature and low temperature (30 K), respectively. When a straight-line relation is obtained in a plot of the left-hand side of Equation (5-1) versus k^2, the value of $\Delta\sigma^2$ can be obtained from the slope. This procedure has been applied to the problem of separating static from thermal disorder in various manganese oxide minerals by Manceau and Combes (1988). The highest value of $\Delta\sigma^2$ (5.1×10^{-3} $Å^2$) was obtained in the mineral chalcophanite ($ZnMn_3O_7 \cdot 3H_2O$), presumably due to a high concentration of vacancies.

5.8 Identity of Neighboring Atoms

Differences in backscattering amplitude and in phase shift exist for atoms of significantly different atomic number (see Figs. 3-4 and 3-5). Because these parameters are separately treated and can be essentially independent of each other in EXAFS analysis, it is possible, in many cases, to distinguish between different neighboring atoms in a coordination shell. This feature of EXAFS analysis is useful in the study of metal complexes in aqueous solutions or in glasses where the first-coordination shell might contain a mixture of different ligands. It is also useful in the study of minerals exhibiting solid solution, or in glasses and melts having variations in short-range ordering. This ability of EXAFS analysis to provide direct and quantitative short-range order information is unique among spectroscopic methods.

Examples of this feature of EXAFS analysis are found in a study of the first coordination shell of Zn complexes in chloride solutions (Brown et al., 1988) and in a study of possible halogen complexes of La, Gd, and Yb in silicate glasses (Ponader and Brown, 1988b). The significant differences in phase shift and backscattering amplitude between oxygen and chlorine (see Fig. 3-6) permitted differentiation between $ZnCl_4$ complexes and $Zn(H_2O)_n$ (n = 5-7) in aqueous chloride solutions and showed that these rare earth elements were not present as Cl complexes in the glasses studied . Another example of this feature is detection of PbO_2Cl_4 complexes in $PbO-PbCl_2$ glasses (Rao et al., 1984). Examples of the detection of short-range ordering in mineral solid solutions by EXAFS analysis will be discussed in the §6.2.2.

5.9 Interpolyhedral Linkages

Although the short-range, one-dimensional nature of EXAFS has been stressed, it is possible in certain favorable cases to extract from EXAFS analysis information about the geometry of interpolyhedral linkages. An excellent example of this is the study of crystalline and glassy GeO_2 (Lapeyre et al., 1983; Okuno et al., 1986). Utilizing first-shell Ge-O distances of 1.74 Å and second-shell Ge-Ge distances of 3.15 Å, the average Ge-O-Ge angle of 130° was derived for both crystal and glass. This finding shows that the average short-range structure of crystalline GeO_2 is maintained in the glassy state. The same procedure has also been used to derive Si-O-Ti (113°) and Zn-O-Si (130°) angles in cordierite glasses (Dumas and Petiau, 1986). When static disorder is large around either absorber involved in the linkage, however, this method is not reliable.

5.10 Bonding Information

Because of the nature of the processes producing XANES, particularly the pre-edge features and the features of the main absorption edge due to bound-state transitions, some information on bonding may be extracted from XANES spectra. For example, the pre-edge of transition metal complexes can be correlated with anti-bonding molecular orbitals involving the metal d and ligand p atomic orbitals (see §4.3 and 4.4). When experiments are designed to take advantage of the natural polarization of synchrotron radiation, polarized single-crystal absorption edge measurements can be made. Such measurements yield valuable information on directional bonds and partially overcome the one-dimensional nature of EXAFS information. For example, the polarized XANES study by Smith et al. (1985) on planar $CuCl_4^{2-}$ complexes in ($CuCl_4$)-bis creatinium and Cu-Cl_2-di-pyridine show dramatically different spectra in the different polarizations due to differences in bonding in and out of the plane of the $CuCl_4$ complex because of different ligands. Significant differences are also seen in the polarized Cu K-XANES of [Cu(dien)(bipyam)](NO_3) (Smith, 1985) (Fig. 5-7). Another example comes from the single crystal XANES study of vanadyl compounds (Templeton and Templeton, 1980) which show strong polarization dependence due to differences in the V-O vanadyl bond and equatorial V-O bonds. The first polarized XANES-EXAFS study of minerals by Manceau et al. (1988) showed significant polarization dependence of iron in biotite and chlorite. Such studies have the potential to yield direct information on specific bond orientation and orbital contributions to these bonds.

Additional information involving differences in electronic structure caused by different second-neighbor ligands can be derived from XANES spectra. For example, the Fe K-XANES of andradite (an insulator) and hematite (with lower band gap) show distinct changes due to differences in Fe second-neighbor bonding (Fig. 4-14) (Calas and Petiau, 1983). A pre-edge

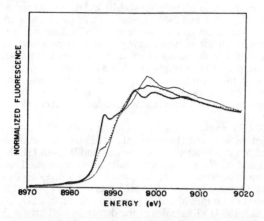

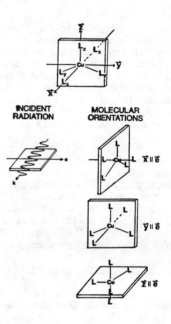

Figure 5-7. Polarized Cu K-XANES spectra of 5-coordinated Cu in [Cu(dien)(bipyam)](NO$_3$). z ∥ e (dark solid line), x ∥ e (light solid line), y ∥ e (dotted line). Molecular orientations are shown at the right. (After Smith, 1985)

feature is observed in the spectrum of hematite at higher energies thought to be caused by multi-electronic effects in the conduction band. This feature is missing from the pre-edge region of Fe-containing insulators.

Electronic structures of sulfides and sulfosalts can be derived from the study of XANES spectra of both cation and anion. The advantage of XANES studies of these structures is the ability of XAS to separate bonding contributions to the projected densities of states from cations and anion. Such studies have been made on chalcopyrite (CuFeS$_2$) and on Fe^{2+}-containing sphalerite (ZnS) (Sainctavit et al., 1986). The three K-edge spectra of Cu, Fe and S in chalcopyrite are shown in Figure 4-12. They are correlated by defining the Fermi level at the first inflection point of the edge: the corresponding local density of states calculated by Hamajima et al. (1981) using an SCF-Xα formalism is also shown and reproduces many features of the experimental spectra.

Bonding information can also be obtained indirectly from temperature-dependent EXAFS measurements as shown by Yang et al. (1986) for arsenic sulfides. Differences in estimated mean-square displacements of As at different temperatures were related to bond stretching force constants for both S nearest neighbors and As second nearest neighbors. The results show that the As-S displacements are lower than As-As displacements with increasing temperature, indicating that As-S bonds have larger force constants, in agreement with Raman measurements. XANES can also be compared to x-ray emission spectra (see e.g., Fischer, 1973) to aid in the assignment of spectral features and to constrain molecular orbital diagrams. This possibility is discussed further in Chapter 14.

VI. MINERALOGICAL AND GEOCHEMICAL APPLICATIONS OF X-RAY ABSORPTION SPECTROSCOPY

6.1 Introduction

Applications of synchrotron-based x-ray absorption spectroscopy in the earth sciences are relatively new, beginning in 1978. Only a few mineralogical studies have been completed to date. However, recent increases in the number of synchrotron facilities around the world and the resulting increase in "beam time" should cause rapid growth in this area. Applications to date have taken advantage of the unique attributes of XAS spectroscopy (its element specificity and the quantitative, short-range structural and chemical information it provides) and the very high x-ray fluxes of synchrotron sources.

The types of mineralogical and geochemical problems that have been addressed with XAS can be divided into the following categories:

(1) structural studies of crystalline materials, including characterization of local structure, oxidation states, and bonding in compositionally complex minerals, and the determination of short-range ordering of cations;

(2) local structure of highly disordered materials such as glasses, gels, solutions, and melts;

(3) identity and structural characterization of dilute species in minerals and glasses;

(4) phase transformations such as nucleation processes;

(5) in-situ studies of surface reactions at water/mineral interfaces;

(6) characterization of local structure in extremely small samples.

New or future applications include high-pressure, high-temperature studies of minerals, with the possibility of real-time observation of transient phenomena such as solid-solid phase transitions or melting. Review of these applications will not be exhaustive, but instead will highlight those studies that demonstrate the advantages and limitations of XAS studies of earth materials.

6.2 XAS studies of minerals

6.2.1 Determination of Local Structure and Cation Oxidation State in Minerals. XAS analysis of local structural details in minerals complements x-ray diffraction analysis of mineral structures, particularly in compositionally complex minerals where more than one element can occupy one type of crystallographic site. An illustration of the type of structural and crystal chemical information that XAS can provide for a compositionally complex mineral comes from the study of Co and Ni in supergene manganese oxides with two-dimensional structures (asbolane and lithiophorite) (Manceau et al., 1987). XAS studies have revealed that these transition elements exhibit different oxidation states and local structure. Co K-edge XANES spectra indicate trivalent, 6-coordinated cobalt in all the phases studied. EXAFS study has shown these ions to have the same local environment as manganese (Fig. 6-1), which suggests that Co is randomly distributed in layers occupied by Mn. The short Co-O bond length (1.92 Å) and the unique XANES spectrum suggest that Co^{3+} occurs as a low-spin octahedral cation, with a subsequent high crystal-field stabilization energy. This suggestion could explain the well known selective up-take of Co by Mn oxides. Unlike Co, Ni exhibits different local surroundings in both lithiophorite and asbolane. In the former, Ni atoms are located in hydrargillite $Al(OH)_3$ layers, and in the latter they occupy sites in separate $Ni(OH)_2$ layers of unknown extent.

The study of Ti in silicates and oxides has been hindered by the lack of definitive information on its formal oxidation states and site geometry. The site geometry of Ti^{4+} cannot be studied directly by optical absorption spectroscopy or most other spectroscopic methods, but can be characterized by XAS as demonstrated by Waychunas (1987). Figure 5-5 shows XANES spectra for a variety of Ti-bearing silicates and oxides. These spectra are characterized by a pre-edge feature whose intensity varies markedly, depending on the distortion of the Ti site (Fig. 5-6) (see §4.3 and 5.6). In addition, the positions of the pre-edge and edge features can be used to estimate the oxidation state of Ti in these phases because of the 1-2 eV shift towards higher energy in going from Ti^{3+} to Ti^{4+}. The main conclusions from this work are (1) that Ti^{3+} is

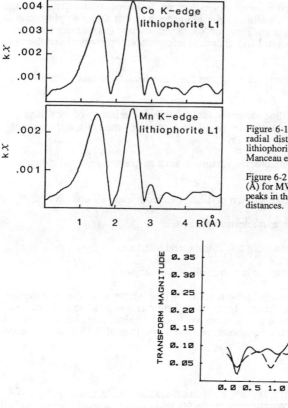

Figure 6-1 (left). Comparison of EXAFS-derived radial distribution functions of Mn and Co for lithiophorite [(Al,Li,Ni)(Mn,Co)O_2(OH)$_2$]. (After Manceau et al., 1987)

Figure 6-2 (below). Fourier transform magnitude vs. R (Å) for MW$_{90}$ (solid line) and MW$_{70}$ (dashed line). The peaks in these rdf's represent different Fe-near-neighbor distances. (From Waychunas et al., 1986b)

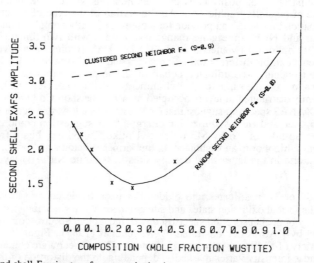

Figure 6-3. Second-shell Fourier transform magnitudes in magnesiowustites (MW) solid solutions versus composition. A short range order (SRO) parameter S is defined as $S = [P(Fe-Fe) - X_{Fe}]/(1-X_{Fe})$, where P(Fe-Fe) is the probability of Fe second-nearest neighbors around the central Fe, and X_{Fe} is the mole fraction of Fe in the solid solution. $S = 0.0$ represents no SRO, whereas $S = 0.9$ represents a high degree of clustering of Fe atoms. The experimental data fall close to the random second-neighbor curve, indicating no significant SRO in these magnesiowustites.

relatively rare, even for samples predicted to have significant Ti^{3+} from previous indirect spectroscopic studies, and (2) that tetrahedral Ti^{4+} occurs only at very small concentration levels in silicates, the majority occuring in 6-coordinated sites with varying degrees of distortion.

Other examples of determination of site geometry and/or oxidation states of elements in minerals include the location of Fe-atoms in natural kaolinites (Bonnin et al., 1982); Fe-site occupancy in Garfield nontronite (Bonnin et al., 1985); Fe-Ni segregation in smectites (Decarreau et al., 1987b); crystal chemistry of chromium in phyllosilicates (Calas et al., 1984b); site occupancies in chromites and Cu-oxidation states in sulfides (Calas et al., 1986); location of calcium in phophates (Harries et al., 1987); modification of the Ca-site geometry in diopside-jadeite solid solution (Davoli et al., 1987); oxidation state and location of copper in vermiculites (Ildefonse et al., 1986); location of Tc, Er, and Lu in epidote (Cressey and Steel, 1988); and characterization of the Cu and Ag sites in argentian tetrahedrites (Charnock et al., 1988).

6.2.2 Short-range order in minerals. One of best uses of EXAFS spectroscopy takes advantage of its ability to distinguish among types of nearest neighbors and to provide quantitative estimates of the extent of short range ordering (SRO) around a selected element in solid solutions. In this application, the inability of the EXAFS probe to detect long-range ordering is a distinct advantage. As has been pointed out earlier, EXAFS is sensitive to interatomic distances, coordination numbers, and the chemical nature of elements within the first few coordination shells surrounding the absorbing atom. The basic assumption made when studying SRO in minerals is that the local structure around the absorbing atom is known. If this structure is not known, the model cannot separate structural from compositional effects resulting in phase cancellation. In addition, this assumption may not be valid when the absorber is a dilute species because local structural relaxation often accompanies element substitutions, and this relaxation is not easy to characterize by spectroscopic or scattering methods other than EXAFS.

An excellent example of the power of EXAFS analysis for determining SRO in oxides is the study of Fe^{2+} substitutions in MgO (Waychunas et al., 1986c). Figure 6-2 shows the Fourier transform magnitude of two magnesiowüstites of compositions $Fe_{0.1}Mg_{0.9}O$ (MW90) and $Fe_{0.3}Mg_{0.7}O$ (MW70). The second-neighbor Fourier transform amplitude of MW90 is significantly larger than that of MW70 because of phase cancellation due to backscattering by Fe and Mg at similar distances. This result indicates that Fe in MW90 has mainly Mg second neighbors. Figure 6-3 plots this second shell amplitude versus mole fraction wüstite for a series of magnesiowüstites quenched rapidly from 1140°C. The data follow a parabolic trend expected for a random arrangement of second-neighbor Fe and Mg atoms and are quite different from the trend predicted for almost complete ordering of Fe in the second-neighbor shell around the absorbing Fe atom. A similar study of Fe^{3+}-doped MgO samples, which were annealed at low temperature, shows evidence for the clustering of Fe^{3+} (Waychunas, 1983).

A more complex solid solution is found in the system diopside ($CaMgSi_2O_6$: Di) – hedenbergite ($CaFeSi_2O_6$: Hd), and EXAFS analysis can be used to obtain SRO information about Fe (Waychunas et al., 1986c). A plot of second neighbor Fourier transform amplitude versus mole percent hedenbergite (Fig. 6-4) shows that the data for samples of Di-Hd solid solution deviate from the trend predicted for a random arrangement of second-neighbor Fe and Mg. This model assumes that Ca and Si remain at the M2 and tetrahedral sites, respectively, and indicates significant clustering of Fe in the edge-sharing chains of M1 octahedra.

SRO studies have also been done on nickel-containing clay minerals using XAS (Manceau and Calas, 1985, 1986). These minerals represent Mg-Ni solid solutions in the 10 Å and 7Å families, and the location of nickel in the most dilute ore-minerals (the so-called "garnierites") has been the subject of debate. This work has shown that nickel is 6 coordinate in a divalent oxidation state, confirming optical absorption spectroscopic studies. The second shell around nickel contains two Si atoms in the tetrahedral layer and six (Ni,Mg) atoms in the octahedral layer. The interatomic distance between the octahedral Ni absorber and octahedral (Ni, Mg) backscattering atoms is equal to b/3 and may be derived from the b cell parameter determined by x-ray diffraction. As in the case of Fe-Mg substitutions in magnesiowüstites and pyroxenes, these studies take advantage of the almost complete cancellation of the total phase

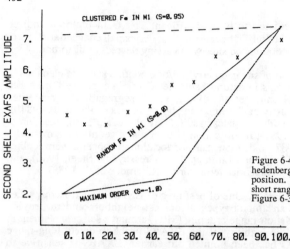

Figure 6-4. Second-shell Fourier transform magnitude of hedenbergite-diopside (Hd-Di) solid solution versus composition. The data indicate some clustering of Fe. The short range order parameter S is defined in the caption of Figure 6-3.

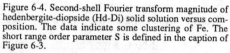

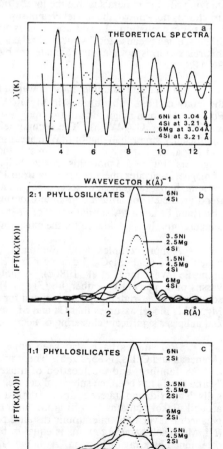

Figure 6-5. Simulation of the second shell around nickel in phyllosilicates as a function of Ni versus Mg content in the octahedral layer. (a) shows the partial EXAFS calculated for 2:1 phyllosilicates. (b) shows the Fourier transforms of these partial EXAFS functions and others representing intermediate compositions. (c) shows Fourier transforms of partial EXAFS functions for 1:1 phyllosilicates over the same range of compositions. (From Manceau and Calas, 1986)

shift when nickel and magnesium backscatterers are simultaneously present in the second coordination shell. This cancellation will make the intensity of the corresponding peak in the Fourier transform very sensitive to the chemical constitution of this shell (Fig. 6-5). The resulting analysis of phyllosilicates of variable Ni content shows strong deviation from a random distribution of cations around nickel (Fig. 6-6), indicating that nickel atoms are segregated in domains, even at the lowest concentration studied (3 mole % Ni). This result suggests that these minerals, derived from supergene transformation of serpen-tinites, have been formed at temperatures too low to favor a random substitution of Ni in the structure.

6.3 Disordered Materials

Disordered natural phases constitute an important class of earth materials and include natural glasses, melts, aqueous solutions, gels and the poorly-crystallized phases that form from them, and metamict minerals that are disordered because of radiation damage. The geochemical behavior of most elements, as well as their phase relations, are controlled in part by the types of sites each element occupies in these disordered phases. Until recently, little has been known about the structural variations present in these materials because few methods can yield structural data on these structurally and chemically complex materials (see chapters on NMR spectrsocopy, this volume). XAS provides one of the few quantitative probes of local structure that is also element specific. Thus it is particularly well-suited for characterization studies of such materials, and significant results have been obtained using XAS. Some of these applications are reviewed below.

6.3.1 Silicate Glasses. One of the major applications of XAS spectroscopy is the study of local structural environments in glasses. Glasses possess little structural order beyond a few angstroms around a cation or anion, i.e., beyond the first coordination shell. Thus the local nature of the structural information provided by XAS, the fact that this information is essentially one dimensional, and the method's element selectivity make XAS well-suited for the study of multicomponent glasses. Several recent reviews of XAS studies of amorphous silicates present reasonably up-to-date treatments of these studies (Calas and Petiau, 1983a; Greaves et al., 1984; Brown et al., 1986b; Calas et al., 1987). Therefore, the discussion below will be limited to only a few representative applications of XAS to glasses.

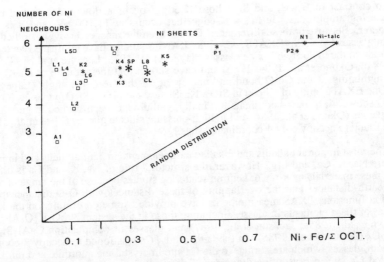

Figure 6-6. Experimentally determined number of nickel second neighbors around a central Ni atom in the octahedral layer as a function of the atomic Ni(Fe) content for various low temperature Mg-Ni phyllosilicates. (From Manceau and Calas, 1986)

Generally speaking, the cations comprising a glass or melt can be divided into two categories: "network formers" and "network modifiers". The first group includes the cations B, Si, Ge, Al, and Ga, whereas the second group includes most of the other cations of the periodic table. Network formers are characterized by high field strengths [= formal charge/d$(M-L)^2$] and directional bonding, which causes them to have well-defined coordination spheres. Thus their local structural environments are realtively easy to define by XAS analysis. The local environments of network modifiers in glasses and melts ranges from well-defined coordination spheres for cations with intermediate field strengths and relatively strong bonds to oxygen, such as Mg^{2+} and the 3d-transition elements, to large, poorly-defined coordination spheres, for cations of low field strength, such as the alkalis and the larger alkaline earths, which form weak bonds with oxygen. Because of the range of distances surrounding these cations in glasses, the site geometry of network modifiers may be difficult to define accurately using XAS, but this method is one of the few capable of providing any distance information at all. When the first shell of ligands surrounding an absorber exhibits strong disorder, as in the case of alkalis in silicate glasses, the second and higher shells will exhibit even greater disorder, making EXAFS analysis of medium-range ordering virtually impossible. In contrast, EXAFS analysis can yield limited information concerning medium-range ordering around network-forming elements like Ge, as was demonstrated earlier (§5.9).

Aluminum in natural silicate melts and in industrial glasses normally acts as a network-forming element. However, for sodium aluminosilicate melts or glasses with peraluminous compositions (where the Al/Na atom ratio exceeds 1.0), some have suggested that a fraction of the Al occurs in octahedral coordination and acts as a network modifier, to explain anomalous changes in physical properties, such as viscosity and electrical conductivity as Al/Na ratio is varied (Hunold and Bruckner, 1980). This possibility has been assessed by XAS study of Al in a series of sodium aluminosilicate glasses using high vacuum, total electron yield detection methods. Such methods are necessary because of the very strong air and matrix absorption problems caused by the low energy of the Al K-edge (1560eV). Analysis of Al K-XANES (Brown et al., 1983) (Fig. 5-3) and Al K-EXAFS (McKeown et al., 1985b) for a series of glasses in the $Na_2O-Al_2O_3-SiO_2$ system (with variable Al/Na and constant Si) clearly shows that Al remains in tetrahedral coordination, even in peraluminous glasses with Al/Na $\leq$ 1.6. Thus another cause besides octahedral aluminum must be sought to explain the effect of Al/Na ratio on the transport properties of these glasses and melts.

In contrast to B, Al, and Si, whose K-edges occur at low energies and have low backscattering amplitudes, Ge has a K-edge that occurs at 11,100 eV, and it is a strong backscatterer. Because of these differences, Ge has been the most extensively studied network former in glasses by XAS. EXAFS as well as XANES studies have confirmed its tetrahedral coordination in the majority of oxide glasses (Sayers et al., 1975; Lapeyre et al., 1983; Greegor et al., 1984b; Okuno et al., 1984, 1986) and have shown that the addition of Na_2O causes a slight lengthening of the Ge-O bond and narrowing of the Ge-O-Ge angle (Lapeyre et al., 1983). One EXAFS study of silicon in SiO_2, $Na_2Si_2O_5$, and $Na_2CaSi_5O_{12}$ glasses (Greaves et al., 1981) showed it to be 4-coordinated. Finally, gallium has been studied in germanate and silicate glasses (Okuno et al., 1984) and in high-pressure silicate glasses (Fleet et al., 1984) and has been found to occupy only 4-coordinated sites.

The most important alkalis and alkaline earth elements in natural melts and in industrial glasses are Na, K, Ca, and Mg. However, the structural role of Na, K, and Ca is difficult to assess in these materials by x-ray or neutron scattering studies because of the range of alkali- or alkaline earth-distances and the overlapping of these distances with O-O distances in radial distribution functions. XAS measurements have provided some constraints on the local site geometries of these elements in glasses, including those in the system $Na_2O-K_2O-Al_2O_3-SiO_2$ (McKeown et al., 1985a; Jackson et al., 1987) and glasses of composition $CaAl_2Si_2O_8$ and $CaMgSi_2O_6$ (Binsted et al., 1985). In the latter study, Ca was found to occupy 7-coordinated sites in both glasses, which are similar to the Ca site in crystalline anorthite and unlike the 8-coordinated Ca site in crystalline diopside. This result is consistent with the similarity in densities of anorthite, anorthite glass, and diopside glass, and the higher density of diopside. A similar study has also been made on Ca in a model basaltic glass by Hardwick et al. (1985).

The structural environments and oxidation states of 3d-transition metal ions in silicate glasses have been studied intensively by many spectroscopic methods, including XAS. Two examples are presented: one involving ferric iron and the other involving tetravalent titanium. XAS studies of other transition elements in glasses are listed in order to provide appropriate references. One of the first of these studies (Brown et al., 1978) examined the structural role of Fe^{3+} in glasses of composition $KFeSi_3O_8$, $NaFeSi_2O_6$, and $NaAl_{.85}Fe_{.15}Si_2O_6$ quenched from 1450° C at one atmosphere pressure. The Fe K-XANES spectra of these glasses is compared with that of crystalline $NaFeSi_2O_6$ (acmite) in Figure 6-7 and show a pre-edge feature that is about 15% of the edge jump. In contrast, the pre-edge of acmite, in which the Fe^{3+} occurs in a somewhat distorted octahedron, is much less intense. This comparison suggests that Fe^{3+} in these glasses is tetrahedrally coordinated, giving rise to a strong 1s→3d transition. This suggestion was confirmed by EXAFS analysis of the glasses, which gave Fe-O distances of 1.87 Å – a value typical of $^{IV}Fe^{3+}$-O. It was further confirmed by ^{57}Fe Mossbauer spectroscopy on the same glass samples which gave I.S. values of 0.04 mm/sec, relative to Fe in Pd, and showed a very small amount (< 5%) of Fe^{2+}. Finally, optical absorption spectroscopy and x-ray radial distribution analysis provided results consistent with the assignment of Fe^{3+} to network-forming, tetrahedral sites in the glasses. The study of Fe^{3+} in sodium disilicate glasses (Calas and Petiau, 1983a) also finds iron to be tetrahedrally coordinated, in contrast with the EXAFS study by Park and Chen (1982), which concluded that Fe^{3+} is octahedrally coordinated. The presence of second and further neighbor contributions to the Fourier transform of Park and Chen's EXAFS spectrum is inconsistent with their experimental spectrum, which shows a single damped sine wave characteristic of contributions from only a single coordination shell.

Ti^{4+} in silicate glasses has been the subject of several XAS studies because its structural role has not been clearly defined by other methods. It is known to occur in 4-, 5-, and 6-coordinated sites in crystalline oxide structures; however, its coordination environment in most glasses is not known. Ti K-XAS studies of TiO_2-SiO_2 glasses (Sandstrom et al., 1980; Greegor et al., 1983) concluded that Ti^{4+} is primarily 4-coordinated, with a small amount being 6-coordinated on the basis of multiple-shell fits to the EXAFS and interpretation of the intensities of the pre-edge feature. They also found that the proportion of 6- to 4-coordinated Ti^{4+} increased with increasing TiO_2 content in these glasses. An interesting structural study of a glass of composition K_2O-TiO_2-$2SiO_2$ (Yarker et al., 1986) combined XAS and neutron scattering data on isotopically substituted samples to determine the structural environment of Ti^{4+}. A two-shell fit of the EXAFS data gave four oxygens at a distance of 1.93 Å and one oxygen at 1.64 Å, which compares well with distances of 1.96 and 1.65 Å, respectively, from neutron diffraction. These data indicate that Ti in this glass is 5-coordinated in a tetragonal pyramid. In an XAS study of multi-component aluminosilicate glasses of cordierite (Dumas and Petiau, 1986) and spodumene composition (Ramos et al., 1985), which used one-shell fits to the data, Ti was found to be 4-coordinated, with an average Ti-O distance of 1.85 Å. This result can be reinterpreted as consistent with the study of Yarker et al. (1986).

Other transition elements in glasses have been investigated using XAS methods, including Fe^{2+} (Calas et al., 1980; Calas and Petiau, 1982; Waychunas et al., 1986b, 1988), Mn^{2+} (Calas and Petiau, 1982), Co^{2+} and Ni^{2+} (Petiau and Calas, 1982), Zn (Pettifer, 1981; Dumas and Petiau, 1986); Zr (Taylor and McMillan, 1982; Waychunas and Brown, 1984; Petiau et al., 1984; Dumas et al., 1985; Brown et al., 1986a; Farges and Calas, 1987). There have also been EXAFS studies of Tc (Antonini et al., 1985), REE (La, Gd, Yb) (Ponader and Brown, 1988a,b), and U (Petiau et al., 1984; Petit-Maire et al., 1986) in silicate and borosilicate glasses. XAS study of the structrural environments of La, Gd, and Yb (at 2000 ppm) in several silicate glasses of differing polymerization prepared with and without fluorine and chlorine (Ponader and Brown, 1988a,b), detected fluorine complexes of Gd and Yb, mixed oxygen-fluorine complexes of La, but no chlorine complexes.

6.3.2 Silicate melts. As with other disordered materials, structural study of multi-component silicate melts is limited to only a few direct methods. The added complications of high-temperature sample containment and control of atmosphere present significant experimental difficulties. The use of synchrotron radiation-based XAS spectroscopy overcomes some of these difficulties. This method was used for the first time by Waychunas et al. (1986b, 1988) to

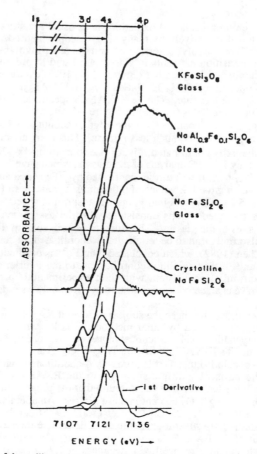

Figure 6-7. Fe K-edges of three silicate glasses and the model compound acmite, showing the first derivative spectra. The presence of intense pre-edge features for the glasses is consistent with Fe^{3+} in tetrahedral coordination. The pre-edge feature of acmite is relatively weak, which is consistent with octahedral Fe^{3+}. (After Brown et al., 1978)

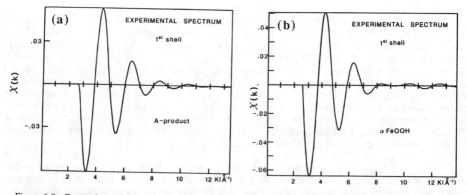

Figure 6-8. Comparison of the partial Fe K-EXAFS functions for the first shell around Fe^{3+}: (a) in a ferric gel produced by complete hydrolysis of ferric nitrate; (b) in goethite. Note that the distribution of Fe-O distances in goethite causes a beat pattern in the EXAFS function for goethite in the k range 8 to 11 $Å^{-1}$. (From Combes et al., 1986)

provide structural data on the coordination geometry of Fe^{2+} in silicate melts of compositions $Na_2FeSi_3O_8$ and $K_2FeSi_3O_8$ at temperatures up to 1200 K and under a vacuum of about 10^{-4} torr. Their EXAFS analysis gave Fe-O distances of 1.94 and 1.96 Å, respectively, for these compositions, and coordination numbers averaging 4. Perhaps the most surprising result of this study was the finding that the Debye-Waller factors for Fe in the melts are no larger than those of Fe in crystalline model compounds, such as $FeAl_2O_4$ (hercynite) at similar high temperatures. This result suggests that the amount of static disorder around the average Fe site in these melts is no greater than that in the well-crystallized oxides studied. The site geometry of Fe in the glasses quenched from these melts was also studied by EXAFS, XANES, and Mossbauer spectroscopies. The resulting Fe-O distances are 1.99 and 2.02 Å, respectively, for the Na and K compositions, and average coordination numbers are similar to those for the melts. The XANES spectra of the melts and glasses are characterized by the presence of strong pre-edge features, consistent with the presence of tetrahedral Fe^{2+}. Finally, the ^{57}Fe Mossbauer spectra gave I.S. values near 0.98 mm/sec relative to metallic Fe, and showed no significant levels of Fe^{3+} or metallic Fe. These results indicate that little structural relaxation occurred around the Fe site during quenching. This study shows that EXAFS spectroscopy has great potential for in-situ studies of various cation environments in multi-component silicate melts.

6.3.3 Hydroxide gels.

Gels are common in low-temperature geological environments and represent an intermediate stage in the formation of many sedimentary and supergene minerals, including silica and aluminum and ferric-iron oxides and oxy-hydroxides. The nature of the resulting crystalline phases strongly depends on the physico-chemical conditions (pH, temperature, activity of the component, etc.) of formation and evolution ("aging") of these gels. Unfortunately, differences among gels at various stages of evolution may not be distinguishable with conventional spectroscopic and diffraction methods because they are x-ray amorphous and the rdf would consist of all possible pair correlations. However, XAS methods are well-suited for the study of such gels because of their element selectivity, which permits derivation of pair correlations involving the element of interest. An Fe K-EXAFS study by Combes et al. (1986) on ferric-iron hydroxide gels has shown them to have short-range order, with shorter Fe-O radial distances (six oxygens at 1.96 Å) than in the derivative crystalline phases (three oxygens at 1.95 Å and three at 2.10 Å) (Fig. 6-8). Fe-O distances in the gels are similar to those encountered in the aqueous solutions (see §6.3.4). The most interesting result is that the local structure around iron is dependent on formation conditions. In gels precipitated from Fe^{2+}-containing solutions, FeO_6 octahedra were found to be edge-shared as in γ-FeOOH; however, in gels precipitated from Fe^{3+}-containing solutions, the local structure around iron is similar to that in goethite (chains of edge-sharing FeO_6 octahedra linked by corners). EXAFS study of the evolution of these gels to hematite at 92°C revealed an intermediate phase which has not been detected by other methods. This intermediate amorphous phase is characterized by the appearance of face-sharing octahedra, as in the structure of hematite; its appearance precedes the formation of hematite nuclei which can be detected by x-ray diffraction. The formation of ferric-silicate precipitates from basic solutions at room temperature and their evolution into the clay mineral nontronite have also been studied through XAS (Decarreau et al., 1987a). A strong local disorder around iron was observed, though a local smectite-like structure appears in "amorphous" precipitates. Although no changes in local order was observed at temperatures of 75 and 100°C, a significant evolution to a more ordered iron environment was observed at a temperature of 150°C.

6.3.4 Aqueous solutions.

Direct structural studies of dilute solutions (< 1 M) containing transition element complexes are difficult for non element-specific methods because of the weak signal from the complexes and the large background contribution from the solvent. These difficulties are largely overcome by XAS spectroscopy, and realization of this fact has led to numerous XAS studies of the local structure of transition element complexes in dilute and concentrated aqueous electrolyte solutions. In one of the first studies of this type, Eisenberger and Kincaid (1975) investigated the structure of Cu complexes in a dilute $CuBr_2$ solution using Cu K- and Br K-EXAFS in the transmission mode. They found Cu-O and Br-O distances of 1.97 and 3.14 Å, respectively. More systematic studies of a variety of aqueous electrolyte solutions at various concentration levels have since been made, including work on dilute (Sandstrom et al., 1977) and concentrated (Sandstrom, 1979; Licheri and Pinna, 1983; Sandstrom, 1984) nickel chloride solutions; highly concentrated $CuBr_2$ solutions (Fontaine et al., 1978); various Mo complexes in aqueous solutions (Cramer et al., 1978); highly concentrated $ZnBr_2$ aqueous solutions (Lagarde et al., 1980); $Sn(ClO_4)_2$ aqueous solutions (Yamaguchi et al., 1982; Yamaguchi et al., 1984); concentrated $CdBr_2$ solutions (Sadoc et al., 1984); $AgNO_3$ solutions (Yamaguchi et al., 1984); dilute $ZnCl_2$ solutions (Parkhurst et al., 1984; Drier and Rabe, 1986; Brown et al., 1988); concentrated Fe^{2+} and Fe^{3+} chloride solutions (Apted et al., 1985); dilute Cr^{3+}, Mn^{2+}, Fe^{3+}, and Ni^{2+} chloride solutions (Garcia et al., 1986b); concentrated copper acetate solutions (Nomura and Yamaguchi, 1986) and manganate and chromate-containing solutions (Garcia et al., 1986a; Bianconi et al., 1985). Most of these studies have examined only the K-edge XAS spectrum of the metal ion in the complex; however, a few have derived interatomic distances and coordination numbers from the EXAFS spectra of both the metal cation and anion (Fontaine et al., 1978; Lagarde et al., 1980; Sandstrom, 1984), which provides pair correlation functions centered on both types of ions and leads to a more accurate description of the local ordering around the metal ion. The added advantage of such studies is that they provide a better opportunity for determining the presence or absence of clusters of cation polyhedra sharing edges or corners. Studies of this type involving bromide complexes (Fontaine et al., 1978; Lagarde et al., 1980; Sadoc et al., 1984) have shown that the solution complexes have local environments similar to those in the crystalline salts. The work of Sandstrom (1984) was the first to measure Cl K-EXAFS and XANES using fluorescence detection. Measurements at the Cl K-edge (2822 eV) are particular challenging because of the relatively low energy of the x-rays, which causes problems of absorption by the solvent. However, such measurements in relatively dilute solutions (to at least 0.001 M) are now quite feasible on wiggler-magnet beam lines.

The study of concentrated ferric and ferrous chloride solutions (Apted et al., 1985) provides a good example of the type of information one can obtain from XAS analysis of electrolyte solutions. Examination of the Fe K-XANES spectrum of the 1.0M $FeCl_3$ (15.0 M Cl^-) solution (Fig. 5-4) shows a strong pre-edge feature at 7113.0 eV, indicative of tetrahedral Fe^{3+}. This feature is also seen in the spectrum of crystalline $FeCl_3$, in which Fe^{3+} is also tetrahedrally coordinated, but it is missing from the spectrum of the 1.0 M $FeCl_3$ (7.8 M Cl^-) solution, indicating the possibility of an octahedral Fe^{3+} complex in the latter. The Fe K-EXAFS results confirm the presence of tetra-chloro complexes in the concentrated chloride solution [d(Fe-Cl) = 2.25Å] and hexa-aquo $[Fe(H_2O)_6^{3+}]$ or trans-$Fe(H_2O)_4Cl_2^+$ complexes (or both) in the less concentrated chloride solution [d(Fe-O) = 2.10 Å]. The study of Zn chloride solutions (Parkhurst et al., 1984; Brown et al., 1988) by fluorescence EXAFS methods examined the types of Zn complexes as a function of concentration (5.6 M to 0.001 M $ZnCl_2$) and Cl:Zn ratio (2:1, 4:1, 10:1, 100:1). The lowest concentration solutions approach the levels of Zn and Cl in natural hydrothermal solutions. The results of this work show that a transition occurs from an inner-sphere tetra-chloro to an outer-sphere aquo complex (with nearest neighbor water molecules) between 5.6 M and 1.0 M. As concentration is lowered from 1.0 M to 0.001 M, the average number of water molecules in the solvation sphere around Zn, increases from 5 to 8. As the Cl:Zn ratio increases, the number of water molecules in the hydration sphere is reduced. A clear extension of this work would involve direct study of metal complexes in hydrothermal solutions at temperature and pressure, where current aqueous thermodynamic data bases rely either on indirect data on complex type or on assumption.

6.3.5 Metamict minerals. Natural metamict materials have received renewed interest recently because of the information they can provide about stable containment, over geologic time periods, of radioactive waste elements such as ^{137}Cs in refractory phases. At radiation doses near the saturation level, these materials are amorphous, and XAS analysis is very useful in the study of local structural environments of certain elements. Several studies have been made on complex Ca-Ti-Nb-Ta oxides of the pyrochlore group by Greegor et al. (1984a, 1986 a,b) and Ewing et al. (1987). Their results generally show that local order beyond the first coordination shell of Ta and Ti is lost with increasing radiation damage. However, this local ordering is found to reappear on annealing the samples. XAS results on metamict materials of several initial structure types show the same type of nearest-neighbor structure around the major network-forming cations, leading to the suggestion that the final local environments of these cations in the metamict state is similar for a wide range of materials (Ewing et al., 1987).

6.4 Nucleation and crystallization processes

Because of its unique attributes, EXAFS spectroscopy can give information on the local organization around an element during the initial stages of nucleation and crystal growth. Such studies have been made on Mg-aluminosilicate glasses (Taylor and McMillan, 1982) and on multi-component glasses of cordierite (Dumas and Petiau, 1984; Dumas et al., 1985; Petiau and Calas, 1985) and spodumene (Ramos et al., 1985) compositions. Ti and Zr atoms are thought to play a nucleating role in cordierite glasses. The Ti K-XANES pre-edges change significantly after heat-treatment of the glasses, and show that this element occupies a 6-coordinated site in the nucleating phase. At the same time, the Ti-O distances in these glasses increase from an average value of 1.82Å to 1.86Å, and the average number of neighbors changes from 4 to 5 (based on a single-shell model). The most significant evolution occurs in the second atomic shell surrounding Ti atoms (Fig. 6-9); this shell consists of light cations (M) such as Si or Al, and the average Ti-M distance increases from 2.90 Å in the initial glass to 3.05 Å in the nucleated material. This distance is consistent with the incorporation of Ti atoms in an Al_2TiO_5 phase. Local order around Zr atoms shows a more complex evolution, characterized by an intermediate phase which appears at 795-800°C, before the formation of tetragonal zirconia, ZrO_2, at 805-850°C. This intermediate phase exhibits a second shell around Zr consisting of (Si, Al) atoms at a distance of 2.80 Å, significantly shorter than the Zr-(Si,Al) distance observed in the initial

Figure 6-9 (right). Evolution of features in Ti K-EXAFS radial distribution functions at various stages of nucleation of cordierite-composition glass. The second-shell feature between 2 and 3 Å changes significantly at different annealing steps, whereas the main feature at ≈1.5 Å changes very little in position. Curve G represents the unannealed cordierite glass, whereas curves A, B, and C represent the glass after successive annealing steps. (From Dumas and Petiau, 1986)

Figure 6-10 (below). Possible structures for selenite adsorbed to α-FeOOH: (a) an outer-sphere, ion-pair adsorption complex, with the first hydration sphere shown as the shaded area; (b) a solid solution of selenite in the oxide phase; (c)-(e) possible inner-sphere complexes on the α-FeOOH surface. Distances determined from the selenium EXAFS are shown for the model structure, (e), that is consistent with the EXAFS data. In all figures, the oxide is shown as the striped area below the line that represents the oxide-water interface. (From Hayes et al., 1987)

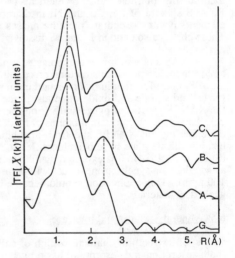

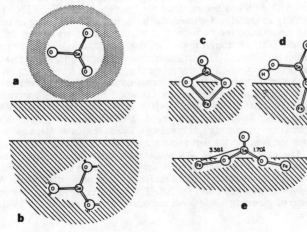

silicate glass (3.32 Å). This variation may be interpreted as a change in the linkage between the ZrO_6 octahedra and the surrounding SiO_4 tetrahedra, which corresponds to an edge-sharing of these polyhedra in the intermediate nucleation species and a corner-sharing arrangement in the initial glass. Thermal treatment above 805°C is accompanied by the disappearance of this second-shell (Si,Al) contribution to the Zr EXAFS and the formation of an intense Zr second-shell contribution characteristic of the local structure in tetragonal zirconia crystals. The presence of these crystals in the heat-treated glasses has been shown by transmission electron microscopy. Thus, evolution of the local structure around Zr and Ti is clearly different during heat treatment of these glasses, although both elements are considered to be efficient nucleating agents.

6.4 Study of dilute species

Although the use of fluorescence detection of XAS makes possible the study of local structure around elements at low concentration levels, only a few applications have been made to materials of geological interest. Several studies of the structural environments of trace elements in glasses and aqueous electrolyte solutions were discussed earlier (§6.3.1 and 6.3.4). In addition, structural environments and chemical states of 3d-transition element impurities in synthetic diamonds have been studied using fluorescence EXAFS methods by Wong et al. (1985). The synthesis process of diamonds in liquid metals and alloys explains the frequent incorporation of dilute transition elements (≈ 0.2 at. %) in the diamonds. Analysis of the K-edge EXAFS spectra of Fe, Co and Ni in diamonds prepared in this way has shown that these elements occur as separate metallic phases which retain the high-temperature fcc structure. These phases also contain 1-2 at. % dissolved carbon.

XAS study of minor components in natural coals, using a helium beam path and sample atmosphere, has provided structural on certain elements in minerals and organic matter (Wong et al., 1984). Sulfur, titanium and vanadium have been shown to be partly associated with organic matter and sulfide and oxide minerals; in the case of sulfur, the XANES spectra were fit using the XANES spectra of pyrite and various heterocyclic organic compounds. Simulated spectra, calculated assuming certain mixtures of model compound spectra, were found to agree well with the experimental XANES spectrum of a bituminous coal. Similarly, K-bearing illite and K-bearing silicate glass have been identified in raw coal and its combustion products, respectively (Spiro et al., 1984). It is expected that the increased use of high intensity insertion devices, such as wigglers, on synchrotron sources will permit studies of even more dilute materials (better than 100 ppm). This application should become increasingly important in trace element studies of geological materials.

6.5 Mineral Surfaces and Interfaces

The structural characterization of solid surfaces and the products of surface reactions (both adsorption and desorption) has proven to be a fruitful application of EXAFS spectroscopy. In fact, the acronym SEXAFS has been coined for "surface EXAFS". SEXAFS studies have been applied to a variety of metal and semiconductor surfaces to characterize surface structure, adsorption sites, and passivation mechanisms (see Stohr (1984) for a review of the method and Citrin (1986) for a relatively thorough review of the applications). All of these applications involve the use of high vacuum ($>10^{-10}$ torr), soft x-ray -vacuum ultraviolet (50-2000 eV) beam lines, because of the desire to study low atomic number elements like oxygen on surfaces at low surface coverages. Several studies involving minerals and silicate glasses have made use of these types of beam lines and the techniques common to SEXAFS spectroscopy (oxygen environments: Brown et al., 1986b; sodium environments: McKeown et al., 1985a; aluminum environments: Brown et al., 1983; McKeown et al., 1985b); however, their objectives were to determine the structural environments and bonding characteristics of these elements in the bulk phases. Such studies are feasible for these low atomic number (Z) elements in low Z matrices because of the 50-100 Å escape depth of photoelectrons from these elements. To date there have been no studies of this type on mineral surfaces, and this appears to be a fruitful area for future work since very little is known about mineral surfaces at the molecular level. Another surface EXAFS spectroscopy method known as "RefEXAFS" has been developed for surface structure studies. It makes use of glancing angle XAS measurements (Bosio et al., 1986) and has been applied in a very limited number of cases (oxidation of Ni metal: Bosio et al., 1984; corrosion of lead phosphate glasses: Greaves, 1986).

Another promising area of research on surfaces is that involving the study of solid/liquid interfaces. Many processes of geological interest occur at such interfaces, but there are few ways to characterize these environments in-situ.. One of the more novel, recent applications of XAS spectroscopy was the in-situ study of cation adsorption complexes at water/mineral interfaces (Roe et al., 1986; Hayes et al., 1987; Brown et al., 1988). Direct structural characterization of such complexes is not generally possible using most spectroscopic or scattering methods because of the low concentration levels of the complexes (coverages are typically less than one monolayer), the interference caused by water, and the need to dry samples for surface-sensitive, high-vacuum spectroscopy or diffraction methods. However, these difficulties can be overcome with XAS. Figure 6-10 is from a recent XAS study of the sorption mechanism of selenium oxyanions on the common ferric iron oxy-hydroxide mineral goethite (Hayes et al., 1987). It illustrates several of the possible geometric arrangements of selenite ions at the goethite/water interface. Separate conventional surface chemistry studies of the sorption of selenite ions at this interface have shown that almost complete uptake can be achieved at pH values less than 6 (Hayes and Leckie, 1987). In-situ Se K-EXAFS spectra of selenate and selenite sorption complexes, measured using fluorescence methods, are shown in Figure 6-11. The spectrum of selenate (upper plot) shows a single, damped sinusoidal wave typical of a single shell of four oxygen backscatterers, whereas the spectrum of selenite (lower plot) shows a distinctive "beat pattern" characteristic of more than one shell of backscatterers at different distances. Fourier transforms of these spectra (Fig. 6-12) show a clear second-shell contribution in the case of selenite but none in the case of selenate. Fitting of the backtransformed spectra using one shell of four oxygens for selenate and two shells for selenite (one shell of three oxygens and a second shell of two irons) gave good fits to the experimental EXAFS. The best-fit models are three oxygens at 1.70 Å and two irons at 3.38 Å around selenium for selenite ions at the goethite/water interface and four oxygens at 1.65 Å around selenium for selenate ions at this interface. These results indicate that selenate is sorbed as an outer-sphere complex (i.e., it is not in direct contact with the goethite surface, but instead is separated from it by a hydration sphere) and that selenite is sorbed as an inner-sphere complex in a bidentate fashion (Fig. 6-10e). This study provides the first molecular-level explanation for the weak binding of selenate and the strong binding of selenite at the goethite/water interface. Applications such as this open up a new research area involving in-situ structural study of reactions at mineral/water interfaces. Further studies of this type should provide useful information about the geochemistry of mineral surfaces, particularly when the next-generation synchrotron source makes feasible real-time, structural studies of chemical reactions on mineral surfaces.

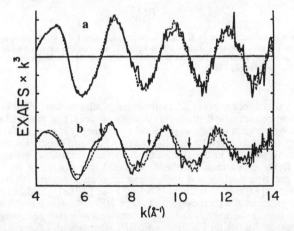

Figure 6-11. (a) Selenium K-EXAFS spectrum of aqueous selenite solution. The dashed line represents the model EXAFS spectrum obtained assuming a single shell of three oxygen backscatterers at a distance of 1.69 Å; (b) Se K-EXAFS spectrum of the selenite/α-FeOOH/water system at 25 mM SeO_3^{2-}. The dashed line is the same as in (a). The arrows indicate beat features due to a second-nearest neighbor Fe ligands at a longer distance than the first shell of oxygen backscatterers. (From Hayes et al., 1987)

502

6.6 XAS Studies of Very Small Samples

Because of the high brilliance of synchrotron x-ray sources, it is possible to carry out XAS studies of extremely small samples, such as those produced in ultra-high pressure diamond anvil cell experiments. One example of this type of study is the investigation of partitioning of Fe within a high pressure silicate perovskite, $Mg_{0.88}Fe_{0.12}SiO_3$, synthesized at 50 GPa and $\approx$ 2000 K (Jackson et al., 1987). Only 30 µg of sample was available from a total of 30 separate synthesis runs under these conditions. This amount precludes many types of spectroscopic characterization; however, XAS study is possible using fluorescence detection techniques. Results from this investigation suggest that Fe^{2+} partitions into the 6-coordinated Si site rather than into the 8- to 12-coordinated Mg site.

6.7 New and Future Applications

Over the next decade, several third-generation synchrotron sources will be constructed in the United States and Europe that will have significant impact on XAS and other studies of earth materials. These machines will utilize wiggler and undulator insertion devices (see §2.1.4) to produce photons ranging from 50 eV to 100 keV in energy with brilliances several orders of magnitude greater than existing sources (see Fig. 2-3). Currently, the PEP ring at Stanford University (Table 2) serves as a test-bed for these new high brilliance-low emittance sources, and extremely low emittances (high brightnesses) have already been achieved using an undulator

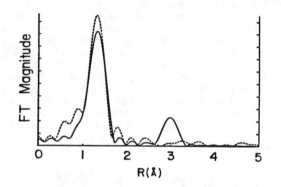

Figure 6-12. Fourier transforms of the Se K-EXAFS spectra of Figure 6-11, uncorrected for phase shift. The solid line represents the rdf of selenite adsorbed to α-FeOOH, whereas the dashed line represents the rdf of selenite in aqueous solution. The second shell Fe contribution is clearly evident in the first case, but it is missing in the second. (From Hayes et al., 1987)

insertion device on this synchrotron radiation source (Coisson and Winick, 1987). The U.S. synchrotron radiation community is planning two new sources, one that will operate in the 7-8 GeV range, producing photons in the x-ray to hard x-ray energy range (1-100 keV), at Argonne National Laboratory (the Advanced Photon Source or APS); and one that will operate in the 1-2 GeV range, producing photons in the soft x-ray to vacuum ultraviolet energy range (50-2000 eV), at Lawrence Berkeley Laboratory (the Advanced Light Source or ALS). The APS will accommodate 34 insertion devices and 35 bending magnets along its 1060 m circumference, providing up to 100 experimental stations (Shenoy *et al.*, 1988). In this ring, positrons will circulate in 1 to 60 bunches at a total current of about 100 mA, with a revolution time of about 3.5 µsec. The European synchrotron radiation community is also planning a major facility (European Synchrotron Radiation Facility -- ESRF) (Buras and Tazzari, 1984). These new sources of x-rays of enormous brilliance will undoubtedly create new opportunities for research in the earth sciences, including new applications of x-ray absorption spectroscopy.

The most obvious research areas that will benefit from these new high brilliance, low emittance x-ray sources are those involving low element concentrations, extremely high pressures and temperatures, and transient phenomena during which rapid data acquisition is needed. Many elements of interest in earth materials (e.g., Ar, Nd, Sm) occur at trace concentration levels (< 1000 ppm), making conventional scattering or spectroscopic studies of their structural environment and bonding virtually impossible. Consequently, little is known about the distributions of these elements in mineral structures (crystallographic sites versus defect sites) or about their crystal chemical behavior (bonding, site partitioning, short range ordering, complex formation, diffusion mechanisms, etc.). Such information is useful in a variety of geological contexts; for example, it is needed to provide an understanding of the "blocking temperatures" of radiogenic elements and their daughter products in minerals that have undergone metamorphism. Moreover, with the gains being made in synchrotron x-ray microprobe analyses of trace elements in minerals at the National Synchrotron Light Source located at Brookhaven National Laboratory, there is a growing need for parallel information on their crystal chemistry. XAS studies of such elements in bulk geologic samples are currently difficult or impossible because of insufficient flux from existing synchrotron x-ray beam lines. The new high brilliance x-ray sources that should become available in the mid 1990's should permit XAS studies of trace elements in minerals, gels, glasses, melts, and electrolyte solutions at concentration levels approaching those in natural systems and at spatial resolution levels of 1 μm^2 or less.

Another research area these new x-ray sources will open up involves reactions, element partitioning, structure, and bonding at mineral surfaces and interfaces. Most chemical reactions involving minerals and fluids occur at these sites of low dimensionality. Furthermore, the strength of minerals and rocks and the transport of elements in earth materials are ultimately related to sites at grain boundaries, surfaces, dislocations, stacking faults, twin boundaries, and domain boundaries. For example, hydrolytic weakening of minerals at crack tips is of great significance in the fracturing and faulting of crustal rocks. Very little is known about these sites of low dimensionality in minerals because few methods have the ability to spatially resolve and isolate such sites or to provide element- specific structure and bonding information for elements in such sites at low concentrations. XAS studies of cations at these types of sites using high brilliance sources will provide the first opportunity to tackle some of these problems. Moreover, these photon sources will permit studies of chemisorption on geologically-relevant single crystals for the first time. Such studies are virtually impossible using conventional spectroscopies because of the typically very low concentration levels of elements at sub-mono-layer coverages and the need for extremely sensitive methods that require high vacuum environments.

A third research area that is well suited to utilize XAS with high brilliance x-rays is the study of earth materials as functions of temperature and pressure. Knowledge of the structure-property relationships of minerals and fluids under extreme conditions is of critical importance in developing models of many geological processes, particularly dynamical processes in the Earth's mantle. As the temperatures and pressures have been pushed to the extreme values typical of those in the Earth's lower mantle and core (T > 3000 K, P > 250 GPa) (see Jeanloz and Heinz, 1984), new information on phase stability, phase transitions, and equations of state has been gained. At these extreme P-T conditions, the need for high brilliance photon sources has increased because of the large pressure and temperature gradients in the DAC (Jeanloz and Heinz, 1984) and the very small sample volume (on the order of picoliters) at the maximum P and T. Both x-ray diffraction and x-ray absorption experiments carried out under such P-T conditions could benefit greatly from the next-generation, high brilliance x-ray sources, which can produce very high x-ray flux in a very small spot size.

One of the few examples from this last research area is the high pressure EXAFS study of FeS_2 (pyrite) by Ingalls et al. (1978) which determined Fe-Fe and Fe-S distances as a function of pressure to ≈ 5 GPa. Other examples on non-geological materials include dispersive XAS studies of solid bromine to pressures of 57.5 GPa (Itie et al., 1986), and on ZnSe and CuBr to pressures of 8.7 GPa (Tranquada and Ingalls, 1984). In addition to studies of the pressure dependence of metal-anion bond lengths, high pressure XAS offers exciting possibilities for studying bonding changes with increasing pressure (e.g., Rohler et al., 1984; Rohler, 1984) as well as phase transitions (Tranquada et al., 1984).

These new sources, when operated in the timing mode (see §2.1.3), will permit new classes of time-resolved XAS experiments (e.g., transient phenomena such as first order solid-solid phase transitions, melting, and higher order transitions such as changes in spin state). Such studies will require rapid detection methods, such as position sensitive detectors employed in dispersive EXAFS studies (e.g., Itie et al., 1986). Development of rapid detection devices for time-resolved synchrotron studies is underway (Clarke et al., 1988). Time-dependent XAS studies have been done on a few materials, including EXAFS study of flash-melted Al films using a single nanosecond pulse of x-rays emitted from a laser-produced plasma (Epstein et al., 1983). Time-resolved EXAFS methods are also used in studies of electrochemical reactions, such as the changes in structure of $Ni(OH)_2$ as it is being electrochemically oxidized in concentrated alkali electrolytes (McBreen et al., 1987). Clearly, much groundwork is needed to fully utilize time-resolved XAS in the study of transient phenomena; however, the potential for major breakthroughs in understanding the kinetics and mechanisms of reactions and phase changes of earth materials is great using XAS.

6.8 Concluding Remarks

X-ray absorption spectroscopy has been developed over the past 15 years into a versatile, element-specific structural probe that has been successfully applied to problems in a variety of disciplines, including the earth sciences. Unlike the other spectroscopic methods discussed in this chapter, EXAFS is capable of providing direct structural information (bond lengths, number and identity of near neighbors, static disorder information) about cation and anion environments in earth materials of all types (crystalline and amorphous solids, gels, liquids [melts and aqueous solutions], and gases). XAS is not restricted to a few elements, like Mossbauer spectroscopy and some of the other methods. In contrast, XAS methods are applicable to most of the elements of interest in earth materials (excluding H) over a range of concentrations (> 1000 ppm). Also in contrast to some of the other methods reviewed in this volume (e.g., XPS and optical-UV spectroscopy), EXAFS spectroscopy requires relatively complex data analysis to extract accurate structural information. For certain mineralogical and geochemical problems (e.g., long-range ordering of cations in minerals; the relationship of thermochemical properties to the structure and dynamics of earth materials), XAS is not applicable. For other problems (coordination number of 3d-transition metals in minerals and silicate glasses), XAS can provide valuable information that complements indirect structural information from other methods (optical-UV absorption, ESR, and Mossbauer methods). The information on bonding from XANES measurements is also directly related to information on orbital energies and compositions provided by x-ray emission spectra, Auger electron spectra, and valence-band x-ray photoelectron spectra. Finally, for certain problems (short-range ordering of cations in minerals; structural environments of certain cations [e.g., alkalis, alkaline earths, rare earths, Ti^{4+}] in amorphous silicates; the structural environment of elements sorbed at mineral/water interfaces or of certain impurity elements [e.g., Ar] in defects in minerals; time-resolved structural studies of transient phenomena such chemical reactions and first-order phase transitions), EXAFS and XANES can provide unique results not duplicated by the other spectroscopic methods discussed in this volume.

Past and future developments of XAS are closely tied to the development of synchrotron radiation sources. It is this aspect of XAS, in particular, that distinguishes it from the other spectroscopic methods reviewed in this volume. In this regard, XAS is not a spectroscopy that can be conveniently carried out in a conventional laboratory in an academic or industrial setting. Instead, those who practice XAS must do so in a few large and complex national facilities that are foreign to most scientists. The use of such facilities requires a team approach because of their complexity and operating schedule (often 24 hours per day). The acquisition of "beam time" on such facilities is still difficult because they are over-subscribed by scientists from a wide range of disciplines. In most cases, time is awarded on a competitive basis following detailed peer- and panel-review of proposals which describe the scientific program requiring beam time. Undoubtedly, the nature of these national laboratories and the proposal process have kept the involvement of earth scientists at a small level compared to the condensed matter physics and materials science communities. However, the rewards can be significant for those mineralogists and geochemists interested in applications of synchrotron radiation, including XAS, to earth science problems, and for those willing to invest the time and energy acquiring the background needed to utilize these facilities. Perhaps the most efficient way for an individual investigator in mineralogy or geochemistry to initiate such studies is to form collaborations with those who are already involved. Such collaborations are welcome. An increase in the level of involvement by earth scientists in synchrotron radiation studies is necesssary for this community to take full advantage of the proposed national and international, third-generation synchrotron facilities (APS, ALS, ESRF) that are now in the final planning stages.

During the ten years following publication of the Eisenberger and Kincaid (1978) Science review of EXAFS entitled "EXAFS: New Horizons in Structure Determinations", many of the new opportunities they discussed have been realized. During the next decade, we expect to observe similar gains in the methods and theory of x-ray absorption spectroscopy and in new applications to earth materials.

VII. ACKNOWLEDGMENTS

The work of Brown and Waychunas reported in this chapter was generously supported by the National Science Foundation, through the Experimental and Theoretical Geochemistry Program of the Earth Sciences Division (Grants EAR 80-16911 and EAR 85-13488) and the Materials Research Division (Materials Resarch Laboratory Program through the Stanford Center for Materials Research). The work of Calas and Petiau was supported by the French CNRS. The collaboration of Brown and Calas on this chapter was initiated during a very pleasant sojourn in the Normandy countryside, which permitted an in-depth study of Norman cuisine and calvados. The preparation of this manuscript was supported by NSF Grant EAR 85-13488 and by CNRS. We thank volume editor Frank Hawthorne for his review of the manuscript, and series editor Paul Ribbe for his patience and tireless efforts in producing the *Reviews in Mineralogy* series.

VIII. REFERENCES

Antonini, M., Merlini, A.E., and Thornley, R.F. (1985) EXAFS structures of technetium in glasses prepared under different redox conditions. J. Non-Cryst. Solids 71, 219-225.

Apted, M.J., Waychunas, G.A., and Brown, G.E., Jr. (1985) Structure and specification of iron complexes in aqueous solutions determined by x-ray absorption spectroscopy. Geochim. Cosmochim. Acta 49, 2081-2089.

Ashley, C.A. and Doniach, S. (1975) Theory of extended x-ray absorption fine structure (EXAFS) in crystalline solids. Phys. Rev. B11, 1279-1288.

Benfatto, M., Natoli, C.R., Garcia, J., and Bianconi, A. (1986) Determination of the third order multiple scattering signal in tertrahedral clusters in liquid solutions. J. Physique 47, C8, 25-29.

Bianconi, A. (1984) XANES - x-ray absorption near edge structure - a probe of local geometrical and electronic structure. unpublished manuscript.

-----, Fritsch, E., Calas, G., and Petiau, J. (1985a) X-ray absorption near-edge structure of 3d transition elements in tetrahedral coordination. The effect of bond-length variation. Phys. Rev. B32, 4292-4295.

-----, Garcia, J., Marcelli, A., Benfatto, M., Natoli, C.R., and Davoli, I. (1985b) Probing higher order correlation functions in liquids by XANES (x-ray absoprion near edge structure). J. Physique 46, C9, 101-106.

Bingham, C., Godfrey, M.D., and Tukey, J.W. (1967) Modern techniques of power spectrum estimation. IEEE Trans. Audio and Electroacoust. 15(2), 56-66.

Binsted, N., Greaves, G.N., and Henderson, C.M.B. (1985) An EXAFS study of glassy and crystalline phases of composition $CaAl_2Si_2O_8$ (anorthite) and $CaMgSi_2O_6$ (diopside). Contrib. Mineral. Petrol. 89, 103-109.

Boland, J.J., Halaka, F.G., and Baldeschwieler (1983) Data analysis in extended x-ray absorption fine structure: determination of the background absorption and threshold energy. Phys. Rev. B28, 2921-2926.

Bonnin, D., Calas, G., Suquet, H., and Pezerat, H. (1985) Site occupancy of Fe^{3+} in Garfield nontronite: a spectroscopic study. Phys. Chem. Minerals 12, 55-64.

-----, Muller, S., and Calas, G. (1982) Le fer dans les kaolins. Etude par spectrométries RPE, Mossbauer, EXAFS. Bull. Minéral. 105, 467-475.

506

Bosio, L., Cortes, R., and Folcher, G. (1986) A laboratory ReflEXAFS spectrometer. J. Physique 47, C8, 113-116.

-----, and Froment, M. (1984) ReflEXAFS studies of protective oxide formation on metal surfaces. In: EXAFS and Near-Edge Structure III, K.O. Hodgson, B. Hedman, and J.E. Penner-Hahn, eds., Springer Proc. Phys. 2, Springer-Verlag, New York, p. 484-486.

Brown, G.E., Jr., Dikmen, F.D., and Waychunas, G.A. (1983) Total electron yield K-XANES and EXAFS investigation of aluminum in amorphous aluminosilicates. Stanford Synchrotron Radiation Lab. Rpt. 83/01, 146-147.

-----, Keefer, K.D., and Fenn. P.M. (1978) Extended X-ray Absorption Fine Structure (EXAFS) study of iron-bearing silicate glasses: iron coordination environment and oxidation state. (abstr.) Abstr. Prog. Geol. Soc. Amer. Ann. Mtg. 10, 373.

-----, Parkhurst, D.A.and Parks, G.A. (1988) Zinc complexes in aqueous chloride solutions: structure and thermodynamic modelling. Geochim. Cosmochim. Acta (to be submitted).

-----, Ponader, C.W., and Keefer, K.D. (1986a) X-ray absorption study of zirconium in Na-Zr silicate glasses. (abstr.) Abstr. Prog. Int'l Mineral. Assoc., 14th Gen. Mtg., Stanford, California, p. 63.

-----, Chisholm, C.J., Hayes, F.K., Roe, A.L., Parks, G.A., Hodgson, K.O., and Leckie, J.O. (1988) In-situ x-ray absorption study of Pb(II) and Co(II) sorption complexes at the γ-Al$_2$O$_3$/water interface. Stanford Synchrotron Radiation Lab. Rpt. 88/01 (in press).

-----, Waychunas, G.A., Ponader, C.W., Jackson, W.E., and McKeown, D.A. (1986b) EXAFS and NEXAFS studies of cation environments in oxide glasses. J.Physique 47, C8, 661-668.

-----, Stohr, J., and Sette, F. (1986c) Near-edge structure of oxygen in inorganic oxides: effect of local geometry and cation site. J. Physique 47, C8, 685-689.

Brown, G.S. and Doniach, S. (1980) The principles of x-ray absorption spectroscopy. In: Synchrotron Radiation Research, H. Winick and S. Doniach, eds., Plenum Press, New York. p. 353-385.

Brown, M., Peierls, R., and Stern, E.A. (1977) White lines in x-ray absorption. Phys. Rev. B15, 738-744.

Buras, B. and Tazzari, S. (eds.) (1984) Report of the European Synchrotron Radiation Facility. c/o CERN, LEP Div., Geneva, Switzerland.

Calas, G., Bassett, W.A., Petiau, J., Steinberg, D., Tchoubar, D., and Zarka, A. (1984a) Mineralogical applications of synchrotron radiation. Phys. Chem. Minerals 121, 17-36.

-----, Brown, G.E., Jr., Waychunas, G.A., and Petiau, J. (1987) X-ray absorption spectroscopic studies of silicate glasses and minerals. Phys. Chem. Minerals 15, 19- 29.

-----, Levitz, P., Petiau, J., Bondot, P., Loupias, G. (1980) Etude de l'ordre local autour du fer dans des verres silicates naturels et synthetiques a l'aide de la spectrometrie d'absorption x. Revue Phys. Appl. 15, 1161-1167.

-----, Manceau, A., Novikoff, A., Boukili, H. (1984b) Comportement du chrome dans les mineaux d'alteration du gisement de Campo Formoso (Bahia, Bresil). Bull. Minéral. 107, 755-766.

Calas, G. and Petiau, J. (1982) Short-range order around Fe(II) and Mn(II) in oxide glasses determined by x-ray absorption spectroscopy in relation with other spectroscopic and magnetic properties. In: The Structure of Non-Crystalline Materials II, P.H. Gaskell, J.M. Parker, and E.A. Davis, eds., Taylor and Francis, London, p. 18-28.

----- (1983a) Structure of oxide glasses: spectroscopic studies of local order and crystallochemistry. Geochemical implications. Bull. Minéral. 106, 33-55.

----- (1983b) Coordination of iron in oxide glasses through high-resolution K-edge spectra: informations from the pre-edge. Solid State Comm. 48, 625-629.

----- (1984) X-ray absorption spectra at the K-edges of 3d transition elements in minerals and reference compounds. Bull. Minéral. 107, 85-91.

----- and Manceau, A. (1986) X-ray absorption spectroscopy of geological materials. J. Physique 47, C8, 813-818.

Carlson, T.A. (1975) Photoelectron and Auger Spectroscopy, Chapter 3. Plenum Press, New York.

Charnock, G.N., Garner, C.G., Patrick, P.A.D., and Vaughan, D.G. (1988) Inves-tigation into the nature of copper and silver sites in argentian tetrahedrites using EXAFS spectroscopy. Phys. Chem. Minerals 15, 296-299.

Citrin, P.H. (1986) An overview of SEXAFS during the past decade. J. Physique 47, C8, 437-472.

Clarke, R., Sigler, P., and Mills, D. (1988) Time-resolved studies and ultrafast detectors: workshop report. Argonne National Lab. Rpt. ANL/APS-TM-2, 51 pp.

Coisson, R. and Winick, H. (eds.) (1987) Workshop on PEP as a Synchrotron Radiation Source. Stanford Synchrotron Radiation Laboratory, Stanford, CA, 558 pp.

Combes, J.M., Manceau, G., and Calas, G. (1986) Study of the local structure in poorly-ordered precursors of iron oxi-hydroxides. J.Physique 47, C8, 697-701.

-----, and Bottero, J.Y. (1988) The polymerization of Fe in ferric solution up to the formation of ferric hydrous oxide gel: a polyhedral approach by XAS. Geochim. Cosmochim. Acta (submitted).

Cook, J.W., Jr. and Sayers, D.E. (1981) Criteria for automatic x-ray absorption fine structure background removal. J. Appl. Physics 52, 5024-5031.

Cox, A.D. and McMilllan, P.W. (1981) An EXAFS study of the structure of lithium germanate glasses. J. Non-Cryst. Solids 47, 257-264.

Cramer, S.P., Hodgson, K.O., Stiefel, E.I., and Newton, W.E. (1978) A systematic x-ray absorption study of molybdenum complexes. The accuracy of structural information from extended x-ray absorption fine structure. J. Amer. Chem. Soc. 100, 2748-2761.

de Crescenzi, M., Antonangeli, F., BVellini, C., and Rosei, R. (1983) Temperature induced asymmetric effects in the surface extended energy loss fine structure of Ni(100). Solid State Comm. 46, 875-880.

Cressey, G. and Steel, A.T. (1988) On EXAFS studies of Tc, Er, and Lu site location in the epidote structure. Phys. Chem. Minerals 15, 304-312.

Crozier, E.D. and Seary, A.J. (1980) Asymmetric effects in the extended x-ray absorption fine structure analysis of solid and liquid zinc. Can. J. Phys. 58, 1388-1399.

Davoli, I., Paris, E., Mottana, A., and Marcelli, A. (1987) Xanes analysis on pyroxenes with different Ca concentration in M2 site. Phys. Chem. Minerals 14, 21-25.

-----, Stizza, S., Bianconi, A., Sessa, V., and Furlani, C. (1983) Sol. State Comm.

Decarreau, A., Bonnin, D., Badaut-Trauth, D., Couty, R., and Kaiser, P. (1987a) Synthesis and crystallogenesis of ferric smectite by evolution of Si-Fe coprecipitates in oxidizing conditions. Clay Minerals 22, 207-223.

-----, Colin, F., Herbillon, A., Manceau, A., Nahon, D., Paquet, H., Trauth-Badaud, D., and Trescases, J.J. (1987b) Domain segregation in Ni-Mg-Fe smectites. Clays Clay Minerals 35, 1-10.

Dreier, P. and Rabe, P. (1986) EXAFS-study of the Zn^{2+} coordination in aqueous halide solutions. J. Physique C8, 809-812.

Dumas, T. and Petiau, J. (1984) Structutal organization around nucleating elements (Ti, Zr) and Zn during crystalline nucleation process in silico-aluminate glasses. In: EXAFS and Near-Edge Structure III. K.O. Hodgson et al., eds., Springer Proc. Phys. 2, Springer-Verlag, New York, p. 311-313.

----- (1986) EXAFS study of titanium and zinc environments during nucleation in a cordierite glass. J. Non-Cryst. Solids 81, 201-220.

-----, Ramos, A., Gandais, M., and Petiau, J. (1985) Role of zirconium in nucleation and crystallization of a $(SiO_2, Al_2O_3, MgO, ZnO)$ glass. J. Mater. Sci. Lett. 4, 129-132.

Eisenberger, P. and Brown, G.S. (1979) The study of disordered systems by EXAFS: limitations. Solid State Comm. 29, 481-484.

----- and Kincaid, B.M. (1975) Synchrotron radiation studies of x-ray absorption spectra of ions in aqueous solutions. Chem. Phys. Lett. 36, 134-136.

----- (1978) EXAFS: new horizons in structure determination. Science 200, 1441-1447.

----- and Lengeler, B. (1980) Extended x-ray absorption fine-structure determination of coordination numbers: limitations. Phys. Rev. B22, 3551-3562.

Epstein, H.M., Schwerzel, R.E., Mallozzi, P.J., and Campbell, B.E. (1983) Flash-EXAFS for structural analysis of transient species: rapidly melting aluminum. J. Amer. Chem. Soc. 105, 1466-1468.

Ewing, R.C., Chakoumakos, B.C., Lumpkin, G.R., and Murakamii, T. (1987) The metamict state. Mat. Res. Soc. Bull. 12, 58-66.

Farges, F. and Calas, G. (1987) XAS study of Zr environment in silicate glasses: relation with Zr-enrichment in magmatic suites. Terra Cognita 7, 415.

Fischer, D.W. (1973) Use of soft x-ray band spectra for determining valence conduction band structure in transition compounds. In: Band Structure Spectroscopy of Metal Alloys. D.J. Fabian and L.M. Watson, eds., Academic Press, London.

Fleet, M.E., Herzberg, C.T., Henderson, G.S., Crozier, E.D., Osborn, M.D., and Scarfe, C.M. (1984) Coordination of Fe, Ga and Ge in high pressure glasses by Mössbauer, Raman and x-ray absorption spectroscopy, and geological implications. Geochim. Cosmochim. Acta 48, 1455-1466.

Fontaine, A., Lagarde, P., Raoux, D., Fontana, M.P., Maisano, G., Migliardo, P., and Wanderlingh, F. (1978) Extended x-ray absorption fine structure studies of local ordering in highly concentrated aqueous solutions of $CuBr_2$. Phys. Rev. Lett. 41, 504-507.

Frick e, H. (1920) K-charicteristic absorption frequencies for the chemical elements magnesium to chromium. Phys. Rev 16, 202-215.

Garcia, J., Benfatto, M., Natoli, C.R., Bianconi, A., Davoli, I., and Marcelli, A. (1986a) Three particle correlation function of metal ions in tetrahedral coordination determined by XANES. Solid State Comm. 58, 595.

-----, Bianconi, A., Benfato, M., and Natoli, C.R. (1986b) Coordination geometry of transition metal ions in dilute solutions by XANES. J. Physique 47, C8, 49-54.

Glenn, G.L. and Dodd, C.G. (1968) Use of molecular orbital theory to interpret x-ray K-absorption spectral data. J. Appl. Phys. 39, 5372-5377.

508

Goulon, J., Tola, P., Brochoin, J.C., Lemonnier, M., Dexpert-Ghys, J., and Guilard, R. (1984) X-ray excited optical luminescence (XEOL): Potentiality and limitations for the detection of XANES/EXAFS excitation spectra. In: EXAFS and Near-Edge Structure III. K.O. Hodgson et al., eds., Springer Proc. Phys. 2, Springer-Verlag, New York, p. 490-495.

Greaves, G.N. (1986) Corrosion studies of glass using conventional and glacing angle EXAFS. J. Physique 47, C8, 819-824

-----, Binsted, N., and Henderson, C.M.B. (1984) The environment of modifiers in glasses. In: EXAFS and Near-Edge Structure III. K.O. Hodgson et al., eds., Springer Proc. Phys. 2, Springer-Verlag, New York, p. 297-301.

-----, Fontaine, A., Lagarde, P., Raoux, D., and Gurman, S.J. (1981) Local structure of silicate glasses. Nature 293, 611-616.

Greegor, R.B., Lyttle, F.W., Chakoumakos, B.C., Lumkin, G.R., and Ewing, R.C. (1986a) Structural investigation of metamict minerals using x-ray absorption spectroscopy. (abstr.) Abstr. Prog. Int'l Mineral. Assoc., 14th Gen. Mtg, Stanford, California, p. 114.

-----, Spiro, C.L., and Wong, J. (1986b) Investigation of the Ta site in alpha-recoil damaged natural pyrochlores by XAS. Stanford Synchrotron Radiation Lab. Rpt. 86/01, 58-59.

-----, Ewing, R.C., and Haaker, R.F. (1984a) EXAFS/XANES studies of metamict materials. In: EXAFS and Near-Edge Structure III. K.O. Hodgson et al., eds., Springer Proc. Phys. 2, Springer-Verlag, New York, p. 343-348.

-----, Kortright, J., Fischer-Colbrie, A., and Schultz, P.C. (1984b) Comparative examination by EXAFS and WAXS of GeO_2-SiO_2 glasses. In: EXAFS and Near-Edge Structure III. K.O. Hodgson et al., eds., Springer Proc. Phys. 2, Springer-Verlag, p. 302-304.

-----, Sandstrom, D.R., Wong, J., and Schultz, P. (1983) Investigation of TiO_2-SiO_2 glasses by-ray absorption spectroscopy. J. Non-Cryst. Solids 55, 27-43.

Grunes. L.A. (1983) Study of the Kedges of 3d transition metals in pure and oxide form by x-ray absorption spectroscopy. Phys. Rev. B27, 2111-2131.

Hamajima, T., Kambara, T., Gondaira, K.I., and Oguchi, T. (1981) Self-consistent electronic structures of magnetic semiconductors by discrete variational $X\alpha$-calculations. III. Chalcopyrite $CuFeS_2$. Phys. Rev. B24, 3349-3353.

Hardwick, A., Whittaker, E.J.W., and Diakun, G.P. (1985) An extended x-ray absorption fine structure (EXAFS) study of the calcium site in a model basaltic glass, $Ca_3Mg_4Al_2Si_7O_{24}$. Mineral. Mag. 49, 25-29.

Harries, J.E., Irlam, J.C., Holt, C., Hasnain, S.S., and Hukins, D.W.L. (1987) Analysis of EXAFS spectra from the brushite and monetite forms of calcium phosphate. Mar. Res. Bull. 22, 1151-1157.

Hayes, T.M. and Boyce, J.B. (1982) Extended x-ray absorption fine structure spectroscopy. Solid State Phys. 37, 173-365.

Hayes, K.F. and Leckie, J.O. (1987) Modeling ionic strength effects on cation adsorption at hydrous oxide/solution interfaces. J. Colloid Interface Sci. 115, 564-572.

-----, Roe, A.L., Brown, G.E., Jr., Hodgson, K.O., Leckie, J.O., and Parks, G.A. (1987) In situ x-ray absorption study of surface complexes: selenium oxyanions on α-FeOOH. Science 238, 783-786.

Heald, S.M., Tranquada, J.M., and Chen, H. (1986) Interface EXAFS using glancing angles. J. de Physique 47, C8, 825-830.

Hertz, G. (1920) Über Absorptionslinien im Röntgenspektrum. Phys. Z. 21, 630-632.

Hunold, K. and Bruckner, R. (1980) Physikalische Eigenschaften und struktureller Feinbau von natrium-aluminosilikat Glasern und Schmelzen. Glass Techn. Ber. 53, 149-161.

Ildefonse, P., Manceau, A., Prost, D., and Toledo-Groke, M.C. (1986) Hydroxy Cu-vermiculite formed by the weathering of Fe-biotites at Salobo, Carajas, Brazil., Clays Clay Minerals 34, 338-345.

Ingalls, R., Garcia, G.A., and Stern, E.A. (1978) X-ray absorption at high pressure. Phys. Rev. Lett. 40, 334-336.

Itie, J.P., Jean-Louis, M., Dartyge, E., Fontaine, A., and Jucha, A. (1986) High pressure XAS on bromine in the dispersive mode. J. Physique C8, 897-900.

Jackson, W.E., Brown, G.E., Jr., and Ponader, C.W. (1987a) EXAFS and XANES study of the potassium coordination environment in glasses from the $NaAlSi_3O_8$- $KAlSi_3O_8$ binary: structural implications for the mixed alkali-effect. J. Non-Cryst. Solids 93, 311-322.

-----, Knittle, E., Brown, G.E., Jr., and Jeanloz, R. (1987b) Partitioning of Fe within high-pressure silicate perovskite: evidence for unusual geochemistry in the lower mantle. Geophys. Res. Lett. 14, 224-226.

Jeanloz, R. and Heniz, D.L. (1984) Experiments at high temperature and pressure: laser heating through the diamond cell. J. Physique C8, 83-92.

Koch, E.E. (ed.) (1983) Handbook on synchrotron radiation, Vols. 1a and 1b. North Holland, New York.

Kossel, W., von (1920) Zum Bau der Rontgenspektren. Z. Phys. 1, 119-134.

Kostroun, V.O., Cehn, M.H., and Crasemann, B. (1971) Atomic radiation transition probabilities to the 1s state and theoretical K-shell fluorescence yields. Phys. Rev. A3, 533-545.

Krause, M.O. (1979) Atomic radiative and radiationless yields for K and L shells. J. Phys. Chem. Ref. Data 8, 307.

Kronig, R. De L. (1931) Zur Theory der Feinstruktur in den Röntgenabsorptionspektren. Z. Phys. 70, 317.

Kunz, C. (ed.) (1979) Synchrotron radiation -- Techniques and Applications. Topics Curr. Phys., Springer-Verlag, New York.

Lagarde, P. (1985) EXAFS studies of amorphous solids and liquids. In: Amorphous Solids and the Liquid State. N.H. March, R.A. Street, and M. Tosi, eds., Plenum Press, New York, p. 365-393..

-----, Fontaine, A., Raoux, D., Sadoc, A., and Migliardo, P. (1980) EXAFS studies of strong electrolytic solutions. J. Chem. Phys. 72, 3061-3069.

Lapeyre, C., Petiau, J., Calas, G., Gauthier, F., and Gombert, F. (1983) Ordre local autour du germanium dans les verres du systeme SiO_2-GeO_2-B_2O_3-Na_2O: etude par spectrometrie d'absorption X. Bull. Minéral. 106, 77-85.

Lear, R.D. and Weber, M.J. (eds.) (1987) Synchrotron Radiation. Energy and Technology Review, Lawrence Livermore National Lab., Nov.-Dec. Issue, p. 33.

Lee, P.A. and Beni, G. (1977) New method for the calculation of atomic phase shifts: application to extended x-ray absorption fine structure (EXAFS) in molecules and crystals. Phys. Rev. B15, 2862-2883.

-----, Citrin, P.H., and Eisenberger, B.M. (1981) Extended x-ray absorption fine structure- its strengths and limitations as a structural tool. Rev. Mod. Phys. 53, 769-806.

----- and Pendry, J.B. (1975) Theory of extended x-ray absorption fine structure. Phys. Rev. B11, 2795.

-----, Teo, B.K., and Simons, A.L. (1977) EXAFS: a new parameterization of phase shifts. J. Amer. Chem. Soc. 99, 3856-3859.

Lenglet, M., Guillamet, R., D'Huysser, A., Durr, J., and Jørgensen, C.K. (1986) Fe, Ni coordination and oxidation states in chromites and cobaltites by XPS and XANES. J. Physique 47, C8, 765-769.

Licheri, G. and Pinna, G. (1983) EXAFS and x-ray diffraction in solutions. In: EXAFS and Near Edge Structure. A. Bianconi, L. Incoccia, and S. Stipcich, eds., Springer Ser. Chem. Phys. 27, Springer-Verlag, New York, 240-247.

Lytle, F.W., Greegor, R.B., and Panson, A.J. (1988) Discussion of x-ray absorption near edge structure: application to Cu in high T_c superconductors, $La_{1.8}Sr_{.2}CuO_4$ and $YBa_2Cu_3O_7$. Phys. Rev. B37, 1550-1562.

-----, Sayers, D.E., and Stern, E.A. (1982) The history and modern practice of EXAFS spectroscopy. In: Advances in X-ray Spectroscopy. C. Bonnelle and C, Mande, eds., Pergamon, Oxford, p. 267.

-----, Sandstrom, D.R., Marques, E.C., Wong, J., Spiro, C.L., Huffman, G.P., and Huggins, F.E. (1984) Measurement of soft x-ray absorption spectra with a fluorescent ion chamber detector. Nucl. Instr. Meth. 226, 542-548.

Manceau, A. and Calas, G. (1985) Heterogeneous distribution of nickel in hydrous silicates from New-Caledonia ore deposits. Amer. Mineral. 70, 549-558.

----- (1986) Nickel-bearing clay minerals. 2. Intracrystalline distribution of nickel: an x-ray absorption study. Clay Minerals 21, 341-360.

----- and Combes, J.M. (1988) Structure of Mn and Fe oxides and oxihydroxides: a topological approach by EXAFS. Phys. Chem. Minerals 15, 283-295.

-----, Llorca, S., and Calas, G. (1987) Crystal chemistry of cobalt and nickel in lithiophorite and asbolane from New-Caledonia. Geochim. Cosmochim. Acta 51, 105-113.

-----, Bonnin, D., Kaiser, P., and Fretigny, C. (1988) Polarized EXAFS spectra of biotite and chlorite. Phys. Chem. Minerals (submitted).

McBreen, J., O'Grady, W.E., Pandya, K.I., Hoffman, R.W., and Sayers, D.E. (1987) EXAFS study of the nickel oxide electrode. Langmuir 1987, 3, 428-433.

McKeown, D.A., Waychunas, G.A., and Brown, G.E., Jr. (1985a) EXAFS and XANES study of the local coordination environment of sodium in a series of silica-rich glasses and selected minerals within the Na_2O-Al_2O_3-SiO_2 system. J. Non-Cryst. Solids 74, 325-348.

----- (1985b) EXAFS study of the coordination environment of aluminum in a series of silica-rich glasses and selected minerals within the Na_2O-Al_2O_3-SiO_2 system. J. Non-Cryst. Solids 74, 349-371.

McMaster, W.H., Nerr del Grande, N., Mallet, J.H., and Hubell, J.H. (1969) Compilation of x-ray cross sections. Rpt. UCRL-50/74, Section 2, Revision 1, Lawrence Radiation Lab., Univ. Calif.

Müller, J.E. and Wilkins, J.W. (1984) Band structure approach to the x-ray spectra of metals. Phys. Rev. B29, 4331-4348.

Natoli, C.R. (1984) Distance dependence of continuum and bound state excitonic resonances in x-ray absorption near-edge structure (XANES). In: EXAFS and Near-Edge Structure III. K.O. Hodgson et al., eds., Springer Proc. Phys. 2, Springer-Verlag, New York, p. 38-42.

510

Nomura, M. and Yamaguchi, T. (1986) Solute structure of copper (II) acetate solutions in liquid and glassy states. J. Physique 47, C8, 619-622.

Novak, G.A. and Gibbs, G.V. (1971) The crystal chemistry of the silicate garnets. Amer. Mineral. 27, 783-792.

Okuno, M., Marumo, F., Sakamaki, T., Hosoya, S., and Miyake, M. (1984) The structure analyses of $NaGaSi_3O_8$, $NaAlGe_3O_8$ and $NaGaGe_3O_8$ glasses by x-ray diffraction and EXAFS measurements. Mineral. J. 12, 101-121.

-----, Yin, C.D., Morikawa, H., Marumo, F., and Oyanagi, H. (1986) A high resolution EXAFS and near-edge study of GeO_2 glass. J.Non-Cryst. Solids 87, 312-320.

Park, J.W. and Chen, H. (1982) The coordination of Fe^{3+} in sodium disilicate glass. Phys. Chem. Glasses 23, 107-108.

Parkhurst, D.A., Brown, G.E., Jr., Parks, G.A., Waychunas, G.A. (1984) Structural study of zinc complexes in aqueous chloride solutions by fluorescence EXAFS spectroscopy. (abstr.) Abstr. Prog. Geol. Soc. Amer. Ann Mtg. 16, 618.

Parratt, L.G. (1959) Electronic band structure of solids by x-ray spectroscopy. Rev. Mod. Phys. 31, 616-645.

Pendry, J.B. (1983) X-ray absorption near edge structure. Comments Solid State Phys. 10, 219-231.

Petiau, J. (1986) La spectrometrie des rayons X. In: Methodes spectroscopiques appliquees aux mineraux. Calas, G., ed., Soc. Fr. Minéral. Cristallogr., p. 200-230.

----- and Calas, G. (1982) Local structure about some transition elements in oxide glasses using x-ray absorption spectroscopy. J. Physique C9, 47-50.

----- (1985) EXAFS and edge structure; application to nucleation in silicate glasses. J. Physique 46, 41-50.

-----, Bondot, P., Lapeyre, C., Levitz, P., and Loupias, G. (1981) EXAFS and near-edge structure of some transition elements and germanium in silicate glasses. In: EXAFS for Inorganic Systems. NERC, p. 127-129.

-----, Dumas, T., and Heron, A.M. (1984) EXAFS and edge studies of transition elements in silicate glasses. In: EXAFS and Near-Edge Structure III. K.O. Hodgson et al., eds., Springer Proc. Phys. 2, Springer-Verlag, New York, p. 291-296.

Petit-Maire, D., Petiau, J., Calas, G., and Jacquet-Francillon, N. (1986) Local structure around actinides in borosilicate glasses. J. Physique 47, C8. 849-852.

Pettifer, R.F. (1981) EXAFS studies of glass. In: EXAFS for Inorganic Systems. NERC, p. 57-64.

Ponader, C.W. and Brown, G.E., Jr. (1988a) Rare earth elements in silicate glass/melt systems: I. Effects of composition on the coordination environment of La, Gd, and Yb. Geochim. Cosmochim. Acta (submitted).

----- (1988b) Rare earth elements in silicate glass/melt systems II. Interactions of La, Gd, and Yb with halogens. Geochim. Cosmochim. Acta (submitted).

Prewitt, C.T., Coppins, P., Phillips, J.C., and Finger, L.W. (1987) New opportunities in synchrotron x-ray crystallography. Science 238, 312-319.

Ramos, A., Petiau, J., and Gandais, M. (1985) Crystalline nucleation process in $(SiO_2-Al_2O_3-Li_2O)$ glasses. J. Physique C8, 491-494.

Rao, K.J., Wong, J., and Rao, G.J. (1984) Determination of the coordination of divalent lead in PbO-$PbCl_2$ glasses by investigations of the x-ray absorption fine structure. Phys. Chem. Glasses 25, 57-63.

Rehr, J.J., Albers, R.C., Natoli, C.R., and Stern, E.A. (1986a) Spherical wave corrections in XAFS. J. Physique 47, C8, 31-35.

----- (1986b) New high-energy approximation for x-ray near edge structure. Phys. Rev. B34, 4350-4353.

Roe, A.L., Schneider, D.J., Mayer, R.J., Pyrz, J.W., Widom, J., and Oue, L., Jr. (1984) X-ray absorption spectroscopy of iron-tyrosinate proteins. J. Amer. Chem. Soc. 106, 1676-1681.

Rohler, J. (1984) High pressure L_{III} absorption in mixed valent cerium. In: EXAFS and Near-Edge Structure III. K.O. Hodgson et al., eds., Springer Proc. Phys. 2, Springer-Verlag, p. 379-384.

-----, Keulerz, K., Dartyge, E., Fontaine, A., Jucha, A., and Sayers, D. (1984) High pressure energy dispersive x-ray absorption of EuO up to 300 kbar. In: EXAFS and Near-Edge Structure III. K.O. Hodgson et al., eds. Springer Proc. Phys. 2, Springer-Verlag, p. 385-387.

Sainctavit, P., Calas, G., Petiau, J., Karnatak, R., Esteva, J.M., and Brown, G.E., Jr. (1986) Electronic structure from x-ray K-edges in ZnS:Fe and $CuFeS_2$. J. Physique 47, C8, 411-414.

-----, Petiau, J., Calas, G., Benfatto, M., and Natoli, C.R. (1987) XANES study of sulfur and zinc K-edges in zincblende: experiments and multiple-scattering calculations. J. Physique 48, C (in press).

Sadoc, A., Lagarde, P., and Vlaic, G. (1984) EXAFS at the Cd K edge: a study of the local order in $CdBr_2$ aqueous solutions. In: EXAFS and Near-Edge Structure III. K.O. Hodgson et al., eds., Springer Proc. Phys. 2, Springer-Verlag, New York, p. 420-422.

511

Sandstrom, D.R. (1979) Ni^{2+} coordination in aqueous $NiCl_2$ solutions: study of the extended x-ray absorption fine structure. J. Chem. Phys. 71, 2381-2386.

----- (1984) EXAFS studies of electrolyte solutions. In: EXAFS and Near-Edge Structure III. K.O. Hodgson et al., eds., Springer Proc. Phys. 2, Springer-Verlag, New York, p. 409-413.

-----, Hodgson, H.W., and Lytle, F.W. (1977) Study of Ni(II) coordination in aqueous solution by EXAFS analysis. J. Chem. Phys. 67, 473-476.

-----, Lytle, F.W., Wei, P.S.P., Greegor, R.B., Wong, J., and Shultz, P. (1980) Coordination of Ti in TiO_2-SiO_2 glass by x-ray absorption spectroscopy. J. Non-Cryst. Solids 41, 201-207.

Sayers, D.E., Lytle, F.W., and Stern, E.A. (1970) Point scattering theory of x-ray K absorption fine structure. Advan. X-ray Anal. 13, 248-271.

-----, Stern, E.A., and Lytle, F.W. (1971) New technique for investigating noncrystalline structures: Fourier analysis of the extended x-ray absorption fine structure. Phys. Rev. Lett. 27, 1204-1207.

----- (1975) Extended x-ray absorption fine structure technique. III-Determination of physical parameters. Phys. Rev. B11, 4836-4845.

Shannon, R.D. and Prewitt, C.T. (1969) Effective ionic radii in oxides and fluorides. Acta Cryst. B25, 925-946.

Shenoy, G.K., Viccaro, P.J., and Mills, D.M. (1988) Characteristics of the 7-GeV advanced photon source: a users guide. Argonne National Laboratory Rpt. ANL-88-9, 52pp. See also 7 GeV Adv. Photon Source Conceptual Design Rpt. ANL-87-15.

Smith, T.A. (1985) Single crystal polarized x-ray absorption edge studies of copper complexes. Stanford Synchrotron Radiation Lab. Rpt. 85/03.

-----, Penner-Hahn, J.E., Berding, M.A., Doniach, S., Hodgson, K.O. (1985) Polarized x-ray absorption edge spretroscopy of single crystal copper (II) complexes. J. Amer. Chem. Soc. 107, 5945-5955.

Spiro, C., Wong, J., Maylotte, D.H., Lamson, S., Glover, B., Lyttle, F.W., and Greegor, R.B. (1984) Nature of potassium impurities in coal. In: EXAFS and Near-Edge Structure III. K.O. Hodgson et al., eds., Springer Proc. Phys. 2, Springer-Verlag, New York, p. 368-370.

Stern, E.A. (1974) Theory of extended x-ray absorption fine structure. Phys. Rev. B10, 3027-3037.

----- and Heald, S.M. (1983) Basic principles and applications of EXAFS. In: Handbook on Synchrotron Radiation, Vol. 1b. E.E. Koch, ed., North Holland, New York, p. 955-1014.

Stohr, J. (1984) Surface crystallography by means of SEXAFS and NEXAFS. In: Chemistry and Physics of Solid Surfaces. V.R. Vanselow and R. Howe, eds., Springer Series in Chem. Phys. Vol. 235, Springer-Verlag, New York, p. 231-275.

Taylor, J.M. and McMillan, P.W. (1982) EXAFS investigation into the role of ZrO -- 2 as a nuclaeting agent in MgO-Al_2O_3-SiO_2 glass. In: Structure of Non-Crystalline Materials II: P.H. Gaskell, J.M. Parker, and E.A. Davis, eds., Taylor and Francis, London, p. 589-596.

Templeton D.H. and Templeton, L.K. (1980) Polarized x-ray absorption and double refraction in vanadyl bisacetylacetonate. Acta Cryst. A36, 237-241.

Teo, B.K. (1986) EXAFS: basic principles and data analysis. In: Inorganic Chemistry Concepts 9, Springer-Verlag, New York, p. 1-349.

----- and Joy, D.C., eds. (1981) EXAFS Spectroscopy. Plenum Press, New York.

----- and Lee, P.A. (1979) Ab initio calculation of amplitude and phase function for Extended X-ray Absorption Fine Structure (EXAFS) spectroscopy. J. Amer. Chem. Soc. 101, 2815-2830.

-----, Simons, A.L., Eisenberger, P., and Kincaid, B.M. (1977) EXAFS: approxi-mations, parameterization, and chemical transferability of amplitude functions. J. Amer. Chem. Soc. 99, 3854-3856.

Tranquada, J.M. and Ingalls, R. (1984) An EXAFS study of the instability of the zincblende structure under pressure. In: EXAFS and Near-Edge Structure III. K.O. Hodgson et al., eds., Springer Proc. Phys. 2, Springer-Verlag, New York, 388-390.

-----, and Crozier, E.D. (1984) High pressure x-ray absorption studies of phase transitions. In: EXAFS and Near-Edge Structure III. K.O. Hodgson et al., eds., Springer Proc. Phys. 2, Springer-Verlag, New York, 374-378.

Vaughan, D. (ed.) (1986) X-ray Data Booklet. Section 2.4: Fluorescent yields for K and L shells, J.B. Kortright, p. 2-19 - 2-20, and Section 2.2: Electron binding energies, G.P. Williams, p. 2-5 - 2-11, Lawrence Berkeley Lab. Pub. 490.

Waychunas, G.A. (1983) Mössbauer, EXAFS and X-ray diffraction study of Fe^{3+} clusters in MgO:Fe and magnesiowüstite $(Mg,Fe)_{1-x}O$ - evidence for specific cluster geometries. J. Mater. Sci. 18, 195-207.

----- (1987) Synchrotron radiation XANES spectroscopy of Ti in minerals: effects of Ti bonding distances, Ti valence and site geometry on absorption edge structure. Amer. Mineral. 72, 89-101.

----- and Brown, G.E., Jr. (1984) Application of EXAFS and XANES spectroscopy to problems in mineralogy and geochemistry. In: EXAFS and Near-Edge Structure III. K.O. Hodgson et al., eds., Springer Proc. Phys. 2, Springer-Verlag, New York, 336-342.

512

-----, and Apted, M.J. (1983) X-ray K-edge absorption spectra of Fe minerals and model compounds: near edge structure. Phys. Chem. Minerals 10, 1-9.

----- (1986a) X-ray K-edge absorption spectra of Fe minerals and model compounds: II EXAFS. Phys. Chem. Minerals 13, 31-47.

-----, Brown, G.E., Jr., Ponader, C.W., and Jackson, W.E. (1986b) High temperature x-ray absorption study of iron sites in crystaline, glassy and molten silicates and oxides. Stanford Synchrotron Radiation Lab. Rpt. 87/01, 139-141.

----- (1988) Evidence from X-ray absorption for network-forming Fe^{2+} in molten alkali silicates. Nature (in press).

-----, Dollase, W.A., and Ross, C.R. (1986c) Determination of short range order (SRO) parameters fron EXAFS pair distribution functions of oxide and silicate solid solutions. J. Physique 47, C8, 845-848.

Will, G. (1981) Energiedispersion und Synchrotron radiation: eine neue Methode und eine neue Strahlenquelle für die Röntgenbeugung. Fortschr. Mineral. 59, 31-94.

Winick, H. (1980) Properties of synchrotron radiation. In: Synchrotron Radiation Research. H. Winick and S. Doniach, eds., Plenum Press, New York, p. 11-25.

----- (1987) Synchrotron radiation. Scientific American 257, 88-99.

----- and Doniach, S., eds. (1980) Synchrotron Radiation Research, Plenum Press, New York, 680 pp.

Wong, J. (1986) Extended x-ray absorption fine structure: a modern structural tool in materials science. Mat. Sci. Eng. 80, 107-128.

-----, Koch, E.F., Hejna, C.I., and Garbauskas, M.F. (1985) Atomic and microstructural characterization of metal impurities in synthetic diamonds. J. Appl. Phys. 58, 3388-3393.

-----, Spiro, C.L., Maylotte, D.H., Lyttle, F.W., and Greegor, R.B. (1984) EXAFS and XANES studies of trace elements in coal. In: EXAFS and Near-Edge Structure III. K.O. Hodgson et al., eds., Springer Proc. Phys. 2, Springer-Verlag, New York, p. 362-367.

Yamaguchi, YT., Lindqvist, O., Boyce, J.B., and Claeson, T. (1984) EXAFS, x-ray and neutron diffraction of electrolyte solutions. In: EXAFS and Near-Edge Structure III. K.O. Hodgson et al., eds., Springer Proc. Phys. 2, Springer-Verlag, New York, p. 417-419.

-----, Claeson, T., and Boyce, J.B. (1982) EXAFS and x-ray difraction studies of the hydration structure of stereochemically active Sn(II) ions in aqueous solution. Chem. Phys. Lett. 93, 528-532.

Yang, C.Y., Paesler, M.A., and Sayers, D.E. (1986) Analysis of bond strengths of arsenic and arsenic chalcogen compounds using the temperature dependence of the EXAFS. J. Physique 47, 391-394.

Yarker, C.A., Johnson, P.A.V., Wright, A.C., Wong, J., Greegor, R.B., Lytle, F.W., and Sinclair, R.N. (1986) Neutron diffraction and EXAFS evidence for TiO_5 units in vitreous $K_2O-TiO_2-2SiO_2$. J. Non-Cryst. Solids 79, 117-136.

ELECTRON PARAMAGNETIC RESONANCE

1 INTRODUCTION

Electron paramagnetism occurs in all atoms, ions, organic free radicals and molecules with an odd number of electrons, color centers (unpaired electrons or holes trapped at lattice defects) and metallic compounds (on account of conduction electrons). Electron Paramagnetic Resonance (EPR) is also called Electron Spin Resonance (ESR) and the two names are equally well justified. It is a spectroscopic technique based upon the resonant absorption of microwaves by paramagnetic substances, tuned by an externally applied magnetic field. It was first discovered by Zavoiski in 1945 but was not extensively applied to materials research until the end of the 1950's. EPR is now a powerful tool which is used in many fields of physics, chemistry and biology.

In mineralogy, EPR studies concern ions with partly-filled inner electron shells (i.e., mainly the d- and f-transition group ions), and color centers caused by natural or artificial irradiation. As with the other spectroscopic techniques, it allows the study of crystalline as well as amorphous compounds: glasses, gels, aqueous solution. However, as it describes the interaction between an electronic spin submitted to the influence of crystal field and an external magnetic field, EPR spectra of oriented single crystals are needed to accurately determine EPR parameters; for well-known centers, the determination (and in some cases the quantitative estimation) can be made using powders on a "fingerprint" basis. This chapter is concerned primarily with the derivation of EPR

1.1 Resonance conditions

The magnetic properties of ions are determined by the nature of the ground state which is usually not completely pure and often contains admixtures from excited states. It is more convenient to analyse the basis of EPR by considering a free ion. If we consider a free electron of spin angular moment S, its magnetic moment μ_s is

$$\mu_s = -g_s \beta S \quad . \tag{1}$$

where g_s is the spectroscopic factor of the free electron and is equal to 2.0023, and ß is the Bohr magneton (ß = $0.927 \cdot 10^{-23}$ JT^{-1} = $0.927 \cdot 10^{-20}$ ergG^{-1}).

Free electrons are aligned at random. If an external magnetic field **B** is applied, a lower energy state occurs in which the electrons are lined up with the magnetic field and a higher energy state corresponds to electrons lined up in opposition to the magnetic field. The degeneracy is lifted, in accordance with the Zeeman effect. The energy of the resulting levels is proportional to the external field and is, for a free electron,

$$E = \pm 1/2 \, g_s \beta B \quad , \tag{2}$$

if the magnetic field is directed along the z (quantization) axis. If an alternating field of frequency ν is applied at right angle to **B**, transitions occur between Zeeman levels, if

$$h\nu = g_s \beta B \quad . \tag{3}$$

514

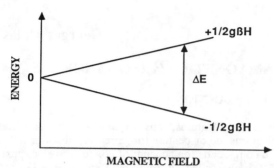

Figure 1. Splitting of a degenerate electronic level (± 1/2) by a magnetic field.

For a multielectronic ion, the g factor can have values distinct from g_S on account of interelectronic interactions; selection rules for electric dipolar transitions, $\Delta S_z = \pm 1$, still hold. For applied magnetic fields of the order of several kG, the convenient radiations are in the microwave range (several GHz). The resonance (Fig. 1) is generally observed when the magnetic field **B** is varied at a fixed frequency ν, as the variation of the microwave frequency is technically not easy to obtain at values greater than some percent of the nominal value (see Part 3 below). The EPR spectrometer measures the energy required to reverse the spin of an electron under externally applied magnetic field. The sample is adjusted in a resonant cavity in such a way that the microwave magnetic (alternating) field is at its maximum. The cavity is placed between the poles of a magnet, and the microwave power is adjusted to avoid saturation effects due to relaxation processes.

1.2 Relaxation processes

It must be pointed out that Zeeman energy levels are not widely separated: 1 to a few cm^{-1} to compare to 10^4 cm^{-1} which is the order of magnitude of optical transitions (see Chapter 7). Consequently, the Boltzman distribution law gives only a small difference in the spin population of ground and excited levels at room temperature. The relaxation processes (see §2.5) have great importance: large values cause saturation phenomena; however, short relaxation times cause a broadening of the EPR absorption line. The excited spins return to the ground state by two distinct processes:

(i) energy is transferred to the phonons of the crystal: this is responsible for the spin-lattice relaxation time, T_1. High T_1 values cause the saturation of the system: the photon absorption diminishes and even disappears if the population of the two levels tends to equal each other. This is sometimes used to separate the contribution of various paramagnetic centers in a complex spectrum, by varying the power of the applied microwave source.

(ii) in non-dilute systems, energy may be transferred to neighbor spins without any loss of energy of the total spin system. This is responsible for the spin-spin relaxation time, T_2.

1.3 Interaction with nuclear spins

The nucleus has a non-zero nuclear spin **I** if there is an odd number of protons and/or neutrons (see Chapter 9). However, the radiations which correspond to Nuclear Magnetic Resonance (NMR) transitions have lower frequencies than is the case for EPR. As a consequence, EPR transitions occur in the GHz range, as NMR transitions do in the MHz range, for similar applied magnetic fields of some kG. The interaction between electronic and nuclear spins is detected in EPR as a perturbation: it results in the so-called hyperfine structure (HFS). For a nuclear spin **I**, each transition leads to $(2I+1)$ hyperfine components which are, to a first approximation, equally separated and of equal probability. The energy separation between these components is referred to as the hyperfine coupling constant. The selection rules correspond to the conservation of the nuclear spin state (ΔS_Z = ±1 and ΔI_Z = 0) in order to keep consistency with the photon angular momentum $\hbar$. If odd nuclear spins exist on the neighboring atoms, each EPR transition is split into components of unequal intensity: it is the <u>superhyperfine</u> (transferred hyperfine) structure (SHFS). Both HFS and SHFS are indirect probes of the nature ("covalency") of the chemical bond in minerals.

1.4 Interelectronic coupling

Indirect coupling of electronic spins occurs in multielectronic ions through the action of <u>spin-orbit</u> coupling (see §2.1.1). It lifts the degeneracy of the electronic levels, even in the absence of an external magnetic field, giving rise to the <u>fine structure</u>. The local symmetry is of primary importance in determining the decomposition scheme. The angular variation of the EPR spectrum and the values of fine structure decomposition are direct probes of the local environment around the paramagnetic center.

1.5 The effective spin Hamiltonian

The <u>Hamiltonian</u> $\mathcal{H}$ is an operator which operates upon a wavefunction; its eigenvalues give the allowed energy levels of the system, if the wavefunction is consistent with Schrödinger equation. The wavefunction of a set of unpaired electrons can be written as the sum of various terms, each term being treated as a perturbation of the preceding ones. The Hamiltonian of an ion in a crystalline environment with an external magnetic field is

$$\mathcal{H} = \mathcal{H}_V + \mathcal{H}_{LF} + \mathcal{H}_{LS} + \mathcal{H}_{SS} + \mathcal{H}_Z + \mathcal{H}_{SI} \quad , \tag{4}$$

where the pure nuclear contributions (nuclear Zeeman effect, nuclear quadrupolar interaction) have been excluded on account of their low energy. The subscript symbols indicate the type of interaction to which the Hamiltonian applies. The first three terms describe the electronic levels involved in optical absorption processes (see Chapter 7): V is the repulsive electron-electron (Coulombian) interaction, LF is the ligand field effect and LS is the spin-orbit interaction. The last three terms correspond to perturbations resulting in closely spaced ($\approx$ some cm^{-1}) levels. They strongly affect EPR spectra: SS is the electron-electron dipolar interaction (fine structure), Z is the Zeeman interaction and SI is the electron spin-nuclear spin interaction (HFS and SHFS). The influence of nuclear Zeeman effect is neglected, if its contribution is smaller than the other terms of the Hamiltonian.

Because of the large energy differences between these levels, as compared to microwave energy, only the lowest energy levels resulting from crystal field or spin-orbit splittings can be studied by EPR. Some of the ground states are orbital singlets, and for these the only degeneracy arises from electron spin. For an orbital degeneracy under the crystal field interaction, some considerations lead to consider only the spin contribution to the state degeneracy. If the ion has an odd number of electrons, the <u>Kramer's theorem</u> applies. This stipulates that there always will be at least a twofold degenerate state in the presence of electric (crystal) fields having any kind of symmetry; this degeneracy can be removed only under an external magnetic field, as it is related to the invariance of the system under time reversal. EPR can thus be observed only if the transition between two levels is permitted <u>and</u> if the separation of the levels is not larger than the applied radiation frequency. Orbital degeneracy can also be lifted in systems with an even number of electrons, under a Jahn-Teller distortion (spontaneous nuclear displacements remove the orbital degeneracy) or spin-orbit interaction: the states arising from these interactions will transform like angular momenta or in a similar manner.

Only spin degeneracy remains: the ground manifold of states has the transformation properties of spin eigenfunctions and we will generally assign a $(2S+1)$ degeneracy to this ground state. For this reason, EPR is often also called ESR, on account of the major influence of electron spin. To interpret the Zeeman splitting of the ground state, it is not necessary to calculate the effects of crystal field on all the energy levels: it is possible to simplify the Hamiltonian (4) by introducing an effective spin Hamiltonian or spin Hamiltonian, which operates upon the effective spin functions and will reproduce the possible behavior of the ground manifold. In particular, the effective spin S is defined such that the number of levels between which EPR transitions are observed is equal to $(2S+1)$. This Hamiltonian consists of spin operators for the electronic and nuclear spins which contribute to the paramagnetism and it also includes the effects of an applied magnetic field. Each of these independent terms must be invariant under the point symmetry of the paramagnetic entity, and the Hamiltonian consists of each of the magnetic fields or operators raised to a certain power. If we restrict it to the quadratic forms, the spin Hamiltonian comprises the last three terms of the full Hamiltonian given in (4), and it can be written as

$$\mathcal{H} = S{\cdot}D{\cdot}S + \beta B{\cdot}g{\cdot}S + S{\cdot}A{\cdot}I \tag{5}$$

where **D**, **g** and **A** are the tensors which couple the vectors indicated.

This spin Hamiltonian will allow us to consider the three major types of interactions which can affect the electronic energy ground states of a paramagnetic center:

(i) the Zeeman splitting by an external magnetic field, and the **g** tensor variations;
(ii) the effect of crystal field (see Chapter 7) which will result in the fine-structure (also called initial splitting or zero-field splitting, ZFS); it practically determines the number and actual position of the observed resonances.
(iii) the interaction between electronic and nuclear unpaired spins corresponding to the hyperfine structure (HFS), which further splits the preceding resonances.

Although the Zeeman interaction is the essential part of this Hamiltonian, the fine and hyperfine structures give EPR spectra their richness, because they are sensitive to the structural and chemical environment of the paramagnetic center.

2 FUNDAMENTALS of EPR

The reader is referred to the numerous works published on this subject, e.g., Low (1960, 1968), Abragam and Bleaney (1969), Goodman and Raynor (1970), Wertz and Bolton (1972), Pake and Estle (1973), Marfunin (1979b), Hervé (1986). Crystal field effects which affect the electronic levels are also described in Orgel (1961), Ballhausen (1962), Burns (1970) and Marfunin (1979a); however, EPR -- as all the other magnetic properties -- is mainly concerned with the nature of the ground state.

2.1 The g TENSOR

2.1.1 Derivation of the Lande factor and the g tensor

As pointed out above, we have to take into account two distinct angular momenta:
(i) the total spin angular momentum S, which has a maximum value of $n/2$ for a n-electron ion. Along a given direction of a quantized space, the S vector can take $(2S+1)$ values between $-S$ and $+S$, in ℏ units ($\hbar = h/2\pi$). The vector modulus is $\sqrt{S(S+1)}$.
(ii) the orbital angular momentum L also contributes to paramagnetism through orbital electron motion. The quantum number L takes integer values between $-L$ and $+L$.

The motion of the electron around the nucleus creates a magnetic field which interacts with the spin through spin-orbit coupling λLS where λ is the spin-orbit coupling constant. For a given atom or free ion, this coupling gives rise to further splitting of the L and S states: they are arranged in accordance with total angular momentum $\mathbf{J}$, obtained for a multielectron atom by vector addition of the components of orbital and spin angular momentum, $\mathbf{J} = \mathbf{L} + \mathbf{S}$. The energy describing Zeeman interaction is defined by the relation

$$E = \text{\ss} g_J \mathbf{J}\mathbf{B} = \text{\ss}(g_L\mathbf{L} + g_S\mathbf{S})\mathbf{B} \quad , \tag{6}$$

where g_J is the Lande factor and g_L and g_S are the gyromagnetic factors concerning orbital and spin moments. The Lande factor is defined by

$$g_J = \frac{3}{2} + \frac{S(S+1) - L(L+1)}{2J(J+1)} \quad . \tag{7}$$

Two limiting cases occur: $S = 0$ and $g = 1$ corresponds to a pure orbital momentum; $L = 0$ and $g = 2$ (in fact, 2.0023) corresponds to a pure spin momentum. When the ion is embedded in a solid, three situations may be recognized, according to the relative importance of crystal field and spin-orbit coupling.

Rare-earth ions. Crystal field effects are small (typically 10^2 cm^{-1}), because the $4f$-subshell is shielded from its surrounding by the outer shells. The spin-orbit coupling (also called Russell-Saunders coupling) holds. It is necessary to consider the total angular momentum $\mathbf{J}$: the J states are generally well-separated, the crystal field splittings are small ($<kT$) and only the ground manifold is populated. Because relaxation time, it is necessary to record EPR spectra at low temperature. For Kramer's ions (ions with an odd number of electrons), the crystal field splitting results in $(J+1/2)$ doublets. Low symmetry crystal fields lead to an anisotropic $\mathbf{g}$ tensor, which reflects the point symmetry at the ion site. g effective values g_{eff} may be calculated from the free ion Lande factor g_J using the wave functions which constitute each doublet. As an example Ce^{3+} ($4f^1$ ion) has a ground state sextuplet $J = 5/2$ and g_{eff} values are: $g_{//} = g_J = 0.86$ and $g_\perp = 2.57 \approx 3\,g_J$, which corresponds to a $|\pm 1/2\rangle$ ground state. g_{eff}-values obtained for non-Kramer's ions (even number of electrons) are more difficult to interpret.

S-state ions (half-filled sub-shells) and free radicals. These correspond to pure spin states; the S-ions have a zero angular momentum $\mathbf{L}$ (symmetric electronic configuration): the angular orbital momentum is "frozen". As $\mathbf{J}$ has no more significance, only $\mathbf{S}$ is subject to an external magnetic field. According to equation (7), $g \approx 2$ and the deviations experimentally observed remain small. Even in anisotropic hosts, it is difficult to detect any deviation of the $\mathbf{g}$ tensor from its isotropic form,

$$\mathbf{g} = g\begin{pmatrix} 1 & 0 & 0 \\ 0 & 1 & 0 \\ 0 & 0 & 1 \end{pmatrix} \quad . \tag{8}$$

S-state ions do not theoretically show any zero-field splitting due to crystal field effects, but such a splitting is observed for cubic and lower symmetry fields.

The transition (non-S) ions. These represent intermediate coupling of $\mathbf{L}$ and $\mathbf{S}$. Here also, $\mathbf{J}$ has no significance. The resonance condition is defined by

$$\mathcal{H} = \beta(L+g_sS)B + \lambda LS \quad . \tag{9}$$

An <u>effective g factor</u>, g_{eff}, is used on a purely experimental basis, describing the resonance condition of equation (3). In this case, the variation of the resonance in the crystallographic coordinates is adequately described using a second rank tensor $\mathbf{g}$, including g_s, λ and the energy difference between ground and excited states (crystal field effects). It can be used to derive, from a first order calculation, a Hamiltonian of the form:

$$\mathcal{H} = \beta \mathbf{B g S} \quad . \tag{10}$$

A special mention has to be made concerning d^3 ions (as Cr^{3+}) in octahedral coordination. The low-lying triplet (t_{2g}) resulting from crystal field splitting is half filled: this gives an effective zero L value. For d^8 ions a similar situation occurs as the e_g doublet is half occupied. Orbital singlet ground states may also be found in d^1 and d^9 ions in distorted octahedral or tetrahedral coordination. All these ions are indeed effective S-state ions at the energy scale of EPR. As S-state ions, they give EPR spectra at room temperature: the absence of nearby excited states gives rise to rather long relaxation times and thus to reasonable linewidths which allow to resolve the experimental spectra. Furthermore the anisotropy of the g-factor is limited by the smaller spin-orbit coupling.

2.1.2 Anisotropy of the g factor

Non-cubic crystal field components induce a perturbation through spin-orbit interaction. The g_{eff} value will depend on the relative orientation of the crystal and the external field **B**. By choosing a system where **g** is diagonal, the energy of levels is

$$E = \beta | \mathbf{B \cdot g} | M_S \quad , \tag{11}$$

where M_S is the magnetic quantum number (projection of **S** along the axis common to **g** and **B**, chosen as quantization axis). E takes integer values between $-S$ and $+S$. If l, m and n are the direction cosines of the magnetic field in the chosen system ($B_x = B\sin\theta\cos\varphi = Bl$; $B_y = B\sin\theta\sin\varphi = Bm$; $B_z = B\cos\theta = Bn$), equation (11) expands as

$$E = \beta \left| (l,m,n) \begin{pmatrix} g_x & 0 & 0 \\ 0 & g_y & 0 \\ 0 & 0 & g_z \end{pmatrix} \right| BM_S = \beta | (g_x l, g_y m, g_z n) | BM_S = g \, \beta B M_S \quad , \tag{12}$$

where g_x, g_y, g_z are **g**-tensor eigenvalues. The anisotropic g factor is defined as

$$g = \sqrt{g_x^2 l^2 + g_y^2 m^2 + g_z^2 n^2} \quad . \tag{13}$$

As for the free electron resonance, the selection rules are $\Delta M_S = \pm 1$. For axial site symmetry, $g_x = g_y = g_\perp$, $g_z = g_{//}$ and equation (13) reduces to

$$g = \sqrt{g_{//}^2 \cos^2\theta + g_\perp^2 \sin^2\theta} \quad . \tag{14}$$

The angular variation of EPR transitions directly gives the orientation of the **g** tensor axes with respect to the crystallographic axes, and it provides direct information on the symmetry of the paramagnetic center. When the symmetry at the site is lower than crystal symmetry, the paramagnetic centers are not magnetically equivalent and give rise to distinct spectra, the symmetry of which is revealed by the angular dependence of g. Figure 2 shows tetragonal and trigonal centers in a cubic matrix. Angular dependence of EPR is the basis for discussing site distortion in crystals and charge compensation processes, through the determination of the local symmetry around the paramagnetic center.

2.1.3 Significance of g values

When the ion is embedded in a solid, the orbital singlet states are characterized by the quenching of kinetic orbital momentum, because the average value of the projection of L along the quantization axis is zero. This quenching is generally partial, and the g factor shows deviations up to several percent from $g_S = 2.0023$. The ground state is allowed to mix with excited states through spin-orbit coupling: the smaller the separation Δ between these states, the higher the deviation of g from 2. First order perturbation calculations give simple expressions for g. In d^1 system, the g value along one of the principal directions is

$$g = g_S - c\lambda/\Delta \ , \tag{15}$$

where Δ is the energy separation between the ground state and the first excited states (i.e., crystal field splitting, as determined by optical absorption spectroscopy: cf. Rossman, this volume), and c is a constant which is usually in the range 1 to 10. The coefficient c is deduced from the "magic pentagon" (Goodman and Raynor, 1970), according to the type of orbitals which are involved in the coupling (Fig. 3). λ is positive when the ground term arises from a subshell less than half full, and is negative for subshells more than half full (electron-hole equivalence): the distinction between both cases is rather easy because the spectra are usually compared to a standart with $g \approx g_S$ (as DPPH, see 3.1). λ allows also to separate paramagnetic defects due to electrons or positive holes: in fluorites, electrons trapped at fluorine vacancies (F centers) have $g = 1.998$ and the hole trapped at interstitial fluorine ions (V_H center) has $g_{//} = 2.0026$ and $g_\perp = 2.039$.

As an example, let us consider the Cu^{2+} ion ($3d^9$) : it exhibits a static Jahn Teller effect in octahedral coordination and occupies strongly elongated coordination polyhedra. If the unpaired d electron occupies the x^2-y^2 orbital, g-values are

$$g_{//} = g_S + \frac{8\lambda}{\Delta E(d_{x^2-y^2} - d_{xy})} \ , \tag{16}$$

$$g_\perp = g_S + \frac{2\lambda}{\Delta E(d_{x^2-y^2} - d_{xz,yz})} \ , \tag{17}$$

where $d_{x^2-y^2}$, d_{xy} and $d_{xz,yz}$ refer to the energy of the x^2-y^2, xy and xz,yz orbitals, respectively. The EPR spectrum of $Cu^{2+}(d^9)$ in diopside glass is given in Figure 4 (Calas and Petiau, 1983). Parallel and perpendicular structures are visible on the same spectrum because of spatial averaging in glasses, with characteristic hyperfine components due to the odd isotopes ^{63}Cu (69.09%) and ^{65}Cu (31.91%) (see §2.2.2). g-values are higher than the free electron value which is in accordance with the above-mentioned assumption: $g_\perp = 2.059$ and $g_{//} = 2.354$. The relation $g_\perp < g_{//}$ confirms that the site is axially elongated.

Principal values of g depend on λ, as well as on crystal field intensity. λ depends on the electronic term considered and is related to the effective nuclear charge and inversely proportional to the cube of the electron-nucleus distance. Some free ion λ values are reported in Table 1. The accuracy of EPR allows us to measure the difference $g - g_S$; this difference remains small because Δ is greater than λ by two orders of magnitude. λ is dependent on unpaired electron localization: λ decreases with electron delocalization on the ligands (degree of "covalency") because the transfer of some of the unpaired spin density onto the ligands reduces the orbital angular momentum (i.e., further quenches the orbital

520

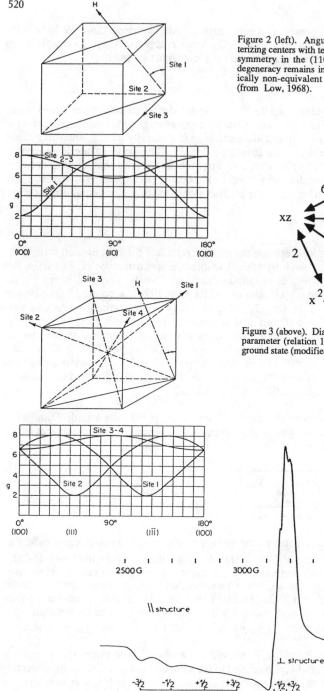

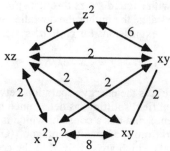

Figure 2 (left). Angular dependence of g-factor characterizing centers with tetragonal (above) and trigonal (below) symmetry in the (110) plane of a cubic lattice. Some degeneracy remains in the plane considered. The magnetically non-equivalent sites are represented at the bottom (from Low, 1968).

Figure 3 (above). Diagram showing the dependence of the c parameter (relation 15) on the type of orbital found in the ground state (modified from Goodman and Raynor, 1970).

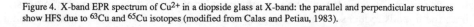

Figure 4. X-band EPR spectrum of Cu^{2+} in a diopside glass at X-band: the parallel and perpendicular structures show HFS due to ^{63}Cu and ^{65}Cu isotopes (modified from Calas and Petiau, 1983).

contribution) and makes the system to behave more like a pure spin system with g values closer to g_s. In Cu-bearing diopside glass, EPR and optical absorption spectra (Calas and Petiau, 1983) allow to determine $\lambda = -600$ cm^{-1}; this value is smaller than free ion value, -830 cm^{-1}, in relation with Cu-O bond covalency (see §4.5.3).

The study of the difference $(g_\perp - g_{//})$ permits the detection of an inversion of the sign of the distortion. An illustrative example may be found for the V^{4+} ion: it occurs in elongated sites in tetragonal (= rutile structure) GeO_2 ($g_{//}=$ 1.963 and $g_\perp =$ 1.921) and in compressed ("vanadyl") octahedra in aqueous complexes ($g_{//} =$ 1.933 and $g_\perp =$ 1.98). With increasing tetragonal compression, $g_\wedge$ value increases and exceeds $g_{//}$ (see §4.5.2).

2.2 The HYPERFINE STRUCTURE (HFS)

The electron senses a magnetic field due to the nucleus when the latter has a nonzero nuclear spin I. This results in a hyperfine coupling tensor $\mathbf{A}$. If the nuclear spin is on one of the surrounding ligands (and in some cases, on the second or further neighbors) the interaction gives rise to the superhyperfine structure (or transferred hyperfine structure, SHFS). The information extracted from HFS and SHFS gives a unique basis for discussing the nature of the absorbing center and/or its environments. It also gives information on the chemical bond between paramagnetic center and ligands.

2.2.1 Determination of hyperfine components

Hyperfine coupling is described, in the presence of a Zeeman effect, by

$$\mathcal{H} = \beta\, \mathbf{B}\cdot\mathbf{g}\cdot\mathbf{S} + \mathbf{S}\cdot\mathbf{A}\cdot\mathbf{I} \ , \tag{18}$$

where $\mathbf{A}$ is the hyperfine tensor. Other terms than this bilinear form (as the nuclear Zeeman term) can contribute to HFS, but they are often neglected. Usually, coupling energy is smaller than Zeeman effect. It is thus possible to expand equation (18) to first order

$$\mathcal{H} = \beta\, |\,\mathbf{B}\cdot\mathbf{g}\,|\,S\zeta + S\zeta' \frac{\mathbf{B}\cdot\mathbf{g}\cdot\mathbf{A}\cdot\mathbf{I}}{|\mathbf{g}\cdot\mathbf{B}|} = g\beta B S\zeta + S\zeta I\zeta' \frac{|\mathbf{B}\cdot\mathbf{g}\cdot\mathbf{A}|}{gB} \ , \tag{19}$$

where ζ and ζ' are the axes along $\mathbf{B}\cdot\mathbf{g}$ and $\mathbf{B}\cdot\mathbf{g}\cdot\mathbf{A}$ respectively, taken as quantization axes. The energy can thus be expressed to first order by

Table 1. Free ion spin orbit coupling constants λ (in cm^{-1}) of some $3d$ ions in a weak octahedral field (ground state terms) (after Abragam and Bleaney, 1969)

Ion	Valency		
	2	3	4
Ti	60	154	—
V	55	104	248
Cr	58	91	164
Mn	—	129	134
Fe	-103	—	129
Co	-178	-145(*)	
Ni	-324	-272(*)	
Cu	-830	-438(*)	

(*) calculated values.

Table 2. Relative intensities of SHFS lines for n equivalent nuclei of spin I.

$I = 1/2$

n =											
1	1	1									
2	1	2	1								
3	1	3	3	1							
4	1	4	6	4	1						
5	1	5	10	10	5	1					
6	1	6	15	20	15	6	1				

$I = 3/2$

n =												
1	1	1	1	1								
2	1	2	3	4	3	2	1					
3	1	3	6	10	12	12	10	6	3	1		

$$E = h\nu = g\beta B M_S + K M_S M_I \quad \text{with} \quad K = |\mathbf{B}\cdot\mathbf{g}\cdot\mathbf{A}| / gB \tag{20}$$

and g is defined as in equation (13). In the orthorhombic system, K expression is

$$\mathbf{B}\cdot\mathbf{g}\cdot\mathbf{A} = B(l,m,n) \begin{pmatrix} g_x & 0 & 0 \\ 0 & g_y & 0 \\ 0 & 0 & g_z \end{pmatrix} \begin{pmatrix} A_x & 0 & 0 \\ 0 & A_y & 0 \\ 0 & 0 & A_z \end{pmatrix} = (A_x g_x l + A_y g_y m + A_z g_z m)B \quad . \tag{21}$$

where l, m and n are the direction cosines of the magnetic field $\mathbf{B}$ (as in equation (12)), and

$$K = \frac{1}{g}\sqrt{A_x^2 g_x^2 l^2 + A_y^2 g_y^2 m^2 + A_z^2 g_z^2 n^2} \quad . \tag{22}$$

The field values which correspond to resonance are, for $\Delta M_S = \pm 1$:

$$B = \frac{h\nu}{g\beta} + M_I \frac{K}{g\beta} \quad . \tag{23}$$

As M_I varies between $-I$ and $+I$, there are $(2I + 1)$ HFS components for each fine structure component. Unlike fine structure components, HFS components are of equal intensity, because HFS levels are almost equally populated on account to their proximity. They are equally separated by $(K/g\beta)$, but this spacing is angle dependent (see equations (13) and (22) for g and K, respectively). However, equation (22) shows that there is no crossing of the position of these transitions, which makes also a clear difference with fine structure transitions. For axial symmetry, equation (22) becomes

$$K = \frac{1}{g}\sqrt{A_{//}^2 g_{//}^2 \cos^2\theta + A_\perp^2 g_\perp^2 \sin^2\theta} \quad , \tag{24}$$

and in cubic symmetry, $K = A$. In fact, HFS components are often unequally separated because second order effects give a dependence of K on the field value. HFS disappears in concentrated systems, because electrons are no more correlated with their nuclei: this contributes to the significant broadening which is observed (see §2.5.3). An example of HFS is shown on the EPR spectrum of Mn^{2+} ($I = 5/2$) in CaO and calcite (Fig. 5): only the stationary central transition $M_S = 1/2\,'M_S = -1/2$ is visible with six HFS components of equal intensity. The local distortion around Mn^{2+} in calcite explains the variations in HFS components intensity as well as the departure from a pure cubic symmetry (anisotropic spectrum). Ten weak "forbidden" transitions ($\Delta I = \pm 1$) occur between the most intense peaks and their relative intensity is also enhanced by distortion effects.

2.2.2 Chemical significance of hyperfine coupling

It is important to consider the kind of orbital to which the resonating electron belongs. In the case of an s orbital (or in a molecule with an orbital of σ-character), the electron density at the nucleus is high (see Hawthorne, this volume): the hyperfine coupling is large and not direction dependent, and this interaction is called the isotropic hyperfine coupling, A_{iso}, or Fermi contact interaction. When the electron belongs to a p, d or f orbital, we have a weaker dipolar interaction: the coupling depends on the direction of the external magnetic field and on the distance between electron and nucleus, resulting in an anisotropic hyperfine coupling, A_{aniso}. Determining these two contributions will consequently allow us to estimate the kind of orbital hybridization.

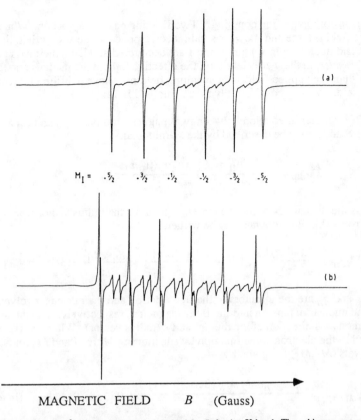

$M_I = \cdot\frac{5}{2} \quad \cdot\frac{3}{2} \quad \cdot\frac{1}{2} \quad \cdot\frac{1}{2} \quad \cdot\frac{3}{2} \quad \cdot\frac{5}{2}$

MAGNETIC FIELD   *B*   (Gauss)

Figure 5. EPR spectra of Mn^{2+} ions in CaO (above) and calcite (below) at X-band. The cubic symmetry at Mn-site in CaO is responsible for the isotropic pattern of HFS and the low intensity of "forbidden" ($\Delta I = \pm 1$) transitions. Distortions effects at the Mn site in calcite explain the more complex spectrum, due to perpendicular and parrallel components of the anisotropic HFS, and the stronger "forbidden" transitions (modified from Angus et al., 1979).

Table 3. Nuclear spins of some of elements giving rise to HFS in minerals.

Isotope	Abundance(%)	Spin	$A_{iso}(G)$	$A_{aniso}(G)$
^1H	100	1/2	508	—
^7Li	92.6	3/2	103	—
^{14}N	99.6	1	557	33.5
^{19}F	100	1/2	17160	1085
^{23}Na	100	3/2	224	687
^{27}Al	100	5/2	983	43
^{29}Si	4.7	1/2	-1218	-61
^{35}Cl	75.5	3/2	1672	102
^{37}Cl	24.5	3/2	1391	85
^{39}K	93	3/2	52	64
$^{51}V^{+4}$	99.8	7/2	—	-105
$^{55}Mn^{+2}$	100	5/2	—	-115
$^{63}Cu^{+2}$	69	3/2	—	-238
$^{65}Cu^{+2}$	31	3/2	—	-250

Anisotropic hyperfine coupling. Two limiting cases may occur. When the external magnetic field is large and A_{aniso} is small, it corresponds to the strong field limit, where the electron and nuclear spin vectors S and I are decoupled and have their axis parallel to this field. However, in the weak field limit the effective field at the nucleus due to the electron is larger than the applied field. The resulting field may have a different direction from the applied field. We will consider only the strong field limit.

The dipolar interaction between two magnetic moments μ_e and μ_n associated with the spins S and I may be described by the Hamiltonian

$$\mathcal{H}_{dip} = - \frac{\mu_e \cdot \mu_n}{r^3} - 3 \frac{(\mu_e \cdot r)(\mu_n \cdot r)}{r^5} \quad , \tag{25}$$

where r is the distance between the two dipoles and r the radius vector from the nucleus to the electron. This Hamiltonian may be written

$$\mathcal{H}_{dip} = -\frac{g_e \beta_e g_n \beta_n}{r^3} \left[I \cdot S - \frac{3(I \cdot r)(S \cdot r)}{r^2} \right] = \frac{g_e \beta_e g_n \beta_n}{r^3} [3\cos^2\theta - 1] \cdot (I \cdot S) \quad , \tag{26}$$

where g_e and g_n are the g-factors of the free electron and the nucleus involved, and β_e and β_n are the nuclear magneton and the Bohr magneton, respectively. θ is the angle between the common axis direction of the dipoles and r (radius vector). When the Hamiltonian (26) is applied to the electron wave function by substituting M_I for I and M_S for S, the energies of the levels (M_I, M_S) are given by

$$E_{M_I, M_S} = \frac{g_e \beta_e g_n \beta_n M_I M_S [3\cos^2\theta - 1]}{<r^3>} = P[3\cos^2\theta - 1] \quad \text{(in ergs)}, \tag{27}$$

where $<r^3>$ indicates the average value of the cube of the electron-nucleus distance. Two particular cases may be considered:

(i) for s electrons, equation (27) averages to zero: there is no contribution of A_{aniso}.
(ii) for electrons in orbitals centered on the nucleus, equation (27) has to be multi-plied by $[3\cos^2\alpha - 1]$, which represents the average direction of S within the orbital (α represents the angle between r and the principal axis of the orbital). In a p orbital, $[3\cos^2\alpha - 1] = 4/5$ and the anisotropic tensor is defined as $A_{aniso} = |4P/5, -2P/5, -2P/5|$.

Isotropic (contact) interaction. It is a purely quantic effect arising from the influence of nuclear magnetism on an electron as it moves at relativistic speed in the immediate vicinity of the nucleus. The isotropic coupling constant A_{iso} is defined by

$$A_{iso} = \frac{8\pi}{3} g_n \beta_n |\psi(0)|^2 \quad \text{(in G)}, \tag{28}$$

where $|\psi(0)|^2$ is the probability of finding the electron (= electron density) at the nucleus location. $A_{iso} = 0$ if the orbital has a node at the nucleus (p, d or f orbitals).

Experimental determination of hyperfine coupling. Hyperfine tensor A is formed by the isotropic and anisotropic contributions: $A = A_{aniso} + A_{iso}u$, where u is the tensor unity. We can measure both components from the angular variation of hyperfine coupling and for an axial symmetry we get $A_{//} = A_{iso} + 2A_{aniso}$ and $A_{\perp} = A_{iso} - A_{aniso}$; hence

$$A_{iso} = \frac{(A_{//} + 2A_{\perp})}{3} \quad \text{and} \quad A_{aniso} = \frac{(A_{//} - A_{\perp})}{3} \quad . \tag{29}$$

Because of the conservation of angular momentum, if momentum is transferred from the microwave photon to the electrons to flip from $-1/2$ to $+1/2$, the nuclear spin cannot change its orientation. Therefore the selection rules for EPR transitions are $\Delta m_S = \pm 1$ and $\Delta m_I = 0$. The transitions with $\Delta m_I = \pm 1$ or $\Delta m_I = \pm 2$ violate these rules and are forbidden in a first approximation; however, distortion of the coordination polyhedron gives them a finite intensity along non-principal axes.

Interaction of electron spin with the nuclear spin of N ligands results in the Super-hyperfine Structure (SHFS) or transferred HFS according to

$$\mathcal{H} = g\beta B \cdot S + \sum_i S \cdot A_i \cdot I_i \quad \text{with } 1 \leq i \leq N \quad . \tag{30}$$

In the case of N identical nuclei of nuclear spin I, it consists (first order approximation) of $2NI+1$ equally spaced lines, the intensity of which is no more constant and obeys the statistical weights of the various configurations of the nuclear spins (Table 3). However, the total intensity of the EPR transitions corresponding to the same $\Delta m_S = \pm 1$ does not change compared to the case where superhyperfine coupling is absent. The contribution of the various ligands is studied with Electron Nuclear DOuble Resonance, ENDOR (see e.g., Abragam and Bleaney, 1969). SHFS is illustrated in Figure 6 by F-centers in fluorite (see §2.1.3) and O_2^- ions in sylvite. The former is a classical defect represented by an electron trapped at an anion vacancy. The contribution of the six F second neighbors (^{19}F has $I = 1/2$) is manifested by seven hyperfine components responsible for the isotropic SHFS (there is no contribution from Ca neighbors). The other example refers to (O_2^-) centers in sylvite, located on anion sites and surrounded by four potassium atoms arranged in a regular tetrahedron. SHFS arises from the odd isotopes ^{39}K and ^{41}K, both with $I = 3/2$.

Finally, it must be pointed out that the intensity of the spectra arising from hyperfine coupling is a function of the natural abundance of the odd isotopes. The lines will further split if various isotopes of odd nuclear spin coexist. Figure 7 compares EPR spectra of Eu^{2+} (with nearly equal proportions of ^{153}Eu and ^{155}Eu isotopes) and the isoelectronic Gd^{3+} in cubic fields of natural fluorites. In this case, the hyperfine coupling constant is different for both Eu isotopes, although I is the same: 36.6 G for ^{151}Eu and 16.2 G for ^{153}Eu. This allows to separate their respective contribution and this low value explains that HFS is only a perturbation of the fine structure (which is the same as isoelectronic Gd^{3+}). In other cases, each isotope presents a distinct value of I. Such an example is presented in Figure 8, concerning Yb^{3+}-substitution in zircon (Ball, 1982). Ytterbium has two odd isotopes which give rise to a resolved HFS: ^{171}Yb ($I = 1/2$) with a 14.3% abundance and ^{173}Yb isotope ($I = 5/2$) with a 16.6% abundance. These HFS lines represent only some percent of the EPR spectrum due to the even isotopes. On account of the number of HFS components of each isotope, the intensity ratio of their EPR spectra is expected to be $^{171}Yb/^{173}Yb = 2.7$, which is observed. In general, only nuclei with A values large enough can be observed: in the opposite case, the HFS and SHFS signals are hidden by the main signal, except in crystals with narrow linewidths (like quartz).

2.2.3 Magnitude of hyperfine interaction

Important deviations are noted when comparing actual and free ion values of

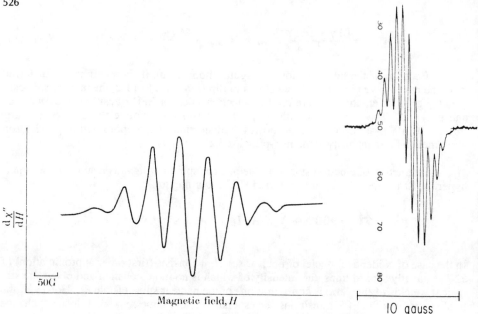

Figure 6. Color centers in halides showing a resolved SHFS at X-band. *Left:* EPR spectrum of F centers in fluorite; the magnetic field is along [110]. The 7-line isotropic HFS is due to the six surrounding ^{19}F nuclei (from Arends, 1964). *Right:* HFS in the EPR spectrum of the O_2^- molecule ion (oriented along [110]), with 13 lines due to the four surrounding ^{39}K nuclei (from Känzig, 1962).

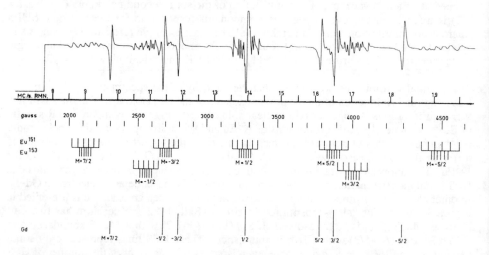

Figure 7. HFS due to Eu^{2+} isotopes in fluorite at X-band; the contribution of both isotopes is separated and the fine structure is similar to isoelectronic Gd^{3+} ions in cubic symmetry in the same sample (from Ryter, 1957).

hyperfine coupling constants (Table 3). One might expect an increase in magnitude of both A_{iso} and A_{aniso}, as cation charge increases, because <r> decreases and valence electrons are more firmly bound. However more important is the nature of the cation-ligand bond, which it is not independent of interatomic distances (Lehmann, 1980b). The formation of a molecular orbital results in transfer of some of the unpaired electron density from the cation to the ligand. This effect has been extensively investigated in EPR spectra of Mn^{2+} and various relationships between A and cation-ligand "covalency" have been proposed.

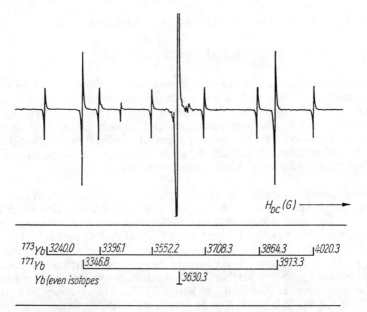

H_{DC} (G) ⟶

^{173}Yb	3240.0	3396.1	3552.2	3708.3	3864.3	4020.3
^{171}Yb		3346.8				3973.3
Yb (even isotopes			3630.3			

Figure 8. HFS of Yb^{3+} ions in tetragonal sites in zircon along [001] at X-band. The contribution of odd and even isotopes is separated below the experimental spectrum. The distinct I value allows to separate contributions from ^{171}Yb, ^{173}Yb and even Yb isotopes (from Ball, 1982).

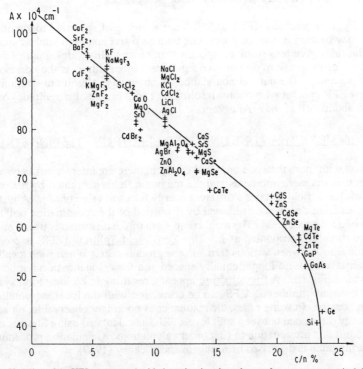

Figure 9. Variation of the HFS parameter A with the ratio c/n, where the covalency parameter c is defined as in relation (36) and n is the number of ligands (from Simanek and Mueller, 1970).

Figure 9 shows the variation of |A| with the parameter $\mathfrak{c}$ defined as

$$\mathfrak{c} = 1 - 0.16(X_{Mn} - X_B) - 0.035(X_{Mn} - X_B)^2 \quad , \tag{31}$$

where X_{Mn} and X_B represent the electronegativity of the manganese and the surrounding anion B, respectively. Note that the value extrapolated for the ionic end member corresponds to the free ion value of |A|, as reported in Table 3, 115 G ($107 \cdot 10^{-4}$ cm^{-1}). |A| does not depend on the majority cation of the host matrix: the local surrounding around Mn^{2+} is the dominant parameter. This property is particularly useful if two distinct possible sites exist in a mineral, because it lets us directly study the partition of the paramagnetic species between both sites and allows to detect unusual coordination polyhedra.

Some examples illustrate electron delocalization by the existence of a SHFS due to next-nearest neighbors: in xenotyme (Fig. 10), Gd^{3+} ions occur in orthorhombic sites in substitution of Y^{3+} ions. Along [001] each HFS component is split into 3 SHFS components. Although A is small (|A| is equal to 4.67 G and 6.09 G for the ^{155}Gd and ^{157}Gd isotopes) SHFS is resolved in the spectra and is due to the two ^{31}P next nearest-neighbors ($I = 1/2$), probably on account of the favored electron delocalization in the P-O bond. Interaction between HFS and SHFS may often result in a rather complex pattern as for Mn^{2+}-containing fluorites (Gehlhoff and Ulrici, 1980). The spectrum (Fig. 11) exhibits six HFS lines due to ^{55}Mn (as in Fig. 5), which are further split into a complex SHFS arising from the eight ^{19}F ligands. As evidenced by the simplicity of the central HFS line (only the nine lines predicted for SHFS are visible on the central transition), fine structure term is nearly zero. This spectrum is strongly different from those which arise from an interaction between HFS and fine structure term (see §2.3.2).

Finally, non-resolved HFS and SHFS are a source of spectral broadening. Major differences may be seen if one compares ions with zero and non-zero nuclear spins. Chlorides and fluorides give less-resolved spectra than oxides. In oxides, major components such as silicon or magnesium do not give significant broadening, as the odd isotopes are only a few percent of the natural abundance. Conversely, aluminum-containing compounds show EPR spectra with poorer resolution (1 to 2 orders of magnitude) on account of the 100% abundance of the ^{27}Al isotope.

2.3 IONS with MORE THAN ONE UNPAIRED ELECTRON - The FINE STRUCTURE

In ions with more than one unpaired electron, there are interactions between the individual electron magnetic moments and the magnetic fields generated by the other electrons. The magnetic field dictates a different energy for each value of M_S. A further complication arises from degeneracy removal by distortion of the coordination polyhedron in the absence of a magnetic field. The axial component of the distortion is usually referred to as D and the rhombic component as E. This Zero-Field Splitting (ZFS) is produced by spin-orbit coupling between various terms in the ground state. When such a splitting occurs, Zeeman levels are no longer equally spaced and the resonances do not occur at the same field value (Fig.12). A fine structure appears, resulting in $2S$ lines for a system with a total spin quantum number S. ZFS can be correlated with the local surrounding of the paramagnetic center. In some cases, the resonance is no longer observable, on account of the great energy difference between split levels. ZFS is calculated using Stevens operators, but we will limit this presentation to first order calculations. Calculation procedure of fine structure Hamiltonian is discussed in Abragam and Bleaney (1969).

2.3.1 The fine structure Hamiltonian

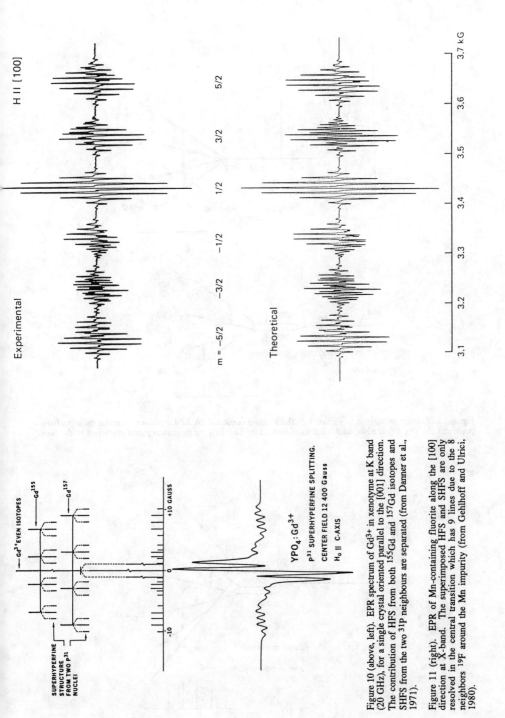

Figure 10 (above, left). EPR spectrum of Gd^{3+} in xenotyme at K band (20 GHz), for a single crystal oriented parallel to the [001] direction. The contribution of HFS from both ^{155}Gd and ^{157}Gd isotopes and SHFS from the two ^{31}P neighbours are separated (from Danner et al., 1971).

Figure 11 (right). EPR of Mn-containing fluorite along the [100] direction at X-band. The superimposed HFS and SHFS are only resolved in the central transition which has 9 lines due to the 8 neighbors ^{19}F around the Mn impurity (from Gehlhoff and Ulrici, 1980).

530

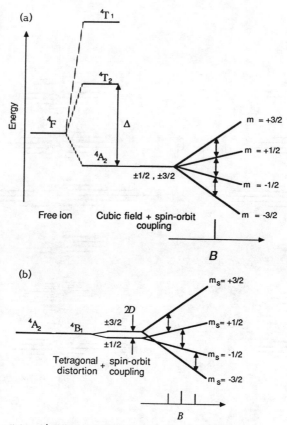

Figure 12. Zeeman splitting of $^4A_2$ level. (a) The splitting in cubic field: EPR transitions occur at the same field position. (b) The effect of an additional axial distortion: ZFS lifts the degeneracy and gives rise to the fine structure.

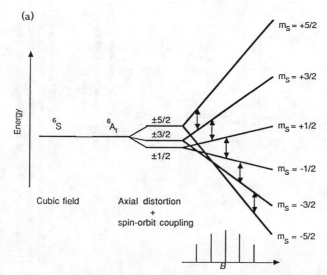

Figure 13. Energy level diagram as a function of external field B for $S = 5/2$. (a) Hyperfine splitting due to a nuclear spin is shown for $I = 5/2$ and (b) [next page] some transitions are indicated (from Bleaney and Ingram in Hawthorne, 1983).

(b)

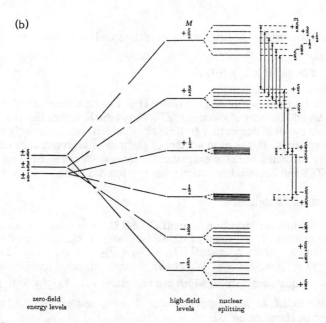

Figure 13 (b) [above; see legend on previous page].

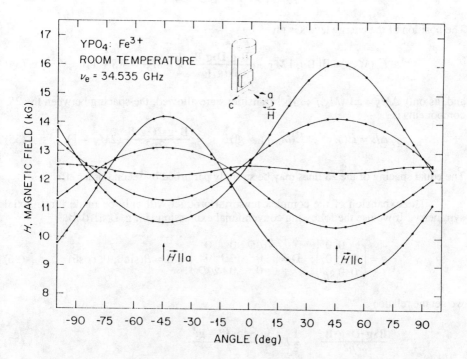

Figure 14. EPR of Fe^{3+} in xenotyme at Q-band; the lines cross at approximately 55° respective to the c axis (from Rappaz et al., 1982). The deviations from the calculated position arise from second-order effects.

We have to consider the first two terms of the spin Hamiltonian of equation (5):

$$\mathcal{H} = \beta B \cdot g \cdot S + S \cdot D \cdot S \quad . \tag{32}$$

ZFS arises from operators which do not involve H or I and are therefore of the type S^a. In systems with an odd number of electrons, ZFS obeys the Kramer's theorem: crystal field splitting leaves a two-fold degeneracy (= Kramer's doublets) which can only be lifted by an external magnetic field. For an even number of electrons, the ground state can be a singlet state too widely separated from the excited states for resonance be observed. The relative magnitude of ZFS and Zeeman interaction gives two limiting cases.

2.3.2 Small zero-field splitting

It occurs when the distortion is small. The $S.D.S$ term can be expanded, giving a rather complex relation. Some simplification arises from the elimination of the skew-symmetric terms, with a zero-trace tensor D $(D_{xx} + D_{yy} + D_{zz} = 0)$ and using the classical relation $S^2 = S(S+1) = S_x^2 + S_y^2 + S_z^2$. We obtain the only diagonal term, using $|M_S\rangle$ as a basis with z as the quantization axis, which can be written $\frac{1}{2}D_{zz}(3S_z^2 - S^2)$. If we choose the quantization axis ζ as the common axis of B and g, and expand the $D_{\zeta\zeta}$ term, first order expansion of Hamiltonian (32) is

$$\mathcal{H} = \beta | B \cdot g | S_\zeta + \frac{B \cdot g \cdot D \cdot g \cdot B}{| B \cdot g |^2} \frac{1}{2}(3S_\zeta^2 - S(S+1)) \quad . \tag{33}$$

The position of the energy levels is given by

$$E(M_S) = \beta | B \cdot g | M_S + \frac{B \cdot g \cdot D \cdot g \cdot B}{| B \cdot g |^2} \frac{1}{2}[3M_S^2 - S(S+1)] \quad , \tag{34}$$

and, as only $\Delta M_S = \pm 1$ $(M_{S-1} \leftrightarrow M_S)$ transitions are allowed, the spacing between the $2S$ components is

$$\Delta E = E(M_S) - E(M_{S-1}) = g\beta B + \frac{3}{2} \frac{B \cdot g \cdot D \cdot g \cdot B}{g^2 B^2}(2M_S - 1) \quad . \tag{35}$$

The equal spacing of the $2S$ lines may be strongly perturbed by second order effects.

The expansion of the complex tensorial product will only be made for an axial symmetry. If we use the following conventional expressions for g, D and B:

$$g = \begin{pmatrix} g_\perp & 0 & 0 \\ 0 & g_\perp & 0 \\ 0 & 0 & g_{//} \end{pmatrix}; \quad D = \begin{pmatrix} -1/3D & 0 & 0 \\ 0 & -1/3D & 0 \\ 0 & 0 & 2/3D \end{pmatrix}; \quad B = B(\sin\theta, 0, \cos\theta) \quad , \tag{36}$$

we get the relation

$$\frac{B \cdot g \cdot D \cdot g \cdot B}{g^2 B^2} = \frac{1}{3}D \frac{3g_{//}^2\cos^2\theta - g^2}{g^2} \quad . \tag{37}$$

Equation (35) may thus be written as

$$\Delta E = g\beta B + \frac{1}{2} D \frac{3g_{//}^2\cos^2\theta - g^2}{g^2} (2M_S - 1) \quad .$$ (38)

In the frequent case where the g factor has only slight angular dependence, the spacing depends on a $(3\cos^2\theta -1)$ term. This gives the following expressions

$$\Delta E = g\beta B + \frac{1}{2} D (3\cos^2\theta - 1)(2M_S - 1); \quad \Delta B = (3\cos^2\theta -1) D/g\beta \quad .$$ (39)

The variation of energy levels as a function of the external magnetic field is shown in Figures 12 ($S = 3/2$) and 13 (= 5/2). Equation (44) cancels for $\theta \approx 55°$, where the lines cross: this is evidenced in Figure 14 which represents the angular dependence of the 5 EPR transitions of substitutional Fe^{3+} in xenotyme with a ZFS of $849 \cdot 10^{-4}$ cm^{-1}. Second order effects shift the transitions from their theoretical position. Line crossing makes a clear difference with hyperfine structure for which no crossing occurs (equation (24) cannot cancel). At $\theta = 0$, the spacing between the fine structure transitions is $D(2M_S - 1)/g\beta$ (in field units); for $\theta = \pi/2$, it is $D(2M_S - 1)/2g\beta$: this allows us to derive ZFS values from oriented single-crystal EPR spectra.   The intensity of the fine structure lines varies as $(S + M_S)(S - (M_S - 1)$. In orthorhombic symmetry, and with a nearly isotropic g factor, there is a more complex angular dependence in $1/3D (3n^2 - 1) + E (l^2 - m^2)$, where l, m and n are the direction cosines of the magnetic field, as defined for equation (12). The anisotropic splitting predicted by (39) is illustrated by EPR spectrum of Gd-containing thorite (Fig. 15). The perpendicular direction gives splittings which are approximately twice lower than the parallel direction. The seven fine structure groups arise from four magnetically nonequivalent orthorhombic sites which are degenerate only for crystal orientations along a and c. D and E parameters are $-718 \cdot 10^{-4}$ and $-475 \cdot 10^{-4}$ cm^{-1} respectively. The interaction between ZFS and HFS may be illustrated by the EPR spectrum of Mn^{2+} in tremolite (Fig. 16). It consists of five groups of lines which obey the relative intensities given above; ZFS value derived from this spectrum is $280 \cdot 10^{-4}$ cm^{-1}, which is high as compared to hyperfine coupling ($|A| \approx 75 \cdot 10^{-1}$ cm^{-1}). These fine structure lines are further split into six equally intense HFS components. In all non-principal axis directions, forbidden transitions ($|\Delta M_S| > 1$) are observed at lower field values.

2.3.3 Strong distortion

The corresponding Hamiltonian is

$$\mathcal{H} = \beta \mathbf{B} \cdot \mathbf{g} \cdot \mathbf{S} + D [S_z^2 - 1/3S (S + 1)] + E (S_x^2 - S_y^2) \quad .$$ (40)

Only the two extreme cases $D = 0$ and $E = 0$ will be considered.

Strong axial distortion. The parameter E is neglected in regard to D. The Kramer's doublets $|\pm M_S\rangle$ are widely separated, as a result of the strong distortion. If we take into account the selection rule $\Delta M_S = \pm 1$, the transitions will be observed in the doublet $|\pm 1/2\rangle$. The energy matrix deduced from the Hamiltonian has the following eigenvalue:

$$E = D\left[\frac{1}{4} - \frac{1}{3}S(S + 1)\right] \pm \frac{1}{2}\beta B \sqrt{g_{//}^2 \cos^2\theta + \left[g_\perp(S + \frac{1}{2})\right]^2 \sin^2\theta} \quad .$$ (41)

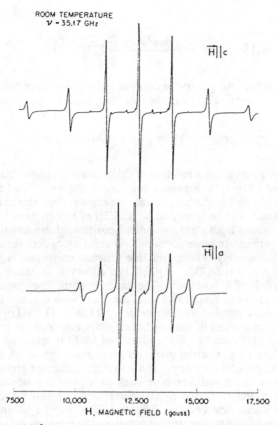

ROOM TEMPERATURE
$\nu = 35.17$ GHz

$\vec{H}\|c$

$\vec{H}\|a$

7500   10,000   12,500   15,000   17,500
H, MAGNETIC FIELD (gauss)

Figure 15. EPR spectra of Gd^{3+} ions in single crystals thorite at Q-band along the a and c directions. Although site symmetry is orthorhombic, the spectra are degenerate along the represented directions and their splitting obey relation (39). Weak lines are due to tetragonal Gd^{3+} ions (remote charge compensation) (from Reynolds et al., 1972).

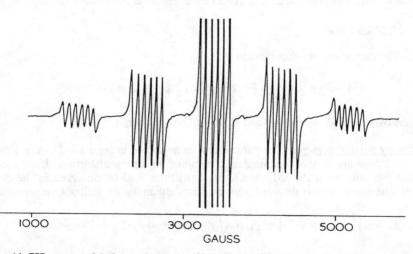

1000   3000   5000
GAUSS

Figure 16. EPR spectrum of single-crystal tremolite at X-band in one of the principal g-tensor directions (in the (001) plane) (from Golding et al., 1972). HFS is a perturbation of the main ZFS spectrum. Only allowed transitions ($\Delta I = 0$) are observed along this direction.

where θ is the angle between the distortion axis (z) and **B**. It is interesting to note that the levels are partly shifted by the same energy. It remains only an interaction similar to a Zeeman term (see equation (17)) by choosing an effective $S' = 1/2$ and effective g values: $g'_{//} = g_{//}$ and $g'_{\perp} = g_{\perp}(S + 1/2)$. There is thus a strong angular dependence which arises from the role played by the distortion axis. This explains the g_{eff} values often encountered for ions in axially distorted environments, e.g., $g_{//} = 2$ and $g_{\perp} = 6$ observed for Fe^{3+} ions ($S = 5/2$).

<u>Strong rhombic distortion</u>. For $D \approx 0$ and $E \gg g\beta B$, a simple classification of the levels according to S_z is no more possible. There is in fact a strong mixing between the states, and it is possible to observe transitions inside each doublet (and not only inside the $|\pm1/2>$ state, as in the axial distortion case). The problem has been extensively studied for Fe^{3+} ions, which often occur in distorted coordination polyhedra of rhombic symmetry: there is an isotropic transition at $g_{eff} = 30/7 = 4.3$, which is effectively observed in many glasses and minerals (Fig. 17). This transition is not indicative of any kind of site as it can be encountered either in octahedral or in tetrahedral coordination: tetrahedral sites in diopside glass and feldspars and octahedral sites in kaolinite and micas. It has been the subject of considerable debate since the pioneering work of Castner et al. (1960) and has been shown to correspond to a transition inside the central Kramer's doublet of the energy system resulting from the rhombic limit of the distortion ($D = 0$ in the Hamiltonian (40)). Figure 18 represents a calculation of g_{eff} for the three crystal field Kramer's doublets considered separately as having effective spins of 1/2 and in the approximation where the fine structure parameters E and D are great respective to the microwave frequency used. The rhombicity of the site is represented by the $\lambda = E/D$ ratio, with values between 0 (pure axial symmetry) and 1/3 (pure rhombic symmetry). $\lambda > 1/3$ corresponds to a more axial symmetry, under an appropriate change of axis. At $\lambda = 0$ and $\lambda = 1$ (pure axial distortion), the effective g-values are 2 and 6 as indicated above. Departures from a pure rhombic distortion is accompanied by the existence of three g values, the position of which has been calculated by various authors (Aasa, 1970): this is illustrated by Figure 17c in which two sites are separated on the basis of distinct intensity of rhombic distortion.

2.3.4 Quadrupolar effects and fine structure in a cubic field

The maximum degeneracy which can remain in a cubic field is four-fold (which corresponds to a $S = 3/2$ ion: ions with $S \geq 2$ show a partial lifting of the degeneracy of the ground state. This includes the S-state ions ($3d^5$ ions such as Fe^{3+} and Mn^{2+}, and $4f^7$ ions such as Gd^{3+} and Eu^{2+}). We will briefly examine a d^5 ion in cubic symmetry: fourth-order terms have to be included in the Hamiltonian:

$$\mathcal{H} = g\beta\mathbf{B}\cdot\mathbf{S} + \frac{1}{6}a\left[S_x^4 + S_y^4 + S_z^4 - \frac{1}{5}S(S + 1)(3S^2 + 3S - 1)\right] \ , \quad (42)$$

where x, y and z are the cubic axes. a is called the <u>cubic fine structure parameter</u> and is related to the intensity of crystal field effects. For $3d^5$ ions in cubic symmetry and weak crystal field, the transitions corresponding to $\Delta M_S = \pm1$ are observed at

$$\Delta E = g\beta B + \frac{a}{120}\left[1 - 5(l^2m^2 + m^2n^2 + n^2l^2)\right]f(M_S) = g\beta B + \frac{aF}{120}f(M_S) \ , \quad (43)$$

where l, m and n are direction cosines of the magnetic field and $f(M_S)$ is a polynomial function of the spin component. On account of the great importance of these ions in mineralogy, we give the expressions relating the field dependence of the various transitions $M_S'M_{S-1}$:

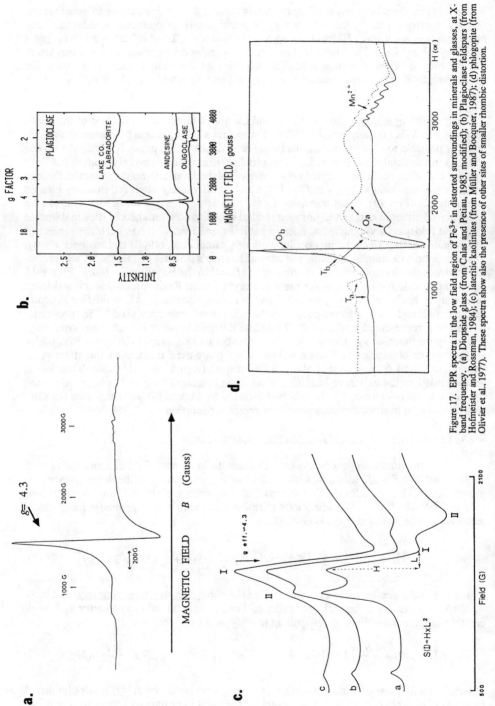

Figure 17. EPR spectra in the low field region of Fe^{3+} in distorted surroundings in minerals and glasses, at X-band frequency. (a) Diopside glass (from Calas and Petiau, 1983, modified); (b) Plagioclase feldspars (from Hofmeister and Rossman, 1984); (c) lateritic kaolinites (from Müller and Bocquier, 1987); (d) phlogopite (from Olivier et al., 1977). These spectra show also the presence of other sites of smaller rhombic distortion.

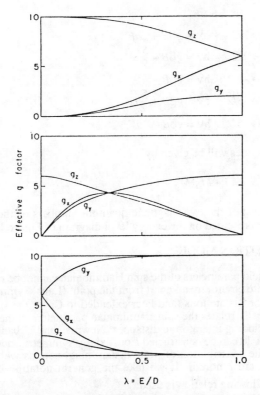

Figure 18. Effective g-values for the Hamiltonian (46) plotted against E/D. The same levels are displayed twice, above and below $\lambda = 1/3$ and a full description of the effects of a rhombic distortion is accounted for by taking only values of λ between 0 and 1/3 (from Wickman et al., 1965).

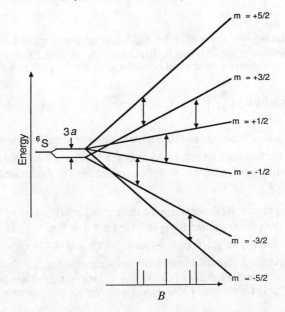

Figure 19. Energy level diagram for Fe^{3+} in a cubic field. The intensity of the five transitions observed is represented at the bottom; their splitting obeys relation (43).

$$5/2 \leftrightarrow 3/2 \quad h\nu = g\beta B + 2aF$$

$$3/2 \leftrightarrow 1/2 \quad h\nu = g\beta B - \frac{5}{2}aF$$

$$1/2 \leftrightarrow -1/2 \quad h\nu = g\beta B$$

$$-1/2 \leftrightarrow -3/2 \quad h\nu = g\beta B + \frac{5}{2}aF$$

$$-3/2 \leftrightarrow -5/2 \quad h\nu = g\beta B - 2aF \quad .$$

The intensity of these lines will be given by

$$I \propto S(S+1) - M_S(M_S - 1) \quad , \tag{44}$$

where M_S is the largest of the two magnetic quantum numbers of the transition. Cubic ZFS of Fe^{3+} ions is represented on the energy level diagram of Figure 19.

2.4 The SUPERPOSITION MODEL

The crystal field parameters of the spin Hamiltonian cannot be related simply to the geometry of immediate environment, apart from the point (Laue) symmetry. An approach has been presented for S-state ions (and later extended to Cr^{3+}), the superposition model (SPM) which empirically relates the spin Hamiltonian parameters to the coordination poly-hedron geometry, including interatomic distances (Newman and Urban, 1975). It assumes that the total crystal field can be constructed from axially symmetric, additive contributions from the ligands. The contribution of more distant neighbors as well as the interaction between the ligands are ignored. If we take the general notation of the crystal field parameters b_n^m, the following relation is used:

$$b_n^m = \sum_i b_n(R_i) \, K_n^m(\theta_i, \varphi_i) \quad , \tag{45}$$

where i runs over the ligands at coordinates R_i, θ_i and φ_i; the functions $K_n^m(\theta_i, \varphi_i)$ are an-gular functions which are tabulated, and $b_n(R_i)$ is an intrinsic radial law defined by compa-rison with a reference distance (the normal bond distance for the paramagnetic ion) R_0 as

$$b_n(R) = b_n(R_0)\left(\frac{R_0}{R}\right)^{t_n} \quad , \tag{46}$$

where t_n is an adjustable exponent. In this notation, the crystal field parameters which have been used in earlier sections can be expressed as $D = 3\,b_2^0$, $E = 1/3\,b_2^2$ and $a = 2\,b_4^0$. The SPM model spatially describes the distortion (relaxation) of the immediate surrounding of the paramagnetic ion.

Second-order crystal field parameters can be considered as giving the major contri-bution to EPR, and the value retained for t_2 is 7 for Fe^{3+} and Mn^{2+}, and $t_2 = 3$ for Cr^{3+}. The K functions are defined as $K_2^0(\theta, \varphi) = \frac{1}{2}(3\cos^2\theta - 1)$ and $K_2^2(\theta, \varphi) = \frac{3}{2}\sin^2\theta\cos 2\varphi$. The curves $b_2(R)$ have been deduced from strain experiments for Fe^{3+} by Sangster (1981) and for Cr^{3+} by Müller and Berlinger (1983); the values obtained are larger than those chosen by Büscher and Lehmann (1986, 1987) and Novak and Vosika (1983). The a-priori knowledge of the $b_2(R)$ function allows us to calculate (for one given direction) the distortion of the coordination polyhedron d_s from ZFS parameters (Büscher et al., 1987):

$$d_s = \frac{1}{2b_2} \left[(3\cos^2\theta - 1)b_2^0 + b_2^2 \sin^2\theta \cos 2\varphi \right] \quad , \tag{47}$$

and to compare it to the distortion deduced from crystal structure determinations:

$$d_c = \frac{1}{2} \sum_i (3\cos^2\theta_i - 1)\left(\frac{R_0}{R_i}\right)^{t2} \quad , \tag{48}$$

where the θ_i are the angles between the cation-ligand bonds and the considered direction. The intrinsic radial law $b_n(R_i)$ and the adjustable exponent t_n are in fact partly correlated and depend on the "normal" M-X bond distance. This explains why the values from various authors or concerning distinct matrices give non-consistent results. SPM has been recently applied to Fe^{3+} and Cr^{3+} in some minerals (Büscher et al., 1987).

2.5 INTENSITY, LINESHAPE and RELAXATION EFFECTS

In order to accurately determine the nature and concentration of paramagnetic centers in minerals, it is necessary to resolve the various signals which can occur at similar field values, to determine the Hamiltonian parameters and to detect departure from the occupancy of a single type, isolated site (distribution effects, correlation between spins). In all these steps, it is necessary to have some information on the probability of the transition, as well as on the expected lineshape of the absorption line. The lineshape gives also some of the fundamental parameters pertaining to the relaxation mechanisms by which the excited spins return to the ground state.

2.5.1 Transition and relaxation processes

There are four ways of observing transitions between two levels: spontaneous emission, induced emission, absorption and non-radiative processes. As the first process is dependent on the cube of the frequency of the radiation, it is not important in EPR spectroscopy (in contrast to optical absorption spectroscopy). Induced emission and absorption are the dominant processes in the microwave region: photons will induce both upward and downward transitions at rates proportional to spin levels population. Absorption is dependent on the population difference between excited and ground states (Δn) and is given by

$$I \approx \Delta n \rho(\nu) \; |<i|S_x|f>|^2 \quad , \tag{49}$$

where $\rho(\nu)$ is the incident energy density and $|i>$ and $|f>$ are the initial and final states, respectively. If these states are eigenfunctions of the same operator, as for a transition along S_z, these two functions will be orthogonal and the resulting transition probability will be zero. This is the reason to set the oscillating magnetic field (induced by the microwave radiation) to be perpendicular to the static, external field.

We can see from equation (49) that if the spin states in both excited and ground states are the same, no transition occurs. As the levels are closely situated, thermal equilibrium has to be taken into account. If only Zeeman effect is considered (as in the transition $|-1/2> \Leftrightarrow |+1/2>$), population difference may be approximated as

$$\Delta n = \frac{Ng\beta B_z}{2kT} \quad , \tag{50}$$

if $\frac{g\beta B_z}{kT}$ is small. T is the absolute temperature and N is the total number of spins.

Equations (49) and (50) show that a significant increase in absorption intensity is expected at low temperature. Equation (49) gives the dependence of transition probability on radiation frequency through the energy density of the radiation, $\rho(\nu)$, which is proportional to ν^2: the sensitivity of EPR spectrometers increases with frequency (see §3.3).

2.5.2 Spin-lattice relaxation time, T_1

As the power of the incident radiation is progressively increased, the observed signal attains a maximum, and further increase in the radiation power produces a decrease of the signal until it totally disappears. This results from the fact that the populations of both ground and excited states become equal, which causes a saturation of the transition. Relaxation processes restore the spin population equilibrium. It is an exponential process with a time constant, the spin-lattice relaxation time, T_1 (also referred to as longitudinal relaxation time), the time in which the system completes $1 - 1/e$ ($= 63\%$) of its return to equilibrium. T_1 describes the energy transfer from the spin system to the lattice (acoustic mode) phonons, in order to satisfy Boltzmann law. It is to the fluctuating magnetic field due to lattice motion that the spin must couple (through spin-orbit coupling), as this allows magnetic energy to be dissipated. More generally, the term lattice refers to the degrees of freedom of the system other than those directly concerned with spin. T_1 is measured from the transient response of the paramagnetic system to the termination or a reversal of the applied external magnetic field. Large T_1 values hinder observation of any EPR signal, as the excited level is immediately saturated and the transition probability is reduced to zero: this occurs in many transition metal ions. The signal lineshape is Lorentzian.

2.5.3 Spin-spin relaxation time, T_2

Even at the ppm level, spins exert on each other a local magnetic field: an electron produces at a distance of 4 Å a local field of 600 G, which is a large perturbation for the system. If there is phase coherence between the spins (in dilute systems), the mean field they mutually exert on each other is the same, and the linewidth is thus reduced. At the same time, the return to equilibrium will act so to randomize the spin system. The corresponding time constant of this process is the spin-spin relaxation time (or transverse relaxation time), T_2, and it will be long. On the contrary, if the group of spins has random phase with respect to each other (as in concentrated systems), a wider field distribution is effectively seen at each spin and results in a faster loss of phase coherence. Broad resonance lines and small T_2 values are observed. In other words, in order to transfer energy from one site to another, the lifetime of this process must be shorter than T_1. Figure 20 shows EPR spectra of Mn^{2+} ions in periclase at two Mn concentrations (de Biasi and Fernandes, 1984). On account of the high hyperfine coupling constant $A = 81 \cdot 10^{-3}$ cm^{-1} and the weak cubic field splitting parameter ($a = 19 \cdot 10^{-3}$ cm^{-1}), the spectrum of single crystals consists of six groups of HFS components ($I = 5/2$) divided into 5 fine structure components; as in each group only the central line is isotropic, powder spectra only show a six line pattern (see §2.3.6). The apparent linewidth of these lines increases by more than a factor of ten with increasing Mn concentration from less than 0.1% to about 1%.

2.5.4 Experimental linewidth

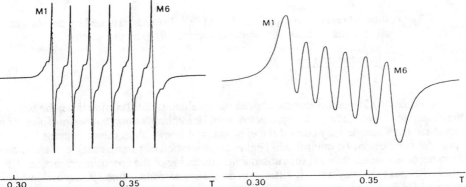

Figure 20. EPR spectrum of Mn^{2+} ions in periclase at X-band at two Mn^{2+} concentrations. The significant broadening is due to spin-spin interactions which cause the broadening of SHFS at high concentrations (from de Biasi and Fernandes, 1984).

Table 4. Representative frequency and magnetic fields at g=2 for various wave bands.

Band	Nominal wavelength (cm)	Frequency (GHz)	Magnetic Field (kG=T)
Q band	0.8	36	12.8
K band	1.5	20	7
X band	3.2	9.3	3.35
S band	10	3	1.07

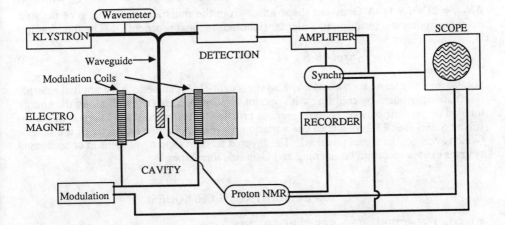

Figure 21. Schematic representation of an EPR spectrometer.

The combined relaxation time $T_r = T_1 T_2/(T_1 + T_2)$ is determined by the shorter of T_1 or T_2. It will determine the experimental linewidth $\Delta B_{1/2}$, by the relation

$$\frac{1}{T_r} = \frac{\pi g \beta \Delta B_{1/2}}{\hbar} \ . \tag{51}$$

Equation (51) means that the shorter the resultant spin relaxation time, the broader the absorption signal (as could be predicted from Heisenberg's uncertainty principle). It is expected that T_1 alone may control the experimental linewidth in dilute systems because spin-spin correlations be minimized. Higher concentration of paramagnetic centers allows T_2 to become smaller than T_1: the subsequent broadening of the spectrum smears out HFS and SHFS and even may make it difficult to detect any resonance line. In favorable cases, narrow EPR lines may be observed in samples of high crystal quality: in calcite, full width of EPR spectra of free radical species as low as 0.015 G has been measured. In concentrated systems, overlapping of electronic orbitals produces an exchange coupling. Increase of exchange frequency with increasing concentration of paramagnetic centers first causes the line to broaden and to shift towards the center of the spectrum (exchange broadening). Further increase of center concentration may result in a narrowing of this line (exchange narrowing) because of strong exchange interaction between neighbor spins.

2.5.5 Lineshape

The expected lineshape of the resonance line is a Lorentzian, if we consider the absorption (or emission) of radiation from an harmonically bound electron (natural lineshape). The equation of a single line may be written as a function of the magnetic field:

$$I(B) = \frac{I_0}{(B-B_o)^2/\Gamma^2 + 1} \ , \tag{52}$$

where B_o and I_0 are the field and absorption intensity at the line center, respectively. Γ is defined as $\Delta B_{1/2} = 2\Gamma$, where $\Delta B_{1/2}$ is the width at half height of the recorded signal (not to be confounded with the peak-to-peak linewidth of signal derivative, which is $\Delta B_{pp} = 2\Gamma/\sqrt{3}$). A Gaussian shape arises from the statistical distribution of the spin magnetic moments, resulting in a dipolar (inhomogeneous) broadening. The function $I(B)$ is

$$I(B) = I_0 \exp(-(B-B')^2/\Gamma^2) \ . \tag{53}$$

Solids generally show such a lineshape, as the local fields (that perturb the applied external field) are distributed by coupling with phonons, resonance occurring all along the energy interval of the distribution. Inhomogeneous broadening is observed as the result of nonresolved SHFS (see §2.2.3), site to site variations or polycrystalline character of the sample (which averages the signal position). The favored random flipping of spins in concentrated systems causes spectrum broadening and Gaussian lineshapes.

3 EXPERIMENTAL TECHNIQUES

3.1 The PRINCIPLE of the SPECTROMETER

An EPR spectrometer is comprised of the following devices:

(i) a radiation source (klystron) in the hyperfrequency range: it gives a mono-chromatic wave, the frequency of which (measured by a wavemeter) can be adjusted in a limited range of frequencies (some percent). EPR spectra are usually recorded at X-band frequency (Table 4), because of the convenience of the electronic components of the spectrometers. Q-band frequency spectrometers are also widely used to get spectra at higher frequency: comparison of measurements made at different frequencies allows separation of frequency dependent components (g factor) from frequency independent (HFS) components.

(ii) an external magnet delivering homogeneous and stable static fields up to 20-25 kG. Electric current intensity inside the electromagnet (and hence the value of the field) is continuously tuned: it is measured at sample location with a NMR spectrometer using the proton resonance. It is checked by comparison with a standard free radical (usually 1,1-diphenyl-2-picrylhydrazyl, DPPH), as the g-value is accurately known (2.0036 ± 0.0002).

(iii) a resonating cavity between the poles of the magnet, where the sample is placed. Resonance inside the cavity produces a standing wave regime. An oscillating field is produced by the electromagnetic wave perpendicular to the static field. The sample is located such that this oscillating field is at a maximum. The yield of the cavity (also called quality factor) is proportional to the ratio of the energy stored inside the cavity to the energy delivered to it. The usually high value of this factor explains the high sensitivity of EPR.

(iv) a detection crystal which analyses the waves transmitted by wave guides. When paramagnetic resonance occurs inside the mineral, energy is absorbed inside the cavity and the detection current inside the detection crystal is modified.

The measurement consists in recording the detected current as a function of mag-netic field. A schematic drawing of a spectrometer working in simple reflection mode is given in Figure 21. The resonance conditions may be expressed in various units. The unit conversions are: $100 \, G = 93.35 \cdot 10^{-4} \, cm^{-1}$; $100 \, MHz = 33.3 \cdot 10^{-4} cm^{-1}$.

3.2 FIELD MODULATION

The EPR signal represents a small change in the overall absorption of the mi-crowave power in the cavity, and has to be amplified. In order to avoid signal perturbations due to noise and baseline shifts, and to enhance broad signals of low intensity, a low amplitude, low frequency (usually 100 kHz) modulation field is superimposed on the static magnetic field. Synchronous detection by a phase detector is used. The a.c. amplification represents the derivative of the $I(B)$ function, $\dfrac{dI(B)}{dB}$. Note that the higher the modulation amplitude, the stronger the derivative signal but the lower the spectral resolution; it is an important experimental parameter, on which depends the quality of EPR spectra.

3.3 SENSITIVITY

Its value is proportional to (Abragam and Bleaney, 1969; Hervé, 1986):

$$N_{min} \propto \frac{V_c T \Delta B}{g^2 \beta^2 S(S+1) B Q \sqrt{P_0}} \qquad (54)$$

where V_c is the volume of the resonance cavity, T the sample temperature, ΔB the linewidth, B the value of the resonance field, Q the yield of the cavity (defined above) and

P_0 the power applied on the cavity. When high powers (a fraction of mW) can be applied without saturating the system, the detection limit can be as low as 10^{11} electron spins in the resonance cavity. Concentrations in the ppb range can be theoretically detected.

By considering equation (54), some experimental adjustments may improve sensitivity. First, low temperature measurements increase the population difference between electronic levels as shown in (50). Low temperature also favors longer relaxation times and thus narrower resonance lines (low ΔB). Saturation of the system (at large T_1 values) hinders the application of high power on the cavity and limits the sensitivity. ΔB limits EPR resolution in glasses and powders where anisotropic g_{eff}-values give a distribution of B: consequently, only the most abundant species and those corresponding to stationary B values are detected. Other factors as dielectric losses must also be taken into account. Finally, higher frequencies improve sensitivity: smaller cavity volume and larger Q and B values give an overall increase by $n^{4.5}$, which compensates the smaller size of the samples. The actual sensitivity of EPR is lower by 2 to 4 orders of magnitude than the theoretical value discussed above, but is still of the order of ppm or better.

3.4 SAMPLES

Low concentrations allow optimum signal strength by keeping to a minimum most line-broadening mechanisms. Provided that the principal directions of **g** and **A** are colinear, single crystal and powder experiments should produce comparable results, except that the latter is unable to give the directions of the tensors principal axes; using powders, the resolution is often poorer and angle-dependent transitions are smeared out. Low temperature measurements are needed in the case of small T_1 values, as in rare-earth ions. In fact, only S-state or effective S-state ions can be studied by EPR at room temperature. Fortunately these concern important minor and trace components, among which are V^{4+}, Cr^{3+}, Mn^{2+}, Fe^{3+}, Cu^{2+}, Gd^{3+}, Eu^{2+} and Mo^{5+}. Synthetic samples may take advantage of isotopic substitution, which gives further informations from HFS and SHFS.

3.5 MEASUREMENT of ABSOLUTE CONCENTRATION

Although it is possible to determine transition probabilities, some variations may come from the sample or the instrument. The basic method is to compare the sample with a known substance containing a determined quantity of a paramagnetic center. All the factors which can affect signal intensity must be kept the same, except the gain, variation of which allows comparisons between the samples. These factors include: temperature, modulation amplitude, time constant, klystron power, filling of the cavity, dielectric constants of the

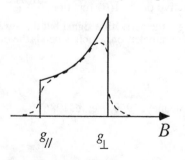
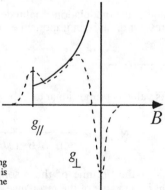

Figure 22. Shape of an EPR spectrum of a center presenting an axial distortion. The derivative (experimetal) spectrum is also reported on the right. The dotted lines represent the actual spectra including broadening effects.

materials. Often the standard is chosen such as to have a simple resonance line, as DPPH. Whatever the standard chosen, several precautions have to be taken when comparing an unknown sample and the reference: the filling of the tube should be the same and the tubes placed in the same region of the cavity; the microwave field must not be distorted by dielectric losses; finally the lineshape and linewidth have to be estimated carefully.

The number of spins is proportional to the double integration of the experimental (derivative) EPR signal. The area A can also be estimated using an expression of the form

$$A = k\Delta B^2 I \quad , \tag{55}$$

where ΔB is the linewidth, I the signal intensity and k a shape factor depending on the type of equation the EPR spectrum resembles (k =1.81 for a natural (Lorentzian) shape and 0.52 for a Gaussian shape). As integrated EPR signal allows quantitative comparison of the standard and the sample, the simplified equation (55) gives less accurate results.

3.6 POWDER SPECTRA

Only powder EPR may be observed in glasses or finely divided minerals (e.g., clays). Powder spectra may be computed, including fine or hyperfine structure; the components which show a pronounced angular dependence smear out and only the stationary features remain. If we consider an axial symmetry, there are more crystallites with axes oriented at $\pi/2$ to the field direction than oriented in this direction: the distribution maximum will occur at $\theta = \pi/2$. Distribution abruptly begins at $\theta = 0$. The mathematical derivation is based on a probability calculation of the crystal axis position in the reference system. A spectrum devoid of fine and hyperfine structures comprises two singular points corresponding to $g_{//}$ and $g_\wedge$ (Fig. 22) and intensity is proportional to $(\sqrt{B^2 - B_\perp^2})^{-1}$ with $B \in \{B_{//}, B_\perp\}$. This relation is characterized by an infinite value for $B = B_\wedge$, at least for zero linewidth. Broadening effects give the experimental shape. The resolution of the spectrum is defined by $\partial = \dfrac{|B_{//} - B_\perp|}{\Delta B}$, where ΔB is the linewidth expected for the center in a single crystal. The larger ∂, the better the spectral resolution. Note that $g_{//}$ and $g_\perp$ values can be obtained from the inflexion points on the absorption curve, which correspond on the derivative curves to the positive and negative maxima for $g_{//}$ and $g_\perp$, respectively. For rhombic distortion, three singular positions are obtained.

In powder spectra, stationary transitions have a higher apparent intensity than transitions which are angular dependent. This is important in the case of the coexistence of various centers or in presence of a distribution of Hamiltonian parameters as in glasses. In the specific case of the rhombic distortion occurring around ferric ions in glasses (see §2.3.3) the effect of a large distribution of the D and E parameters has been studied by Brodbeck (1980). It has been found by this author that only a small fraction of Fe^{3+} ions contributes to the resonance at $g_{eff} \approx 4.3$, although it is the major feature on the spectrum (see above Fig. 17a). Site rhombicity distribution favors the stationary transition inside the middle Kramer's doublet and produces a peak at $g_{eff} = 4.3$.

4   RECENT MINERALOGICAL APPLICATIONS of EPR

On account of its great sensitivity and because of interactions between paramagnetic ions, only paramagnetic centers embedded in diamagnetic matrices can be detected, restricting EPR to the "white" minerals, i.e., the minerals devoid of iron. Among the groups which can be studied by EPR are: quartz, pure or nearly pure Na-Ca-Mg-Al end-members of the various groups of silicates, Ca-Mg carbonates and sulfates, complex oxides (perovskites, spinels). We will not discuss the use of EPR to study some aspects of the supergene geochemistry of transition elements: biogeochemistry (association with humic acids from soils or kerogen from sediments: see Premovic et al., 1986), speciation in natural waters (Carpenter, 1983) or adsorption processes (see the review of Pinnavaia, 1982).

4.1 MINOR and TRACE COMPONENTS in MINERALS

In addition to a basic understanding of the mechanisms which govern trace element (actually ions) partitioning in natural systems, EPR spectroscopy gives us a way to distinguish between different generations/locations of a mineral. Various processes may lead to modification around a site occupied by a cation impurity in a crystal. Differences of ionic radii or of bond type between trace and major components often give unexpected behavior. More difficult to predict is the charge compensation mechanism by which heterovalent substitution can occur, although it is manifested by a lower symmetry than the unrelaxed, non-substituted site.

4.1.1 Determination of site occupancy

Trace element incorporation in multisite minerals gives rise to an important crystal chemical problem: which site is actually occupied? This cannot be solved by refining the crystal structure, nor can it be determined by electron microprobe analysis. Spectroscopic techniques are the appropriate tool, and among these EPR. Various parameters can give information needed to locate an element in crystal structure: the orientation of the axes of the Hamiltonian, the values of the E and D parameters, the value of HFS constant A, the origin of SHFS; however, it must be pointed out that EPR is only sensitive to Laue symmetry, i.e., it does not reflect the presence or absence of an inversion center.

The spinel family has been extensively studied. The EPR spectrum of Mn^{2+} in aluminous spinels is particularly interesting (Stombler et al., 1972; Shaffer et al., 1976), because the two end members $MgAl_2O_4$ and $MnAl_2O_4$ have direct and partly inverse (13%) structures, respectively. The variation in the linewidth of the EPR transitions, as manganese concentration increases, shows progressive incorporation of manganese into the octa

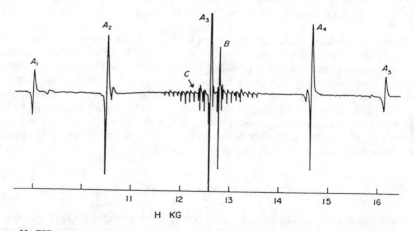

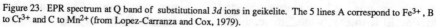

Figure 23. EPR spectrum at Q band of substitutional $3d$ ions in geikelite. The 5 lines A correspond to Fe^{3+}, B to Cr^{3+} and C to Mn^{2+} (from Lopez-Carranza and Cox, 1979).

hedral sites, and the magnetic exchange mechanisms between A and B sites may be observed. By a simple perturbation analysis, Shaffer et al. (1976) were able to derive the value of the fine structure parameter D which was interpreted as due to a "layering" of alternate divalent and trivalent cations in octahedral coordination.

The ilmenite group contains $3d$-ions as common impurities (Fig. 23). These elements show contrasted behavior in $MgTiO_3$ (Lopez-Carranza and Cox, 1979; Haider and Edgar, 1980). Divalent Co, Ni and Mn ions occupy Mg sites, in accordance with the solid solutions in which these divalent ions replace Mg. Two sites are observed for Cr^{3+} and Mn^{4+} ions, associated with Mg and Ti atoms, but with some uncertainty as whether they are substitutional or interstitial. Fe^{3+} ions occupy both Mg and Ti sites, on the basis of the orientation of the paramagnetic centers, but substitution at Mg-site is predominant and the ratio between both types of location depends on synthesis conditions. The two-site occupancy is in accordance with the existence of a solid solution between ilmenite/geikeilite structures and haematite. Haider and Edgar (1980) performed oxidation-reduction treatments and showed Fe^{2+}/Fe^{3+} and Fe^{3+}/Fe^{4+} interplay at Mg and Ti sites, respectively.

Finally, sites with only minor differences may be distinguished by EPR. Spectra of minor amounts of Mn^{2+} or Fe^{3+} substituting calcium in calcite show distinct ZFS in the two non-equivalent Ca sites. Site occupancy depends on the ion: Mn^{2+} ions do not show any site preference, but Fe^{3+} ions site occupancy ratio between both sites may attain a value of 10 (Marshall and Reinberg, 1963).

4.1.2 Local distortion around the substituting element

In forsterite, Mn^{2+} and Gd^{3+} ions replace Mg at the M2 site, with pseudo-symmetry axes (those describing an undistorted octahedron) close to those of the non-substituted site (Gaite, 1980). However, in contrast to Mn^{2+}, Gd^{3+} occupies strongly distorted octahedra: this is accounted for by the difference in both ionic charge and radius between Mg and Gd^{3+}. Fe^{3+} occupies M1, M2 and Si sites, but site preference is dependent on synthesis conditions (Niebuhr, 1975); at M2 sites, the distortion is smaller for Fe^{3+} than for Gd^{3+} ions, because the former have a smaller ionic radius (Gaite and Hafner, 1984). Similarly, the distortion at the Si site is not important. The absence of significant distortion rules out a charge-balanced substitution mechanism. Finally, Cr^{3+} ions are located at M1 and M2 sites with some preference for M1 (Rager, 1977). Superhyperfine interactions occur also in Cr-Al-doped crystals (Bershov et al., 1983): Figure 24 shows the EPR spectrum of a synthetic crystal of forsterite with the resonance lines of Fe^{3+} and Cr^{3+}. The fine structure due to Cr^{3+} is further split into a 5-line pattern: one arises from the even isotopes and is located at the center of a quartet attributed to the HFS due to the ^{53}Cr isotope ($I = 3/2$, with a natural abundance of 9.54%). Another spectrum arises from Cr-Al pair centers: it is manifested by the presence of a 6-lines pattern, each line being at the center of a HFS due to ^{53}Cr. The sextet has the significance of a SHFS due to ^{27}Al ($I = 5/2$) which has been confirmed by ENDOR. The relative concentration of this Cr-Al center is around 0.25 of total Cr. Although the majority of the Cr^{3+} ions are replacing Mg^{2+} ions without any short-range charge compensation, the presence of HFS and SHFS indicates some local charge compensation for the Cr-Al centers. If Cr^{3+} occurs at the M1 site, this charge compensation is obtained with Al^{3+} atoms located in the nearest (edge-sharing) Si sites; for Cr^{3+} at M2, only a next-nearest compensation seems compatible with observed SHFS anisotropy.

In simple oxide minerals, such as rutile or cassiterite, Fe^{3+} ions are found occurring in distorted environments (Calas and Cottrant, 1982; Thorp et al., 1986), on account of nearby charge compensation (see §4.1.5). In natural cassiterites, there exist three

548

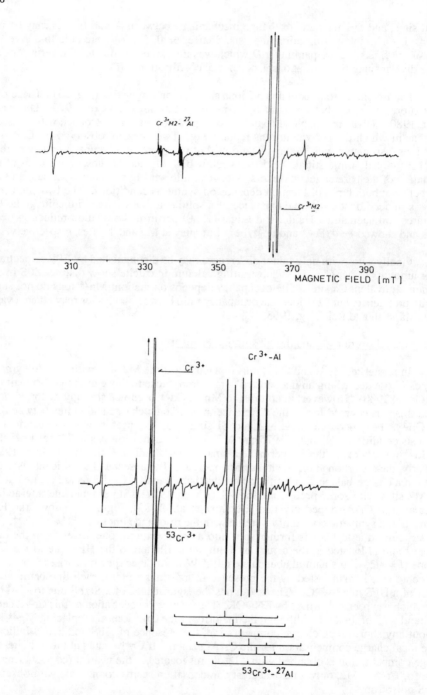

Figure 24. EPR spectrum of Cr^{3+} in forsterite at X-band, showing the transitions representative of substitutional Fe^{3+} (in M2 sites) and Cr^{3+} (M1 sites). The complex HFS and SHFS are expanded on the right: four HFS lines due to ^{53}Cr and six SHFS transitions due to $Cr^{-27}Al$ pairs. Each SHFS line is at the center of a weak quartet representing the minority HFS due to the ^{53}Cr nuclei (from Bershov et al., 1983).

substitutional sites and one described as interstitial. Of the three substitutional centers, only one preserves the local symmetry of the Sn^{4+} site, in relation with a non-local charge compensation mechanism. The others paramagnetic centers show resonances near $g_{eff} \neq$ 4.3 and their orientation and distortion has been recently determined (Dusausoy et al., 1988). Site occupancy is related to the paragenesis and location of the samples (Izoret et al., 1985).

4.1.3 Application of the superposition model (SPM)

Recent studies have applied the SPM to minerals (Lehmann, 1980a; Prissok and Lehmann, 1986; Büscher et al., 1987). The angular variation of the distortion is deduced by comparing the local structure of the non-substituted (unrelaxed) site and that derived interatomic distances calculated by SPM. In andalusite, Fe^{3+} occupies both 5- and 6-coordinated Al sites (Holuj and Jesmanowicz, 1978), but Cr^{3+} is only in 6-fold coordination (Vinokurov et al., 1962; Holuj et al., 1966). Application of SPM (Büscher et al., 1987) shows that Fe^{3+} retains the axis orientations of the AlO_6 octahedron, but a strong distribution of interatomic distances is found. On the contrary, chromium ions show considerable deviation in axes orientation. Five-coordinated iron shows a strong deviation in the distortion pattern, as compared to crystallographic data. In zoisite, only the largest Al site is occupied by Fe^{3+} and Cr^{3+}. Figure 25 compares the distortion patterns deduced from SPM and calculated at the monoclinic Al site by applying the relation (48). A correct agreement is found for substituting Fe^{3+} concerning the orientation of principal axes and distortion magnitude. There is no relaxation during substitution process, on the contrary to Cr^{3+} for which considerable differences are found between experimental and calculated patterns. However an accurate description of site geometry cannot be obtained if the size mismatch is too important between host and substituting ions.

4.1.4 Unusual coordination

In (synthetic) powellite $CaMoO_4$ and wulfenite $PbMoO_4$, Fe^{3+} occupies the Mo-site rather than the Ca(Pb) site, as shown by the orientation of the Hamiltonian axes, the large zero-field splitting and rhombic distortion (Friedrich and Karthe, 1979). Conversely, in scheelite $CaWO_4$, some Fe^{3+} occurs as interstitials (Golding et al., 1978). There is insufficient data to know if this arises from the growth history or from intrinsic local structure differences. In $AlPO_4$ (berlinite), V^{4+} occupies tetrahedral sites (de Biasi, 1980) and a small fraction (0.3%) of Cr^{3+} ions has been found in 4-fold coordination (Henning et al., 1967). In silicate minerals, some unusual coordinations are also found: tetravalent Nb and V replace silicon in zircon and thorite (Greenblatt, 1980; Di Gregorio et al., 1982), and tetrahedral V^{4+} is found in cristobalite (de Biasi, 1983a). Finally, tetrahedral complexes often stabilize high oxidation states: CrO_4^- in xenotyme and apatite, MnO_4^- in barite and MnO_4^{2-} in powellite.

Unusual ligands may also be determined: in some Mn^{2+}-bearing fluorites, disappearance of the superhyperfine coupling with the ^{19}F nuclei has been interpreted as the replacement of the 8 F nearest neighbors by 4 oxygens (Gehlhoff and Ulrici, 1980). Another example is given by Mn^{2+} in scheelite which exists mainly in 8-coordination, substituting regularly for Ca; some spectra show a strong axial distortion (Marfunin et al., 1966). Figure 26 reports the detail of one component of the HFS: the five SHFS components show the presence of $(MnF_4)^{2-}$ groups substituting a tungstate group. We have no indication of the geochemical significance as the existence of fluorine-complexing in hydrothermal systems in the latter case. These few examples point out that the incorporation processes of a trace element strongly depend on its chemical state.

550

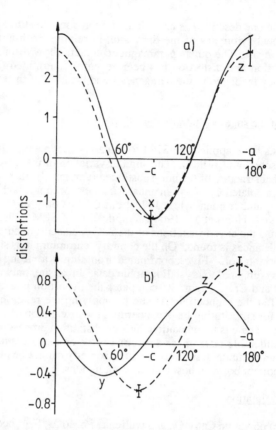

Figure 25. Relaxation effects around Fe^{3+} (top) and Cr^{3+} ions (bottom) in zoisite. The calculations of distortion patterns by SPM (full lines) are compared to unsubstituted (unrelaxed) Al site distortion pattern (dotted lines). The difference among both diagrams is explained by considerable relaxation around substituted Cr^{3+} (from Büscher et al., 1987).

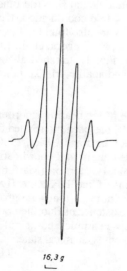

16,3 g

Figure 26. SHFS of Mn in scheelite: it causes the splitting of each of the six HFS transitions similar to those reported in Figure 5 and has been interpreted as arising from $(MnF_4)^{2-}$ complexes (modified from Marfunin et al., 1966).

4.1.5 Charge compensation processes

Charge compensation is usually indirectly discussed through relaxation effects during substitution. In spodumene, the substitution of Mn^{2+} rises the problem of a distinct charge as compared to the two possible sites occupied by Al^{3+} and Li^+. Mn^{2+} enters only the Al site, retaining the same symmetry. Distortion patterns deduced from application of SPM show that medium range, random charge compensation holds in the crystal studied (Vassilikou-Dova and Lehmann, 1987).

Perovskites ABO_3 (A = 8- and B = 6-coordinated) have been the subject of detailed studies on charge compensation during heterovalent substitution (see the review of Müller, 1981). On the basis of relative ionic radii, $3d$-ions are expected to occur at B sites and rare-earth ions at A sites. Charge compensation processes are still obscure for rare-earth ions (Boldu et al., 1987). The EPR spectrum of Fe^{3+} in $BaTiO_3$ shows two centers: a cubic one, with a low cubic fine structure parameter and another with an anisotropic g factor. The former is due to regular substitution at the B site, the original symmetry being retained and the latter to a nearby vacancy at an oxygen position of the B octahedron. ZFS (D) is higher than the EPR frequency , giving an anisotropic signal (see §2.3.3). g-values are those predicted for axially distorted d^5 ion, with $g_{//} = 2 = g_z$ and $g_\wedge = 6$. The geometry of the Fe^{3+}- and Mn^{2+}-vacancy pair has been calculated using SPM (Siegel and Müller, 1979a). The Fe^{3+} (Mn^{2+}) ion can move along the cation-vacancy axis, with a subsequent contraction of the four equatorial oxygens. SPM shows that the reduction of ionic radius (estimated as 4%) is confirmed by a lower coordination number (5 instead of 6).

In natural cassiterite, Fe^{3+} ions show various environments (see §4.1.2) (Dusausoy et al., 1988). One center confirms the reaction: $2Sn^{4+} \Rightarrow Fe^{3+} + Nb^{5+}$, as demonstrated in synthetic, intentionally co-doped cassiterites. Another center is interpreted as substitutional Fe^{3+} and results from the mechanism: $Sn^{4+} + O^{2-} \Leftrightarrow Fe^{3+} + OH^-$. In fluorites, Gd^{3+} ions also exhibit different local symmetries. Cubic symmetry is the most common in natural samples and demonstrates remote charge compensation (Bill and Calas, 1978). Other possible local symmetries are tetragonal (interstitial F^-), trigonal (O^{2-} replacing F^-) and rhombic (alkali replacing Ca nearest neighbors) and have been observed only in small amounts in natural fluorites (Chatagnon and Galland, 1982; Méary et al., 1984). The existence of different couplings shows the complexity of incorporation of minor components in minerals during heterovalent substitution. Natural fluorite and cassiterite show distinct charge compensation processes. This can be related to the type of excess charge, Fe^{3+}/Sn^{4+} versus TR^{3+}/Ca^{2+} and explained in terms of structural relaxation around the chemical defect (Stoneham, 1987). Synthetic (non-hydrothermal) samples show the reverse of what is observed in natural minerals in absence of intentional co-doping: cassiterites grown from vapor phase have only Fe^{3+} ions at regular Sn^{4+} sites, as thermally equilibrated fluorites grown from melt show only REE ions at tetragonal sites. This illustrates the influence of the growth history on the incorporation of minor components in mineral structure.

4.2 DETERMINATION of OXIDATION STATES

It relies on EPR analysis of a paramagnetic ion, the other oxidation state being obtained by difference with the element concentration in the mineral. The advantage is that, once a calibration curve has been obtained the analysis is fast and non-destructive and its sensitivity allows the study of small samples (10-50 mg). In contrast to wet chemical titration, there are no interactions between the various redox couples and it is possible to determine oxidation states of trace elements. However, ionizing radiations can induce oxidation or reduction of some trace components as in Ti-containing quartz (Okada et al., 1971) or Tm-containing fluorites (Chatagnon et al., 1982): oxido-reduction conditions during crystal growth cannot be deduced from EPR data. A limitation exists in case of site distribution (as in glasses): some ions may not be analysed by EPR, because their spectrum is smeared out by the variations in the fine-structure parameters or they have distinct relaxation times. EPR was used extensively during the lunar studies, to determine the presence of exotic oxidation states in lunar samples (see Weeks, 1973).

4.2.1 Third row transition elements

Fe^{2+}/Fe^{3+} ratios in feldspars have been measured by Hofmeister and Rossman (1984) using EPR and optical absorption spectroscopy. The proportionality of the Fe^{3+} response given by both methods is verified, although a distinct titration curve is found for potassium feldspars and plagioclases. This indicates a distinct distribution of site symmetry around substituted Fe^{3+}. Because the resonance at $g_{eff} = 4.3$ is stationary, only the sites with strong rhombic distortion are detected (actually those with a high D/n ratio and $E/D ≅ 1/3$: see §2.3.3).

Direct titration of Fe^{3+} in silicate glasses by EPR (Schreiber et al., 1979; Schreiber et al., 1980, 1984; Fudali et al., 1987; Calas and Miché, 1988) rises the same problem concerning the influence of distribution effects because only the sites with rhombic distortion are usually determined by EPR. The same limitation should also apply to Cr^{3+}, the EPR analysis of which also relies on a peak describing the strong local distortion at the site (for Cr^{3+}, this peak occurs at $g_{eff} ≈ 5.2$ instead of 4.3 in a $S = 5/2$ system): in this case also, a distribution of the local geometry would lead to deviations in the actual redox values obtained. Schreiber (1976) did not observe such deviations and obtained a linear relation between wet chemical titration and EPR signal intensity. EPR studies show that glasses quenched from melts synthesized at high oxygen fugacity contain significant amounts of Cr^{5+}, an exotic oxidation state, which is not detected by optical absorption spectroscopy. In fact both techniques are complementary, as EPR does not allow detection of Cr^{2+} or $(CrO_4)^{2-}$ groups, only detected by optical absorption spectroscopy. The Ti^{3+}/Ti^{4+} redox equilibrium has also been examined. Ti^{3+} gives a broad resonance at $g = 1.95$, already followed as a function of the oxygen fugacity in basaltic glasses (Bell et al., 1974) and sodium silicate glasses (Iwamoto et al., 1983). Signal intensity increases as measurement temperature decreases, due to larger T_1, but at constant temperature, the EPR signal intensity is proportional to the absorbance of the octahedral Ti^{3+} ion. Mutual redox reactions combining various transition elements have also been studied (see e.g., Schreiber et al., 1978, 1980; Schreiber, 1980; Schreiber and Balazs, 1981).

4.2.2 Rare earth elements

Only Gd^{3+} and Eu^{2+} ions are detected at room temperature. The determination of Eu^{2+}/Eu^{3+} ratios in glasses is easier than that of redox equilibria in $3d$-elements, because of smaller dependence of distribution effects. Eu^{2+}/Eu^{3+} ratios are determined in glasses of the same composition containing either Gd or Eu and analysed by neutron activation analysis. Gadolinium only occurs as Gd^{3+}, which allows scaling of the EPR measurements. Eu^{2+} is analysed by EPR and Eu^{3+} is obtained by difference with total Eu. Comparison of the spectra of Gd^{3+} and Eu^{2+} limits instrumental errors and the sensitivity of the method allows to study small samples (50 mg). The Eu^{2+}/Eu^{3+} equilibrium is dependent on oxygen fugacity as well as on melt composition (Morris and Haskin, 1974; Morris et al., 1974; Lauer and Morris, 1977). The same procedure has been applied to analyse the Eu^{2+} content of natural fluorites by reference to Gd assumed to exist as Gd^{3+} (Méary et al., 1984, 1985). The linear calibration curves of cubic Gd^{3+} vs total Gd ensures that all the gadolinium is incorporated in the cubic sites of the fluorite structure. In most fluorites, a positive Eu anomaly is observed on the normalized REE patterns and correspond to low Eu^{2+}/Eu^{3+} ratios. Higher ratios are responsible for the strong negative Eu anomaly in normalized REE patterns of some samples. Other unusual valence states are also present in fluorites: Tm^{2+} (unstable in the presence of water) is likely to be related to natural irradiation of the mineral (Chatagnon et al., 1982), a case similar to Sm^{2+}, detected in natural green fluorites by optical absorption spectroscopy (Bill and Calas, 1978).

4.3 PARAMAGNETIC RADIATION DEFECTS in MINERALS

Electrons or positive holes are formed by high-energy radiation (α-, γ- or x-rays, electrons, neutrons) and are trapped at structural or chemical defects. On account of their mutual annihilation, they cannot reach important concentrations and remain usually at the ppm level, unless irreversible modifications are attained (as colloidal calcium in purple natural fluorites: Bill and Calas, 1978). Their relation with coloration, geothermometry and dating will be discussed in §4.4, 4.7 and 4.8. Defect centers in minerals have been reviewed by Marfunin (1979b) and Vassilikou-Dova and Lehmann (1987).

Quartz is by far the most studied mineral, on account of its technological and geological importance. The EPR characteristics of defects in natural and synthetic quartz have been recently reviewed (Weil, 1984). Two types of centers can be recognized: impurity-associated defects and oxygen-vacancy trapped electron centers. The Al-associated center is common and is the first paramagnetic center discovered in quartz (Griffiths et al., 1954); it shows HFS arising from ^{27}Al and (weak) from ^{17}O. It is described as an electron hole trapped at an oxygen surrounding the substituted Al. Of the two types of oxygens found in quartz (Si-O = 1.6101 Å and Si-O = 1.6145 Å: Le Page et al., 1980), only the one associated with the longer bond traps the hole in one of the $2p$ levels. Other impurity-associated holes include atomic H as well as less frequent centers (Ge, P, Ti, Fe and atomic Cu and Ag). These are the origin of the coloration mechanisms of quartz (see §4.4). The family of oxygen-vacancy defects is usually referred to as E centers. The best known is the E'_1 center (Weeks, 1956), an oxygen vacancy having an unpaired electron localized in the sp^3 hybrid orbital extending from an adjacent silicon into the vacancy on the short-bond side (Fig. 27). The spectrum is characterized by narrow linewidths (0.1 G), because the near absence of isotopes with $I \neq 0$, and shows a weak HFS arising from ^{29}Si neighbors (4.7% natural abundance). E'_1 centers occurs also in vitreous silica (Griscom, 1980a), and, in both cases, it is believed to be a major feature decorating the defect structure of the material. The thermal stability of the Al-associated and E'_1 centers has been studied by Jani et al (1983) and Jani and Halliburton (1984). Figure 28 shows the results of pulse anneals of electron-irradiated quartz: E'_1 centers appear only after heating the crystal above 200°C and are annealed near 425°C, at higher temperature than the Al-centers which anneal out at 300°C. These results suggest that the formation of the E'_1 centers by electron irradiation is a two-step process: a diamagnetic precursor is converted into a paramagnetic defect by heating the crystal above 200°C. The release of holes trapped by Al-centers when heating is a possible source for E'_1 centers through their recombination with one of the electrons of the diamagnetic precursor, but the migration of interstitial alkali ions during irradiation seems also to play a role. Neutron irradiation causes ionic displacements and allows direct observation of E'_1 centers. The high thermal stability of the E'_1 centers ensures permanence of the paramagnetic defect over geologic time: it is thus possible to trace anomalous irradiation of the mineral in relation to possible uranium sources (Chatagnon, 1986) and - jointly with thermoluminescence- to decipher the various formation conditions in quartz (Kostov, 1986). Serebrenikov et al. (1982) have also reported the occurrence of E_1 centers in shock-metamorphosed quartz, associated with the induced dislocations.

554

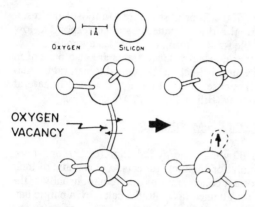

OXYGEN
VACANCY

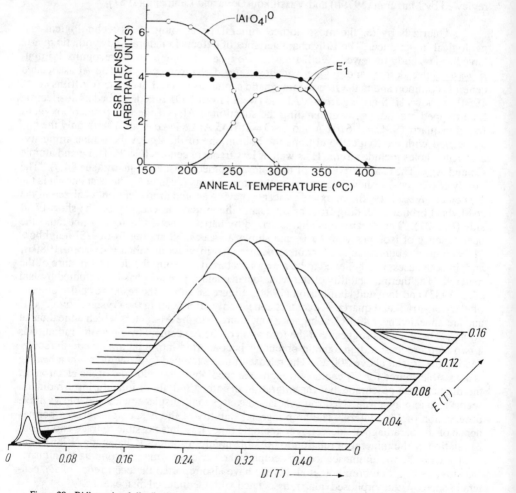

Figure 27. Model of the E'_1 center, a hole trapped by a neutral oxygen vacancy (from Griscom, 1980a).

Figure 28 (below). Thermal stability of paramagnetic defects in quartz after irradiation with 1.7 MeV electrons. The concentration of Al-associated and E'_1 centers is measured after 15 min annealing steps. The filled circles represent the evolution of E'_1 centers in a sample previously heated at 300°C (from Jani et al., 1983).

$|AlO_4|^0$

E'_1

ESR INTENSITY
(ARBITRARY UNITS)

ANNEAL TEMPERATURE (°C)

E (T)

D (T)

Figure 29. Bidimensional distribution pattern of D and E parameters describing EPR spectra of Mn^{2+} ions in oxide glasses. The two gaussian functions are scaled to have the same height (from Kliava, 1986).

4.4 RELATION to COLOR

We will give only some indications on the color induced by irradiation in quartz and feldspars. Here also, quartz is one of the most studied minerals. The various colors are related to specific defect centers, mostly paramagnetic: smoky quartz is due to the Al-center; rose quartz can be produced by irradiation which creates a defect associated with both Al and P (Maschmeyer and Lehmann, 1983b); Ti^{3+} is found in most natural rose quartz and is not affected by irradiation. Fe^{4+} ions (S = 2) are the origin of the color of amethyst (Cox, 1976, 1977). In morions and citrine, other types of hole centers are found, mostly related to Al (Samoilovich, 1970; Maschmeyer and Lehmann, 1980, 1983a). In feldspars, several X-ray-induced hole centers have been detected by Speit and Lehmann (1982); as in quartz, there are holes trapped on oxygens as well as centers related to impurities (Ti,Pb). Smoky feldspar and amazonite have been investigated by Hofmeister and Rossman (1985a,b). The origin of anomalous dichroism or pleochroism may also be studied by EPR: in amethyst, it has been related to unequal occupancy of Si sites by Fe^{4+} ions (Cox, 1977).

4.5 SILICATE GLASSES

EPR and optical absorption spectroscopy have been applied to the study of paramagnetic ions and radiation-induced paramagnetic defects in silicate glasses (see the reviews of Wong and Angell, 1976 and Griscom, 1980a). Distribution effects give significant broadening and make spectra analysis difficult; furthermore, because of distinct angle dependence and relaxation times of the paramagnetic centers, some sites may not be seen and an incorrect picture of the behavior of the element is obtained. In ions with a non-zero nuclear spin, HFS allows to study chemical bond.

4.5.1 Coordination of ions

The EPR spectrum of Fe^{3+} in a synthetic basaltic glass at X-band (Fig. 17a) shows two major resonances: at low field, the g = 4.3 signal denotes a strong rhombic distortion without any indication of cation coordination (see §2.3.3); at g = 2, a broad signal is attributed to isolated (slightly distorted) and coupled Fe^{3+} ions. Mn^{2+} ions show the same type of spectrum, although the g = 2 resonance is more intense than the g = 4.3 one, and exhibits a 6-HFS line pattern. The integrated spectra show that resonance of Fe^{3+} ions at $g_{eff} = 4.3$ concerns only a minor portion of the total intensity (Griscom, 1980a). SPM has been applied to Fe^{3+} and Mn^{2+} by Brodbeck and Buckrey (1981) and Kliava (1982, 1986) which has also discussed the statistical site-to-site fluctuations of the local environments. SPM suggests two sites, with only a small overlap between the distribution functions of the EPR parameters. This is represented on the joint probability densities of D and E parameters of Mn^{2+} ions in phosphate glasses (Kliava, 1982, 1986). Figure 29 shows this joint probability density and illustrates the distribution of E and D: a higher degree of short-range disorder is found in sites with $g_{eff} = 4.3$ as compared to the ones with $g_{eff} = 2$ This bimodal joint probability function found by Kliava (1982) is distinct from the unimodal representation derived from Mössbauer spectra of Fe^{2+} in silicate glasses (Levitz et al., 1980). SPM allows to quantify the radial and angular distortion. Low values of mean distortion are found, some few percent of the bond length and several degrees for angular distortion from the regular polyhedron (octahedron or tetrahedron). Baiocchi et al. (1980) have shown that raising the temperature diminishes the intensity of the resonance at g = 4.3. The rhombic distortion which gives rise to this signal seems to be produced during cooling. A recent reinterpretation of the EPR spectra of Gd^{3+} and Eu^{2+} in sodium silicate glasses with SPM (Brodbeck and Iton, 1985) suggests that they are incorporated at sites with high coordination numbers and a moderate distortion: a single site distribution gives them a network modifier role.

4.5.2 Molecular complexes

Some transition metal ions occur as discrete molecular complexes in extreme oxidation states (CrO_4^{2-}, VO_4^{3-},...), that are not paramagnetic and are evidenced only with optical absorption spectroscopy. Lower oxidation states can also give rise to discrete complexing in glasses and such complexes have been identified by EPR: VO^{2+} (vanadyl), CrO^{3+} and MoO^{3+}. All these groups have a d^1 configuration, which makes the interpretation of EPR spectra easier (Calas, 1982; Calas and Petiau, 1986). These species can be interpreted as a contraction of one cation-oxygen bond, resulting in a discrete covalent bonding and a strongly axial spectrum. The spectrum of tetravalent V in diopside glass (Fig.30) exhibits a well-resolved HFS ($A_{//} = 164\cdot10^{-4}$cm^{-1} and $A_\wedge = 71\cdot10^{-4}$ cm^{-1}) because the 100% abundance of the ^{51}V isotope ($I = 7/2$). The relation $g_\wedge = 1.955 > g_{//} = 1.937$ indicates a strong axial compression: the center is in fact better explained as a VO^{++} molecular ion. HFS is less resolved at high field because site distribution and second order effects, as observed for other cations in glasses (Imagawa, 1968). However in fully polymerized silica glass, 4-coordinate V^{4+} has been found (Fritsch et al., 1987).

4.5.3 Evolution of the EPR spectra with glass composition

The dependence of EPR spectrum on glass composition has been shown for VO^{++} and Cu^{2+} ions in silicate glasses (Hosono et al., 1979; Kawazoe et al., 1980) and in glasses relevant to geological compositions (Calas, 1982; Calas and Petiau, 1983). Cu^{2+} ions show an increase of Cu-O (π) covalency in polymerized glasses and a decrease of crystal field strength. EPR spectra of VO^{2+} ions show a continuous evolution as a function of glass composition: $g_{//}$ = 1.942 and 1.935, $g_\wedge$ = 1.995 and 2.005, $|A_{//}|$ =160·10^{-4} and 173·10^{-3} cm^{-1} and $|A_\perp|$ = 61·10^{-4} and 70·10^{-4} cm^{-1} for glasses of basaltic and rhyolitic composition, respectively. There is an increase in the $g_{//}$ -$g_\perp$ difference from the former to the latter. According to (19) and (20), this indicates a progressive axial compression of the coordination polyhedron, corresponding to greater π-bonding. A similar evolution holds for hyperfine coupling; it indicates less covalent bonding with the surrounding oxygens (not only that implied in the vanadyl grouping) in polymerized glasses. Both structural and chemical changes are related to the evolution of effective charges of the surrounding oxygens: in polymerized compositions, the strong polarization of the oxygens prevents them from donating electron density to central V^{4+}, the charge of which remains high and allows stronger vanadyl (π) bonding with a non-polarized oxygen. This gives a high stability to this vanadyl-group, and could explain the strong decrease of mineral/liquid partition coefficients with increasing oxygen fugacity (Irving, 1978).

4.5.4 Nucleation and crystallization processes

Paramagnetic centers in nucleating phases. EPR is used to monitor nucleation processes in glasses containing transition metal ions. In some cases, it helps, as an indirect method, in determining redox equilibria which influence phase precipitation (as nucleation of cuprite in silicate glasses: Duran et al., 1984). EPR is also used to monitor phase precipitation through paramagnetic tracers. The behavior of chromium during glass nucleation has been investigated by Durville et al. (1984) in a glass of cordierite composition. EPR spectra show the fast incorporation of minor Cr^{3+} ions in spinel $MgAl_2O_4$ crystals, after as few as 10 min at 950°C. At lower temperature (900°C), the Cr^{3+} ions first cluster together in antiferromagnetic phases like magnesiochromite $MgCr_2O_4$. The concentration of these Cr-rich nuclei decreases after further heating as they transform in the more dilute $MgAl_2O_4$ spinel. EPR of Fe^{3+} has enabled to monitor spinel crystallization in aluminosilicate glasses (Stryjak and McMillan, 1978). On the contrary to Cr^{3+} ions only a part of Fe^{3+} is incorporated in spinel, the other Fe^{3+} ions remaining in the glass network. Other

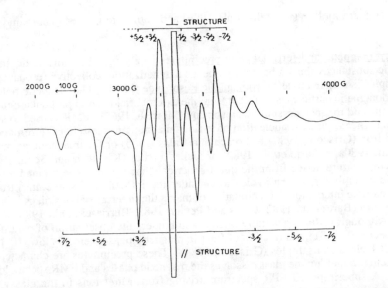

Figure 30. X-band EPR spectrum of vanadyl ions in a diopside glass (after Calas and Petiau, 1983, modified).

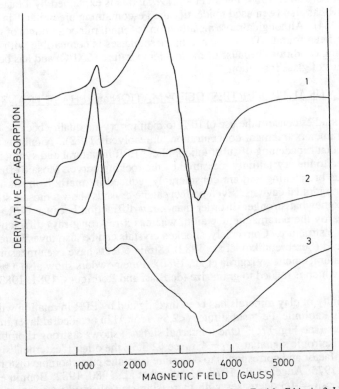

Figure 31. X-band EPR spectra of natural glasses. 1. Basaltic glass from Erta'ale, Ethiopia; 2. Lipari obsidian; 3. Pantellerite from Boïna, Ethiopia. The broad signal at $g_{eff} = 2$ arises from FMR of magnetite. The weak signal at $g_{eff} \approx 4.3$ is due to Fe^{3+} in glassy matrix. Weak HFS from diluted Mn^{2+} is visible at $g = 2$ in the basaltic glass. (from Calas and Petiau, 1983).

studies concern mullite (Reisfeld et al., 1986) and cordierite glasses (Thorp and Hutton, 1983).

Ferromagnetic and ferrimagnetic precipitates in glasses. Ferromagnetic and ferrimagnetic substances show a bulk spontaneous magnetization; collective resonance of unpaired spins is referred to as Ferromagnetic Resonance (FMR). The magnetic field inside the ferromagnetic particles is different from the external field and in particular depends on their shape and magnetocrystalline anisotropy (Morrish, 1965). FMR is used to study, in silica and silicate glasses, nucleation of various magnetic phases either ferromagnetic as metallic iron (Griscom, 1980a,b; Fritsch and Calas, 1985) or ferrimagnetic as magnetite (Griscom, 1980a,b; Auric et al., 1982) or nickel ferrites (Komatsu and Soga, 1984). A simplifying feature arises from the quasi-spherical shape of the fine-grained, dispersed magnetic precipitates, which avoids to consider the influence of shape anisotropy and particle-particle interactions. Precipitation of magnetite in terrestrial obsidians is shown in FMR spectra (Bart et al., 1982; Calas and Petiau, 1983; Burriesci et al., 1983; Griscom, 1984). Although FMR is generally intense and hinders the observation of paramagnetic dilute species (Fig. 31), the fraction of Fe^{3+} in magnetic phases remains low, 10-15% for Mount St Helens ash samples (Griscom, 1984). These precipitates are characteristic of slow cooling: remelting the glass dissolves the magnetic phases and FMR spectra loose intensity. A superimposed EPR spectrum arising from Mn^{2+} ions in the glassy matrix proves that they are not incorporated in magnetic clusters (Calas and Petiau, 1983). On the contrary, tektites do not show any FMR spectra unless they are annealed in an oxidizing environment (Griscom, 1984; Fudali et al., 1987): this is explained by a faster quenching. Finally, FMR has also been used to identify the Fe-containing precipitates in lunar glasses (Griscom, 1984). Although almost metallic iron is found, minor amounts of magnetitelike phases have been found. The existence of these phases is compatible with equilibration under reducing conditions because of the low temperature (600°C) and has been explained by the kinetics of glass formation.

4.6 STRUCTURAL PROPERTIES, DEFORMATION and PHASE TRANSITIONS

Few studies concern the use of EPR to monitor crystal quality because the influence of many parameters (element concentration, unresolved SHFS). Nienhaus et al. (1986) have shown that broadening of substitutional Fe^{3+} EPR in natural and synthetic amethysts can be related to the crystallinity measured by the rocking curve. Mosaic effects cause inhomogeneous broadening and are considerably reduced in quartz from fissures as compared with amigdoidal cavities. Even perfect calcite crystals show a mosaic structure: EPR indicates a somewhat stronger disorder than x-ray diffraction because EPR measures the deviation nearby the paramagnetic cation, whereas x-ray (or g-ray) diffraction gives the overall mosaic structure. Conversely, shocked crystals can be also investigated by the E'_1 centers induced (Serebrennikov et al., 1982). Strain effects have been measured in Mn^{2+}-containing natural calcite (Vizgirda et al., 1980). Fine powders show also a broadening of EPR spectra which is related to grain size (de Biasi and Rodrigues, 1981, 1983).

Crystallinity of clay minerals has been investigated by EPR in relation with formation conditions. In kaolinite, Fe^{3+} substitutes (<2.5%) for Al in octahedral layer in two paramagnetic centers (see Fig. 17c). One ("internal signal") shows a strong rhombic distortion and gives an isotropic signal at $g_{eff} = 4.3$ (see §2.3.3); the other ("external" signal) has an anisotropic g-factor and corresponds to departure from the pure rhombic distortion (Meads and Malden, 1975; Angel et al., 1978; Mestdagh et al., 1980, 1982; Bonnin et al., 1982). The signal at $g_{eff} = 4.3$ decreases with increasing crystallinity, as measured by X-rays diffraction or infrared spectroscopy (Hall, 1980; Mestdagh et al., 1980; Brindley et al., 1986). The relative intensities of "internal" and "external" signals is related to layer stacking disorder and regular stacking of greater crystallinity layers, respectively. Muller and Bocquier (1987) have proposed a tracing of laterite formation conditions on the basis of the successive *in situ* transformations of the kaolinites from the various horizons.

Phase transitions in the perovskite family have been extensively studied by EPR (Müller et al., 1986). Temperature dependence of the Gd^{3+} EPR spectrum in $BaTiO_3$ is shown in Figure 32 (Rimai and DeMars, 1962). Modifications of the ZFS D parameter, responsible for the 7-line pattern in the tetragonal phase are visible below phase transition temperature, showing local modifications previous to (first order) phase transition. Fe^{3+}-containing $BaTiO_3$ polymorphs (Müller, 1981) show a nearly constant cubic fine structure parameter a, which indicates that the geometry of the oxygen coordination polyhedron remains unchanged around impurity Fe^{3+}; there is no simple relation between D and the electric polarization, neither in direction nor in intensity variation at the phase transition (in contrast to what is found from EPR experiments on Gd^{3+}-containing crystals). SPM calculations (Siegel and Müller, 1979b) show that Fe^{3+} remains at the center of the octahedron and does not significantly participate in the collective off-center motion of the Ti^{4+} ions during ferroelectric phase transition. The consequence is that the transition temperature is reduced on doping (1% doping results in a 15 K decrease). A recent EPR study of Cr^{3+} ions in $BaTiO_3$ phases (Müller et al., 1985) has shown that (like Fe^{3+}) they remain centered in the oxygen cage.

4.7 GEOTHERMOMETRY

Intersite distribution coefficients vary as a function of temperature and can be determined by EPR. Since the pioneering study of Schindler and Ghose (1970), there has been much work on the relation between crystallization/equilibration temperature of dolomites and distribution of Mn^{2+} among Ca- and Mg-sites defined by the distribution coefficient $K = (Mn^{2+}$ in Mg sites$)/(Mn^{2+}$ in Ca sites$)$. The EPR spectrum of Mn^{2+} in dolomite (Fig. 33) shows two sites: one has a nearly perfect cubic symmetry (Ca-site) and the other a strong axial distortion (Mg-site). D values illustrate this difference: $-9.5\ 10^{-4}$ and $-143 \cdot 10^{-4}\ cm^{-1}$ for Ca- and Mg-sites, respectively, and the relaxation around Mn at the Ca-site is shown by SPM calculations (Prissok and Lehmann, 1986). K varies in a broad range, from 5 to above 250. Contradictory results have been obtained on K evolution with increasing temperature: both increasing (Angus et al., 1984) and decreasing (Lloyd et al., 1985; Prissok and Lehmann, 1986) evolutions have been reported. However, the evolution with increasing K was found in samples from various geological locations and formation conditions of sedimentary dolomites are known to induce variations of K from 5 to 70 (Lumsden and Lloyd, 1984).

Studies based on thermal annealing of radiation defects are less common (see §4.3). Lloyd and Lumsden (1987) have shown that the decrease of the EPR signal resulting from natural irradiation in dolomites is not a first-order process. Thermal annealing kinetics may be used to study the thermal gradient in a specific geological location, provided laboratory calibration gives a time scale independent of the irradiation dose received by the samples. Yellow color of fluorite is related to O_3^- (Bill, 1982): thermal bleaching studies give minimal cooling rates in hydrothermal systems (Calas and Touray, 1972).

4.8 EPR DATING

The basic principle of EPR dating is the same as for thermoluminescence: if a mineral is subject to irradiation from radioisotopes (U, Th, K families), the defect concentration is assumed to increase linearly with time if the system is closed. It is necessary to establish the relation between EPR signal intensity and the radiation dose using artificial

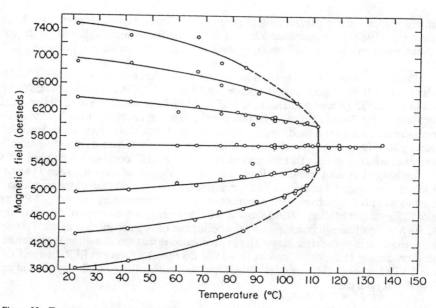

Figure 32. Temperature dependence of EPR spectrum of Gd^{3+} in perovskite $BaTiO_3$. With the decreasing temperature, the tetragonal $\Rightarrow$ cubic first order phase transition is accompanied by the collapse of the ZFS spectrum (from Rimai and de Mars, 1962).

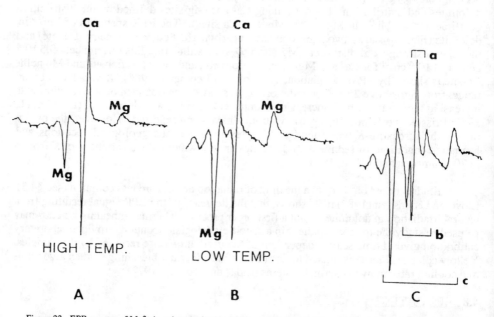

Figure 33. EPR spectra of Mn^{2+} ions in substitution of Mg and Ca in dolomites, in the vicinity of the high-field transition ($M_I = -5/2$). The anisotropic spectrum in Mn at Mg sites is clearly separated from that in Ca site. Sample A corresponds to a higher temperature (820°C) than sample B (210°C). The effect of additional calcite is illustrated by sample C, and contributes to the intermediate feature, b (see Fig. 5); a and b are the isotropic contribution from Mn^{2+}(Ca) and anisotropic contribution of Mn^{2+}(Mg) in dolomite (modified from Lloyd et al., 1985).

irradiation. Extrapolation to zero intensity gives the accumulated dose (AD) since time 0 (closure of the system). Time may be derived if the annual dose rate is known (from the concentrations in U, Th and K and assessing influence of cosmic rays). The age is calculated from the ratio between AD and the annual dose rate. The linearity of the relation AD vs time requires that there is no saturation of the available hole/electron traps. EPR dating has been tested only in recent ($<10^6$ years) minerals, mainly calcite (Hennig and Grün, 1983; Nambi, 1985). EPR dating concern speleothems, shells (Ikeya and Ohmura, 1981), foraminifera (Sato, 1982) and corals (Ikeya and Ohmura, 1983). Bone phosphates have been dated (Ikeya and Miki, 1980; Grün et al., 1987), but uranium uptake by teeth enamel or dentine will change the measured age. Dating recent volcanic rocks has been proposed (Shimokawa et al., 1984). Finally, dating of recent faults via the in-filling quartz has also been proposed (Ikeya et al., 1982): strain associated with fault formation causes electron/hole traps to be cleared out, but it is not known wether or not the process is complete.

4.9 RECENT REFERENCES on EPR STUDIES of MINERALS

Apart the references quoted above, we report here some references which were published posteriorly to the two reviews concerning EPR application in mineralogy (Low, 1968; Marfunin, 1979b), which the reader is asked to refer to for a more complete bibliography. A recent review has been also recently published by Vassilikou-Dova and Lehmann (1987) in which EPR data are presented according to the kind of paramagnetic ions and defects. For each group of minerals we will quote first the references on trace components and second the data on color centers.

Nesosilicates: $3d$ ions in forsterite (see §4.1.2) and in aluminosilicates (see §4.1.3); location of REE (Ball, 1982) and of Nb^{4+} in zircon (Di Gregorio et al., 1980); location of Fe^{3+}, Cr^{3+} and Mn^{2+} in garnets (Novak and Vosika, 1983; Rager et al., 1984); radiation centers in zircon (Danby and Hutton, 1980).

Ring silicates: EPR of Cr^{3+} in emerald (Edgar and Hutton, 1978; Martirosyan et al., 1980, 1983); relation with color of natural beryl (Blak et al., 1982; Lehmann, 1983); EPR of Cu^{2+} in dioptase (Reddy et al., 1985) and plancheite (Sharma et al., 1982).

Inosilicates: relaxation around Fe^{3+} and Mn^{2+} ions in spodumene (Gaite et al., 1985); oxidation states and sites occupied by chromium in diopside (Schreiber, 1977a); location of Fe^{3+} and Mn^{2+} in tremolite (McGavin et al., 1982).

Phyllosilicates: extensive work concern clay minerals (see the review of Pinnavaia, 1982; see §4.6) and micas (see the review of Rossman, 1984); EPR of Fe^{3+} in nontronite (Bonnin et al., 1985) and of $3d$-ions in halloysite (Chaikum and Carr, 1987); determination of radiation-induced centers in kaolinites (Cuttler, 1980 and 1981; Muller and Calas, 1988) and phlogopite (Avesov et al., 1980, 1982).

Tectosilicates: location of Fe^{3+} in quartz (Rager and Schneider, 1986); paramagnetic defects in quartz have been the subject of numerous studies (see the review of Weil, 1984; see §4.3, 4.4 and 4.6); EPR analysis of Fe^{3+} in feldspars (see §4.2.1); variation of Fe^{3+} EPR parameters with plagioclase composition (Scala et al., 1978); location of Mn^{2+} in sodalite (Hassib, 1980); sulfur-associated centers in lazurite (Galimov et al., 1981); location of Fe^{3+} in scolecite (Hutton and Scala, 1976) and of Eu^{2+} in zeolites (Iton et al., 1983); radiation defects in feldspars (Hofmeister and Rossman, 1984, 1985a,b; Matyash et al., 1982) and sodalite (Pizani et al., 1985).

Phosphates: location of Gd^{3+} in monazites (Rappaz et al., 1981) and of Fe^{3+} in xenotyme (Rappaz et al., 1982); EPR of 4-coordinate ions: Cr^{5+} in apatite (Pifer et al., 1983) and xenotyme (Greenblatt et al., 1981), Mn^{5+} and V^{4+} in chlorapatite and xenotyme (Greenblatt, 1980); radiation defects in hydroxyapatite (Close et al., 1981), berlinite (Requardt and Lehmann, 1985a,b,c), and brazilianite (Landrath and Lehmann, 1987).

Tungstates and molybdates: EPR of rare earth ions in scheelites (Caruba et al., 1983); ZFS of Mn^{2+} in scheelite (Biederbick, 1980); EPR of Tb^{3+} in scheelite (Shlenkin, 1980) and powellite (Kurkin and Shlenkin, 1978), of Mo^{5+} in powellite (Friedrich and Karthe, 1980), of Fe^{3+} in wulfenite (Friedrich and Karthe, 1979), of 4-fold coordinated Mn^{5+} in powellite (Greenblatt, 1980); radiation defects in scheelite (Cord et al., 1981).

Sulfates and arsenates: local relaxation around Fe^{3+} and Cr^{3+} in alums (Büscher and Lehmann, 1986); EPR of 4-fold coordinated Mn^{5+} in barite (Greenblatt, 1980); radiation defects in rhombic sulfates (Danby et al., 1982; Ryabov et al., 1983), gypsum (Albuquerque and Isotani, 1982), celestite (Bernstein, 1979) and astrakhanite (Khazanov et al., 1978); a review of the correlation between paramagnetic defects and thermolumines-cence of anhydrite and other minerals was published by Nambi (1985).

Carbonates: reinvestigation of Mn^{2+} EPR in calcite (Marshall et al., 1979) and aragonite (White et al., 1977); applications of EPR of Fe^{3+} and Mn^{2+} in dolomite and cal-cite (see §4.7 and 4.8); EPR study of phase transitions of calcium carbonates (Barnett et al., 1985; Barnett and Nelson, 1986); shock-induced defects in calcite (Vizgirda et al., 1980); EPR of defects related to calcite thermoluminescence (Calderon et al., 1984).

Oxides and hydroxides: EPR of Fe^{3+} and Mn^{2+} in periclase (de Biasi and Fernandes, 1984; Boldu et al., 1984) and corundum powders (de Biasi and Rodrigues, 1985); Cr^{3+}-vacancy pairs in periclase (Xiao-Pengh He and Mao-Lu Du, 1987); cation distribution in spinels (see §4.1.1): Cr^{3+} (Krygowska-Doniec and Kwiatkowska, 1980), and Fe^{3+} (de Biasi, 1983b); location of Cr^{3+} ions in alexandrite (Edgar et al., 1985; Troup et al., 1982; Forbes, 1983), of Fe^{3+} in cassiterite (see §4.1.2 and 4.1.5) and rutile (Thorp et al., 1986); location of Nb^{4+}, Mo^{5+} and W^{5+} in rutile and cassiterite (Madacsi et al., 1982) and of Mo^{5+} and V^{4+} in brookite (Grunin and Razumeenko, 1980); crystal chemistry of $3d$ and $4f$ ions in perovskites: see §4.1.5 and 4.6; location of $3d$ ions in ilmenite group: see §4.1.1; EPR of Mn^{2+} in manganite (Glaunsinger et al., 1979); defects in zincite (Schallenberger and Hausmann, 1981; Neuman, 1981, Wada et al., 1982).

Sulfides: EPR of Mn^{2+} in sphalerite (Kreissl and Gehlhoff, 1984) and pyrite (Srinivasan and Seehra, 1982) and of Cu^{2+} in digenite (Walczak et al., 1986) and multi-component sulfides (Bente, 1987).

Halides: EPR of $3d$-ions (see the review of Gehlhoff and Ulrici, 1980) and rare-earth ions in fluorite (Chatagnon et al., 1982; see §4.1.5 and 4.2.2), of Mn^{2+} in sellaite (Remme et al., 1985) and of Fe^{3+} and Cr^{3+} in chiolite (Jacoboni et al., 1981; Leblé et al., 1982); determination of Mn^{2+}-F distances in fluoroperovskites and other fluorides (Barriuso and Moreno, 1984);), color centers in fluorites (Bill and Calas, 1978).

Diamond: Two-dimensional imaging of paramagnetic centers in natural and irradiated diamonds (Hoch and Day, 1979; Zommerfeld and Hoch, 1986); nitrogen-associated paramagnetic defects (Shcherbatova et al., 1978; Loubser and Van Wyk, 1978; Loubser et al., 1982; Van Wyk, 1982; Van Wyk and Loubser, 1983).

REFERENCES

Aasa, R. (1970) Powder line shapes in the electron paramagnetic resonance spectra of high-spin ferric complexes. J. Chem. Phys. 52, 3919-3930.

Abragam, A. and Bleaney, B. (1969) Electron paramagnetic resonance of transition ions. Clarendon Press.

Albuquerque, A.R.P.L. and Isotani, S. (1982) The EPR spectra of X-ray irradiated gypsum. J. Phys. Soc. Japan 51, 1111-1118.

Angel, B.R. and Vincent, W.E.J. (1978) Electron spin resonance studies of iron oxides associated with the surface of kaolins. Clay Mineral. 26, 263-272.

Angus, J.G., Raynor, J.B. and Robson, M. (1979) Reliability of experimental partition coefficients in carbonate systems: evidence for inhomogeneous distribution of impurity ions. Chem. Geol. 27, 181-205.

-----, Beveridge, D. and Raynor, J.B. (1984) Dolomite thermometry by electron spin resonance spectroscopy. Chem. Geol. 43, 331-346.

Arends, J. (1964) Color centers in additively colored CaF_2 and BaF_2. Phys. Status Solidi 7, 805-815.

Auric, P., Dang, N.V., Bandyopadhyay, A.K. and Zarzycki, J. (1982) Superparamagnetism and ferrimagnetism of the small particles of magnetite in a silicate matrix. J. Non-Cryst. Solids 50, 97-106.

Avesov, A.D., Gasanov, E.M., Novozhilov, A.I. and Samoilovich, M.I. (1982) Some structure defects in neutron-irradiated fluorphlogopite. Phys. Status Solidi A73, K67- K68.

-----, Vakhidov, S.A., Gasanov, E.M., Novozhilov, A.I. and Samoilovich, M.I. (1980) Some radiation-optical properties of neutron-irradiated fluorphlogopite. Phys. Status Solidi A61, K171-K173.

Ball, D. (1982) The paramagnetic resonance of Nd^{3+} and Yb^{3+} in zircon structure silicates. Phys. Status Solidi A111, 311-320.

Baiocchi, E., Monetenero, A., Momo, F. and Sotgiu, A. (1980) High temperature ESR study of Fe(III) in lead silicate glass. J. Non-Cryst. Solids 37, 134-147.

Ballhausen, C.J. (1962) An introduction to ligand field theory. McGraw Hill.

Barberis, G.E., Calvo, R., Maldonado, H.G. and Zarate, C.E. (1975) ESR spectra and linewidths of Mn^{2+} in calcite. Phys. Rev. B12, 853-860.

Barnett, J.D., Nelson, H.M. and Tyagi, S.D. (1985) High-pressure EPR study of the calcite-$CaCO_3$ (II) phase transition near 1.6 GPa. Phys. Rev. B31, 1248-1257.

----- and Nelson, H.M. (1986) Preliminary EPR study of a solid-solid critical point in $CaCO_3$ near 200°C and 1.5GPa. Phys. Rev. B33, 4464-4470.

Barriuso, M.T. and Moreno, M. (1984) Determination of the Mn^{2+}-F^- distance from the isotropic superhyperfine constant for $[MnF_6]^{4-}$ in ionic lattices. Phys. Rev. B29, 3623-3631.

Bart, J.C.J., Burriesci, N., Cariati, F., Cavallaro, S., Giordano, N. and Petrera, F. (1982) Nature and distribution of iron in volcanic glasses: Mössbauer and EPR study of Lipari pumice. Bull. Minéral. 105, 43-50.

Bell, P.M., Mao, H.K. and Weeks, R.A. (1974) Crystal-field transition and electron paramagnetic resonance of trivalent titanium in iron-free magnesium-aluminum-calcium-silicate glass synthesized under controlled oxygen fugacity. Carnegie Inst. Wash. Year Book 73, 492-500.

Bente, K. (1987) Stabilization of Cu-Fe-Bi-Pb-Sn sulfides. Mineral. Petrol. 36, 205-217.

Bernstein, L.R. (1979) Coloring mechanisms in celestite. Amer. Mineral. 64, 160-168.

Bershov, L.V., Gaite, J.M., Hafner, S.S and Rager, H. (1983) Electron paramagnetic resonance and ENDOR studies of Cr^{3+} – Al^{3+} pairs in forsterite. Phys. Chem. Minerals 9, 95-101.

Biasi, R.S. de (1980) ESR of V^{4+} ions in α-cristobalite $AlPO_4$. J. Phys. C 13, 6235-6238.

----- (1983a) ESR and optical spectroscopy of V^{4+} ions in α-cristobalite SiO_2. J. Magnetism Magnetic Mater. 31-34, 653-654.

----- (1983b) Cation distribution and crystal anisotropy in $MgFe_2O_4$. J. Mater. Sci. Lett. 2, 363-365.

----- and Fernandes, A.A.R. (1984) Measurement of small concentrations of Cr and Mn in MgO using electron spin resonance. J. Amer. Ceram. Soc. 67, C173-C175.

----- and Rodrigues, D.C.S. (1981) Influence of chromium concentration and particle size on the ESR linewidth of Al_2O_3:Cr^{3+} powders. J. Mater. Sci. 16, 968-972.

----- (1983) Influence of iron concentration and particle size on the ESR linewidth of Al_2O_3:Fe^{3+} powders. J. Mater. Sci. Lett. 2, 210-212.

----- (1985) Measurement of small concentrations of Cr and Mn in α-Al_2O_3 using electron spin resonance. J. Amer. Ceram. Soc. 68, 409-412.

Biederbick, R., Hofstaetter, A., Scharman, A. and Born, G. (1980) Zero-field splitting of Mn^{2+} ions in scheelites. Phys. Rev. B21, 3833-3838.

Bill, H. (1982) Origin of the coloration of yellow fluorites: the O_3^- center structure and dynamical aspects. J. Chem. Phys. 76, 219-224.

----- and Calas, G. (1978) Color centers, associated rare-earth ions and the origin of coloration in natural fluorites. Phys. Chem. Minerals 3, 117-131.

Blak, A.R., Isotani, S. and Watanabe, S. (1982) Optical absorption and electron spin resonance in blue and green natural beryl. Phys. Chem. Mineral. 8, 161-166.

Boldu, J.L., Abraham, M.M. and Boatner, L.A. (1987) EPR characterization of Er^{3+} and Yb^{3+} in single crystals of synthetic perovskite ($CaTiO_3$). J. Chem. Phys. 86, 5267-5272.

-----, Muñoz, E.P., Chen, Y. and Abraham, M.M. (1984) EPR powder pattern analysis for cubic sites of Fe^{3+} in MgO. J. Chem. Phys. 80, 574-575.

Bonnin, D., Calas, G., Suquet, H. and Pezerat, H. (1985) Site occupancy of Fe^{3+} in Garfield nontronite: a spectroscopic study. Phys. Chem. Minerals 12, 55-64.

----- and Calas, G. (1982) Le fer dans les kaolins. Etude par spectrométries RPE, Mössbauer, EXAFS. Bull. Minéral. 105, 467-475.

Brindley, G.W., Kao, C.C., Harrison, J.L., Lipsicas, M. and Raythartha, R. (1986) Relation between structural disorder and other characteristics of kaolinites and dickites. Clays Clay Minerals 34, 239-249.

Brodbeck, C.M. (1980) Investigations of g-value correlations associated with the g=4.3 ESR signal of Fe^{3+} in glass. J. Non-Cryst. Solids 40, 305-313.

----- and Bukrey, R.R. (1981) Model calculations for the coordination of Fe^{3+} and Mn^{2+} ions in oxide glasses. Phys. Rev. B24, 2334-2342.

----- and Iton, L.E. (1985) The EPR spectra of Gd^{3+} and Eu^{2+} in glassy systems. J. Chem. Phys. 83, 4285-4299.

Burns, R.G. (1970) Mineralogical applications of crystal field theory. Cambrige Universtiy Press.

Burriesci, N., Giordano, N., Cariati, F., Petrera, M. and Bart, J.C.J. (1983) Characterization of iron distribution in various pumice grades and the derived zeolites. Bull. Minéral. 106, 571-584.

Büscher, R. and Lehmann, G. (1986) Zero-field splitting and local relaxation of Fe^{3+} and Cr^{3+} in alums. Chem. Phys. Lett. 124, 202-205.

----- (1987) Correlation of zero-field splittings and site distortions. IX. Fe^{3+} and Cr^{3+} in ß-Ga_2O_3. Z. Naturforsch. 42a, 67-71.

-----, Such, K.P. and Lehmann, G. (1987) Local relaxation around Fe^{3+} and Cr^{3+} in Al sites in minerals. Phys. Chem. Minerals 14, 553-559.

Calas, G. (1982) Spectroscopic properties of transition elements in glasses of geological interest. J. Physique C9, 311-314

----- and Cottrant, J.F. (1982) Cristallochimie du fer dans les cassitérites bretonnes. Bull. Minéral. 105, 598-605.

----- and Miché, C. (1988) Ferrous-ferric equilibria in the Na_2O-Al_2O_3-SiO_2 system (in preparation).

----- and Petiau, J. (1983) Structure of oxide glasses: spectroscopic studies of local order and crystallochemistry. Geochemical implications. Bull. Minéral. 106, 33-55.

----- (1986) Etudes spectroscopiques de la structure des verres et de la nucléation cristalline. In "Méthodes d'étude spectroscopiques des minéraux", G. Calas Ed., 529-547, Soc.Franç. Minéralogie et Cristallographie.

----- and Touray, J.C. (1972) An upper limit for the crystallization temperature of yellow fluorite. Modern Geol. 3, 209-210.

Calderon, T., Aguilar, M., Jaque, F. and Coy-Ill, R. (1984) Thermoluminescence from natural calcites. J. Phys. C 17, 2027-2038.

Carpenter, R. (1983) Quantitative electron spin resonance (ESR) determinations of forms and total amounts of Mn in aqueous environmental samples. Geochim. Cosmochim. Acta 47, 875-885.

Caruba, P., Iacconi, P., Cottrant, J.F. and Calas, G. (1983) Thermoluminescence, fluorescence and Electron Paramagnetic Resonance properties of synthetic hydrothermal scheelites. Phys. Chem. Minerals 9, 223-228.

Castner, T., Newell, G.S., Holton, W.C. and Slichter, C.P. (1960) Note on the paramagnetic resonance of iron in glass. J. Chem. Phys. 32, 668-673.

Chaikum, N. and Carr, R.M. (1987) Electron spin resonance studies of halloysites. Clay Minerals 22, 287-296.

Chatagnon, B.(1986) La résonance paramagnétique électronique du centre E'1 dans le quartz. Aspect fondamental et intérêt en géologie et en prospection minière. Thèse d'Etat, INPL, Nancy.

----- and Galland, D. (1982) R.P.E. de la fluorite naturelle: inventaire des ions lanthanides et autres impuretés paramagnétiques. Bull. Minéral. 105, 37-42.

-----, Gloux, P. and Méary, A. (1982) L'ion paramagnétique Tm^{2+} dans la fluorite. Mineral. Deposita 17, 411-422.

Close, D.M., Mengeot, M. and Gilliam, O.R. (1981) Low-temperature intrinsic defects in X-irradiated hydroxyapatite synthetic single crystals. J. Chem. Phys. 74, 5497-5503.

Cord, B., Hofstatter, A. and Scharman, A. (1981) EPR investigations on the dynamic behaviour of hole centres in $CaWO_4$. Phys. Status Solidi B106, 499-504.

Cox, R.T. (1976) ESR of an S=2 centre in amethyst quartz and its possible identification as the d^4 ion Fe^{4+}. J. Phys. C 9, 3355-3361.

----- (1977) Optical absorption of the d^4 ion iron (4+) in pleochroic amethyst quartz. J. Phys. C 10, 4631-4643.

Cuttler, A.H. (1980) The behaviour of a synthetic ^{57}Fe-doped kaolin. Mössbauer and electron paramagnetic resonance studies. Clay Minerals 15, 429-444.

----- (1981) Further studies of a ferrous iron-doped synthetic kaolin. Dosimetry of X-ray-induced defects. Clay Minerals 16, 69-80.

Danby, R.J., Boas, J.F., Calvert, R.L. and Pilbrow, J.R. (1982) ESR of thermoluminescent centres in $CaSO_4$ single crystals. J. Phys. C, 15, 2483-2493.

----- and Hutton, D.R. (1980) A new radiation defect centre in natural zircon. Phys. Status Solidi B98, K125-K128.

Danner, J.C., Ranon, U. and Stamires, D.N. (1971) Hyperfine, superhyperfine and quadrupole interactions of Gd^{3+} in YPO_4. Phys. Rev. B3, 2141-2149.

Di Gregorio, S., Greenblatt, M., Pifer, J.H. (1980) ESR of Nb^{4+} in natural zircon. Phys. Status Solidi B101, K147-K150.

----- and Sturge, M.D. (1982) An ESR and optical study of V^{4+} in zircon-type crystals. J. Chem. Phys. 76, 2931-2937.

Duran, A., Fernandez Navarro, J.M., Garcia Sole J. and Agullo-Lopez, F. (1984) Study of colouring process in copper ruby glass by optical and EPR spectroscopy. J. Mater. Sci. 19, 1468-1475.

Durville, F., Champagnon, B., Duval, E., Boulon, G., Gaume, F., Wright, A.F. and Fitch, A.N. (1984) Nucleation induced in a cordierite glass by Cr^{3+} : a study by small angle scattering, electron paramagnetic resonance and laser spectroscopy. Phys. Chem. Glasses, 25, 126-133.

Dusausoy, Y., Ruck, R. and Gaite, J.M. (1988) Study of the symmetry of Fe^{3+} sites in SnO_2 by electron paramagnetic resonance. Phys. Chem. Minerals 15, 300-303.

Edgar, A. and Hutton, D.R. (1978) Exchange-coupled pairs of Cr^{3+} in emerald. J. Phys. C 11, 5051-5063.

-----, Spaeth, J.M. and Troup, G.J. (1985) EPR and optical fluorescence spectra of Cr^{3+} ions in alexandrite. Phys. Status Solidi A88, K175-178.

Forbes, C.E. (1983) Analysis of the spin Hamiltonian parameters for Cr^{3+} in mirror and inversion symmetry sites of alexandrite ($Al_{2-x}Cr_xBeO_4$). Determination of the relative site occupancy by EPR. J. Chem. Phys. 79, 2590-2595.

Friedrich, M. and Karthe, W. (1979) Structure and kinetics of Fe^{3+}- doped $PbMoO_4$. Phys. Status Solidi B94, 451-456.

----- (1980) EPR investigation of Mo^{5+} in $CaMoO_4$. Phys. Status Solidi B97, 113-117.

Fritsch, E., Babonneau, F., Sanchez, C. and Calas, G. (1987) Vanadium incorporation in silica glasses. J. Non-Cryst. Solids 92, 282-294.

----- and Calas, G. (1985) Thermal darkening in silica glass: a spectroscopic study. J. Non-Cryst. Solids 72, 175-179.

Fudali, R.F., Darby Dyar, M., Griscom, D.L. and Schreiber, H.D. (1987) The oxidation state of iron in tektite glass. Geochim. Cosmochim. Acta, 51, 2749-2756.

Gaite, J.M. (1980) Pseudo-symmetries of crystallographic coordination polyhedra. Application to forsterite and comparison with some EPR results. Phys. Chem. Minerals 6, 9-17.

-----, Bookin, A.S. and Dritz, V.A. (1985) Local distortion of the spodumene structure around isolated cations using EPR and the superposition model. Phys. Chem. Minerals 12, 145-148.

----- and Hafner, S.S. (1984) Environment of Fe^{3+} at the M2 and Si sites of forsterite obtained from EPR. J. Chem. Phys., 80, 2747-2751.

Galimov, D.G., Tarzimanov, K.D. and Yafaev, N.R. (1981) On the nature of colour centres in ultramarine and blue alkali borate glasses containing sulphur. Phys. Status Solidi, A64, 65-71.

Gehlhoff W. and Ulrici, W. (1980) Transition metal ions in crystals with the fluorite structure. Phys. Status Solidi, B102, 11-59.

Glaunsinger, W.S., Horowitz, H.S. and Longo, J.M. (1979) Preparation and magnetic properties of manganite. J. Sol. St. Chem., 29, 117-120.

566

Golding, R.M., Kestigian, M. and Tennant, C.W. (1978) EPR of high-spin Fe^{3+} in calcium tungstate, $CaWO_4$. J. Phys. C, 11, 5041-5048.

-----, Newman, R.H., Rae, A.D. and Tennant, W.C. (1972) Single crystal ESR study of Mn^{2+} in natural tremolite. J. Chem. Phys., 57, 1912-1918.

Goodman, B.A. and Raynor, J.B. (1970) Electron spin resonance of transition metal complexes. In "Advances in Inorganic Chemistry and Radiochemistry", 13, 135-162.

Greenblatt M. (1980) Electron spin resonance of tetrahedral transition oxyanions (MO_4^{n-}) in solids. J. Chem. Educ., 57, 546-551.

-----, Pifer, J.H., McGarvey, B.R. and Wanklyn, B.M. (1981) Electron spin resonance of Cr^{5+} in YPO_4 and YVO_4. J. Chem. Phys., 74, 6014-6017.

Griffiths, J.H.E., Owen, J. and Ward, I.M. (1954) Paramagnetic resonance in neutron-irradiated diamond and quartz. Nature 173, 439-442.

Griscom, D.L. (1980a) Electron spin resonance in glasses. J. Non-Cryst. Solids 40, 211-272.

----- (1980b) Ferromagnetic resonance of fine grained precipitates in glass: a thumbnail review. J. Non-Cryst. Solids 42, 287-296.

----- (1984) Ferromagnetic resonance of precipitate phases in natural glasses. J. Non-Cryst. Solids, 67, 81-117.

Grün, R., Schwarcz, H.P. and Zymela, S. (1987) Electron spin resonance dating of tooth enamel. Can. J. Earth Sci. 24, 1022-1037.

Grunin, V.S. and Razumeenko, M.V. (1980) Electron spin resonance of Mo^{5+} and V^{4+} ions in brookite (TiO_2). Sov. Phys., Solid State 22, 1449-1450.

Haider, A.F.M. (1982) Spectroscopic studies of Fe^{3+} ions in $MgTiO_3$. Indian J. Phys. 56A, 16-22.

----- and Edgar, A. (1980) ESR study of transition metal ions in magnesium titanate. J. Phys. C 13, 6329-6350.

Hall, P.L. (1980) The application of electron spin resonance spectroscopy to studies of clay minerals. I. Isomorphous substitution and external surface properties. Clay Minerals 15, 321-335.

Halliburton, L.E., Jani, M.G. and Bossoli R.B. (1984) Electron spin resonance and optical studies of oxygen vacancy centers in quartz. Nucl. Instr. Methods B1, 192-197.

Hassib, A. (1980) Electron spin resonance of Mn^{2+} in sodalite. Phys. Chem. Minerals 6, 31-36.

Hawthorne, F. (1983) Quantitative characterization of site-occupancies in minerals. Amer. Mineral. 68, 287-306.

Hennig, G.J. and Grün, R. (1983) ESR dating in Quaternary geology. Quar. Sci. Rev. 2, 157-238.

Henning, J.C.M., Liebertz, J. and Van Stapele, R.P. (1967) Evidence for Cr^{3+} in four-coordination: ESR and optical investigations of Cr-doped $AlPO_4$ crystals. J. Phys. Chem. Solids 28, 1109-1114.

Hervé, A. (1986) La résonance paramagnétique électronique. In "Méthodes d'étude spectroscopiques des minéraux", G. Calas Ed., 313-389, Soc. Franç. Minéralogie et Cristallographie.

Hoch, M.J.R. and Day, A.R. (1979) Imaging paramagnetic centres in diamond. Solid State Comm. 30, 211-213.

Hofmeister, A.M. and Rossman, G.R. (1984) Determination of Fe^{3+} and Fe^{2+} concentrations in feldspar by optical absorption and EPR spectroscopy. Phys. Chem. Minerals 11, 213-224.

----- (1985a) A spectroscopic study of irradiation coloring of amazonite: structurally hydrous, Pb-bearing feldspar. Amer. Mineral. 70, 794-804.

----- (1985b) A model for the irradiative coloration of smoky feldspar and the inhibitive influence of water. Phys. Chem. Minerals 12, 324-332.

Holuj, F. and Jesmanowicz, A. (1978) EPR of Fe^{3+} in andalusite and kyanite at V-band and the pair spectra. Phys. Status Solidi A48, 191-198.

----- , Ther, J.R. and Hedgecock, N.E. (1966) ESR spectra of Fe^{3+} in single crystals of andalusite. Can. J. Phys. 44, 509-523.

Hosono, H., Kawazoe, H. and Kanazawa, T. (1979) Behavior of VO^{2+} as a spectroscopic probe for characterization of glasses. J. Non-Cryst. Solids 33, 125-129.

Hutton, D.R. and Scala, C.M. (1976) An EPR study of Fe^{3+} in scolecite. Phys. Status Solidi B75, K167-168.

Ikeya, M. and Miki, T. (1980) Electron spin resonance dating of animal and human bones. Science 207, 977-979.

----- and Tanaka, K. (1982) Dating of a fault by electron spin resonance on intrafault material. Science 215, 1392-1393.

----- and Ohmura, K. (1981) Dating of fossil shells with electron spin resonance. J. Geol. 89, 247-251.

----- (1983) Comparison of ESR ages of corals from marine terraces with ^{14}C and $^{230}Th/^{234}U$ ages. Earth Planet. Sci. Lett. 65, 34-38.

Imagawa, H. (1968) ESR studies of cupric ion in various oxide glasses. Phys. Status Solidi 30, 469-478.

Irving, A. (1978) A review of experimental studies on crystal-liquid trace-element partitioning. Geochim. Cosmochim. Acta 42, 743-771.

Iton, L.E., Brodbeck, C.M., Suib, S.L. and Stucky, G.D. (1983) EPR study of europium ions in type A-zeolite. The general classification of the EPR spectra of S-state rare-earth ions in disordered polycrystalline or glassy matrices. J. Chem. Phys. 79, 1185-1196.

Iwamoto, N., Hikada, H. and Makino, Y. (1983) State of Ti^{3+} ion and Ti^{3+}-Ti^{4+} redox reaction in reduced silicate glasses. J. Non-Cryst. Solids 58, 131-141.

Izoret, L., Marnier, G. and Dusausoy, Y. (1985) Caractérisation cristallochimique de la cassitérite des gisements d'étain et de tungstène de Galice, Espagne. Can. Mineral. 221-231.

Jacoboni, C., Leblé, A. and Rousseau, J.J. (1981) Détermination précise de la structure de la chiolite $Na_5Al_3F_{14}$ et étude par RPE de $Na_5Al_3F_{14}$:Cr^{3+}. J. Solid State Chem. 36, 297-304.

Jani, M.G., Bossoli, R.B. and Halliburton, L.E. (1983) Further characterization of the E'_1 center in crystalline SiO_2. Phys. Rev. B27, 2285-2293.

------ and Halliburton, L.E. (1984) Point defects in neutron-irradiated quartz. J. Appl. Phys. 56, 942-946.

Känzig, W. (1962) Paraelasticity, a mechanical analog of paramagnetism. J. Phys. Chem. Solids 23, 479-499.

Kawazoe, H., Hosono, H., Kokumai, H. and Kanazawa, T. (1980) Structural distribution and rigidity of the glass network determined by EPR of Cu^{2+}. J. Non-Cryst. Solids 40, 291-303.

Khasanov, R.A., Khasanova, N.M., Nizamutdinov, N.M. and Vinokurov, V.M. (1978) Tetrahedral paramagnetic centers in single crystals of Zn-astrakhanite. Sov. Phys. Cryst. 23, 303-306.

Kliava, J. (1982) Superposition model analysis of the short-range ordering of Mn^{2+} in oxide glasses. J. Phys. C 15, 7017-7029.

------ (1986) EPR of impurity ions in disordered solids. Phys. Status Solidi B134, 411-455.

Komatsu, T. and Soga, N. (1984) ESR and Mössbauer studies of the precipitation processes of various ferrites from silicate glasses. J. Mater. Sci. 19, 2353-2360.

Kostov, R.J. (1986) Electron paramagnetic resonance and thermoluminescence of quartz from the Rhodope Massif (Bulgaria). Bulgarian Geol. Soc. Monogr. 6, 150-157.

Kreissl, J. and Gehlhoff, W. (1984) EPR investigations of ZnS:Mn and ZnSe:Mn. Phys. Status Sol. A81, 701-7.

Krygowska-Doniec, M. and Kwiatkowska, J. (1980) Electron spin resonance of Cr^{3+} in $MgCr_2O_4$ normal spinel. Phys. Status Solidi B102, K41-K43.

Kurkin, I.N. and Shlenkin, V.I. (1978) Phase relaxation of Er^{3+} ions in $CaMoO_4$ single crystals. Sov. Phys., Solid State 20, 2163-2164.

Landrath, K.D. and Lehmann, G. (1987) The structures of two hole centers in the mineral brazilianite $NaAl_3(PO_4)_2(OH)_4$. Z. Naturforsch. 42a, 572-576.

Lauer, H.V. and Morris, R.V. (1977) Redox equilibria of multivalent ions in silicate glasses. J. Amer. Ceram. Soc. 60,443-451.

Leblé, A., Rousseau, J.J., Fayet, J.C. and Jacoboni, C. (1982) EPR investigations on $Na_5Al_3F_{14}$: Fe^{3+}. A validity test of the superposition model. Solid State Comm. 43, 773-776.

Lehmann, G. (1980a) Correlation of zero-lied splittings and site distortions. II Application of the superposition model to Mn^{2+} and Fe^{3+}. Phys. Status Solidi B99, 623-633.

------ (1980b) Variation of hyperfine splitting constants for $3d^5$-ions with interatomic distances. J. Phys. Chem. Solids 41, 919-921.

------ (1983) Optical absorption and electron spin resonance in blue and green natural beryl: a comment. Phys. Chem. Minerals 9, 278.

LeMarshall, J., Hutton, D.R., Troup, G.J. and Thyer, J.R.W. (1971) A paramagnetic resonance study of Cr^{3+} and Fe^{3+} in sillimanite. Phys. Status Solidi A5, 769-773.

Le Page, Y., Calvert, L.D. and Gabe, E.J. (1980) Parameter variation in low-quartz between 94K and 298K. J. Phys. Chem. Solids 41, 721-725.

Levitz, P., Bonnin, D., Calas, G. and Legrand, A.P. (1980) A two-parameter distribution analysis of Mössbauer spectra in non-crystalline solids using general inversion method. J. Phys. E 13, 427-432.

Lloyd, R.V., Lumsden, D.N. and Gregg, J.M. (1985) Relationship between paleotemperatures of metamorphic dolomites and ESR determined Mn(II) partitioning ratios. Geochim. Cosmochim. Acta 4, 2565-2568.

------ and Lumsden, D.N. (1987) The influence of radiation damage line in ESR spectra of metamorphic dolomites: a potential paleothermometer. Chem. Geol. 64, 103-108.

Lopez-Carranza, E., and Cox, R.T. (1979) ESR study of Fe^{3+} and Mn^{2+} ions in trigonal sites in ilmenite structure $MgTiO_3$. J. Phys. Chem. Solids 40, 413-420.

Loubser, J.H.N. and Van Wyk, J.A. (1978) Electron spin resonance in the study of diamond. Rep. Progr. Phys. 41, 1201-1248.

------ and Welbourn, C.M. (1982) Electron spin resonance of a tri-nitrogen centre in Cape yellow type Ia diamonds. J. Phys. C 15, 6031-6036.

Low, W. (1960) Paramagnetic resonance in solids. Solid St. Phys. Suppl 2. Academic Press.

------ (1968) Electron spin resonance. A tool in mineralogy and geology. Adv. Electronics Electr. Phys. 24, 51-108.

Lumsden, D.N. and Lloyd, R.V. (1984) Mn (II) partitioning between calcium and magnesium sites in studies of dolomite origin. Geochim. Cosmochim. Acta 48, 1861-1865.

Madacsi, D.P., Stapelbroek, M., Bossoli, R.B., Gilliam, O.R. (1982) Superhyperfine interactions and their origins for nd^1 ions in rutile-structure oxides. J. Chem. Phys. 77, 3803-3809.

Marfunin, A.S. (1979a) Physics of minerals and inorganic compounds. Springer Verlag.

------ (1979b) Spectroscopy, luminescence and radiation centers in minerals. Springer Verlag.

------, Bershov, L.V. and Mineeva, R.M. (1966) La résonance paramagnétique électronique de l'ion VO^{2+} dans le sphène et l'apophyllite et de l'ion Mn^{2+} dans la trémolite, l'apophyllite et la scheelite. Bull. Soc. fr. Soc. Minéral. Crist. 89, 177-183.

Marshall, S.A., Marshall, T., Zager, S.A. and Zimmerman, D.N. (1979) Electron spin resonance absorption spectrum of divalent manganese in single crystal calcite. Phys. Stat. Solidi B91, 275-282.

------ and Reinberg, A.R. (1963) Paramagnetic resonance absorption spectrum of trivalent iron in single-crystal calcite. Phys. Rev. 132, 134-142.

Martirosyan, R.M., Manvelyan, M.O., Menatsakanyan, G.A. and Sevastianov, V.S. (1980) Spin-lattice relaxation of Cr^{3+} ions in emerald. Sov. Phys., Solid State 22, 563-565.

------ (1983) Width of an ESR line of Cr^{3+} in emerald. Sov. Phys, Solid State 25, 904-905.

Maschmeyer, D. and Lehmann, G. (1983a) New hole centers in natural quartz. Phys. Chem. Minerals 10, 84-88.

------ (1983b) A trapped-hole center causing rose coloration of natural quartz. Z. Krist. 163, 181-196.

------, Nieman, K., Hake, H., Lehman, G. and Rauber, A. (1980) Two modified smoky quartz centers in natural citrine. Phys. Chem. Minerals 6, 145-156.

Matyash, I.V., Bagmut, N.N., Litovchenko, A.S. and Proshko, V.Ya. (1982) Electron paramagnetic resonance study of new paramagnetic centers in microcline-perthites from pegmatites. Phys. Chem. Minerals 8, 149-152.

McGavin, D.G., Palmer, R.A., Tennant, W.C. and Devine, S.D. (1982) Use of ultrasonically modulated electron resonance to study S-state ions in mineral crystals: Mn^{2+} and Fe^{3+} in tremolite. Phys. Chem. Minerals 8, 200-205.

Meads, R.E. and Malden, P.J. (1975) Electron spin resonance in natural kaolinites containing Fe^{3+} and other transition metal ions. Clay Minerals 10, 313-345.

Méary, A., Galland, D., Chatagnon, B. and Diebolt, J. (1984a) A study of Gd, Ce and Eu impurities in fluorite deposits by electron paramagnetic resonance and neutron activation analysis. Phys. Chem. Minerals 10, 173-179.

-----, Touray, J.C., Galland, D. and Jebrak, M. (1985) Interprétation de l'anomalie en europium des fluorines hydrothermales. Données de la résonance paramagnétique électronique: application au gîte de fluorine de Montroc (Tarn). Chem. Geol. 48, 115-124.

Mestdagh, M.M., Herbillon, A.J., Rodrique, L. and Rouxhet P.G. (1982) Evaluation du rôle du fer sur la cristallinité des kaolinites. Bull. Minéral. 105, 457-466.

-----, Vielvoye, L. and Herbillon, A.J. (1980) Iron in kaolinite. The relationships between kaolinite crystallinity and iron content. Clay Minerals 15, 1-14.

Morris, R.V. and Haskin, L.A. (1974) EPR measurement of the effect of glass composition on the oxidation state of europium. Geochim. Cosmochim. Acta 38, 1435-1445.

-----, Biggar, G.M. and O'Hara, M.J. (1974) Measurement of the effect of temperature and partial pressure of oxygen on the oxidation states of europium. Geochim. Cosmochim. Acta 38, 1447-1459.

Morrish, A.H. (1965) Physical principles of magnetism. John Wiley.

Muller, J.P. and Bocquier (1987) Textural and mineralogical relationships between ferruginous nodules and surrounding clayey matrices in a laterite from Cameroon. In: Proc. Int'l Clay Conf., Denver, 1985, Schultz, L.G., van Olphen, H. and Mumpton, F.A. Eds., 186-196.

----- and Calas, G. (1988) Tracing kaolinites through their defect centers (in prep.)

Müller, K.A. (1981) Paramagnetic point and pair defects in oxide perovskites. J. Physique 42, 551-557.

----- and Berlinger, W. (1983) Superposition model for six-coordinated Cr^{3+} in oxide crystals. J. Phys. C 16, 6861-6874.

----- (1986) Microscopic probing of order-disorder versus displacive behavior in BaTiO3 by EPR. Phys. Rev. B34, 6130-6136.

----- and Albers, J. (1985) Paramagnetic resonance and local position of Cr^{3+} in ferroelectric BaTiO3. Phys. Rev. B32, 5837-5844.

Nambi, K.S. (1985) Scope of electron spin resonance in thermally stimulated luminescence studies and its chronological applications. Nucl. Tracks 10, 113-131.

Neuman, G. (1981) On the defect structure of zinc-doped zinc oxide. Phys. Status Solidi 105b, 605-612.

Newman, D.J. and Urban, W. (1975) Interpretation of S-state ion EPR spectra. Adv. Phys. 24, 793-844.

Niebuhr, H.H. (1975) Electron spin resonance of ferric iron in forsterite, Mg2SiO4. Acta Cryst. A31, 274.

Nienhaus, K., Stegger, P., Lehmann, G. and Schneider, J.R. (1986) Assessment of quality of quartz crystals by EPR and g-diffraction. J. Cryst. Growth 74, 391-398.

Novak, P. and Vosika, L. (1983) Superposition model for zero-field splitting of Fe^{3+} and Mn^{2+} ions in garnets. Czech. J. Phys. B33, 1134-1147.

Okada, M., Rinneberg, H., Weil, J.A. and Wright, P.M. (1971) EPR of Ti^{3+} centers in α-quartz. Chem. Phys. Lett. 11, 275-276.

Olivier, D., Vedrine, J.C. and Pezerat, H. (1977) Application de la RPE à la localisation des substitutions isomorphiques dans les micas: Localisation du Fe^{3+} dans les muscovites et les phlogopites. J. Solid St. Chem. 20, 267-279.

Orgel, L. (1966) An introduction to transition-metal chemistry. Methuen.

Pake, G.E. and Estle, T.L. (1973) The physical principles of electron paramagnetic resonance. Frontiers in Physics. Benjamin.

Pifer, J.H., Ziemski, S. and Greenblatt, M. (1983) Effect of anion substitution on the electron spin resonance of Cr^{5+} in calcium phosphate apatite. J. Chem. Phys. 78, 7038-7043.

Pinnavaia, T.J. (1982) Electron spin resonance studies of clay minerals. In: Developments in Sedimentology 34, 139-161, Elsevier.

Pizani, P.S., Terrile, M.C., Farach, H.A. and Poole, C.P. (1985) Color centers in sodalite. Amer. Mineral. 70, 1186-1192.

Premovic, P.L., Pavlovic, M.S. and Pavlovic, N.Z. (1986) Vanadium in ancient sedimentary rocks of marine origin. Geochim. Cosmochim. Acta 50, 1923-1931.

Prissok, F. and Lehmann, G. (1986) An EPR study of Mn^{2+} and Fe^{3+} in dolomites. Phys. Chem. Minerals 13, 331-336.

Rager, H. (1977) Electron spin resonance of trivalent chromium in forsterite, Mg2SiO4. Phys. Chem. Minerals 1, 371-378.

------ and Schneider H. (1986) EPR study of Fe^{3+} centers in cristobalite and tridymite. Amer. Mineral. 71, 105-110.

------, Zabinski, W. and Amthauer, G. (1984) EPR study of South African grossularites ("South African Jades"). N. Jb. Miner. Mh. 433-443.

Rappaz, M., Abraham, M.M., Ramey, J.O. and Boatner, L.A. (1981) EPR characterization of Gd^{3+} in the monazite-type rare-earth orthophosphates: LaPO4, CePO4, PrPO4, NdPO4, SmPO4 and EuPO4. Phys. Rev. B23, 1012-1030.

------, Ramey, J.O., Boatner, L.A. and Abraham, M.M. (1982) EPR investigations of Fe^{3+} in single crystals of the zircon-structure orthophosphates LuPO4, YPO4 and ScPO4. J. Chem. Phys. 76, 40-45.

Reddy, K.M., Jacob, A.S. and Reddy, B.J. (1985) EPR and optical spectra of Cu^{2+} in dioptase. Ferroelectrics, Lett. Sect. 6, 103-112.

Reisfeld, R., Kisilev, A., Busch, A. and Ish-Shalom, M. (1986) Spectroscopy and EPR of chromium(III) in mullite transparent glass-ceramics. Chem. Phys. Lett. 129, 446-449.

Remme, S., Lehmann, G., Recker, R. and Wallrafen, F. (1985) EPR and luminescence of Mn^{2+} in MgF2 single crystals. Sol. State Comm. 56, 73-75.

Requardt, A. and Lehmann, G. (1985a) A hole center causing coloration in AlPO4 and GaPO4. Phys. Status Solidi 127B, 695-701.

----- (1985b) An O^{2-} radiation defect in AlPO4 and GaPO4. J. Phys. Chem. Solids 46, 107-112.

----- (1985c) Impurity-related electron centers in α−berlinite (AlPO4). Phys. Status Solidi 127B, K117-120.

Reynolds, R.W., Boatner, L.A., Finch, C.B., Chatelain, A. and Abraham, M.M. (1972) EPR investigations of Er^{3+}, Yb^{3+} and Gd^{3+} in zircon-structure silicates. J. Chem. Phys. 56, 5607-5625.

Rimai, L. and de Mars, G.A. (1962) Electron paramagnetic resonance of trivalent gadolinium ions in strontium and baryum titanates. Phys. Rev. 127, 702-709.

Rossman, G. R. (1984) Spectroscopy of micas. Rev. Mineral. 13, 145-181.

Ryabov, I.D., Bershov, L.V., Speranskiy, A.V. and Ganeev, I.G. (1983) Electron paramagnetic resonance of PO_3^{2-} and SO_3^- radicals in anhydrite, celestite and barite: the hyperfine structure and dynamics. Phys. Chem. Minerals 10, 21-26.

Ryter C. (1957) Résonance paramagnétique électronique dans la bande de 10.000 mC/s de l'europium et du gadolinium soumis à un champ cristallin cubique. Helv. Phys. Acta 30, 353-373.

Samoilovich, M.I., Tsinober, L.I. and Kreiskop, V.N. (1970) Features of the smoky color of natural quartz: morions. Sov. Phys. Cryst. 15, 438-440.

Sangster, M.J.L. (1981) Relaxations and their strain derivatives around impurity ions in MgO. J. Phys. C 14, 2889-2898.

Sato, T. (1982) ESR dating of planktonic foraminifera. Nature 300, 518-521.

Scala, C.R., Hutton, D.R. and McLaren, A.C. (1978) NMR and EPR studies of some chemically intermediate plagioclase feldspars. Phys. Chem. Minerals 3, 33-44.

Schallenberger, B. and Hausmann, A. (1981) Über Eigenstörstellen in Zinkoxid. Z. Phys. B 44, 143-153.

Schindler, P. and Ghose, S. (1970) Electron paramagnetic resonance of Mn^{2+} in dolomite and magnesite and Mn^{2+} distribution in dolomites. Amer. Mineral. 55, 1889-1895.

Schreiber, H.D. (1977a) On the nature of synthetic blue diopside crystals: the stabilization of tetravalent chromium. Amer. Mineral. 62, 522-527.

----- (1977b) Redox states of Ti, Zr, Hf, Cr and Eu in basaltic magmas: an experimental study. Proc. Lunar Sci. Conf. 8th, 1785-1807.

----- (1980) Properties of redox ions in glasses: an interdisciplinary perspective. J. Non-Cryst. Solids 42, 175-184.

----- and Balazs, G.B. (1981) Mutual interactions of Ti, Cr and Eu redox couples in silicate melts. Phys. Chem. Glasses 22, 99-103.

----- and Haskin, L.A. (1976) Chromium in basalts: experimental determination of redox states and partitioning among synthetic silicate phases. Proc. Lunar Sci. Conf. 7th, 1221-1259.

-----, Lauer, H.V. and Thanyasiri, T. (1980) The redox state of cerium in basaltic magmas: an experimental study of iron-cerium interactions in silicate melts. Geochim. Cosmochim. Acta 44, 1559-1612.

-----, Minnix, L.M. and Balazs, G.B. (1984) The redox state of iron in tektites. J. Non-Cryst. Solids 67, 349-359.

-----, Thanyasiri, T. and Lauer, H.V. (1979) Redox equilibria of iron in synthetic basaltic melts. Lunar Planet. Sci. 10, 1070-1072.

-----, Lach, J.J. and Legere, R.A. (1978) Redox equilibria of Ti, Cr and Eu in silicate melts: reduction potentials and mutual interactions. Phys. Chem. Glasses 19, 126-139.

Serebrennikov, A.I., Valter, A.A., Mashkovtsev, R.I. and Scherbakova, M. Ya. (1982) The investigation of defects in shock-metamorphosed quartz. Phys. Chem. Minerals 8, 153-157.

Shaffer, J.S., Farach, H.A. and Poole C.P. (1976) Electron spin resonance study of manganese-doped spinel. Phys. Rev. B13, 1869-1875.

Sharma, K.B.N., Reddy, B.J. and Lakshman, S.V.J. (1982) Absorption spectra of Cu^{2+} in shakkutite and plancheite. Phys. Lett.,92A, 305-308.

Shcherbatova, M. Ya, Nadolinnyi, V.A. and Sobolev, E.V. (1978) The N3 center in natural diamonds from ESR data. J. Struct. Chem. 19, 261-269.

Shimokawa, K., Imai, N. and Hirota, M. (1984) Dating of a volcanic rock by electron spin resonance. Chem. Geol. 46, 365-373.

Shlenkin, V.I. (1980) Phase relaxation of Tb^{3+} ions in $CaWO_4$ single crystals. Sov. Phys., Solid State 22, 520-2.

Siegel, E. and Müller, K.A. (1979a) Structure of transition-metal-oxygen vacancy-pair centers. Phys. Rev. B19, 109-120.

----- (1979b) Local position of Fe^{3+} in ferroelectric $BaTiO_3$. Phys. Rev. B20, 3587-3595.

Simanek, E. and Müller, K. (1970) Covalency and hyperfine structure constant A of iron group impurities in crystals. J. Phys. Chem. Solids 31, 1027-1040.

Speit, B. and Lehmann, G. (1982) Radiation defects in feldspars. Phys. Chem. Minerals 8, 77-82.

Srinivasan, G. and Seehra, M.H. (1982) Temperature dependence of the ESR spectra of Mn^{2+} in iron pyrite (FeS_2). Solid State Comm. 42, 857-859.

Stombler, M.P., Farach, H.A. and Poole, C.P. (1972) Electron Spin Resonance of manganese-substituted spinel. Phys. Rev. B6, 40-45.

Stoneham, A.M. (1987) Quantitative modelling of defect processes in ionic crystals. Phys. Chem. Minerals 14, 401-406.

Stryjak, A.J. and McMillan, P.W. (1978) Microstructure and properties of transparent glass-ceramics. J. Mater. Sci. 13, 1805-1808.

Thorp, J.S. and Hutton, W. (1983) Radiation-induced paramagnetic centres in glass-ceramics derived from the MgO-Al_2O_3-SiO_2-TiO_2 system. J. Phys. Chem. Solids 44, 1039-1047.

-----, Eggleton, H.S., Egerton, T.A. and Pearman, A.J. (1986) The distribution of iron centres in Fe-doped rutile powders. J. Mater. Sci. Lett. 5, 54-56.

Troup, G.J., Edgar, A., Hutton, D.R. and Phakey, P.P. (1982) 8mm wavelength EPR spectra of Cr^{3+} in laser-quality alexandrite. Phys. Status Solidi A71, K29-31.

Van Wyk, J.A. (1982) Carbon-13 hyperfine interactions of the unique carbon of the P2 (ESR) or N3 (optical) centre in diamond. J. Phys. C 15, L981-L983.

----- and Loubser, J.H.N. (1983) Electron spin resonance of a di-nitrogen centre in Cape yellow type Ia diamonds. J. Phys. C 16, 1501-1506.

Vassilikou-Dova, A.B. and Lehmann, G. (1987) Investigations of minerals by electron paramagnetic resonance. Forschr. Mineral. 65, 173-202.

Vinokurov, V.M., Zaripov, M.M., Stepanov, V.G., Pol'skii, Y.E., Cherkin, G.K. and Shekun, L.Y. (1962) EPR of trivalent chromium in andalusite. Sov. Phys., Solid State 3, 1797-1800.

Vizgirda, J., Ahrens, T.J. and Fun-Dow Tsay (1980) Shock-induced effects in calcite from Cactus Crater. Geochim. Cosmochim. Acta 44, 1059-1069.

Wada. T., Kikuta, S., Kiba, M., Kiyozumi, K., Shimojo, T. and Kakehi, M. (1982) Electron spin resonance and cathodoluminescence in ZnO. J. Cryst. Growth 59, 363-369.

Walczak, J., Debinski, H., Murasco, B., Kuriata, J. and Sadlowski, L. (1986) EPR study of copper (I) sulfide. Phys. Status Solidi A97, 291-295.

Weeks, R.A. (1956) Paramagnetic resonance of lattice defects in irradiated quartz. J. Appl. Phys. 27, 1376-1381.

----- (1973) Paramagnetic resonance spectra of Ti^{3+}, Fe^{3+} and Mn^{2+} in lunar plagioclases. J. Geophys. Res. 78, 2393-2401.

Weil, J.A. (1984) A review of electron spin resonance and its applications to the study of paramagnetic defects in crystalline quartz. Phys. Chem. Minerals 10, 149-165.

Wertz J. and Bolton, J. (1972) Electron spin resonance. Elementary theory and practical application. Mc Graw Hill.

White, L.K., Szabo, A., Carkner, P. and Chasteen, N.D. (1977) An electron paramagnetic resonance study of manganese (II) in the aragonite lattice of a clam shell, Mya arenaria. J. Phys. Chem. 81, 1420-1424.

Wickman, H.H., Klein, M.P. and Shirley, D.A. (1965) Paramagnetic resonance of Fe^{3+} in polycrystalline ferrichrome A. J. Chem. Phys. 42, 2113-2117.

Wong, J. and Angell, C.A. (1976) Glass structure by spectroscopy. M. Dekker.

Xiao Pengh He and Mao-Lu Du (1987) An EPR study of the structure of the Cr^{3+}-vacancy centre in MgO. Phys. Status Solidi B144, K51-K56.

Yeung, Y.Y. and Newman, D.J. (1986) Superposition-model analyses for the Cr^{3+} $^4A_2$ ground state. Phys. Rev. B34, 2258-2265.

Zommerfeld, W. and Hoch, M.J.R. (1986) Imaging of paramagnetic defect centers in solids. J. Mag. Res. 67, 177-189.

Chapter 13

Michael F. Hochella, Jr.

AUGER ELECTRON AND X-RAY PHOTOELECTRON SPECTROSCOPIES

1. INTRODUCTION

This chapter deals with the two most widely used surface sensitive spectroscopies today, x-ray photoelectron spectroscopy (XPS) and Auger electron spectroscopy (AES). Since these two methods were introduced for general laboratory use around 1970, they have shown remarkable applicability to scientific investigations in many disciplines, and a number of excellent reviews already exist in the literature (for XPS, see e.g. Riggs, 1975, Briggs and Seah, 1983, Feldman and Mayer, 1986, and references therein; for AES, see e.g. Joshi et al., 1975, Riviere, 1983, and Briggs and Seah, 1983, and references therein). However, a guide to the fundamentals and use of these techniques has never been written for geologists despite their increasing role in geochemical research. Therefore, the intent of the author is to present an understandable, relatively condensed chapter on XPS and AES, with explanations, procedures, examples, and applications of these techniques specifically slanted towards the geosciences. After additional introductory comments in this section, we will discuss the fundamental principles, spectral interpretation, instrumentation, experimental procedures, chemical quantification, and applications of XPS and AES in geochemical research. A concluding section discusses advantages and disadvantages of the techniques, and compares them with other surface spectroscopies. The level of this coverage is aimed at a broad range of users, from the novice, for which reading this chapter in sequence would be most appropriate, to the user with up to moderate experience who has specific questions or is looking for useful references.

1.1 Surfaces, near-surfaces, and why we study them

Considering only the solid state, what do we mean by the words "surface" and "near-surface"? It is generally acceptable in the field of surface science to consider only the top layer of atoms, that is the top monolayer, as the true surface of a solid. Although there are techniques which are sensitive *only* to the top layer of atoms, e.g., ion scattering spectroscopy (ISS), scanning tunneling and atomic force microscopy (STM and AFM, respectively), and high resolution electron energy loss spectroscopy (HREELS), we will see later that XPS and AES are generally not capable of such analyses. Instead, XPS and AES typically obtain their information from the top several monolayers of the solid, and this we define as the *near-surface*. Going deeper into the solid, instruments such as the electron microprobe analyze to depths of approximately 0.3 to 3 microns in oxides depending on the electron beam accelerating voltages used. This translates into thousands to tens of thousands of monolayers, and analyses this deep are generally regarded as being representative of the bulk. As we will see later, one of the problems in materials characterization is getting chemical information from between the near-surface as measured by XPS or AES and the deep bulk as measured by the electron microprobe. Nevertheless, this chapter will deal almost exclusively with near-surfaces.

Why are near-surfaces important to study? When solids chemically interact with their environment, the interactions are generally dictated by the chemistry and structure of the surface and near-surface, not the bulk. Therefore, a complete understanding of critically important geochemical processes, such as fluid-solid interactions of all sorts, depend on knowing the surface properties of minerals and other important earth materials. Advances in these studies will allow us to make inroads towards a much more satisfying understanding of the controls on groundwater chemistry, element partitioning at mineral/aqueous solution interfaces, mineral and rock weathering, and natural heterogeneous catalytic reactions which may influence processes such as ore deposit formation.

1.2 General descriptions of XPS and AES

Spectroscopic techniques can be systematically classified by considering what is used to stimulate the material of interest and what is detected in response. Leaving details for later, x-ray photoelectron spectroscopy uses x-rays as the stimulator, and *photoelectrons* are detected in response. The energies of these photoelectrons are analyzed for elemental and chemical state information. For Auger spectroscopy, either x-rays or electrons can be used to generate *Auger electrons* whose energies are typically used for elemental identification only. It is important to note that in this chapter (and generally elsewhere), the term "AES" implies the use of an electron beam for Auger electron stimulation, whereas "x-ray induced AES" is specifically stated if x-rays are used. This is simply because Auger spectroscopy is most commonly done with electron stimulation. However, x-ray stimulated Auger spectra have a number of interesting uses as described throughout this chapter.

Because the basis of XPS and AES is the measurement of the energies of photo- and Auger electrons, respectively, these techniques are classified as *electron spectroscopies*. The relationship between these electron spectroscopies and the most commonly used *x-ray spectroscopies* (x-ray fluorescence (XRF) and electron probe microanalysis (EPMA)) are shown in Table 1. The comparison is continued in Table 2, showing the important attributes of each technique. Although many details may differ between electron and x-ray spectroscopies, the most critical difference comes from the fact that x-rays can travel through solids without energy loss (although attenuation is severe), whereas electrons can only travel extremely short distances through solids before undergoing energy loss. This single principle allows XRF and EPMA to be used as bulk characterization techniques, and allows XPS and AES to be used as techniques sensitive to the near-surface.

1.3 Historical perspective

Where did XPS and AES come from, and how did they evolve into the tools that we use today? A very brief overview of the evolution of these techniques is given below. Excellent and more complete historical accounts which detail the development of surface sensitive spectroscopies can be found in, e.g., Auger (1975), Jenkin et al. (1977), and Briggs and Seah (1983).

The long and intriguing history of XPS can be traced to the turn of the century, where a number of classic discoveries by some of the great men of science added pieces to a puzzle which would eventually develop into a practical spectroscopy more than five decades later. It was Hertz in 1887 who coined the word "photoelectric" to help describe unexplainable phenomena which occurred during his historic experiments with electromagnetic waves. After the discovery of x-rays by Rontgen in 1896 and the electron by Thomson in 1897, Einstein in 1905 showed how the photoelectric effect as described by Hertz could be explained. His formalism involved Planck's quantum theory of radiation, and it was Einstein's conception of the photon upon which the photoelectric effect was finally understood at its most fundamental level. A number of studies quickly followed in which the first attempts were made to measure the yield and energies of photoelectrons, although the work of Einstein was rarely mentioned. The most fundamental relationship describing photoelectrons, however, could not come about before the discovery of the nuclear atomic and fundamental atomic structure. This was achieved between 1910 and 1915 in the Manchester University laboratories of Rutherford. It was Rutherford (1914) himself who made the first attempt at expressing the fundamental photoelectron relationship

$$E_K = h\nu - E_B \quad , \tag{1}$$

where E_K is the kinetic energy of a photoelectron, $h\nu$ the energy of the incident photons, and E_B the energy by which the photoelectron had been bound to its parent atom.

After World War I, photoelectron research continued, but it was eventually overshadowed by the more rapid success of x-ray spectroscopy. The principle problem with photoelectron spectroscopy was the lack of a high energy resolution electron

Table 1. Basic x-ray and electron spectroscopies

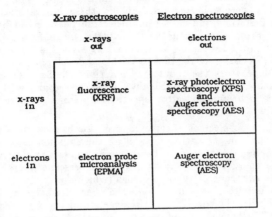

	X-ray spectroscopies	Electron spectroscopies
	x-rays out	electrons out
x-rays in	x-ray fluorescence (XRF)	x-ray photoelectron spectroscopy (XPS) and Auger electron spectroscopy (AES)
electrons in	electron probe microanalysis (EPMA)	Auger electron spectroscopy (AES)

Table 2. Comparison of x-ray and electron spectroscopies

	XRF	EMPA	XPS	AES
Sample Preparation	- grinding - pellet pressing or flux-fusion	- cutting - polishing - coating	none if surface is clean	
Typical diameter of analysis area	20-30 mm	several μm	several mm	several μm
Minimum area of analysis area	---	1-2μm	~150μm	~0.03μm
Typical depth of analysis	few 10's of μm	few μm	10 - 50 A	10-50A
Elements that can be detected	Be to U*	Be to U*	Li to U	Li to U
Detection limit in best case	10^{-9} g	10^{-15} g	10^{-11} g	10^{-11} g
Data reduction for quantitative analysis	fairly well understood		not well understood, but improving	
Mapping Capability	no	yes	no	yes
Speciman damage	All are 'nondestructive' techniques, although incident electron beams can cause chemical and/or structural modifications			

* Detection of elements lighter than Na requires a thin-window or windowless detector.

spectrometer. Despite a number of efforts, this was not to come until well after World War II in the Uppsala, Sweden laboratories of Kai Siegbahn. The first high resolution photoelectron spectra were collected on NaCl in these laboratories in 1954. In this experiment, for the first time, individual photoelectron peaks could be seen and binding energies measured. From then until the late 1960's, Siegbahn's group singlehandedly developed the entire groundwork for XPS, climaxing with the publication of a book on the subject in 1967 (Siegbahn et al., 1967). With the development of ultra-high vacuum systems for general laboratory use about the same time, the first commercial XPS instruments became available around 1970. Finally, in 1981, Siegbahn won the Nobel Prize in physics for his work in bringing electron spectroscopy from the dreams of Einstein and Rutherford into reality.

The history of Auger electron spectroscopy is closely related to the history of photoelectron spectroscopy. In the early 1920's, after the emission of photoelectrons from a number of elements had been crudely determined, additional electrons were observed which seemed to be energetically related to x-ray fluorescent photons. In 1922, Pierre Auger confirmed the existence of these electrons (which were later to bear his name) by bombarding Ar atoms with x-rays in a cloud chamber. He showed that these electrons were always ejected with the same energy irrespective of the x-ray energy used to eject them (see Auger, 1975, for a review of his early work). Much later, Ruthemann (1942), Hillier (1943), and Lander (1953) were among the first to show that Auger electrons could also be stimulated by electron beam bombardment. However, as with photoelectron spectroscopy, further advances in Auger spectroscopy would have to wait for the development of high resolution electron energy analyzers. After these analyzers became available (see above), it was Harris (1968a,b) who first demonstrated electron stimulated AES as we know it today. To allow Auger spectra to be analyzed more easily, he was also the first to differentiate the direct analyzer signal. Palmberg et al. (1969) described the first practical AES instrument which made use of a cylindrical mirror electron energy analyzer (CMA), combining acceptable energy resolution with a high collection efficiency. Scanning Auger microscopy (SAM), which is electron stimulated AES using a highly focused and scanning electron beam, was first realized on modified scanning electron microscopes (SEM's) fitted with CMA's (see Wells and Bremer, 1969, and MacDonald, 1970). The refinement of SAM instrumentation has dramatically increased the popularity of Auger spectroscopy. It has found wide ranging applications in many fields, and has established itself as the second most commonly used surface analytic technique after XPS.

2. FUNDAMENTAL PRINCIPLES

The following section describes the physical origin (mechanism of ejection) and fundamental properties of photo- and Auger electrons. This provides the entire basis for the characteristics and unique applicability of photoelectron and Auger spectroscopies. The attenuation lengths of electrons traveling through solids, which dictates the depth of analysis of these techniques, is also discussed in this section. However, before beginning these discussions, atomic energy level nomenclature, which will be used to name principal spectral lines, will be briefly reviewed.

2.1 Nomenclature

The conventional naming of the principal spectral lines in photoelectron and Auger electron spectra comes directly from classical quantum number formalism as well as the so-called *j-j coupling scheme*. The overall naming system for the first four principal atomic shells is shown in Table 3, along with the number of electrons populating each sublevel. We start with the *principal quantum number*, n, which takes on the values 1,2,3,4..., and are also designated K,L,M,N..., respectively. Next, we use the *orbital angular momentum quantum number l, which for each level n, can take on the values 0,1,2...n-1 (designated s,p,d,f etc., respectively). Adding the spin momentum quantum number*, s, which takes on values of ±1/2, the *total* electron angular momentum is the

Table 3. X-ray and spectroscopic nomenclatures for the first four principal atomic levels

Principal shell	Total electrons in shell	Quantum numbers (j-j coupling)			X-ray nomenclature	Spectroscopic nomenclature	Electrons each level
		n	l	j			
K	2	1	0(s)	1/2	K	$1s_{1/2}$	2
L	8	2	0(s)	1/2	L_1	$2s_{1/2}$	2
		2	1(p)	1/2	L_2	$2p_{1/2}$	2
		2	1(p)	3/2	L_3	$2p_{3/2}$	4
M	18	3	0(s)	1/2	M_1	$3s_{1/2}$	2
		3	1(p)	1/2	M_2	$3p_{1/2}$	2
		3	1(p)	3/2	M_3	$3p_{3/2}$	4
		3	2(d)	3/2	M_4	$3d_{3/2}$	4
		3	2(d)	5/2	M_5	$3d_{5/2}$	6
N	32	4	0(s)	1/2	N_1	$4s_{1/2}$	2
		4	1(p)	1/2	N_2	$4p_{1/2}$	2
		4	1(p)	3/2	N_3	$4p_{3/2}$	4
		4	2(d)	3/2	N_4	$4d_{3/2}$	4
		4	2(d)	5/2	N_5	$4d_{5/2}$	6
		4	3(f)	5/2	N_6	$4f_{5/2}$	6
		4	3(f)	7/2	N_7	$4f_{7/2}$	8

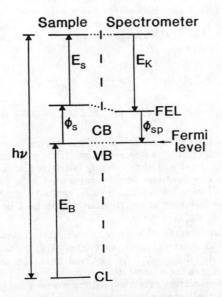

Figure 1. Energy level diagram for x-ray photoemission and measurement. In this case, the sample is assumed to be a conductor in good electrical contact with the spectrometer. This diagram can be used to help visualize the origin of Eqns. 2-4. CB, VB, and CL refer to conduction band, valence band, and core level, respectively. All other symbols are defined in the text, along with an explanation of the diagram.

vector sum of the angular and the spin momenta. In the j-j coupling scheme, this summation is simply considered for individual electrons and is termed j where $j = l + s$. Therefore, j takes on values of 1/2, 3/2, 5/2, 7/2, etc. for the various sublevels as shown in Table 3. Unique combinations of the n, l, and j quantum numbers results in a description of the subatomic levels observed from all photoelectron lines and most Auger lines.

Each subatomic level can be named according to either the "x-ray" or "spectroscopic" nomenclature as listed in Table 3. For example, the level described by $n = 3$, $l = 2$, and $j = 5/2$ would be designated equivalently as either M_5 or $3d_{5/2}$. By convention, Auger peaks are named after the three levels that are involved in the Auger transition (see Fig. 3 and accompanying text) according to the x-ray nomenclature. On the other hand, it is conventional to name photoelectron peaks simply by naming the level from which the electron was ejected in terms of the spectroscopic nomenclature. The photoelectron and Auger ejection processes are described in detail in the next two sections.

It should be emphasized that, although the nomenclature which results from the simple j-j coupling model is the most common in use today and is fully adequate for naming all possible photoelectron lines, it is not adequate for naming all possible Auger transitions. This requires the L-S coupling scheme (also called Russell-Saunders coupling), or a combination of the j-j and L-S formalisms. However, this quantum description is not needed for everyday use of Auger spectroscopy, as the additional lines that are described are only seen with very high resolution electron energy analyzers. Therefore, a further description will not be given here. The interested reader is referred to Briggs and Riviere (1983) for a discussion of this subject and additional references.

2.2 Origin and energies of photoelectrons

Electrons can be ejected from their parent atom if they undergo an interaction with a photon whose energy exceeds that of their bound state, their so-called *binding energy*. Therefore, the energy of this ejected electron, now a photoelectron, should be equal to the energy of the photon minus the electron binding energy. This statement has already been expressed as Eqn. 1, but in this section we will add an additional term to the equation for completeness as follows:

$$E_K = h\nu - E_B - \phi_{sp} \quad , \tag{2}$$

where E_K, $h\nu$, and E_B are as defined earlier, and ϕ_{sp} is the *work function* of the spectrometer defined below.

The relative probability that a photon will eject a certain electron from a certain element is called the *photoionization cross-section*. This probability is not only a function of the atomic energy level and the element, but it is also a function of the photon energy. Photoionization cross-sections can be calculated at a particular photon energy using, for example, the Hartee-Fock-Slater atomic model (e.g., Scofield, 1976, Yeh and Lindau, 1985), and these numbers are particularly important in chemical quantification (see section 5.3).

In order to better understand the meaning and relationships of the factors used in Eqn. 2, it is convenient to construct an energy level diagram as shown in Figure 1. In constructing this schematic diagram, it is assumed that the sample from which photoelectrons originate is a conductor and that it is in good electrical contact with the instrument. The case of insulating materials will be covered next. Referring to Figure 1, our point of reference for all energy measurements, by convention, is the *Fermi level*, which is always located between the valence and conduction bands. An electron at the Fermi level is defined as having a binding energy of zero, i.e., $E_B = 0$. Therefore, during the ejection process, E_B is expended by exciting the electron from its bound state to the Fermi level. However, additional work is required to remove the electron completely from the material, and this is called the *sample work function*, or ϕ_s. Once outside the material, the electron is at the *free electron level*, and in the absence of outside interferences (e.g., stray magnetic fields), this is also called the *vacuum level*. At this point, the electron has kinetic energy E_s. In an actual XPS instrument, before this single photoelectron arrives at an electron multiplier in order to be counted, it passes through an electron spectrometer (i.e., an electron energy analyzer). During this passage, a small amount of work is necessarily done on the photoelectron, and this is called the *spectrometer work function*, or ϕ_{sp}. At this point, the photoelectron has kinetic energy E_K.

In order to obtain an absolute relationship that includes the terms defined above, it is necessary to consider the relative position between the Fermi level of the sample and that of the instrument. Because in this example the sample can conduct electrons, and because it is in good electrical contact with the instrument, the Fermi levels of the sample and instrument will be aligned as shown schematically in Figure 1. Therefore, we can write

$$E_s + \phi_s = E_K + \phi_{sp} \quad , \tag{3}$$

and

$$h\nu = E_B + E_s + \phi_s \quad . \tag{4}$$

Combining Eqns. 3 and 4, we have again derived Eqn. 2, the fundamental photoelectron equation.

There are several important factors to consider from the preceding discussion. First, E_B is conveniently referenced to the Fermi level of the *sample* only because it matches the Fermi level of the spectrometer. Second, the work function of the sample, ϕ_s, does not come into play, i.e., we do not have to know what it is for a routine measurement of E_B. Finally, E_B as measured for all conducting samples in good electrical contact with the same instrument can be directly compared. This will not be the case for insulators as we will now see.

For insulating samples, the sample surface is necessarily electrically insulated from the spectrometer. Therefore, the Fermi level of the sample floats with respect to the spectrometer, and measuring the binding energy of an element in the sample as before will not result in a true E_B. The situation now is as presented in Figure 2; the Fermi levels of the sample and the spectrometer are different, and a new term, ϕ_e, has been added. ϕ_e represents the energy of the static surface charge which is a balance between external electrons from some instrumental source which flood the insulating sample in response to the positive surface charge buildup due to the loss of photoelectrons. The top of this energy band provides a reference, because it must coincide with the free electron level of the sample. Therefore, according to Figure 2, we can write the expression

$$E_B = h\nu - E_K - (\phi_s - \phi_e) \quad . \tag{5}$$

Thus, E_B now depends on the work function of the sample, ϕ_s, and the static surface charge, ϕ_e. Although it is possible to obtain these values, they are generally not known from sample to sample. Without taking into account ϕ_s and ϕ_e, one only measures an *apparent* E_B, and E_B's measured from different insulating samples cannot be directly compared. Herein lies the biggest problem with XPS measurements on insulators (or conductors which are electrically insulated from the spectrometer).

Fortunately, there are a number of techniques for which one can indirectly obtain absolute E_B's from insulating samples. These techniques all involve adding some standard element or compound, with well-known E_B's, to the insulator. Then, after collecting a spectrum, the shift of a standard E_B between its known and observed values is measured and this shift is used to offset all peaks from the insulating sample. This is called *static charge referencing*, and specific methods and standards for this will be

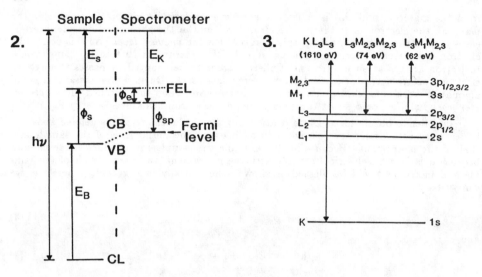

2.

Sample   Spectrometer

E_S E_K

FEL

ϕ_e

ϕ_S CB ϕ_{sp} Fermi level

$h\nu$ VB

E_B

CL

3.

K L_3L_3 $L_3M_{2,3}M_{2,3}$ $L_3M_1M_{2,3}$
(1610 eV)   (74 eV)   (62 eV)

$M_{2,3}$ 3$p_{1/2,3/2}$
M_1 3s

L_3 2$p_{3/2}$
L_2 2$p_{1/2}$
L_1 2s

K   1s

Figure 2. Energy level diagram for x-ray photoemission and measurement. In this case, the sample is assumed to be an insulator. This diagram can be used to derive Eqn. 5. CB, VB, CL refer to conduction band, valence band, and core level, respectively. All other symbols are defined in the text associated with Figures 1 and 2, along with an explanation of the diagram.

Figure 3. Schematic representation of the three most probable Auger transitions for Si, along with their names and the energies of the resulting Auger electrons. The downward arrows represent the filling of a core vacancy, and the upward arrows represent the ejection of an electron (the Auger electron) using the energy released in the de-excitation event. Subatomic levels are shown in both x-ray and spectroscopic nomenclatures (see Table 3). However, Auger peaks are named using the x-ray nomenclature of the three subatomic levels involved in the transition as shown. From Hochella et al. (1986a).

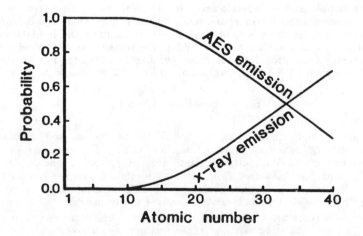

Figure 4. Probability of x-ray vs. Auger electron emission as a function of the atomic number of the emitting element when the initial core vacancy is in the K shell. See text for explanation. Modified from Siegbahn et al. (1967).

discussed in section 5.2.

2.3 Origin and energies of Auger electrons

As should be apparent from the previous section, the energies of photoelectrons depend on the energy of the photons which eject them (Eqn. 2). This is not the case for Auger electrons, in that their energies are *independent* of the stimulating source. As explained in section 3.1.1, this fundamental difference can sometimes be very useful in distinguishing between photo- and Auger peaks in the same spectrum.

Figure 3 schematically shows the three Auger transitions that result in the most intense spectral lines for Si. An Auger electron can be produced when an atom undergoes an inner-shell ionization stimulated either by an incident photon or electron. The resulting hole will be filled with an electron from a higher atomic level. This de-excitation process results in the emission of an x-ray photon *or*, if the created photon is internally recombined, in the ejection of another electron, the *Auger* electron (upward arrows, Fig. 3). Thus, the emission of a photon or an Auger electron are directly competing processes, and the summed probabilities for their occurrence in any particular case must equal unity.

Because these two possible mechanisms form the basis for x-ray fluorescence analysis (e.g., as used in XRF and EPMA) and Auger electron spectroscopy, it is fundamentally important to know which process is more likely to occur for each element. As shown in Figure 4 for atoms in which the initial vacancy is in the K shell, the probability of Auger electron emission is much more likely for lighter elements than photon emission. This is also true when the initial vacancy is in higher principal shells of the heavier elements. When all of the probability curves for photon and Auger electron emission are overlapped, it is apparent that x-ray fluorescence analysis of the elements heavier than approximately atomic number 10 should be somewhat favorable, but that the Auger analysis of all elements should be favorable. This is misleading on two accounts. First, as explained in section 2.4.2, H and He do not have enough electrons to undergo an Auger transition, and therefore they cannot be detected in Auger spectroscopy. Second, although all other elements can be detected by AES, the relative sensitivities for different elements varies by up to a factor of 100. A good rule of thumb is to remember that at a primary beam energy of 3 keV (this is the beam energy used most commonly for analysis of insulating surfaces), elements with atomic numbers between 30 and 39 and greater than 55 will have higher detection limits than the other detectable elements. Charts of the relative sensitivities of all elements up to atomic number 79 are given in Davis et al. (1976). On the other hand, the detection limit for AES in the best case is as good or better than XRF (Table 2).

The kinetic energy of an Auger electron is dependent on the energy available from the de-excitation process and the binding energy of the electron which is ejected. This is expressed as

$$E(VXY) = E(V) - E(X) - E(Y') , \qquad (6)$$

where E(VXY) is the energy of the Auger electron, and V, X, and Y are the three subatomic levels involved in the Auger process. E(V) and E(X) are the energies of the subatomic levels from which the hole is created and the filling electron originates, respectively. E(Y) is the energy of the level from which the Auger electron comes, and E(Y') is the energy of this level with the atom in an Auger transition imposed (doubly ionized) final state. Although these three energy levels can be measured or approximated (Shirley, 1972), it should be apparent that binding energy information is not directly or easily available from an Auger electron. Therefore, Auger energies are always reported and used in terms of kinetic energies.

2.4 Some important characteristics of photoelectrons and Auger electrons

2.4.1 Stimulation of photo- and Auger electron ejection. As mentioned above, Auger electrons can be ejected by *either* photons or electrons, whereas photoelectrons are *only* stimulated by photons. This important fundamental consideration can be easily rationalized by remembering that Auger electrons are the result of an electronic transition, irrespective of how the first vacancy in the Auger ejection process was created. On the other hand, photoelectrons depend on their characteristic energy by the complete transfer of energy from a photon to a bound electron as inferred in Eqn. 2. Herein lies the reason why "photoelectron-like" electrons are *not* generated by electron bombardment, as electron-electron collisions rarely result in a complete transfer of momenta. Therefore, the only characteristic electrons which are stimulated from a sample by electron bombardment are Auger electrons.[1] Likewise, *both* photo- and Auger electrons are stimulated by photons (x-rays). In addition, by convention and as mentioned previously, the term AES implies that the Auger electrons are stimulated by electron bombardment. When x-rays are used, this should be specified as x-ray stimulated or x-ray induced AES.

2.4.2 Elemental detection and identification. As described above, the origins of photoelectrons and Auger electrons are fundamentally different. Therefore, XPS and AES should not necessarily have the ability to detect the same elements. For all practical purposes, however, the only elements that they *cannot* detect are H and He. For XPS, the reason is that the photoionization cross-sections for H and He 1s orbitals are extremely small. For AES, the three-electron Auger process precludes the detection of H and He. In addition, it would seem that the three electrons of Li are not in the proper configuration for an Auger transition to occur (i.e., the 2s to 1s transition does not provide enough energy to eject the remaining 1s electron). However, in condensed matter, characteristic Li Auger lines occur due to weakly bound electrons which populate molecular orbitals in which Li participates.

Because the configurations and energies of the electrons of each element are unique, the photoelectron and Auger electron spectrum for each element is unique. Therefore, there is no ambiguity in identifying any detectable element with either XPS or AES. Fortunately, due to relative photoionization cross-sections (for XPS) or Auger transition probabilities (for AES), most elements do not have more than 4 or 5 intense lines in a spectrum. This helps minimize line overlap for materials with complex compositions.

2.4.3 Depth of analysis. Referring only to solids, the depth of analysis for XPS and AES depends solely on the distance that an electron in the energy range of photo- or Auger electrons (generally 0-2 keV) can travel through the sample without encountering an energy loss event. The energy loss mechanism for electrons in solids usually involves inelastic collisions with other electrons. For inorganic materials, the average distance that an electron in this energy range can travel through a solid without undergoing energy loss is very short, usually less than a few 10's of angstroms and in certain cases less than 10 angstroms. The exact distance that any given electron will travel depends on the energy of the electron and the material through which it is traveling. This characteristic of free electron travel through condensed matter is what makes XPS, AES, and related electron techniques near-surface sensitive.

Electron attenuation lengths vs. kinetic energy are shown in Figure 5. The attenuation length is defined as the distance at which the probability of electron escape from a solid without the occurrence of an inelastic scattering event drops to e^{-1} (i.e., 1/2.718 or 36.8%). The term *escape depth* has exactly the same meaning as attenuation length, except that it is only defined for electrons traveling *normal* to the surface. Therefore, the attenuation length times the cosine of the angle between the direction of electron travel and the sample surface normal is equal to the escape depth. Attenuation lengths (and escape depths) are experimentally measurable, and are usually determined by

---

[1] So-called "secondary" electrons which make scanning electron microscopy (SEM) possible are also ejected by electron bombardment. However, these electrons are not characteristic in that they do not result from specific and fixed ejection mechanisms as do photo- or Auger electrons.

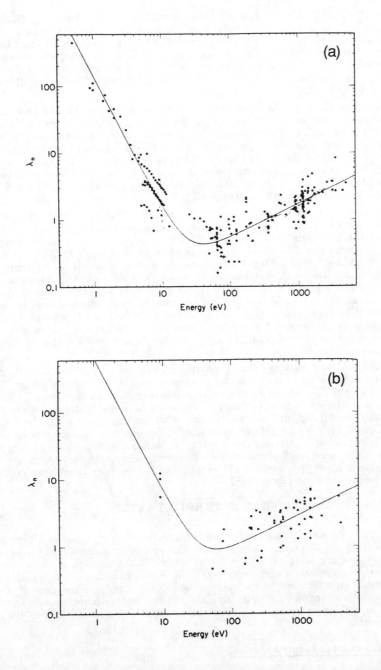

Figure 5. Compilation of measured attenuation lengths in nanometers (λ_n) vs. electron energy for pure elements (a) and inorganic compounds (b). The solid curves are empirical least squares fits to the data. From Seah and Dench (1979).

measuring the attenuation of electrons traveling through thin films (see, e.g., Hochella and Carim, 1988). It is interesting to note from Figure 5 that electrons seem to have maximum interaction with solids at energies around 50 eV. At energies above and below this, interaction decreases and attenuation length goes up. For the energies of the most commonly observed photo- and Auger electrons (between 50 and 2,000 eV), attenuation lengths in metals are generally between 5 and 15 A, and for oxides (including silicates), between 15 and 30 A.

Considering now the depth of analysis in terms of the escape depth, the probability of electron escape falls off as an exponential of the depth of origin. More specifically, 63% (i.e., $1 - e^1$) of the signal originates from less than one times the escape depth, 86% (i.e., $1 - e^2$) is from less than two times the escape depth, 95% (i.e., $1 - e^3$) is from less than three times the escape depth, and so on. Therefore, nearly a quarter of the signal (23%) comes from between one and two times the escape depth, and 9% from between two and three times the escape depth. To help reflect this situation, three times the escape depth is often called the *information depth*.

From the preceeding discussion, it should be obvious that the depth of analysis varies in a single XPS or AES spectrum depending on the kinetic energy of the electrons which produced the line of interest. Nevertheless, it is important to consider whether there are any characteristic differences between the surface sensitivity of XPS vs. AES. In general terms, for any given material, the surface sensitivity depends on the energy of the detected electrons, not on the mechanism by which they were ejected. However, for a number of elements common in silicates, the most intense (and therefore useful) Auger electron lines are at lower kinetic energy than the most intense photoelectron lines for the same element. In these cases, AES may be significantly more surface sensitive than XPS. For example, it has been estimated by Hochella et al. (1988) that for labradorite, the information depth associated with the most intense Al and Si AES lines is 4 times less than the most intense XPS lines (≈ 20 A as opposed to ≈ 80 A).

Finally, it should be mentioned that the term *inelastic mean free path* is often used in the literature and can be confused with attenuation length and escape depth. The inelastic mean free path is defined as the average distance that an electron of a given energy will travel between successive inelastic collisions in some material. Mean free paths are normally calculated, and the relationship between them and escape depths are discussed in Powell (1984) and references therein. For surface characterization and chemical quantification, it is attenuation lengths, or escape depths, which are needed. However, there are a number of problems with measuring escape depths. Accurate calculations of inelastic mean free paths could eventually lead to more reliable estimations of attenuation lengths (e.g., see Penn, 1987).

3. SPECTRAL INTERPRETATION

In this section, we will discuss the fundamental features which occur in all electron spectra due to incident x-rays or electrons and the secondary features which appear according to special circumstances. We will also introduce the concept of the chemical shift and a related factor, the Auger parameter. The fundamental spectral features are used for elemental identification in the near-surface region, and, in the case of x-ray induced spectra, for valence band characterization. The secondary features and chemical shifts are used to acquire information on the structural and/or oxidation state of the elemental constituents.

3.1 Interpretation of x-ray induced spectra

3.1.1 Elemental lines. A monoatomic sample of a heavier element provides a particularly good example of elemental lines and other fundamental features seen in all x-ray induced electron spectra. Figure 6 shows a spectrum of a pure Ag sample generated from incident non-monochromatic Mg x-rays. It is immediately apparent that one can actually "see" the electronic structure of Ag, at least from the 3s through 4d levels which

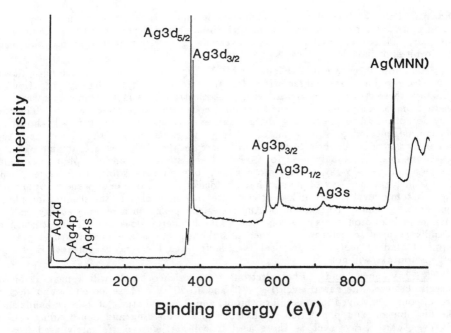

Figure 6. X-ray photoelectron and Auger spectra of Ag stimulated with a nonmonochromatic Mg x-ray source. The relatively small peaks visible to the low binding energy side of the Ag3d$_{5/2}$, Ag3p$_{3/2}$, and Ag3p$_{1/2}$ lines are x-ray satellite lines. See the text for a further description of the photoelectron lines in this spectrum. The four Auger lines collectively labeled Ag(MNN) are exactly equivalent to and specifically named in the electron stimulated Auger spectrum for Ag shown in Figure 11.

fall in the energy range plotted. (Imagine the excitement of Siegbahn and others collecting these spectra for the first time in the 1950's!) The x-axis in Figure 6 could be plotted in terms of kinetic energy, which is actually being measured in the electron spectrometer, or binding energy, which is calculated from Eqn. 2 with hν = 1256.6 eV, the energy of the most intense Mg x-ray line (K $\alpha_{1,2}$), and ϕ_{sp} equal to 4.75 eV. In the literature, nearly all XPS spectra are plotted in terms of binding energy, and therefore Figure 6 and all subsequent XPS spectra in this chapter are plotted in this way. Unfortunately, there is no convention as to whether the binding energy scale should increase to the left or right. Because a number of XPS spectra in this chapter are taken from the literature, both scale directions are used here and the reader is reminded to check this direction in all cases to avoid confusion.

As explained above (section 2.1), the j-j coupling scheme predicts that all subatomic levels with l > 0 (i.e. p,d,f, etc.) should show pairs of photopeaks. This is commonly referred to as *spin-orbit splitting*. Note that in Figure 6, spin-orbit splitting is apparent for the Ag 3p and 3d levels, but not for the 4p and 4d levels. This apparent inconsistency is simply due to a lack of spectral resolution in this survey scan over a large binding energy region. Narrow high-resolution scans of the 4p and 4d regions would show the expected splitting. Furthermore, it can be shown theoretically that the magnitude of this splitting should decrease as l increases for a given principal shell (K,L,M,N, etc.). As an example of this, Figure 6 shows that the separation between the Ag 3p$_{1/2}$ and 3p$_{3/2}$ sublevels is approximately 30 eV, whereas the splitting between the Ag 3d$_{3/2}$ and 3d$_{5/2}$ sublevels is only approximately 6 eV. It can also be shown that, for given principal and sublevel quantum numbers (n and l), the spin-orbit splitting increases with increasing atomic number. In addition, the relative intensities of spin-orbit split pairs can be predicted according to the number of electrons in each sublevel of the pair as shown in Table 3. Therefore, p level spin-orbit split lines will have an intensity ratio of 1:2, d pairs 2:3, f pairs 3:4, and so on, with the higher j level being the more intense of the pair. These area ratios can be convenient in determining the atomic sublevel of spin-orbit split lines in unknown samples.

Also note in Figure 6 that four Ag MNN Auger lines appear. These lines, like the photoelectron lines, are characteristic for and can be used to identify Ag. However, although these Auger lines appear with photoelectron lines in the same spectrum, they result from a completely different ejection mechanism as explained earlier. Therefore, they do not relate to the binding energy scale even though they are typically shown with photopeaks plotted only in terms of binding energy. However, the photo- and Auger peaks always relate directly to the kinetic energy scale. The reason that binding energy scales are most commonly used for x-ray induced spectra is because photopeak kinetic energies change as x-ray sources are changed, but photopeak binding energies do not. This is just another way of saying that photoelectron binding energies are a function of the subatomic level from which they originate *and* the energy of the stimulating photon. On the other hand, the position of Auger electron peaks on a binding energy scale will change with photon energy because their kinetic energy does not. This situation can actually be used to determine whether an unknown line is a photo- or Auger peak. In terms of binding energy, Auger peaks will shift when the x-ray wavelength is changed, but photopeaks will not.

3.1.2 Valence bands. Photoelectrons from valence bands will appear at binding energies between the Fermi level ($E_B = 0$) and 15-20 eV. These loosely bound electrons are, of course, involved in molecular orbitals, and a careful study of valence bands shows the distribution and population of electrons in the bonding and non-bonding orbitals. However, hard x-rays such as those used to collect Figure 6 are rarely used to study valence bands. These shallow atomic levels have a very low photoionization cross-section for photons in this energy range. On the other hand, they have a very high cross-section for photons in the ultra-violet range. As a result, ultra-violet photoelectron spectroscopy (UPS) has evolved into a distinct field of its own. Typical photon sources for UPS are He, Ne, and Ar gas discharge lamps, although tunable synchrotron light sources have also become available. The photon excitation energies typically used are less than 100 eV (as opposed to energies well over 1000 eV for x-ray sources), and therefore only the most shallow core levels and the valence band are normally collected. The extensive subject of UPS and valence band spectroscopy is beyond the scope of this chapter. An excellent review of the field is given by DeKock (1977).

3.1.3 Background. The form of the background of the Ag spectrum in Figure 6, consisting of a series of steps which rise toward higher binding energy, is characteristic of x-ray induced spectra. The steps appear to occur at peak positions in the spectrum, with large steps following large peaks and small steps following small ones. These background features are a direct result of characteristic electrons which have been ejected by normal photoelectron or Auger processes, but which subsequently have encountered random and most likely multiple energy loss events before finally escaping the sample. Because these once characteristic electrons have lost kinetic energy, they appear at higher binding energies and result in a background which steps up on the high binding energy side of each peak. Naturally, the most intense lines provide the most characteristic electrons for these random and multiple energy losses, and therefore the steps are largest after these lines.

Occasionally, this distinctive background from x-ray induced spectra can be used to recognize photoelectron peaks. For example, Fe 2p photopeaks can be very broad and difficult to recognize due to multiple-splitting (see section 6.1). However, a slight rise in the background, starting at the position of the Fe $2p_{3/2}$ photopeak, is characteristic of small amounts of Fe in the near-surface region of the sample.

3.1.4 X-ray satellite lines. The photopeaks labeled in Figure 6 are induced by the most intense characteristic x-ray line coming from the Mg x-ray source, namely K $\alpha_{1,2}$ (due to the $L_{2,3}$ to K transition). However, the Mg x-ray tube which was used to stimulate the spectrum in Figure 6 is a nonmonochromatic source, and there are many other characteristic Mg x-ray wavelengths coming off the anode besides the principal K $\alpha_{1,2}$ line. Because of this, photoelectrons with different kinetic energies will be generated from the same subatomic level in the sample. Then, when the photoelectron spectrum is plotted in terms of binding energy based on the energy of the most intense x-ray line (for Mg anodes, $h\nu = 1256.6$ eV for K $\alpha_{1,2}$), extra photopeaks around the principal ones will appear. These extra lines are called *x-ray satellites*.

Specifically, Mg and Al x-ray anodes are by far the most commonly used to stimulate photoelectron spectra. From these sources, the most intense satellite lines are generated by K α_3 and K α_4 x-rays (resulting from $L_{2,3}$ to K transitions in a multiply ionized atom). These x-rays have higher energies than K $\alpha_{1,2}$ but far less intensity, and as a result relatively small photopeaks generated by them will appear at *lower* binding energies when the binding energy scale is based on the energy of K $\alpha_{1,2}$ x-rays. It is the K $\alpha_{3,4}$ x-rays which have generated the visible satellite lines for the principal photopeaks in Figure 6. Although many other characteristic x-rays from Mg and Al anodes are hitting the sample, they do not have sufficient intensity to create generally noticeable x-ray satellite peaks. The relative positions and intensities for K α_3 and K α_4 induced satellite lines with respect to the main photopeaks generated by K $\alpha_{1,2}$ for both Mg and Al anodes are presented in Table 4.

X-ray satellite lines can be eliminated by using a monochromatic x-ray source, but only at the expense of decreased x-ray intensity and signal to noise ratio. Monochromatic x-ray sources have a number of other advantages and disadvantages discussed in section 4.3.2. Finally, it is important to remember that only photopeaks will have x-ray satellites, whereas x-ray induced Auger lines will not due to the nature of their excitation.

3.1.5 Multiplet splitting.

Multiplet splitting, not to be confused with spin-orbit splitting, can occur for core level photopeaks when the atom has unpaired valence electrons. For example, after photoionization in an s level, the remaining unpaired electron in that s level and unpaired electrons in the valence level can have the same direction of spin (parallel or coupled spin) or opposite directions of spin (anti-parallel or uncoupled spin). This results in two slightly different final state energies, and the s level photopeak will be split into two components. The multiplet-splitting of p and higher sublevels is more complicated due to orbital-angular momentum coupling. For example, 2p sublevels of first row transition metals in high-spin or paramagnetic states exhibit considerable line broadening due to complex multiplet splitting phenomena.

For geologically relevant materials, perhaps the most important examples of multiplet splitting occur for Fe in paramagnetic minerals. For example, the Fe 2p lines for hematite and fayalite are shown in Figure 7. As a result of multiplet splitting, these peaks are broad (4-6 eV) and asymmetric, and the hematite spectrum even shows additional weak lines approximately 8 eV to the high binding energy side of the principal Fe 2p lines. However, all of these features have been predicted by the calculated multiplet structure of free ferrous and ferric ions (Gupta and Sen, 1974). Despite the complexity of these line shapes and background, spectra of this sort can still be used to determine the oxidation state of the iron in the near-surface of minerals (see section 6.1).

3.1.6 Shake-up lines.

When photoelectrons are emitted, the parent atom does not always immediately return to an ionic ground state. The ion may instead return to an excited state. In the case of *shake-up lines*, this occurs as a result of a valence electron being excited to a higher unfilled valence level during the photoelectron emission process. Therefore, photoionization occurs simultaneously with photoexcitation, and the energy for the latter process is not available to the photoelectron. This results in a reduction of its kinetic energy, or equivalently an increase in the apparent binding energy. Like multiplet splitting, shake-up lines are most likely to occur in paramagnetic compounds. Examples of shake-up lines for copper oxide and copper sulfate are shown in Figure 8.

3.1.7 Energy loss lines.

When electrons loose discrete amounts of energy due to some interaction mechanism before leaving the sample, *energy loss lines* are observed to the high binding energy side of photo- and Auger electron peaks. This energy loss mechanism can be very well defined, as with *plasmon losses* in conductors where discrete

588

Table 4. Relative energies and intensities of $K\alpha_3$ and $K\alpha_4$ x-ray satellite photopeaks

Anode	Displacement relative to principal photopeak (eV)*		Intensity relative to principal photopeak (%)	
	$K\alpha_3$	$K\alpha_4$	$K\alpha_3$	$K\alpha_4$
Mg	-8.4	-10.2	8.0	4.1
Al	-9.8	-11.8	6.4	3.2

* In terms of binding energy.

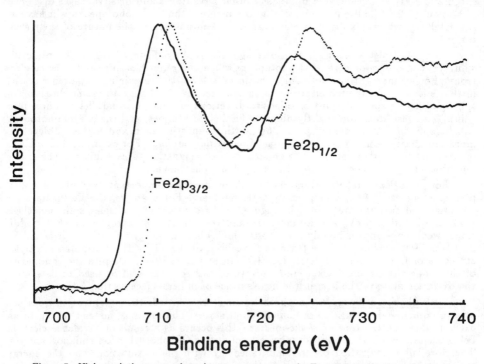

Figure 7. High resolution x-ray photoelectron spectra over the Fe 2p region for hematite (dotted curve) and fayalite (solid curve). The unusually wide and asymmetric line shapes for both, as well as the weak lines approximately 8 eV above the main Fe 2p lines for hematite, are due to multiplet splitting. The shift in peak position between the main Fe 2p lines for hematite and fayalite are characteristic of ferric and ferrous iron, and therefore this can be used to determine iron oxidation states in the near surface regions of materials. See section 6.1 for additional examples.

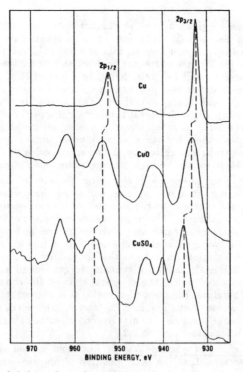

Figure 8. Examples of shake-up lines observed in the Cu 2p spectra of metallic Cu, CuO, and CuSO$_4$. The main Cu 2p lines are displaced due to the chemical shift effect (section 3.1.9). Note that in these spectra, the binding energy is increasing to the left. From Wagner et al. (1979a).

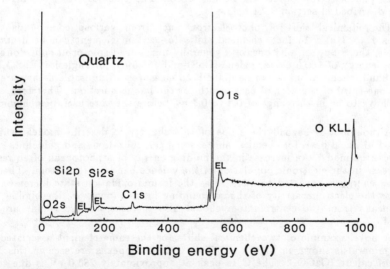

Figure 9. X-ray photoelectron spectrum of quartz showing three energy loss features (labeled EL). Also note the x-ray satellite lines to the low binding energy side of major peaks (see section 3.1.4) and the adventitious carbon line (see section 3.1.8). The x-ray induced oxygen KL$_{2,3}$L$_{2,3}$ Auger line is labeled O KLL in this spectrum.

amounts of energy can be given up to the characteristic oscillations of conduction electrons. Plasmon loss features provide for an important part of conductor surface analysis, but because they are so rarely seen in minerals, the interested reader is referred to Briggs and Riviere (1983) and references therein, and no further discussion of discrete surface and bulk plasmons will be given here. In contrast, the energy loss mechanism may not be well defined and is best described as a continuum of discrete losses resulting in broad and generally weak energy loss lines. This is exactly what is seen in most silicate minerals and many other important mineral groups. For example, energy loss features in the quartz x-ray induced electron spectrum are shown in Figure 9. It is important not to mistake these for other primary or secondary spectral features. In general, energy loss lines can be recognized by their very broad width (10 eV or more at half height) and position 15-20 eV to the high binding energy side of major peaks.

3.1.8 Photopeaks from adventitious elements. Because of the reactivity of surfaces, it is not surprising that most survey spectra will show the presence of *adventitious elements* (i.e., elements not inherent to the sample). This can occur with time even if the sample is prepared in vacuum. The most commonly observed adventitious elements which appear on the surfaces of earth materials are carbon and oxygen. Nitrogen is only occasionally present. (Hydrogen is also expected, but it is not detectable by XPS.) The usual source of this contamination is from the air and/or residual gases in the vacuum. Because most earth materials naturally contain oxygen, small amounts of adventitious oxygen usually go unnoticed unless the chemical shift of the contaminant oxygen line is particularly large. Therefore, the C 1s photopeak at approximately 285 eV is by far the most commonly seen adventitious line in XPS (and AES) spectra (see Fig. 9). This peak can be used for a relatively rough but highly convenient static charge referencing technique on insulating surfaces (see section 5.2).

3.1.9 Chemical shifts of photopeaks. It has been known since the early work of Siegbahn and his associates that the exact position of photoelectron and Auger lines can be indicative of the chemical and/or structural environment of the element, as well as its formal oxidation state. This *chemical shift* phenomenon has been studied extensively for most elements in a wide range of materials and has allowed XPS and AES to be used for more than just near-surface elemental identification. In this section, only the chemical shift of photoelectron peaks will be discussed. The chemical shift of Auger peaks is briefly described in section 3.2.1 below.

The chemical shift of photoelectron lines from various compounds commonly reaches 1 to 3 eV. In fact, chemical shifts for certain element can be dramatic. For example, the S $2p_{3/2,1/2}$ photopeak (a generally unresolved spin-orbit split doublet) has a binding energy of 160.5 eV for galena (PbS) and 168.5 eV for anglesite ($PbSO_4$). On the other hand, chemical shifts can be difficult to measure and unusable if they are less than about one-third of the normal peak width for the line in question. Therefore, shifts generally have to be in the range of 0.3 to 0.7 eV before they are useful as chemical indicators.

Although it is beyond the scope of this chapter to describe models which give a detailed physical basis for chemical shifts, some generalizations and guidelines concerning them can be made. An increase in the binding energy of a photopeak often results from a decrease in the electronic population of the valence bands of that atom. Therefore, the binding energy is expected to increase as the formal oxidation state increases. In addition, as the electronegativity of the surrounding ligands increases, the binding energy of photopeaks from the central atom should also increase. These trends are generally obeyed.

A novel example of two chemical shifts for the same element associated with the same mineral is shown in Figure 10. Two C 1s photopeaks are seen in this XPS spectrum of calcite ($CaCO_3$). The C 1s peak at approximately 285.0 eV is due to adventitious carbon on the surface of the calcite which is commonly seen after air exposure. This carbon is most likely in the form of hydrocarbons, i.e., $(CH_2)_n$. In addition, the C 1s peak at approximately 289.7 eV is due to the CO_3^{2-} groups in the carbonate mineral itself. Ion sputtering will decrease the intensity of the 285.0 eV line, whereas tilting the

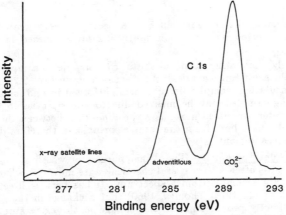

Figure 10. The chemical shift of C 1s photopeaks for calcite. The peak at approximately 289.7 eV is due to the carbon in the carbonate groups of the calcite. The peak at approximately 285.0 eV is adventitious carbon on the calcite surface due to air exposure. The latter line can be eliminated by low energy ion sputtering for short times.

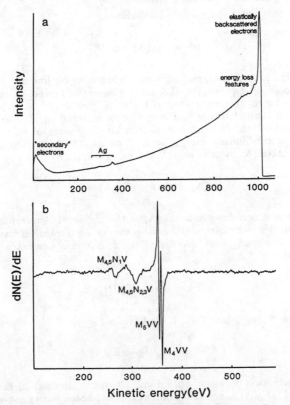

Figure 11. (a) Electron induced Auger spectra of Ag using a 1 kV beam. Four Ag MNN Auger peaks, equivalent to those in the x-ray induced spectra (Fig. 6), are marked. See text for a full description of this spectrum. (b) High resolution differentiated spectrum over the Ag MNN region of the spectrum in (a). The 4 Auger lines have been labeled according to the three subatomic levels involved in the Auger transition as explained in section 2.1 and Figure 3. The "V" in the label indicates the valence level. The position of each Auger line (by convention) is taken as the point of greatest negative excursion of the differentiated peak.

sample with respect to the entrance slit of the electron energy analyzer will increase its intensity. Angle resolved XPS and ion sputtering are discussed in sections 5.5.1 and 5.5.2, respectively.

Unfortunately, comparing chemical shifts from sample to sample requires the measurement of absolute binding energies. As discussed in section 2.2, measuring absolute energies from insulating samples in not straightforward, and small chemical shifts between insulating samples may be masked due to uncertainties in the measurements. The Auger parameter, discussed next, uses the *difference* between the chemical shifts of photoelectron and Auger lines from the same element, and therefore does not depend on absolute energy measurement.

3.1.10 The Auger parameter. The Auger parameter was first described by Wagner (1972, 1975a,b) and is often a very useful chemical state indicator when other evidence of chemical state is not available from a spectrum. It has been well established (Wagner, 1975b; Kowalczyk et al., 1974; Thomas, 1980) that a change in the Auger parameter of any element is directly proportional to the change in the extra-atomic relaxation (also referred to as polarization energy) for that element in the compound being measured. This relaxation involves the screening of the final state ion in the Auger electron emission process by electrons of neighboring atoms (not necessarily next nearest neighbors) and/or by electrons in the conduction band.

The Auger parameter, α, is defined as follows:

$$\alpha = E_K(A) - E_K(P) \quad , \tag{7}$$

where $E_K(A)$ is the kinetic energy of an x-ray induced Auger line and $E_K(P)$ is the kinetic energy of a photoelectron line for the same element observed in the same spectrum. Because x-ray induced spectra are almost always plotted in terms of binding energy, Eqn. 7 is more convenient to use if it is rewritten in terms of binding energy. The fundamental equation which relates the measured kinetic energy of an x-ray induced Auger line with its *apparent* binding energy, as seen on an XPS spectrum plotted in terms of binding energy, takes the form of Eqn. 2 and can be written

$$E_K(A) = h\nu - aE_B(A) - \phi_{sp} \quad , \tag{8}$$

where (A) signifies Auger lines and $aE_B(A)$ is the apparent Auger binding energy. The fundamental equation which relates binding energy of photoelectrons to the measured kinetic energy has already been given in Eqn. 2, but it is rewritten here explicitly for photoelectrons as

$$E_K(P) = h\nu - E_B(P) - \phi_{sp} \quad . \tag{9}$$

Substituting Eqns. 8 and 9 into Eqn. 7, we have

$$\alpha = E_B(P) - aE_B(A) \quad . \tag{10}$$

However, there are two problems with Eqn. 10. First, if used in this form, α can be negative. Second, $aE_B(A)$ is not independent of the photon excitation energy as is $E_B(P)$. Therefore, in this form, α will change for different x-ray sources. In order to keep the Auger parameter positive and independent of photon excitation energy, the modified Auger parameter, α' , is commonly used where

$$\alpha' = \alpha + h\nu = E_B(P) - aE_B(A) + h\nu \quad . \tag{11}$$

Because the Auger parameter is simply based on the difference in energy between an Auger peak and a photoelectron peak for the same element in the same XPS spectrum, absolute peak positions (and therefore static charge corrections) are not needed. It is possible to measure peak positions to within 0.05 eV in high energy resolution spectra, and therefore Auger parameters can be accurate to within 0.1 eV or less, even when measured in different laboratories using different instruments. Examples of the use of Auger parameters is given in the text which accompanies Figures 28 and 34 below.

3.2 Interpretation of electron induced spectra

As explained in section 2.4.1 above, Auger peaks will be the only element-specific signals produced in electron induced spectra. In this section, the shape and chemical shift of Auger lines as well as other characteristic features of electron induced spectra will be discussed.

Figure 11 is an Auger spectrum of pure Ag using a 1 kV primary electron beam. The differences between this spectrum and the one stimulated with x-rays (Fig. 6) exhibit the fundamental differences between electron and x-ray induced electron spectroscopies. In Figure 11a, a gently sloping background is sandwiched between a broad hump at low kinetic energy and a relatively sharp and intense peak at high kinetic energy. The high energy peak is at the energy of the primary electron beam and represents the electrons in the primary beam which have been elastically scattered from the sample into the electron energy analyzer. These electrons are often referred to as *elastically backscattered* electrons. The weak lines at slightly lower kinetic energies are energy loss lines which originate exactly as described above for these lines in x-ray induced spectra (section 3.1.7). Most of the sloping background is due to the *inelastically backscattered* electrons from the primary beam and secondary electrons (i.e., electrons from the sample) which have been ejected by the primary beam. Finally, the broad hump at low energy is named the *secondary electron* peak. This is obviously an unfortunate choice of terms, but it has stuck for historical reasons. These "secondary" electrons are those with energies less than 50 eV (by definition), and they are mostly valence and/or conduction electrons ejected by interaction with the energetic primary electrons. It is these "secondary" electrons which are used to form secondary electron images in SEM instruments.

Superimposed on the gently sloping background of Figure 11a are relatively weak and broad peaks (marked on figure) which are characteristic of Ag and are in fact the Ag MNN series of Auger peaks equivalent to those shown in Figure 6. Herein lies one of the problems associated with electron induced Auger spectroscopy. Because the inelastically backscattered electrons create a relatively intense background, and because the Auger lines are generally small compared to this background, the signal-to-noise ratio for AES is often poor. In order to allow for easier detection of relatively weak and/or broad lines, it is common practice to differentiate the direct spectrum as shown in Figure 11b for the Ag MNN portion of the spectrum shown in Figure 11a. This can be done directly while collecting the data by modulating the electron energy analyzer voltage and sending the signal through a lock-in amplifier, or indirectly by collecting the signal in a normal scan, storing it, and finally differentiating it with a computer. The former method was used almost exclusively in the early days of Auger spectroscopy. However, with the availability of inexpensive computers, the latter method is by far the more commonly used method today.

3.2.1 Chemical shift and shapes of Auger lines. The chemical shift of Auger lines occurs for the same reasons and in the same directions as for photoelectron lines (discussed above). However, as first reported by Wagner and Biloen (1973), the chemical shifts of Auger lines are often substantially greater than for photoelectron lines for the same element. The reasons for this lie in the fact that Auger transitions involve three atomic sublevels (as opposed to one for photoemission) and result in a doubly ionized final state (as opposed to a singly ionized final state for photoemission). A theoretical explanation as to why this results in larger chemical shifts can be found in Wagner and Biloen (1973) and Wagner (1975b). However, the potentially greater chemical shift for Auger lines is somewhat offset by the fact that Auger lines are typically broader than photoemission lines. Also, Auger lines involving shallow atomic levels tend to group into a closely spaced series of peaks, making for potentially complex line overlaps. In addition, the shape of electron induced Auger lines can be very sensitive to even slight amounts of beam damage in the near-surface region. It is for these reasons that, historically speaking, the chemical shift phenomenon for photoemission lines has been more extensively studied and utilized than for Auger lines.

Despite the potential drawbacks presented above, the shape of Auger lines are potentially very sensitive to changes in the chemistry and atomic structure of compounds. This should be the case particularly when the Auger transition involves one or two valence levels. As a result, one of the most diagnostic Auger lines for minerals should be the O $KL_{2,3}L_{2,3}$ peak (also called the O KVV line, V signifying valence level). This is also by far the most intense oxygen Auger line. For silicates, Wagner et al. (1982) suggested that the x-ray induced O KVV line becomes more asymmetric as the ratio of alumina to silica tetrahedra increases. However, Hochella and Brown (1988) have shown for a much larger series of silicate minerals and glasses that this relationship does not hold, and that the change in the O KVV Auger line is surprisingly and disappointingly small through a series of silicate minerals. However, there are a number of compounds in which Auger line shapes (both in direct and differentiated forms) have been shown to be diagnostic of composition and/or structure (e.g., see the review by Briggs and Riviere, 1983). It is likely that in the future this will also be found to be true for certain geologic materials.

4. INSTRUMENTATION

As discussed in the Introduction of this chapter, at least fifty years separated the time between the conception of the theoretical framework for photo- and Auger electron spectroscopy and building the first successful electron spectrometers. This hiatus was not due to a lack of interest, but occurred because this instrumentation proved to be exceedingly difficult to develop. Commercial systems have only been available for general laboratory use since about 1970, and they originally had very few features and little flexibility. Fortunately, highly sophisticated instruments which are relatively easy to operate and maintain have recently come on the market.

The major components of modern XPS and AES equipment include vessels and pumps capable of ultra-high vacuum, x-ray and electron sources, an electron energy analyzer, electronics and computer, and accessories. The most commonly used accessories include sample introduction and transfer devices, sample manipulators with tilt capabilities, ion guns, secondary electron detectors, and low energy electron flood guns. In addition, many systems have in-vacuum fracturing devices, gas inlet systems, and heating and cooling stages. More sophisticated systems may have the components necessary for other surface sensitive techniques such as low energy electron diffraction (LEED), ultra-violet photoelectron spectroscopy (UPS), secondary ion mass spectroscopy (SIMS), high resolution electron energy loss spectroscopy (HREELS), and ion scattering spectroscopy (ISS).

Although the design concepts and workings of many of the components mentioned above are fascinating, they are complex and space does not allow for a full discussion here. Nevertheless, it is *necessary* that users of these techniques have some understanding of vacuum systems and electron energy analyzers, along with certain accessories. With this knowledge, users will have the ability to choose appropriate experiments, collect spectra properly, and fully appreciate the research possibilities and limitations of these techniques. Therefore, the most important aspects of the design philosophy and function of modern XPS/AES systems are covered below. The only major components which will not be described are the x-ray and electron beam sources; these components are in widespread use in many common instruments and some knowledge is assumed.

Table 5. Time for monolayer contamination formation on a surface at various pressures and 20°C

	Pressure (torr)	Surface collision frequency ($cm^{-2}sec^{-1}$)	Monolayer formation time (sec)*
atmosphere	760	10^{23}	10^{-9}
lower limit of:			
medium vacuum	10^{-3}	10^{17}	10^{-3}
high vacuum	10^{-6}	10^{14}	10^{0}
ultrahigh vacuum	10^{-9}	10^{11}	10^{3}

* Assuming a sticking coefficient of unity, i.e., all collisions result in attachment.

4.1 The need for ultra-high vacuum and how it is obtained

All modern surface analytic systems operate under ultra-high vacuum (UHV) conditions. UHV conditions are defined by pressures between 10^{-9} and 10^{-12} torr (760 torr = 1 atm). To put this in perspective, the best rotary vacuum pumps (not including turbomolecular pumps) can pull vacuums on the order of 10^{-2} torr, and the more sophisticated pumps in conventional electron microprobes, TEM's and SEM's keep these instruments in the 10^{-5} to 10^{-6} torr range.

Surface analytic instruments must operate under at least medium vacuums so that the probability of interaction between electrons from the sample and residual gas molecules in the vacuum remains low. The kinetic theory of gases predicts that the mean free path of particles in a vacuum of 10^{-5} torr is on the order of a few meters (see, e.g., Moore et al., 1983), thus providing a good chance for most emitted electrons to make an uninterrupted journey from the sample surface to the end of the electron energy analyzer (typically less than 1 meter).

However, the restriction of electron mean free path in the instrument vacuum only provides a minimum vacuum condition needed for practical surface analysis. Much higher levels of vacuum are needed for sample surface cleanliness. In fact, UHV conditions are needed in order to keep sample surfaces relatively clean for extended periods of time. This can be rationalized again with simple gas kinetic theory. Air under room conditions has on the order of 10^{19} molecules/cm^3, and surface collision frequency is such that a monolayer of contamination will form on a clean surface in only 10^{-9} sec! (assuming that every atom or molecule that hits the surface bonds to it, an assumption that is not unreasonable for the first monolayer). Referring to Table 5, UHV conditions are needed before monolayer contamination time is on the same order as experiment duration.

However, for many surface analytic experiments and measurements which are useful in geochemistry, maintaining an atomically clean surface is not necessary. Also, most mineral surfaces which have been exposed to air will immediately develop at least one monolayer of contamination as predicted, but the thin contaminant layer usually passivates the highly active clean surface and remains thin. Then, after putting the sample into UHV conditions, further growth of the contaminant layer with time will be minimized. Practical aspects of surface cleanliness will be discussed in more detail in section 5.1 below.

How is an ultra-high vacuum obtained? It is first necessary to have pumps which can function efficiently at very low pressures. However, it is also necessary to have vessels which can be sealed properly and heated to at least 100°C. It can be shown that in a vessel without leaks and with UHV pumps, the rate limiting step in obtaining UHV conditions is the release of adsorbed gas molecules from the inner-vessel walls. By "baking" the system, or heating it up to between 100 and 200°C for 24 to 48 hours, the desorption of these gases into the vacuum, where they are then subject to pumping, is greatly accelerated. Therefore, for all practical purposes, it is necessary that the vessel and all attached components be bakeable while maintaining structural and functional integrity. As a result, UHV vessels and components are typically made of stainless steel, with strong glass ports (windows) and dense ceramic parts wherever insulators are needed. All joints are precision welded and flanges are sealed with metal (usually copper) gaskets. When the metal flanges are bolted together, the gaskets between them are deformed under high mechanical stress and provide extremely strong and durable metal-to-metal seals. These UHV seals can be broken and resealed repeatedly, although a new gasket is generally required each time.

There are three major categories of UHV pumps, namely ion-sputter pumps (often called ion pumps), UHV vapor diffusion pumps, and cryopumps. Space only allows for a very brief description of each type. *Ion-sputter pumps* work on the principle of gas molecule ionization and trapping. The ionization is the result of high energy electrons traveling between a Ti cathode and a stainless steel anode. Magnets are appropriately positioned in order to maximize the electron flight paths. Ionized gases are accelerated toward and collide with the Ti cathode, implanting themselves or reacting with fresh Ti. In either case, they are effectively removed from the vacuum. *UHV vapor diffusion pumps* work on the principle of momentum transfer between a stream of moving vapor and residual gases in the vacuum. The vapor is continually condensed and reboiled to repeat the cycle. Silicones or polyphenyl ethers, both having extremely low vapor pressures, are fluids generally used for this application. *Cryopumps*, as suggested by the name, condense gases onto metal plates or in traps at very low temperatures. These pumps have closed circuit He refrigeration systems which cool outer plates in the pumping core to 30 to 50 K, condensing the major components in air. The inner core has a cryosorbant material (usually charcoal or a zeolite) cooled to 10-20 K in order to trap H, He, and Ne.

Each of the three UHV pumps mentioned above has distinct attributes, and each is in common use today. Ion-sputter pumps have no fluids or moving parts, and they are essentially maintenance free. UHV vapor diffusion pumps are relatively inexpensive and can pump all gases efficiently. Cryopumps are very clean and provide high pumping speeds for easily condensed gases.

In UHV systems, additional pumping systems are needed to rough pump sample entry locks, forelines, or the whole system when it has been up to air. The most commonly used pumps for this purpose are conventional vane mechanical pumps, turbomolecular pumps which operate via a high speed turbine, and sorption pumps which operate with a liquid nitrogen cooled zeolite. In addition, titanium sublimation pumps (TSP's) are used in most UHV systems as an aid to trap reactive gases. These pumps work by the sublimation of Ti from a hot Ti-W filament, resulting in the deposition of thin films of Ti on the inner walls of the stainless steel TSP vessel. Fresh (unreacted) Ti is extremely reactive with water vapor, O_2, CO, H_2, and other relatively reactive gases. The pumping is achieved by the process of chemisorption, or, as it is sometimes called, "gettering", where the formation of stable titanium compounds occurs.

Finally, pumping speed depends not only on the pumps themselves, but also on the vessel design. Because low pressure gases move by molecular flow (as opposed to viscous flow), the transport of gas molecules from the analytic chamber to the pump throat is a statistically random process. Therefore, the diameter, length, and path of the plumbing which connects the pump to the analytic chamber is critical. In addition, an ion gauge (a devise for measuring low pressures) within the pump may give a pressure reading which is much lower than the true pressure in the analytic chamber. Commercial systems are usually, but not always, designed with these factors as a top priority. Therefore, it is important to be aware of these design factors. More reading on this subject is available in the text by Moore et al. (1983).

4.2 Electron energy analyzers

Electron energy analyzers are the heart of any electron spectroscopic instrument. It is the ability of these devices to accurately measure the energy of electrons which make XPS and AES a reality. There are many analyzers which can achieve high energy resolution (see review by Roy and Carette, 1977), but only two basic designs are in use today, and the important factors concerning both are briefly discussed below.

4.2.1 Concentric hemispherical analyzer (CHA).
As shown in Figure 12, a CHA is made of two concentric metallic hemispheres with an entrance slit and retarding grid at one end and an exit slit and electron multiplier (detector) at the other. A voltage is applied to each hemisphere, for example V_1 to the outer hemisphere and V_2 to the inner. It follows that if ΔV, the voltage difference, is sufficiently large, a high energy electron entering the analyzer may follow a hemispherical path, not strike either surface, and enter the electron multiplier. If ΔV is sufficiently small, the same could be true for a low energy electron. At a certain ΔV, the energy that an electron must have in order to pass through the analyzer without colliding with it or the exit slit is called the *pass energy*, and it is related to ΔV by a constant as follows:

$$PE = H\Delta V \quad , \tag{12}$$

where PE is the pass energy and H is a geometric factor related to the separation of the hemispheres.

At this point, it is important to consider the advantages and disadvantages of high and low pass energies. To demonstrate this, a hypothetical example will be given in which it is assumed that the energy window (resolution) of the analyzer is 1% of the electron energy. For measuring high energy electrons, we would want the PE to be very large (i.e., ΔV is very large), but electrons with energies of 999 and 1,000 eV could not be distinguished from one another (their relative difference in energy is only 1 eV or 0.1% of the electron energy). On the other hand, one can imagine that the count rate would be excellent with such a wide window (1% of 1000 eV or 10 eV). On the other hand, for measuring low energy electrons, we would want the PE to be very small (i.e., ΔV is very small). Electrons with energies of 9 and 10 eV would be easily separated. In fact, with the 1% window, we could separate electrons at 9.9 and 10 eV. But the resolution here is essentially so good (the window so small), that the count rate suffers. To compromise between good energy resolution and a reasonable count rate, a single intermediate PE is desirable for *all* electrons coming from the sample. In order to do this, a *retarding grid* at the entrance slit of the analyzer is used. Depending on the desired PE, the voltage on the grid can be adjusted to accelerate or decelerate the electrons which one wants to pass through the analyzer. The kinetic energy of any electron that makes it through the analyzer is then

$$E_K = (-R) + PE + \phi_{sp} \quad , \tag{13}$$

where R is the voltage applied to the retarding grid. As a practical example, assume that a scan of electrons with energies from 10 to 1000 eV is desired at a pass energy of 100 eV, and that the spectrometer work function is 5 eV. Eqn. 13 indicates that R will increase from 95 volts at the beginning of the scan (10 eV) to -895 volts at the end of the scan (1000 eV). Therefore, the "retarding" grid will accelerate electrons by as much as 95 volts at the beginning of the scan and decelerate electrons by as much as 895 volts at the end, but at all times keeping the pass energy constant for the electrons for which transmission through the analyzer is desired. Because the PE stays constant throughout the scan, this mode of analyzer operation is call *constant analyzer energy* (CAE) mode. CAE is the analyzer mode under which nearly all XPS spectra are collected when using CHA analyzers. Larger PE's are used for wide survey scans, where resolution can be sacrificed for better counting rates, whereas smaller PE's are used for narrow scans,

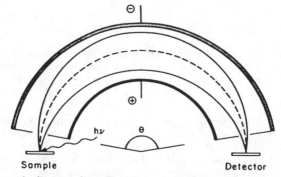

Figure 12. Schematic diagram of a concentric hemispherical analyzer (CHA). The retarding grid (not shown) would normally be positioned just before the entrance slit to the analyzer. The voltages applied to the inner and outer hemispheres in this case are positive (+) and negative (-) as depicted; however, both hemispheres may have a positive or negative voltage. In every case, ΔV is the voltage difference between the two hemispheres. The dashed curve represents the median trajectory of an electron successfully traversing the analyzer to the detector. The solid trajectories represent electrons which enter the analyzer off the analyzer axis (i.e. the line centering and normal to the entrance slit). This condition obviously increases the detected signal, but decreases the energy resolution. See text for further details. From Hercules (1979).

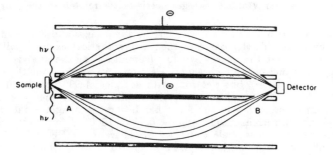

Figure 13. Schematic diagram of a cylindrical mirror analyzer (CMA) consisting of concentric inner and outer cylinders. The electron trajectories shown are analogous to those shown and explained in Figure 12. See text for further details. From Hercules (1979).

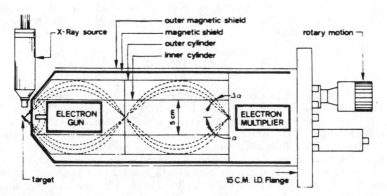

Figure 14. Schematic diagram of a double-pass cylindrical mirror analyzer (DPCMA). This analyzer has two CMA's in series and a cone shaped front end with fringe field rings to smooth and flatten the field gradients near the entrance of the analyzer. From Palmberg (1975).

where the study of line shapes requires the best energy resolution at the expense of lower counting rates.

The mode for CHA analyzers most commonly used for electron induced Auger spectra collection is called *constant retard ratio* (CRR) mode. In this mode, the PE is not constant, but varies with E_K. The retard ratio, RR, is defined as

$$RR = \frac{PE + (-R)}{PE} . \tag{14}$$

For a scan from 10 to 1000 eV with $\phi_{sp} = 5.0$ eV and RR = 4, Eqns. 13 and 14 indicate that at 10 eV the PE is 1.25 eV and R is -6.25 volts. At 1000 eV the PE is 248.8 eV and R is -746.7 volts. If the same scan is performed now at RR = 10, at 10 eV the PE is 0.5 eV and R is -4.5 and at 1000 eV the PE is 99.5 eV and R is -895.5. The first thing to notice is that during a scan in CRR mode, the PE changes such that the energy resolution is best at low E_K and the count rate is best at high E_K. This is desirable for AES data collection, where the probability of line overlap is greatest at low E_K, and Auger peaks generally have weaker intensities at high E_K. By increasing RR, the same conditions apply, but the PE remains smaller across the scan. Therefore, smaller RR's are used for wide survey scans, whereas larger RR's are used for narrow scans where better energy resolution is desired.

4.2.2 Cylindrical mirror analyzer (CMA). The general design of the CMA is shown in Figure 13 along with the trajectories of electrons successfully passing through it. In this case, there are two coaxially arranged cylinders. The inner cylinder is typically kept at earth potential, whereas the outer cylinder has a negative voltage which is ramped to scan the desired energy range. One of the original problems with the CMA design was that electrons entering the analyzer saw asymmetric field gradients, and thus underwent severe aberrations. This problem was overcome by making the end of the CMA cone shaped (see Fig. 14) and adding fringe field (or terminating) rings. The ring electrodes smooth out the abrupt termination of the fields at the end of the CMA, assuring smoothly changing field gradients around the entrance to the analyzer.

The CMA design shown in Figure 13 has no retarding grids, so that its energy resolution is variable, and on average not especially good. Also, the geometry of this analyzer is such that it has a very well defined focal point, and movement of the sample from this point results in an apparent shift in the measured energy of the electrons. (In actual practice of electron induced Auger spectroscopy when using the CMA, this focal point is found by moving the sample in and out until the measured energy of the elastically backscattered peak matches the accelerating voltage of the electron gun.) On the other hand, the entrance area into the analyzer is very large and its transmission is very high. Therefore, this particular analyzer is well suited for electron stimulated Auger spectroscopy, where collection efficiency and good transmission characteristics are more important than energy resolution.

In order to make this type of analyzer suitable for XPS, it is necessary to add retarding grids to the front end and to add another CMA to the back end, making this into what is called a *double-pass* CMA (Fig. 14). This design gives the energy resolution demanded by XPS spectroscopists, but with the retarding grid grounded, the analyzer is also well suited for AES.

4.2.3 General comments on electron energy analyzers. A few concluding remarks concerning CHA and CMA analyzers are in order. By far the most commonly used analyzer for XPS is the CHA. These analyzers have a number of advantages which are particularly well-suited to XPS, including the possibility of adding an input lens between the sample position and the entrance slit of the analyzer. This opens up much needed space around the sample position in which to put excitation sources and accessories. The addition of a lens also allows excellent energy resolution while maintaining high luminosity (i.e., good analyzer transmission coupled with a large acceptance area on the sample). In addition, the acceptance solid angle of the CHA can be very small, and it is therefore ideal for angle-resolved XPS (discussed in the next section).

On the other hand, the most common analyzer used for electron induced AES is the CMA. The CMA provides remarkable collection efficiency (because of the large acceptance cone) and has excellent transmission characteristics. All of this is achieved at moderate (and adequate) energy resolution. Therefore, good quality spectra can be collected in significantly shorter times which is often critical because of surface electron beam damage (see section 5.4.2). In addition, the CMA design allows for the possibility of mounting the electron gun coaxially (i.e., inside) the CMA as shown in Figure 14. On rough surfaces, this greatly reduces "shadowing" effects which are often a problem with side-mounted electron guns.

The evolution of electron energy analyzers continues today, both with the continued optimization of the standard designs described above, and with new designs. The most important directions for future analyzer development are in the areas of improved energy resolution and collection and counting efficiency. The goal, of course, is to allow for high energy resolution data collection in much shorter periods of time. The time factor is not just for convenience, but more importantly for the minimization of sample damage due most commonly to the impinging electrons in electron stimulated Auger spectroscopy.

4.3 Selected accessories

4.3.1 Fast entry locks. In the early days of surface analysis, sample entry was far from convenient. In some cases, the entire instrument was opened to the air to introduce a sample, and pumpdown times could be considerable (at least several hours for medium vacuum and 36-48 hours for UHV). Most instruments today are equipped with UHV compatible fast entry locks. These systems allow samples to be inserted into the instrument in minutes, and UHV conditions can be restored in equally short periods of time. Fast entry locks are generally small vessels attached to the main system through a gate valve that is capable of holding at least high vacuum conditions when open to air on one side. The other end of the fast entry lock has a simple door or hatch, capable of holding medium vacuums, through which the samples are loaded from air. After the samples are loaded and the outer door is closed, the entry lock is evacuated to low to medium vacuum range. The inner gate valve is then opened, allowing the samples to be inserted into the UHV side of the system. When the gate valve is again closed, UHV conditions are quickly restored if the samples are clean and UHV compatible.

4.3.2 X-ray monochromaters. The continuous spectrum from the x-ray tube is often monochromatized to provide a single x-ray wavelength for the collection of XPS spectra. The monochromatic source is produced by one or more single crystals which diffract only the K $\alpha_{1,2}$ line from the anode. Their are several advantages to using a monochromatic source for XPS. First, the photopeaks are considerably narrower allowing for less overlap of closely spaced peaks and better characterization of line shapes. The width of photopeaks result from a convolution of the source width, the analyzer resolution, and the inherent width of the core level. The widths of unmonochromatized Al and Mg K $\alpha_{1,2}$ lines are 0.85 and 0.70 eV, respectively. After diffraction from a monochromatizing crystal, the width for both lines is approximately 0.2 eV. Thus, for example, the full width at half maximum for the Ag $3d_{5/2}$ line is reduced from approximately 1.1 to 0.6 eV by switching from a non-monochromatic to a monochromatic Al x-ray source.[1] Additional advantages of using monochromatic x-rays include the elimination of x-ray satellite lines from the spectra and high energy Bremsstrahlung from the x-ray source which can damage delicate samples. Unfortunately, monochromatic sources have at least two disadvantages. First, x-ray diffraction is not an efficient process, and a good deal of the intensity from the Kα line is lost in the monochromatizing process. Second, charging problems on insulating samples are much more common with monochromatic

---

[1]These line widths are reported for illustration purposes only. Actual results will depend on instrument type, quality, set-up, and performance.

sources. Non-monochromatic sources can be placed very close to the sample, and the x-rays pass through an ultra-thin aluminum window which provides a soft electron source to help neutralize and stabilize surface charge. Monochromatic sources usually rely on a separate electron flood gun to help handle the charging problem on insulating surfaces, and these can be difficult and/or time-consuming to adjust to the optimum settings.

4.3.3 Ion guns. The ion gun is one of the most important and useful accessories on surface analytical instruments. They provide the ability to ion-sputter the sample for cleaning and depth chemical profiling (discussed in section 5.5.2). Ion guns (excluding liquid metal ion guns) work by ionizing a gas, and then accelerating and focusing this beam of ions onto the sample. Ionization can be achieved by simple electron bombardment ionization or by the combined action of magnetic and electrostatic fields on the gas in a confined space (cold-cathode discharge). Modern ion guns typically used with surface analytical equipment can generate ion beams with energies up to 10 kV and with ion current densities of up to a few hundred $\mu A/cm^2$. For most guns, the ion beam can be focused to much less than 1 mm in diameter (in some cases below 0.1 mm) and the beam can be rastered.

5. PRACTICAL ASPECTS OF XPS AND AES

This section covers the most important aspects of preparing for and actually performing XPS and AES measurements on mineral and glass surfaces. This includes sample preparation, both outside and within the UHV environment of the instrument, collecting spectra and identifying elemental lines, quantifying the chemical data, understanding the adverse affects of electron beams on surfaces, neutralizing surface charge, and performing chemical depth profiling. The problems associated with surface analysis of insulating materials are emphasized.

5.1 Sample preparation

Preparation techniques are quite variable depending on the material itself, the history of the sample, the information actually needed in the surface analysis, and the vacuum level that you intend to maintain in your instrument. Sample preparation can also continue once the sample is in the instrument by way of fracturing, heating, cooling, ion sputtering, etc. However, as discussed below, there are a number of steps that are generally followed for all samples for which surface analysis will be performed.

In general, the sample, its holder, and the material which secures the sample to the holder must be ultra-high vacuum compatible. Therefore, one should beware of organics, polymers, and extremely porous materials, although there are certainly exceptions. Also, all materials which are to be introduced to the vacuum should only be handled with clean gloves or clean demagnetized tools. Tools and sample holders (the latter usually made of stainless steel or aluminum) can be most effectively degreased by washing with a strong detergent, rinsing in distilled water, and finally boiling in acetone and then ethanol for several minutes. Cleaned articles should only be stored in degreased glass or stainless steel containers and never touched with hands.

The "cleanliness" of the sample is a far more complex factor to deal with compared to the accessory cleaning described above. For example, it is intuitively obvious that the standard degreasing procedure used for a sample *may* destroy the surface chemistry that one is attempting to study. Generally speaking, if a mineral or rock chip lacks obvious gross contamination (e.g., kerosene from sawing) and has not been handled with bare hands, it will be UHV compatible. Exceptions include rocks with very high porosity and permeability, or minerals with loosely bound water of hydration.

Attaching samples to sample holders is another important step which, because of UHV restrictions, should be carefully considered. Attachments which are mechanical (metal clips, screws, etc.) are generally the best because their cleanliness can be assured. If cements are used, special UHV-compatible silver epoxy or methanol-based "dag" are the best options. The dag, a colloidal suspension of graphite particles in methanol, is particularly useful because it can be painted onto a sample holder, and after a sample is set or a powder sprinkled onto the wet dag, it can be completely dried within 30 minutes if placed in a 100°C oven. The adhesion, although not nearly as strong as epoxy, is generally adequate. When the mount is no longer needed, the sample can be removed from the holder by hand and the dag dissolved with methanol. One other sample mounting media which can be particularly convenient is indium foil. Indium is a very malleable metal, and it can be used to attach odd-shaped samples or to mount powders.

As mentioned previously, important aspects of sample preparation can be performed after the sample is inserted into the instrument. These might include heating, ion sputtering, or fracturing. These processes are generally used separately or in combination to obtain atomically clean and ordered surfaces. For example, light ion sputtering can be used to clean contamination from a surface followed by heating to reorder or recrystallize the ion damaged surface. Surfaces exposed by fracture under UHV conditions are also particularly useful for basic studies in which a "virgin" surface is needed.

5.2 Identification of elemental lines

XPS and AES spectra are generally collected in one of two ways, commonly referred to as survey and narrow scans. *Survey scans* are collected over a wide electron energy range with the analyzer set for relatively low energy resolution but high transmission. Most, if not all, of the elemental identification can be accomplished from these scans. *Narrow scans* are taken from small energy regions with the analyzer set for high energy resolution so that line shapes can be studied and line areas and exact positions can be obtained.

The identification of elemental lines in XPS and AES spectra is generally straightforward using handbooks such as Wagner et al. (1979a) for XPS and Davis et al. (1976) for AES. These handbooks have convenient charts and tables for identifying not only elemental lines, but also for identifying satellites and certain energy loss lines. In addition, many modern systems now have computer-aided line identification software.

When identifying lines, it is also helpful to remember the following. In x-ray induced spectra, photopeaks are generally narrower than Auger peaks and often occur in pairs (spin-orbit splitting). Spin-orbit split lines have well defined intensity ratios which vary depending on the atomic sublevel (explained above in section 3.1.1). Therefore, the intensity ratio allows the sublevel to be identified which, when combined with position, can be used to more readily identify the element. Once a photopeak or photopeak pair has been tentatively identified, look for the remaining peaks for that element, keeping in mind the expected intensity ratios. In addition, consider all the potential sources of lines in XPS and AES spectra as described in sections 3.1 and 3.2 above.

Finally, when collecting spectra from insulating surfaces, the peaks will generally be charge shifted. In x-ray induced spectra, surface charging will cause peaks to occur at *lower* kinetic energy (higher binding energy) than expected. This is because the loss of electrons on an insulating surface causes the surface to charge positively. For electron induced spectra, insulating surfaces can conceivably charge positively or negatively (see section 5.4.1). Typically, however, surface charging will be negative and will cause peaks to occur at a *higher* kinetic energy than expected. In all cases (except inhomogeneous samples on which differential charging across a surface can develop), every peak will be drifted by the same amount. A correction can be made by identifying a line, measuring its drift compared to its expected position, and applying this correction to all lines. In XPS spectra, drifts of 4 to 8 eV are common on mineral and glass surfaces. For AES, drifts can be considerably higher.

Static charge referencing techniques are commonly used in conjunction with XPS to make a precise charge shift in order to obtain accurate binding energies for elements in insulators. A number of these techniques are described in Swift et al. (1983). However, only the two most commonly used techniques, the *adventitious carbon method* and the gold dot method, are briefly described here. The adventitious carbon method, reviewed by Swift (1982), depends on the presence of contaminant carbon on the surface of the sample, typically from air exposure (see section 3.1.8). The position of the adventitious C 1s line is generally considered to be 284.8 eV, and all peaks are simply adjusted according to this line. The advantage of this technique is that adventitious carbon is commonly present and easily measured on sample surfaces. No sample manipulation or modification is needed. The disadvantages are that the carbon line can shift, presumably depending on the interaction of the adventitious carbon with the particular chemistry of the surface to which it is attached. Adventitious C 1s binding energies have been reported from 284.6 to 285.2 eV. Therefore, peak positions derived via this method could conceivably be in error by as much as 0.6 eV. The *gold dot method* (see, e.g., Stephenson and Binkowski, 1976; Swift et al., 1983) requires that a small spot of gold be deposited on the surface of the sample. Usually this is done with a thin film evaporator mounted in the surface analytic instrument. The Au $4f_{7/2}$ binding energy is generally agreed to be 84.0 eV, and all lines are adjusted accordingly. This method has two principal disadvantages. First, a gold evaporator is needed. Second, the gold binding energies are know to shift depending on the thickness of the thin film laid down and the composition of the sample (Kohiki and Oki, 1985). However, although more testing will be required, this technique may provide the most accurate and reproducible binding energies for silicate surfaces as long as the thicknesses of the gold films can be carefully controlled.

One of the more recent ideas concerning static charge referencing involves the implantation of Ar into the near-surface of the sample using a conventional ion sputter gun (Kohiki et al., 1983a,b). The spectral energies are then referenced to the Ar $2p_{3/2}$ line (242.3 eV). Although this technique is theoretically sound and has been successfully tested, Hochella et al. (1988) have shown that even low levels of Ar implantation can result in significant structural and chemical modifications of mineral surfaces. Therefore, the Ar implantation method of charge referencing is not recommended for earth materials.

5.3 Chemical analysis

5.3.1 Chemical quantification in XPS. It is fairly straight forward to use photoelectron peak intensities to obtain semi-quantitative to quantitative analyses of chemically homogeneous near-surface regions of materials. XPS chemical analysis is simplified by the fact that surface modification during data collection is generally very slight or nonexistent, and one does not have to be concerned with backscattering phenomena as in AES quantification. What is given below is a general outline of what is involved in the quantification of XPS data with emphasis on the practical aspects of these measurements. One general assumption in this treatment is that the near-surface chemistry is homogeneous. The picture becomes considerably more complicated when the surface is covered with thin overlayers, including overlayers due to contamination. For this condition, and for a more lengthy treatment of XPS quantification, Briggs and Seah (1983) is recommended.

From first principles, the intensity, I, of a photoelectron peak from an element in a homogeneous sample depends on a number of factors as follows:

$$I \propto FAn\sigma\theta\lambda yT \ , \tag{15}$$

where F is the x-ray flux of the wavelength which generated the peak of interest, A is the area on the sample surface from which photoelectrons are being collected, n is the number of atoms of the measured element in the analytic volume, σ is the photoionization cross-section for the sampled orbital for the excitation energy used, θ is the angle of tilt of the sample with respect to the analyzer entrance direction, λ is the attenuation length of the photoelectrons in the sample, y is the probability for the photoelectron to undergo an energy loss event besides electron-electron collision (e.g., a plasmon loss), and T is the transmission function (i.e. the detection efficiency) of the analyzer at the energy which is being measured. All factors in Eqn. 15 except I and n are sometimes conveniently grouped together and referred to as S, the *atomic sensitivity factor*. Therefore,

$$n \propto I/S \qquad (16)$$

and

$$c_x = k \frac{n_x}{\sum_i n_i} \times 100 = k \frac{I_x/S_x}{\sum_i I_i/S_i} \times 100 \quad , \qquad (17)$$

where c_x is the concentration of element x in *atomic percent*, and k is a proportionality constant.

There are various ways in which Eqn. 15 can be considerably simplified. For example, what is commonly needed in surface analysis is the ratio of two elements in the near-surface region of a sample. Because one is generally working with a ratio of intensities collected at the same time from the same sample, F, A, and θ all cancel. Further, if the two photopeaks of interest are reasonably close in kinetic energy (usually within a few hundred eV), y, λ, and T will be approximately constant and also cancel. Therefore, $S \approx \sigma$ and

$$\frac{n_a}{n_b} = \frac{I_a}{I_b} \cdot \frac{\sigma_b}{\sigma_a} \quad . \qquad (18)$$

Photoionization cross-sections can easily be obtained from calculated tabulations such as those found in Scofield (1976) or Yeh and Lindau (1985). This method of quantification is very simple and generally results in semi-quantitative analyses.

Another way in which chemical quantification using XPS can be simplied is by empirically deriving a set of sensitivity factors using ones own instrument and standards. The standards must be homogeneous with well-known bulk chemistry. Ideally, they are then broken in the XPS instrument under vacuum and analyzed. Alternatively, they can be broken outside the instrument in the presence of an inert gas or nitrogen and transferred into the instrument without air exposure. The atomic sensitivity factor is then obtained from Eqn. 16 using the measured I and the known n. The unknown sample must also have a clean and homogeneous near-surface region, and the instrument conditions and sample geometry must be identical to those used for the standard measurement. Obviously, it is best to use relatively flat surfaces. The only two factors which contribute to S, and which do not factor out in this case, are y and λ. However, if the sample and unknown are similar materials (as, for example, if they were both silicates), y and λ can be assumed to be constant. Quantification using this approach is relatively easy and results in good semi-quantitative to quantitative analyses. Unfortunately, a 'permanent' set of standards, as is typically used with the electron microprobe, cannot be used for surface sensitive spectroscopies. Over time, surfaces become contaminated even under UHV conditions. Atomically cleaning the sample by ion sputtering could easily result in a compositional change in the standard near-surface (section 5.5.2).

5.3.2 Chemical quantification in electron stimulated AES. Chemical quantification via electron stimulated AES (see, e.g., Holloway, 1978; Powell, 1980; Prutton, 1982; Seah, 1983) is generally more difficult than with XPS for a number of reasons. First, the signal to noise ratio in electron stimulated AES is generally poor (see section 3.2), and therefore measured Auger peak intensities are not as precise (reproducible) as measured photopeak intensities. Second, because the excitation source is an electron beam, one must worry about near-surface damage in the form of electron stimulated desorption and absorption, beam heating, and field induced elemental migration (see section 5.4.2), as well as anisotropic Auger emission from different crystallographic orientations (e.g., Armitage et al, 1980; LeGressus et al., 1983) and rough surfaces (deBernardez et al.,

1984; Wehbi and Roques-Carmes, 1985; Hochella et al., 1988). Third, the fundamental principals of chemical quantification with AES is more complicated than with XPS because of the additional factors of backscattering and fluorescense (see Ichimura et al., 1983, and below). Despite these problems, it is still possible to obtain good semi-quantitative analyses from mineral surfaces.

The intensity of an Auger signal, I, depends on the following factors:

$$I \propto I_p I_b n \sigma_e \theta \lambda y T (1 - \omega) \quad , \tag{19}$$

where n, θ, λ, y, and T are as defined for Eqn. 15, I_p is the electron flux from the primary beam, I_b is the electron flux due to backscattered primary electrons, σ_e is the ionization cross-section for the initial core vacancy needed to initiate the Auger decay process of interest, and ω is the fluorescence yield generated from this initial core vacancy.

Because of the complex interactions of high energy electrons in solids and the competition of x-ray fluorescence and Auger electron emission, practical surface chemistry quantification using AES is most readily achieved with the use of standards from which sensitivity factors can be derived. Sensitivity factors derived from pure elements or simple compounds, like those available in Davis et al. (1976), can be used for qualitative analysis, but sensitivity factors derived from materials similar to the unknown are best for semi-quantitative to quantitative analyses. Sensitivity factors are derived using the same procedures used for deriving XPS sensitivity factors (see previous section).

As an example of sensitivity factor use in Auger spectroscopy, repeated analyses of quartz surfaces at 3 keV give sensitivity factors for O and Si of 0.50 and 0.36, respectively, using the O $KL_{2,3}L_{2,3}$ and Si $L_3M_{2,3}M_{2,3}$ lines (Hochella et al., 1986b). When these same sensitivity factors are used to calculate the O/Si ratio from O and Si Auger intensities collected from orthoclase, albite, wollastonite, diopside, forsterite, titanite, and kyanite standards, the results deviate from that expected as shown in Figure 15. The deviation becomes progressively greater as the actual O/Si ratio increases. Hochella et al. (1986b) suggest that at least part of this is due to changes in the ionization cross section and the probability of the Auger decay process for these two Auger transitions. Whatever the cause, the fact that this deviation occurs at all is disturbing. This suggests that Auger sensitivity factors needed for quantitative analysis of silicates can only be derived from standards which are quite similar to the unknown unless fundamental parameters can be derived which can account for these deviations. Obviously, much more work will be required before routine quantitative analysis of silicates using AES is possible.

5.4 Avoiding electron induced surface charging and damage

5.4.1 Electron induced charging. The obvious problem with using electron stimulated AES on minerals is charging of the surface. This can be a very serious problem and prohibit data collection, although often it can be avoided. It has been know for some time that, during electron bombardment, insulating surfaces can remain neutral or even charge positively under certain conditions (Bruining, 1954). Charge neutralization will occur when the number of incident electrons equals the number leaving the surface. Therefore, at neutral conditions, the following equation will hold:

$$I_p = I_\gamma + I_\delta + I_A \quad , \tag{20}$$

where I_p is the primary electron beam current, I_γ is the electron current produced by primary backscattered electrons (both elastic and inelastic scattering) along with electrons ejected from the sample with energies over 50 eV and not including characteristic Auger electrons, I_δ is the current produced by the "secondary" electrons (defined in section 3.2), and I_A is the total Auger electron current. As can be seen from Figure 11, I_A is small compared to I_γ and I_δ, and for the purposes of these arguments, it can be ignored.

606

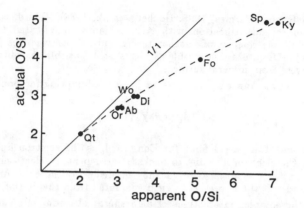

Figure 15. Apparent oxygen to silicon ratios, calculated using the sensitivity factors derived from the O $KL_{2,3}L_{2,3}$ and the Si $L_3M_{2,3}M_{2,3}$ lines from quartz, for eight silicate mineral standards and plotted against the actual oxygen to silicon ratios. The 1/1 line (solid) is drawn to show the deviation from ideality. The standards, obtained from the C.M. Taylor Corporation, are quartz (Qt), orthoclase (Or), albite (Ab), wollastonite (Wo), diopside (Di), forsterite (Fo), titanite (Sp, for sphene), and kyanite (Ky). From Hochella et al. (1986b).

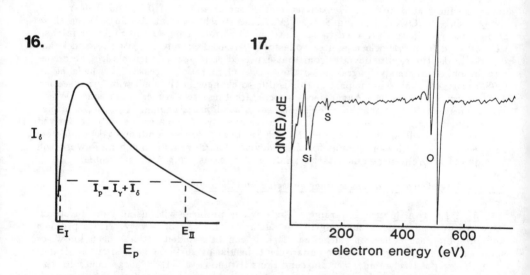

Figure 16. Secondary electron emission (I_δ) as a function of primary electron beam energy (E_p) for an insulator. E_I and E_{II} are the crossover voltages where the primary beam current (I_p) is equal to the backscattered electron current (I_γ) plus the secondary electron current (I_δ); no charging will occur here. From Hochella et al. (1986a).

Figure 17. Differentiated Auger electron spectrum of quartz stimulated with a rastered 3 kV electron beam. The effective beam current density over the area of analysis was approximately 2 amps/cm², and the sample was exposed to the beam, including before and during data collection, for approximately 5 minutes. The split negative excursion of the Si Auger line is due to the chemical shift of Si in SiO_2 (lower energy line) and metallic Si (higher energy line). Metallic Si is generated in the near surface of the quartz due to electron stimulated desorption of O_2. The sulfur signal is due to a submonolayer quantity of SO_2 surface contamination.

Neutralization will therefore depend on the characteristics and dependencies of I_γ and I_δ. I_γ is known to be independent of the primary beam energy, E_p (e.g. Goldstein et al., 1981). However, it has been experimentally demonstrated that I_δ is dependent on E_p (e.g., Dawson, 1966) as shown in Figure 16. Therefore, by adjusting E_p, there should be conditions for which Eqn. 20 is satisfied. The beam energy for which insulating surfaces remain neutral is commonly called a "crossover voltage". Unfortunately, the crossover voltages may be at beam energies which are too low to efficiently excite the Auger transitions of interest. In this case, tilting the sample with respect to the primary electron beam effectively allows the crossovers to occur at higher voltages. This is because tilting the sample allows the primary electron interactions to occur closer to the surface, thus increasing the backscattered and secondary emission efficiencies (Kanter, 1961; Newbury et al., 1973).

Other factors which indirectly influence surface charging include surface contamination and roughness. We have found that it is generally more difficult to suppress charging on rougher surfaces. This observation can be rationalized from the discussion above. As a surface becomes more irregular, differential charging can develop as the beam intersects the surface at many angles, most of which are not appropriate for the beam conditions used. The role of surface contamination and its effects on charging are not as clear. However, we have seen in a number of cases in which charging is easier to control on mineral surfaces that have slight contamination due to air exposure compared to the same surface without any contamination.

5.4.2 Electron induced damage. Another problem with electron stimulated AES of mineral surfaces is the chemical changes which can occur due to electron bombardment. As the electrons in the primary beam hit the sample, the most likely stopping mechanism is multiple inelastic collisions with valence electrons. This process has a number of consequences, including electron stimulated desorption and adsorption (ESD and ESA) (e.g. Carriere and Lang, 1977; Pantano and Madey, 1981) and electric field or thermally induced ion migration (e.g. Gossink et al., 1980; Yau et al., 1981; Storp, 1985), all of which can change the chemistry of the near surface. For example, the change in the surface chemistry of quartz and calcite due to ESD effects are shown in Figures 17 and 18. In these cases, the electronic configuration of the quartz and calcite is sufficiently disrupted by the bombarding electrons such that O_2 and CO_2 are ejected into the vacuum. In addition, electron bombardment causes the surface temperature to increase, and it has been shown that the heat at the surface is proportional to the beam power and inversely proportional to the thermal conductivity of the sample. Temperatures as high at 1100°C have been measured on quartz surfaces due to electron beam bombardment.

Surface damage due to the electron beam can be avoided (or at least minimized) by following the suggestions of Levenson (1982). To prevent possible electron stimulated adsorption of residual gases in the vacuum onto the sample surface, the vacuum should be maintained at the lowest possible level, and at least in the 10^{-10} torr range. The sample should be in good thermal contact with the sample holder for heat dissipation. The beam energy and the beam current density should be kept as low as possible while still giving acceptable signal-to-noise ratios in a reasonable time. In order to effectively reduce the beam current density without changing the beam conditions, the beam can be rastered over an area, or defocused, or both. In all cases, reproducibility studies should be made on test samples. Change in Auger spectra with time is a sure sign of electron stimulated damage, and if this is the case, the beam conditions, area of raster, and/or the time of data collection per area should be modified to avoid these problems.

5.5 Depth profiling

Up to this point, we have been concerned only with the average chemistry within the depth of analysis of XPS and AES. Depth profiling techniques allow us to look at the distribution of elements as a function of distance from the surface, both within the range of the depth of analysis (via angle resolved analysis) and beyond this range (via ion sputtering). These techniques have added a third dimension to surface analysis, and

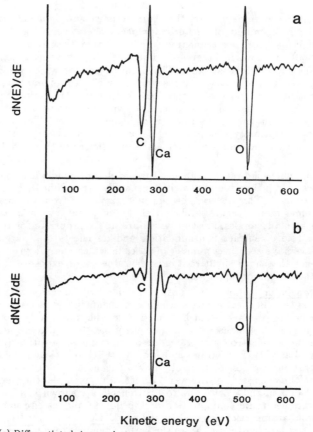

Figure 18. (a) Differentiated Auger electron spectrum of calcite stimulated with a rastered 3 kV electron beam. The effective beam current density over the area of analysis was less than 0.1 amps/cm^2, and the sample was exposed to the beam for less than 3 minutes. (b) Differentiated Auger electron spectrum of calcite using the same beam conditions and time as in (a), but with the beam fixed (unrastered); this results in a beam current density of over 10 amps/cm^2. The C line is nearly eliminated and the O line is significantly reduced. Independent experiments and calculations have shown that this desorption is not due to electron beam heating, and it therefore seems likely to be some sort of electron stimulated desorption, although the exact mechanism is not as yet clear.

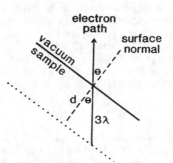

Figure 19. Schematic diagram showing the origin of Eqn. 21 relating the effective depth of analysis (measured perpendicular to the surface) to the angle of electron ejection (which is towards the entrance slit of the electron energy analyzer).

have many important applications in surface science. The use of these techniques in surface geochemistry is just starting to be explored.

5.5.1 Angle resolved surface analysis. As explained previously (section 2.4.3), the depth of analysis of XPS and AES depends on the distance that electrons can travel through materials without undergoing energy loss. This distance does not depend on the origin of the electron, but only on its energy and the material through which it is traveling. Therefore, if one has an instrument which can collect photo- or Auger electrons emerging from the sample surface as a function of angle, one can effectively dictate the depth of analysis of the measurement. If we consider the information depth (3 times the escape depth) as the effective depth of analysis (see section 2.4.3), it is obvious that

$$d = 3\lambda\cos\theta \quad , \tag{21}$$

where d is the depth of analysis, λ is the attenuation depth normal to the surface (escape depth), and θ is the angle between the sample normal and the direction of signal collection. This situation is shown in Figure 19. Therefore, when θ equals 0, the depth of analysis is at its maximum, and it decreases as θ increases. It is obvious that, for Eqn. 21 to be valid, the area of analysis should be flat or nearly so. It is also easy to see that the sampling depth resolution will improve as the angular collection solid angle decreases (good angle resolving instruments have collection solid angles below approximately 6°). Obviously, CHA's are much better suited for this type of work than CMA's. However, some CMA's have internal apertures which can limit the angle of collection and give relatively crude (but often useful) angle resolved data.

Although a thorough discussion will not be given here, it is interesting to note a special case in angle resolved analysis which can be used to measure the thickness of a very thin overlayer, such as a thin precipitate, on a substrate, such as a mineral surface. In this case, it can be shown (Baer et al., 1970) that

$$x = -\lambda_f \cos\theta \left[\ln(1 - \frac{I_{f,x}}{I_{f,\infty}})\right] \quad , \tag{22}$$

where x is the thickness of the overlayer, λ_f is the attenuation length of the electrons that are being measured, $I_{f,x}$ is the intensity of the signal emanating from the element of interest in the film of thickness x, and $I_{f,\infty}$ is the signal emanating from the same element in a layer of infinite thickness compared to the depth of analysis. Therefore, by measuring $I_{f,x}$ and $I_{f,\infty}$, and by knowing λ_f and θ, the thickness of the overlayer can be determined. The method assumes that the overlayer is continuous and uniform, and that the element whose photoelectrons are measured is either not present in the underlying substrate or that there is a chemical shift such that the intensity from that element in the substrate and overlayer can be separated.

Although angle resolved analysis can be quantitative as described above and as shown in, for example, Pijolat and Hollinger (1981) and Bussing and Holloway (1985), it is very commonly used in a semi-quantitative to qualitative manor. For example, survey spectra of a sample are collected at θ values of 0 and 70°. If the ratio of all elements within the two spectra are the same, then the sample is considered chemically homogeneous (i.e. not layered) within the information depth. Otherwise, variations in the chemistry with depth within the information depth are generally apparent.

5.5.2 Ion sputtering. If chemistry as a function of depth is needed beyond the information depth of electron spectroscopy, there are a number of options from which to choose, both destructive and non-destructive. The most popular non-destructive depth profiling techniques include Rutherford backscattering (RBS) and nuclear reaction analysis (NRA), while the most used destructive techniques are either XPS or AES coupled with ion sputtering or secondary ion mass spectroscopy (SIMS). Because this chapter is limited to electron spectroscopies, a discussion of only XPS and AES coupled with ion sputtering is presented below. However, it should be noted that RBS and NRA generally have lower depth resolutions and are not suited for the analysis of as many elements as XPS or AES coupled with ion sputtering or SIMS. On the other hand as discussed below, the destructive nature of ion sputtering creates its own problems, and RBS and NRA analysis may be the method of choice when these problems cannot be tolerated.

Ion sputtering is often described as playing billiards with atoms, and this simple analogy goes a long way in helping to understand a number of observed characteristics of sputtering (see Carter and Colligon, 1968, for an extensive review and analysis of the ion bombardment of solids). Ion sputtering occurs when near-surface atoms or atomic fragments are ejected from a material into a vacuum via energetic bombardment of ions or atoms. Ion sputtering is most commonly performed with charged ions accelerated in the electrostatic field of an ion gun (see section 4.3.3). Assuming that the energy of the bombarding ions is above the threshold to cause physical sputtering, a number of possibilities can occur when these ions collide with a solid surface. Impinging ions can be scattered back into the vacuum after one or more collisions, or they can implant themselves into the target material. In either case, collisions with target atoms can cause these atoms, perhaps along with neighboring atoms, to be ejected into the vacuum either as neutral or charged species. However, this ejection process is inefficient, and on average, most of the energy of the bombarding ion is converted to heat in the target. The rate of sputtering depends on the following: (1) the energy of the bombarding ions; (2) the atomic number of the bombarding element; (3) the ion current density of the bombarding ions; (4) the angle of impact of the bombarding ions; (5) the target composition; and (6) the crystallographic orientation of the target with respect to the ion beam. Therefore, sputtering rates are difficult to know unless a system is specifically calibrated. However, one aspect of sputtering that simplifies things is that under a given set of sputtering conditions (ion energy, current density, and impact angle), sputtering yields for most materials rarely differ by more than a factor of 10, and they are often within a factor of 2 or 3. Even within this limited range, there is no clear trend between sputtering rate and average atomic number of the target. To give a very rough idea of what typical sputtering rates can be, modern ion guns operating between 1 and 4 keV accelerating potential over an area of up to a few square millimeters will sputter away amorphous SiO_2 at a rate in the range of 10 to 100 angstroms per minute.

Unfortunately, chemical depth profiling, that is alternating XPS or AES analyses with periods of ion sputtering, is complicated by a number of factors. First, because XPS and AES will be analyzing to a depth of several angstroms to several 10's of angstroms, a measured chemical profile through a chemical boundary will not have the shape of the true profile unless the composition changes very slowly with depth (with respect to the depth of analysis). The shape of the true profile can be recovered from the measured profile by deconvoluting the analytic broadening function from the data (see, e.g., Hofmann, 1980, 1983). Related to this is the fact that physical sputtering creates surface roughening, also reducing the depth resolution of the technique. Second, the incoming ions often change the crystal structure of the near-surface region. Most surfaces which have ordered arrangements of atoms become disordered under ion bombardment, and crystalline materials most often become amorphous to the depth of ion penetration. Third, the chemistry of the near-surface can be modified, both due to the addition of small amounts of the sputtering ion by implantation and because of differential sputtering. *Differential sputtering* is the phenomena in which certain constituents, or elements, of the target material will sputter faster than others, thus changing the chemistry in the near-surface region. Thus, unless these potential problems are taken into account, misinterpretation of sputter depth profiles can easily be made.

A number of studies have explored sputter damage problems in many classes of materials (Naguib and Kelly, 1975; Storp and Holm, 1977; Hofmann and Thomas, 1983; Storp, 1985, and references therein) including minerals (Yin et al., 1972, 1976; Braun et al., 1977; Tsang et al., 1979; Coyle et al., 1981; Fischer et al., 1983; Ajioka and Ushio, 1986). Most recently, Hochella et al. (1988) showed that moderate doses of argon ions (10^{17} ions/cm^2) at only 1 keV accelerating potential can result in significant structural and chemical changes in silicate, sulfate, and carbonate mineral surfaces within 1 minute of exposure. For example, under these conditions, the sodium in albite sputters preferentially relative to all other major constituents including oxygen, and silicon sputters slightly more rapidly than aluminum. Carbon sputters most rapidly for calcite, followed by oxygen and then calcium. With continued sputtering, the near-surface chemistry quickly reaches a steady state as the elements which sputter more rapidly become depleted from the upper monolayers. Therefore, the sputtering rate of these elements is reduced, matching that of the elements which sputter more slowly but which are now relatively concentrated at the surface. It is changes in chemistry from this steady state which indicate a *real* change in chemistry with depth.

Due to the many problems with ion sputtering as described above, monitoring chemistry as a function of depth with this technique cannot realistically be done when only subtle changes are present. For example, it would be difficult or impossible to document an oxidation state change with depth using this technique. However, when major chemical changes with depth are present, depth profiling with ion sputtering coupled with either XPS or AES has been very successful in materials science. Applications to earth materials are very likely in the near future.

6. APPLICATIONS

In this section, we will look at various aspects of XPS and AES work on geological materials, concentrating on examples that show the capabilities of the techniques. These contributions can be grouped into four areas: (1) the oxidation state of near-surface atoms; (2) the adsorption of species on mineral surfaces; (3) the leaching and weathering of mineral surfaces; (4) the atomic structure of minerals and glasses.

6.1 Studies of the oxidation state of near-surface atoms

In studies involving adsorption, chemical exchange, and corrosion (weathering) of surfaces, it is often particularly important to determine the oxidation state of near-surface atoms. It has been well-known since the early days of XPS that photopeaks from the same element in different oxidation states can have different binding energies, energy loss features, and spin-orbit splittings. In addition, a quick look through an XPS data reference book, such as Wagner et al. (1979a), shows that the various common oxidation states of most elements should be discernible with XPS. On the other hand, AES is rarely used for oxidation state studies for reasons which have already been given (section 3.2.1).

Excellent examples of determining oxidations states within near-surface regions of minerals can be found in Dillard et al. (1981) for Pb, Murray et al. (1985) for Mn, and Myhra et al. (1988) for Ti. Many other examples can be found in the references listed in section 6.2. To show the utility of oxidation state evaluation using XPS in some detail, three examples involving S, Fe, and Co on mineral surfaces are given below.

In a study of gold deposition on sulphide surfaces at low temperatures, Jean and Bancroft (1985) collected high resolution XPS spectra in the S 2p region for a number of sulfides before and after surface oxidation treatments. Their results for pyrrhotite ($Fe_{1-x}S$) are shown in Figure 20. The oxidized pyrrhotite gives a S 2p signal at roughly 168 eV representing S^{6+} (spectrum A), whereas the cut and polished phyrrhotite surface gives two S peaks, one at approximately 168 eV and the other at approximately 162 eV representing S^{2-} (spectrum B). It is know that when a fresh sulfide surface is exposed by fracturing in a vacuum, only the 162 eV line is visible. Therefore, the chemical shift covering sulfur oxidation states is roughly 6 eV. Even though the S 2p photopeaks are relatively wide (the $2p_{3/2,1/2}$ spin orbit split lines are not easily resolved), the large chemical shift makes the oxidation states easily identifiable.

The use of XPS to distinguish between ferrous and ferric iron has been extensively studied (Brundle et al., 1977; McIntyre and Zetaruk, 1977, and references therein). Perhaps the most comprehensive study is by McIntyre and Zetaruk (1977). They have

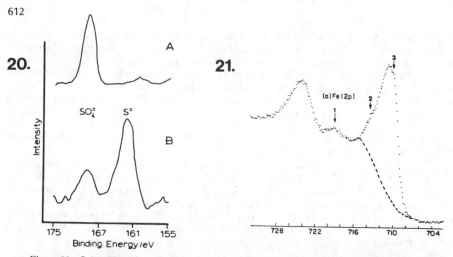

20.

21.

Figure 20. S 2p x-ray photoelectron spectra of (A) an oxidized pyrrhotite surface, prepared at 500°C for 5 hours in moist air, and (B) a cut and polished pyrrhotite plate without heat treatment. The chemical shift between S^{6+} and S^{2-} is approximately 6 eV. Notice that the binding energy is increasing to the left. From Jean and Bancroft (1985).

Figure 21. Fe 2p region of the X-ray photoelectron spectrum of hematite showing the background (dashed curve) according to the widely used background estimating algorithm of Barrie and Street (1975). Notice that the binding energy is increasing to the left. From McIntyre and Zetaruk (1977).

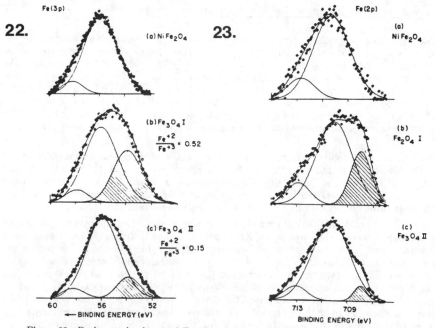

22.

23.

Figure 22. Background subtracted Fe 3p x-ray photoelectron spectra for (a) $NiFe_2O_4$, (b) pure Fe_3O_4, and (c) partially oxidized Fe_3O_4. The cross-hatched areas are assigned to the contribution from ferrous iron, and the two additional curves are assigned to ferric iron. See text for further details. From McIntyre and Zetaruk (1977).

Figure 23. Background subtracted Fe 2p x-ray photoelectron spectra for (a) $NiFe_2O_4$, (b) pure Fe_3O_4, and (c) partially oxidized Fe_3O_4. The fit curves are assigned as in Figure 22 for ferrous and ferric iron. From McIntyre and Zetaruk (1977).

shown that the Fe 2p and 3p photopeaks can be used to determine Fe oxidation states in near-surface regions with reasonable accuracy, even for mixed valence phases such as Fe_3O_4. For the best analysis of these lines, it is first necessary to subtract the background using an algorithm like that first suggested by Barrie and Street (1975). An example of this estimated background is depicted in Figure 21 for the Fe $2p_{3/2}$ line of Fe_2O_3. Each point of the background is proportional to the integrated area under the peak from that point to the low binding energy side of the peak. Such background subtraction routines can be found in most XPS data reduction packages. Once the background is subtracted, detailed line shape analysis can be done.

For phases containing Fe as a major constituent, the Fe 3p line is probably the best for determining ferric/ferrous ratios, despite the fact that the spin-orbit splitting generally cannot be resolved. Figure 22 shows the background subtracted Fe $3p_{3/2}$ lines for $NiFe_2O_4$ (a), Fe_3O_4 (b), and partially oxidized Fe_3O_4 (c). Two Gaussian-Lorenzian lines best fit the Fe $3p_{3/2}$ photopeak from $NiFe_2O_4$, this peak serving as a standard for the ferric component for the same line from Fe_3O_4 because the two compounds are isostructural. The Fe $3p_{3/2}$ line for unoxidized Fe_3O_4 can then be partially fit by including the ferric lines fit to the peak for $NiFe_2O_4$ and only allowing their intensity to vary. The remainder of the peak is due to ferrous iron, and the ferrous to ferric peak area ratio is very close to 0.50 as expected. The partially oxidized Fe_3O_4 shows a reduced ferrous component, also as expected. Therefore, even for this case with mixed valence states, the Fe $3p_{3/2}$ line can be curve fit to give reasonable results. In addition, as shown in this example, it is generally found that the Fe $2p_{3/2}$ lines for ferrous compounds have binding energies approximately 1.5 to 3 eV lower than for ferric compounds.

Although the spin-orbit split Fe 2p photopeaks (the most intense XPS lines for Fe) are difficult to work with because of a sharply rising background and a large line width due to multiplet splitting, McIntyre and Zetaruk (1977) have shown that Fe oxidation states can also be estimated using the Fe $2p_{3/2}$ line. The use of this line may lead to an oxidation state evaluation which is less accurate than when the Fe 3p line is used, but it is nevertheless important for cases of low Fe concentrations because it is approximately 10 times more intense than the Fe 3p line. Although McIntyre and Zetaruk (1977) have shown that the Fe $2p_{3/2}$ peak shapes from purely ferrous and ferric compounds can be rationalized by fitting them with 3 or 4 narrow lines according to the prediction of multiplet splitting for iron by Gupta and Sen (1974), it is easier to fit this line as was done for the Fe 3p line in Figure 22. The results for the fitting of the Fe $2p_{3/2}$ for $NiFe_2O_4$, Fe_3O_4, and partially oxidized Fe_3O_4 are shown in Figure 23. The estimated ferrous/ferric ratio for Fe_3O_4 in this case is 0.32, significantly below the expected value of 0.50; this error may be due to an additional ferrous component hidden under the ferric component. However, the change in the ferrous/ferric ratio between Fe_3O_4 and partially oxidized Fe_3O_4 as measured using this line is very similar to the change measured using the Fe 3p line (Fig. 22). Apparently, the Fe $2p_{3/2}$ line can be used to accurately measure the relative change in oxidation states between samples, but can only give a qualitative estimate of the actual ferrous/ferric ratio.

An excellent example of the use of XPS to determine the oxidation state of iron in more complex minerals can be found in Stucki et al. (1976), who measured the ferrous/ferric ratio in nontronite (a dioctahedral smectite clay with ferric iron as the dominant octahedral cation) and nontronites which had been reduced by hydrazine and dithionite. XPS spectra of the Fe $2p_{3/2}$ photopeaks of the three materials are shown in Figure 24. The XPS spectrum for unreduced nontronite only shows evidence for ferric iron (based simply on a relatively high binding energy of the Fe $2p_{3/2}$ line), and as the clay is reduced, the Fe $2p_{3/2}$ line develops a distinct shoulder 2 to 3 eV lower in binding energy, approximately the same separation seen by McIntyre and Zetaruk (1977) for ferrous and ferric iron in simple oxides. In this case, Stucki et al. (1976) showed that the areas under the ferrous and ferric components of the Fe $2p_{3/2}$ photopeaks agree to within approximately 10% of the ferrous/ferric ratio determined colorimetrically from the bulk samples.

More recently, a number of workers have used XPS to determine the oxidation state of iron in the near-surface region of minerals which have been in contact with

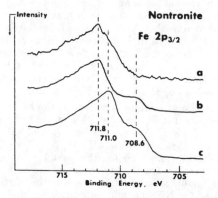

Figure 24. Fe 2p x-ray photoelectron spectra for (a) unaltered nontronite, (b) hydrazine-reduced nontronite, and (c) dithionite-reduced nontronite. As the ferric iron in nontronite is progressively reduced, the spectral contribution from ferrous iron (in these spectra at 708.6 eV) increases as shown. From Stucki et al. (1976).

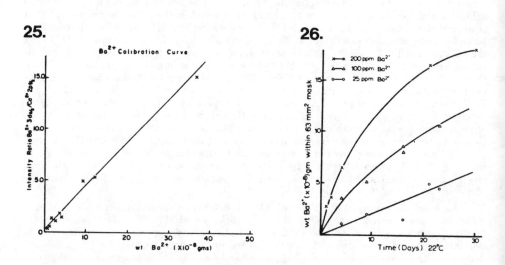

Figure 25. Calibration for XPS Ba analyses on calcite surfaces plotted as the known amount of Ba on the surface (in terms of weight in the area of analysis) vs. the Ba $3d_{5/2}$/Ca $2p_{1/2}$ peak area ratio. See text for details. From Bancroft et al. (1977).

Figure 26. Sorption plot for Ba (in terms of weight per area) on calcite surfaces as a function of time and initial Ba concentration in solution. Amounts of Ba sorbed on calcite surfaces are derived from the calibration plot shown in Figure 25 after measurement of the Ba $3d_{5/2}$ and Ca $2p_{1/2}$ intensities on each surface. See text for further details. From Bancroft et al. (1977).

aqueous solutions in natural and laboratory environments. Berner and Schott (1982) show that pyroxenes and amphiboles taken from natural weathering environments show only ferric iron in the near-surface region. Schott and Berner (1983) further suggest that bronzite and fayalite surfaces, in the presence of dissolved O_2, become hydrous ferric oxides on the outermost layers, with a ferric protonated or hydroxylated silicate just below. At low pH or under anoxic conditions, Fe is leached from the near-surface region, but the oxidation state of the remaining iron remains ferrous. Finally, White et al. (1985) and White and Yee (1985) suggest that the oxidation of near-surface iron in augite, biotite, and hornblende may be related to the reduction of iron in solution. The latter case is a good example of how surface analytic and chemical information can help understand the processes that are occurring in solutions in contact with the solid surface.

So far in this analysis of oxidation states using XPS, we have concentrated primarily on peak position, with higher oxidation states corresponding to higher binding energies. Occasionally, if the peak shift with oxidation state is not great enough, or if the absolute binding energies are hard to measure (as can be the case for insulators), other information in the XPS spectra can be used, with or without absolute peak positions, to determine oxidation state. For example, Murray and Dillard (1979) have shown that during the adsorption of cobalt onto MnO_2 surfaces, cobalt(II) in solution is oxidized to cobalt(III). To determine this oxidation state change, it is possible to use the peak position of the Co $2p_{3/2}$ line as well as the presence or absence of satellite lines and the separation of the Co $2p_{3/2}$ and Co $2p_{1/2}$ photopeaks. For example, the Co 2p photopeaks for Co(II) have associated energy loss lines 5 to 6 eV away (see Fig. 8). For Co(III) oxides, these features are greatly reduced or altogether missing. In addition, the Co $2p_{3/2}$ - Co $2p_{1/2}$ separation is 16.0 ±0.2 eV for Co(II) and 15.0 ±0.2 eV for Co(III) (see McIntyre and Tewari, 1977, and references therein). Thus in this case, it is possible to combine three factors in order to make the best determination of the oxidation state of Co.

6.2 Studies of sorption reactions on mineral surfaces

One of the most important factors to consider at rock-water interfaces is that of sorption reactions, including adsorption, ion exchange, and surface-limited precipitation. For example, a number of recent studies (e.g., see Hayes et al., 1987; Martin and Smart, 1987; and references therein) have shown that one of the critical processes which dictates the migration of components in natural groundwaters is that of sorption on mineral surfaces in both rocks and soils. When specifically considering the migration of radionuclides or other toxins in groundwater, the practical importance of such studies becomes clear. In addition, phenomena such as ore deposit formation may also be affected by such processes.

XPS and AES are ideally suited to aid in the study of sorption reactions for a number of reasons. First and foremost, of course, is that they are inherently surface sensitive. However, XPS is particularly versatile because it can detect every element of geochemical interest except H, it can be used to estimate surface coverage of sorbed species or thickness of thin precipitate films, and it can provide important information on the chemical state of the substrate surface before reaction, and both the substrate and sorbed species after reaction. Electron stimulated AES has until now been underused in this area of research. However, it is becoming apparent that sorption reactions do not occur homogeneously across mineral surfaces, and the ability of AES to analyze surfaces with considerably better spatial resolution than XPS will promote much more work with this technique. The major drawback of XPS and AES in sorption studies is the fact that measurements must be made under vacuum. Therefore, if the sorbed species is sensitive to vacuum, as might be the case for a hydrated surface complex, other techniques with the ability to study surface complexes *in situ* (i.e., at the water-solid interface), such as x-ray absorption spectroscopy, must be considered.

A large number of XPS-based studies concerning the sorption of metal complexes in solution onto mineral surfaces have been done. Such studies include Cr, Ni, and Cu on chlorite, illite, and kaolinite (Koppelman and Dillard, 1977; Koppelman et al., 1980), Co on illite (Dillard et al., 1981) and goethite (Schenck et al., 1983), Pb on montmorillonite

(Counts et al., 1973), and Ti, Co, Ni, Cu, and Pb on manganese and ferromanganese oxides (Murray and Dillard, 1979; Dillard et al., 1981, 1982, 1984). In addition to these, several other studies are described in more detail below.

An excellent early example of an XPS-based study concerning metal sorption on mineral surfaces is that of Bancroft et al. (1977); this study examines the sorption of Ba^{2+} on calcite as a function of Ba concentration in solution and time of solution/crystal exposure. In order to physically calibrate the Ba XPS signal from the calcite surface, known quantities of Ba in solution were evaporated onto single calcite cleavage surfaces and then the Ba $3d_{5/2}$ XPS signal measured. Because the Ba coverage in this work was the equivalent of less than a monolayer in most cases, the Ba $3d_{5/2}$ intensity was normalized by dividing by the Ca $2p_{1/2}$ signal. A calibration curve of the Ba $3d_{5/2}$ signal as a function of the amount of Ba actually present on the surface is shown in Figure 25. The actual sorption experiments were done by placing single crystals of calcite in 500 ml of $CaCO_3$ saturated solution with between 25 and 200 ppm Ba added. After various lengths of time, the crystals were removed from solution, lightly rinsed, and analyzed with XPS. A sorption plot obtained in this way is shown in Figure 26. These results clearly show that the amount of Ba associated with the calcite surface is directly proportional with the amount of time the crystal is in contact with solution and the starting amount of Ba in solution. In addition, the solution chemistry (as monitored by atomic adsorption spectroscopy) showed that the amount of Ba leaving solution was equivalent to the amount of Ca entering solution, indicating that the mechanism of Ba sorption may be an exchange process. Thus, XPS was used successfully to directly monitor the sorption of Ba on calcite surfaces directly, and was able to detect surface concentrations to below 10^{-9} g/cm^2. In addition, these workers used the same basic technique to study the uptake of Ni^{2+}, La^{3+}, and Ba^{2+} on MnO_2 surfaces (Brule et al., 1980). They again found that the uptake was characteristic of a sorption rather than a precipitation process, and that affinity for sorption was much higher for Ba than for Ni or La.

A good example of detailed line shape analysis and the use of Auger parameters is given by Dillard and Koppelman (1982) in a study of the adsorption of aqueous Co(II) onto kaolinite surfaces from pH 3 to 10. The amount of Co adsorbed on kaolinite increases dramatically with increasing pH between pH 6 and 8, and the purpose of the study was to examine the chemical nature of the adsorbed Co species over this range. The Co 2p photopeaks of the adsorbed Co on kaolinite at pH 6, 7, and 10, as well as for pure $Co(OH)_2$, are shown in Figure 27. The intense and broad satellite feature to the high energy side of the Co $2p_{3/2}$ and $2p_{1/2}$ lines and the separation of the spin-orbit split lines (16 eV) clearly indicate that the adsorbed species is Co(II) in all samples (see section 6.1 for details of this oxidation state assignment). However, the peak to satellite ratios, peak widths, and peak positions suggest that there are important differences between the adsorbed species over the pH range. The Co 2p photopeaks for the pH 7.5 - 10 samples are nearly identical in every respect, but are slightly different than the spectrum taken from $Co(OH)_2$. This led Dillard and Koppelman to conclude that the adsorbed species from pH 7.5 to 10 are very similar to each other and to a cobalt hydroxide complex. As the pH decreases below 7.5, the spectra of the adsorbed Co complex show a greater and greater satellite intensity compared to the principal photopeak (see spectra at pH 6 and 7 in Fig. 27). Dillard and Koppelman (1982) interpret this as an increase in the covalency of ligand to Co bonding, although the actual complex at these lower pH's is not known. A plot of the modified Co Auger parameters (calculated using the Co $2p_{3/2}$ and Co L_3VV lines (Fig. 28) for the Co surface complexes formed at various pH's and several standard compounds) gives further support to the idea that the complexes which form above pH 7.5 are very similar and are $Co(OH)_2$-like, and that at lower pH's, unspecified changes occur in the complex. However, the decrease in the kinetic energy in the Co L_3VV line with decreasing pH of sorbate formation (see Fig. 28) is also indicative of more covalent Co-ligand bonding (Wagner et al., 1979a,b).

Our last example of a sorption study concerns the XPS examination of the adsorption of phosphate, sulfate, and selenite on goethite surfaces (Martin and Smart, 1987). The purpose of this study was to examine the surface speciation and coverage of the anionic species as a function of pH. For Se at pH 3 and 12, the adsorbed complex was

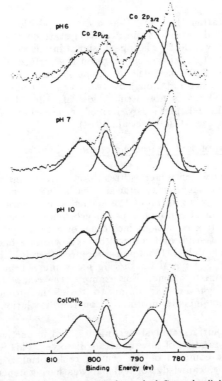

Figure 27. Co 2p x-ray photoelectron spectra for sorbed Co on kaolinite surfaces prepared at pH 6, 7, and 10, along with the Co 2p spectrum from pure Co(OH)$_2$. The observed spectra are dotted, and the solid curves are fitted components. The broad and relatively intense peaks to the high binding energy side of the narrow spin orbit split Co 2p lines are energy loss satellites, characteristic of Co(II) in many compounds. For these spectra, the binding energy scale is uncorrected for charge shift. See text for further details. From Dillard and Koppelman (1982).

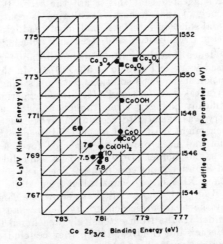

Figure 28. Modified Co Auger parameters calculated using the Co 2p$_{3/2}$ and Co L$_3$VV lines for Co sorbed onto kaolinite at several pH's (solid circles with pH numbers) and several standard compounds. All Auger parameters marked with solid circles are from Dillard and Koppelman (1982); those marked with solid squares are from Wagner et al. (1979b). See text for further details. From Dillard and Koppelman (1982).

found to contain Se^{4+} rather than Se^{6+} (based on the position of the Se $3d_{5/2}$ line), and the adsorption process did not affect the oxidation state on the goethite surface (ferric). An appropriate XPS-derived X/Fe photopeak signal intensity ratio (where X = P, S, or Se) was used to estimate the percent coverage on the surface. These results were found to be consistent with coverage assumed by replacement of two A-type hydroxyls (those singly coordinated to surface Fe^{3+} ions) by two oxygens belonging to the anionic complex. Thus, although XPS was not used to identify the adsorption site, it was completely consistent with models previously derived from earlier infrared spectroscopic studies on this system (e.g., see Atkinson et al., 1974).

6.3 Studies of the alteration and weathering of mineral surfaces

The mechanisms by which minerals break down are as important and interesting as the mechanisms by which they form. Indeed, weathering is one of the most critical factors in the geochemical cycling of all of the elements in the rock forming minerals, and weathering is one of the most common types of geochemical reactions because it conceivably occurs under all surface conditions including, of course, ambient conditions. Therefore, it is no wonder that studies of mineral breakdown date back more than a century (see Helgeson et al., 1984, Velbel, 1986, and references therein). Further, an understanding of the fundamentals of mineral weathering has a much broader impact in the earth sciences than on geochemistry alone. For example, the complete understanding of any rock-water interaction, whether pertaining to toxins in groundwater or ore-forming hydrothermal fluids, must include information on the dissolution of mineral surfaces by the fluid.

In the last decade, surface analysis techniques, and in particular XPS, have added a great deal to our understanding of how minerals dissolve. This is probably no more evident than for the most common of the crustal rock forming minerals, the feldspars. Before XPS was available, knowledge of feldspar weathering depended almost entirely on information gathered on the solution side of the mineral-water interface. From this, it was generally believed that a chemically altered layer, up to hundreds or thousands of angstroms thick, was formed in the early stages of dissolution, and that diffusion of feldspar constituents through this layer controlled the dissolution rate of the mineral (see, e.g., Paces, 1973). A number of XPS measurements on both artificially and naturally weathered feldspars (Petrovic et al., 1976; Berner and Holdren, 1977, 1979; Holdren and Berner, 1979) have suggested that this is not the case. It is now thought (as discussed in more detail below) that mineral dissolution is controlled by reactions that occur right at the mineral-solution interface, and most readily at high energy sites such as surface inhomogeneities, irregularities, and structural defects.

The first direct evidence that deep altered layers do not form on dissolving silicates came from the XPS-based work of Petrovic et al. (1976). In this study, sanidine grains were exposed to dissolution in aqueous solutions with pH's ranging from 4 to 8 at 82°C and 1 atm. XPS spectra taken before and after these experiments (lasting up to 377 hours) showed no significant change. Even though analytic uncertainty in the XPS-derived surface chemistry was taken into account, Petrovic et al. (1976) estimated that any depleted layer which might exist could not extend deeper than about 17A from the surface, far thinner than had been previously expected for a leached layer. This work was followed by Holdren and Berner (1979), who this time did dissolution experiments with albite at pH 6 and 8 under room conditions. The basic results were the same; Figure 29 shows some of their XPS survey scans taken from albite powder both before and after dissolution at pH 6 for 241 hours. All peaks show the same relative height to one another, indicating that the near-surface chemistry before and after dissolution is very similar. Along with similar results from extensively weathered feldspars recovered from several different soils (Berner and Holdren, 1979), they concluded, as did Petrovic et al. (1976), that the mechanism controlling feldspar dissolution is surface controlled reactions.

More recently, similar results have been obtained from fayalite and various pyroxenes and amphiboles (Schott et al., 1981; Berner and Schott, 1982; Schott and Berner,

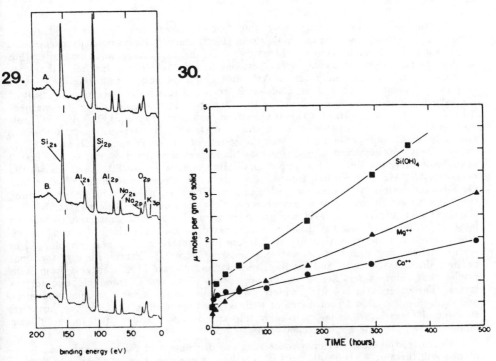

Figure 29. X-ray photoelectron spectra of albite prepared as follows: (A) ground but otherwise untreated starting material; (B) ground and treated in buffered pH 6.0 aqueous solution for 241 hours; and (C) ground and etched in 5% HCl and 0.09 N H_2SO_4. The spectra are very similar, indicating little or no change in near-surface chemistry of the three albites. Note that the binding energy is increasing to the left. From Holdren and Berner (1979).

Figure 30. Release of Ca^{2+}, Mg^{2+}, and $Si(OH)_4$ (in terms of μ moles per gram of solid) vs. time of dissolution at pH 6 and 20°C. See text for details. From Schott et al. (1981).

Table 6. Summary of surface composition results, using XPS, for the dissolution of etched tremolite. Elemental ratios given refer to ratios of peak areas (from Schott et al., 1981).

pH	Temp. °C	Time (days)	Mg2s/Si2p	Ca2p$_{3/2}$/Si2p
		TREMOLITE		
	starting material		0.46	0.81
6	20	9	0.45	0.69
6	20	16	0.43	0.68
6	20	40	0.42	0.68
6	60	16	0.43	0.67
6	60	24	0.41	0.66
1	20	9	0.39	0.66
1	20	16	0.38	0.64
1	20	24	0.37	0.63
1	60	16	0.36	0.61
1	60	24	0.35	0.61

1983), although in these studies very thin leached layers were positively identified. Schott et al. (1981) and Berner and Schott (1982) showed for various pyroxenes and amphiboles, which had been weathered both in the laboratory and naturally, that alkaline/alkaline-earth depleted layers between 5 and 20A form. In their laboratory experiments, it was apparent from the XPS measurements, and by following the changes in solution chemistry, that these depleted layers form relatively quickly as a result of incongruent dissolution, after which the thin depleted layer stabilizes due to congruent dissolution. A good example of this is shown for tremolite $(Ca_2Mg_5Si_8O_{22}(OH)_2)$ in Table 6 and Figure 30. For instance, at pH 6 and 20°C, the Mg/Si near-surface ratio shows a relatively slight and gradual drop over the course of the experiment (960 hours = 40 days). On the other hand, the Ca/Si ratio undergoes a sharp drop during the first 9 days before leveling out. This agrees qualitatively with Figure 30, which shows the release of Ca, Mg, and $Si(OH)_4$ into solution during this same experiment. Notice that Ca is initially released at a higher rate than Mg. However, after several hundred hours, the Mg, Ca, and Si release rates have reached near congruent proportions (Ca:Mg:Si = 2.5:5.2:8). At any rate, no evidence has been found for a thick protective or leached layer which might control the dissolution rate via diffusion constraints. Instead, all evidence points only to very thin depleted layers (on the order of several monolayers) in which surface and near-surface reactions control dissolution rates. Finally, for iron-bearing silicates (bronzite and fayalite), Schott and Berner (1983) suggest from XPS results that solution weathering in the presence of dissolved O_2 produces two surface layers. The outer layer is a hydrous ferric oxide, whereas the inner layer (just above the undisturbed mineral) is a protonated or hydroxylated ferric silicate. As before, however, they found these layers to be exceedingly thin and do not consider them to be diffusion inhibiting and protective in the usual sense.

Recently, the basic concepts concerning the dissolution of minerals have had to be refined still further. Certain dissolution experiments such as those of Chou and Wollast (1984) have again suggested the presence of a highly altered and relatively thick surface layer through which certain components must diffuse. However, this evidence for a thick and highly altered layer does not agree with the XPS results presented above. In order to explain both the solution and near-surface data, Berner et al. (1985) presented a simple model which suggests that altered surface layers on dissolving silicates may not be uniform or continuous, but discontinuous and of varying thicknesses. Circumstantial evidence for this comes from the fact that silicates generally do not dissolve uniformly, but instead have areas of enhanced dissolution typically referred to as etch pits. Berner et al. (1985) suggested that relatively deep dissolution may occur at high energy sites on the surface, such as a step, or where solid state defects intersect the surface. In areas of relatively low energy, altered layers may not exist or would be very shallow (a few monolayers). Therefore, XPS analysis, which covers a relatively large surface area and analyzes to a depth of only several 10's of angstroms, would show an altered layer that was, on the average, very thin. However, the solution results would suggest, on the average, a much deeper altered layer.

What is obviously needed to test this idea is a surface analytic tool with high spatial resolution. Small spot XPS can currently obtain spatial resolution of approximately 150 μm. On the other hand, Hochella et al. (1986a,b) have recently shown that scanning Auger microscopy (SAM) can semi-quantitatively analyze the surfaces of non-conducting materials with spatial resolution slightly better than 1 μm. This technique was used to analyze the surfaces of labradorite altered under hydrothermal conditions in an attempt to test the theory of Berner et al. (1985). The results of this study (Hochella et al., 1988) show that the dissolution of labradorite under hydrothermal conditions is not uniform and is in fact highly complex. In general, the surfaces of the reacted grains are incongruently modified to a depth of at least 20 to 30A with a relative enrichment of Al and a depletion of Na, Ca, and Si compared to the starting composition. Locally, some portions of the reacted labradorite surfaces have had Na and Ca almost entirely removed (Figs. 31 and 32). Because the spatial resolution used in this particular study was in the 2 to 10 μm range to prevent surface damage induced by the primary electron beam, surface chemical variations at individual surface defects or other high energy sites could not

Figure 31. Low magnification photomicrograph of labradorite grains altered under hydrothermal conditions. Arrows a, b, and c point towards areas of Auger analysis which are magnified in Figure 32(a), (b), and (c), respectively. From Hochella et al. (1988).

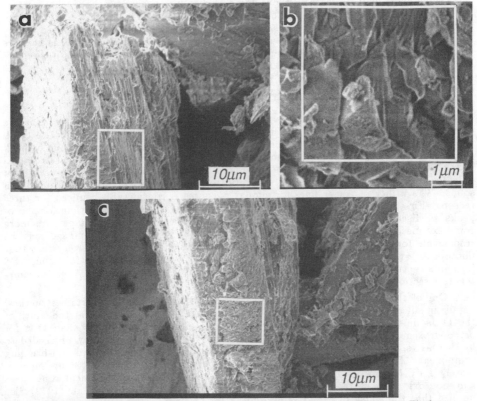

Figure 32. High magnification photomicrographs of the areas denoted in Figure 31. The boxes in each photomicrograph denote the areas of Auger surface analysis. The measured surface chemistry in the area denoted in (a), an area showing polysynthetic twinning but otherwise relatively unperturbed, is the same as most of the surface analyses of nearby grains. The measured surface chemistry in the area denoted in (c), an area on the same grain showing a roughened texture, is markedly different in that the near-surface Ca is nearly entirely removed. From Hochella et al. (1988).

be detected. However, these results definitely show that feldspar dissolution can be highly non-uniform in both depth and lateral extent and support the position of Berner et al. (1985).

Although space does not allow for the discussion of additional studies which have used XPS and AES to study mineral surface alteration and dissolution, several others are listed here for completeness. Thomassin et al. (1977) used XPS to study the dissolution kinetics of chrysotile, and Zingg and Hercules (1978) and Perry et al. (1984) used XPS to characterize various alterations of lead sulfide surfaces. Mossotti et al. (1987) studied the alteration of carbonate building stone in acid rain environments using XPS. Several other studies have used XPS to determine the surface alteration of minerals and rocks at temperatures above ambient, including feldspars (Fung et al., 1980), basalts (Crovisier et al., 1983), and titanate minerals (Myhra et al., 1984). Perry et al. (1983) have used AES to study HF- and HF/H_2SO_4-treated feldspar surfaces, and Mucci et al. (1985) and Mucci and Morse (1985) used similar techniques to study the surface reactions of carbonates with seawater. Bisdom et al. (1985) have used SAM to study the surface chemistry of highly weathered materials in soils. Finally, Remond et al. (1981, 1982, 1983, 1985) have used XPS and SAM to investigate alteration of mineral surfaces during the process of polishing; this affects the optical properties (color and reflectance) of minerals in reflected light.

Much more will be learned about mineral dissolution as SAM and other surface analytic techniques are applied to these problems. Recently, for example, Petit et al. (1987) have used nuclear reaction analysis (NRA) to depth profile for hydrogen in the near-surface region of a weathered and unweathered diopside. Studies such as this are critical in understanding the true depth of alteration, as H^+ and OH^- substitution are expected to be present and a key factor in dissolving silicates.

6.4 Studies of the atomic structure of minerals and glasses

We have seen in the preceding sections that XPS can, in many cases, determine the oxidation state of an element in the near-surface region of a solid. It is logical to think that if the oxidation state of that element remains constant, any chemical shift or peak shape change for that element from material to material will be due to structural changes which affect its environment. Therefore, the chemical shift should be useful not only for oxidation state information, but also for deriving atomic structural details in solids. The utility of Auger lines for similar purposes is considerably reduced for three reasons. First, although the chemical shift of Auger lines is often greater than for photo-peaks as discussed earlier, Auger peaks are much broader and their shapes are more complex. Second, there is always the possibility that electron beam damage may be responsible for changes in peak shapes and/or positions. Third, by the very fact that photopeaks are easier to work with, there is a relatively large data base for using XPS lines in structural studies and a very small one for Auger lines. However, Auger parameters (see section 3.1.10) can have structural significance as shown below.

One of the first studies which showed that XPS was sensitive to certain structural details in silicate minerals was that by Yin et al. (1971). They showed that the width of the O 1s line for olivine, which has only non-bridging oxygens, is less than that for orthopyroxene, which has both bridging and non-bridging oxygens.[1] They proceeded to fit two peaks to the orthopyroxene O 1s line representing the bridging and non-bridging components of the photopeak, and found these components to be separated by approximately 1 eV (Fig. 33). This is a chemical shift stimulated by environment differences and not oxidation state change. Shortly after this study, Adams et al. (1972) also implied that the width of the O 1s photopeak for a mineral might be dependent on the

---

[1]Non-bridging oxygens are defined as those which are bonded to no more than one tetrahedrally coordinated cation; bridging oxygens are bonded to two, thus "bridging" two tetrahedra. Rarely, bridging oxygens can be shared by three tetrahedra forming a trimer. Mullite has trimers consisting of 2 Al and 1 Si tetrahedra.

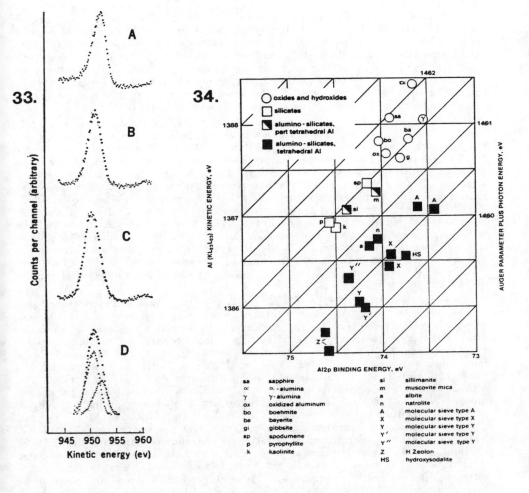

Figure 33. Oxygen 1s photoelectron spectra for (A) Fe$_2$O$_3$, (B) olivine, (C) orthopyroxene, and (D) the same orthopyroxene line with fitted bridging and non-bridging components. See text for details. From Yin et al. (1971).

Figure 34. Compilation of modified Al Auger parameters (measured using the Al 2p and Al KL$_{2,3}$L$_{2,3}$ line positions) for a number of Al and O containing inorganic compounds. See text for details. From Wagner et al. (1982).

number of oxygen bonding environments in that structure.

Despite this initial success, later studies showed that the structure induced chemical shift was not going to be as useful as was hoped. For example, Anderson and Swartz (1974) showed that the Al 2p photopeak for sillimanite, with aluminum in both fourfold and sixfold coordination, was essentially identical in position and shape to that for kyanite, with aluminum in sixfold coordination only. The same observations were made by Fenn and Brown (1974). Urch and Murphy (1974) and Urch and Webber (1976) report a correlation between the binding energies of the Al 2s and 2p photopeaks and the Al-O bond length in a number of minerals, but the overall chemical shifts are relatively small (less than 1 eV) and the scatter in the data is significant, making the correlation relatively weak. Urch and Webber (1976) report that no such correlation exists between Si 2p binding energies and Si-O bond lengths, but they have shown how it is possible to calculate a potential at the Si atomic site which does relate to the Si 2p binding energy.

In addition, there has been some interest in the shape of the x-ray induced O $KL_{2,3}L_{2,3}$ Auger line for indicating variation in oxygen bonding environments in silicates (e.g., Wagner et al., 1982, and Hochella and Brown, 1988). This intense oxygen Auger line involves two final state holes in valence levels, and one might predict that its shape would be sensitive to changes in the average bonding configuration around oxygen in a structure. In fact, Wagner et al. (1982) suggested that the O $KL_{2,3}L_{2,3}$ line shape was characteristic of the silicate structure, and related its shape specifically to (Si,Al)-O-(Si,Al) linkages. However, this analysis was based only on spectra collected from sheet and framework silicates. Hochella and Brown (1988) have subsequently shown that the O $KL_{2,3}L_{2,3}$ line shape is essentially constant for quartz, albite, nepheline, jadeite, diopside, and forsterite. Therefore, the O $KL_{2,3}L_{2,3}$ line shape is not always sensitive to the average oxygen environment in silicates. Instead, Hochella and Brown (1988) have suggested that the O 1s photopeak may be a useful indicator of the number of chemically distinct oxygen environments in silicates (see below).

Wagner et al. (1981, 1982) and West and Castle (1982) have had more general success exploring the use of XPS as a structural probe using Auger parameters. Because Auger parameters result from the energy difference between two lines from the same element in the same spectrum, charge referencing is not necessary, and measurements are usually reproducible to ± 0.05 eV even on insulators. An example of the use of aluminum Auger parameters, calculated with the Al 2p and Al $KL_{2,3}L_{2,3}$ line positions, is shown in Figure 34 for 21 aluminosilicates and aluminum oxides and hydroxides. The aluminosilicates with aluminum in tetrahedral coordination have Al Auger parameters grouped between 1460 and 1461 eV. The silicates with aluminum partially or fully in octahedral coordination have Auger parameters between 1461.0 and 1461.5 eV, and aluminum oxides and hydroxides generally have Auger parameters above 1461.5 eV. This means that the polarizability or extra-atomic relaxation of aluminum (the ability of electrons from neighboring atoms to screen final state holes caused by Auger emission from aluminum) is lowest when it is tetrahedrally coordinated in silicates, and that it increases with octahedral coordination. This same general result was also observed by West and Castle (1982) using the Al 1s line instead of the Al 2p as above. They were also able to relate these polarizabilities to those deduced from refractive index measurements.

Perhaps the most recent interest in using XPS as a structural probe for silicates has been in the area of glass structure analysis. Major articles in this field include Nagel et al. (1976), Bruckner et al. (1976, 1978a,b, 1980), Kaneko and Suginohara (1977, 1978), Jen and Kalinowski (1979), Lam et al. (1980), Smets and Lommen (1981), Veal et al. (1982), Kaneko et al. (1983), Puglisi et al. (1983, 1984), Smets and Krol (1984), Onorato et al. (1985), Tasker et al. (1985), Goldman et al. (1986), and Hochella and Brown (1988). The most interest with XPS in this role has been in the Na_2O - SiO_2 and Na_2O - Al_2O_3 - SiO_2 systems. It has been shown in most of these studies that clearly resolvable features in the O 1s photopeak can be attributed to bridging and non-bridging oxygens. Recent work by Onorato et al. (1985), Tasker et al. (1985), and Goldman (1986) have all shown that O 1s photopeaks collected from peralkaline glasses in the Na_2O - Al_2O_3 - SiO_2 system can be most appropriately fit with three components, a symmetric bridging oxygen

(Si-O-Si), an asymmetric bridging oxygen (Si-O-^{IV}Al), and a non-bridging oxygen. Using their fitting approach, the overall bridging to non-bridging ratios derived from the O 1s spectra match exceptionally well with the same ratios calculated from the ideal structural models for each glass composition, and it also predicts correctly the reversals in physical properties such as density and viscosity (Riebling, 1964, 1966) which occur near the SiO_2 - $Na_2O \cdot Al_2O_3$ join.

An example of the fitting scheme now used for O 1s photopeaks in the Na_2O - Al_2O_3 - SiO_2 system is shown in Figures 35 and 36. Figure 35 shows the charge shifted O 1s spectra of a series of glasses with constant silica content and varying Al/Na ratio. When Al/Na = 1, the O 1s peak is slightly asymmetric towards higher binding energy. As the Al/Na ratio decreases from 1, a shoulder develops to the low binding energy side, and the major peak shifts to higher binding energy. Each spectral envelope is fit with three Gaussian peaks, allowing position and intensity to vary; however, the peak widths are fixed based on bridging and non-bridging component widths for glasses along the Na_2O - SiO_2 join. The results are shown in Figure 36. In this figure, the Si-O-Si and Si-O-^{IV}Al components are labeled BO1 and BO2, respectively, and the non-bridging component is labeled NBO. The sum of the components is indistinguishable from the observed photopeaks.

Unfortunately, not all silicates show O 1s photopeaks that have clearly resolvable features that can be assigned to different oxygen species. Hochella and Brown (1988) have looked at a number of mineral and rock composition glasses, comparing their O 1s, O 2s, and O $KL_{2,3}L_{2,3}$ peaks with those of several mineral standards. Although the components of the O 1s lines in these spectra were too close in energy to give resolved peaks (or even distinct shoulders), the peak width (full width at half maximum or FWHM) in all cases was proportional to the number of chemically distinct oxygen environments present (in the case of the minerals) or thought to be present (in the case of the glasses) in the structure. However, more work is needed to determine the applicability of this finding to a much wider range of silicate materials.

All in all, the use of XPS as an atomic structural probe is slowly growing, especially for amorphous materials. Further developments will come with increasing interest in the structure (instead of just the chemistry) of mineral surfaces, and the structural effects on weathering, ion exchange, adsorption, and so on.

7. CONCLUSIONS

In this chapter, we have explored the fundamental principals, instrumentation, practical aspects and methods, and applications of XPS and Auger spectroscopies from a mineralogical perspective. The purpose of this concluding section is not to summarize all of this information, but instead to carefully compare the XPS and AES techniques, first examining their similarities and then their respective advantages. Knowing this, it becomes clear as to which technique one should apply to a given surface problem. This will be followed by a brief statement on future developments in XPS and AES instrumentation, and conclude with a basic comparison of XPS and AES with other important surface analysis techniques.

7.1 Similarities in XPS and AES

The most obvious similarity between XPS and AES, and one of their many very important characteristics, is that they can detect all elements except H and He. In addition, although the limit of detection for any given element is likely to be different, the best *overall* detection limit is about the same for both. Regarding the depth of analysis, there is a common misconception that AES is more surface sensitive than XPS. However, as discussed in section 2.4.3, the depth of analysis depends solely on the energy of the electrons to be analyzed and the material through which they pass before entering the vacuum. Therefore, in general, AES is no more surface sensitive than XPS. It is true that there are a number of common elements, such as Si, for which the kinetic

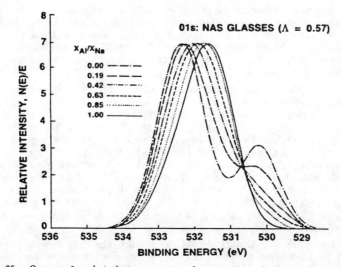

Figure 35. Oxygen 1s photoelectron spectra for a series of glasses in the system Na₂O – Al₂O₃ – SiO₂. X_{Al}/X_{Na} is the mole fraction of alumina to soda with the mole fraction of silica constant at 0.67. Note that the binding energy is increasing to the left. See text for details. From Tasker et al. (1985).

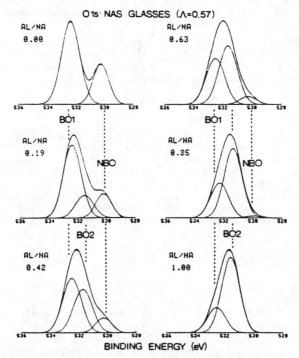

Figure 36. Oxygen 1s photoelectron spectra (the same as those shown in Fig. 35) showing a 3-component Gaussian fit. The components labeled BO1 and BO2 are thought to represent Si-O-Si and Si-O-^{IV}Al linkages, respectively; NBO is the nonbridging oxygen component. From Tasker et al. (1985).

energy of the line commonly measured in AES analysis (Si $L_3M_{2,3}M_{2,3}$, 74 eV) is much lower than the line commonly used in XPS analysis (Si 2p, 1383 eV for Al Kα x-rays). In these specific cases, AES would be more surface sensitive than XPS. Finally, the two techniques are similar in that semi-quantitative analysis can be done with about the same ease (using appropriate sensitivity factors). However, as mentioned below, improving on semi-quantification is generally easier with XPS than with AES.

7.2 XPS advantages over AES

The major advantage that XPS has over AES is that chemical state information is readily available from XPS, whereas this same information from AES is only available in certain cases. There are several reasons for this. First, although chemical shifts for Auger lines are often greater than equivalent shifts for photopeaks (section 3.2.1), Auger lines are generally much broader (up to 4 or 5 times) and their shapes cannot be as easily characterized because of the inherently poor signal to noise ratios. Second, and probably most important, is the fact that electron beam damage in AES is always possible, and one must be concerned with the possibility that the features in the Auger lines which one is using for chemical state information are due to beam damage rather than inherent to the material's surface. Third, precisely because of the first two problems, there is relatively limited published work on the use of AES as an oxidation state indicator, and practically no work on its use as a structural state tool. By comparison, the data base for such studies with XPS is large, making it much easier to quickly interpret oxidation and structural state information in XPS spectra.

Another advantage that XPS enjoys over AES is that sample charging is far less of a problem. As pointed out above (section 5.4.1), there are ways to prevent charging of insulating surfaces in AES, but there is always the possibility that conditions will not allow neutralization of charging, and data collection will not be possible on certain areas of certain samples. With XPS, positive charging on insulating surfaces due to exiting photoelectrons can be easily stabilized or neutralized with an external source of low energy electrons to flood the surface. The most convenient source is the x-ray tube end-window of a non-monochromatic source. In addition, special low energy electron flood guns are also commonly used in conjunction with monochromatic x-ray sources. In standard XPS systems, it is rare when data cannot be collected due to charging problems, and charge stabilization or neutralization generally takes little to no effort.

Two additional advantages that have already been thoroughly discussed above will only be briefly mentioned here. The first concerns quantitative analysis. AES quantitative analysis from first principles is complicated mostly by the fact that one must deal with the complex interactions of the primary electrons with the near-surface region (section 5.3.2). However, both electrons and x-rays are subject to channelling (or diffraction) by a crystalline material, so at least this complication is shared. The second additional advantage is that it is generally unlikely that the excitation source for XPS will modify the surface of any earth materials. In the case of polymers and certain other organics, the high energy white radiation from non-monochromatic x-ray sources has been known to cause surface damage. However, instances of x-ray induced surface damage to silicates, sulfides, sulfates and carbonates have never been reported.

7.3 AES advantages over XPS

There is only one major advantage that AES enjoys over XPS, and that is spatial resolution. In fact, at the present time, AES has far better spatial resolution than any other surface analytic technique except perhaps for SIMS (secondary ion mass spectroscopy) instruments fitted with liquid metal ion sources. For AES, analysis has been demonstrated on metallic surfaces in areas as small as 300A in diameter (Venables et al., 1976). In addition, the spatial resolution of AES rivals that of most bulk microprobes. The use of AES as a high-resolution "microprobe" for geologic materials has already been explored in Hochella et al. (1986a). Table 7 compares AES with the most popular microprobes currently in use, including the electron microprobe (EPMA), and scanning

Table 7. Commonly used microanalysis techniques compared with SAM

	EPMA	SEM/EDS	STEM/EDS	SAM
lightest detectable element	Be[*]	Be[*]	Be[*]	Li
best lateral spatial resolution for chemical analysis	1-2μm[**]	1-2μm[**]	0.01μm	0.05μm
depth resolution for chemical analysis	1-3μm[**]	1-3μm[**]	sample thickness	0.003μm[***]
quantification	quant.	semi-quant.	semi-quant.	semi-quant.
sample charging problems	none	few	none	can be significant
sample preparation	polish and sput. coat	sput. coat	ultra-thin sample	high vacuum compatible

[*] Detection of elements lighter than Na requires a thin-window or windowless x-ray detector.
[**] Estimated for oxides with beam voltages in the range 10-15 KeV.
[***] Estimated for oxides.

and scanning-transmission electron microscopy with energy dispersive analysis (SEM/EDS and STEM/EDS, respectively).

In short, it should be apparent that except in unusual circumstances, XPS is the method of choice for surface analysis of earth materials unless high spatial resolution is needed. However, spatial resolution is such an absolute necessity for certain surface problems that AES has maintained its tremendous popularity despite the disadvantages and shortcomings it has in comparison with XPS.

7.4 Further developments in XPS and AES

XPS and AES instrumentation, quantification techniques, and experimental procedure are still a fair way from reaching a mature stage. Significant improvements will continue to occur in the foreseeable future. One improvement which will affect both AES and XPS, but AES to a greater extent, will be more efficient electron energy analyzers and detectors. New designs are currently being developed which will significantly reduce data collection times. Although this may just improve the convenience of using XPS, it is critical in AES where most electron beam damage is time dependent. However, electron beam damage is also dependent on beam current density, and more efficient analyzers and detectors would also allow the use of lower beam current densities at standard counting times.

Other advances which will only affect XPS instrumentation are higher flux x-ray sources (including rotating anode type) and higher spatial resolution. The present limit on XPS spatial resolution, whether using a fine x-ray beam or a very narrow extraction lens, is about 150 μm. It is possible that XPS spatial resolution can be significantly improved beyond this level with further development of the axial magnetic field photoelectron spectromicroscope described by Beamson et al. (1981).

Naturally, it is also expected that chemical quantification techniques for both XPS and AES will continue to improve. This not only includes an enlarged data base and understanding of sensitivity factors, but it also involves advances in the understanding of the fundamental principles which affect signal intensities.

7.5 Comparison with other surface analytic techniques

Although XPS and AES are presently by far the most commonly used techniques for studying the surfaces of materials, they are by no means the only "important" ones for this purpose. There has been an explosion in the number of surface sensitive spectroscopies available for use in the last two decades, and in addition a number of bulk spectroscopies are now sensitive enough to study certain aspects of surfaces. Each technique generally has its own set of advantages which makes it particularly well suited for certain materials and/or the study of certain aspects of surface chemistry. However, it is well beyond the scope of this chapter to list and describe all of these techniques. The purpose of this section is to very briefly mention a few of the other spectroscopic techniques which are presently or potentially important for studying mineral surfaces and the chemical reactions that occur there.

Ultra-violet photoelectron spectroscopy (UPS) is closely related to XPS except for the excitation source (see, e.g., DeKock, 1977, for a detailed discussion of UPS). UPS uses an inert gas discharge lamp which emits very narrow (in energy) and low energy resonance lines. For example, when He is used in the discharge lamp, the dominant emission lines occur at 21.2 and 40.8 eV. This is only enough energy to eject electrons from the valence and shallowest core levels. Therefore, UPS is used to study bonding in materials by mapping out the density of states in bonding and nonbonding orbitals and shallow core lines. It should be noted that although x-rays could be used to map valence bands, UV light is much preferred for two reasons. First, UV linewidths are much narrower than x-ray linewidths (including monochromatized x-rays); this results in much improved energy resolution in the spectra. Second, the photoionization cross-section for valence electrons is much higher for light in the UV range compared to the x-ray range. Therefore, spectral intensities are orders of magnitude greater when using UV light as the excitation source.

An extremely important analytic technique which can be used to study mineral surfaces (as well as the bulk) is secondary ion mass spectroscopy (SIMS) (see, e.g., Metson et al., 1985, and Nesbitt et al., 1988). For this technique, a primary ion beam bombards the material to be analyzed. The secondary (sputtered) ions are then mass analyzed with either a quadrupole or magnetic sector mass analyzer. The technique is very sensitive, with ppm detection for most elements and ppb for some. In addition, all elements can be detected. Although several tens of angstroms of the surface can be removed during the course of an analysis, SIMS can also be operated in the "static" mode whereby the ion beam is rastered over a large area. In this mode of operation, high spatial resolution is lost, but an analysis can be made with only a few monolayers of erosion. A closely related technique which is still in the development stage, but which should be very important in the future, is surface analysis by laser ionization (SALI) (Becker and Gillen, 1984, 1985). This technique takes advantage of the fact that most of the atoms and molecules stimulated from surfaces by ion or electron beams (or even heat) are neutral. SALI operates by universal ionization of these neutrals by a pulsed and intense laser light which passes approximately 1 mm above and parallel to the sample surface. The atomic or molecular ions are then mass analyzed by time-of-flight mass spectrometry. The detection sensitivity for this technique is astounding, with ppm to ppb detection *in each monolayer* for all masses. Also, at least in principle, chemical quantification should be much more straight foward with SALI than with SIMS.

Although not a true spectroscopy, it is important to include low energy electron diffraction (LEED) in this discussion. LEED can be used to study the structure of crystalline surfaces (or surfaces with ordered sorbed species). In this method, a focused electron beam impinges upon a sample at right angles; electron energies used are below a few hundred electron volts. If the surface is atomically ordered, the electrons that are elastically scattered form a diffraction pattern which can be viewed on a hemisphereical screen above the sample. The pattern is due to the constructive and destructive interference of the electron waves backscattered from the surface. The pattern is mostly dependent on the two-dimensional arrangement of atoms on the surface, but features in the pattern can also result from the atomic arrangement 2 or 3 monolayers deep.

Several spectroscopies known for bulk analysis have also been applied to surfaces. These include, among others, conversion electron Mossbauer spectroscopy, electron spin resonance spectroscopy, Fourier transform infrared spectroscopy, and nuclear magnetic resonance spectroscopy. However, perhaps the most revolutionary example of a bulk technique now also being used to study the surface of geologic materials is x-ray absorption spectroscopy (XAS) used in conjunction with synchrotron radiation. If the element for which the absorbtion edge studied is specifically localized on surfaces, the state and atomic environment of that element can be carefully probed. In this regard, the extended x-ray adsorption fine structure (EXAFS) part of the XAS spectrum can be used to obtain relatively accurate interatomic distances and coordination numbers, and also to identify the ligands of the x-ray absorbing element. Geochemically pioneering work using this technique has recently been published by Hayes et al. (1987). They have used XAS to directly determine the structure of sorbed selenate and selenite complexes at the goethite-water interface.

The future of mineral surface geochemistry looks very bright. This is due to the incredible instrumental advances which have occurred over the last decade, not only with the introduction of new techniques, but with sweeping improvements of the more established methods such as XPS and AES which still form the analytical backbone of surface analysis. The instrumental advances have not only expanded well established fields of study, such as mineral weathering, but they have created new ones, such as the study of geochemical heterogenous catalytic reactions. As mineral surface geochemistry continues to grow as one of the newer and more exciting subfields of geochemistry, it will depend more and more on surface spectroscopies to lead the way.

ACKNOWLEDGMENTS

The author sincerely appreciates collaboration with and assistance and advice from the following individuals: at Stanford, Gordon Brown, Ray Browning, Kris Butcher, Altaf Carim, Lisa Dennis, Steve Didziulis, Carrick Eggleston, Heather Ponader, and Eric Winterburn; at Perkin-Elmer, Arthur Turner and David Harris; and at the USGS (Menlo Park), Jim Lindsay and Victor Mossotti. The author is also indebted to the NSF through Stanford's Center for Materials Research, the Perkin-Elmer Corporation (Physical Electronics Division), the Chevron Oil Field Research Company, and the USGS for either direct or indirect financial assistance.

REFERENCES

ADAMS I., THOMAS J.M. and BANCROFT G.M. (1972) An ESCA study of silicate minerals. Earth Planet. Sci. Lett. 16, 429-432.

AJIOKA T. and USHIO S. (1986) Characterization of the implantation damage in SiO_2 with x-ray photoelectron spectroscopy. Appl. Phys. Letters 48, 1398-1399.

ANDERSON P.R. and SWARTZ W.E., Jr. (1974) X-ray photoelectron spectroscopy of some aluminosilicates. Inorg. Chem. 13, 2293-2294.

ARMITAGE A.F., WOODRUFF D.P. and JOHNSON P.D. (1980) Crystallographic incident beam effects in quantitative Auger electron spectroscopy. Surface Sci. 100, L483-L490.

ATKINSON R.J., POSNER A.M., PARFITT R.L. and SMART R.ST.C. (1974) Infrared study of phosphate adsorption on goethite. J. Chem. Soc., Faraday I., 70, 1472-1479.

AUGER P. (1975) The Auger effect. Surface Sci. 48, 1-8.

BAER Y., HEDEN P.F., HEDMAN J., KLASSON M. and NORDLING C. (1970) Determination of the electron escape depth in gold by means of ESCA. Solid State Comm. 8, 1479-1481.

BANCROFT G.M., BROWN J.R. and FYFE W.S. (1977) Quantitative x-ray photoelectron spectroscopy (ESCA): Studies of Ba^{2+} sorption on calcite. Chem. Geol. 19, 131-144.

BANCROFT G.M., BROWN J.R. and FYFE W.S. (1977) Calibration studies for quantitative ESCA of ions. Anal. Chem. 49, 1044-1048.

BARRIE A. and STREET F.J. (1975) An Auger and x-ray photoelectron spectroscopic study of sodium metal and sodium oxide. J. Electron Spectrosc. Related Phenom. 7, 1-31.

BEAMSON G., PORTER H.Q. and TURNER D.W. (1981) Photoelectron spectromicroscopy. Nature

290, 556-561.

BECKER C.H. and GILLEN K.T. (1984) Surface analysis by nonresonat multiphoton ionization of desorbed or sputtered species. Anal. Chem. 56, 1671-1674.

BECKER C.H. and GILLEN K.T. (1985) Can nonresonant multiphoton ionization be ultrasensitive?. J. Optical Soc. Amer. B2, 1438-1442.

BERNER R.A. and HOLDREN G.H., Jr. (1977) Mechanism of feldspar weathering: Some observational evidence. Geology 5, 369-372.

BERNER R.A. and HOLDREN G.R., Jr. (1979) Mechanism of feldspar weathering. II. Observations of feldspars from soils. Geochim. Cosmochim. Acta 43, 1173-1186.

BERNER R.A. and SCHOTT J. (1982) Mechanism of pyroxene and amphibole weathering. II. Observations of soil grains. Amer. J. Sci. 282, 1214-1231.

BERNER R.A., HOLDREN G.R., Jr. and SCHOTT J. (1985) Surface layers on dissolving silicates. Geochim. Cosmochim. Acta 49, 1657-1658.

BISDOM E.B.A., HENSTRA S., KOOISTRA M.J., VAN OOIJ W.J. and VISSER T.H. (1985) Combined high-resolution scanning Auger microscopy and energy dispersive x-ray analysis of soil samples. Spectrochimica Acta 40B, 879-884.

BRAUN P., FARBER W., BETZ G. and VIEHBOCK F.P. (1977) Effects in Auger electron spectroscopy due to the probing electrons and sputtering. Vacuum 27, 103-108.

BRIGGS D. and SEAH M.P. (1983) Practical Surface Analysis by Auger and X-ray Photoelectron Spectroscopy. Wiley, New York, 533 pp.

BRIGGS D. and RIVIERE J.C. (1983) Spectral interpretation. In: Practical Surface Analysis by Auger and X-ray Photoelectron Spectroscopy (D. Briggs and M.P. Seah, eds.). Wiley, New York, 87-139.

BRUCKNER R., CHUN H.-U. and GORETZKI H. (1976) Discrimination between bridging and non-bridging oxygen in sodium silicate glasses by means of x-ray induced photoelectron spectroscopy (ESCA). Glastechn. Ber. 49, 211-213.

BRUCKNER R., CHUN H.-U. and GORETZKI H. (1978a) XPS measurements on alkali silicate and soda aluminosilicate glasses. Japan. J. Appl. Phys. 17, 291-294.

BRUCKNER R., CHUN H.-U. and GORETZKI H. (1978b) Photoelectron spectroscopy (ESCA) on alkali silicate- and soda aluminosilicate glasses. Glastechn. Ber. 51, 1-7.

BRUCKNER R., CHUN H.-U., GORETZKI H. and SAMMET M. (1980) XPS measurements and structural aspects of silicate and phosphate glasses. J. Non-Cryst. Solids 42, 49-60.

BRUINING H. (1954) Physics and applications of secondary electron emission. Pergamon Press, London.

BRULE D.G., BROWN J.R., BANCROFT G.M. and FYFE W.S. (1980) Cation adsorption by hydrous manganese dioxide: A semi-quantitative x-ray photoelectron spectroscopic (ESCA) study. Chem. Geol. 28, 331-339.

BRUNDLE C.R., CHUANG T.J. and WANDELT K. (1977) Core and valence level photoemission studies of iron oxide surfaces and the oxidation of iron. Surface Science 68, 459-468.

BUSSING T.D. and HOLLOWAY P.H. (1985) Deconvolution of concentration depth profiles from angle resolved x-ray photoelectron spectroscopy data. J. Vac. Sci. Tech. A, 3, 1973-1981.

CARRIERE B. and LANG B. (1977) A study of the charging and dissociation of SiO_2 surfaces by AES. Surface Sci., 64 , 209-223.

CARTER G. and COLLIGON J.S. (1968) Ion bombardment of solids. Elsevier, New York, 446p.

CHOU L. and WOLLAST R. (1984) Study of the weathering of albite at room temperature and pressure with a fluidized bed reactor. Geochim. Cosmochim. Acta 48, 2205-2218.

COUNTS M.E., JEN J.S.C. and WIGHTMAN J.P. (1973) An electron spectroscopy for chemical analysis study of lead adsorbed on montmorillonite. J. Phys. Chem. 77, 1924-1926.

COUNTS M.E., JEN J.S.C. and WIGHTMAN J.P. (1973) An electron spectroscopy for chemical analysis study of lead adsorbed on montmorillonite. J. Phys. Chem. 77, 1924-1926.

COYLE G.J., TSANG T., ADLER I. and BEN-ZVI N. (1981) XPS studies of ion-bombardment damage of transition metal sulfates. J. Electron Spectros. Related Phenom. 24, 221-236.

CROVISIER J.L., THOMASSIN J.H., JUTEAU T., EBERHART J.P., TOURAY J.C. and BAILLIF P. (1983) Experimental seawater-basaltic glass interaction at 50°C: Study of early developed phases by electron microscopy and X-ray photoelectron spectrometry. Geochim. Cosmochim. Acta 47, 377-387.

DAVIS L.E., MACDONALD N.C., PALMBERG P.W., RIACH G.E. and WEBER R.E. (1976) Handbook of Auger electron spectroscopy, 2nd edition. Eden Prairie, Minnesota, Physical Electronics

632

Division.

DAWSON P.H. (1966) Secondary electron emission yields of some ceramics. J. Appl. Phys. 37, 3644-3645.

DEBERNARDEZ L.S., FERRON J., GOLDBERG E.C. and BUITRAGO R.H. (1984) The effect of surface roughness on XPS and AES. Surface Sci. 139, 541-548.

DEKOCK R.L. (1977) Ultraviolet photoelectron spectroscopy of inorganic molecules. In: Electron Spectroscopy: Theory, Techniques, and Applications (C.R. Brundle and A.D. Baker, eds.). Academic Press, New York.

DILLARD J.G., KOPPELMAN M.H., CROWTHER D.L., SCHENCK C.V., MURRAY J.W. and BALISTRI-ERI L. (1981) X-ray photoelectron spectroscopy (XPS) studies on the chemical nature of metal ions adsorbed on clays and minerals. In: Adsorption from Aqueous Solutions (P.H. Tewari, ed.). Plenum, New York, 227-240.

DILLARD J.G. and KOPPELMAN M.H. (1982) X-ray photoelectron spectroscopic (XPS) surface characterization of cobalt on the surface of kaolinite. J. Colloid Interface Sci. 87, 46-55.

DILLARD J.G., CROWTHER D.L. and MURRAY J.W. (1982) The oxidation states of cobalt and selected metals in Pacific ferromanganese nodules. Geochim. Cosmochim. Acta 46, 755-759.

DILLARD J.G., CROWTHER D.L. and CALVERT S.E. (1984) X-ray photoelectron spectroscopic study of ferromanganese nodules: Chemical speciation for selected transition metals. Geochim. Cosmochim. Acta 48, 1565-1569.

FELDMAN L.C. and MAYER J.W. (1986) Fundamentals of Surface and Thin Film Analysis. North-Holland, New York, 352 pp.

FENN P.M. and BROWN G.E., Jr. (1974) X-ray induced electron emission spectroscopy of crystalline and amorphous silicates (abstr.). EOS 55, 1202.

FISCHER H., GOTZ G. and KARGE H. (1983) Radiation damage in ion-implanted quartz crystals. II. Annealing behaviour. Physica Status Solidi 76, 493-498.

FUNG P.C., BIRD G.W., MCINTYRE N.S., SANIPELLI G.G. and LOPATA V.J. (1980) Aspects of feldspar dissolution. Nuclear Tech. 51, 188-196.

GOLDMAN D.S. (1986) Evaluation of the ratios of bridging to nonbridging oxygens in simple silicate glasses by electron spectroscopy for chemical analysis. Phys. Chem. Glasses 27, 128-133.

GOLDSTEIN J.I., NEWBURY D.E., ECHLIN P., JOY D.C., FIORI C. and LIFSHIN E. (1981) Scanning electron microscopy and x-ray microanalysis. Plenum Press, New York, 673 pp.

GOSSINK R.G., VAN DOVEREN H. and VERHOEVEN J.A.T. (1980) Decrease in the alkali signal during Auger analysis of glasses. J. Non-Crystal. Solids 37, 111-124.

GUPTA R.P. and SEN S.K. (1974) Calculation of multiplet structure of core p-vacancy levels. Phys. Rev. B, 10, 71-77.

HARRIS L.A. (1968a) Analysis of materials by electron-excited Auger electrons. J. Appl. Phys. 39, 1419-1427.

HARRIS L.A. (1968b) Some observations of surface segration by Auger electron emission. J. Appl. Phys. 39, 1428-1431.

HAYES K.F., ROE A.L., BROWN G.E., Jr., HODGSON K.O., LECKIE J.O. and PARKS G.A. (1987) In situ x-ray adsorption study of surface complexes: Selenium oxyanions on α-FeOOH. Science 238, 783-786.

HELGESON H.C., MURPHY W.M. and AAGAARD P. (1984) Thermodynamic and kinetic constraints on reaction rates among minerals and aqueous solutions. II. Rate constants, effective surface area, and the hydrolysis of feldspar. Geochim. Cosmochim. Acta 48, 2405-2432.

HERCULES D.M. (1979) ESCA and Auger spectroscopy. American Chemical Society, Washington, D.C., 152 pp.

HILLIER J. (1943) On microanalysis by electrons. Phys. Rev. 64, 318.

HOCHELLA M.F., Jr., HARRIS D.W. and TURNER A.M. (1986a) Scanning Auger microscopy as a high-resolution microprobe for geologic materials. Amer. Mineral. 71, 1247-1257.

HOCHELLA M.F., Jr., TURNER A.M. and HARRIS D.W. (1986b) High resolution scanning Auger microscopy of mineral surfaces. Scanning Electron Microsc. 1986/II, 337-349.

HOCHELLA M.F., Jr., PONADER H.B., TURNER A.M. and HARRIS D.W. (1988) The complexity of mineral dissolution as viewed by high resolution scanning Auger microscopy: Laboradite under hydrothermal conditions. Geochim. Cosmochim. Acta (in press).

HOCHELLA M.F., Jr. and CARIM A.H. (1988) A reassessment of electron escape depths in silicon and thermally grown silicon dioxide thin films. Surface Sci. (in press).

HOCHELLA M.F., Jr., LINDSAY J.R. and MOSSOTTI V.G. (1988) Sputter depth profiling in mineral surface analysis. Amer. Mineral. (in review).

HOCHELLA M.F., Jr. and BROWN G.E., Jr. (1988) Aspects of silicate surface structure analysis using X-ray photoelectron spectroscopy (XPS). Geochim. Cosmochim. Acta (in press).

HOFMANN S. (1980) Quantitative depth profiling in surface analysis: A review. Surface Interface Anal. 2, 148-160.

HOFMANN S. (1983) Depth profiling. In: Practical Surface Analysis by Auger and X-ray Photoelectron Spectroscopy (D. Briggs and M.P. Seah, eds.), Wiley and Sons, New York, 141-179.

HOFMANN S. and THOMAS J.H., III (1983) An XPS study of the influence of ion sputtering on bonding in thermally grown silicon dioxide. J. Vac. Sci. Tech. B, 1, 43-47.

HOLDREN G.R., Jr. and BERNER R.A. (1979) Mechanism of feldspar weathering - I. Experimental studies. Geochim. Cosmochim. Acta 43, 1161-1171.

HOLLOWAY P.H. (1978) Quantitative Auger electron spectroscopy - Problems and propects. Scanning Electron Microsc. 1978/I, 361-374.

HUNTRESS W.T., Jr. and WILSON L. (1972) An ESCA study of lunar and terrestrial materials. Earth Planet. Sci. Letters 15, 59-64.

ICHIMURA S., SHIMIZU R. and LANGERON J.P. (1983) Backscattering correction for quantitative Auger analysis. III. A simple functional representation of electron backscattering factors. Surface Sci. 124, L49-L54.

JEAN G.E. and BANCROFT G.M. (1985) An XPS and SEM study of gold deposition at low temperatures on sulphide mineral surfaces: Concentration of gold by adsorption/reduction. Geochim. Cosmochim. Acta 49, 979-987.

JEN J.S. and KALINOWSKI M.R. (1979) An ESCA study of the bridging to non-bridging oxygen ratio in sodium silicate glass and the correlations to glass density and refractive index. J. Non-Cryst. Solids 38/39, 21-26.

JENKIN J.G., LECKEY R.C.G. and LIESEGANG J. (1977) The development of x-ray photoelectron spectroscopy: 1900-1960. J. Electron Spectros. Related Phenom. 12, 1-35.

JOSHI A., DAVIS L.E. and PALMBERG P.W. (1975) Auger electron spectroscopy. In: Methods of surface analysis (A.W. Czanderna, ed.). Elsevier, Amsterdam.

KANEKO Y. and SUGINOHARA Y. (1977) Fundamental studies on quantitative analysis of O^0, O^-, and O^{2-} ions in silicate by x-ray photoelectron spectroscopy. Nippon Kinzoku Gakkai J. 41, 375-380.

KANEKO Y. and SUGINOHARA Y. (1978) Observation of Si 2p binding energy by ESCA and determination of O^0, O^-, O^{2-} ions in silicate. Nippon Kinzoku Gakkai J. 42, 285-289.

KANEKO Y., NAKAMURA H., YAMANE M., MIZOGUCHI K. and SUGINOHARA Y. (1983) Photoelectron spectra of silicate glasses containing trivalent cations. Yogyo-Kyokaishi. 91, 321-324.

KANTER H. (1961) Energy dissipation and secondary electron emission in solids. Physical Rev. 121, 677-681.

KOHIKI S., OHMURA T. and KUSAO K. (1983a) A new charge-correction method in x-ray photoelectron spectroscopy. J. Electron Spectrosc. Related Phenom. 28, 229-237.

KOHIKI S., OHMURA T. and KUSAO K. (1983b) Appraisal of a new charge correction method in x-ray photoelectron spectroscopy. J. Electron Spectrosc. Related Phenom. 31, 85-90.

KOHIKI S. and OKI K. (1985) An appraisal of evaporated gold as an energy reference in x-ray photoelectron spectroscopy. J. Electron Spectrosc. Related Phenom. 36, 105-110.

KOPPELMAN M.H. and DILLARD J.G. (1977) A study of the adsorption of Ni(II) and Cu(II) by clay minerals. Clays and Clay Minerals 25, 457-462.

KOPPELMAN M.H., EMERSON A.B. and DILLARD J.G. (1980) Adsorbed Cr(III) on chlorite, illite, and kaolinite: An x-ray photoelectron spectroscopic study. Clays and Clay Minerals 28, 119-124.

KOWALCZYK S.P., LEY L., MCFEELY F.R., POLLAK R.A. and SHIRLEY D.A. (1974) Relative effect of extra-atomic relaxation on Auger and binding-energy shifts in transition metals and salts. Phys. Rev. B, 9, 381-391.

LAM D.J., PAULIKAS A.P. and VEAL B.W. (1980) X-ray photoelectron spectroscopy studies of soda aluminosilicate glasses. J. Non-Cryst. Solids 42, 41-48.

LANDER J.J. (1953) Auger peaks in the energy spectra of secondary electrons from various materials. Physical Rev. 91, 1382-1387.

LEGRESSUS C., DURAUD J.P., MASSIGNON D. and LEE-DEACON O. (1983) Electron channeling effect on secondary electron image contrast. Scanning Electron Microsc. 1983/II, 537-542.

LEVENSON L.L. (1982) Electron beam-solid interactions: Implications for high spatial resolution Auger electron spectroscopy. Scanning Electron Microsc. 1982/III, 925-936.

MACDONALD N.C. (1970) Auger electron spectroscopy in scanning electron microscopy: Potential measurements. Appl. Phys. Letters 16, 76-80.

MARTIN R.R. and SMART R.St.C. (1987) X-ray photoelectron studies of anion adsorption on goethite. Soil Sci. Soc. Am. J. 51, 54-56.

MCINTYRE N.S. and TEWARI P.H. (1977) Comments on "Adsorption of Co(II) at the oxide-water interface". J. Colloid Interface Sci. 59, 195-196.

MCINTYRE N.S. and ZETARUK D.G. (1977a) X-ray photoelectron spectroscopic studies of iron oxides. Anal. Chem. 49, 1521-1529.

METSON J.B., BANCROFT G.M. and NESBITT H.W. (1985) Analysis of minerals using specimen isolated secondary ion mass spectrometry. Scanning Electron Microsc. 1985(II), 595-603.

MOORE J.H., DAVIS C.C. and COPLAN M.A. (1983) Building Scientific Apparatus. Addison-Wesley, London.

MOSSOTTI V.G., LINDSAY J.R. and HOCHELLA M.F., Jr. (1987) Alteration of limestone surfaces in an acid rain environment by gas-solid reaction mechanisms. Materials Performance 26, 47-52.

MUCCI A., MORSE J.W. and KAMINSKY M.S. (1985) Auger spectroscopy analysis of magnesian calcite overgrowths precipitated from seawater and solutions of similar composition. Amer. J. Sci. 285, 289-305.

MUCCI A. and MORSE J.W. (1985) Auger spectroscopy determination of the surface-most adsorbed layer composition on aragonite, calcite, dolomite, and magnesite in synthetic seawater. Amer. J. Sci. 285, 306-317.

MURRAY J.W. and DILLARD J.G. (1979) The oxidation of cobalt(II) adsorbed on manganese dioxide. Geochim. Cosmochim. Acta 43, 781-787.

MURRAY J.W. and DILLARD J.G. (1979) The oxidation of cobalt(II) adsorbed on manganese dioxide. Geochim. Cosmochim. Acta 43, 781-787.

MURRAY J.W., DILLARD J.G., GIOVANOLI R., MOERS H. and STUMM W. (1985) Oxidation of Mn(II): Initial mineralogy, oxidation state and ageing. Geochim. Cosmochim. Acta 49, 463-470.

MYHRA S., SAVAGE D., ATKINSON A. and RIVIERE J.C. (1984) Surface modification of some titanate minerals subjected to hydrothermal chemical attack. Amer. Mineral. 69, 902-909.

MYHRA S., WHITE T.J., KESSON S.E. and RIVIERE J.C. (1988) X-ray photoelectron spectroscopy for the direct identification of titanium in hollandites. Amer. Mineral. (in press).

NAGEL S.R., TAUC J. and BAGLEY B.G. (1976) X-ray photoelectron study of soda-lime-silica glass. Solid State Comm. 20, 245-249.

NAGUIB H.M. and KELLY R. (1975) Criteria for bombardment-induced structural changes in non-metallic solids. Radiation Effects 25, 1-12.

NESBITT H.W., METSON J.B. and BANCROFT G.M. (1988) Quantitative major and trace element whole rock analyses by secondary ion mass spectrometry using the specimen isolation technique. Chem. Geol. (in press).

NEWBURY D.E., YAKOWITZ H. and MYKLEBUST R.L. (1973) Monte Carlo calculations of magnetic contrast from cubic materials in the scanning electron microscopy. Appl. Phys. Letters 23, 488-490.

ONORATO P.I.K., ALEXANDER M.N., STRUCK C.W., TASKER G.W. and UHLMANN D.R. (1985) Bridging and nonbridging oxygen atoms in alkali aluminosilicate glasses. J. Amer. Ceram. Soc. 68, C148-C150.

PACES T. (1973) Steady-state kinetics and equilibrium between ground water and granitic rock. Geochim. Cosmochim. Acta 37, 2641-2663.

PALMBERG P.W., BOHN G.K. and TRACY J.C. (1969) High sensitivity Auger electron spectrometer. Appl. Phys. Letters 15, 254-255.

PALMBERG P.W. (1974) Combined ESCA/Auger system based on the double pass cylindrical mirror analyzer. J. Electron Spectros. Related Phenom. 5, 691-703.

PANTANO C.G. and MADEY T.E. (1981) Electron beam damage in Auger electron spectroscopy. Applications Surface Sci. 7, 115-141.

PENN D.R. (1987) Electron mean-free-path calculations using a model dielectric function. Phys. Rev. B, 35, 482-486.

PERRY D.L., TSAO L. and GAUGLER K.A. (1983) Surface study of HF- and HF/H_2SO_4-treated feldspar using Auger electron spectroscopy. Geochim. Cosmochim. Acta 47, 1289-1291.

PERRY D.L., TSAO L. and TAYLOR J.A. (1984) The galena/dichromate solution interaction and the nature of the resulting chromium(III) species. Inorganica Chimica Acta 85, L57-L60.

PETIT J-C., MEA G.D., DRAN J-C., SCHOTT J. and BERNER R.A. (1987) Mechanism of diopside dissolution from hydrogen depth profiling. Nature 325, 705-707.

PETROVIC R., BERNER R.A. and GOLDHABER M.B. (1976) Rate control in dissolution of alkali feldspar - I. Study of residual feldspar grains by x-ray photoelectron spectroscopy. Geochim. Cosmochim. Acta 40, 537-548.

PIJOLAT M. and HOLLINGER G. (1981) New depth-profiling method by angular-dependent x-ray photoelectron spectroscopy. Surface Sci. 105, 114-128.

POWELL C.J. (1980) Quantification of surface analysis techniques. Appl. Surface Sci. 4, 492-511.

POWELL C.J. (1984) Inelastic mean free paths and attenuation lenghts of low-energy electrons in solids. Scanning Electron Microsc. 1984, 1649-1664.

PRUTTON M. (1982) How quantitative is analysis in the scanning Auger electron microscope?. Scanning Electron Microsc. 1982, 83-91.

PUGLISI O., MARLETTA G. and TORRISI A. (1983) Oxygen depletion in electron beam bombarded glass surfaces studied by XPS. J. Non-Cryst. Solids 55, 433-442.

PUGLISI O., TORRISI A. and MARLETTA G. (1984) XPS investigation of the effects induced by the silanization on real glass surfaces. J. Non-Cryst. Solids 68, 219-230.

REMOND G., HOLLOWAY P.H. and GRESSUS C. (1981) Electron spectroscopy and microscopy for studying surface changes of mechanically prepared pyrite and quartz. Scanning Electron Microsc. 1981/I, 483-492.

REMOND G., HOLLOWAY P.H., HOVLAND C.T. and OLSON R.R. (1982) Bulk and surface silver diffusion related to tarnishing of sulfides. Scanning Electron Microsc. 1982/III, 995-1011.

REMOND G., PICOT P., GIRAUD R., HOLLOWAY P.H. and RUZKOWSKI P. (1983) Contribution of electron spectroscopies to x-ray spectroscopy applied to the geosciences. Scanning Electron Microsc. 1983/IV, 1683-1706.

REMOND G., HOLLOWAY P.H., KOSAKEVITCH A., RUZAKOWSKI P., PACKWOOD R.H. and TAYLOR J.A. (1985) X-ray spectrometry, electron spectroscopies and optical micro-reflectometry applied to the study of ZnS tarnishing in polished sulfide ore specimens. Scanning Electron Microsc. 1985/IV, 1305-1326.

RIEBLING E.F. (1964) Structure of magnesium alumino-silicate liquids at 1700°C. Can. J. Chem. 42, 2811-2821.

RIEBLING E.F. (1966) Structure of sodium aluminosilicate melts containing at least 50 mole % SiO_2 at 1500°C. J. Chem. Phys. 44, 2857-2865.

RIGGS W.M. (1975) Surface analysis by x-ray photoelectron spectroscopy. In: Methods of surface analysis (A.W. Czanderna, ed.). Elsevier, Amsterdam.

RIVIERE J.C. (1983) Auger techniques in analytical chemistry. The Analyst 108, 649-684.

ROY D. and CARETTE J-D. (1977) Design of electron spectrometers for surface analysis. In: Topics in Current Physics, Vol. 4, Electron Spectroscopy for Chemcial Analysis (H. Ibach, ed.), Springer-Verlag, Berlin, 13-58.

RUTHEMANN G. (1942) Elektronenbremsung an rontgenniveaus. Nafurwissenscheften 30, 145.

RUTHERFORD E. (1914) The connexion between the β and γ ray spectra. Phil. Mag. 28, 305-319.

SCHENCK C.V., DILLARD J.G. and MURRAY J.W. (1983) Surface analysis and the adsorption of Co(II) on goethite. J. Colloid Interface Sci. 1983, 398-409.

SCHOTT J., BERNER R.A. and SJOBERG E.L. (1981) Mechanism of pyroxene and amphibole weathering - I. Experimental studies of iron-free minerals. Geochim. Cosmochim. Acta 45, 2123-2135.

SCHOTT J. and BERNER R.A. (1983) X-ray photoelectron studies of the mechanism of iron silicate dissolution during weathering. Geochim. Cosmochim. Acta 47, 2233-2240.

SCHOTT J. and BERNER R.A. (1983) X-ray photoelectron studies of the mechanism of iron silicate dissolution during weathering. Geochim. Cosmochim. Acta 47, 2233-2240.

SCOFIELD J.H. (1976) Hartree-Slater subshell photoionization cross-sections at 1254 and 1487 eV. J. Electron Spectros. Related Phenom. 8, 129-137.

SEAH M.P. and DENCH W.A. (1979) Quantitative electron spectroscopy of surfaces: A standard data base for electron inelastic mean free paths in solids. Surface Interface Anal. 1, 2-11.

SEAH M.P. (1983) A review of quantitative Auger electron spectroscopy. Scanning Electron

Microsc. 1983, 521-536.

SHIRLEY D.A. (1972) Relaxation effects on Auger energies. Chem. Phys. Letters 17, 312-315.

SIEGBAHN K., NORDLING C.N., FAHLMAN A., NORDBERG R., HAMRIN K., HEDMAN J., JOHANSSON G., BERMARK T., KARLSSON S.E., LINDGREN I. and LINBERG B. (1967) ESCA: Atomic, Molecular, and Solid State Structure Studied by Means of Electron Spectroscopy. Almqvist and Wiksells, Uppsala.

SMETS B.M.J. and LOMMEN T.P.A. (1981) The incorporation of aluminum oxide and boron oxide in sodium silicate glasses, studied by x-ray photoelectron spectroscopy. Phys. Chem. Glasses 22, 158-162.

SMETS B.M.J. and KROL D.M. (1984) Group III ions in sodium silicate glass. Part 1. X-ray photoelectron spectroscopy study. Phys. Chem. Glasses 25, 113-118.

STEPHENSON D.A. and BINKOWSKI N.J. (1976) X-ray photoelectron spectroscopy of silica in theory and experiment. J. Non-Cryst. Solids 22, 399-421.

STORP S. and HOLM R. (1977) ESCA investigations of ion beam effects on surfaces. Surface Sci. 68, 459-468.

STORP S. (1985) Radiation damage during surface analysis. Spectrochimica Acta 40B, 745-756.

STUCKI J.W., ROTH C.B. and BAITINGER W.E. (1976) Analysis of iron-bearing clay minerals by electron spectroscopy for chemical analysis (ESCA). Clays and Clay Minerals 24, 289-292.

SWIFT P. (1982) Adventitious carbon - the panacea for energy referencing?. Surface Interface Anal. 4, 47-51.

SWIFT P., SHUTTLEWORTH D. and SEAH M.P. (1983) Static charge referencing techniques. In: Practical Surface Analysis by Auger and X-ray Photoelectron Spectroscopy (D. Briggs and M.P. Seah, eds.). Wiley, New York, 437-444.

TASKER G.W., UHLMANN D.R., ONORATO P.I.K. and ALEXANDER M.N. (1985) Structure of sodium aluminosilicate glasses: X-ray photoelectron spectroscopy. J. Physique C8, 273-280.

THOMAS T.D. (1980) Extra-atomic relaxation energies and the Auger parameter. J. Electron Spectrosc. Related Phenom. 20, 117-125.

THOMASSIN J.H., GONI J., BAILLIF P., TOURAY J.C. and JAURAND M.C. (1977) An XPS study of the dissolution kinetics of chrysotile in 0.1 N oxalic acid at different temperatures. Phys. Chem. Minerals 1, 385-398.

TSANG T., COYLE G.J. and ADLER I. (1979) XPS studies of ion bombardment damage of iron-sulfur compounds. J. Electron Spectrosc. Related Phenom. 16, 389-396.

URCH D.S. and MURPHY S. (1974) The relationship between bond lengths and orbital ionisation energies for a series of aluminosilicates. J. Electron Spectrosc. Related Phenom. 5, 167-171.

VEAL B.W., LAM D.J., PAULIKAS A.P. and CHING W.Y. (1982) XPS study of CaO in sodium silicate glass. J. Non-Cryst. Solids 49, 309-320.

VELBEL M.A. (1986) Influence of surface area, surface characteristics, and solution composition on feldspar weathering rates. Geochemical Processes at Mineral Surfaces (J.A. Davis and K.F. Hayes, eds.), ACS Symposium Series 323, 615-634.

VENABLES J.A., JANSSEN A.P., HARLAND C.J. and JOYCE B.A. (1976) Scanning Auger electron microscopy at 30 nm resolution. Philosophical Magazine 34, 495-500.

WAGNER C.D. (1972) Auger lines in x-ray photoelectron spectroscopy. Anal. Chem. 44, 967-973.

WAGNER C.D. and BILOEN P. (1973) X-ray excited Auger and photoelectron spectra of partially oxidized magnesium surfaces: The observation of abnormal chemical shifts. Surface Sci. 35, 82-95.

WAGNER C.D. (1975a) Auger parameter in electron spectroscopy for the identification of chemical species. Anal. Chem. 47, 1201-1203.

WAGNER C.D. (1975b) Chemical shifts of Auger lines, and the Auger parameter. Disc. Faraday Soc. 60, 291-300.

WAGNER C.D., RIGGS W.M., DAVIS L.E., MOULDER J.F. and MUILENBERG G.E. (1979a) Handbook of x-ray photoelectron spectroscopy. Eden Prairie, Minnesota, Physical Electronics Division.

WAGNER C.D., GALE L.H. and RAYMOND R.H. (1979b) Two-dimensional chemical state plots: A standardized set for use in identifying chemical states by x-ray photoelectron spectroscopy. Anal. Chem. 51, 466-482.

WAGNER C.D., SIX H.A. and JANSEN W.T. (1981) Improving the accuracy of determination of line energies by ESCA: Chemical state plots for silicon - aluminum compounds. Appl. Surface Sci. 9, 203-213.

WAGNER C.D., PASSOJA D.E., HILLERY H.F., KINISKY T.G., SIX H.A., JANSEN W.T. and TAYLOR J.A. (1982) Auger and photoelectron line energy relationships in aluminum-oxygen silicon-oxygen compounds. J. Vac. Sci. Tech. 21, 933-944.

WEHBI D. and ROQUES-CARMES C. (1985) Surface roughness contribution to the Auger electron emission. Scanning Electron Microsc. 1985/I, 171-177.

WELLS O.C. and BERNER C.G. (1969) Improved energy analyser for the scanning electron microscope. J. Scientific Instrum. (J. Phys. E) 2, 1120-1121.

WEST R.H. and CASTLE J.E. (1982) The correlation of the Auger parameter with refractive index: An XPS study of silicates using Zr $L\alpha$ radiation. Surface Interface Anal. 4, 68-75.

WHITE A.F., YEE A. and FLEXSER S. (1985) Surface oxidation - reduction kinetics associated with experimental basalt - water reaction at 25°C. Chem. Geol. 49, 73-86.

WHITE A.F. and YEE A. (1985) Aqueous oxidation-reduction kinetics associated with coupled electron-cation transfer from iron-containing silicates at 25°C. Geochim. Cosmochim. Acta 49, 1263-1275.

YAU Y.W., PEASE R.F.W., IRANMANESH A.A. and POLASKO K.J. (1981) Generation and applications of finely focused beams of low energy electrons. J. Vac. Sci. Tech. 19, 1048-1052.

YEH J.J. and LINDAU I. (1985) Atomic subshell photoionization cross sections and asymmetry parameters: $1 < Z < 103$. Atomic Data and Nuclear Data Tables, 32, 1-155.

YIN L., GHOSE S. and ADLER I. (1972) Investigation of a possible solar-wind darkening of the lunar surace by photoelectron spectroscopy. J. Geophys. Res. 77, 1360-1367.

YIN L., TSANG T. and ADLER I. (1976) On the ion-bombardment reduction mechanism. Proceedings of the Lunar Science Conference 7, 891-900.

YIN L.I., GHOSE S. and ADLER I. (1971) Core binding energy difference between bridging and nonbridging oxygen atoms in a silicate chain. Science 173, 633-635.

ZINGG D.S. and HERCULES D.M. (1978) Electron spectroscopy for chemical analysis studies of lead sulfide oxidation. J. Physical Chem. 82, 1992-1995.

Chapter 14

LUMINESCENCE, X-RAY EMISSION AND NEW SPECTROSCOPIES

LUMINESCENCE SPECTROSCOPY

Luminescence processes comprise phenomena which respond to incident radiation or particles with the emission of visible light in excess of that produced by thermal black body radiation. Visible light emission processes due to thermal radiation are termed incandescence. In practice we distinguish among the types of luminescence by reference to the kind of incident radiation or particles, and by the kinetics of the emission process. Thus cathodoluminescence (CL) is a response to incident high-energy electrons, as in an electron microprobe. In similar respect, photoluminescence is a response to photons, chemiluminescence is a response to chemical radicals, candoluminescence is a response to gas phase radicals (Ivey, 1974), and radioluminescence is a response to high-energy particles. The kinetic distinction holds that photoluminescence emission that occurs faster than 10^{-8} s is called fluorescence, whereas slower emission is called phosphorescence, but this is entirely arbitrary. A more meaningful distinction is based on differences in the luminescence mechanism to be described below. Along these lines, thermoluminescence (TL) is identical to phosphorescence, except that the usual mode of TL spectrum collection uses temperature rather than time as independent variable.

Luminescence spectroscopy is a constantly evolving and highly promising methodology. Many new techniques in luminescent analysis are appearing almost daily, along with improved theoretical models for the understanding of emission and energy transfer processes. For example, the application of tunable lasers has particularly enhanced luminescence studies, both in resolution (such as fluorescence line-narrowing methods (Macfarlane, 1987; Selzer, 1981)) and sensitivity. The promise of the methodology derives from the many types of information that can be derived, at least in principle, from mineralogical samples. These include trace element, defect and interstitial identities (down to the ppb range in some cases) the nature of the sites at which these species reside, the interaction and physical separation of the species, and their electronic and thermal properties. All of this information can be used to determine the chemical, thermal and deformational history of the material. The subject development here proceeds on a fairly basic level, without recourse to much theory. The aim is to provide a working knowledge of mineralogical luminescence processes without getting bogged down in excessive complexity. The reader is encouraged to follow up references to detailed works in the literature, although a true appreciation for some of the material may require considerable depth in the theory of solid-state physics. Applications described here make no effort to cover the tremendous body of work presently available. Instead, representative work is described.

Reviews of luminescence in minerals generally are aimed at the mineral collector (Robbins, 1983; Gleason, 1960), but there are several reviews aimed at the mineralogist, e.g., Walker (1985) and Geake and Walker (1975). However, many reviews of luminescence in solids are available, e.g., Curie (1963), Williams (1966), Johnson (1966), White (1975) and Leverenz (1968). An interesting historical perspective is given by Harvey (1957) of all types of luminescence from ancient times up to the twentieth century. An overview of luminescence and related areas of mineralogical spectroscopy by Marfunin (1979) gives the most detailed treatment in a mineralogical context, though most of the references are from the Russian literature. Reviews of CL in minerals by Nickel (1978) and Remond (1977), and in inorganic solids by Garlick (1966) are notable. Very thorough works on TL are those of McKeever (1985) and McDougall (1968). Barnes (1958) has compiled observations of visible and infrared luminescence in an extensive list of minerals.

TYPES OF LUMINESCENCE SPECTRA AND MODES OF EXCITATION

Luminescence emission may be from the bulk of the material itself, or from isolated impurities, clustered defects, exsolved phases, included phases, or grain boundary contaminants. In general, the emission process involves the transition of a electron from an excited state to one lower in energy. As the electronic structure of an atom is related to its particular quantum states

640

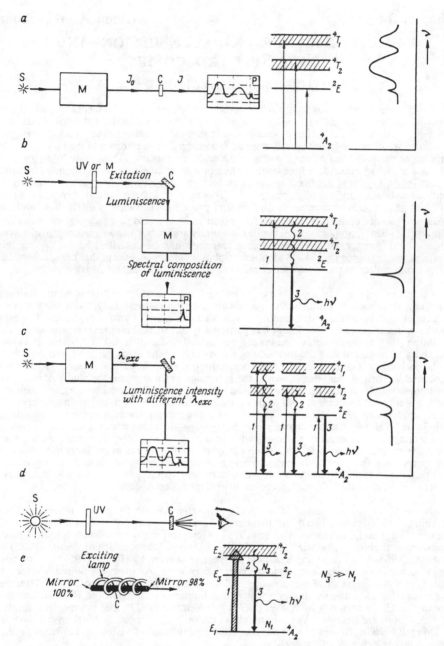

Figure 1. (a) Absorption experiment (S = source, C = crystal, P = plot of spectrum, M = monochromator). The observed transitions are in the Cr^{3+} ion in α–Al_2O_3. (b) Emission spectrum. The monochromator analyzes the fluorescence and phosphorescence output. Emission is from only a single narrow band. (c) Excitation spectrum. The integrated luminescence intensity is measured as a function of varied excitation energy. In the case shown all absorption transitions excite luminescence via the ^2E->$^4A_2$ transition. A second monochromator is usually used to separate specific emission bands for study. (d) Visual observation of luminescence, as with a UV lamp. (e) The emission of a ruby laser. Excitation into the $^4T_2$ level by flash lamp pumping, followed by radiationless decay to level ^2E and subsequent emission. [From Marfunin, 1979]

and the symmetry and local fields at its site, analysis of the luminescence from a sample can reveal the type of atom and the structural site responsible for the luminescent activity. The analysis of this process is the realm of luminescence spectroscopy, which includes emission spectroscopy of the visible light emission, and excitation spectroscopy of the light required to excite the emission. The relation of these techniques to absorption spectroscopy is shown in Figure 1. The emission spectrum will only show lines or bands (broad lines as opposed to energy bands within the context of band structure) corresponding to electronic transitions which emit light. The excitation spectrum resembles the absorption spectrum but will only contain bands which represent transitions connected with the emission. Thus the excitation spectrum is generally a type of subset of the absorption spectrum. The intensities of lines and bands seen in the absorption and excitation spectrum differ as well. The absorption line intensity is related only to the oscillator strength of the single transition identified, whereas the excitation line intensity is related to the oscillator strength of the absorption transition, the efficiency of internal electronic processes which transfer energy to the final excited state from which emission occurs, and to the efficiency of the emission transition. The reader is referred to the chapter on optical spectroscopy for a discussion of the probability of electronic transitions and oscillator strengths.

Photons in the visible range have energies from 1.6 to 3.25 eV, which correspond to typical electronic energy level separations of transition metal and rare earth impurities in insulating and chalcogenide materials. Hence these elements are most commonly associated with luminescence and coloration in minerals.

Luminescence may be produced by a wide variety of processes. In this section we mainly consider ultraviolet-photon and electron stimulated luminescence, as these are the types most likely to be used by earth scientists. Most UV-excited luminescence is produced using low pressure mercury discharge lamps, in which the gas pressure is optimized to preferentially excite one of the mercury ion electronic transitions. The commonly used transitions emit short-wave ultraviolet at 2537 Å, and long-wave ultraviolet at 3650 Å. The shorter wave emission does not penetrate regular glass appreciably, hence fused silica lamp envelopes must be used. Other sources of UV are xenon and hydrogen discharge lamps. Laser excitation in the UV has also been utilized with the advantage of very selective excited state generation and high excitation energy density. Cathodoluminescence is commonly observed under 10-20 keV electron bombardment with the electron microprobe beam, though this usually allows only qualitative (by eye or photograph) evaluation. A similar acceleration voltage is used in devices dedicated to cathodoluminescence excitation, such as the Luminoscope (Nuclide Corporation), which uses a cold cathode gas discharge. Other devices have been described by Zinkernagel (1978) and Ramseyer (1982) that used harder vacuum and have no ionized gas. Such gas produces ion bombardment which can create radiation damage in the samples, and there is also a luminescence from the gas discharge itself. Direct electron stimulation has several advantages over UV stimulation. The energy density is much greater than for UV photons as the penetration depth of 15 keV electrons is on the order of microns, whereas UV photons penetrate from a fraction to several mm (Fig. 2). Additionally, in the case of the microprobe beam, focusing laterally to a few microns is possible, so that single mineral grains, inclusions, or grain boundaries can be excited for spectroscopy. Thus cathodoluminescence is much brighter than UV luminescence and is able to discern spatial variations on a relatively fine scale. Because of the excitation energy density, many more minerals are observed to cathodoluminesce than luminesce under UV.

ENERGY LEVEL DIAGRAMS AND SIMPLE MODELS

FOR FLUORESCENCE CENTERS

Luminescence can be activated by various electronic excitation schemes. The most simple type involves absorption and emission of light by a single ion on a particular crystallographic site, largely independent of the rest of the host structure. The best examples of this type of luminescent activity are minerals activated by lanthanide rare earth ions, e.g., apatite, fluorite and zircon, and by the actinide rare earths, notably via the uranyl ion $(UO_2)^{2+}$. In these minerals, the luminescence is often characteristic of the type of rare earth, and may be largely unaffected by variations in mineral structure. This is so because the electronic energy levels responsible for much of the absorption and emission of light are mainly of shielded 4f or 5f orbital

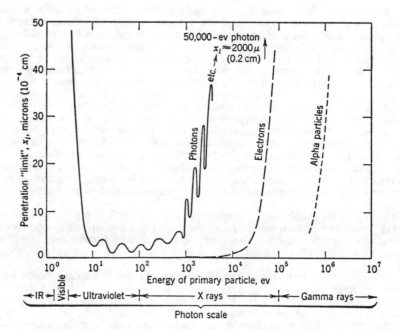

Figure 2. Penetration depths of various radiation sources in a hypothetical optically transparent solid of density 4 g cm⁻³. The irregular fluctuations in the photon curve are due to photoelectric absorption edges. [From Leverenz, 1968]

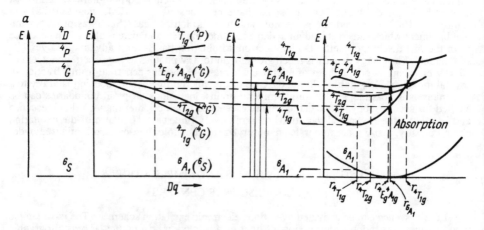

Figure 3. Electronic energy level diagram and configurational coordinate model for Mn²⁺ in octahedral oxygen coordination. (a) Ground and three excited states for a d⁵ electronic configuration in the absence of a crystal field. (b) Splitting of states in an octahedral crystal field of varying strength Dq. (c) Energy level diagram for fixed field Dq value, with possible absorption transitions. (d) Configurational coordinate diagram of variation of energy levels at fixed Dq with lattice vibrations. Lowest temperatures (minimum lattice vibrations) are at the bottom of parabolic-like curves. At higher temperatures the vibrations effectively spread the energy of the band, smearing the transition energies over a range. [From Marfunin, 1979]

character and, as such, are not involved in bonds with nearby atoms. Hence variations in the symmetry and strength of the crystal field results only in relatively subtle changes in the relative energy (and thus emission) from these levels. For example, the emission spectrum of the uranyl ion in many uranium minerals, in silicate glass, and in aqueous solution are very similar (Jorgensen, 1979; Hoffman, 1970; Blasse, 1968). The emission of the uranyl ion is greenish-yellow when observed, almost independent of its environment in a material.

Electronic levels in a rare earth may also be of mixed character, e.g., 4f-3d. These levels will be much more sensitive to variations in site geometry and local crystal field than 4f levels, and lead to variations in associated emission bands for the same RE ion in different structures. Other ions that activate luminescence, viz. the first row transition-metal ions, have 3d electronic levels that are quite sensitive to site geometry. As the luminescence due to these ions usually involves such 3d levels, the luminescence spectrum is often characteristic of the ion site.

A typical example is the Mn^{2+} ion in octahedral coordination in silicates, oxides, carbonates and phosphates. This activator configuration is responsible for much of the orange-red luminescence observed in minerals. The electronic structure of isolated Mn^{2+} is shown in Figure 3a. The term symbols refer to differing spin and angular momentum multiplicities of the 3d electrons in the ground (S) and three excited states. Figure 3b depicts the splitting of these states in an octahedral crystal field as a function of field strength. For a given field strength the energy levels can be specified as in Figure 3c. As the energy levels shift in energy with variation in interatomic distance, the complete energy level diagram must include this dependence which allows evaluation of the effects of bond vibrations (Fig. 3d). A typical absorption process is indicated by the vertical arrow. This represents absorption of a photon (or energy from some other source) which excites an electron in the ground $^6A_1$ state creating the $^4T_{1g}$ excited state. The x-axis in the drawing represents interatomic distance in a general way, and is called the configurational coordinate. The fact that electronic transitions are so much faster than crystal lattice vibrations allows the assumption that the transition path is strictly vertical (Franck-Condon principle). Although decay of the excited state could conceivably occur instantaneously back along the same arrow to the ground state, in reality an excited state has a lifetime of at least 10^{-8} s (it is longer, about 10^{-3} s, in Mn^{2+} because the transition is spin and symmetry forbidden). This allows time for the excited state to relax toward a lower energy with a reduction in the Mn-O distance. This contraction excites a lattice vibration which carries off the energy lost by the excited electronic state. Emission can now occur from the lowest energy point on the excited state curve back to the ground state. These processes are illustrated in Figure 4, with the wiggly line indicating the vibrational relaxation process in excited and ground states. The emission transition is smaller in energy than the absorption process, a general consequence in luminescent processes. The emitted light is thus of longer wavelength than the absorbed light, a fact first documented by the British physicist George Stokes, and now called Stokes' shift. Ion-lattice interaction is indicated by the relative state energy curvature in Figures 3d and 4. The larger this interaction, the larger is the Stokes shift and, assuming that emission occurs from more than one vibrational substate within the excited state, the broader the emission band. The measure of the ion-lattice coupling is the Huang-Rhys factor or S parameter (Walker, 1985). Rare earth ions can have very small lattice interactions so that their emission transitions can be virtually the same energy as their excitations and have extremely narrow line widths.

Absorption transitions can also occur into the other excited states in the diagram. However, where lines cross, higher energy states can revert to lower energy states, and emission may not occur from the particular state excited. The probability of transitions to the excited states varies with the symmetry of the ground and excited states, and the selection rules of the transition. Hence particular photon energies may be very poor at producing an excited state, whereas others easily cause the transition. In an example like Mn^{2+}, the ground to ^4E_g transition is not the most likely, so that light in this part of the spectrum will not cause the most efficient production of luminescence. Higher energy ultraviolet will excite transitions of higher probability to higher energy excited states, resulting in increased luminescent response.

The general possibilities are shown in Figure 5. The typical Stokes shift is seen in Figure 5a. Figure 5b shows a symmetrical configurational coordinate model which gives rise to nearly identical absorption and emission energies (zero Stokes shift). This case gives rise to strong

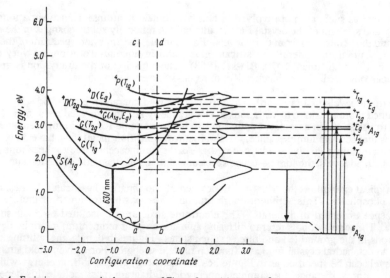

Figure 4. Emission processes in the system of Figure 3, modeled on Mn^{2+} in calcite. Additional crystal field states are shown. Absorption into any level occurs by transitions between the a-b window, representing lowest vibrational excitation (window widens with increasing temperature). Note that the curvature of the levels interacts with the window width (and hence temperature) to determine absorption spectrum band widths. Levels above T_{1g} intersect the T_{1g} curve and can relax to T_{1g} without luminescent emission. The final emission occurs after T_{1g} relaxes to its lowest vibrational energy (wiggly arrow). Additional relaxation of A_{1g} to its lowest vibrational state is the final process. The Stokes shift is a consequence of the difference in the lowest energy points for the A_{1g} and T_{1g} curves. [From Marfunin, 1979]

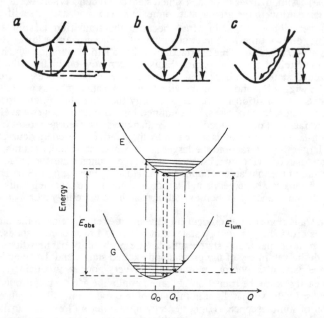

Figure 5. Possibilities for energy exchange between the excited and the ground state. (a) Stokes shift. (b) Zero Stokes shift (also occurs if the excited state curve is flat). (c) Relaxation of excited state through thermal quenching. [From Marfunin, 1979] (d) Configurational coordinate diagram with detailed vibrational sublevels. Short dashed lines represent transitions without the generation of lattice phonons (no phonon transitions---pure electronic transitions). Long dashed lines represent the limits of energy of the emission and absorption transitions, resulting in the largest possible Stokes shift. Solid lines represent transitions from the lowest vibrational states. [From Walker, 1985]

self-absorption in the material, weakening the emission. A similar situation results from a flat energy versus bond length dependence. A system may have interlevel crossings that quench luminescence (Fig. 5c) because the excited state energy is removed entirely by lattice vibrations. Figure 5d shows a more detailed picture of the vibrational sublevels. The two solid lines indicate transitions from the center coordinate positions of the two lowest vibration levels of the excited states. A small Stokes shift is produced. The two long dashed lines indicate the widest possible energy separation. This transition would produce the largest Stokes shift. The short dashed lines indicate no-phonon (also called zero-phonon) transitions, i.e., purely electronic transitions. In actual cases, there are usually many vibrational-electronic transitions in a band, with the band width proportional to the curvature of the energy level.

The analogous configurational coordinate diagram for Mn^{2+} in tetrahedral coordination would look much the same as that of Figures 3 and 4. The energy levels are in the same order as in the octahedral case (true only for d^5 systems, otherwise the levels are reversed in order), but with a smaller crystal field splitting. The splitting is 4/9ths that of the octahedral system in the ideal case. This smaller splitting increases the energy of the $^4T_1$ level (called $^4T_{1g}$ in the octahedral case) relative to the ground $^6A_1$ state ($^6A_{1g}$ in the octahedral case), shifting the emission to higher energy. The characteristic emission color is green. Mn^{2+} luminescence may not always be simply related to coordination, however, as pairing of green-emitting centers in silicate glass may produce Mn-Mn interaction and red luminescence (Turner and Turner, 1970), and the effect of bond distance and site distortion may overide coordination in determining the ground to first excited state splitting.

Luminescence from non-cubic single crystals is polarized due to symmetry constraints on the possible electromagnetic interactions between excited and ground states. The case is identical to the group theoretical analysis of the allowed bands and symmetries of absorption spectra. Crystals may contain strain which distorts the local symmetry and partially allows emission with a forbidden polarization component. Thus luminescence is sensitive to the elastic properties of crystals. If strain alters bond distances, the energy of the emission will shift as well.

ACTIVATORS, SENSITIZERS AND LUMINESCENCE QUENCHING

The Mn ion discussed above is an example of an activator. Small amounts of Mn, doped into six-coordinated sites in calcite or silicate minerals, will create a weak red-orange luminescence. The weakness is due to the spin-forbidden nature of the excitation transitions. Many other ions can serve as activators, but the most important are transition metals and rare earths with partially filled d and f shells, respectively. The concentrations of activators necessary to produce observable luminescence can be minute, even down below the ppm range. Hence in principle, luminescence analysis can serve as a form of trace element analysis. The emission spectra created by various activators in synthetic sphalerite and in natural fluorites are shown in Figure 6.

This picture is unfortunately complicated by the interaction of activators. Two activators, simultaneously present in a mineral may interact with one another indirectly or directly. In indirect interaction, the emission from one activator is partially absorbed by the second activator, creating emission in the latter. This process has been called cascade fluorescence. In such a case, the excitation (absorption) and emission spectrum of each activator is not very strongly altered and can be compared with spectra of the isolated activators. In direct interaction, the excited state of one activator transfers its energy to the other activator. No (or reduced) emission occurs from the original activator (in this case called a Sensitizer), but the second activator is pumped into an excited state from which emission occurs. This process is seen most notably with Pb^{2+} and Mn^{2+} in calcite and in many of the spectacularly luminescent minerals from Franklin, New Jersey. The Pb^{2+} has strong absorption in the ultraviolet, and by itself is an activator of luminescence in ore minerals (cerussite, anglesite) and silicates (feldspars). However, if located near a Mn^{2+} activator, there is a high probability of energy transfer between the two ions, creating the excited state in the Mn ion.

As the Mn^{2+} transitions are spin-forbidden, the absorption process in the absence of the sensitizer would produce weak emission because the excited state is sparsely populated. If

646

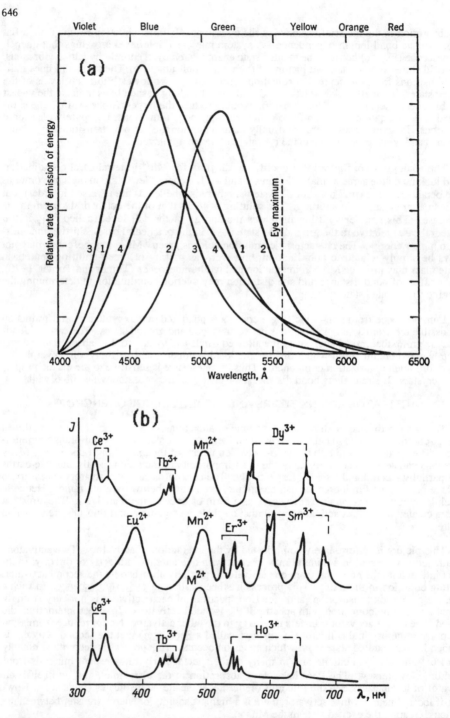

Figure 6. Emission spectra of various activators. (a) In synthetic sphalerite. 1. Interstitial Zn. 2. Cu+ (0.01 wt %). 3. Ag+ (0.008 wt %). 4. Au+ (0.002 wt %). Mn²⁺ activation produces a band at about 5910 Å. [From Leverenz, 1968] (b) In natural fluorites with varying concentrations of Mn²⁺ and RE activators. [From Marfunin, 1979]

energized by energy transfer from the Pb ion, which has strong absorption, the excited Mn^{2+} state is well-populated and the emission is intense. The necessary condition of activator-sensitizer proximity gives rise to a concentration dependence on the emission intensity. The probability of energy transfer varies with r^{-6} to r^{-8}, where r is the interactivator separation, and with the lifetime of the excited state in the sensitizer. In general, only a fraction of an atom percent of Pb is necessary to sensitize most of the Mn in such luminescent materials. The excitation spectra for synthetic calcite doped with Mn^{2+} alone and with additional sensitizers are shown in Figure 7. The absorption bands responsible for the luminescence are centered near 2500 Å in the case of Pb and Tl sensitizers, and this explains the strong luminescence of such doped calcites to short-wave ultraviolet. Rare earth sensitization shifts the absorption bands responsible for emission to about 3100 Å. Ce^{3+} may be a common sensitizer for Mn^{2+} in many natural calcites, and it may also assume an activator role (Blasse and Aguilar, 1984).

The direct transfer of energy between ions can be further classified as resonant or nonresonant, depending on whether the energy transferred to the activator exactly matches a ground to excited state transition energy, or is larger or smaller respectively. In the latter case, the proper energy for excitation of the activator is adjusted by creation or annihilation of a lattice vibration.

Luminescence quenching occurs by several processes. Essentially all excited states can have their energy dissipated to the lattice by various types of thermal quenching. This is the fundamental reason why luminescence is always limited at high temperatures. Thermal quenching refers to the increasing probability of activating new lattice vibration modes as temperature increases, some of which will interact with an excited state to produce a nonradiative transition. Ions that have large charge-transfer absorption tails overlapping the emission peaks of activators will absorb this emission energy and terminate luminescence. This quenching is not specific and will affect any type of activator or activator-sensitizer combination. The most common agent of this sort, called a killer or poison, is Fe^{3+}. This ion has exceptionally strong charge-transfer bands, and will quench Mn^{2+} emission in many minerals with a threshold of only about 0.1%. However, luminescence can sometimes be observed in Fe^{3+} minerals if the emission is in the red to infrared part of the spectrum. Fe^{3+} can in some cases (feldspar, spinels) actually be an activator on this basis (White et al., 1986). Fe^{2+}, Ni^{2+} and Co^{2+} normally have much weaker charge-transfer bands, and thus can co-exist with activators up to a concentration of several percent. The effect of Ni^{2+} concentration in Ag-activated synthetic sphalerite and Mn-activated synthetic willemite is shown in Figure 8.

Self-quenching of luminescence can occur if there is an unfavorable interaction of activators or sensitizers at high concentrations. This does not occur with Mn^{2+} and WO_4 in $MnCO_3$ or $CaWO_4$, respectively, but is observed in most rare earth compounds. The quenching is due to energy transfer between nearby ions of the same type, where the excitation of the recipient ion is into a state which can decay nonradiatively. This process requires close energy agreement between the transitions in the donor and receiving ion. Such interaction causes rare earth activators to have maximum efficiency at a moderate concentration level, perhaps less than one percent. If the energies of the transitions are not similar, transfer of energy is less likely and "concentration quenching" occurs at much larger activator concentrations. Another type of self-quenching is self-absorption of the emission. This is common in rare earth activation, but also occurs in transition activation, as with Cr^{3+} in ruby.

FLUORESCENCE, PHOSPHORESCENCE AND THERMOLUMINESCENCE

By popular acceptance, fluorescence is light emission that occurs promptly following excitation and decays quickly once excitation has stopped. Phosphorescence is pictured as emission which continues for a considerable time, sometimes even for years (Millson and Millson, 1964), after excitation has ceased. The difficulty with this picture is that rather different processes can produce emission with about the same time dependence. A more practical distinction is one that separates the two in terms of processes which are temperature independent and temperature dependent. Thus, fluorescence is due to electronic transitions within single ions, or energy transfer between closely associated ions in a structure. Phosphorescence involves energy transfer associated with energy bands (as in band theory), and the interaction

648

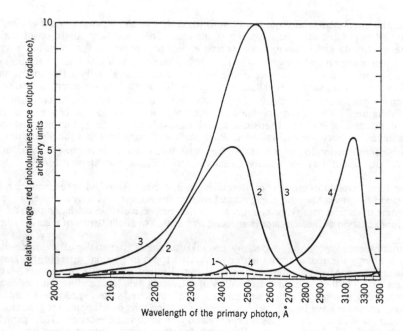

Figure 7. Emission spectra of synthetic calcite activated by Mn^{2+} (1), and sensitized by Tl (2), Pb (3) and Ce (4). [From Leverenz, 1968]

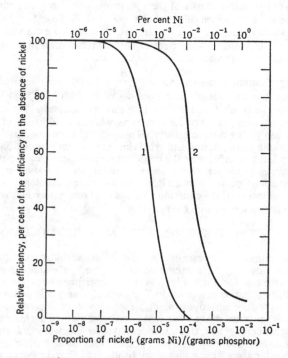

Figure 8. Quenching action of Ni^{2+} in 1. Synthetic sphalerite activated by Ag^+ and excited by CL. 2. Synthetic willemite activated by Mn^{2+} and excited by short wave UV. [From Leverenz, 1968]

with <u>electron traps</u> which prevent electron transfer until suitable thermal stimulation is provided. To illustrate this, Figure 9 shows a fluorescence process in a hypothetical tetrahedrally coordinated ion, such as Mn^{2+} in willemite. The steps in the luminescence process begin with a stable ground state (a). This is stimulated to an excited state by a UV photon or high energy electron (b). The excited state adjusts structurally by losing heat as lattice vibrations (c). Emission follows (d), then the ground state relaxes with further dissipation of vibrational energy. The cartoons depicting band structure indicate that the conduction band (shaded) is never involved, but only the band gap energy levels of the Mn ion. The structural cartoons show the expansion of the electron cloud of the Mn ion upon excitation, and the accompanying lattice expansion. These changes are reversed after emission occurs. Note that the configurational coordinate diagrams introduce the notion of a varying probability distribution in the wavefunctions of the electronic-vibrational states. The highest probability tends to be at the limits of the energy level curves, and thus transitions are most likely to these coordinate positions, rather than at the exact low point of the energy level curve.

Figure 10 similarly illustrates the processes for one type of thermally stimulated phosphorescence. In this case, the excitation moves an electron from the activator (X_1) into the conduction band (a). An impurity ion (X_2) with an energy level close to the conduction band, but within the band gap, acts as an electron trap which binds the electron if it loses some thermal energy (b). The electron remains trapped on the impurity until sufficient thermal stimulation excites it back into the conduction band (c). The electron then can lose thermal energy to become bound to the original type of activator ion, and then produces emission with accompanying relaxation and heat liberation as it settles into the stable ground state (d). Both the activator and electron-trap impurity center behave similarly in this scheme.

In the opposite situation the activator can accept an electron from the valence band, producing a positively charged hole which can then be trapped by an impurity with an energy level near the top of the valence band. Thermal stimulation in this case acts to free the hole. Other possibilities include the presence of both types of activators, as well as activators with differing types of charge compensation. These include local charge compensation such as anion vacancies which can trap electrons, and charge-compensating pairs of activators (ZnS: Cu^+, Ga^{3+}), and nonlocal charge compensation via the production of holes and electrons in the valence and conduction bands, respectively. These scenarios can lead to complex luminescent behavior and emission processes, including activator-to-band and activator-to-activator transitions. Some of the situations are pictured in Figure 11. The reader should consult the literature on chalcogenide phosphor materials for a thorough explanation of the mechanisms and kinetics, e.g., Shionoya (1966) and Williams (1968).

Processes involving trapping are temperature sensitive, with the dependence related to the depth (energy) of the trapping level. The greater the depth, the more thermal stimulation is necessary to free the electron. Hence deep traps lead to long-decay phosphorescence, and shallow traps to short decays. For any given trapping depth the higher the ambient temperature, the shorter the decay rate. It is possible to probe trapping levels, and hence the type of impurity states in phosphorescent minerals, by scanning the excitation spectrum with IR-visible light. The depth of the levels will be directly reflected in the energy of the bands in the excitation spectrum. This operation of using low-energy light to produce a higher-energy emission is an example of a stimulated emission process. It is the basis of the ruby laser.

Where phosphorescence is normally studied at a given temperature, we can also study the emission as sample temperature is ramped up. The phosphorescence is then called thermoluminescence. Ramping up temperature increases the mean of the Boltzmann distribution of thermal energy available in a solid, so that this experiment is similar (though of much lower resolution) to the excitation spectral scan. The function obtained (emission intensity versus temperature) is called a <u>glow curve</u> (example in Fig. 12). At each maximum in the glow curve, it is possible to record the complete emission spectrum. Thus information can be obtained about the depth of the trapping level and the associated activators and sites producing the luminescence.

Thermoluminescence is also associated with trapping levels not populated by the excitation of activators, but rather by radiation damage. In the case of radiation-produced trapped electrons

650

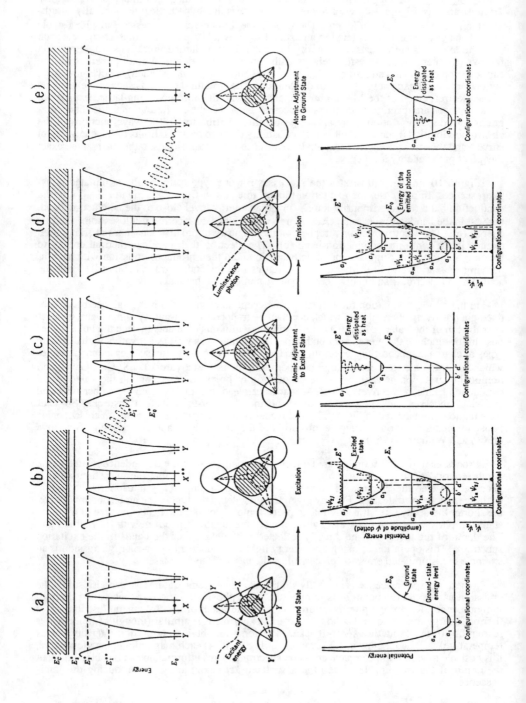

Figure 9. Band, structural and configurational coordinate model for fluorescence activated by a tetrahedrally coordinated cation. (a) Ground state. Electron localized on cation X (Mn^{2+}). Anions indicated by Y. E_c^* = conduction band, E_1^{***}, E_2^* and E_3^* are excited states of the Mn^{2+} ion. (b) Excitation raises the electron to the E_1^{**} state. The excited activator ion expands in size. The transition is not from the lowest vibrational state of E_0, but from the highest probability position of the wave function. (c) Anions relax outward from the excited cation. The relaxation produces a vibration and loss of energy as heat. The excited state E_1^{**} moves to a lower energy vibrational state. (d) Emission occurs spontaneously from cation by E_1^* to E_0 transition. As in adsorption, the transition is not from the bottom of the E_1^* curve, but from the highest probability position in the E_1^* and E_0 wavefunctions. The lowest part of the diagram shows the product $\Psi_1 \Psi_2 = \Psi$ ground Ψ excited , which is proportional to the transition probability. The Mn^{2+} ion contracts in size. (e) Contraction of the anions about the manganese ion with loss of vibrational energy as heat to return to the original configuration in (a). [From Leverenz, 1968]

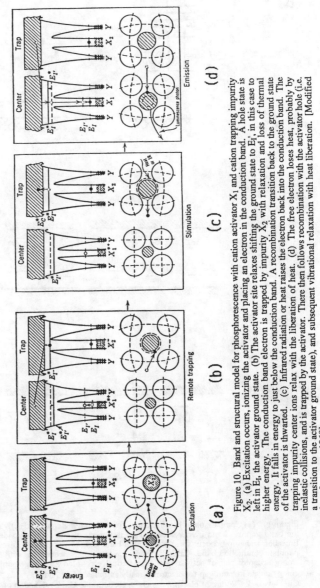

(a) Excitation (b) Remote trapping (c) Stimulation (d) Emission

Figure 10. Band and structural model for phosphorescence with cation activator X_1 and cation trapping impurity X_2. (a) Excitation occurs, ionizing the activator and placing an electron in the conduction band. A hole state is left in E_1, the activator ground state. (b) The activator site relaxes shifting the ground state to E_1', in this case to higher energy. The conduction band electron is trapped by impurity X_2 with relaxation and loss of thermal energy. It falls in energy to just below the conduction band. A recombination transition back to the ground state of the activator is thwarted. (c) Infrared radiation or heat raises the electron back into the conduction band. The trapping impurity center ions relax with the liberation of heat. (d) The free electron loses heat, probably by inelastic collisions, and is trapped by the activator. There then follows recombination with the activator hole (i.e. a transition to the activator ground state), and subsequent vibrational relaxation with heat liberation. [Modified from Leverenz, 1968]

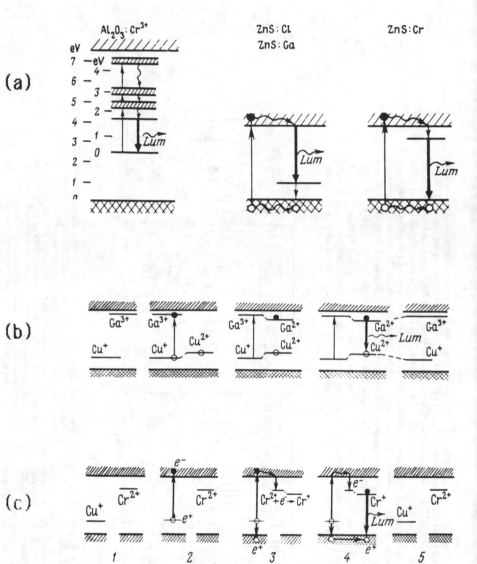

Figure 11. (a) Comparison of band schemes for different materials and activators. $Al_2O_3:Cr^{3+}$. All transitions in the band gap, no band involvement. This is typical of silicate and oxide activation. ZnS:Cl or ZnS:Ga. Emission transitions occur from conduction band to activator levels. Direct band gap ionization necessary to populate conduction band. ZnS:Cr. Emission transitions occur from activator levels to the valence band. (b) Donor-acceptor pair of activators, electrostatically coupled for charge compensation. ZnS:Cu,Ga. 1. Starting energy levels in band gap for electrostatically associated Cu^+ and Ga^{3+}. 2. Cu^+ ionized to Cu^{2+}, relaxation of Cu^{2+} level. 3. Capture of electron by Ga^{3+}, reduction to Ga^{2+}, relaxation of Ga^{2+} level. 4. Recombination with luminescent emission. 5. return to original state by thermal relaxation. (c) $ZnS:Cu^+Cr^{2+}$ coactivators. 1. Cu^+,Cr^{2+} Activator levels in band gap. 2. Excitation-ionization of Cu^+ to Cu^{2+}, free electron produced in conduction band. 3. Free electron from conduction band captured by Cr^{2+}. Transfer of hole state from Cu^{2+} to valence band (this is the same as capture of an electron from the valence band by Cu^{2+} with production of a hole). 4. Recombination of electron on Cr^{2+} with free hole in valence band. 5. Relaxation to initial states. [After Marfunin, 1979]

in minerals, excitation of thermoluminescence will empty the traps, and no additional trapping will occur since the source of radiation is no longer present. Radiation-produced thermoluminescent activity is of considerable geological interest as mechanical deformation of minerals, recrystallization during metamorphism or heating during magma emplacement or hydrothermal activity can affect the trapping levels and thus the appearance of the thermoluminescence spectrum. Thermoluminescence seems to be a widely studied process, but is in need of increased critical evaluation as to the nature of the trapping centers and the mechanism by which they are influenced by petrologic processes.

APPLICATIONS

Trace element analysis and crystal chemistry

As noted earlier, an advantage of luminescent techniques is their sensitivity to extremely small concentrations of activators and sensitizers. This advantage is mitigated by the difficulty of attaching significant quantitative concentrations to intensity measurements without standards of nearly identical constitution. Another advantage is sensitivity to variations in site geometry, charge compensation and activator valence. This opens up the possibility of site-occupancy determination and perhaps site geometry characterization, both of which are important concepts in the description of short-range order, site partitioning and geothermometry. Unfortunately, this advantage is also mitigated by the difficulty of unraveling luminescence data, especially if specially prepared single crystals are not used as model compounds. The following section consists of brief reviews of work performed on minerals that commonly display luminescence activity.

Apatites. In their survey of CL, Smith and Stenstrom (1965) found a large variety of colored emissions from apatites of differing origins. Some of the variations were apparently due to bulk compositional differences, but others were related to chemical variations during crystallization, resulting in color variations in crystal rims and growth zones. As apatites are known to concentrate rare earth ions, it is likely that the various responses were due to fluctuations in the rare earth distribution in these samples. Portnov and Gorobets (1969) made similar observations from a wider set of rock types mainly from the Soviet Union. Mariano and Ring (1975) made detailed spectral CL measurements on apatites from six locations with europium concentrations of from 130 to 500 ppm by weight. Their work indicated that Cl is a good tool for determining the relative concentrations of several rare earths, and the proportion of the europium valence states. Divalent europium was prevalent in carbonatite and igneous apatites. Other important activators are samarium and dysprosium. Growth zonation and other petrological features of the luminescence were not considered.

Knutson et al. (1985) examined the short-wave ultraviolet-excited luminescence of apatite crystals from the Panasqueira tin-tungsten deposits in Portugal. They were able to take photographs of sections of the apatite crystals specific to the separate luminescence emission bands. These photographs demonstrated remarkable zonation. Statistical correlations allowed connection of the neutron activation analyses with the observed luminescence intensities from each of the bands. The results showed that the stronger emission bands were due to cerium, Eu^{2+} and rare-earth-sensitized Mn^{2+}. Over relatively short sampling distances (less than ten meters), the luminescent zonation, and hence the rare earth zonation, of apatites are identical. This indicated a complex crystallization history for the Panasqueira ore body with many localized hydrothermal cells.

Roeder et al. (1987) examined the rare earth element distribution in a suite of 14 apatites from a variety of geologic environments. They were able to identify Sm^3, Dy^{3+}, Tb^{3+}, Eu^{3+}, Eu^{2+} and Mn^{2+} bands in their CL spectra by comparison with spectra of earlier workers. The apatites had a wide range of Ce/Y ratios, indicative of LREE/HREE enrichment variations. The CL spectra had well-resolved Sm^{3+} and Dy^{3+} features (Fig. 13) whose peak height ratios were roughly compatible with the Ce/Y ratios. Thus CL may be useful as a convenient method for determining relative REE fractionation. Eu^{3+}/Eu^{2+} ratios could also be roughly estimated from the CL spectra, but because of the variability of the Eu^{2+} band positions, this could not be easily

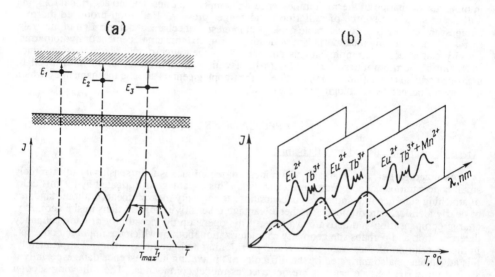

Figure 12. Thermoluminescence Glow Curves. (a) Relation of trapping level depth to temperature of emission. Temperature becomes a direct measure of stimulation energy. (b) Emission spectra can be collected at each thermoluminescence peak, thus allowing association of trapping levels with specific activators and luminescence mechanisms. [After Marfunin, 1979]

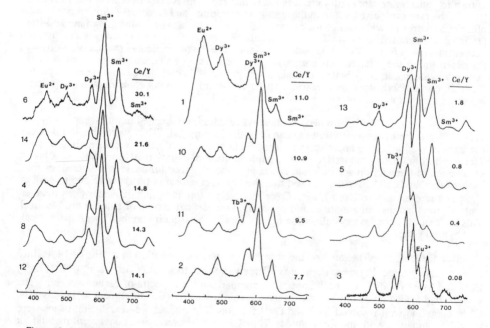

Figure 13. Cathodoluminescence spectra of apatites from a variety of localities. The Ce/Y ratios are indicated. [From Roeder et al., 1987]

done quantitatively. Eu^{2+} transitions involve valence electron states that may interact strongly with the crystal lattice (Palilla and O'Reilly, 1968). Thus broad emission bands and site-specific spectra are to be expected. Llallagua (Bolivia) apatite had a definite positive Eu anomaly of about one hundred times the other rare earths. By comparison with glass Eu standards equilibrated under differing oxygen fugacity conditions (to produce varied trivalent/divalent Eu ratios), Roeder et al. (1987) concluded that most of the Eu in this apatite was trivalent.

Cassiterite. Hall and Ribbe (1971) examined the Cl spectrum of cassiterites of hydrothermal origin. They found considerable zonation with respect to W and Ti luminescence activation, and also a dependence on Fe content; Fe tended to quench the blue-green W emission, but apparently sensitized the yellow Ti emission. Ti^{4+}, Fe^{3+} and W^{4+} were assumed to occupy the Sn position in the structure. Substitution of Si^{4+} and W^{4+} for Sn^{4+} were found to affect the Ti^{4+} emission wavelength. One way in which this can be explained has been suggested by Fonda (1957). A larger ion substituent in a crystal will cause an expansion of the structure, resulting in weaker crystal fields and smaller crystal field splitting. Smaller splitting will increase the ground to excited state energy separation, thus shifting the emission to shorter wavelengths. Substitution of smaller ions has the opposite effect. This rationale seems to work for the much smaller Si^{4+} ion substituting for Sn^{4+}, shifting the Ti^{4+} emission from yellow to orange. However, the substitution of the slightly smaller W^{4+} ion for Sn^{4+} creates a shift to shorter rather than longer wavelengths. As W^{4+} is also an activator in cassiterite, the picture is probably more complex than simple structure expansion or contraction. Additional work seems necessary to identify the activator, sensitizer and quenching agent relationships.

Fluorite. Fluorite was the mineral examined by Stokes (1852) in his original work on luminescence. Its name was subsequently used to describe the phenomenon. Due to its large Ca site and ability to tolerate substantial trivalent substitutions, fluorite is a reservoir for many of the rare earths, as well as Mn^{2+}. There is a considerable amount of work available on fluorite luminescence, perhaps the most impressive being the compilation by Prizbram (1956). More recent studies have been summarized by Marfunin (1979). Most of the rare earths which activate strong visible luminescence in fluorite have been studied in synthetic samples. Thus the individual rare earth spectra are available for comparison. Correlation of the color and luminescence of fluorites with the particular rare earth activators has been very successful, but little attention has been paid to quantitative concentration estimates from UV or CL excitation measurements. The nature of the rare earth defect centers in fluorite is of interest to solid state physicists because of the laser applications, but this has not been addressed in any detail in the mineralogical literature. Fluorites often also demonstrate cascade fluorescence, in addition to various radiationless energy transfers between rare earth ions.

Fluorite may also have trapped electron defects such as F-centers and related larger clusters, which give rise to luminescent activity. The response of fluorites to irradiation is largely due to variations in the type and density of such defects. McDougall (1970) has attempted to utilize TL techniques to determine relative temperatures of crystallization of fluorites. The assumption made is that the original temperature is related to the concentration of specific defects which later produce the TL response. Blanchard (1966, 1967) examined thermoluminescence in a variety of natural and synthetic rare-earth doped fluorites.

Calcite and other carbonates. Calcite is abundant and of such importance in sedimentary processes that it has been studied in some detail by luminescence spectroscopy. Much of the CL results has been compiled by Nickel (1978). Medlin (1968) has thoroughly investigated the luminescence of calcite, dolomite and magnesite. Most calcite luminescence is due to Mn^{2+} with varying amount of rare earth or heavy-metal sensitization. Extremely small amounts of Pb and Mn can create strong orange-red Mn emission in calcite. Because the two cation sites in dolomites are of differing size, Sommer (1972) found that two Mn luminescence bands could be determined. The relative intensity of these bands should presumably afford a measure of Mn^{2+} site partitioning. However, Wildeman (1970) determined by ESR that most of the Mn in dolomite resides in the smaller Mg site. By the crystal field splitting criterion this should require Mn emission in calcite to have higher energy. Spectra of dolomite and calcite obtained by Graves and Roberts (1972) and Medlin (1968) confirm this, and do not show separate bands due to Mn occupation of both sites in dolomite. (Note that all of this logic concerning site size assumes that sites do not relax appreciably to accommodate a substituting element. This is untrue, and Mn sites in any of the carbonates may be much less different in size than expected (see Dollase (1975) for distance-least-squares simulations of substituents in solid solutions).

Gies (1975) studied some 350 calcite and other carbonate samples under short-wave ultra-violet excitation, and correlated the various types of emissions with detailed trace element and rare earth analyses. One set of calcites had broad emission in the near-UV and blue and slight emission in the red-orange. These emissions added to a white luminescence attributed to structural defects. A second set of calcites displayed the strong orange-red emission typical of Mn activation with Pb or (possibly) Zn as sensitizer. A third set of calcites were activated by rare earths and displayed a flesh-colored emission. The emission spectrum showed many sharp features, suggesting activation by several rare earths. Cerium activation in calcites has been shown by Blasse and Aguilar (1984) to be a source of red luminescence formerly thought to be a variety of Mn-activated emission. Rare earth emission in calcites remains rather ill-characterized at present.

Sommer (1972) has shown that the emission from the Mn transition in trigonal and orthorhombic carbonates is strongly dependent on the mean interatomic distances, in accordance with Fonda's (1957) result. Gies (1975) also observed the short-wave UV emission from smithsonite, aragonite, strontianite, cerussite and witherite. Besides shifts in the emission band position similar to Sommer's (1972) results, Gies found that Pb-rich aragonites had uniform Pb-activated luminescence. This is an indication that exsolution of Pb-rich carbonate had not occurred. Most workers find a strong relationship between Fe^{2+} in carbonates and the lumines-cence intensity. Mariano and Ring (1975) observed strong Eu^{2+} emission in strontianite from several carbonatites. When the strontianite occurred with apatite, it was noted that the typical Eu^{2+} emission from the apatite was missing due to the strong partitioning of Eu^{2+} into the strontianite. Calcite with small amounts of trapped uranyl ion is fairly common. It is unclear if the uranyl unit occupies a particular structural position. The emission spectrum is similar to that for uranyl ion in other crystals and in solution (Jorgensen, 1979).

Feldspars. Walker (1985) has reviewed cathodoluminescence studies on the feldspars. Plagioclase emission consists of three bands, one near the infrared, one in the green, and one in the blue. The overall emission color depends on the relative strength of these three bands (Fig. 14). The near-infrared band is due to the presence of tetrahedral Fe^{3+}. The excitation spectrum of the iron luminescence allows estimation of the ligand field parameter Dq and the Racah parameter B. The latter is smaller that that for Fe^{3+} in octahedral coordination, suggesting the tetrahedral site assignment. In principle, it might be feasible to decide among the possible tetra-hedral sites from the spectral variations, but this was not possible with available data.

Because of the small Dq value, usually a characteristic of large sites, the green emission is believed to be from Mn activation on the calcium sites. As tetrahedral Mn^{2+} also produces a small Dq value and a green emission, it is important to note the opposing effects of ligand number and bond distance on Dq. Studies on synthetic anorthites with varying Mn concentra-tions show that the green emission is correlated with Mn content up to a critical self-quenching value (Telfer and Walker, 1978). The activation of the blue emission was not assigned, although it was attributed to the same sources as the blue emission observed in quartz. How-ever, Mariano and Ring (1975) determined that the blue emission in feldspars is due to Ti^{4+} activation. The other main possibility, the blue Eu^{2+} emission, was examined in a synthetic anorthite. This emission had a similar color, but different band shape and decay rate from most feldspar emissions. Also, the concentration of Eu^{2+} necessary to detect emission appeared larger than that usually seen in luminescent feldspars.

Walker (1985) indicates that the Mn emission band is less prevalent in the K-rich feldspars. This may be due to the difficulty of charge compensating the Mn^{2+}-K substitution. The tetrahedral ferric band is apparently observed in all types of feldspars. A recent study by White et al. (1987) includes an absorption spectrum for Fe^{3+} in orthoclase. This spectrum is essentially identical to the excitation spectrum observed by Walker in bytownite (Fig. 15). Marfunin (1979) also notes the luminescence of Pb^{2+} in amazonites and other feldspars. The emission is in the UV near 3000 Å.

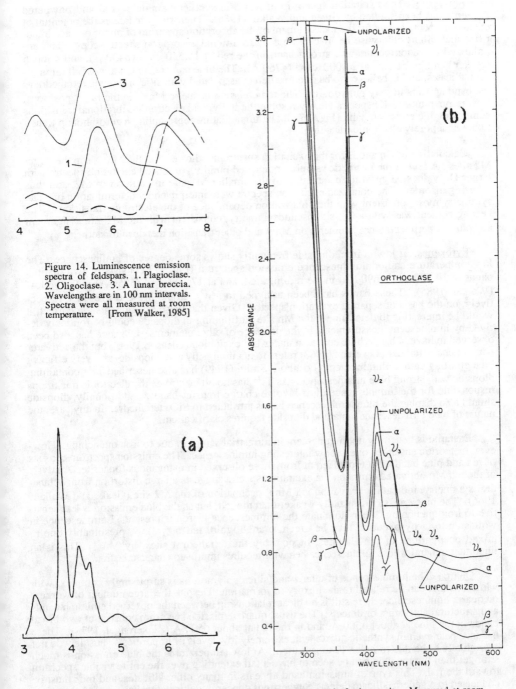

Figure 14. Luminescence emission spectra of feldspars. 1. Plagioclase. 2. Oligoclase. 3. A lunar breccia. Wavelengths are in 100 nm intervals. Spectra were all measured at room temperature. [From Walker, 1985]

(a)

(b)

ORTHOCLASE

ABSORBANCE

ν_1

ν_2

ν_3

ν_4 ν_5

ν_6

UNPOLARIZED

WAVELENGTH (NM)

Figure 15. (a) Excitation spectrum of the infrared Fe^{3+} emission band of a bytownite. Measured at room temperature. Wavelength in 100 NM intervals. [From Walker, 1985] (b) Absorption spectrum of Fe^{3+} in orthoclase. Measured at room temperature. [From White et al., 1986]

Forsterite. The excitation spectrum of Mn^{2+} in synthetic single-crystal and powdered forsterite has been studied by Green and Walker (1985). Their study indicates the potential of excitation spectroscopy for the measurement of the absorption spectrum of minor or trace impurities that might otherwise be inaccessible by traditional absorption spectroscopy. The unpolarized emission occurs as a strong band in the red at 15,550 cm^{-1} (6430 Å), and a much weaker band in the blue at 21,000 cm^{-1} (4760 Å). The blue band occurs in almost all luminescent silicates, and is believed to be due to a defect associated with silicon-oxygen tetrahedra as the band appears in very pure quartz. The red emission is due to Mn^{2+} in the Mg^{2+} positions. Low temperature (5 K) spectra clearly resolve the individual electronic-vibrational transitions composing the emission band (Fig. 16), including a feature representing a zero-phonon transition, i.e., a purely electronic transition.

Calculation of Dq and B for the excitation spectrum indicated that the Mn^{2+} is in the larger M2 sites. A few of the synthetic samples, quenched rapidly, produced excitation spectra with larger Dq values, suggesting occupation of the smaller M1 site in addition to M2, and thus inferring disorder. The polarized excitation spectra were interpreted by determination of a site symmetry most consistent with the polarization dependence of the electronic transitions. The best agreement was with a C_{2v} site pseudosymmetry, consistent with the M2 site assignment and other experimental measurements on Mn^{2+} and other transition metals in forsterite.

Pyroxenes. If low in iron, diopside frequently shows some degree of luminescence. The low temperature cathodoluminescence emission spectrum (Fig. 17) consists of two distinct bands at 14,500 cm^{-1} (6900 Å) in the red-infrared, and at 17,100 cm^{-1} (5850 Å) in the yellow (Walker, 1985). These bands have been assigned to Mn^{2+} in the M1 and M2 sites, respectively, on the basis of separate excitation spectra. Given that these assignments are correct, it would be interesting to determine if the Mn site partitioning in diopside could be routinely derived by luminescent measurements in the presence of other elements. Diopside has also been observed to have a bright blue emission under UV excitation (Gleason,1960), but the activator was not characterized. Leverenz (1968) refers to titanium-activated diopside as a very efficient blue emitting cathodoluminescent phosphor. Smith (1949) has also described Ti^{4+} containing diopside with strong blue luminescence. As Ti^{4+} has no 3d electrons, the electronic transitions responsible for the luminescence must involve charge-transfer bands. Additionally, diopside should also display the blue luminescence bands attributed to SiO_4 tetrahedra. In any case, the nature of blue luminescence in natural diopside requires clarification.

Enstatite is generally not luminescent in terrestrial samples, due to iron quenching. However, meteoritic enstatites often show interesting luminescence. The emission spectrum consists of red and blue bands, not too different from those observed in manganous forsterite. Analysis of the red Mn-activated emission via excitation spectra suggests a large distortion from octahedral symmetry, indicating activation via Mn^{2+} occupation of the M2 site (Geake and Walker, 1966). The blue emission is usually weaker than the red, but as the blue emission is less sensitive to iron quenching, it can dominate the luminescence if iron is present. Luminescence in wollastonite, a pyroxenoid, may be green, orange, gold, red or yellow, presumably due to mixed occupations of the Ca sites, or possibly the tetrahedral sites, by Mn^{2+}. This is an important mineral whose luminescence behavior requires improved characterization.

Quartz. The luminescence of quartz and silica polymorphs is surprisingly complex, with variations that can be related to the history of formation. Despite the large number of workers on quartz luminescence, there still is no clear relationship between the observed emission bands and structural activators or defects. The usual quartz emission spectrum consists of two broad bands in the blue at about 4500 Å, and in the red at about 6500 Å (Zinkernagel, 1978). These bands appear even in synthetic quartz with extremely low levels of impurities. Thus they do not seem likely to be due to Al or Ti substitution. At low temperatures the blue emission is much stronger than the red, and can be seen to have a tail extending over the entire visible spectrum toward the red. This type of emission band appears in many other silicates, and thus must be connected to a generic silicate defect. Suggestions concerning this defect have included trapped electron-hole pairs on "dangling" sp^3 bonds, various types of other oxygen vacancies, oxygen associated hole centers, and various defects related to radiation damage, as well as many kinds

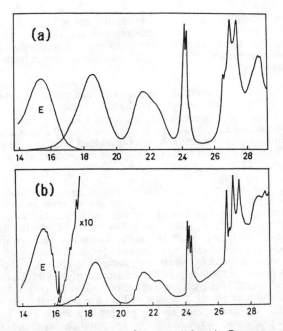

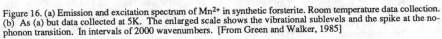

Figure 16. (a) Emission and excitation spectrum of Mn^{2+} in synthetic forsterite. Room temperature data collection. (b) As (a) but data collected at 5K. The enlarged scale shows the vibrational sublevels and the spike at the no-phonon transition. In intervals of 2000 wavenumbers. [From Green and Walker, 1985]

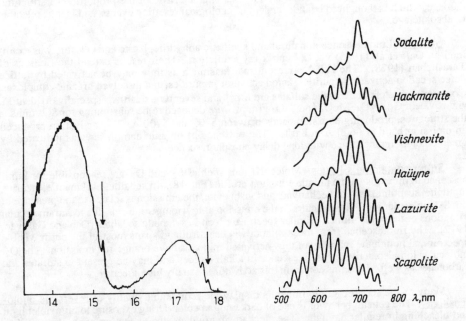

Figure 17 (left). Luminescence emission spectrum of Mn^{2+} in diopside at 5 K. Arrows indicate the no-phonon lines at 17,800 (M2 site) and 15,250 (M1 site) wavenumbers. [From Walker, 1985]

Figure 18 (right). Emission spectra for S_2^- molecular activation in aluminosilicates. [From Marfunin, 1979]

of cation impurities. The blue luminescence is readily excited by cathodoluminescence, but cannot be induced with UV except below room temperature. The red emission band has been assigned to sodium, hydrogen and hydroxide impurities, to non-bridging oxygens, oxygen vacancies and to cation impurities. There is obviously much work to be done to sort out all of the suggested mechanisms.

Zircon. Natural and doped synthetic zircons have been studied in some detail by Iacconi and Caruba (1984) by thermoluminescence methods. Most of the emission observed is explained by Dy^{3+} activation in the naturals, which gives rise to sharp bands at 4800, 5800, 6620 and 7600 Å. Variations in the luminescent color observed is probably due to small amounts of other rare earths. Emission from electron-hole pairs trapped by silicate tetrahedra is suggested as the cause of the band at 3800-4000 Å. Cathodo-luminescence of zircon is also reported by Gorz et al. (1970).

Several efforts have been made to use thermoluminescence of zircon as an age-determination tool. The basic approach, which is also used with several other minerals, is to assume that zircon acts as a perfect dosimeter in recording the effects of included radioactive impurities since the date of crystallization. The integrated dose is assumed to be simply related to the thermoluminescent intensity. This type of analysis assumes that standardization techniques based on the exposure of model samples to very high dose rates for short intervals are equivalent to geologically produced low dose rates over extremely long intervals, i.e., that the process is dose-rate independent. A further problem is possible thermal resetting of the TL during the life of the zircon. Vaz and Senftle (1971) suggest another technique based on the comparison of TL from laboratory-annealed and unannealed zircons from the same source that are then exposed to an intense ^{60}Co source. The unannealed zircons presumably display TL induced by the new gamma ray dose in proportion to the amount of structural damage, and thus electron traps, created by included radioactive elements during the zircon history. The annealed zircons are presumed to be recrystallized so that the gamma ray dose indicates the TL consistent with minimum structural damage. The difference is thus characteristic of the proportion of structural damage, which is thought to be independent of geological resetting events, and is thus relatable to absolute age.

Sulfates. Luminescence in natural and synthetic anhydrite, barite and celestite was examined by Gaft et al. (1985) via UV and x-ray excitation. Their results extend the findings of Tarashchan (1978). The sulfates are unusual inasmuch as they may be activated by lattice defects, cation activator impurities, anion activator impurities, and dissolved organic complexes. Common activators in all of the sulfates are Mn^{2+} and several rare earths, especially divalent Eu and trivalent Gd. VO_4, MoO_4 and TiO_4 groups were deduced to be substituting for SO_4 units in the structures, producing emission bands between 5200 and 5900 Å. Ti^{3+} activation produced an emission band at 6600 Å in barite. The presence of organic complexes was indicated by green-colored luminescence with long delay phosphorescence.

Sulfur-containing silicates. Molecular ions such as S_2^- and O_2^- are responsible for luminescence in minerals with a unique emission profile (Fig. 18). In sodalite, hackmanite, lazurite, tugtupite, scapolite, cancrinite, hauyne, and vishnevite, the emission is related to the presence of the S_2^- molecular ion. The structure in the emission at low temperatures is due to intramolecular vibrations of the sulfur-dimer. Similar spectra are seen in synthetic sodalites (Deb and Gallivan, 1972). The fluorescence from these minerals is beautiful and spans most of the entire visible spectrum, although the majority of thus-activated minerals have orange-red emission. The O_2^- molecular ion also activates luminescence in alkali halides and may be present in sulfates and carbonates as well. The emission from this activator is usually blue-green.

Many of the sodalite-group minerals display the remarkable quality of reversible photosensitivity. The general behavior is one of darkening in color during exposure to ultraviolet light and bleaching in regular daylight. The mechanism involves the excitation of an electron on a S_2^{2-} ion creating the S_2^- ion which absorbs strongly at 4000 Å, and the trapping of an electron on a Cl vacancy site. The latter is an F-center which absorbs strongly at about 5300 Å. These two bands create the pink-purple color. Lower energy light or thermal stimulation can free the

F-center trapped electrons which then can combine with the S_2^- ion to form the uncolored S_2^{2-} ion. This process is particularly well exemplified in hackmanite (Kirk, 1955).

Diamonds. Diamonds are usually luminescent and may display emission colors in many regions of the visible spectrum. These emissions appear mainly to be related to nitrogen impurities and platelets. However, other impurity centers, donor-accepter pair transitions, and other electronic and structural defects also give rise to luminescent activity. Davies (1977) and Dean (1965, 1966) have reviewed some of the processes and defect centers involved in diamond luminescence. Emission varies among the types of diamonds, and affords some degree of source identification. The luminescence consists of much fine structure superimposed on broad bands. Some well characterized emission features include a broad band at 6900 Å associated with a zero-phonon line at 6374 Å and due to single nitrogen impurities coupled with carbon vacancies (NV band), a band at 5900 Å due to donor-acceptor transitions (nitrogen donors, Al acceptors; NAl band), a broad band at 5300 Å associated with sharp lines at 5034 (H3) and 4961 Å (H4) due to nitrogen aggregates interacting with radiation damage-produced centers, sharp strong lines at 4914 and 4843 Å due to radiation damage, a broad band at 4590 Å, possibly due to a closely associated donor-acceptor pair, and a sharp line at 4153 Å due to nitrogen triad impurities (N3). The relative intensity of these lines and bands varies among the various types of diamonds and with their radiation exposure history. The 5900 Å band is common only in synthetic diamonds. Marfunin (1979) notes that diamond luminescence has been used by several Russian investigators as an indication of the geologic conditions of formation. Susuki and Lang (1981) were able to relate fine details of cathodoluminescence to growth zonation in diamond.

Molybdates and tungstates. Two of the best known luminescent minerals are scheelite, $CaWO_4$, and powellite, $CaMoO_4$, which form a complete solid solution series. The bright fluorescence emission is in the blue near 4300 Å for scheelite, and the yellow for powellite. Hence the color of emission can serve to indicate the Mo-W ratio in ore if this solid solution is the principle economic phase. Prospecting for Mo-W ores was greatly extended by the use of ultraviolet lamps to detect these ore minerals after World War II. Other tungstates and molybdates are also luminescent but few such minerals are of economic consequence. A large body of work exists on synthetic tungstate and molybdate phosphors (see Leverenz, 1968).

Uranium minerals. Most of the luminescent uranium minerals owe their activity to the uranyl ion, $(UO_2)^{2+}$. Frondel (1958) compiled an exhaustive list of uranium minerals including a complete description of luminescence properties. The most spectacular species include andersonite, autunite, schroeckingerite, uranocircite and zippeite. The U^{6+} ion may also activate luminescence in uranium minerals, as it does in many solids (Blasse et al., 1979), although this has not been verified. Uranyl ion activation of luminescence in calcite, other carbonates, sulfates, opal, and silicates is rather common. Lower valence states of uranium may also activate luminescence and are found in natural minerals. However, intervalence charge-transfer or other strong absorption mechanisms obscure or quench any luminescence in these species.

Glasses. Because glasses are believed to be reasonable analogs of melt structure, luminescence of particular ions in glasses can be studied to provide information on ion site partitioning in melts. Unfortunately, natural glasses cannot be studied because of the usual high iron concentration. Divalent Mn-doped glasses have been the most commonly examined, but there is disagreement as to the meaning of the emission wavelengths and intensities (Bingham and Parke, 1965; Turner and Turner, 1970). The emission is divided into two bands at about 6060 and 5260 Å, equivalent to red-orange and green emission, respectively. As has been discussed earlier, the red emission is consistent with an octahedral ligand field in crystals, whereas the green emission is found for tetrahedral fields or very large sites which also have relatively small crystal field splitting (see Feldspar section above). The green emission is thus ambiguous and could be assigned to network-forming sites or large network-modifying sites.

The magnetic exchange coupling of nearby magnetic ions can induce additional electronic state splittings which may reduce the ground to excited state transition energy, and thus shift the emission toward the red. This may be a possibility with Mn^{2+} because of the observed emission color change with increasing Mn^{2+} concentration. The green emission band weakens and the red-orange strengthens, an indication that the red band could be due to Mn^{2+}- Mn^{2+} interactions. Detailed study of the excitation spectra at low temperatures may help to clarify the situation. Phosphate glasses, which presumably contain divalent cations only in octahedral coordination, show only the red emission. The luminescence from Ni^{2+} and Cr^{3+} have also been studied. Details of glass luminescence, including rare earth activation which is important for glass laser materials, have been reviewed by Wong and Angell (1976).

Petrology

Because of its ubiquitous nature in rocks and sediments, the luminescence of quartz has been studied carefully in an attempt to relate variations in emission or thermoluminescence to geologic history. Unfortunately, many of the results seem to be contradictory or at least irreproducible. Some of these problems may be due to variations in the excitation methods. Perhaps worse, the mechanisms of luminescence in quartz are not well understood, so that constraints cannot be placed on how geologic factors produce particular changes in emission.

Despite these difficulties, Zinkernagel (1978) was able to compare most of the luminescence observations in quartz, and arrived at three general types of overall emission: (1) violet, (2) brown, and (3) nonluminescent (see Fig. 19). The violet category could be further subdivided on the basis of the blue component, into relatively blue or relatively red types. The brown category could be subdivided into three subdivisions based on the shape of the dominant red emission band, and the sharpness of the weak blue band. Overall, Zinkernagel found that this classification coincided with specific modes of quartz occurrence. Igneous rocks contained only the violet type (24 different samples). The samples with the greatest blue emission had the highest temperature of crystallization, had the most rapid cooling rates, and occurred in volcanic rocks. Plutonic quartz, also crystallized at high temperatures but presumably cooling more slowly, had much less blue component in the emission. Quartz from contact metamorphosed metasediments (five samples) near the high temperature contact had emission generally similar to quartz from the volcanic rocks. With increasing distance from the contact, the quartz emission graded toward the brown category. In addition, the quartz grains had, in some cases, a complex luminescence petrography with patchy red and blue dominant areas. These effects may be due to temperature irregularities or chemical inhomogeneities within the grain. As the sediment luminescence was presumably not of the violet type to start with, the temperature of the contact has reset the luminescence to a new "grade."

Only brown luminescence was observed in quartz from metasediments free of matrix. Other metasediments showed differences in luminescence between quartz cement and grain nuclei. Quartz from schistose rocks showed a red-brown emission in one case (staurolite-amphibolite facies) and a dark brown emission in another (Precambrian biotite schist). The violet emission of igneous rocks is reset by regional metamorphism into the brown emission category, indicating that the presumed temperature effect is reversible. Authigenic quartz has a brown type of emission with the smallest blue component of all luminescing quartz. It was possible to easily see the authigenic overgrowths in detrital quartz grains by the strong contrast between violet core and brown rim CL. The occurrence of nonluminescent quartz was restricted to diagenetic origin without any metamorphism.

An exception to the observed trends in occurrence versus luminescence is found in hydrothermally derived quartz. Initial CL excitation produces a bottle-green emission which shifts to brown emission with long term electron bombardment. Most grains also show intensive zonation. Attempts to model hydrothermal quartz luminescence by the examination of variously produced synthetic samples resulted in the discovery of still additional complexity in the synthetic's luminescence. Zinkernagel's work shows conclusively that there is a preserved temperature effect on the luminescence of quartz. Uses of this effect include the determination of the source rocks from the luminescence of detrital grains, and the identification of diagenetic alteration such as pressure solution. Additional applications, such as the relation of quartz luminescence to deformational history, may be possible if the luminescence mechanisms are established.

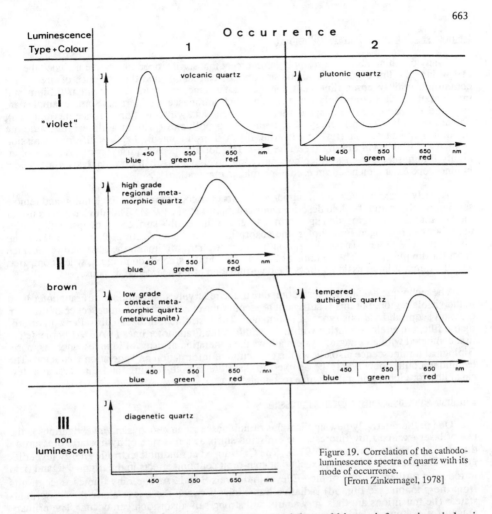

Figure 19. Correlation of the cathodo-luminescence spectra of quartz with its mode of occurrence.
[From Zinkernagel, 1978]

Feldspars hold some promise of yielding source and thermal history information via luminescence analysis, but the results are fragmentary. Kastner (1971) summarized her own results on authigenic feldspars together with those of Smith and Stenstrom (1965), and concluded that authigenic feldspars and albites from low grade metamorphic rocks show no characteristic luminescence. Conversely, alkali feldspars from pegmatites and igneous rocks had a distinctive CL response. The general concept of luminescent activity increasing and varying with increasing temperature of formation also appears to hold. The solubility of activators or the rate of production of defect states as a function of temperature may account for this.

In calcites the luminescence is related largely to the Mn^{2+} and Fe^{2+} concentrations, with considerable zonation possible. Recrystallization or other large chemical change in carbonate rocks may reset the luminescence, resulting in more uniform emission over a large area (Sommer, 1972). Conversely, inhomogeneities tend to indicate preservation of the initial composition. Luminescence in calcite has generally been used as a petrography tool, with the shift in the emission peak serving to indicate Mg content. Dissolution and recrystallization features are often revealed by dramatic changes in CL coloration.

It has been postulated that fluorite thermoluminescence can be used to determine the temperature of crystallization because of the varying dependence of particular defect concentrations and their associated trapping levels on temperature (McDougall, 1970). Age determination of fluorites by thermoluminescent dosimetry methods has not proven reliable (Blanchard, 1966).

Remote sensing and mineral exploration

Many ore minerals are so strongly luminescent that sorting operations have included the use of UV lamps. Trace quantities of these minerals can also signal the presence of ore bodies containing mainly nonluminescent species. Lead-zinc, manganese, tungsten, lithium and uranium ore bodies commonly contain many such luminescent minerals. Some examples are anglesite and cerussite (lead), sphalerite, hydrozincite and willemite (zinc), scheelite (tungsten), powellite (molybdenum), eucryptite, petalite and pollucite (lithium and cesium), autunite (uranium) and cassiterite (tin). Sphalerite, if low in iron content, displays differing emission from Ag, Cu and Mn doping (see Fig. 6a). In addition, petroleum hydrocarbons are often highly luminescent, even at trace concentration levels. Thus drill core analysis via UV and luminoscope excitation has been used in petroleum exploration.

A fairly recent application of luminescence spectroscopy to mineral exploration and remote sensing is the Fraunhofer line descriminator (Hemphill et al., 1984). This device makes use of the fact that the solar emission spectrum that reaches the earth's surface is not a smooth continuum, but has hydrogen Fraunhofer line absorption from the sun's atmosphere throughout the visible and UV range. These absorption lines create sharp minima in the observed reflectance spectra from the earth. The sampling of these gaps in the smooth continuum, when compared with the direct solar emission, allows the determination of luminescence response.

The technique has limited resolution due to the relatively small number of Fraunhofer lines in the visible region, but this is sufficient to identify the characteristic luminescence of uranium ores, oil seeps and various types of pollutants. The instrument utilizes Fabry-Perot interferometric filters to attain resolution within the width of the Fraunhofer lines (1 Å). Airborne detection combined with an imaging system allows the generation of maps of luminescence response. Biological luminescence also allows the collection of information on vegetation character. The technique could conceivably be adapted to satellite remote-sensing, and thus create a new capability for global mineral resource reconnaissance.

Luminescence as a sensor for high pressure

One of the relatively new applications of luminescence in geophysical investigations is the use of laser-excited ruby fluorescence emission shifts as a measure of hydrostatic pressure in diamond anvil cells (Mao et al., 1978). The Cr^{3+}-doped corundum electronic energy level diagram is that shown in Figure 1, except that spin-orbit coupling effects in the trigonal ligand field introduce a further splitting of the ^2E level into levels E and 2A. The ruby fluorescence results from no-phonon (see Fig. 5d) radiative transitions between these levels and the $^4A_2$ ground state. The transitions are thus very sharp, and strong self-absorption can occur. The primary reason for the energy shift upon the application of pressure is the increase in crystal field splitting as bond distances shorten, although this cannot account for all of the shift (Berry et al., 1984). Increased splitting reduces the transition energy, and hence shifts the emission toward longer wavelengths.

A second effect of pressure is a change in the lifetime of the excited states due to suppression of nonradiative transition processes from the excited states. If a nonradiative energy transfer mechanism is suppressed, the lifetime of an excited state increases, and the radiative transition linewidth decreases. In ruby the nonradiative transitions are due to two-phonon Raman processes that become more probable with increasing temperature (Sato-Sorenson, 1986). This creates a problem for experimentalists: increasing pressure narrows the emission lines, but increasing temperature broadens the lines. Increasing temperature also produces a shift in the emission to longer wavelengths due to the same Raman process (Powell et al., 1966), hence the temperature shift must be known in order to calculate pressure from the observed emission shift. For these reasons there is a search for other fluorescent systems which retain narrow emission lines at high temperatures and pressures but retain emission shifts for pressure calibration.

X-RAY EMISSION SPECTROSCOPY

Of all the spectroscopies dealt with in this volume, x-ray emission spectroscopy is probably the most common in the geological sciences. Electron microprobe analysis (electron-induced x-ray emission) and x-ray fluorescence analysis (x-ray-induced x-ray emission) are the mainstays of rock and mineral composition determination. These techniques have evolved since their origination. Microprobe-type analysis is now done with other electron beam apparatus, such as the transmission electron microscope (TEM) and the scanning electron microscope (SEM). Energy analysis is now commonly performed with so-called energy dispersive detectors, rather than only with single energy (crystal diffraction or wavelength dispersive) spectrometers. Other related new techniques are on the horizon. Proton and particle induced x-ray emission (PIXE) offers advantages over electron microprobes for many elements at low concentrations. Synchrotron radiation induced x-ray emission combines the advantages of XRF measurements with collimated, polarized and extremely intense x-ray sources.

The field of x-ray emission spectroscopy can be divided into two categories: the analysis of gross and fine emission spectrum structure (these categories are sometimes referred to as spectrometry and spectroscopy, respectively). For elemental analysis we are generally concerned only with gross x-ray structure, as the energy separation of individual element emission lines is usually much greater than the line width. Fine structure studies require high resolution instrumentation, but can reveal the subtle shifts, intensity changes and additional structure in x-ray spectra which contain information on formal valence, electronic bonding configuration, coordination and effective charge. Both areas are dependent on the signal/background ratio characteristic of each excitation method.

Analysis of the gross structure of x-ray emission spectra for basic chemical analysis will not be considered in any detail here. Many texts are available that cover the techniques and theory, for example Muller (1972), Birks (1969, 1971), Reed (1975) and Anderson (1973). Instead, we will consider the various modes of x-ray fluorescence excitation, how these differ in the quality of elemental spectra produced, and new developments in elemental analysis. The rest of this section will describe applications in fine structure analysis.

Excitation of x-ray emission

If photons or other particles with sufficient energy strike an atom, it is possible to excite one of the atomic electrons into a higher energy state in the atom, or to eject it from the atom entirely. This process thus empties one of the electronic states of the atom. This state will then be filled very rapidly by another electron, which drops into the empty state from some higher energy situation. The difference in energy between the original and final state of this electron can be emitted as a new photon with energy characteristic of the particular energy level spacings in the atom. This process is shown in Figure 20, where excitation of K-shell (or 1s) electrons results in various types of x-ray photons. As the energies of these photons are characteristic of particular atoms, such x-ray emission can be used to determine chemical composition. This is the basis of x-ray fluorescence analysis and electron microprobe analysis, where Bragg spectrometer systems or energy-dispersive detector systems are used to collect x-ray emission spectra emitted by a sample under intense x-ray or electron excitation. If protons or particles are used to excite the x-ray photons the process is called PIXE, for Proton (or Particle) Induced X-ray Emission. Figure 20 also shows other possible results of inner shell electronic stimulation of atomic electrons, viz. Auger and photoelectron emission. These processes are treated in a separate chapter elsewhere in this volume.

X-ray-induced x-ray fluorescence (XRF). This mode of x-ray excitation is the oldest and still one of the most utilized. X-rays are stimulated by use of both the bremsstrahlung and characteristic radiation from a high atomic number (typically tungsten) anode x-ray tube. The bremsstrahlung spectrum is continuous, with highest energy equivalent to the acceleration voltage applied to the tube. It is produced by the electrons from the cathode of the x-ray tube which undergo decelerations due to the strong electrostatic fields near the nuclei of the anode target. The bremsstrahlung spectrum is thus dependent on anode atomic number; higher Z materials having more efficient x-ray production. The characteristic spectrum consists of the x-ray fluorescent emission directly excited by the incoming electrons and indirectly excited by the bremsstrahlung.

Excitation of x-ray emission in samples via x-ray tubes has two major disadvantages, but a single great advantage. The x-ray beam cannot be focussed easily to a small diameter, thus

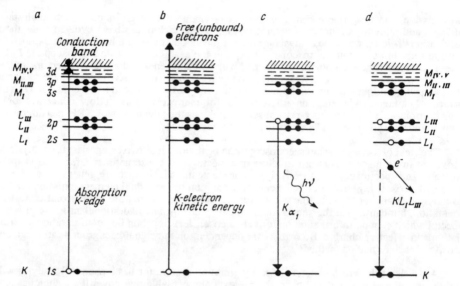

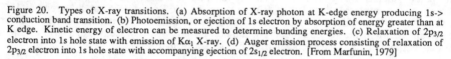

Figure 20. Types of X-ray transitions. (a) Absorption of X-ray photon at K-edge energy producing 1s-> conduction band transition. (b) Photoemission, or ejection of 1s electron by absorption of energy greater than at K edge. Kinetic energy of electron can be measured to determine bunding energies. (c) Relaxation of $2p_{3/2}$ electron into 1s hole state with emission of $K\alpha_1$ X-ray. (d) Auger emission process consisting of relaxation of $2p_{3/2}$ electron into 1s hole state with accompanying ejection of $2s_{1/2}$ electron. [From Marfunin, 1979]

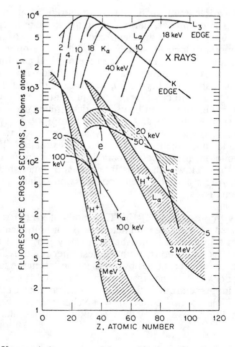

Figure 21. Fluorescence X-ray emission cross sections as a function of atomic number for the production of $L\alpha$ and $K\alpha$ X-rays. Proton excitation indicated by hatched pattern and $^1H^+$ symbol, electron excitation by e. [From Sparks, 1980]

microprobe-like lateral resolution is not possible. Also the x-ray flux obtainable from a sealed x-ray tube results in relatively low excitation intensity compared to other methods. Offsetting these limitations is the relative availability and low cost of in-laboratory XRF apparatus.

In order to compare x-ray excitation with the other techniques, several topics must be considered: x-ray fluorescence cross sections, the excitation energy density in the sample, and the magnitude and effects of background level in the excited spectrum. The cross sections for x-ray fluorescence are a measure of the probability of generating an x-ray of a given type for each impingent x-ray photon or other particle. For x-rays these probabilities vary dramatically as a function of energy, with enhanced probability at energies just above characteristic absorption edges. This enhanced excitation probability is of major importance to x-ray absorption spectroscopy, and is discussed in that chapter of this volume. In general, however, the x-ray fluorescence cross sections for x-ray photons are some 10-100 times that of electrons, and some 5-1000 times that of protons for elements of $Z \geq 10$ and typical electron and proton energies (Sparks, 1980). This is shown in Figure 21, where fluorescence cross sections for the indicated x-ray emission lines are contrasted. This comparison is directly applicable for samples thin enough so that absorption effects are negligible. For thick samples with complete absorption the total fluorescent yield is a better comparison (Fig. 22), and gives the maximum number of fluorescence events per incident particle.

The excitation density is related to the depth of penetration of the x-ray photons or charged particles in the sample. Because, in general, x-rays penetrate much further into geological materials than do electrons, the volume of sample contributing to x-ray emission can be much larger in x-ray excitation. This requires that XRF samples be remade into homogeneous thick masses for optimum results.

The x-ray emission spectrum background with x-ray excitation is due to incident x-ray scattering and bremsstrahlung from both photoelectron ejection and Auger electron emission. The effect of scattering depends on the detector and geometry of the analysis instrument, but can be minimized with proper design. The bremsstrahlung emission is much weaker than from proton or electron excitation. This leads to high signal/background ratios for x-ray excited x-ray fluorescence. Figure 23 compares the theoretical fluorescence spectrum for incident x-rays of two characteristic energies, and for a continuous band of energies. The optimal type of x-ray excitation spectrum appears to consist of a few x-ray characteristic emission lines, spanning the spectral energy range required.

X-ray excitation produces minimal heating in the sample compared to charged-beam excitation methods. For materials of average $Z = 6$, the energy dissipation for equivalent x-ray emission is 20 times less than for electrons. At $Z = 20$ this ratio increases to 2000 times (Kirz et al., 1978).

Synchrotron x-ray induced x-ray fluorescence (x-ray microprobe). The radiation from synchrotron sources is highly linearly polarized in the plane of the synchrotron ring. Thus excitation of x-ray fluorescence in a target will result in little scattered radiation entering a detector at right angles to the incident beam direction (Fig. 24). Hence synchrotron radiation (SR) may produce a reduced background compared to x-ray tube sources.

SR is also well collimated because of the highly defined orbit of the radiating electron or positron beam. This allows the development of x-ray optics to focus the x-ray beam to a small point, perhaps on the order of microns. With the use of specialized synchrotron insertion devices, such as undulators, the direct synchrotron beam may be reduced to tens of microns in size. In either case, the focussing capability of SR promises a future true x-ray microprobe.

SR intensity is from four to ten orders of magnitude larger than that from x-ray tubes. This provides a sensitivity to the ppb level in most cases. An example of this is shown in Figure 25, where x-ray fluorescence from 4.6×10^9 cadmium atoms in a 0.45 mm^2 sample (2.2 ppm) is easily seen over background. If all of the Cd were present on the surface of the sample, this would be equivalent to 0.08% of a monolayer! The minimum detection limit in this case was estimated at 2.7×10^8 cadmium atoms (130 ppb).

668

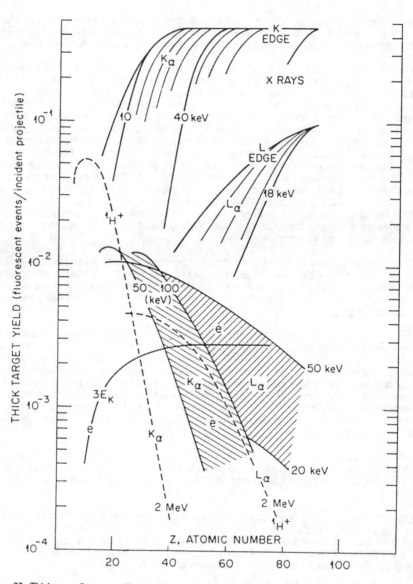

Figure 22. Thick target fluorescent X-ray yield as a function of atomic number. Electron excited yield indicated by hatched region. [From Sparks, 1980]

Despite lower excitation density and energy dissipation, SR intensity will lead to sample heating. However, this problem will still be much smaller than with particle beam excitation.

Electron-induced x-ray fluorescence (Electron Microprobe Analysis, EMA). The x-ray fluorescence cross sections for electron stimulation (Figs. 21 and 22) indicate the relative ineffi- ciency of electrons as x-ray excitation sources. This inefficiency is slightly offset by the greater excitation density relative to x-ray excitation, because electrons of 10-30 keV energies penetrate only a few microns into geological materials. However, the relative background is much greater than for x-ray excitation due to a much larger bremsstrahlung production. In effect, a sample

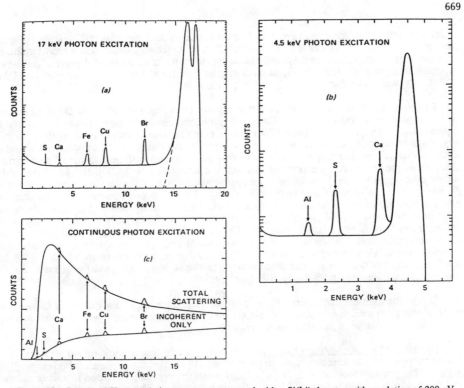

Figure 23. Calculated X-ray emission spectra as measured with a Si(Li) detector with resolution of 200 eV (independent of energy). Model sample consists of 250 nanograms of Al, S, Ca, Fe, Cu or Br on 25 mg/cm² carbon substrate, an approximation of a biological sample. (a) Excitation by 17 KeV X-rays. (b) Excitation by 4.5 KeV X-rays. (c) Excitation by continuous X-ray spectrum (analogous to continuous synchrotron radiation excitation). [From Jaklevic, 1976]

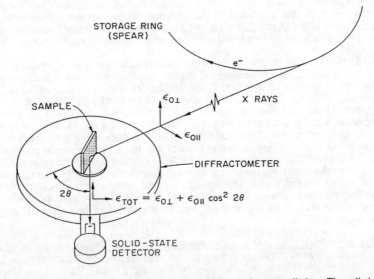

Figure 24. Geometry of X-ray fluorescence spectrometer used with synchrotron radiation. The radiation is almost completely linearly polarized with the electric vector in the ring plane. Consequently, at $2\theta = 90°$, there is very little direct beam scatter into the detector. SPEAR is an electron and positron storage ring at the Stanford synchrotron radiation laboratory (SSRL). [From Sparks and Hastings, 1975]

under electron beam bombardment acts like an anode in an x-ray tube. The atoms of each element in the sample, besides their characteristic x-ray emission, contribute a bremsstrahlung spectrum. The summation of these individual spectra constitutes much of the overall emission spectrum background. Thus the signal/background ratio for electron excitation may be 100 to 1000 times worse than for x-ray excitation (Sparks, 1980).

Electron microprobe analysis is nevertheless successful because of the methods for dealing with background corrections. The ability to obtain lateral resolution down to about 1 micron in typical instruments, and to perhaps 100 x 100 Å domains in analytical TEMs, is the principle advantage of electron beam excitation. This advantage is limited to thin samples. Analysis of thick specimens requires larger acceleration voltages to create sufficient depth of penetration. Because of the large cross section for electron scattering, the beam will spread appreciably, limiting the possible lateral resolution. A major weakness is the difficulty in estimating accurate fluorescence cross sections due to the complexity of the fluorescence excitation process via electron bombardment. The cross sections for electrons actually change continuously as the electron slows down in the sample. Hence the cross sections actually represent integrals of complex functions over a range in energy. The cross sections are thus not as well known as for x-ray excitation, and this affects correction procedures necessary for quantitative analysis of combinations of elements (the usual case). Quantitative analysis errors are roughly $\pm 1\%$ for XRF (Muller, 1972) given similar composition standards, but about $\pm 6\%$ for electron microprobe analysis (Beaman and Solosky, 1972) at best.

Due to the high excitation density, electron excitation results in dissipation of large quantities of energy in a relatively small sample volume. This causes heating with the possible loss or migration of volatile or mobile constituents.

Excitation by channelling electrons. Charged particle beams incident on a crystal may scatter cooperatively in directions parallel to particular crystallographic planes called channelling planes. This phenomenon occurs with electrons when the crystal is oriented in such a manner that the Bragg condition is satisfied for the channelling planes. The electrons are then scattered repeatedly along the thickness of the crystal. At incident beam angles less then the Bragg angle, the electron intensity is maximized along planes having the highest projected electron densities. At incidence angles above the Bragg angle, the channelling occurs between these planes (Fig. 26). This sets up the possibility of analyzing the contents of the planes or the sites between them by examining the x-ray emission during channelling in each beam orientation, and comparing it with that produced in the absence of channelling. This is the basis of the ALCHEMI technique (Atom Location by CHannelling-Enhanced MIcroanalysis) (Spence and Tafto, 1983), which can be employed with a transmission electron microscope (TEM).

ALCHEMI can be used with many types of crystals to obtain information about site partitioning and occupancies. Mineralogical studies employing the technique have examined site occupancies in olivines (Smyth and Tafto, 1982; McCormick et al., 1987), orthoclase (Tafto and Buseck, 1983), spinel (Tafto, 1982), ilmenite and dolomite (McCormick and Smyth, 1987), and garnet (Otten and Buseck, 1987). The technique is restricted by certain structural requirements, viz. the sites of interest must be in alternating planes, the planes must have significantly different projected electron densities so that channelling is strong enough, and a few types of atoms must occupy only one type of site to serve as internal standards (Otten and Buseck, 1987). Within estimated error, ALCHEMI results agree with most previous site occupation studies done with spectroscopic or x-ray diffraction methods. The quantitative applicability of the technique is still developing, but will always vary according to the details of channelling in each structure. The newest application is a form called axial ALCHEMI, where the channelling is directed between columns of atoms rather than planes (Otten and Buseck, 1987).

Proton (particle) induced x-ray fluorescence (PIXE). Proton excitation of x-ray fluorescence (Johansson et al., 1970) is more efficient than electron excitation for low Z elements, as shown by the excitation efficiencies in Figures 21 and 22. More importantly, the bremsstrahlung continuum x-ray background produced by proton bombardment is much smaller than that produced by electrons. This is due to the greater mass of protons over electrons, so that the protons are much less affected by nuclear charges. The ratio of bremsstrahlung

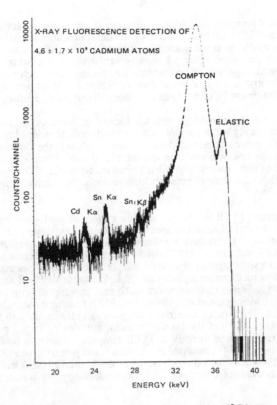

Figure 25. X-ray emission spectrum from a resin bead loaded with 4.6 x 10⁹ Cd atoms excited with 37 KeV synchrotron radiation. Counting time 2200 s. [From Sparks, 1980]

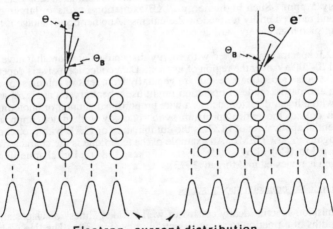

Figure 26. Distribution of electron beam current in electron channeling down crystal planes. Beam incidence angle smaller than Bragg angle (left), and larger than Bragg angle (right). [From Otten and Buseck, 1987]

background of 3 MeV protons to 20-40 keV electrons is about 1/100 (Jacklevic, 1976). The net effect of these factors is that proton-induced x-ray emission is substantially more sensitive than electron-induced emission to small quantities of sample, with the sensitivity increasing toward smaller Z. Neither technique is as potentially sensitive as synchrotron radiation induced emission. The overall signal/background ratios expected from each type of excitation as a function of Z and matrix average Z are shown in Figure 27. Experimental arrangements and background contributions in PIXE analysis have been discussed by Richter (1984).

PIXE analysis has the advantage of beam focussing, and spot sizes of a few microns are possible without sacrificing beam current. Excitation density is reduced relative to electron excitation because of the deep penetration. A related problem is the fluorescence contributions from underlying grains not detectable on the sample surface. This is much less of a problem with EMA because of the usual shallow (1 m) penetration depth. PIXE excitation also creates sample heating to about the same degree as electrons for a comparable x-ray fluorescence emission (Kirz et al., 1978).

PIXE applications. PIXE has been used in a variety of geological applications, many of which have served as a test of the technique's capabilities. Benjamin et al.(1984) were able to determine the presence of fractionation effects in trace levels of actinides and lanthanides in meteoritic minerals. Roeder et al. (1987) used PIXE analysis to measure rare earth concentrations in apatites, and also compared rare earth analyses produced by electron microprobe, PIXE and neutron activation methods on a glass standard. Quantitative PIXE analyses were claimed down to tens of ppm, whereas electron microprobe results were difficult below a few hundred ppm (Roeder, 1985). Annegarn et al. (1984) used PIXE to determine trace levels of platinum group metals in ores. They preconcentrated 50 gram samples of ore into NiS buttons. These were then treated with HCl and the insoluble metal sulfides collected on filter paper. The collected solids were then used directly as PIXE samples. Detection limits of about 0.5 ppb were attained. Carlsson (1984) examined the accuracy and precision of PIXE analyses on thick geological standard samples including those from the United States Geological Survey. He found a standard deviation in the analyses of 1.6%, neglecting pulse statistics, and accuracies of 2-3%.

Durocher et al. (1988) used proton beams of 40 MeV energy to carry out PIXE analyses of rare earth K-shell emission in a Durango, Mexico apatite sample. The greater energy spacing of the characteristic K-emission spectra of the rare earths enabled much improved resolution of individual element lines (Fig. 28) compared to L emission from a similar sample excited by 3 MeV protons (Rogers et al., 1984) (Fig. 29). These workers estimated their detection limit to be approximately 5 ppm. Such high-energy PIXE excitation results in larger bremsstrahlung background and reduced utility for the low Z elements. Another disadvantage is the difficulty of fine-focussing such high energy particle beams.

Most PIXE spectra are collected with energy-dispersive detectors that have energy resolution of from 120-200 eV over the region of interest. Data collection periods depend on the beam current and the pulse repetition rate, but are generally from tens of minutes to several hours in duration. An alternative mode of operation might use a bent crystal analyzer as a wavelength sorter along with a linear detector such as a wire proportional counter or photodiode array. This system could collect the entire spectrum synchronously with much improved resolution. Another potential application of PIXE is the combination of PIXE with proton channeling in a manner analogous to ALCHEMI. An example of the technique is the characterization of dilute Co atoms in a Mo (Co) alloy single crystal (Ecker, 1984). Ion channelling for structure characterization is reviewed in Morgan (1973).

High-resolution x-ray emission spectroscopy

X-ray emission spectra ideally consist of well-defined "lines" of Gaussian or Lorentzian shape with widths of a fraction to several electron volts (Fig. 30). The line width is inversely proportional to the sum of the lifetimes of the initial and final electronic states, which in the case of an inner shell x-ray transition, and unlike most optical and luminescence transitions, are both excited states with short lifetimes. The initial state is an electron vacancy or "hole" in an atomic or molecular orbital produced by one of the excitation processes. The filling of this hole

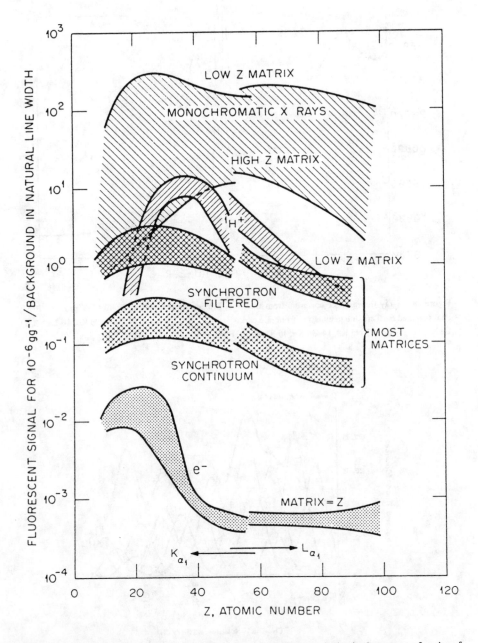

Figure 27. Signal/Background (S/B) ratios for various modes of X-ray excitation in elements as a function of atomic number. [From Sparks, 1980].

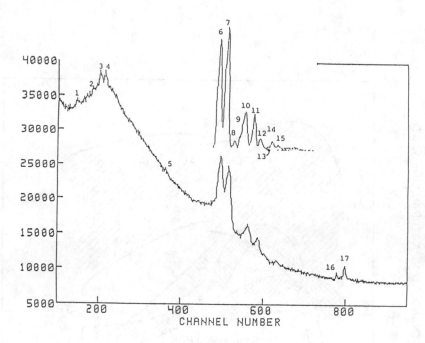

Figure 28. X-ray fluorescence spectrum from Durango, Mexico apatite excited by 40 MeV protons (PIXE). Peaks labeled 6-15 are due to emission from rare earths. (6-La Kα, 7-Ce Kα, 8-Pr Kα, 9-Nd Kα, 10-La Kβ, 11-Ce Kβ, 12-Sm Kα, 13-Nd Kβ, 14-Gd Kα, 15-Tb Kα). [From Durocher et al., 1988]

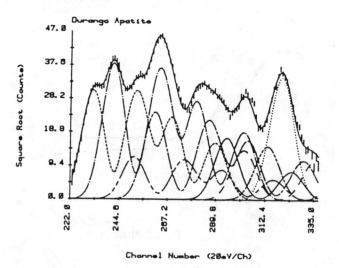

Figure 29. X-ray emission spectrum of Durango, Mexico apatite using 2.8 Mev protons for excitation (PIXE). Only the portion of spectrum containing rare earth L edge X-ray emission is shown. Each rare earth has four major emission lines (as compared to two on the K edge). The overlap of lines is a serious problem for L edge emission. [From Rogers et al., 1984]

by an electron from a higher shell orbital generates the x-ray. Usually, the spectrometers used for x-ray emission analysis do not have sufficient resolution to resolve the true width of x-ray lines or details of fine structure, and this is unnecessary for elemental analysis work. Typical energy-dispersive detectors (intrinsic Ge) have energy resolution of $\Delta E/E = 2 \times 10^{-2}$ at best, whereas curved crystal spectrometers of simple design have resolution of about 2×10^{-3}. In contrast, resolution of 10^{-4} or better is needed to resolve 1 eV features in a 10 keV K x-ray emission spectrum.

Changes in the chemical state of the atom have the largest effect on the energy of the valence electrons in the outer orbitals. Thus any x-rays produced by the transitions of these electrons to inner orbitals will carry information about the atom valence, type of bonding, coordination number, and so forth. These variations in x-ray energy are a very small fraction of the total x-ray energy in the case of transitions into the K-shell, but are proportionally larger for transitions to more outer orbitals. Thus given a fixed spectrometer resolution, analysis of L-shell spectra may reveal more detail than K-shell spectra.

High-resolution x-ray emission spectroscopy has had two differing applications in mineralogy. Initial studies sought to connect variations in discrete ion valence, coordination or bond length to specific details of the spectra. This type of study continues up to the present, but is not always successful due to the complexity of factors giving rise to single feature intensity and energy shifts. The other application is the construction of molecular orbital energy level diagrams for all of the possible bonding, anti-bonding and non-bonding states in a mineral. Work by Fischer and Baun (1968), Fischer (1970, 1971), Urch (1970), and Glen and Dodd (1968) demonstrates the power of x-ray emission spectra in testing molecular orbital calculations and elucidating bonding interactions and physical properties.

In the following sections, the nomenclature and basic physical processes of x-ray emission in minerals are briefly described. A number of specific applications of x-ray emission spectroscopy are then reviewed.

X-ray energy level diagrams. In the simplest approximation, one can assume that x-ray transitions are entirely like ideal optical electronic transitions, i.e. the relevant transition-moment operator considers only pure electric dipole interactions, the transition involves only one electron and its initial and final orbital states (hydrogen atom approximation), and there is little adjustment in an atom in response to the production of the initial excited state (frozen orbital model). These assumptions all fail to varying degrees, particularly for shorter wavelength x-rays, but they are a useful starting point. Direct application of the quantum mechanical selection rules (unpolarized case: $\Delta l = \pm 1$; $\Delta m = 0, \pm 1$) immediately allows us to specify the allowed and forbidden x-ray transitions (Rooke, 1974). For example, for Δl: np→ns, nd→np allowed; np→np, ns→ns, nd→nd, ns→nd forbidden. The resulting picture can be cast as an energy level diagram with the possible one-electron emission transitions indicated as so-called diagram lines (Fig. 31).

Diagram lines. Diagram lines have specific designations. Transitions which fill an electron hole in the K-shell (1s state) are K x-rays. If the infalling electron is from the 2p state (L-shell), it is a Kα x-ray. As the 2p state is divided into $2p_{1/2}$ and $2p_{3/2}$ states by spin-orbit coupling (the subscripts are equal to j, the total angular momentum), Kα x-rays are further subdivided into Kα_1 and Kα_2, respectively. Kβ x-rays are produced by electrons from the M, N or higher shells, and there are a variety of numeric subscripts designating various spin-orbit coupling split states. In general, the subscript indicates the relative intensity of a given type of line. In similar fashion, transitions from higher energy levels to a hole in the L-shell (2s, 2p states) are L x-rays, transitions to a hole in the M-shell (3s, 3p, 3d states) are M x-rays, and so forth. Energy tables for diagram lines have been complied by Bearden (1967).

Another way of speaking of emission spectra is in terms of bands and edges. Thus spectra consisting of transitions from the valence orbitals (or valence band) in solids to the $L_{II,III}$ edge (2p states) are said to be $L_{II,III}$ band spectra. Such transitions might consist of $M_V \rightarrow L_{III}$ ($3d_{5/2} \rightarrow 2p_{3/2}$) or L$\alpha_1$, $M_{IV} \rightarrow L_{III}$ ($3d_{3/2} \rightarrow 2p_{3/2}$) or L$\alpha_2$, $M_{IV} \rightarrow L_{II}$ ($3d_{3/2} \rightarrow 2p_{1/2}$) or L$\beta_1$, and so on. Because of partial failure of our original assumptions, some lines are also observed which

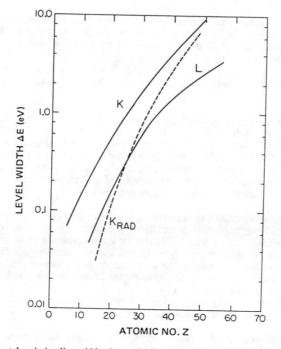

Figure 30. Experimental emission line widths due to the finite lifetime of the core hole state as a function of atomic number. K = K shell emission, L = L shell emission. The dashed line indicates the portion of the width for K X-ray transitions due to fluorescence emission processes. The rest is due to Auger type processes. These two relaxation processes determine the excited state lifetime. [From Parratt, 1959]

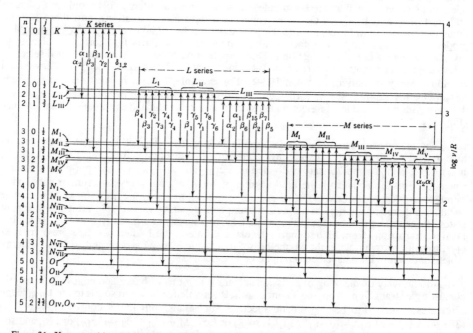

Figure 31. X-ray transition diagram and diagram lines. [From Klug and Alexander, 1974]

break the selection rules. These are usually quadrupole-allowed transitions, $\Delta l = 0, \pm 2$, such as $K\beta_4$ (4d→1s), and $K\beta_5$ (3d→1s).

Satellites. Besides the diagram lines, many additional features appear in actual x-ray emission spectra. These may be called non-diagram lines or satellites. The term satellite is appropriate because these additional features are always associated with particular diagram lines. They arise from spin interactions between a hole state and another hole in the atom (i.e., a multiply ionized atom), from interactions between a hole state and a partially unfilled orbital as in 3d transition metal complexes, from crossover and other molecular orbital transitions (see next section), and from other interactions. The intensity of satellite lines is usually very weak compared to the diagram lines (Aberg, 1967). However, excitation by charged particles produces much more multiply ionized atoms than does x-ray photon excitation, and a corresponding increase in this type of satellite intensity. In some cases, such satellite lines can even exceed the diagram lines $K\alpha_{1,2}$ in intensity, as shown in Figure 32 (Knudson et al., 1971). Satellite intensity can be very sensitive to the details of electronic structure. Thus it would be interesting to see if high resolution PIXE analysis might reveal new bonding information about mineralogical solids by virtue of enhanced multiple ionization satellite lines.

Satellites have various designations. For example, $K\alpha_{3,4}$ are multiple-ionization satellites of $K\alpha_{1,2}$; whereas $K\beta'$, the origin of which is still open to question, is a satellite of $K\beta_{1,3}$. Double ionization satellites may have particular sensitivity to the number of unpaired electrons in an unfilled d-shell, and are thus especially useful in valence, spin-state and bonding determinations.

X-ray emission transition probabilities. Given the dipole and frozen orbital approximation noted above, a relatively simple transition moment integral formulation can be composed for a given pair of bonded atoms in a structure. Beginning with the general form of the transition moment integral, the transition intensity, I, is given (Rooke, 1974) by

$$I = C \times v^3 \left[\int \Psi_i \cdot r_j \exp(ik \cdot r) \, \Psi_f \, d\tau \right]^2 ; \qquad (1)$$

where C is a constant defined as $4\pi^2 e^2/hc^3$, and $r_j \exp(ik \cdot r)$ is the transition moment operator. r_j is the component of r in the direction of the electric vector polarization. Expansion of the operator yields the terms due to dipole and other multipole moments:

$$\exp(ik \cdot r) = 1 + ik \cdot r + (ik \cdot r)^2/2! + (ik \cdot r)^3/3! + \ldots \qquad (2)$$

In the dipole approximation all terms except the first are considered zero. In fact, the second term (quadrupole moment) is usually nonzero but small. The higher terms, octapole and hexadecapole, are generally negligible. In the dipole approximation we then have:

$$I = C \times v^3 \left[\int \Psi_i \cdot r_j \, \Psi_f \, d\tau \right]^2 . \qquad (3)$$

This formulation is sufficient for isolated atoms, where the initial and final wavefunctions would be well-represented by atomic orbital wavefunctions, and the diagram line energy level picture would be basically adequate. However, in the case of a typical insulating or semiconducting mineral, we must consider bonding interactions between pairs of atoms, usually a cation M, and an anion O. Proper representation of this interaction in the transition moment integral requires replacement of the final-state wavefunction by molecular orbital wavefunctions of the M-O system. The initial state may be approximated by the inner atomic orbital on the metal atom having the electron hole, as it has minimal involvement in bonding. The simplest way to approximate the molecular orbital is by a linear combination of atomic orbitals (LCAO):

$$I = C \times v^3 \left[\int \Phi_i \cdot r_j \, (\Sigma \, \alpha_{rf} \, \Phi_f) \, d\tau \right]^2 . \qquad (4)$$

Here only atomic orbital wavefunctions (Φ) are used, and the α_{rf} are the molecular orbital coefficients over the set of r atomic orbitals. If, for example, the initial electron hole is in a 1s

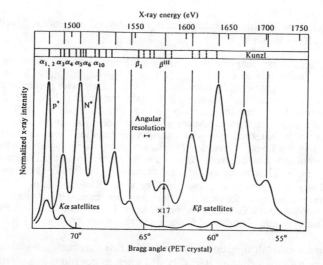

Figure 32. Kα emission spectrum of Al metal excited by protons (p+) and nitrogen ions (N+). The heavier nitrogen ions excite additional ionizations within the Al atoms, creating an enhanced satellite emission structure. [From Knudson et al., 1971]

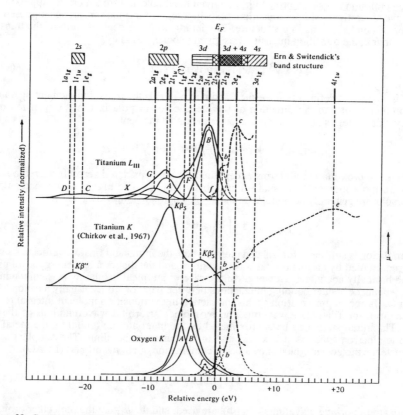

Figure 33. Correlation of Ti L$_{III}$, Ti K and O K X-ray emission (solid lines) and absorption (dashed lines) spectra of TiO with band structure (Ern and Switendick (1965) and molecular orbital scheme. [From Fischer, 1973]

atomic-like orbital on the M atom, and the final state is composed of a linear combination of s and p atomic orbitals from both M and O atoms, we get by expanding (Urch, 1985):

$$I = C \times \nu^3 \left[\int \Phi_{M1s} \cdot r_j \, \alpha_{Mns} \, \Phi_{Mns} \, d\tau + \int \Phi_{M1s} \cdot r_j \, \alpha_{Mnp} \, \Phi_{Mnp} \, d\tau \right.$$

$$\left. + \int \Phi_{M1s} \cdot r_j \, \alpha_{Ons} \, \Phi_{Ons} \, d\tau + \int \Phi_{M1s} \cdot r_j \, \alpha_{Onp} \, \Phi_{Onp} \, d\tau \right]^2 . \tag{5}$$

The value of the first integral is zero, since ns→1s transitions are forbidden in the dipole approximation. The second integral is finite since np→1s is an allowed transition. The 3rd and 4th integrals indicate "cross over" transition probabilities, where an electron is transferred from either an s or p orbital on the anion into the M 1s core hole. If these latter integrals are assumed to be small compared to the second integral, the intensity reduces to

$$I = C \times \nu^3 \times (\alpha_{Mnp})^2 \times \left[\int \Phi_{M1s} \cdot r_j \, \Phi_{Mnp} \, d\tau \right]^2 , \tag{6}$$

where the molecular orbital coefficient is taken outside of the integral.

If several molecular orbitals in the M-O bonding system have p character, then instead of the single "diagram line" situation, several transitions to the M1s state will be observed. The intensity of each transition will be proportional to the square of the relevant molecular orbital coefficient. Thus the amount of p character in each state can be determined from the x-ray line intensity. This example features the use of the M K x-ray emission spectrum since it examines M1s transitions, however O K and M L x-ray spectra could be examined as well. O K spectra would have intensities determined largely by O 2p character in the molecular orbitals, and M L spectra would analogously probe M ns and M nd character. It is conceivable that the total atomic orbital components of a molecular orbital could be determined from such a set of spectral measurements.

In order to estimate the complete set of atomic orbital contributions to molecular orbitals, the differing x-ray spectra from a single M-O system must be aligned to a common energy scale, and the component features assigned with a molecular orbital energy level diagram. Figure 33 depicts such a diagram for Ti O (NaCl structure). This energy alignment is possible because the spectra all probe the same set of molecular orbital energy level spacings, even though the absolute energy regions of each type of spectrum may be quite different. The alignment can only be accomplished if the ionization energies of the initial electronic state is known. These can be obtained from core-level XPS spectra (see the chapter on XPS spectroscopy in this volume). For example, in the case of the M-O system discussed above, the M L x-ray emission spectra may show major lines at energies E_1 and E_2, and the O K x-ray emission spectra lines at energies E_3 and E_4. If from XPS spectra the ionization energies for the 1s core levels of the M and O atoms are E_{Ms} and E_{Os}, respectively, then the final states have energies of $E_{Ms}-E_1$, $E_{Ms}-E_2$, $E_{Os}-E_3$ and $E_{Os}-E_4$. If the magnitude of $E_{Ms}-E_1$ is equal to $E_{Os}-E_4$, then the M L emission line at energy E_1 and the O K emission line at energy E_4 originate from electronic transitions to the same molecular orbital. The relative intensity of each line then indicates the atomic orbital contributions. As in Figure 33, the number of molecular orbital states in even a simple structure can be quite large. Such a construction provides a self-consistency test for molecular orbital calculations of mineral structures (Tossell, 1973, 1975; Sherman, 1984).

As seen by this example, the type of emission spectrum predicted from simple atomic orbital energy diagrams and diagram lines differs substantially in complexity and information from that actually observed in minerals. Single diagram lines representing simple (e.g. 2p→1s) transitions on a single atom are replaced by molecular orbital transitions whose number, type and intensity depend on the ligand field symmetry, metal-ligand distance, spin interactions and ligand identity. Transitions due to crossovers, a type of charge transfer, can also occur, and emphasize the interaction between metal and ligand due to chemical bonding. The transitions will also vary in energy due to changes in bond length and formal ion valence.

<u>Spectrometers</u>. The resolution necessary for x-ray emission fine structure measurements depends on the energy region explored. Below 1000 eV, resolution of $10^{-3} \, \Delta E/E$ is sufficient

for most features; from 1 to 10 keV, 5 x 10^{-4} is needed. For higher energy K-spectra, 10^{-4} or better is required. The four commonly used crystal spectrometers utilize a single flat crystal, single curved (Johann) crystal, single curved and ground (Johansson) crystal, or a double-flat crystal geometry.

The single flat-crystal spectrometer is the simplest, but requires narrow slits and/or use of a high order Bragg reflection to achieve the highest resolution (Thomsen, 1974), thus limiting the intensity. The improved resolution of higher order Bragg reflections is due to their narrower rocking curves (more severe diffraction conditions) compared to low order reflections. A Johann crystal of the same type used with a Rowland circle geometry can have similar resolution to the flat crystal, but as much as 100 times the intensity. A Johansson crystal is still better by about a factor of three in intensity with similar resolution (Lu and Stern, 1980). In practice, the resolution of focussing crystal spectrometers is largely controlled by aberrations in the crystal bend and surfacing. The radius of the spectrometer must also be consistent with the inherent resolution of the crystal.

There are several differing geometries for a double crystal spectrometer. The crystals may be cut symmetrically or asymmetrically and oriented in parallel or antiparallel alignments. In general, the resolution will be about three times better than a single flat crystal, but the precise geometry, the order of the Bragg reflection used, and the slit arrangements determine ultimate performance (Azaroff, 1974). Most high resolution spectrometers used with synchrotron radiation sources have parallel double crystal x-ray optics. For very soft x-rays, it is necessary to use diffraction gratings instead of Bragg crystals. The geometry of grating spectrometers has been reviewed by Cuthill (1974).

Applications to mineralogy and geochemistry

Iron minerals. The x-ray emission spectrum of iron minerals, particularly iron oxides, has been studied by several workers in an attempt to correlate valence and coordination with spectral features. Nefedov (1966) and Koster and Mendel (1970) noted variations in the intensity of the Kβ satellites relative to the Kβ diagram lines as a function of iron formal valence (Fig. 34). In particular, as shown by Urch and Webber (1977), the intensity of the Kβ' satellite increases as the number of 3d unpaired electron spins increases. The Kβ' satellite is about 15 eV lower in energy than the Kβ$_{1,3}$ doublet ($2p_{3/2,1/2} \to 1s$), and thus is likely due to a perturbation to the 2p states which splits the Kβ$_{1,3}$ line. The mechanism for this is incompletely understood. It probably involves spin interactions between the 2p electron hole in the final state and the 3d shell (Tsutsumi et al., 1976; Sherman, 1984). However, Tossell (1974) has interpreted the experimental data in terms of exciton formation, which is consistent with the energies predicted by his molecular orbital calculations. In high spin configurations, Fe^{3+} ions have five unpaired electrons, and Fe^{2+} only four, and this leads to a small but reproducible variation in Kβ' intensity. Variations between high and low spin compounds are much larger. The Kβ" satellite (about 15 eV lower in energy than the Kβ$_5$ line) is also affected by the number of unpaired 3d spins, but in the opposite sense. This feature decreases in intensity as the number of unpaired 3d spins increases. It has been assigned to an O 2s $\to$ Fe 1s crossover transition, in analogy with Mn spectra (Tsutsumi et al., 1976). Presumably the O 2s state must have some degree of p-character mixing, or else this transition is allowed only via quadrupole coupling. The Kβ spectral lines also shift in energy with formal cation valence. Koster and Mendel (1970) found a 1.1 eV increase in the Kβ$_5$ energy in going from FeO to Fe_2O_3. The strong Kβ$_{1,3}$ doublet showed a smaller increase of 0.4 eV. The satellite lines also shifted, but because of their relative weakness and the effect of background tails due to stronger nearby lines, the observations are probably much less reliable.

Koster and Rieck (1970) examined a large number of iron oxides and spinels, and also found that iron valence affected the Kβ$_5$ line position, although the energy shift was not linearly related to the average valence. This line was apparently insensitive to coordination, as Fe^{2+} in chromite and wustite had the same Kβ$_5$ energy. The Kβ$_1$ line seemed sensitive to both valence and coordination, with the coordination change from octahedral to tetrahedral giving a larger positive energy shift (0.5 eV) then the change from Fe^{2+} to Fe^{3+} (0.4 eV).

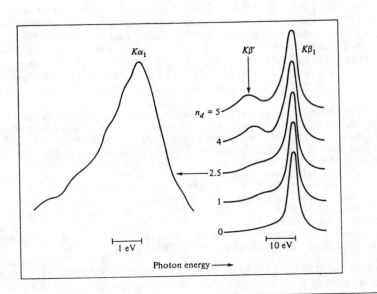

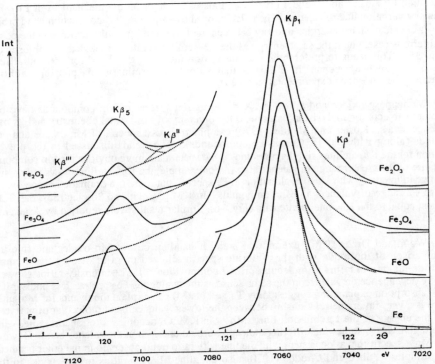

Figure 34. Top: Intensification of the Kβ' satellite as a function of the number of unpaired 3d electrons in iron compounds. Satellite structure in the Kα emission line is also observed, but unresolved. [From Nefedov, 1966] Bottom: Fe K X-ray emission spectrum for a series of oxides of with varying Fe valency. [From Koster and Mendel , 1970]

Narbutt (1980) examined Kα and Kβ spectra of iron in a suite of 54 samples, including 23 minerals. The Kβ line positions of the oxides agree reasonably well with those of Koster and Rieck (1970) and Koster and Mendel (1970). However, there are effects on the $K\beta_1$ energies of the silicates due to coordination, ligand symmetry and overall chemistry which overwhelm any valence effect. For example, Fe^{2+} in hedenbergite and Fe^{3+} in andradite have the same $K\beta_1$ energies, and Fe^{2+} in the dodecahedral (X) site in almandine has 0.26 eV higher energy than Fe^{3+} in hematite. Additional work is obviously required to understand these variations.

Using an electron microprobe, Albee and Chodos (1970) examined the iron $L\alpha_{1,2}$ and $L\beta_1$ line intensities in a variety of minerals, including hematite-ilmenite and magnetite-ulvospinel solid solutions. They found a considerable dependence of the intensity ratio of these lines on average iron valence. However, basic theory would suggest no such intensity change, as the transitions are due to electron holes in the $2p_{3/2}$ and $2p_{1/2}$ subshells, which have a 2:1 probability of vacancy production regardless of the number of unpaired 3d electrons. Other workers (e.g., O'Nions and Smith, 1971) have observed similar valence effects, but because the intensity ratio is very strongly dependent on the acceleration voltage and emission take-off angle due to self absorption effects, it is difficult to produce quantitative results. In principle, spectra can be corrected for these effects, which are much less of a problem with K spectra at typical excitation energies. Self-absorption effects can be used to determine L_{III} absorption edge positions from L band emission spectra (Dodd and Ribbe, 1978).

Tossell et al. (1974) calculated molecular orbital energies for ferric and ferrous octahedral Fe-O clusters via SCF Xα methods. Their results generally agree well with the Kβ spectra measured by Koster and Mendel (1970), and also shed light on the mechanism of $L\alpha_{1,2}/L\beta_1$ intensity variation due to valence. The calculation shows that the $7a_{1g}$ conduction band orbitals (Fe 4s character) have energies midway between the $L\alpha_{1,2}$ and $L\beta_1$ emission lines in the octahedral ferrous case, but at the same energy as the $L\beta_1$ emission line for the octahedral ferric case (Fig. 35). This leads to preferential self absorption into the $7a_{1g}$ band for octahedral ferric minerals. Such a mechanism indicates that valence determination is possible, although extremely dependent on experimental technique.

Manganese oxides and silicates. X-ray emission spectra of Mn compounds have been popular subjects for analysis, in part due to the diversity of Mn crystal chemistry and its many valence states. Koster and Mendel (1970) examined the Kβ spectra of Mn oxides and compounds (along with Ca, Ti, Cr and Fe compounds). The trends discussed in the previous section for the Kβ satellites and diagram lines in Fe spectra are more obvious for Mn compounds because of the greater range of valence states. For example, the Kβ" satellite shows only a small change in intensity with valence alteration in Fe spectra, but there is a dramatic change over the +7 to +2 range of valences in Mn compounds. In $KMnO_4$, with no 3d electrons, the Kβ" is almost equal to the $K\beta_5$ line in intensity (Fig. 36). Similar results were reported by Asada et al. (1975).

Wood and Urch (1976) examined Kβ and L band spectra of Mn oxides and fluorides. Comparison of the oxide with the fluoride spectra allowed them to discuss the increased fluoride ionicity in terms of molecular orbital composition. Lα/Lβ intensity ratios showed a strong dependence on valence in the same sense as the analogous Fe spectra. It is unclear if the self-absorption explanation suggested by Tossell (1974) can explain the results for Mn, unless there is a very favorable relationship between the lowest valence band molecular orbital levels and the energy of the Lβ emission line. Mn^{7+} in $KMnO_4$ occupies a tetrahedral site, which would give rise to somewhat differing molecular orbital energies in the valence band. However, the Mn^{7+} spectrum was consistent with the intensity ratio trend observed in the other octahedral Mn samples. Albee and Chodos (1970) also examined Lα/Lβ ratios in Mn oxides in their electron microprobe study.

Urch and Wood (1978) examined the spectra of Mn minerals, focussing on use of the $K\beta'/K\beta_{1,3}$ intensity ratio as a measure of Mn valence. Their results should be relatively free of self-absorption effects, and appear to be insensitive to bond lengths and coordination environ-

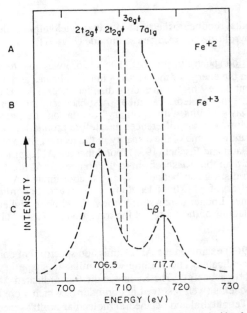

Figure 35. Experimental Fe L X-ray emission spectrum for hematite compared with calculated energies for empty molecular orbitals responsible for Lα absorption bands in ferrous and ferric oxides. A- ferrous octahedral cluster calculation. B- ferric octahedral cluster calculation. C- experimental spectrum. [From Tossell et al., 1974]

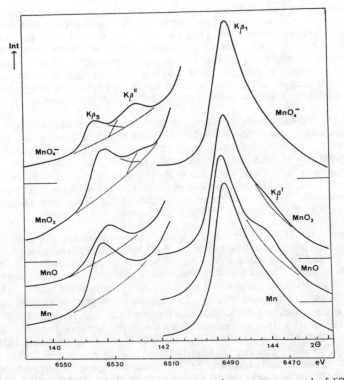

Figure 36. Manganese Kβ X-ray emission spectra from a series of manganese compounds of differing valence. [From Koster and Mendel, 1970]

ments. Mn^{2+} in rhodonite, rhodochrosite and Mn acetate yield almost identical spectra, as does Mn^{4+} in psilomelane, pyrolusite and manganese dioxide (Fig. 37).

Si, Al and O emission spectra in silicates. Day (1963) was able to detect a small effect of coordination number on the energy of the Al $K\alpha_{1,2}$ line in silicates and aluminates. Octahedral Al emission occurs about 0.2 eV higher in energy than tetrahedral Al. Day utilized "secondary" excitation, i.e., x-rays from a sealed x-ray tube. A subsequent attempt by Baun and Fischer (1965) to reproduce Day's results with "primary" excitation, i.e., electron excitation, was largely unsuccessful. However, the differences in satellite production between the two excitation methods may account for some of the disparity in the results, since the $K\alpha$ satellite structure is difficult to resolve. Baun and Fischer (1965) found that the Al $K\beta$ emission spectrum showed a much larger variation than the $K\alpha$ spectrum to differences in Al bonding. However, the $K\beta$ emission also failed to distinguish between 4- and 6-coordinated Al. Wardle and Brindley (1971) determined the effect of Al-O bond length for both 6- and 4-coordinated Al on the energy of the $K\alpha_{1,2}$ emission line. Longer bonds shifted the Al $K\alpha$ line to higher energies. This trend is consistent with calculations of the energy of the transition by Tossell (1973) and may explain Day's results.

Dodd and Glen (1969) examined the Al $K\beta$ emission spectrum of corundum, pyrope, beryl and several framework silicates. The $K\beta$ line was similar in shape in all of the framework silicates, but shifted subtly in energy. The $K\beta$ line in the 6-coordinated Al samples appeared to have two distinct components (well resolved in corundum), which were assigned to $3p(\sigma)$ and $3p(\pi) \rightarrow 1s$ transitions. Tetrahedral and octahedral molecular orbital schemes have been calculated for Al-O clusters by Tossell (1973, 1975). Both coordinations should give rise to three $K\beta$ emission lines, two of which are close in energy, so that small differences in orbital contributions (or bond length) may be the cause of the differences in the observed spectra.

Si emission spectra have been examined in some detail, with attempts to correlate line intensity, energy and width to details of SiO_4 tetrahedra polymerization in silicates. Koffman and Moll (1966) surveyed a large suite of silicates, but found no energy dependence of the $K\alpha$ emission with Si-O bond length. Tossell (1973) reassessed these workers' data and found a relationship between their Si $K\alpha_{3,4}$ energies and the type of tetrahedral linkage. Specifically, the Si $K\alpha_{3,4}$ energy increases in going from isolated tetrahedra to chain silicates to framework silicates. Accompanying this energy shift was a change in the effective Si charge, such that Si is most positive in SiO_2 and decreases as the degree of polymerization decreases.

Si $K\beta$ spectra also show a distinct energy dependence on bond length (White and Gibbs, 1967) and degree of polymerization (Tossell, 1973) with the same sense as the $K\alpha_{3,4}$ results. The shift in energy is particularly large in going from quartz to stishovite. Tossell (1973) also predicted a dependence of the Si $K\beta$ emission band (there are many associated t_2 "like" states giving rise to the $3p \rightarrow 1s$ Si $K\beta$ emission due to the low symmetry of most Si sites) on the Si-O-Si bond angle. This, in addition to the known difference in Si-O bond length depending on the bridging or non-bridging nature of the oxygen, indicated that Si $K\beta$ spectra might be used to identify the proportion of non-bridging oxygen in a silicate mineral or glass.

Hogarth and Urch (1976) found a dependence of the separation of the Si $K\beta_1$ and $K\beta'$ satellite with the type of ligand bonding to the Si. Si-O bonds gave a separation of about 15 eV, Si-F about 20 eV and Si-C about 8 eV. Thus the separation appears to be a measure of the effective charge on the Si (largest with the most ionic ligand) and hence, the relative covalency of the bond. Urch (1985) discusses the Si $K\beta$ spectrum in terms of the atomic orbital contributions to the various t_2 molecular orbitals. The $K\beta_1$ line is assigned to a (mixed Si 3p-O 2p) $4t_2 \rightarrow$ (Si 1s) $1a_1$ transition, and the $K\beta'$ line to a (mainly O 2s with some Si 3p) $3t_2 \rightarrow$ (Si 1s) $1a_1$ transition. The small Si 3p contribution to the $K\beta'$ emission is the reason for its low relative intensity. In the quartz emission spectrum (Fig. 38) these features occur at about 1832 and 1817 eV, respectively. A small shoulder at about 1835 eV may be due to a $5t_2 \rightarrow 1a_1$ transition, where the $5t_2$ state has a small amount of Si 3p character. Figure 38 demonstrates how the various emission spectra of quartz can be used to estimate these atomic contributions. If it is assumed that crossover transitions contribute little intensity, then the Si $K\beta$ spectra will have

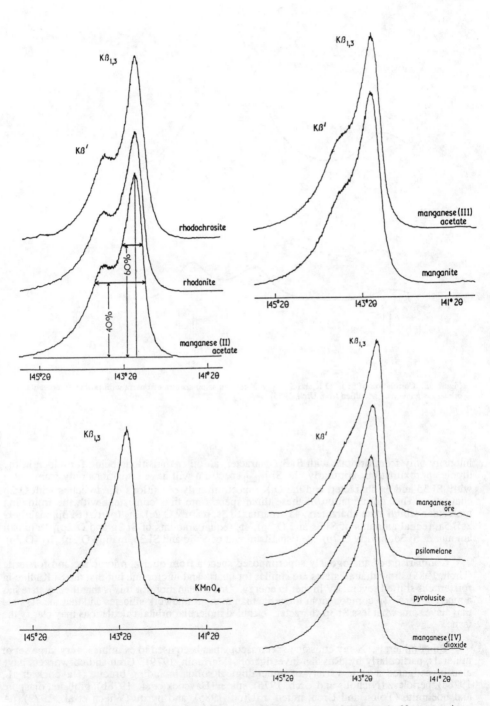

Figure 37. Manganese Kβ X-ray emission spectra from manganese compounds and minerals. Note comparison with Figure 16 (energy scale reversed). [From Urch and Wood, 1978]

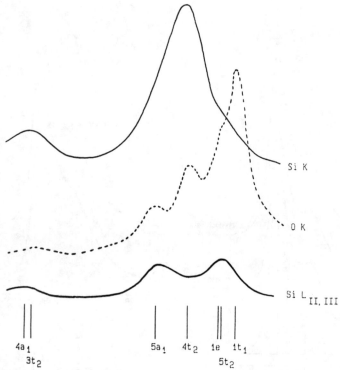

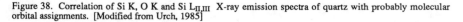

Figure 38. Correlation of Si K, O K and Si L$_{II,III}$ X-ray emission spectra of quartz with probably molecular orbital assignments. [Modified from Urch, 1985]

intensity only from orbitals with Si 3p character, all other transitions being forbidden in the dipole approximation. Similarly, the Si L$_{II,III}$ spectrum will have transitions only from states with Si 3d and 3s character, and the O Kα spectrum only transitions due to states with O 2p character. The combination of these three spectra are thus consistent with the following molecular orbital state characters: 4a$_1$ (partial Si 3s, mainly O 2s), 3t$_2$ (partial Si 3p, mainly O 2s), 5a$_1$ (equal amounts of Si 3s and O 2p), 4t$_2$ (equal amounts of Si 3p and O 2p), 1e (small amount of Si 3d, mostly O 2p), 5t$_2$ (small amount of Si 3d and Si 3p, mainly O 2p), 1t$_1$ (O 2p).

Comparison of analogously superimposed spectra from quartz, microcline and forsterite (Urch, 1985) show that all states are similar for quartz and microcline but that the Si Kβ line in forsterite is shifted several eV higher in energy. One explanation for this is that the 5t$_2$ state has a much larger Si 3p component in forsterite than in the framework silicates and that the 4t$_2$ state may have somewhat less Si 3p character. Detailed molecular orbital calculations may clarify the issue.

Other minerals. X-ray emission spectroscopy has been used to examine a very large set of minerals, particularly by Russian investigators (Marfunin, 1979). Urch and co-workers have also investigated a wide variety of mineralogical solids including brucite (Haycock et al., 1978a), periclase (Nicholls and Urch, 1975), spinel (Haycock et al., 1978b), gibbsite, diaspore and boehmite (Tolon and Urch, noted in Urch, 1985), and pyrite (Wiech et al., 1972). A growing volume of work has also been done with photoemission spectroscopy, which probes the core ionization levels directly, and can be used to align the energy axis of the various emission spectra.

Glasses. The Si Kα and Kβ emission spectra of several compositional series of glasses have been investigated. Dodd and Glen (1970) examined Li_2O-Al_2O_3-SiO_2 glasses and found a progressive shift of the $K\beta_1$ peak to higher energies as silica content decreased. Their spectrometer resolution was not good enough to resolve any changes in line structure. However, these results suggested a lengthening and weakening of the Si-O bond. In Na_2O-SiO_2 glasses, Wiech et al. (1976) and Alter and Wiech (1978) observed a systematic change in the Kβ spectrum with the appearance of a second feature about 4 eV higher in energy than the main $K\beta_1$ peak. The new feature appears to be consistent with the $5t_2 \rightarrow 1a_1$ transition in quartz discussed earlier. The intensity of the feature and its correlation with the amount of Na_2O suggest that it is due to silicon associated with non-bridging oxygen, i.e., with nonpolymerized SiO_4 tetrahedra (Brüchner et al., 1978).

In addition to the intensification of the $5t_2 \rightarrow 1a_1$ feature with increasing alkali content, an overall shift of the Si Kα spectrum to lower energy occurs. Such a shift is consistent with the Si $K\alpha_{3,4}$ satellite results observed by Koffman and Moll (1966), i.e., the glass network depolymerizes with increasing alkali and this results in progressively smaller positive charge on the Si. This logic suggests that the oxygens associated with the isolated SiO_4 tetrahedra carry a larger negative charge compared to bridging oxygens. It is this enhanced negative charge on the non-bridging oxygens which is balanced by the alkali ions. It is interesting to note that the energy shift occurs without any measurable broadening in the Kα emission band, suggesting a structure wide variation rather than a mixing of two types of Si atoms. Brüchner et al. (1978) also examined silica glasses with other alkali ions, and found that there was a shift in the O 1s XPS spectrum to lower energy according to the sequence Li, Na, K, Cs. They concluded that increasing ionicity of the charge-compensating cation resulted in a larger negative charge (charge density) on the non-bridging oxygens.

An assortment of alkali and alkali earth-silica glasses was studied by DeJong et al. (1981). They analyzed their Kβ emission spectra in terms of Q distribution theory (Engelhardt et al., 1975), and also did CNDO/2 molecular orbital calculations on small molecular units to emulate the effect of alkalis on the Q_3 and Q_4 units. The results of their calculations, coupled with the experimental spectra, led to the conclusion that the spectra were mainly able to sense the difference between Q_4 (Si in fully polymerized SiO_4 tetrahedra) and Q_R (all of the other possible types of Si). This was due to the relatively large energy shift created by the production of the first non-bridging oxygen coordinated to a silicon atom. Some inference as to the nature of the Q_R distribution was made on the basis of total spectral band width. The alkali-silicate glasses appeared to have strictly bimodal spectra, whereas the alkali earth silicate glasses produced broader Si Kβ bands, particularly in the case of the barium-rich glass. Unfortunately, DeJong et al. (1981) had much lower resolution spectra than those presented by Alter and Wiech (1978). Hence their CNDO/2 calculations are not sufficiently tested by the experimental spectra. Improved resolution in Si Kβ spectra, as well as new Si L and O K spectra, should help in deciphering the bonding changes in these glasses.

NEW SPECTROSCOPIES

In this section we consider two spectroscopic methods which have been known for some time, but have seen substantial renewed interest and development in the last decade. Both RBS and EELS have seen increasing use in materials-related disciplines because of their high spatial resolution capabilities, and their applicability to thin film technologies. Neither has been used in many geological applications, but this situation will certainly change with the growing awareness of the importance of surface chemistry and fine details of mineral structure to geochemical processes.

Rutherford backscattering spectroscopy (RBS)

RBS owes its name to the original type of experiments that Rutherford devised to probe atomic structure. In these early studies, a beam of particles was directed on thin foils and the intensity of scattering as a function of scattering angle was observed. The backscattering occurred only from near the vicinity of the atom nuclei, allowing an estimation of the nuclear dimensions. Such backscattering is essentially elastic, so that the energy of the scattered particles does not change appreciably if the film is very thin. However, if the film is thicker than several monolayers of atoms, the energy of the backscattered particle is reduced by electrostatic interactions with atomic electrons. This energy loss effect allows measurement of the depth of the backscattering process in the material. In addition, the energy of backscattering is dependent on the mass of the backscattering atom, so that in a given backscattering direction, the particles will have an energy range characteristic of the atomic number of the scatterers. These two properties give RBS its utility for depth profiling on a near atomic resolution scale, yet it can also be used for bulk chemical analysis.

Theory. The process of RBS is shown in Figure 39 (Gossmann and Feldman, 1987). Particles of mass M_1, atomic number Z_1 and energy E_0 impinge on a sample with a small concentration of impurity atoms on the surface. The bulk material has mass M_2 and atomic number Z_2, the impurity atoms mass M_3 and atomic number Z_3. A detector collects the backscattering particles over a small angular range and analyses their energy. The energy of the backscattered particles, assuming only elastic interactions, is given by:

$$E_1 = \left[\frac{(M_2^2 - M_1^2 \sin^2\theta)^{1/2} + M_1 \cos\theta}{M_2 + M_1} \right]^2 E_0 \tag{7}$$

At $\theta = 90°$, this simplifies to $E_1 = [(M_2-M_1)/(M_2+M_1)]E_0$. Thus the energy of the backscattered particle depends on the masses of both the incident particle and the backscatterer.

In Figure 39, the mass of the bulk material, M_2, is smaller than that of the impurity, M_3. Hence observation of scattering at a given angle would detect higher energy particles scattering from M_3 atoms, and lower energy particles scattering from M_2 atoms. Since the M_3 atoms are only near the surface of the sample, their energy is well defined. However, the M_2 atoms produce scattering over a range of energies, corresponding to the energy loss of particles scattered at various depths in the sample. This picture is quantitative. The difference in energy between the M_3 peak and the beginning of the M_2 band is representative of the mass difference M_2-M_3. Similarly, the energy width of the M_2 band is representative of the thickness of the sample probed by the particle beam.

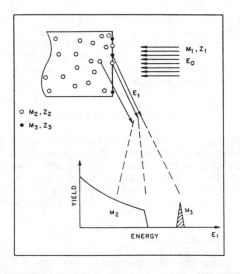

Figure 39. Geometry of Rutherford Backscattering. The incident particles have mass M_1, energy E_0 and atomic number Z_1. The sample in this drawing consists of a thin layer of impurities on top of a substrate. The impurities have mass M, the substrate atoms have mass M_2. A solid state detector intercepts the scattered beams in the direction shown, yielding the spectrum at the bottom. [From Gossmann and Feldman, 1987]

The energy loss of the scattered (and penetrating) particles is related to the stopping power of the material. This is characterized by the rate of energy loss with depth, or dE/dx. For scattering at any given angle the variation in energy with thickness is

$$\delta E_1/\delta\tau = E_1/E_0 \left(dE/dx\right)_{E0} - 1/\cos q \ (dE/dx)_{E1} \ . \tag{8}$$

According to this relation, the depth resolution of the RBS technique is dependent on the geometry of scattering via the scattering angle, θ, the energy resolution of the detector, δE_1, and the rate of energy loss, dE/dx. In practice, the parameters are such that 1 MeV is an optimal energy for light particle scattering due to its relative invariance from material to material, and the fact that the ion scattering at these energies is almost purely elastic. With such energy, thickness resolution of about 2 nm is possible using standard solid state energy-dispersive detectors. Research is continuing on types of detectors that will enable substantially better depth resolution (van der Veen, 1985). Resolution in the lateral directions is limited by the particle beam optics, but is generally on the order of microns. It is conceivable that substantially narrower beams could improve lateral resolution by a factor of 10 or more. Detailed treatment of the physics of the ion backscattering process in materials can be found in Chu et al. (1978).

Applications. RBS works best on samples of low relative average mass having heavy impurities. In this application it is ideal for thin semiconductor film analysis, where many of the problems surround characterization of the depth of diffusion (or ion implanted) layers. An example of RBS analysis of a thin high temperature superconductor film is shown in Figure 40. In this example the incident particles are 1.7 MeV He[7] ions, and the RBS spectrum shows characteristic "edges" at high energy (channel number) due to scattered particles from the top of the thin film. Similar "edges" occur at the low energy side of the scattering band due to scattering from the bottom edge of the film. The width of each elemental scattering region is the same, but shifted in energy due to the mass differences. Because of its low mass, the oxygen scattering from the top of the film does not appear until about channel number 150, and thus its scattering profile is somewhat buried under the low energy scattering band from the substrate. Depth resolution is on the order of 5 nm.

In geochemistry, the most obvious application of RBS would be in the measurement of diffusion profiles with resolution at least 100 times better than an electron microprobe. Another area of promise is the examination of mineral surface chemistry changes due to weathering and other surface reactions. Unlike Auger spectroscopy, several hundred nanometers of sample can be probed, and this is done simultaneously.

RBS is an ideal technique to combine with PIXE, since RBS is not particularly sensitive to trace amounts of impurities, nor is it especially useful for identifying unknown chemical elements (especially if there are very many of them). It is mainly a spatial, rather than a chemical, analysis tool. Fortunately, accelerators which can produce particle beams for RBS are usually ideal sources for PIXE analysis. The combination of these two methodologies in the same instrument could provide an extremely potent system for geochemical analysis.

Electron energy loss spectroscopy (EELS)

Earlier in this chapter we considered the interaction of electrons with a sample in the production of characteristic and bremsstrahlung x-ray emission. Many other interactions can occur, among them, Auger electron emission, plasmon and exciton production, electron single and multiple scattering (elastic and inelastic), and excitation of inner shell electrons. One way to measure these phenomena is to record the energy spectrum of the transmitted electron beam. The various interactions appear in this spectrum as energy losses from the original unaffected beam energy. This type of spectroscopy is called electron energy loss spectroscopy, or EELS, and has been studied for more than 40 years. However, over the last decade there has been an explosion of interest in the technique, particularly with the development of low-cost high-resolution spectrometer systems which can be attached to standard transmission electron microscopes.

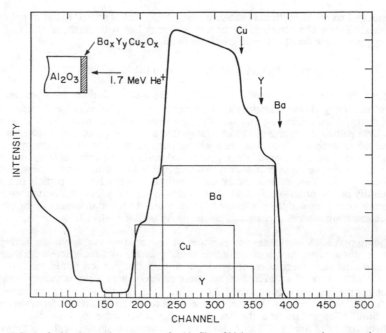

Figure 40. Rutherford backscattering spectrum of a thin film of high temperature ceramic superconductor. The channel scale represents scattered particle energy increasing to the right. The various sections of the scattering band due to the different elements in the sample are shown by the rectangles. [From Gossmann and Feldman, 1987]

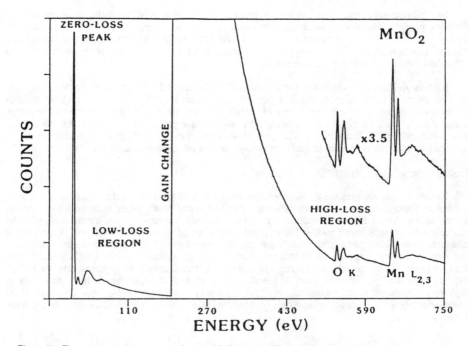

Figure 41. Electron energy loss spectrum from an MnO_2 sample. [From Rask et al., 1987]

Much of the current interest in EELS is tied to its ability to record the electron analog of the x-ray absorption spectrum. The edges which are observed may be treated much as EXAFS and XANES spectra, to determine chemical and structural details of individual regions on a sub-micron scale. EELS may also be combined with channelling for site partitioning and occupation determination.

In this section, we consider the basic details of EELS analysis, and some recent work on mineralogical samples. Excellent reviews of EELS techniques, theory, and applications have been written by Egerton (1986), Colliex (1984), and Joy (1979).

The EELS spectrum. An example of an EELS spectrum is shown in Figure 41. The spectrum can be divided up into three energy regions for discussion: (i) the very low energy loss region within 1 eV or so of the zero-loss peak; (ii) the low-loss region with energy loss of up to 50 eV; (iii) the high energy loss region. For a sufficiently thin specimen, there is little chance that an electron will be involved in two interaction events during its passage through the sample. In this case, the features observed in the spectrum are each due to separate types of interactions. If the specimen is thick enough such that the probability of multiple events is large, features due to the summed energy loss of the two (or more) events appear. The region near the zero loss peak is mainly due to the non-resolvable inelastic electron scattering which activates phonon modes in the sample. The features in the low loss region are in part due to plasmon formation. Plasmons are quantized collective oscillation modes of the valence electrons in a solid. They are easily excited in materials with weak binding energy for the outer electrons, such as alkali metals; in oxides and silicates they give rise to much weaker features due to the larger binding energy. Another constituent of low loss structure are excitations of individual valence and other outer-shell electrons. The low loss region thus contains much information about the valence band structure of a material.

The high energy loss region contains features due to excitations of inner shell electrons. Thus the edges, near-edge fine structure, and extended fine structure corresponding to x-ray absorption occur in this region. The background of the energy loss spectrum has a smooth logarithmic decay with increasing energy loss, reflecting the decreasing probability of higher energy inelastic interactions. The signal/noise ratio of the background also degrades with increasing energy loss due to degrading counting statistics.

EELS applications. The main use of EELS in thin film technology has been for the chemical analysis of low atomic number elements. These elements have absorption edges (or they might better be called energy loss edges) at relatively small energy losses where the signal/noise is good enough for quantitative analysis. With proper background subtraction and correction for any multiple scattering, the area under the edges is proportional to the product of the appropriate ionization cross section (e.g., 1s) and the incident beam intensity. The latter is often approximated by the zero-loss peak intensity.

An angle limiting aperture is usually placed below the specimen to limit the angular portion of the transmitted electron beam which reaches the energy loss spectrometer (Fig. 42). The angle β is chosen to exclude most of the elastically scattered electrons, but accept most of the inelastically scattered electrons below a certain energy loss. The elastic scattering produces a spreading of the electron beam in the sample (dotted area in Fig. 42) which limits spatial resolution for x-ray emission analysis. This effect worsens for increasing average atomic numbers and thicker specimens. Ideal EELS samples are thus rather thin, on the order of 1000 Å. The EELS spectrum originates from the hatched area in Figure 42, and thus is generated by a smaller region than that which produces the x-ray emission. Auger emission occurs from the top few hundred Ångströms of the specimen.

A thorough analysis of the high energy loss spectrum of Mn oxides and minerals with varying Mn oxidation states has been done by Rask et al. (1987). They found significant well-resolved shifts in the positions of the Mn L_2 and L_3 edges, and the O K edge as a function of Mn valence state. The Mn L edges shifts to higher energy, whereas the O K edge shifts to lower energies with increasing oxidation state. These trends with valence change are very similar to those observed in the L_2, L_3 and O K edge x-ray absorption spectra of the transition metal oxides and compounds (Fischer, 1971). This study demonstrates the possibility of spatial

692

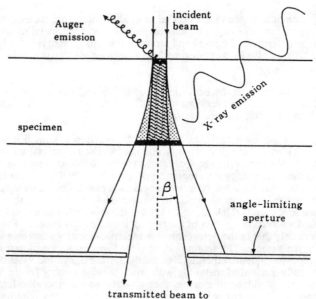

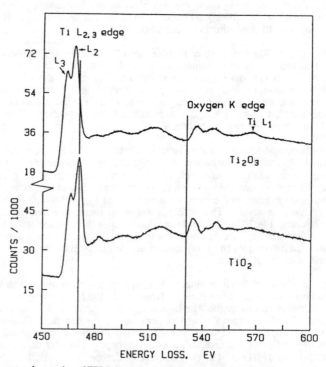

Figure 42. Excitation regions in a thin transmission electron microscope sample. The hatched region is that which contributes to the EELS spectrum. The dotted region represents the spread in the electron beam due to elastic scattering inside the sample. Auger emssion occurs only from the near surface region. [From Egerton, 1986]

Figure 43. High energy loss region of EELS spectrum of TiO_2 and Ti_2O_3. Note the opposite energy shifts in the Ti and O edges with the valence change. [From Otten and Buseck, 1987]

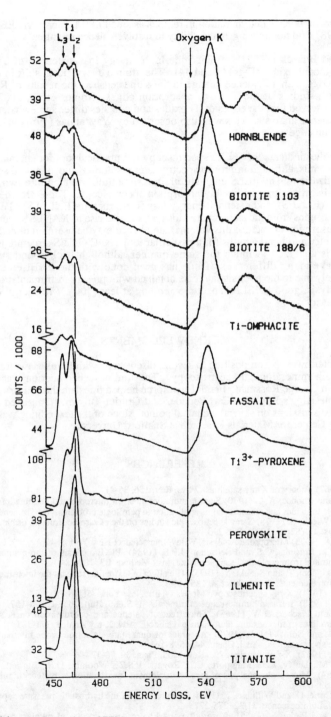

Figure 44. High energy loss regions of EELS spectra of titanium minerals and model compounds. [From Otten and Buseck, 1987]

694

transition element valence determination in regions as small as 0.1 micron via EELS. However, Rask et al. (1987) did not investigate the ability to resolve valence mixtures.

Otten and Buseck (1987) examined the high energy loss spectrum of Ti in a suite of minerals and model oxides (Figs. 43 and 44). They found a significant shift in the Ti L edges and the O K edge with Ti valence change in the same sense as the results of Rask et al., and consistent with studies of the Ti x-ray absorption edges (Grunes et al., 1982). Otten and Buseck also observed that the edges of Ti oxides and silicates differ somewhat, adding a complication to valence analysis. This was also observed by Waychunas (1987) in a study of Ti K x-ray absorption edges.

These studies indicate that EELS spectroscopy is a practical tool for chemical and valence analysis in minerals, although improved spectrometer resolution, and signal/noise enhancement will aid quantization. The EELS edge spectra are sometimes divided into two categories for further analysis, in analogy with x-ray absorption spectroscopy. These are energy loss near-edge structure (ELNES), and extended energy loss fine-structure (EXELFS). Features close to the edge have origins from the same interactions which give rise to XANES structure (discussed in the x-ray absorption chapter of this volume), and hence may be useful in directly determining the site symmetry and bond lengths for a particular ion. The EXELFS structure is analogous to EXAFS and is analyzed in much the same manner, although background subtraction and corrections have major differences. Little has been done with the EXELFS spectrum from minerals, largely due to the poor signal/noise obtained with present instrumentation. For useful structural analysis, good EXELFS spectra extending at least 400 eV in energy loss above the edge are necessary.

ACKNOWLEDGMENTS

This chapter owes its composition to several discussions with Frank Hawthorne concerning new methods in mineralogical spectroscopy. The luminescence section was improved by comments from George Rossman. Frank Hawthorne helped translate the author's American text into respectable King's English. Carl Ponader and Gordon Brown, Jr. assisted the author via useful general discussions on several technical points. Much of the text editing was facilitated by the staff of the Center for Materials Research at Stanford University.

REFERENCES

Aberg, T. (1967) Theory of X-ray satellites. Phys. Rev. 156, 35-41.

Albee, A.L. and Chodos, A.A. (1970) Semiquantitative electron microprobe determination of Fe^{2+}/Fe^{3+} and Mn^{2+}/Mn^{3+} in oxides and silicates and its application to petrologic problems. Am. Mineral. 55, 491-501.

Alter, A. and Wiech, G. (1978) X-ray spectroscopic studies on the electronic structure of binary glasses. Jap. J. Appl. Phys. 17 (Suppl. 17-2), 288-290.

Anderson, C.A. (1973) Microprobe Analysis. Wiley, Interscience: New York.

Annegarn, H.J., Erasmus, C.S. and Sellschop, J.P.F. (1984) PIXE analysis of the platinum group elements preconcentrated from geological samples. Nucl. Inst. Methods B3, 181-184.

Asada, E., Takiguchi, T. and Suzuki, Y. (1975) The effect of Oxidation state on the intensities of $K\beta5$ and $K\beta''$ of 3d transition elements. X-ray Spect. 4, 186-189.

Azaroff, L.V. (1974) X-ray spectroscopy. McGraw-Hill: New York, 560 p.

Barnes, D.F. (1958) Infrared luminescence of minerals. U.S.G.S. Bull. 1052-C, 71-157.

Baun, W.L. and Fischer, D.W. (1965) The effect of valence and coordination on K series diagram and nondiagram lines of magnesium, aluminum and silicon. Adv. X-ray Anal. 8, 371-383.

Beaman, D.R. and Solosky, L.F. (1972) Accuracy of quantitative electron probe microanalysis with energy dispersive spectrometers. Anal. Chem. 44, 1598-1610.

Bearden, J.A. (1967) X-ray wavelengths. Rev. Mod. Phys. 39, 78-124.

Benjamin, T.M., Duffy, C.J., Maggiore, C.J., Rogers, P.S.Z., Woolum, D.S., Burnett, D.S. and Murrell, M.T. (1984) Microprobe analysis of rare earth element fractionation in meteoritic minerals. Nucl. Inst. Methods B3, 677-680.

Berry, D.E., Curie, D. and Williams, F. (1984) Interpretation of the hydrostatic pressure dependence of the ruby R-lines. J. Luminescence 31/32, 275-277.

Bingham, K. and Parke, S. (1965) Absorption and Fluorescence spectra of divalent manganese in glasses. Phys. Chem. Glasses 6, 224-232.

Birks, L.S. (1971) Electron probe microanalysis. Second edition. Wiley, Interscience: New York.

Birks, L.S. (1969) X-ray spectrochemical analysis. Second edition. Wiley, Interscience: New York.

Blanchard, F.N. (1967) Thermoluminescence of synthetic fluorite. Am. Mineral. 52, 371-379.

Blanchard, F.N. (1966) Thermoluminescence of fluorite and age of deposition. Am. Mineral. 51, 474-485.

Blasse, G. (1968) Fluorescence of Uranium-activated compounds with rocksalt lattice. J. Electrochem. Soc. 115, 738-742.

Blasse, G. and Aguilar, M. (1984) Luminescence of natural calcite ($CaCO_3$). J. Luminescence 29, 239-241.

Blasse, G., Bleijenberg, K.C. and Krol, D.M. (1979) The luminescence of hexavalent uranium in solids. J. Luminescence 18/19, 57-62.

Brüchner, R. , Chun, H.U. and Goretzki, H. (1978) XPS-Measurements on alkali silicate-and soda aluminosilicate glasses. Japan J. Appl. Phys. 17, (Suppl. 17-2), 291-294.

Carlsson, L.E. (1984) Accuracy and precision in thick-target PIXE-PIGE analysis determined with geological standards. Nucl. Inst. Methods B3, 206-210.

Chu, W-K., Mayer, J.W. and Nicolet, M-A. (1978) Backscattering Spectrometry. Academic Press: New York.

Colliex, C. (1984) Electron energy-loss spectroscopy in the electron microscope. In Cosslett, V.E. and Barer, R., eds. Advances in optical and electron microscopy. Academic Press: London 9, 65-177.

Curie, D. (1963) Luminescence in Crystals. John Wiley: New York, 221 p.

Cuthill, J.R. (1974) Grating spectrometers and their application in emission spectroscopy. In Azaroff, L.V., ed. X-ray spectroscopy. McGraw-Hill: New York, 560p.

Davies, G. (1977) The optical properties of diamond. Chemistry and Physics of Carbon 13, 1-144.

Day, D.E. (1963) Determining the coordination number of aluminum ions by X-ray emission spectroscopy. Nature 200, 649-651.

Dean, P.J. (1966) Lattices of the diamond type. In Goldberg, P., ed. Luminescence of Inorganic Solids. Academic Press: New York, p. 120.

Dean, P.J. (1965) Bound excitons and donor-acceptor pairs in natural and synthetic diamond. Phys. Rev. 139, A588-A602.

Deb, S.K. and Gallivan, J.B. (1972) Photoluminescence of O_2^- and S_2^- ions in synthetic sodalites. J. Luminescence 5, 348-360.

DeJong, B.H.W.S., Keefer, K.D., Brown, G.E. Jr. and Taylor, C.M. (1981) Polymerization of silicate and aluminate tetrahedra in glasses, melts and aqueous solutions-III. Local silicon environments and internal nucleation in silicate glasses. Geochim. et Cosmochim. Acta 45, 1291-1308.

Dodd, C.G. and Glen, G.L. (1969) A survey of chemical bonding in silicate minerals by X-ray emission spectroscopy. Am. Mineral. 54, 1299-1311.

Dodd, C.G. and Glen, G.L. (1970) Studies of chemical bonding in glasses by X-ray emission spectroscopy. J. Am. Ceram. Soc. 53, 322-325.

Dodd, C.G. and Ribbe, P.H. (1978) Soft X-ray spectroscopy of ferrous silicates. Phys. Chem. Minerals 3, 145-162.

Durocher, J.J.G., Halden, N.M., Hawthorne, F.C. and McKee, J.S.C. (1988) PIKXE and Micro-PIKXE analysis of minerals at E_p = 40 MeV. Nucl. Inst. Methods (in press).

Ecker, K.H. (1984) Proton induced X-ray excitation applied to lattice location studies in molybdenum (cobalt) by channeling. Nucl. Inst. Methods B3, 283-287.

Egerton, R.F. (1986) Electron energy-loss spectroscopy in the electron microscope. Plenum Press: New York, 410 p.

Englehardt, G., Zeigan, D., Jancke, H., Hoebbel, D. and Weiker, W. (1975) Zur abhangigkeit der struktur der silicatanionen in wassrigen natriumsilicatlosungen vom Na:Si verhaltnis. Z. Anorg. Allg. Chem. 418, 17-28.

Ern, V. and Switendick, A.C. (1965) Electronic band structure of TiC, TiN, and TiO. Phys. Rev. A137, 1927-1936.

Fischer, D.W. and Baun, W.L. (1968) Band structure and the titanium $L_{2,3}$ X-ray emission and absorption spectra from pure metal, oxides, nitride, carbide and boride. J. Appl. Phys. 39, 4757-4776.

Fischer, D.W. (1970) MO interpretation of the soft X-ray $L_{II,III}$ emission and absorption spectra of some titanium and vanadium compounds. J. Appl. Phys. 41, 3561-3569.

Fischer, D.W. (1971) Soft X-ray band spectra and molecular orbital structure of Cr_2O_3, CrO_3, CrO_4^{-2} and $Cr_2O_7^{-2}$. J. Phys. Chem. Solids 32, 2455-2480.

Fischer, D.W. (1973) Use of soft X-ray band spectra for determining valence conduction structure in transition metal compounds. In Fabian, D.J. and Watson, L.M., eds. Band Structure spectroscopy of metals and alloys. Academic Press: London.

Fonda, G.R. (1957) Influence of activator environment on the spectral emission of phosphors. J. Opt. Soc. Amer. 47, 877-880.

Frondel, C. (1958) Systematic mineralogy of uranium and thorium. U.S.G.S. Bull. 1064, 400 p.

Gaft, M.L., Bershov, L.V., Krasnaya, A.R. and Yaskolko, V.Ya. (1985) Luminescence centers in anhydrite, barite, celestite and their synthesized analogs. Phys. Chem. Minerals 11, 255-260.

Garlick, G.F.J. (1966) Cathodo- and Radioluminescence. In Goldberg, P., ed. Luminescence of Inorganic Solids. Academic Press: New York, 765 p.

Geake, J.E. and Walker, G. (1975) Luminescence of minerals in the near- Infrared. In Karr, C., ed. Infrared and Raman spectroscopy of lunar and terrestrial minerals. Academic Press: New York, 375 p.

Geake, J.E. and Walker, G. (1966) Geochim. Cosmochim. Acta 30, 929-937.

Gies, H. (1975) Activation possibilities and geochemical correlations of photoluminescing carbonates, particularly calcites. Mineral. Deposita 10, 216-227.

Gleason, S. (1960) Ultraviolet guide to minerals. Van Nostrand: Princeton, 244 p.

Glen, G.L., and Dodd, C.G. (1968) Use of MO theory to interpret X-ray K-absorption spectral data. J. Appl. Phys. 39, 5372-5377.

Gossmann, H-J. and Feldman, L.C. (1987) Materials Analysis with High Energy Beams Part 1: Rutherford Backscattering. MRS Bulletin 12, 26-28.

Gorz, H., Bhalla, R.J.R.S.B., and White, E.W. (1970) Detailed cathodoluminescence characterization of common silicates. Penn. State Univ. MRL Spec. Pub. 70-101, 62-70.

696

Graves, W.E. and Roberts, H.H. (1972) Thermoluminescence spectral shifts of some naturally occurring calcium carbonates. Chem. Geol. 9, 249-256.

Green, G.R. and Walker, G. (1985) Luminescence excitation spectra of Mn^{2+} in synthetic forsterite. Phys. Chem. Minerals 12, 271-278.

Grunes, L.A., Leapman, R.D., Wilker, C.N., Hoffman, R. and Kunz, A.B. (1982) Oxygen K near-edge fine structure: An electron-energy-loss investigation with comparisons to new theory for selected 3d transition metal oxides. Phys. Rev. B 25, 7157-7173.

Hall, M.R. and Ribbe, P.H. (1971) An electron microprobe study of luminescence centers in cassiterite. Am. Mineral. 56, 31-45.

Harvey, E.N. (1957) A history of luminescence from the earliest times until 1900. Philadelphia, Memoirs Am. Philosoph. Soc. Vol. 44.

Haycock, D.E., Kasrai, M., Nicholls, C.J. and Urch, D.S. (1978a) The electronic structure of magnesium hydroxide (Brucite) using X-ray emision, X-ray photoelectron, and Auger spectroscopy. J.C.S. Dalton, 1791-1796.

Haycock, D.E., Nicholls, C.J., Webber, M.J., Urch, D.S. and Kasrai, M. (1978b) The electronic structure of magnesium dialuminum tetraoxide (Spinel) using X-ray emission and X-ray photoelectron spectroscopies. J.C.S. Dalton, 1785-1790.

Hemphill, W.R., Theisen, A.F. and Tyson, R.M. (1984) Laboratory analysis and airborne detection of materials stimulated to luminesce by the sun. J. Luminescence 31/32, 724-726.

Hoffman, M.V. (1970) Fluorescence and energy transfer in $SrZnP_2O_7:UO_2^{+2}$. J. Electrochem. Soc. 117, 227-232.

Hogarth, A.J.C.L. and Urch, D.S. (1976) A study of the bonding in some five and six co-ordinate compounds of silicon based on Si $K\beta_{1,3}$ emission spectra. J.C.S. Dalton, 794-798.

Iacconi, P. and Caruba, R. (1984) Trapping and emission centers in X-irradiated natural zircon. Characterization by thermoluminescence. Phys. Chem. Minerals 11, 195-203.

Ivey, H.F. (1974) Candoluminescence and radical-excited luminescence. J. Luminescence 8, 271-307.

Jacklevic J.M. (1976) Proceedings of the energy research and development administration X- and Gamma-ray symposium. Ann Arbor, Michigan pp. 1-6.

Johansson, T.B., Akselsson, R. and Johansson, S.A.E. (1970) X-ray analysis: Elemental trace analysis at the 10-12 g level. Nucl. Inst. Methods 84, 141-143.

Johnson, P.D. (1966) Oxygen Dominated lattices. In Goldberg, P., ed. Luminescence of Inorganic solids. Academic Press: New York, p. 287.

Jorgensen, C.K. (1979) Excited states of the uranyl ion. J. Luminescence 18/19, 63-68.

Joy, D.C. (1979) The basic principles of electron energy loss spectroscopy. In: Hren, J.J., Goldstein, J.I. and Joy, D.C., eds. Introduction to analytical electron microscopy. Plenum Press: New York, p. 223.

Kastner, M. (1971) Authigenic feldspar in carbonate rocks. Am. Mineral. 56, 1403-1442.

Kirk, R.D. (1955) The luminescence and tenebrescence of natural and synthetic sodalite. Am. Mineral. 40, 22-31.

Kirz, J., Sayre, D. and Dilger, J. (1978) Comparative analysis of X-ray emission microscopies for biological specimens. Ann. N.Y. Acad. Sci. 306, 291-305.

Knudson, A.R., Nagel, D.J., Burkhalter, P.G. and Dunning, K.L. (1971) Aluminum X-ray satellite enhancement by ion-impact excitation. Phys. Rev. Lett. 26, 1149-1152.

Knutson, C., Peacor, D.R. and Kelly, W.C. (1985) Luminescence, color and fission track zoning in apatite crystals of the Panasqueira tin-tungsten deposit, Beira-Baixa, Portugal. Am. Mineral. 70, 829-837.

Koffman, D.M. and Moll, S.H. (1966) Effect of chemical combination on the K X-ray spectra of silicon. Adv. in X-ray Anal. 9, 323-328.

Koster, A.S. and Mendel, H. (1970) X-ray Kβ emission spectra and energy levels of compounds of 3d transition metals-I. Oxides. J. Phys. Chem. Solids 31, 2511-2522.

Koster, A.S. and Rieck, G.D. (1970) Determination of valence and coordination of iron in oxidic compounds by means of the iron X-ray fluorescence emission spectrum. J. Phys. Chem. Solids 31, 2505-2510.

Leverentz, H.W. (1968) An introduction to luminescence of solids. Wiley: New York, 569 p.

Lu, K.Q. and Stern, E.A. (1980) Johann and Johansson focussing arrangements: Analytical analysis. AIP Conference Proc. 64 (Laboratory EXAFS facilities-1980), 104-108.

Mao, H.K., Bell, P.M., Shaner, J.W. and Steinberg, D.J. (1978) Specific volume measurements of Cu, Mo, Pd, and Ag and calibration of the ruby R_1 fluorescence pressure guage from 0.06 to 1 Mbar. J. Appl. Phys. 49, 3276-3283.

Macfarlane, R.M. (1987) Optical spectral linewidths in solids. In Lasers, Spectroscopy and new ideas. Springer series in optical sciences 54, 335 p.

Mariano, A.M. and Ring, P.J. (1975) Europium activated cathodoluminescence in minerals. Geochim. Cosmochim. Acta 39, 649-660.

Marfunin, A.S. (1979) Spectroscopy, Luminescence and Radiation Centers in Minerals. Springer-Verlag: Berlin, 352 p.

McCormick, T.C. and Smyth, J.R. (1987) Minor element distributions in ilmenite and dolomite by electron channeling-enhanced X-ray emission. Am. Mineral. 72, 778-781.

McCormick, T.C., Smyth, J.R. and Lofgren, G.E. (1987) Site occupancies of minor elements in synthetic olivines as determined by channeling-enhanced X-ray emission. Phys. Chem. Minerals 14, 368-372.

McDougall, D.J. (1968) Thermoluminescence of geological materials. Academic Press: London, 678 p.

McDougall, D.J. (1970) Relative concentrations of lattice defects as an index of the temperature of formation of fluorite. Econ. Geol. 65, 856-861.

McKeever, S.W.S. (1985) Thermoluminescence of solids. Cambridge U.P.: Cambridge, 376 p.

Medlin, W.L. (1968) The nature of traps and emission centers in rock materials. In McDougall, D.J., ed. Thermoluminescence of geological materials. Academic Press: London, 678 p.

697

Millson, H.E. and Millson, H.E., Jr. (1964) Duration of phosphorescence. J. Opt. Soc. Am. 54, 638-640.

Morgan, D.V. (1973) Channelling. Wiley: New York, 486 p.

Muller, R.O. (1972) Spectrochemical analysis by X-ray fluorescence. Plenum Press: New York.

Narbutt, K.I. (1980) X-ray spectra of iron atoms in minerals. Phys. Chem. Minerals. 5, 285-295.

Nefedov, V.I. (1966) Multiplet structure of the $K\alpha_{1,2}$ and $K\beta$-β' lines in the X-ray spectra of iron compounds. J. Struct. Chem. (USSR) (English Transl.) 7, 672-677.

Nicholls, C.J. and Urch, D.S. (1975) X-ray emission and photoelectron spectra from magnesium oxide: A discussion of the bonding based on the unit Mg_4O_4. J.C.S. Dalton, 2143-2148.

Nickel, E. (1978) The present status of cathode luminescence as a tool in sedimentology. Minerals Sci. Eng. 10, 73-100.

O'Nions, R.K. and Smith, D.G. (1971) Investigations of the $L_{II,III}$ X-ray emission spectra of Fe by electron microprobe, Part 2. The Fe $L_{II,III}$ spectra of Fe and Fe-Ti oxides. Am. Mineral. 56, 1452-1463.

Otten, M.T. and Buseck, P.R. (1987) The determination of site occupancies in garnet by planar and axial ALCHEMI. Ultramicroscopy 23, 151-158.

Otten, M.T. and Buseck, P.R. (1987) The oxidation state of Ti in hornblende and biotite determined by electron energy-loss spectroscopy, with inferences regarding the Ti substitution. Phys. Chem. Minerals 14, 45-51.

Palilla, F.C. and O'Reilly, B.E. (1968) Alkaline earth halophosphate phosphors activated by divalent europium. J. Electrochem. Soc. 115, 1076-1081.

Parratt, L.G. (1959) Electronic band structure of solids by X-ray spectroscopy. Rev. Mod. Phys. 31, 616-645.

Portnov, A.M. and Gorobets, B.S. (1969) Luminescence of apatite from different rock types. Dokl. Akad. Nauk. SSSR Seriya Geologischeskaya 184, 110-114. (in Russian)

Powell, R.C., DiBartolo, B., Birang, B. and Naiman, C.S. (1966) Temperature dependence of the widths and positions of the R and N lines in heavily doped ruby. J. Appl. Phys. 37, 4973-4978.

Prizbram, K. (1956) Irradiation colours and luminescence.-A contribution to mineral physics. Pergamon Press Ltd: London, 327 p.

Ramseyer, K. (1982) Workshop-Conference on the geological applications of cathodoluminescence, University of Manchester.

Rask, J.H., Miner, B.A. and Buseck, P.R. (1987) Determination of manganese oxidation states in solids by electron energy-loss spectroscopy. Ultramicroscopy 21, 321-326.

Reed, S.J.B. (1975) Electron microprobe analysis. Cambridge University Press: U.K.

Remond, G. (1977) Applications of cathodoluminescence in mineralogy. J. Luminescence 15, 121-155.

Richter, F.W. (1984) Experimental arrangements for PIXE analysis. Nucl. Inst. Methods B3, 105-113.

Robbins, M. (1983) The collector's book of fluorescent minerals. Van Nostrand Reinhold: New York, 289 p.

Roeder, P.L., MacArthur, D., Ma, X-P., Palmer, G.R. and Mariano, A.N. (1987) Cathodoluminescence and microprobe study of rare-earth elements in apatite. Am. Mineral. 72, 801-811.

Rogers, P.S.Z., Duffy, C.J., Benjamin, T.M. and Maggiore, C.J. (1984) Geochemical applications of nuclear microprobes. Nucl. Inst. Methods B3, 671-676.

Rooke, G.A. (1974) Theory of emission spectra. In X-ray spectroscopy. Azaroff, L.V., ed. McGraw-Hill: New York: 560p.

Sato-Sorenson, Y. (1986) Measurements of the lifetime of the ruby R_1 line under high pressure. J. Appl. Phys. 60, 2985-2987.

Selzer, P.M. (1981) Laser Spectroscopy of solids. In Yen, W.M. and Selzer, P.M., eds. Topics in Applied Physics. Springer-Verlag, Berlin: 49, Chapter 4.

Sherman, D.M. (1984) The electronic structures of manganese oxide minerals. Am. Mineral. 69, 788-799.

Shionoya, S. (1966) Luminescence of lattices of the ZnS type. In Goldberg, P., ed. Luminescence of inorganic solids. Academic press, New York: p.206.

Smith, A.L. (1949) Some new complex silicate phosphors containing calcium, magnesium and beryllium. J. Electrochem. Soc. 96, 287-296.

Smith, J.V. and Stenstrom, R.C. (1965) Electron-excited luminescence as a petrologic tool. J. Geol.73, 627-635.

Smyth, J.R. and Tafto, J. (1982) Major and minor element site occupation in heated natural forsterite. Geophys. Res. Lett. 9, 1113-1116.

Sommer, S.E. (1972) Cathodoluminescence of carbonates, I. Characterization of cathodoluminescence from carbonate solid solutions. 2. Geological applications. Chem. Geol. 9, 257-284.

Sparks, C.J. (1980) X-ray fluorescence microprobe for chemical analysis. In Synchrotron Radiation Research, eds. Winick, H. and Doniach, S. Plenum Press: New York, 754p.

Sparks, C.J. and Hastings, J.B. (1975) X-ray diffraction and fluorescence at the Stanford Synchrotron Radiation Project. Oak Ridge National Laboratory Report ORNL-5089.

Stokes, G.Q. (1852) On the change of refrangibility of light. Phil. Trans. Roy. Soc. London A142, 463.

Suzuki, S. and Lang, A.R. (1981) Visible emission from crystal growth-sector boundaries: A new luminescence phenomenon. J. Luminescence 26, 47-52.

Taftø, J. (1982) The cation distribution in $(Cr,Fe,Al,Mg)_3O_4$ spinel as revealed from the channelling effect in electron induced X-ray emission. J. Appl. Cryst. 15, 378-381.

Taftø, J. and Buseck, P.R. (1983) Quantitative study of Al-Si ordering in an orthoclase feldspar using an analytical transmission electron microscope. Am. Mineral. 68, 944-950.

Tarashchan, A.N. (1978) Luminescence of Minerals. Naukova dumka: Kiev.

Telfer, D.J. and Walker, G. (1978) Ligand field bands of Mn^{2+} and Fe^{3+} luminescence centres and their site occupancy in plagioclase feldspars. Mod. Geol. 6, 199-210.

Thomsen, J.S. (1974) High Precision X-ray spectroscopy. In X-ray spectroscopy. Azaroff, L.V., ed. McGraw-Hill: New York: 560p.

Tossell, J.A. (1973) Molecular orbital interpretation of X-ray emission and ESCA spectral shifts in silicates. J. Phys. Chem. Solids 34, 307-319.

698

Tossell, J.A., Vaughan, D.J. and Johnson, K.H. (1974) The electronic structure of rutile, wustite, and hematite from molecular orbital calculations. Am. Mineral. 59, 319-334.

Tossell, J.A. (1975) The electronic structures of Mg, Al and Si in octahedral coordination with oxygen from SCF Xα MO calculations. J. Phys. Chem. Solids 36, 1273-1280.

Tsutsumi, K., Nakamori, H. and Ichikawa, K. (1976) X-ray Mn Kβ emission spectra of manganese oxides and manganates. Phys. Rev. B 13, 929-933.

Turner, W.H., and Turner, J.E. (1970) Absorption spectra and concentration-dependent luminescence of Mn^{2+} in silicate glasses. J. Am. Cer. Soc. 53, 329-335.

Urch, D.S. (1985) X-ray spectroscopy and chemical bonding in minerals. In Berry, F.J. and Vaughan, D.J., eds., Chemical Bonding and Spectroscopy in Mineral Chemistry. Chapman and Hall: London, 325p.

Urch, D.S. (1970) The origin and intensities of low energy satellite lines in X-ray emission spectra: a molecular orbital interpretation. J. Phys. C Solid State Phys. 3, 1275-1291.

Urch, D.S. and Weber, S. (1977) Fe $K\beta_{1,3}$ X-ray emission spectra from complexes which contain ferric iron in unconventional spin states. X-ray Spect. 6, 64-65.

Urch, D.S. and Wood, P.R. (1978) The determination of the valency of manganese in minerals by X-ray fluorescence spectroscopy. X-ray Spect. 7, 9-11.

van der Veen, J.F. (1985) Ion beam crystallography of surfaces and interfaces. Surf. Sci. Rep. 5/6, 199-288.

Vaz, J.E. and Senftle, F.E. (1971) Geologic age dating of zircon using thermoluminescence. Mod. Geol. 2, 239-245.

Walker, G. (1985) Mineralogical applications of Luminescence techniques. In Berry, F.J. and Vaughan, D.J., eds. Chemical Bonding and Spectroscopy in Mineral Chemistry. Chapman and Hall: London, p. 103.

Wardle, R. and Brindley, G.W. (1971) The dependence of the wavelength of AlKα radiation from aluminosilicates on the Al-O distance. Am. Mineral. 56, 2123-2128.

Waychunas, G.A. (1987) Synchrotron radiation XANES spectroscopy of Ti in minerals: Effects of Ti bonding distances, Ti valence, and site geometry on absorption edge structure. Am. Mineral. 72, 89-101.

White, W.B. (1975) Luminescent materials. Trans. Am. Crystallogr. Assoc. 11, 31-49.

White, E.W. and Gibbs, G.V. (1967) Structural and chemical effects on the Si Kβ X-ray line for silicates. Am. Mineral. 52, 985-993.

White, W.B., Masako M., Linnehan, D.G., Furukawa, T. and Chandrasekhar, B.K. (1986) Absorption and luminescence of Fe^{3+} in single-crystal orthoclase. Am. Mineral. 71, 1415-1419.

Wiech, G., Koppen, W. and Urch, D.S. (1972) X-ray emission spectra and electronic structure of the disulphide anion. Inorg. Chim. Acta 6, 376-378.

Wiech, G., Zopf, E., Chun, H.-U. and Brückner, R. (1976) X-ray spectroscopic investigation of the structure of silica, silicates and oxides in the crystalline and vitreous state. J. Non-Cryst. Solids 21, 251-261.

Wildeman, T.R. (1970) The distribution of Mn^{2+} in some carbonates by electron paramagnetic resonance. Chem. Geol. 5, 167-177.

Williams, F. E. (1966) Theoretical basis for solid-state luminescence. In Goldberg, P., ed. Luminescence of inorganic solids. Academic Press: New York, p. 2.

Williams, F .E. (1968) Theoretical aspects of point and associated luminescent centers. In Proc. Int. Conf. Luminescence, Budapest 1, 113-123.

Wong,J. and Angell, C.A. (1976) Glass Structure by spectroscopy. Marcel Dekker: New York, 864 p.

Wood, P.R. and Urch, D.S. (1976) Valence-band X-ray emission spectra of manganese in various oxidation states. J.C.S. Dalton Trans., 2472-2476.

Zinkernagel, U. (1978) Cathodoluminescence of quartz and its application to sandstone petrology. Contrib. Sediment. 8, 1-69.